PHYSICAL BIOCHEMISTRY

PHYSICAL BIOCHEMISTRY

Applications to Biochemistry and Molecular Biology

Second Edition

David Freifelder
Brandeis University

W. H. Freeman and Company
New York

Project Editor: *Judith Wilson*
Copy Editor: *Diana Steeds*
Designer: *Christy Butterfield*
Cover Design: *Sharon Smith*
Production Coordinator: *Bill Murdock*
Illustration Coordinator: *Cheryl Nufer and Richard Quiñones*
Artists: *Donna Salmon, Georg Klatt*
Compositor: *Syntax International*
Printer and Binder: *The Maple-Vail Book Manufacturing Group*

Library of Congress Cataloging in Publication Data

Freifelder, David Michael, 1935–
Physical biochemistry.

Includes bibliographies and index.
1. Biological chemistry—Technique. 2. Molecular biology—Technique. 3. Biophysics—Technique. I. Title.
[DNLM: 1. Biochemistry—Methods. 2. Molecular biology—Methods. QU 4 F862p]
QH345.F72 1982 574.19′283 81-19521
ISBN 0-7167-1315-2 AACR2
ISBN 0-7167-1444-2 (pbk.)

Printed in the United States of America

12 13 14 15 16 VB 9 9 8 7 6 5 4 3 2 1 0

CONTENTS

Preface to the Second Edition vii

Preface to the First Edition ix

Chapter 1 *Characterization of Macromolecules* 1

Part One

DIRECT OBSERVATION 39

Chapter 2 *Light Microscopy* 40

Chapter 3 *Electron Microscopy* 72

Part Two

GENERAL LABORATORY METHODS 117

Chapter 4 *Measurement of pH* 118

Chapter 5 *Radioactive Labeling and Counting* 129

Chapter 6 *Autoradiography* 169

Chapter 7 *Membrane Filtration and Dialysis* 193

Part Three

SEPARATION AND IDENTIFICATION OF MATERIALS 215

Chapter 8 Chromatography 216

Chapter 9 Electrophoresis 276

Chapter 10 Immunological Methods 323

Part Four

HYDRODYNAMIC METHODS 361

Chapter 11 Sedimentation 362

Chapter 12 Partial Specific Volume and the Diffusion Coefficient 455

Chapter 13 Viscosity 470

Part Five

SPECTROSCOPIC METHODS 493

Chapter 14 Absorption Spectroscopy 494

Chapter 15 Fluorescence Spectroscopy 537

Chapter 16 Optical Rotary Dispersion and Circular Dichroism 573

Chapter 17 Nuclear Magnetic Resonance and Electron Spin Resonance 603

Part Six

MISCELLANEOUS METHODS 653

Chapter 18 Ligand Binding 654

Chapter 19 Miscellaneous Methods 685

Answers to Problems 723

Index 753

PREFACE TO THE SECOND EDITION

Five years have passed since the first edition of *Physical Biochemistry*. During this time, I have had the opportunity to use the book as a text for my own course. This has been quite a valuable experience because I have had direct input from my students, who had need to read more carefully than did I or any of the people who helped me with the first edition. My students took an interest in the book, told me about the sections that were unclear, and suggested improvements. In addition, W. H. Freeman and Company has provided me with copies of all published reviews. I have taken these seriously and have incorporated many of the reviewers' suggestions. However, the main reason for this revision is that many new techniques have been developed or have become major tools in the past five years. For example, in the first edition, agarose-gel chromatography of DNA was worthy of a paragraph, whereas now this technique is more widely used than centrifugation in the study of DNA.

In order to prepare this revised edition, I have consulted people who are experts in the use of each of the techniques described in the first edition; they have informed me of the latest and most important developments. Laboratory supply companies have been significant sources of information. Bio-Rad Laboratories has provided me with many figures.

The result of my efforts is a larger book containing a great many new techniques. The following subjects are totally new in the second edition: helix-coil transitions; c_0t analysis; description of buffer systems; metal-ion electrodes; high-speed autoradiography; dissociating and associating systems; new techniques of immunoelectrophoresis; Southern, Northern, and Western transfers; light scattering; ligand binding; calculation of concentrations from interference diagrams; electron spin resonance spectroscopy; an explanation of voltage versus current control in electrophoresis; high-performance liquid chromatography; DNA sequencing; fluctuation spectroscopy; and the Beckman Airfuge. Sections describing Raman spectroscopy, NMR spectroscopy, affinity chromatography, hydroxyapatite chromatography, structure of macromolecules, viscoelasticity, electron microscopy, scintillation counting, and gas chromatography have been expanded considerably.

Gerald Fasman, Martin Gellert, Chris Miller, Alfred Redfield, Serge Timasheff, Helen Van Vunakis, and Bruno Zimm have provided me with general advice, photographs, and useful information, and I owe them my thanks. Frederick W. Dahlquist, Jonathan Greer, Ivan I. Kaiser, Jon Robertus, and Carl W. Schmid, Jr. reviewed the second edition, and I appreciate their comments. My indefatigable secretary, Mildred Kravitz, joined me in typing the revised manuscript. I thank my children, Joshua and Rachel, whose enthusiasm and pride in having an author-father gave me the desire to continue.

January 1982 DAVID FREIFELDER

PREFACE TO THE FIRST EDITION

Modern biochemistry and molecular biology are concerned with the functions of biological systems. A century ago, the only means of study was by direct observation of such systems at work. Today, much more sophisticated and detailed observations can be made through electron microscopy and the specialized microscopic techniques that have been developed in the past forty years.

Scientists realized in the late nineteenth century that something was to be gained by studying the chemistry of cells. For decades thereafter, biochemists relied on the chemical methods available to them, and, indeed, great advances in understanding were achieved. Probably the most significant improvement in chemical technique was the development of the use of radioisotopic tracers; this vastly increased the sensitivity of detection, and the number of kinds of biological molecules that could be identified. As it became necessary to have more sensitive methods for separating the various components of a biochemical reaction, chromatography and electrophoresis became routine procedures.

When the attention of physicists and physical chemists was directed toward biology (perhaps because of the ability of living cells to create order even though the laws of physics state that there is a tendency toward disorder in the universe), the techniques of physics and physical chemistry—hydrodynamics, spectroscopy, scattering, and diffraction—entered the biological field.

A significant advance in biochemistry was the recognition that biological systems contain not only the small molecules with which organic chemistry is concerned, but also giant molecules, the *macromolecules*, whose molecular weights we now know can be at least 100 billion times the mass of a hydrogen atom. The importance of macromolecules to biological systems lies in the specificity that they confer both on biological reactions and in forming structural units. It is probably fair to say that in the past twenty years the greatest effort in biochemistry and molecular biology has been to characterize and understand macromolecules and their interactions with one another. This has required sophisticated methods both for separation and purification and for detailed observation of small parts of the molecules. Hence, the greater part of this book deals with techniques for the characterization of macromolecules with which to find answers to the following questions.

1. What is the precise structure at the atomic level of a macromolecule or an aggregate of macromolecules?

2. What properties of a macromolecule determine its structure and what forces participate in stabilizing it?

3. If a macromolecule binds other molecules (either small molecules or other macromolecules), what is their structure, what is the number of binding sites, and what are the physical constants (e.g., dissociation constants) for the binding?

4. Where is a particular macromolecule located within a cell or a small unit like a virus?

It should be noticed that the greatest concern is with the determination of physical parameters and structure. This derives from the (correct) belief that these characteristics determine biological function. For example, it has been said that a protein with a specialized function (e.g., an enzyme) can be thought of as an active site consisting of a few amino acids held together by α helices, β sheets, β turns, and random coils so that the structure reflects the requirements of the protein for its activity.

At present, the precise structure (at the atomic level) of a macromolecule can be determined only by X-ray diffraction analysis, and no technique has had a greater impact on the study of macromolecules than this one. (Consider, for example, the impact on genetics of the determination of the structure of DNA.) By knowing a few precise structures, not only can we establish rules for determining structure, but also we have reference molecules to study using other techniques, thus explaining how to interpret the data so obtained. X-ray diffraction, unfortunately, is a very difficult and complex technique, and often years are required to determine the structure of a protein or a macromolecular complex. Traditionally, books of this sort contain general descriptions of the theory of X-ray diffraction. However, this book does not for two reasons: (1) it is very unlikely that any but a few readers will ever use the technique or even have an occasion to attempt to unravel an X-ray diffraction pattern, and (2) it is virtually impossible, in less than an entire book, to explain the principles and analytical methods so that they can be thoroughly understood. However, throughout the book, reference is made to information gained by such analysis.

Students wishing to understand modern biological thinking can find excellent texts that give biological facts and theories as we know them today. However, well-read students who attempt to read current scientific literature (i.e., journals and review articles) quickly discover that a great deal of technical information is needed. To obtain this information, they can turn to textbooks of biophysical chemistry and find excellent theoretical analyses of the available methods, together with detailed mathematical derivations describing the physics underlying the methods. Alternatively, there are manuals available that describe procedures for the use of various instruments and techniques.

In many years of teaching physical biochemistry, I have come to realize that a student who has labored through mathematical derivations rarely has achieved sufficient understanding of the techniques to read, understand, and judge the work presented in current scientific literature. Furthermore, many students lack the mathematical sophistication required for obtaining any information at all from this standard approach. For these reasons, I teach physical biochemistry in a way that completely avoids mathematical derivations. In presenting an equation, I make clear what assumptions have been made in its derivation and what conditions must be satisfied before the equation is usable. Techniques are described in detail, but mostly with words, and many examples are given as a teaching device. I feel that the approach has been successful, and this book is designed in accord with it. Derivations are left to those instructors who feel that they are necessary to round out understanding. The single aim of this book is to enable the student to read and understand the current literature.

The book is directed at advanced undergraduates or beginning graduate students who have a general knowledge of chemistry and physics. For a few of the techniques presented (especially in the sections on spectroscopic methods), a more extensive background in physics is required, and the student may have to draw on the instructor's knowledge of such techniques for a thorough grasp of them. In the course of writing this book, I had the help of many active researchers as reviewers and was pleased to discover that the book is informative to such people, filling them in on techniques developed since graduation.

I would like to thank the following people whose aid was invaluable in achieving correctness and clarity: Elliot Androphy, Dan Alterman, Carol Orr, and Jon Tumen, four undergraduates at Brandeis University, and Robert Suva, a Brandeis graduate student, who combed the manuscript for ambiguities and inaccuracies; Andrew Braun, who read the entire manuscript in search of flaws in presentation; Richard Mandel, without whom I could not have written the chapters on spectroscopy; Alfred Redfield, Helen Van Vunakis, Lawrence Levine, Robert Baldwin, Bruno Zimm, Ross Feldberg, Sherwin Lehrer, Inga Mahler, and Serge Timasheff, who read the specialized chapters; Phil Hanawalt, Paul

Schimmel, and Peter Von Hippel, who reviewed the manuscript submitted for publication. Further thanks go to all of the people who gave me data and allowed the use of their illustrations and photographs. I also want to thank Mildred Kravitz and Barbara Nagy, who typed thousands of pages to bring the manuscript into its final form. My final debt is to my wife, Dorothy, who made use of her extraordinary skills as proofreader, scientist, editor, and logician, requiring perfection of me and of this book as she read the page proofs.

February 1976 DAVID FREIFELDER

PHYSICAL BIOCHEMISTRY

1

Characterization of Macromolecules

For roughly half a century the aim of biochemistry has been to assemble a complete catalog of the chemical *reactions* occurring in living cells. The motivation for this great effort was the belief that a significant number of the biological properties of cells could be understood in terms of the reactions in which covalent bonds are formed or broken. Indeed, from the great collection of biological reactions that has been obtained, we now understand in some detail how energy is generated by chemical degradation, how biological molecules are interconverted, and how giant molecules—the *macromolecules*—are assembled from amino acids, nucleotides, sugars, and lipids.

In the past thirty years, it has become apparent that the physical *interactions* between molecules—that is, those that do not form or break covalent bonds—are at least as important as the chemical *reactions*. For example, the *regulation* of chemical reactions (i.e., the degree to which they are allowed to occur) is accomplished both by physical changes in the structure of macromolecules and by variation in the availability of active sites on macromolecules resulting from the noncovalent binding of both small and large molecules. Furthermore, the large macromolecular aggregates found either in cells or in organisms (i.e., membranes, cell walls, chromosomes, tendons, hair, etc.) derive many of their special properties from noncovalent physical interactions. Therefore, chemical reactions are only half the story; clearly, to understand a complex biological system, knowledge of the physical properties of the constituent molecules is essential. The attainment of this knowledge is the goal of *physical biochemistry*. The application of the information so obtained to biological systems is the foundation of the modern discipline called *molecular biology*. A large part of this book describes methods for characterizing macromolecules. Because the language describing macromolecules is generally unfamiliar to the student of biochemistry, the terminology and concepts used in considering the properties and shapes of macromolecules and the transitions between various forms will be explained first.

Basic Terminology

The term *macromolecule* means a large molecule but by convention the word is reserved for a polymer. The number of monomers per polymer is called the *degree of polymerization*. The lower limit of the degree of polymerization that would require the use of the term macromolecule is undefined; most biochemists would not use the term for a tetramer but would do so if the molecule consisted of twenty monomers. The poorly defined word oligomer is usually used for a macromolecule that consists of more than one monomer but that is not large enough to be called a polymer. The range in size of macromolecules is very great. For example, the smallest naturally occurring RNA molecule and the largest known DNA molecule have molecular weights of 2.5×10^4 and 2×10^{10} respectively.

A polymer consisting of a single type of monomer is called a *homopolymer*. If the monomer is Q and there are n monomers per polymer, the polymer may be denoted poly(Q) or occasionally polyQ_n or $(Q)_n$; poly(Q) is the most common notation. Sometimes the identity of the monomer is ambiguous. For example, polyethylene, $H{-}(CH_2CH_2)_n{-}H$, which is made by polymerizing ethylene ($CH_2{=}CH_2$), contains no double bonds and could just as well be called polyethane or polymethane. This confusion is especially prevalent in biochemistry because biochemical dimers often have special names. Thus the polysaccharide amylose can be viewed as a polymer of either α-D-glucose or of α-D-maltose, since the latter is the name of the α-D-glucose dimer.

If two or more different monomer types are found in a macromolecule, it is called a *copolymer*. Protein molecules, which are linear sequences of amino acids, and nucleic acids, which are linear sequences of nucleotides, and polysaccharides, which are polymers of sugars, are examples of biological copolymers. An *alternating copolymer* contains two monomers A and B and has the chemical sequence A-B-A-B-A-B . . . Many biological polymers consist of chemically distinct substances; some examples of such complex polymers are glycoproteins (sugars attached to proteins), lipoproteins (lipids attached to proteins) and lipopolysaccharides (lipids linked to polysaccharides). There are also complex aggregates such as nucleoproteins (nucleic acids and proteins held by noncovalent bonds), phages and viruses (a particular class of nucleoproteins), and membranes (a system of interacting proteins, lipids, lipoproteins, and lipopolysaccharides).

In the following section the chemical structures of proteins and polynucleotides are described.

Polypeptides and Polynucleotide Chains

The components of proteins and polypeptides are the *amino acids*. The chemical structures of the common amino acids are shown in Figure 1-1 in which the amino acids are grouped to indicate their usual locations in proteins. *Polar* amino acids are those that interact significantly with water (i.e., they are *solvated*). They are also called the *hydrophilic* amino acids. Polar residues are either charged, like aspartic acid or lysine, or they are neutral but have strong dipoles, like serine and asparagine. These amino acids can form hydrogen bonds between themselves or with water (Figure 1-2).

Because of this strong interaction with water, polar amino acids tend to be on the surfaces of proteins, thereby maximizing contact with water. Frequently amino acids carrying opposite charges (e.g., on the negative carboxyl and positive amino groups) tend to interact with one another to form bonds and are therefore often in close proximity. The *nonpolar* amino acids are not charged nor easily solvated by water, and therefore tend to be internal, minimizing contact with water. They are also called *hydrophobic* amino acids. The sulfhydryl (SH) group of the amino acid cysteine can combine with the SH of another cysteine to form a *disulfide bridge* (—S—S—).

Amino acids polymerize by forming a covalent bond, called a *peptide bond*, between the carboxyl group of one and the amino group of another. The resulting structure of a polypeptide or protein is shown in Figure 1-3, in which R_1 (like R_2 and R_3) represents the *side chain*, or distinguishing group, of an amino acid. The amino acid sequence of a polypeptide is, by convention, written with the amino terminus at the left; thus, alanine-glycine-tyrosine is NH_2-alanine-glycine-tyrosine-COOH.

Four of the component bases of nucleic acids are depicted in Figure 1-4. The bases consist of relatively hydrophobic rings to which are attached polar groups that interact by means of hydrogen bonding to form the base pairs indicated. A base is covalently coupled with a sugar (deoxyribose for DNA, or ribose for RNA) to which a phosphate group is attached at the 5′-carbon; the structures thus formed are polymerized by means of phosphodiester bonds between the 3′-carbon of one sugar unit and the 5′-carbon of the adjacent sugar to form a nucleic acid, as shown in Figure 1-5.

Naturally occurring polynucleotide strands are always terminated by one 3′-OH group and one 5′-*P* (phosphate) group. The terms 3′ end and 5′ end are used to distinguish the two termini. By convention single-

Polar charged amino acids (tend to be on protein surface)

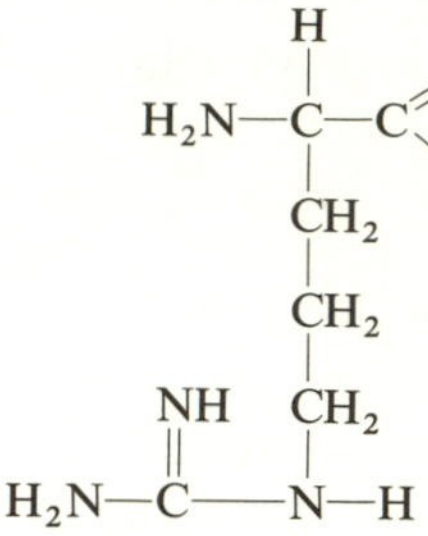

Arginine

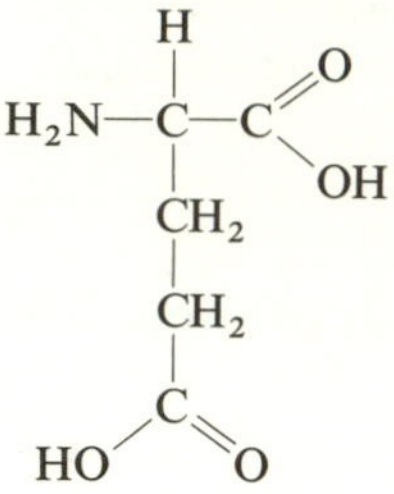

Glutamic acid

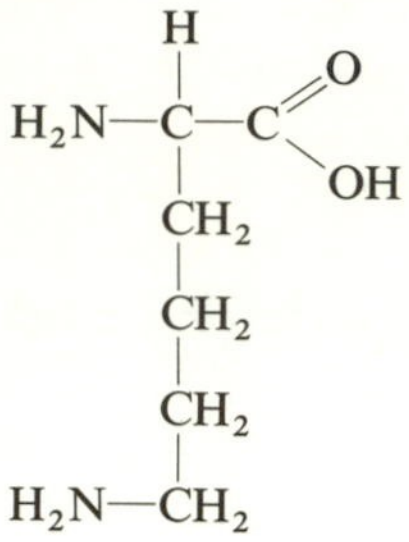

Lysine

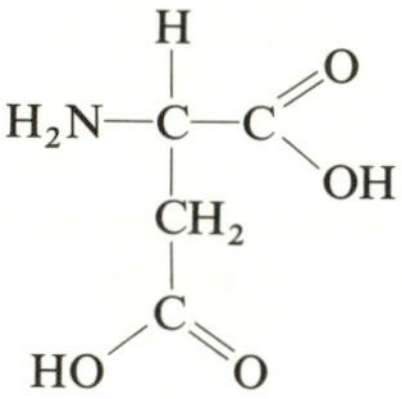

Aspartic acid

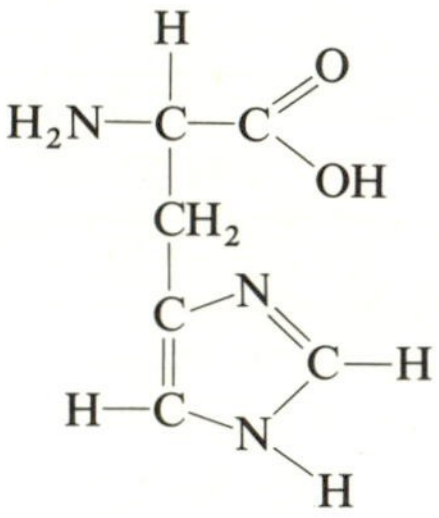

Histidine

Polar uncharged amino acids (tend to be on protein surface)

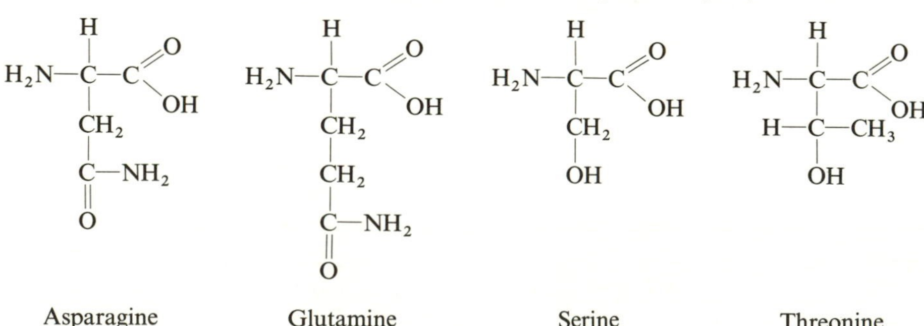

Figure 1-1
The amino acids and their chemical structures.

Nonpolar amino acids (tend to be internal)

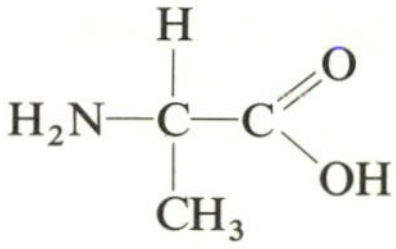

Alanine

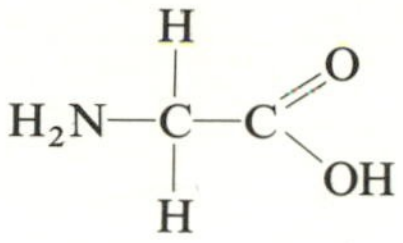

Glycine

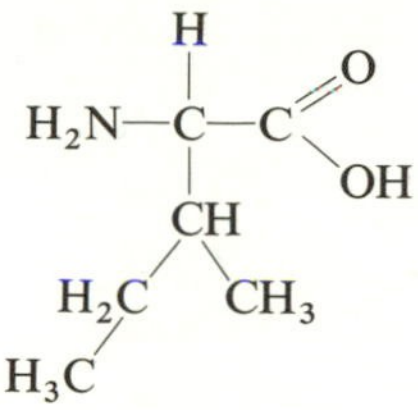

Isoleucine

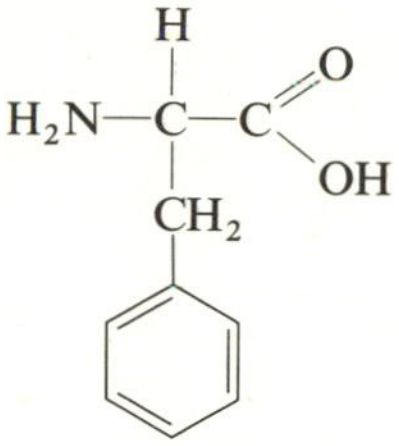

Phenylalanine

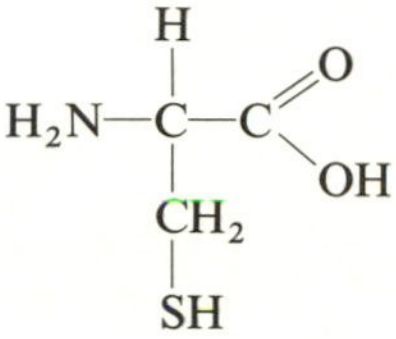

Cysteine

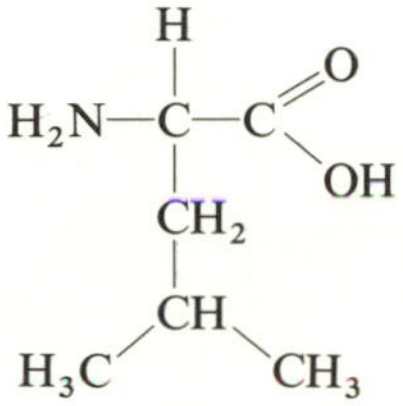

Leucine

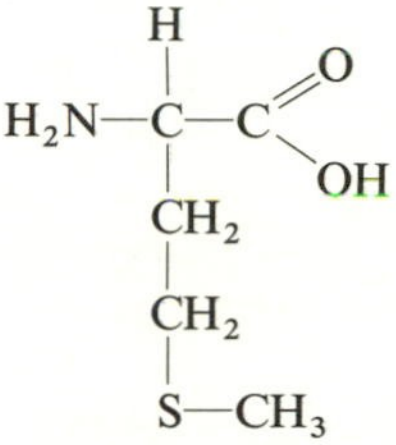

Methionine

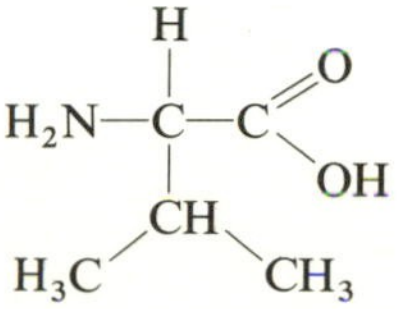

Valine

Amino acids equally frequently internal and external

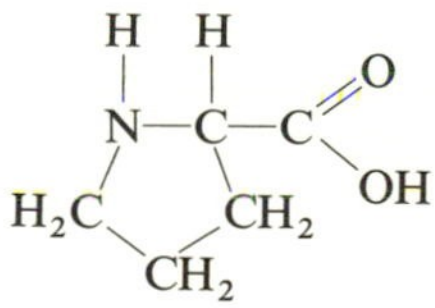

Proline

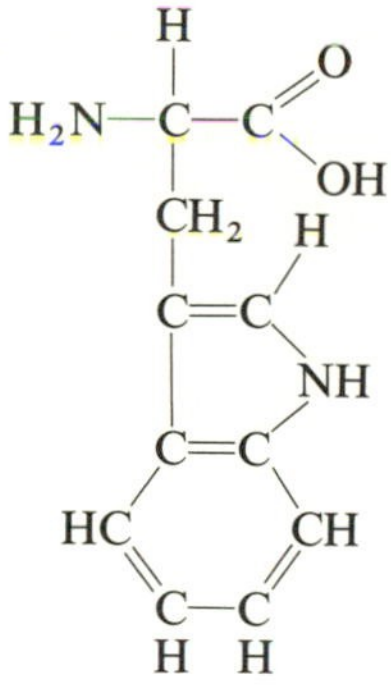

Tryptophan (nonpolar)

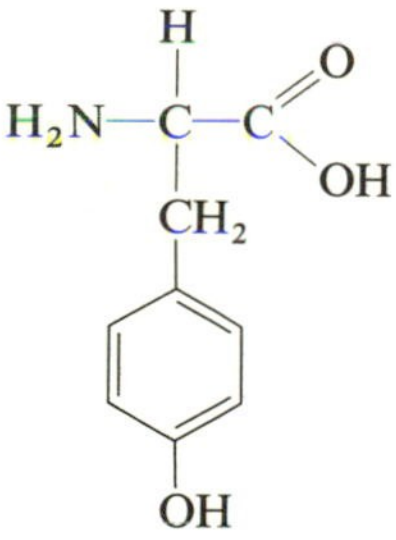

Tyrosine (polar)

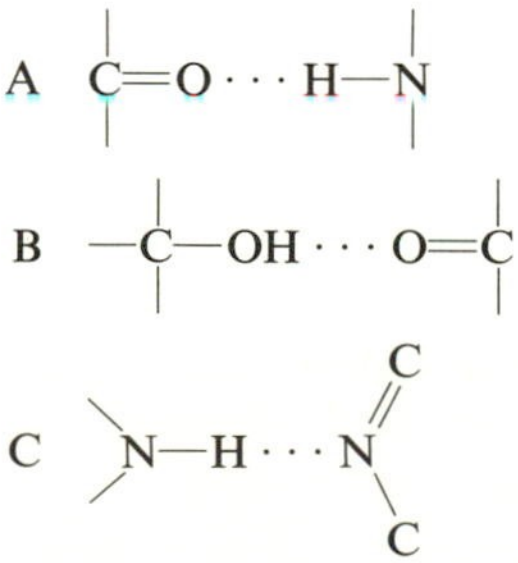

Figure 1-2
Structures of three types of hydrogen bonds (indicated by three dots): (A) the type found in proteins and nucleic acids; (B) a weak bond found in proteins; (C) the type found in DNA.

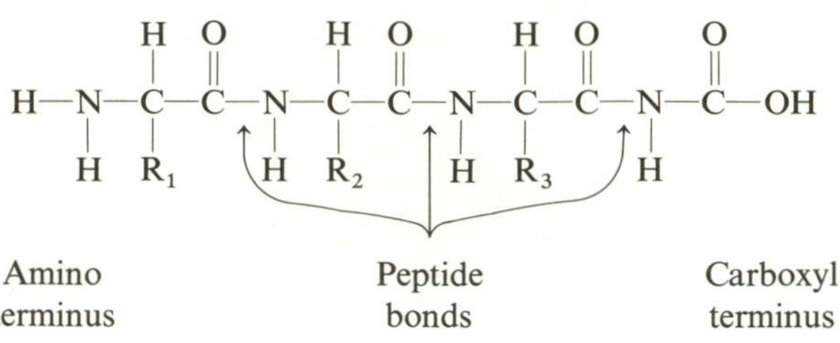

Figure 1-3
Structure of a polypeptide chain, showing amino and carboxyl termini, peptide bonds, and the locations of the side chains (R_1, R_2, and R_3).

deoxyribose deoxyribose

Thymine Adenine

deoxyribose deoxyribose

Cytosine Guanine

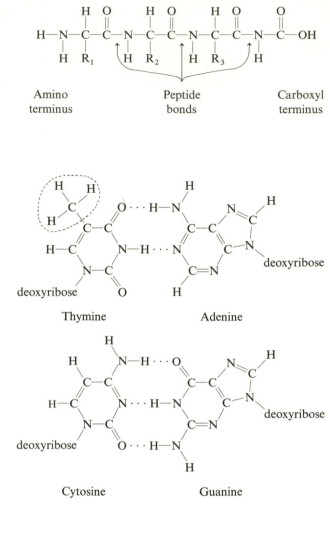

Figure 1-4
The two common base pairs of DNA. If the encircled methyl group were replaced by a hydrogen, the result would be uracil.

base base base

5′-Phosphoryl end | Sugar | Phosphate | Sugar | Phosphate | Sugar | 3′-OH end

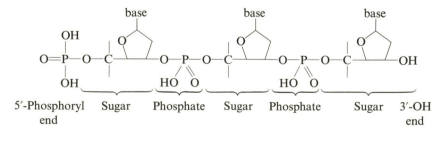

Figure 1-5
Structure of a single polynucleotide chain. The sugar (ribose or deoxyribose) and phosphate moieties alternate, a phosphate always connecting the 3′- and 5′-carbon atoms. At physiological pH the phosphates are negatively charged.

stranded polynucleotides are always written so that the 5′-*P* terminus is at the left; thus, A-T-C (adenine-thymine-cytosine) means 5′A-T-C3′; a common notation for such a trinucleotide is pApTpC. In double-stranded (duplex) DNA the two strands are *antiparallel*. That is, at each end of the duplex structure one polynucleotide strand terminates with a 5′-*P* group and the other strand ends with a 3′-OH group, as shown below:

$$\begin{array}{l} \ \ P5'\text{\_\_\_\_\_\_\_\_\_\_\_\_}3'\text{OH} \\ \text{HO}3'\text{\_\_\_\_\_\_\_\_\_\_\_\_}5'P \end{array}$$

The two chains coil about each other in forming the double helix. This arrangement places the polar sugar and phosphate backbone groups on the outside of the structure in contact with water. The polar groups on the bases form hydrogen bonds with each other. Bases from each chain complement one another along the helix forming the base pairs A·T and G·C. There is some evidence that these interbase hydrogen bonds are stronger than the hydrogen bonds formed with water because they are in such an apolar environment.

Polymer Structures

The *primary structure* of a polymer consisting of different monomer types refers to the monomer sequence—for example, the amino acid sequence of a protein and the base sequence of a polynucleotide. Such sequences can be determined by chemical analysis, using many of the separation procedures described in Chapters 8 and 9. The actual chemical methods, however, will not be discussed in this book.

Because of the interactions between various amino acid side chains and between nucleic acid bases and because of the relative degree of interaction of different molecules with water, biological polymers are rarely fully extended, linear chains; instead, they fold to form complex three-dimensional structures. Frequently, monomeric units at some positions along the chain engage in specific interactions with other units close by or at some distance along the linear sequence. An extensive pattern of such interactions (mainly hydrogen bonds) that forms an identifiable substructure in the polymer is called *secondary structure*. The principal forms are α helices and β sheets in proteins (see below) and double helices in nucleic acids. Although secondary structure relates certain portions of the linear sequence, it carries little, if any, information about their orientations in space. *Tertiary structure* is the description of all of the interactions of all of the atoms of a macromolecule in three-dimensional space. Many biological polymers interact with one another to form complex structures such as multisubunit proteins, viruses, membranes, filaments, and so forth. This is sometimes called the *quaternary structure*.

The peptide bond is planar (Figure 1-6), which allows one to consider it as a unit when assessing protein structure. On the other hand, all bonds involving the α-carbon are flexible and allow a wide variety of possible orientations. Each α-carbon is bonded to two planar peptide groups (one toward the amino end and one toward the carboxyl end). The orientation of the peptide groups with respect to one another can be described by two angles, usually designated ϕ and ψ. The entire protein backbone, then, can be described by a list of pairs of angles ϕ, ψ, one pair for each amino acid.

The phosphodiester bonds of nucleic acids are also flexible (Figure 1-7). However, because the bases consist of planar, strongly hydrophobic ring systems surrounded by only a few charged groups, they tend to stack one above the other (see Chapter 16), thereby minimizing contact with water. This tends to increase the rigidity of the structure, even in a single-stranded polynucleotide.

A linear polymer that has free rotation about all bonds in the chain and has no interaction of side groups is called a *random coil* (Figure 1-8). It does not have a unique three-dimensional structure or size because it is continually being distorted by Brownian motion. Its size can be described by an average value—the *radius of gyration*, which will be discussed shortly. Strictly speaking, a random coil is the structure that would exist if each monomer could be at *any* angle with respect to the adjacent monomer. Of course such a perfectly random structure cannot exist because no bond is perfectly flexible and because no two distant monomers can occupy the same space. However, there is a tendency to consider a molecule to be a random coil if it is *fairly* flexible and if there are neither attractive nor repulsive forces (other than complete overlap) between any pair of monomers. This is an extremely rare occurrence,

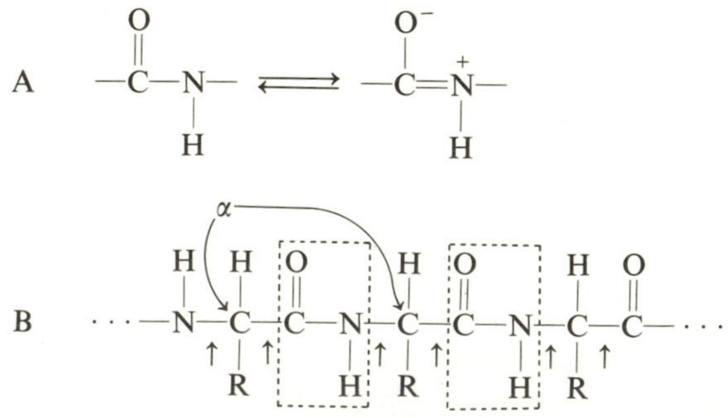

Figure 1-6
A. Resonance structure of peptide bond, showing the rigidity conferred by partial double-bond characteristics. This is why the peptide bond is planar. B. Part of a polypeptide chain: arrows point to bonds about which there is free rotation; the rigid peptide units are inside the boxes.

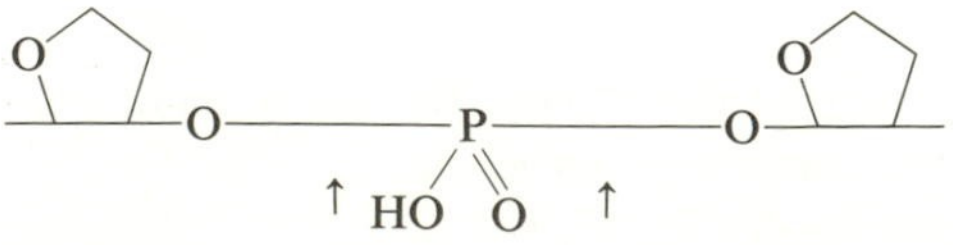

A —C(=O)—N(H)— ⟷ —C(O⁻)=N⁺(H)—

B ···—N(H)—C(H)(R)—C(=O)—N(H)—C(H)(R)—C(=O)—N(H)—C(H)(R)—C(=O)—··· (α)

Figure 1-7
A phosphodiester bond; the points of possible rotation are indicated by the arrows.

O —O—P(HO)(=O)—O— O

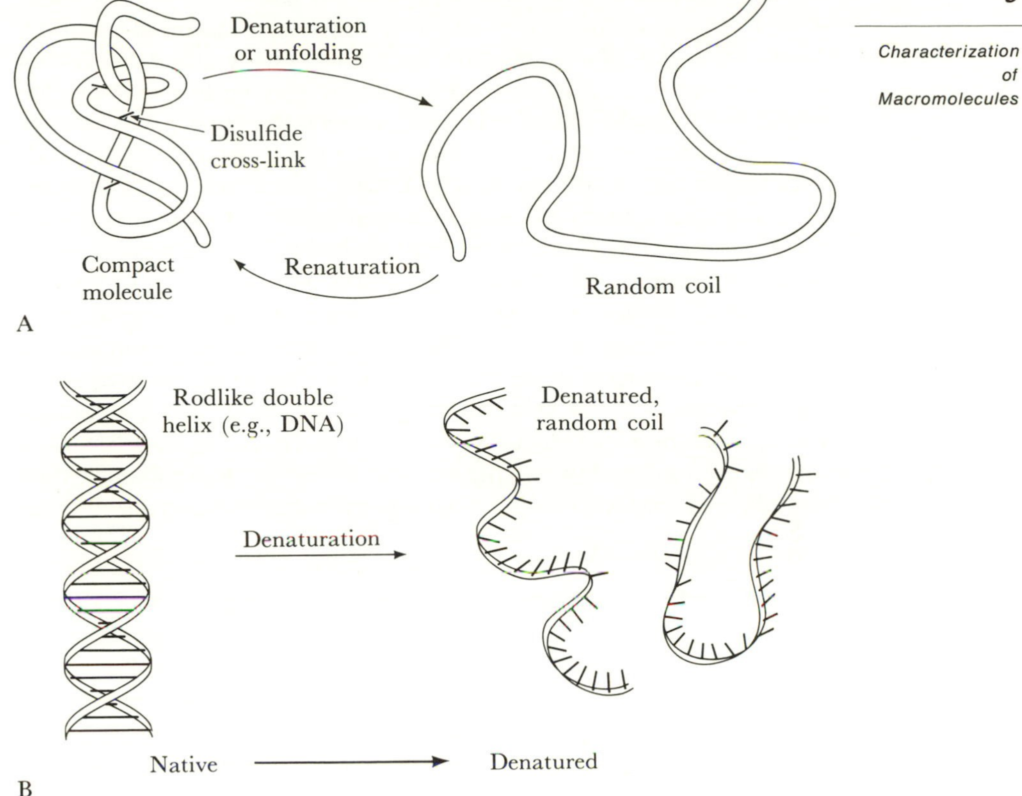

Figure 1-8
Several forms of macromolecules and how they are interconverted.

if it happens at all, because in biological macromolecules there tend to be very significant attractive interactions between the monomers. A protein in which all hydrogen bonds are broken but a *few* disulfide bonds remain is sometimes called a *near-random* coil.

Biological macromolecules are not usually random coils but have a complex orientation in space (tertiary structure), which is vital for activity of the molecule. The tertiary structure is governed by many forces and interactions, such as hydrogen bonding, hydrophobic interactions, van der Waals forces, charged attraction (ionic bonds), and charge repulsion, each acting between the backbone and the side-chain atoms. The term *side-chain interactions* is often used when referring to interactions that do not include the backbone. If the interactions maintain a generally spherical or ellipsoidal shape, the protein is *globular*; if the molecule is extended and rodlike, it is a *fibrous* protein. Often the interactions reduce the volume of the molecule to less than that of a random coil; in that case, the molecule is said to be *compact* (Figure 1-8).

Many polymers have a *helical structure* (Figure 1-9). A helix arises when monomeric units of the structure are related by a constant rotation about some axis plus a constant translation along that axis. Such a structure can fit in a cylinder. A simple analogy is the structure that would be obtained if a stack of playing cards were pinned together with a straight pin at one corner but each card were rotated at a fixed angle with respect to the one below. The α helix of proteins is one example of this structure. In an α helix a hydrogen bond is present between the C=O group of one peptide group and an NH group in a peptide group three residues along the chain (Figure 1-10). The NH group of the first peptide unit is also hydrogen-bonded with the C=O group three residues *back* along the chain. Thus, every peptide unit is engaged in two hydrogen bonds. Note that the side chains do not participate in the formation of the α helix. Other helical structures are also known. For instance, in polynucleotides, the planar nucleotide bases are stacked one above another but slightly rotated. In DNA, two polynucleotide strands, each extended by this stacking, hydrogen bond to one another

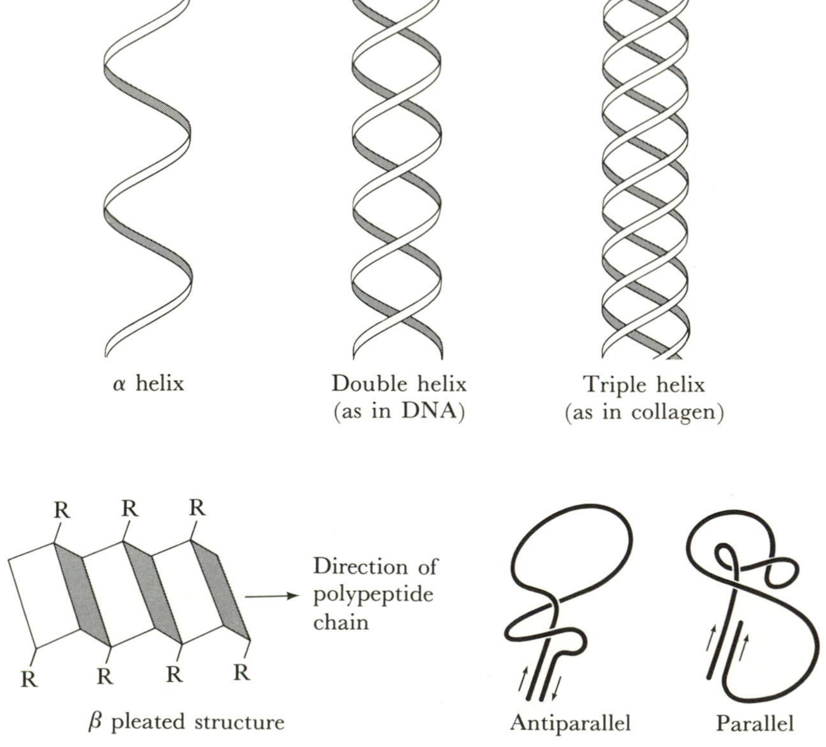

Figure 1-9
Several conformations of macromolecules.

to form a *double-stranded helix* (Figure 1-9). Some molecules (e.g., the protein collagen) can form a *triple-stranded helix*. Other types of single helices are also present in some proteins.

Helical molecules (single-stranded and multiple-stranded) are examples of *extended* or *rodlike* molecules.

A common structure in proteins is the *β structure*, in which two sections of a polypeptide chain (or in some cases two different chains) are aligned side-by-side and held together by hydrogen bonds. To maximize the number of hydrogen bonds, the polypeptide chains are pleated as shown in Figure 1-9. In this structure, the plane of the pleat contains the peptide group and the side chains are located alternately

A

O H H O H H O H H O H H O H H O H H O H H
C—N—C—C—N—C—C—N—C—C—N—C—C—N—C—C—N—C—C—N—C—
−3 R −2 R −1 R 0 R 1 R 2 R 3 R

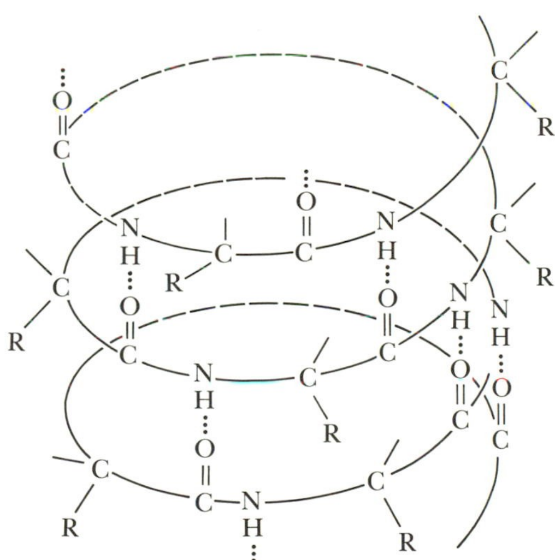

B

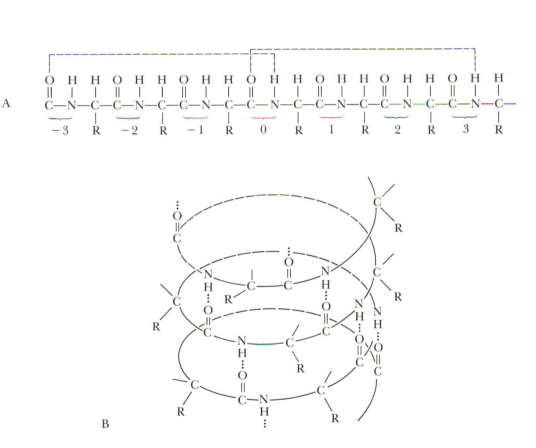

Figure 1-10
A. The two hydrogen bonds (dashed lines) in which a particular peptide group (numbered zero) is engaged in an α heliz. The peptide groups are indicated by braces; they are numbered using positive and negative numbers to denote direction along the helix; the positive direction is from the NH_2 to the COOH terminus. B. An α helix showing how the hydrogen bonds stabilize the structure. The dots represent the hydrogen bonds. The H atoms that are not in hydrogen bonds are omitted for clarity. Note that the C═O bonds point upward and the N—H bonds downward along the helix.

above and below the plane of the sheet. The figure shows only one of the strands of the β structure; normally, a second strand would be adjacent to the one shown. The alignment of the two strands may be such that the adjacent chains are running either in the same direction (*parallel*) or in the opposite direction (*antiparallel*), as shown in Figure 1-9. Such regions of a protein are called parallel and antiparallel β-*pleated sheets* respectively.

Cross-Links

Certain chemical and physical agents produce covalent bonds between two nucleotides. These are called *cross-links* and can be either intrastrand or interstrand: intrastrand cross-links prevent the strand from forming a random coil; interstrand cross-links prevent strand separation. A covalent bond between a polynucleotide and another molecule (e.g., a protein) is also called a cross-link. Proteins contain natural cross-links, the disulfide bonds (Figure 1-11).

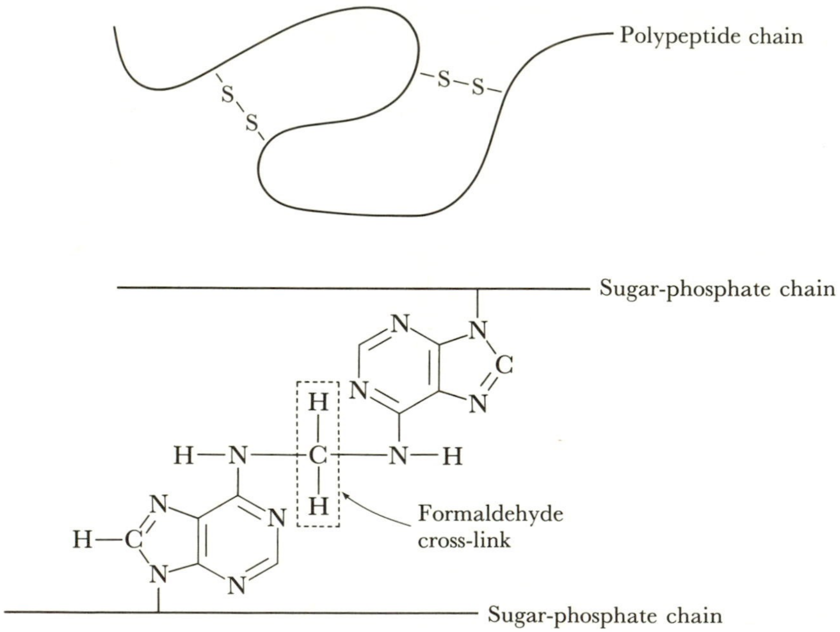

Figure 1-11
Two types of cross-links: (top) a disulfide cross-link (connecting two cysteines) within a polypeptide chain; and (bottom) a methylene bridge produced by formaldehyde reacting with the amino groups of two adenines in a double-stranded polynucleotide.

Characterization of Macromolecules

Many numerical values are used to describe macromolecules. One of the most important is the molecular weight M. Often all of the molecules in a sample of a biological macromolecule do not have the same value of M. Such a sample is said to be *polydisperse*; it is *monodisperse* if all molecules have the same value of M.

The molecular weight of a polydisperse sample is necessarily an average value. Two major types of averages, the number-average molecular weight M_n and the weight-average molecular weight M_w are considered because some experimental techniques yield M_n whereas others yield M_w.

M_n is defined as

$$M_n = \frac{\sum n_i M_i}{\sum n_i} = \sum f_i M_i \qquad (1)$$

in which n_i is the number of molecules and f_i is the fraction of the total number of molecules having molecular weight M_i.

M_w is defined as

$$M_w = \frac{\sum w_i M_i}{\sum w_i} \qquad (2)$$

in which w_i is the weight of all molecules having molecular weight M_i.

Since $w_i = n_i M_i$, the equations can also be written:

$$M_n = \frac{\sum w_i}{\sum \frac{w_i}{M_i}} \quad \text{and} \quad M_w = \frac{\sum n_i M_i^2}{\sum n_i M_i} = \frac{\sum f_i M_i^2}{\sum f_i M_i} \qquad (3)$$

The difference between M_n and M_w is illustrated in the following examples.

Example 1-A

☐ What are the values of M_n and M_w for a collection of molecules of which $\frac{1}{4}$ of the molecules have molecular weight of 5×10^4, $\frac{1}{2}$ of 2×10^5, and $\frac{1}{4}$ of 10^7?

For M_n, the relevant numbers are $f_1 = \frac{1}{4}, M_1 = 5 \times 10^4, f_2 = \frac{1}{2}, M_2 = 2 \times 10^5, f_3 = \frac{1}{4}$ and $M_3 = 10^7$. Thus

$$M_n = \tfrac{1}{4}(5 \times 10^4) + \tfrac{1}{2}(2 \times 10^5) + \tfrac{1}{4}(10^7) = 2.6 \times 10^6$$

For M_w, using equation (3),

$$M_w = [\tfrac{1}{4}(5 \times 10^4)^2 + \tfrac{1}{2}(2 \times 10^5)^2 + \tfrac{1}{4}(10^7)^2]/2.6 \times 10^6$$
$$= 9.6 \times 10^6$$

The value of M_w is greater than M_n because it is strongly affected by the heavier molecules. More specifically, M_w is an average over the mass fraction of molecules whereas M_n is an average over the number fraction of molecules. Since heavier molecules contribute more mass than lighter molecules, M_w is more sensitive to the heavier molecules whereas M_n is more sensitive to lighter molecules. This is further illustrated by Examples 1-B and 1-C.

Example 1-B

☐ What are the values of M_n and M_w for a sample of a molecule whose true molecular weight is 2×10^5 but is contaminated with an impurity that is 2% by weight and has a molecular weight of 300?

The sample is 98% by weight of the molecule of interest. Using equation (3), $\sum w_i = 0.98 + 0.02 = 1$ and $w_1 = 0.98$, $M_1 = 2 \times 10^5$, $w_2 = 0.02$ and $M_2 = 300$. Thus

$$M_n = 1\bigg/\left[\frac{0.98}{2 \times 10^5} + \frac{0.02}{300}\right] = 1.4 \times 10^4$$

Using equation (2)

$$\begin{aligned} M_w &= [(0.98)(2 \times 10^5) + (0.02)(300)]/(0.98 + 0.02) \\ &= 1.96 \times 10^5. \end{aligned}$$

Note that M_n is very sensitive to the presence of the low-molecular-weight impurity whereas M_w is nearly the same as the true value of M.

Example 1-C

☐ What are the values of M_n and M_w for a sample whose true molecular weight is 2×10^5 when there is a high-molecular-weight impurity that is 2% by weight and has a molecular weight of 10^7?

In this case:

$$\begin{aligned} M_n &= 1/[(0.98/2 \times 10^5) + (0.02/10^7)] = 2.02 \times 10^5 \\ M_w &= [(0.98)(2 \times 10^5) + (0.02)(10^7)]/(0.98 + 0.02) \\ &= 3.96 \times 10^5 \end{aligned}$$

In contrast with Example 1-A, the high-molecular-weight contaminant affects M_w but not M_n.

These examples show the following: *A contaminant of low molecular weight makes M_n significantly lower than the true M and has only a small effect on M_w. A high-molecular-weight contaminant raises the value of M_w and has virtually no effect on M_n.*

The significance of these averages is that some techniques for measuring M yield M_n (e.g., osmometry and freezing-point depression) whereas other techniques (e.g., equilibrium centrifugation and light scattering) yield M_w.

The dimensions of a macromolecule are also of importance. If the molecule is rigid and has a regular structure, such as a rod or a sphere, dimensions such as length and radius are meaningful. Indeed it is reasonable to state the length of a DNA molecule; for DNA, the mass per unit length is constant so that the length is proportional to the molecular weight. However, most rodlike molecules are somewhat flexible and are constantly being bombarded by solvent molecules so that their shape varies from one instant to the next. Thus, the dimensions of a rod may be continually changing with time and some type of average is required. Dimensions that are averaged over a long time interval could be defined but, because experimentally a large number of molecules is examined at once, it is more reasonable to take an average over the population at a particular instant. The two parameters that are used are the *end-to-end distance* and the *radius of gyration*. These will be discussed shortly. The problem of defining dimensions also exists for a rigid molecule whose shape is not a simple geometric figure, as, for example, a highly folded protein molecule. For such a molecule, the radius of gyration is also a useful parameter. It is also often useful to describe a molecule in terms of the smallest sphere or ellipsoid that can contain the molecule.

Macromolecules are so large that they take up a great deal of space. This has important effects on the properties of solutions of macromolecules. In the usual theory of solutions developed to describe solvent-solute interactions, an assumption is made that the solute and solvent molecules have roughly the same size, yet a macromolecule may be thousands (even hundreds of thousands) of times larger than a solvent molecule. Because molecules are continually rotating, their effective volume is even greater than their real volume. The volume parameter that describes this is called the *excluded volume*. It is the excluded volume effect that is the major cause of non-ideality of solutions of macromolecules. In performing measurements with solutions of macromolecules, it is usually necessary to extrapolate all measurements to zero concentration in order to eliminate the non-ideality resulting from the excluded volume effect.

The *end-to-end distance*, *h*, is the average separation between the two ends of a linear molecule. For a rigid rod it equals the length of the molecule but for a flexible molecule, its value depends upon the molecular weight and the flexibility. For a random coil (that is, a perfectly flexible molecule), *h* can easily be calculated using the solution to a problem in statistics called the *random walk*. In the one-dimensional random walk a person is constrained to walk in a straight line but the choice to take a step forward or backward is random. If N steps of length l are taken from the origin, after N random steps, the walker is, on the average, $N^{\frac{1}{2}}$ paces from the origin. If the problem is extended to three dimensions, the result is the same. This analysis is equivalent to determining the end-to-end distance of a random coil as long as the bond angles are assumed to be capable of having any value. Thus, we can write for the end-to-end distance h

$$h = N^{\frac{1}{2}}l \tag{4}$$

This can be rewritten

$$h = (Nl)^{\frac{1}{2}}l^{\frac{1}{2}} = L^{\frac{1}{2}}l^{\frac{1}{2}} \tag{5}$$

in which $Nl = L$, the total length; L is called the *contour length* of the molecule. Since L is proportional to the molecular weight of the molecule, h is also proportional to $M^{\frac{1}{2}}$.

Of course, the angle between the two monomers is not random when they are linked in a polymer. However, with little error we can assume that it is random and consider l to be the *effective segment length* of a hypothetical truly random coil. If this length is just the length of the monomer, then N is the degree of polymerization. This effective length can then be compared to the actual linear dimensions of the monomer. If they are similar, the macromolecule is flexible and if they differ by a large value, the macromolecule must be rigid. If a molecule is rigid, it will of course be very long and thin.

The value of h can be determined from sedimentation measurements, although not very accurately; l is then calculated from equation (4). For highly flexible organic polymers, such as polyethylene, it is found that l is 2–3 Å (the length of the $—CH_2—$ group); however, for very rigid molecules, such as double-stranded DNA, l is about 1000 Å, which is much greater than the length of the sugar-phosphate unit. Thus, DNA is a very stiff molecule.

The radius of gyration, R_G, is the root-mean-square average of the distances of all parts of a molecule from its center of mass. That is,

$$R_G^2 = \frac{\sum m_i r_i^2}{\sum m_i} \tag{6}$$

in which m_i is the mass of the ith element at a distance r_i from the center of mass. The actual physical significance of R_G in a geometric sense is not obvious but this average appears as a direct measurement in the analysis of the angular dependence of the scattering of light from a solution of macromolecules (see Chapter 19). For simple geometric shapes and for the random coil, R_G can be calculated. A few values are tabulated below.

Sphere	$\sqrt{3/5}r$	r = radius of sphere.
Rod	$\sqrt{1/12}L$	L = length of rod.
Random coil	$\sqrt{N/6}l$	N = number of units of length l.

The value of R_G can be used to estimate the shape of a molecule by comparing the measured value with that expected for a sphere. This

Table 1-1
Radii of Gyration of Selected Molecules.

Molecule	M	$\bar{v}$	$R_{G,obs}$ (Å)	$R_{G,sphere}$ (Å)	$R_{G,obs}/R_{G,sphere}$
Lysozyme	13,930	0.70	14.3	12.2	1.17
Myoglobin	16,890	0.74	16.0	13.2	1.21
tRNA	26,600	0.53	21.7	13.8	1.57
Myosin	493,000	0.73	468	45.2	10.4
DNA	4×10^6	0.55	1170	74	15.8

calculation requires writing r, the radius of a sphere, in terms of measurable quantities. The volume of a sphere is $4\pi r^3/3$ or $M\bar{v}/N_A$ in which M is the molecular weight, $\bar{v}$ is the partial specific volume (the measurement of which is described in Chapter 12) and N_A is Avogadro's number. Thus $r = (3M\bar{v}/4\pi N_A)^{\frac{1}{3}}$ and $R_{G,sphere} = \sqrt{3/5}(3M\bar{v}/4\pi N_A)^{\frac{1}{3}}$. Values of $R_{G,observed}$, and $R_{G,sphere}$ for three proteins (lysozyme, myoglobin, and myosin) and two nucleic acids (tRNA and DNA) are shown in Table 1-1. The data shown in this table indicate that lysozyme and myoglobin are not strictly spherical but that their overall shape is not very nonspherical. Myosin and DNA, on the other hand, are very nonspherical. The values of R_G for these molecules would suggest that they are rodlike, as is indeed the case. Some idea of the flexibility of the molecules can also be gained by comparing $R_{G,observed}$ to the value for a rigid rod. For DNA, the mass per unit length of DNA is known from X-ray diffraction studies to be 200 molecular weight units per ångström. Thus a molecule for which $M = 4 \times 10^6$ has a length of 2×10^4 Å and $R_{G,rod} = \sqrt{1/12} \times (2 \times 10^4) = 5774$ Å, roughly five times greater than the observed value. Hence DNA is not extended as much as a rigid rod so that it must be somewhat flexible. It is nonetheless a very stiff molecule (because of the base stacking and the duplex structure) because $R_{G,observed}$ is much greater than that of a sphere; the same conclusion can be drawn for myosin.

Intramolecular and Intermolecular Forces and Subunits

The three-dimensional structure of a macromolecule is determined by three factors: the allowable bond angles, the interaction between the components of the macromolecule (that is, the monomers), and the interaction between the solvent and the components. The solvent interactions are of two types: *solvation* or solvent binding, which is an attraction between the components and the solvent molecules, and the *hydrophobic interaction*, which is a solute-solute interaction as a consequence either of the inability to interact with the solvent or an avoidance interaction. The basic rule of the hydrophobic interaction is that *if a collection of molecules is unable to be solvated, the molecules will instead stick close to one another in order to minimize contact with the solvent.* This effect can be seen quite clearly in the nucleic acids and the proteins. In the first case, the nucleotide bases interact weakly with water (that is, they are only sparingly soluble) so that, to minimize contact with water, they stack one above another. This is one of the factors that makes even the single-stranded polynucleotides somewhat rigid. For proteins the hydrophobic interaction may result in stacking of ring side chains, as in the case of phenylalanine and, more significantly, of clustering of the very nonpolar amino acids such as leucine, isoleucine, and valine. Hydrophobic interactions are also responsible, in part, for the structure of biological membranes. It is important to note that hydrophobic interactions are not directional. That is, interacting groups are brought together but, except for the stacking of rings, the cluster usually lacks the angular constraints imposed by covalent bonds and hydrogen bonds.

The principal positive interactions in a macromolecule are hydrogen bonds, ionic bonds, and the van der Waals attraction. Hydrogen bonds account for the double-helix structure of DNA, for the α helix and β structure of proteins, and for certain other interactions. Ionic bonds are found in proteins, joining pairs of oppositely charged amino acids, and in nucleoproteins, joining the positively charged amino acids, arginine and lysine, with the negatively charged phosphates in nucleic acids. The van der Waals attraction is a weak force that exists between all molecules. It is effective only at very small distances. If two regions of a macromolecule have complementary shapes, the regions can approach one another closely and the van der Waals force can be quite strong. Thus the van der Waals force is responsible in part for interactions between two regions of a molecule that can fold in such a way that complementary surfaces are produced.

All of the interactions serve to give every macromolecule a unique three-dimensional structure suited to its biological function. The surface of a macromolecule may also be capable of interaction with the surface of another macromolecule, usually to perform some function. One example just given is that of a nucleoprotein, in which ionic bonds stabilize the interaction. It is very common among proteins designed to aggregate or self-assemble that the interaction is through complementary hydrophobic patches on the surface. These patches clearly do not help the solubility of the individual proteins but two molecules having such patches can reduce unfavorable contact with water by pairing with one another in such a way that some of the patches are removed from solvent contact. This is the reason that hemoglobin is a tetramer consisting of four interacting polypeptide chains.

There is a third cause of pairing of macromolecules. In the complex folding of a protein, for instance, there may be generated two surface regions whose three-dimensional shapes are complementary, as shown

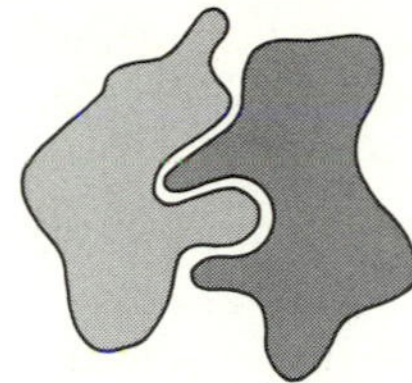

Figure 1-12
An example of two proteins interacting in regions with complementary shapes.

in Figure 1-12. The very weak van der Waals forces between two amino acids are quite insufficient to form a stable pair which, at ordinary temperatures, can resist thermal disruption. However, when the shapes are complementary, these weak forces can add up to make quite a strong attraction. This shape effect is not only responsible for the pairing of proteins and other macromolecules but also for the powerful binding of small molecules to a macromolecule, such as in the specific binding of a substrate molecule to an enzyme.

These surface interactions are responsible for the widespread phenomenon of the existence of *subunits* of proteins. That is, most proteins whose molecular weight is above 50,000 consist of several polypeptide chains joined together. Sometimes these chains are identical but they may also be of two or more types—for instance, hemoglobin consists of two α chains and two β chains. The large multifunctional proteins may contain several different chains as, for example, *Escherichia coli* RNA polymerase, which consists of five different subunits.

These interactions are also responsible for the structure of large macromolecular aggregates such as viruses, chromosomes, and membranes.

In the following section we examine how changes in the environment of a macromolecule that alter the strength of these interactions affect the structure of the macromolecule.

The Concept of Native and Denatured Structures

The term *native* structure is commonly used but is difficult to define. It can mean any of the following: (1) the structure of a macromolecule as it exists in nature; (2) the structure of a macromolecule as isolated, if it retains enzymatic activity; (3) the form of a macromolecule that has no biological activity but possesses secondary structure.

Denatured is an equally vague term usually meaning a form of a macromolecule that has less secondary structure than that which is called native. For proteins, it usually means a random (or near-random) coil. For double-stranded (native) DNA, the term specifically means single-stranded DNA that may or may not have intrastrand hydrogen bonds and may be randomly aggregated with one or more different single strands. If some of the native structure of a molecule is lost, the molecule is considered *partially denatured.*

A transition from an ordered to a disordered structure is often called a *helix-coil transition*, (even if the native state is not helical), although the term denaturation is used with equal frequency. Helix-coil transitions are usually detected by monitoring a change in some physical property of the molecule—for example, intrinsic viscosity, optical density, sedimentation coefficient, and so forth. The usual agents used to induce a helix-coil transition are temperature, pH, salt concentration, and chemical denaturants such as urea and guanidium chloride for proteins, and formamide, formaldehyde, and ethylene glycol for nucleic acids (Figure 1-13). The most common agent is temperature, so that denaturation is often called *melting*.

A helix-coil transition can be either *non-cooperative* or *cooperative*, a distinction best described by the following example. Consider a protein whose structure is determined by an ionic interaction between a single pair of amino acids A and B, one positively charged and one negatively charged. At any temperature, equilibrium exists between the molecules in which A and B are paired (the "helical" form) and those in which A and B are unpaired (the "coil"). At a temperature at which the pairing is very stable most of the molecules are native whereas at some high temperature most, if not all, are denatured. There will be a temperature range in which the equilibrium changes rapidly; that is, if a graph were to be made of the fraction of unpaired A and B versus temperature, it would show a rapid rise in this temperature range. Since there is only a single pair per molecule, the temperature range is very narrow and the

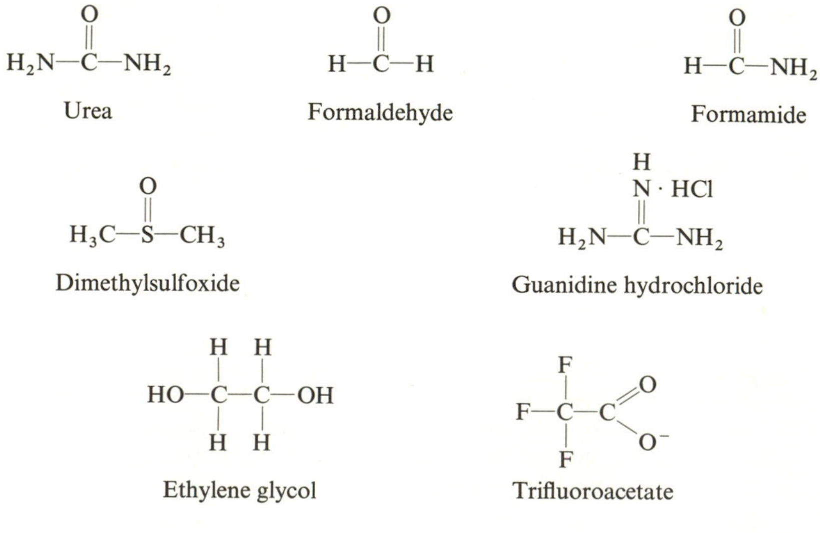

$H_2N{-}C({=}O){-}NH_2$

Urea

$H{-}C({=}O){-}H$

Formaldehyde

$H{-}C({=}O){-}NH_2$

Formamide

$H_3C{-}S({=}O){-}CH_3$

Dimethylsulfoxide

$H_2N{-}C({=}NH \cdot HCl){-}NH_2$

Guanidine hydrochloride

$HO{-}CH_2{-}CH_2{-}OH$

Ethylene glycol

$F_3C{-}C({=}O){-}O^-$

Trifluoroacetate

Figure 1-13
Chemical formulas for several common denaturants. Note that many contain groups that can form hydrogen bonds.

transition is said to be *sharp*. This is not necessarily true for a molecule whose shape is determined by a very large number of pairs because there are two different kinds of transitions. In a *non-cooperative* transition, the probability of any pair existing is independent of the existence of any other pair; in a *cooperative* transition, the probability of a pair existing depends on the existence of other pairs. In general all pairs do not have the same strength, because they are either chemically different or in different locations in the molecule. Thus in a non-cooperative transition, as the temperature is raised, pairs will be disrupted at different temperatures so that the helix-coil transition will not be sharp. In a cooperative transition there is much greater difficulty in disrupting the first pair—because it is stabilized by the existence of all other pairs—than in breaking the last pair, which is stabilized only by its own intrinsic binding energy. Thus the first pair that is disrupted is broken at a higher temperature than if it were the *only* pair; this is also true of the second, third, etc. pairs, but not true of the last pair. However, once the first pair is finally disrupted, the second pair is less stable; when this one is disrupted then the third is weakened, and so forth. Therefore, *when there is cooperativity, a transition is much sharper than when there is no cooperativity.*

Non-cooperative transitions are rare in biological macromolecules. Consider a protein, a large fraction of whose amino acids are hydrogen-bonded. At any particular temperature the molecule is being bombarded by solvent molecules, which has the effect that at any instant various regions of the molecule are moving with respect to one another. The motion is not very great because the regions are constrained by many hydrogen bonds. As the temperature increases, the molecule is bombarded with greater force. The rigidity imposed by the hydrogen bonds still maintains the structure until a temperature is reached at which one hydrogen bond breaks. This allows the regions held together by this hydrogen bond to move more freely with respect to one another, which thereby strains nearby hydrogen bonds. A smaller increment in energy is now needed to break these bonds, after which the molecule again increases in flexibility making it even more susceptible to disruption. This process continues by addition of small increments of energy until the molecule is totally disrupted. This is a somewhat idealized picture in that the process is not necessarily completely progressive. That is, at a particular temperature *several* regions might be so weakened that the conversion to the disrupted form may be occurring *simultaneously* in several parts of the molecule.

A similar situation exists for a DNA molecule. The structure of the double-stranded helix is maintained by two factors, namely, (1) the tendency for the hydrophobic bases within a single strand to stack one above the other with the planes of their rings parallel and (2) the hydrogen bonds between the strands, which hold the two strands together. The strength of a hydrogen bond, which is a weak bond in any case, is affected by the angle made by the two component groups. Thus hydrogen bonding in a polynucleotide is weak unless the bases are stacked, because the stacking provides the orientation that is necessary for a large number of hydrogen bonds to form simultaneously. At a particular temperature, bombardment of the molecule by solvent molecules tends to break the hydrogen bonds and to alter the relative orientation of the bases. This is very difficult though because, in order to break one hydrogen bond, the adjacent bonds have to be strained in order to tip the plane of one base with respect to an adjacent base; the first would have to tip also with respect to its other neighbor. Thus, there is an enormous stabilization resulting from the stacking so that the huge DNA molecule typically undergoes a helix-coil transition at temperatures 30–40°C above the values usually encountered for proteins. Note also that the base pairing causes the bases to project inward from the sugar-phosphate backbone. This encourages the bases to stack so that hydrogen bonding and base stacking are synergistic.

The hydrogen bonds that are most susceptible to disruption are those at the ends of a DNA molecule because the terminal base pair is stabilized only by one pair of stacked bases. Thus, as the temperature increases, the hydrogen bonds at the ends of the double helix are the first to break. This breakage destabilizes the next pair and denaturation proceeds progressively inward. In addition to the termini, sequences rich in adenine·thymine (A·T) base pairs also denature early in the transition. These base pairs have two hydrogen bonds and have intrinsically lower stability than guanine·cytosine (G·C) pairs, which have three hydrogen bonds. An internal region at which base pairs are disrupted is called a *bubble*. Further breakage of hydrogen bonds occurs preferentially at the ends of bubbles with high A·T content because these regions are nearly equivalent to the ends of the molecule (Figure 1-14). Thus, denaturation of DNA proceeds both by enlargement of bubbles and progressive opening of the helix from the ends.

Helix-coil transitions are frequently described by the temperature at which the transition is 50% complete.* This temperature is designated T_m or the *melting temperature* (Figure 1-15). The value of T_m depends on the means of detecting the transition. For example, T_m for the change of absorbance of a DNA solution need not be the same as the value of T_m for the change of viscosity of the solution, as shown in the figure. The melting temperature is sometimes called the midpoint of the transition; however, it is important to realize that it is not the temperature midway between the temperatures at which the transition stops and starts. In fact, the melting curves are not usually symmetric on the temperature axis about the value of T_m.

In the discussion so far, the helix-coil transition has been considered as a conversion from an ordered to a disordered state. The opposite

* Other parameters that might be varied, such as the pH, the ionic strength, and the concentration of a denaturant, are sometimes used but the temperature is by far the most common.

point of view—that of the coil-to-helix transition can also be taken. In this way, it can be seen that cooperativity strengthens the ordered state and provides a means for achieving it. Consider a protein in which many amino acids are paired. There may be alternate modes of pairing and a molecule with any of these pairs might be quite stable; however, it might also lack biological activity. One of the combinations of pairs will be thermodynamically the most stable one. If it is assumed that the molecule with the correct configuration has strong cooperative interactions and that those with the incorrect configuration do not, the role of cooperativity in the creation of the correct configuration becomes clear. This can be seen by examining the transition of a newly synthesized (unordered) protein molecule to the correct configuration. Since all of

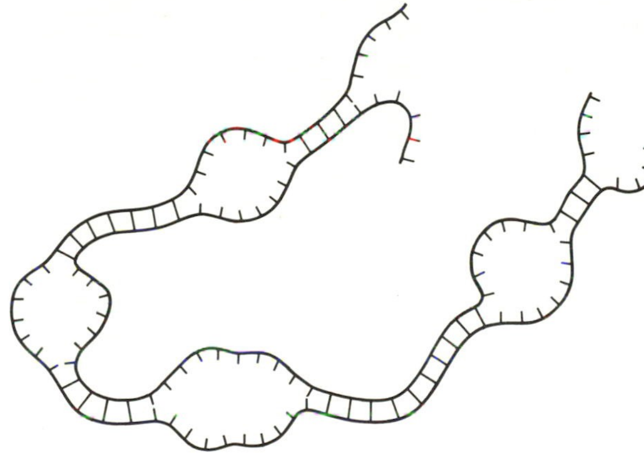

Figure 1-14
A partially denatured DNA molecule showing bubbles and frayed ends.

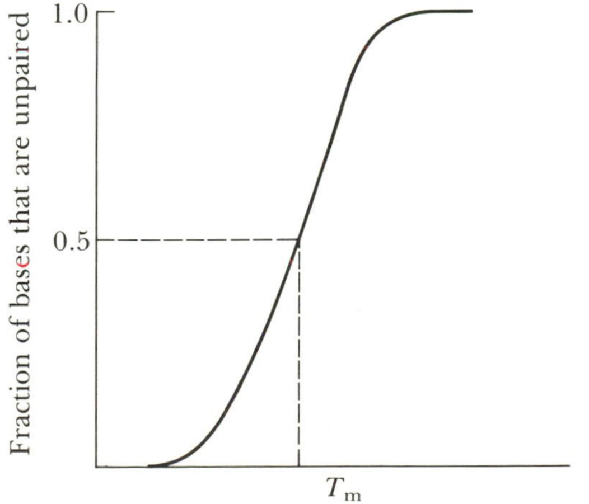

Figure 1-15
A helix-coil transition for DNA showing the melting temperature T_m.

the amino acid side-chain interactions are weak and easily disrupted by thermal vibrations, the first amino acid pair that forms may have little stability and break spontaneously. Another pair may form a moment later and this too will break. However, if the interactions that maintain the correct structure are cooperative, then even though the making and breaking of pairs just described will still occur, there will be a different outcome of the formation of the first correct pair. This is because this pair, once formed, increases the probability of the formation of a second correct pair; furthermore, once the second correct pair has formed, the first correct one will be stabilized. In this way, the probability of forming successive correct pairs increases and the stability of all pairs becomes greater. This effect of cooperativity is even easier to see in the formation of double-stranded DNA from two complementary single strands. That is, a base pair is stable only if the adjacent pair can hydrogen-bond and the two pairs are stable only if the next-to-nearest neighbors can bond.

The examples just given demonstrate two important reasons why biological macromolecules have evolved with cooperative interactions. First, the stability of an ordered structure, in which there are cooperative interactions, is greater in the sense that it is more resistant to thermal disruption. If stability is of utmost importance, then why, one might ask, did the molecules evolve with weak bonds rather than strong bonds such as covalent bonds? The answer is the second reason for cooperativity; that is, in carrying out its biological function and in adapting to the varied conditions of nature, a molecule frequently must change its configuration. This would not be a simple matter if the structure were stabilized entirely by covalent bonds. To be changeable, weak bonds are necessary but in order to provide strength, several of the weak bonds must act together.

In the following an example is given of a helix-coil transition that occurs in nature by reducing the effectiveness of the cooperative interactions that stabilize the native structure.

Example 1-D ☐ Effect of helix-destabilizing proteins on the DNA helix.

An essential step in the replication of DNA molecules within living cells is the separation of the two polynucleotide strands. This is accomplished in part by the helix-destabilizing proteins. At any temperature, there will be temporary breakage of hydrogen bonds in small regions owing to collisions with solvent molecules. This phenomenon is called *breathing* of DNA. Sometimes a helix-destabilizing molecule, which binds tightly to several adjacent bases, enters the transient bubble, binds to one strand of the DNA molecule, and thereby prevents reformation of the hydrogen bonds. This distortion of the helix weakens the stacking interaction (the source of the cooperative stabilization) so that the bubble fluctuates in size. The helix-destabilizing protein that is bound undergoes a small change in shape that enables it to bind also to a second identical protein molecule. Thus another helix-destabilizing molecule, which is by itself capable of binding to the bases in the fluctuating bubble, now has a greater probability of binding because it can bind both to the bases and to the bound protein molecule. That is, the binding of a helix-destabilizing protein is also cooperative. Furthermore the second helix-destabilizing molecule binds *adjacent to the bound molecule.* This further enlarges the bubble by breaking down stacking and, if there are enough bound molecules, the process continues until the two polynucleotide strands are totally separate.

The principal reason for studying a helix-coil transition is that it provides information about the structure of a particular macromolecule. The rules used in these studies are the following.

Rule I. If a reagent lowers the T_m, it must reduce or eliminate some factor that stabilizes the native form. If T_m is increased, the effectiveness of that factor is enhanced.

Rule II. If a change in pH produces a helix-coil transition, the pH at which the transition is 50% complete is near the pK of a component of the molecule that is required for stability of the native structure. The interaction invariably alters the pK of the group.

Rule IIIA. If T_m increases with increasing ionic strength, the molecule contains groups of like charge that repel one another and reduce stability whereas if T_m decreases, there are unlike charges that attract and enhance stability.

Rule IIIB. If T_m increases with increasing ionic strength and it can be shown that there are no groups having like charge, then hydrophobic interactions are a major feature of the structure. This is because nonpolar residues are less keen to go into a salt solution than they are into water, so that salt keeps the nonpolar residues together and requires a higher temperature to break down structure.

Rule IIIA is the usual explanation for the salt effect on the denaturation of DNA whereas rule IIIB is very often the case for proteins.

The use of these rules can be seen in a few examples.

Example 1-E

☐ Effect of sodium trifluoroacetate ($NaCF_3COO$) on the stability of DNA.

Figure 1-16 shows helix-coil transitions for DNA in 0.01 M, 0.1 M, and 4 M $NaCF_3COO$. The values of T_m are 75°C, 82°C, and 66°C respectively. This is to be compared to similar curves for 0.01 M, 0.1 M and 4 M NaCl, for which the values of T_m are 72°C, 85°C, and 99°C respectively. For both salts the value of T_m for 0.1 M is greater than that for 0.01 M. This is because, at low ionic strength, the negatively charged phosphate groups of the two strands repel one another and weaken the interaction. These charges are partly neutralized in the 0.1 M solutions. (This is an example of rule III.) In the 4 M NaCl the T_m is greater than at 0.1 M because the electrostatic repulsive force is incompletely eliminated at 0.1 M. However, in 4 M $NaCF_3COO$ the T_m drops enormously indicating that between the concentrations of 0.1 M and 4 M a stabilizing force has been eliminated. $NaCF_3COO$ is one of a class of reagents that increases the solubility of weakly polar organic molecules in water and thereby reduces hydrophobic interactions. This decreases the stacking tendency of the bases and thereby reduces the thermal stability of the DNA.

Other examples of helix-coil transitions can be found throughout this book (e.g., Figures 14-15, 14-20, 16-11, 17-12, and 17-13.)

Renaturation (Figure 1-8) refers to the reformation of the native structure from a denatured form. Actually the structure of a renatured molecule depends on the criterion used to assay restoration of the native configuration. For DNA, it specifically means reformation of a double-stranded helix from separated polynucleotide strands. For proteins, the

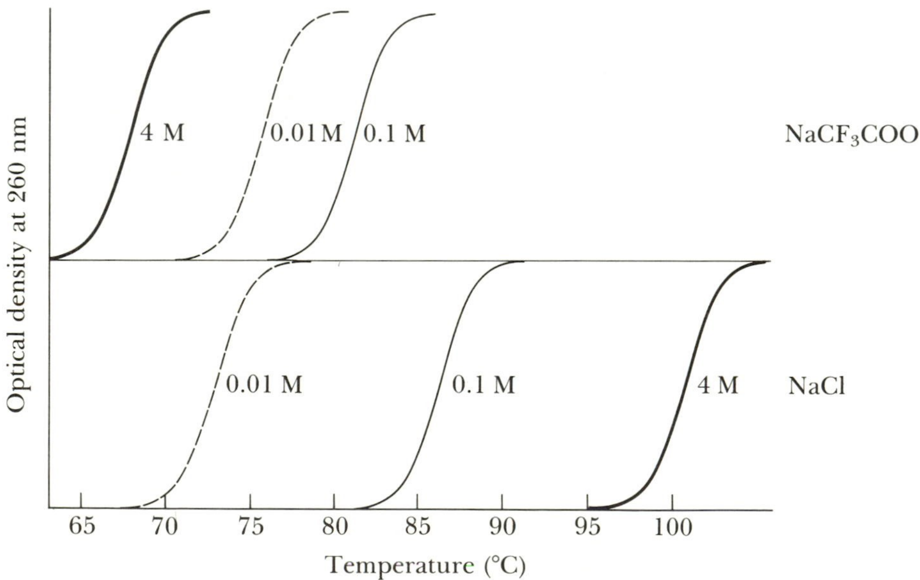

Figure 1-16
Melting curves for a DNA sample dissolved in various concentrations of sodium trifluoroacetate ($NaCF_3COO$) and NaCl. T_m increases as the NaCl concentration increases yet decreases above a concentration of $NaCF_3COO$ of 1.5 M.

term is less definite. Renaturation will be discussed in greater detail in the succeeding section.

If a structure consists of subunits, separation of the subunits is occasionally called denaturation but more often *dissociation* (Figure 1-17). The latter term is used, for example, to describe a protein consisting of

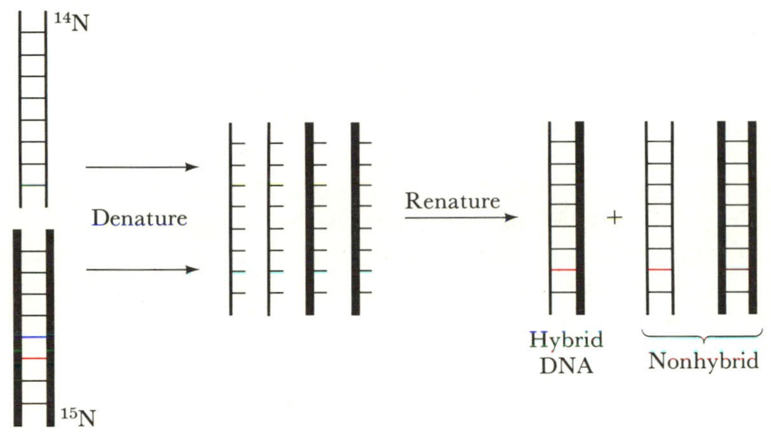

A. Formation of hybrid DNA

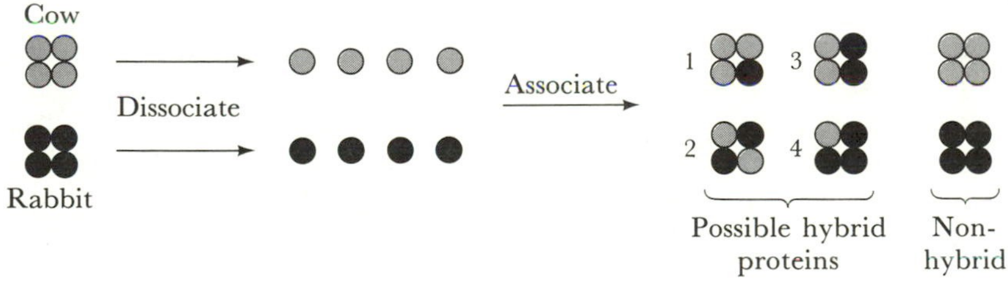

B. Formation of hybrid proteins

Figure 1-17
Hybridization of nucleic acids and proteins. (A) Two DNA molecules labeled with either ^{14}N or ^{15}N are denatured to produce single strands, which are then renatured. In some cases, strands having the same label renature to form the original DNA. However, hybrid DNA consisting of one strand labeled with ^{14}N and the other labeled with ^{15}N can also form. If equal amounts of ^{14}N- and ^{15}N-labeled single strands are mixed, then by random association the renatured mixture will be 25% [^{14}N]DNA, 25% [^{15}N]DNA, and 50% (hybrid) [^{14}N$\cdot$^{15}N] DNA. (B) Formation of hybrid proteins by dissociation of subunits and reassociation. Two proteins of the same type, one from a cow and the other from a rabbit, can reassociate to form four possible hybrid types, which differ by the number and arrangement of subunits; types 2 and 3 have the same number of each subunit, but they are arranged differently.

several polypeptide chains that has been reduced to a mixture of the separated chains. In this case, the individual subunits are not necessarily denatured by the dissociation process. On the other hand, separation of two strands of DNA is always called denaturation.

The reconstruction of a subunit-containing structure from subunits derived from *different* larger units is called *hybridization.* The mechanism may be renaturation or reassociation. For example, if DNA whose strands are labeled with the isotope nitrogen-14 is denatured and mixed with single-stranded (denatured) DNA labeled with nitrogen-15 and then subjected to renaturation, double-stranded DNA is formed, some of which consists of one ^{14}N and one ^{15}N strand (Figure 1-17). Such a $[^{14}N \cdot {}^{15}N]$DNA is an example of *hybrid DNA.* Similarly, a double-stranded polynucleotide containing one DNA and one RNA strand is another kind of hybrid—a DNA·RNA hybrid. Proteins can also form hybrids. For example, if a protein that contains four subunits is obtained from a cow, dissociated, and mixed with the dissociated form of a similar rabbit protein, the subunits may reassociate in such a way that hybrid structures are formed, as shown in Figure 1-17. Hybridization is discussed in greater detail in Chapter 19.

The Coil-Helix Transition—Renaturation

Consider a sample of double-stranded DNA dissolved in a solution of low ionic strength; for example, 0.01 M NaCl. In this solution most of the phosphate groups (which are totally dissociated) of the deoxyribose-phosphate chain are not shielded by a Na^+ counterion and hence the full negative charge is available for electrostatic interaction. Thus the phosphate groups repel one another and, as a result, a DNA molecule is less stable in 0.01 M NaCl than in a solution in which almost all of the phosphates are neutralized, for example, 0.5 M NaCl. This causes the T_m to be lower in 0.01 M NaCl than in 0.5 M NaCl. If a DNA sample in 0.01 M NaCl is first heated to the temperature at which strand separation occurs and then returned to room temperature, the mutual repulsion of the negatively charged phosphates keeps the strands separate.

If the DNA is in 0.5 M NaCl, the situation is quite different. Once a sample that contains strands that have been separated by heating is cooled, there is no charge repulsion so that hydrogen bonds can reform both within a single strand (in*tra*strand hydrogen bonding) and between two different strands (in*ter*strand hydrogen bonding). However, in a very long and complex molecule containing tens of thousands of bases, it is not likely that hydrogen-bond formation will result in reconstitution of the original double-stranded helix. This is because there are many short complementary base sequences that can form base pairs while the adjacent non-complementary bases remain unpaired. This is called *random hydrogen-bond formation.* The important point about these paired regions is that they are short and often contain intervening unpaired bases. Hence, these short double-stranded tracts will not be very stable and they will be disrupted at temperatures much lower than T_m. This has the following consequence. If a sample of DNA that contains either random intermolecular or intramolecular hydrogen bonds is reheated to a few degrees below T_m, all of these weakly paired regions will break and the sample will consist entirely of separated single strands. However, at a temperature below the T_m, the double-stranded state is more stable than complementary single strands. Therefore, as long as no activation energy is required, the single strands ought to form double-stranded DNA again. This occurs in the following way. Two strands collide and form randomly hydrogen-bonded base pairs; because they are not stable at the renaturation temperature, these paired bases rapidly unpair. However, sometimes a collision leads to base pair formation at such a position in the molecule that each adjacent pair of bases can also form hydrogen bonds, that is, at what might be called the "correct" position. In this case, a cooperative interaction occurs and the entire double-stranded molecule is "zipped up" again. Note that renaturation does not occur at room temperature because the process has an activation energy, which is simply the energy required to break the random hydrogen bonds. A few degrees below T_m, there are no randomly hydrogen-bonded base pairs so that there is no activation energy and renaturation can occur.

There are three requirements for renaturation to occur. (1) There must be complementary single strands in the sample. (2) The ionic strength of the solution of single-stranded DNA must be fairly high so that the bases can approach one another; operationally, this means >0.2 M. (3) The DNA concentration must be high enough for intermolecular collisions to occur at a reasonable frequency. The first two requirements determine whether renaturation *can* occur; the third requirement only affects the rate. Further discussion of the uses of renaturation is presented in Chapter 19.

Linear and Circular Polynucleotide Molecules

Some single-stranded or double-stranded DNA molecules (e.g., the single-stranded DNA from *E. coli* phage ϕX174 or the double-stranded DNA from *Pseudomonas* phage PM-2) have no terminal nucleotides because each nucleotide is joined covalently by means of a phosphodiester bond to the adjacent one. Such a molecule is *circular*. If there are terminal nucleotides, the molecule is *linear* (Figure 1-18). If both polynucleotide strands of double-stranded DNA are circular, the DNA is a *covalent* or *closed circle*; if there are one or more single-strand interruptions, the circle is *open* or *nicked*. Some linear DNAs (e.g., from *E. coli* phage λ) contain short, complementary single strands at each end of the molecule (i.e., strands whose base sequences are such that the strands can be joined by the standard adenine·thymine and guanine·

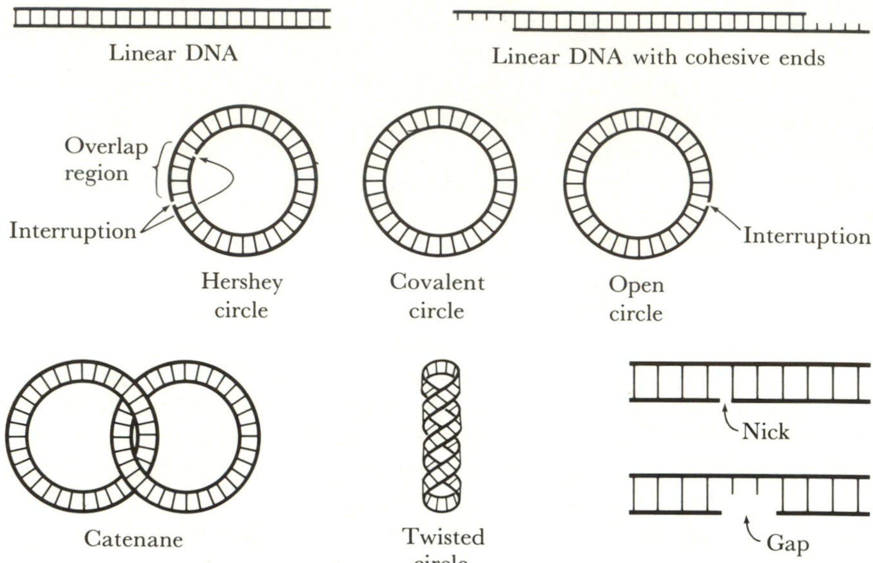

Figure 1-18
Various conformations of DNA.

cytosine base pairs to form double-stranded DNA). These short strands are called *cohesive ends* or *sticky ends.* When these cohesive ends join, the resulting structure is a *Hershey circle.* If a covalent circle is twisted (imagine taking a linear double-stranded DNA and twisting it before joining the ends but making sure that each strand is continuous), it is called (interchangeably) a *twisted circle*, a *superhelix*, or a *supercoil.* The number of twists per unit of molecular weight is the *superhelix density.* Note that if the hydrogen bonds of a superhelix are broken, the individual circular strands cannot be physically separated; however, if one of the circles contains a physical interruption, the double-stranded circle can separate into two single-stranded molecules, one linear and the other circular.

In a double-stranded polynucleotide, a broken phosphodiester bond is called a *nick.* If one or more nucleotides are missing, the result is a *gap.*

Supercoiling of DNA is very widespread in nature and is essential for many stages of DNA replication, RNA synthesis, and DNA recombination. Its role in each of these processes is probably due, in part, to the fact that a supercoiled DNA molecule has extensive regions of unpaired bases. To see why this is so, the means of generating a supercoiled molecule from a non-supercoiled circle must be examined.

Consider a nicked circle that is about to be converted to a covalent circle. All bases are paired and the number of base pairs per turn of the helix is the usual number, namely ten. Such a molecule is *relaxed* and can lie on a flat surface. Prior to sealing the nick, the nicked strand could be rotated about the continuous strand. There are two possible directions of rotation; namely, to unwind the helix (*underwinding*) and to wind the helix more tightly (*overwinding*). In either case, if the rotated strand is released before sealing, it would rotate in a way such that the initial state of winding would be restored. Thus an underwound or overwound helix is strained. If sealing occurs before releasing the winding, the resulting covalent circle will also be under strain. This distortion of the standard structure can be accommodated in two ways (Figure 1-19).

1. The total number of turns in the strained helix is readjusted to the number for the unstrained helix. This is accomplished by taking up the excess or deficient turns into twisting of the helix itself; that is, the molecule becomes supercoiled. The sense of the supercoiling of the underwound molecule is opposite to that of the overwound molecule; an underwound supercoil is said to have *negative superhelicity* and an overwound molecule has *positive superhelicity*.

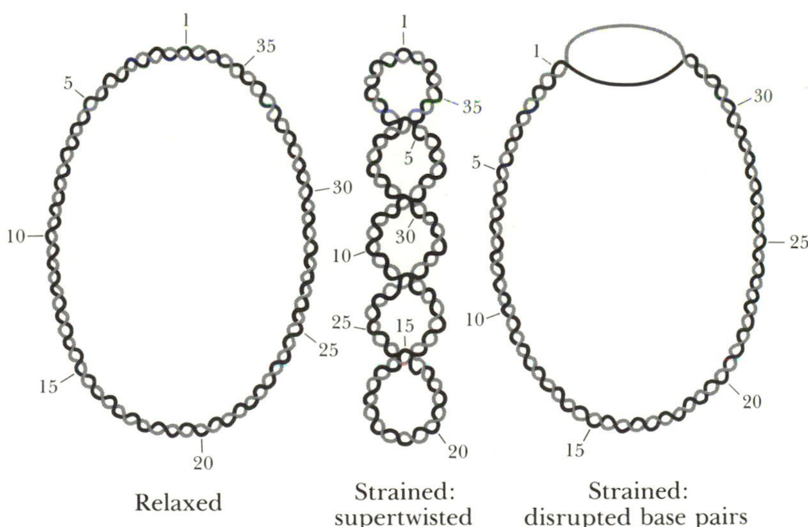

Figure 1-19
Three forms of covalent circles. The left panel shows the non-supercoiled form. In the center is the underwound, twisted circle. At the right is an underwound circle that is not twisted but shows an extensive single-stranded region. The numbers indicate complete turns of the helix. [From *DNA Replication* by A. Kornberg. W. H. Freeman and Co. Copyright © 1980. Adapted from a diagram supplied by Dr. M. Gellert.]

2. If the molecule is underwound, the strain can be moved to a single section of the molecule, which would then be single-stranded. This allows the remainder of the molecule to have the normal number of base pairs per helical turn. Alternatively there could be several single-stranded regions in different parts of the molecule.

Since the DNA molecule is a dynamic structure, it would be expected that the molecule would alternate between these two states passing through many intermediate stages.

All naturally occurring supercoils are negatively twisted with a superhelix density of approximately 0.05 twist per 10 base pairs.

Relation Between Structure and Function

The main reason for determining the structure of a macromolecule is to understand its biological function. There are numerous examples in which the relation between structure and function is immediately obvious. For instance, the so-called structural proteins, namely, those that are responsible for maintaining particular cellular structures, are usually fibrous and not very flexible. Examples are myosin, one of the two main proteins in muscle, and collagen, the protein of which tendon is composed.

Collagen is an especially good example of structure suited to function. The job of collagen in tendon is to provide tensile strength in a long fiber. To begin with, the molecule must be extended rather than spherical; this is accomplished in the following way. Collagen contains a large amount of the amino acid proline. This molecule does not have a normal α-amino group (it is really an *imino* acid) and fails to form a typical peptide bond. The bond it forms is much less flexible and thereby helps to maintain the protein molecule in a form that is more extended than a random coil. The extended single strands are held together by interstrand hydrogen bonds; the result, a *triple-stranded helix*, is the first stage in generating tensile strength. Additional strength is obtained by side-to-side aggregation of the triple helices. Collagen is able to aggregate side-by-side because it has numerous positive and negative amino acids arranged along the chain. These charged groups interact by ionic bonds. Having so many charges of both signs on a single chain, a collagen strand would have a tendency to fold back on itself, since this in*tra*-molecular interaction would occur more easily than in*ter*molecular aggregation. This is avoided because the molecule is a triple helix and hence so rigid that it does not bend. In order to generate a long fiber (as is needed in tendon), the side-to-side aggregates must also be able to join longitudinally. This is accomplished not by means of an end-to-end interaction but staggering the triple helices by one-fourth the length of the molecule (Figure 1-20). The final strengthening of the fiber is obtained by chemical modification of proline and lysine to yield hydroxyproline and hydroxylysine. These hydroxylated amino acids can cross-link from

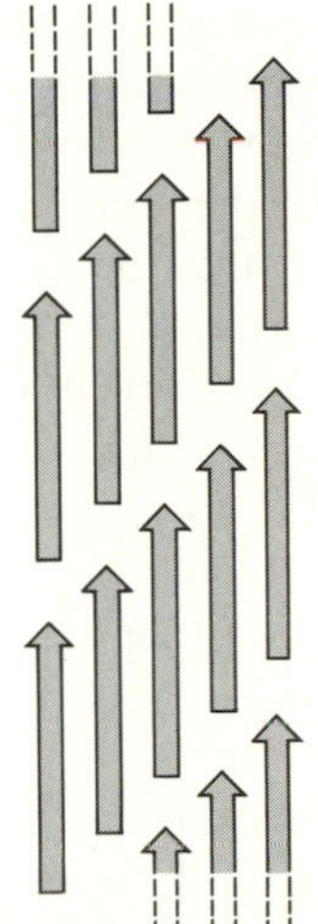

Figure 1-20
A model of collagen showing the triple-stranded helices in the quarter-stagger array.

strand to strand. Thus, various structural features of the molecule combine to form a complex macromolecular array that has a great deal of tensile strength.

Certain features of the structure of enzymes are also in accord with their catalytic role. For example, the specificity of binding of a substrate molecule by an enzyme requires that its binding site have a well-defined and constant shape; that is, the binding site of the enzyme must be relatively stiff. On the other hand, the regulation of activity of an enzyme is most easily accomplished by altering the shape of the binding site; therefore, the enzyme must not be completely rigid. In accord with these tenets, a typical enzyme is a highly folded, globular protein that contains a large number of hydrogen-bonded and interacting amino acids, providing a somewhat rigid and compact structure. Some of the amino acid interactions with a protein are usually interdependent so that when a regulatory molecule binds to a site in the enzyme molecule, the effect can propagate and alter the shape of the substrate-binding site. It must be pointed out that not all enzymes are regulated in this way; many regulated enzymes make use of at least two separate polypeptide chains, one with an active site and one with a regulatory site.

Determination of the Structure of Macromolecules

A variety of physical techniques is required to determine the structure of a macromolecule. Before study can begin, the molecule must be isolated and purified; the major techniques for accomplishing this are chromatography, electrophoresis, and sedimentation, which are described in Chapters 8, 9, and 11, respectively. Some information can also be gained from impure samples but ultimately purification is required. The determination of the chemical structure mainly requires chemical techniques but often chromatographic and electrophoretic analyses play an important role. For example, the amino acid composition of a protein is determined by acid hydrolysis followed by chromatographic separation of the amino acids that are released. Many physical properties, such as the dimensions of the molecule, the molecular weight, the location of individual monomers, and the three-dimensional configuration, are determined by a variety of optical and hydrodynamic measurements. It is also possible to determine certain chemical information, such as whether a particular amino acid is on the surface of a highly folded protein or buried deep within the molecule. This can be done using radioactive labeling, fluorescence spectroscopy, or nuclear magnetic resonance, to name just a few techniques. Identification of the amino acids in the binding site of an enzyme is a major goal of these types of studies.

If the complete three-dimensional structure of any macromolecule is to be worked out, the method of X-ray diffraction is needed. Why then is X-ray diffraction analysis not always done first? The main reason is that this technique, although very powerful, is very difficult, expensive, and extremely time-consuming. Furthermore, the task of the X-ray diffractionist is simplified if many structural features of the molecule are already known from prior studies with the other techniques listed above. For example, interpretation of the structure is nearly impossible without the amino acid sequence and mechanistic models of enzyme action require prior identification of the amino acids in the active site.

SELECTED REFERENCES

Anfinsen, C. B. 1973. "Principles That Govern the Folding of Protein Chains." *Science (Wash. DC)* 181:223–230.

Bailey, J. L. 1967. *Techniques of Protein Chemistry*. American Elsevier.

Bloomfield, V. A., D. M. Crothers, and I. Tinoco. 1974. *Physical Chemistry of Nucleic Acids*. Harper & Row.

Cantoni, G. L., and D. R. Davies, eds. 1971. *Procedures in Nucleic Acid Research*, vol. 2. Harper & Row.

Cantor, C. R., and P. R. Schimmel. 1980. *Biophysical Chemistry. Part I: The Conformation of Biological Macromolecules*. W. H. Freeman and Company.

Chou, P. Y., and G. D. Fasman. 1978. "Empirical Predictions of Protein Conformation." *Annu. Rev. Biochem.* 47:251–276.

Cohn, W. E., ed. 1976. *Progress in Nucleic Acid Research and Molecular Biology*. Academic Press. Two volumes per year containing a fine set of review articles.

Cold Spring Harbor Laboratory. 1972. "Structure and Function of Proteins at the Three-Dimensional Level." *Cold Spring Harbor Symp. Quant. Biol.* 36.

Dickerson, R. E., and I. Geis. 1969. *The Structure and Action of Proteins*. Harper & Row. An excellent book for the beginner.

Fasman, G. D., ed. 1976. *CRC Handbook of Biochemistry and Molecular Biology., Section A: Proteins. Section B: Nucleic Acids*. CRC Press.

Flory, P. 1953. *Principles of Polymer Chemistry*. Cornell University Press. A classic theoretical book.

Gutfreund, H., ed. 1974. *Chemistry of Macromolecules*, vol. 1. Butterworths. Contains many review articles.

Haschemeyer, R. H., and A. E. V. Haschemeyer. 1973. *Proteins: A Guide to Study by Physical and Chemical Methods.* Wiley.

Mandelkern, L. 1972. *An Introduction to Macromolecules.* Springer-Verlag.

Neurath, H., and R. L. Hill, eds. 1975. *The Proteins.* Academic Press. Contains numerous articles on many aspects of protein structure.

Schultz, G. E., and R. H. Schirmer. 1979. *Principles of Protein Structure*, Springer-Verlag.

Tanford, C. 1961. *Physical Chemistry of Macromolecules.* Wiley. A great theoretical text.

Tanford, C. 1970. "Protein Denaturation." *Adv. Protein Chem.* 24:1–95.

Ts'O, P. O. P, ed. 1975. *Basic Principles of Nucleic Acid Chemistry.* Academic Press.

Von Hippel, P. H., and J. D. McGhee. 1972. "DNA-protein Interactions." *Annu. Rev. Biochem.* 41:231–300.

Wold, F. 1971. *Macromolecules: Structure and Function.* Prentice-Hall.

PROBLEMS

1–1. A polynucleotide, poly(dA-dT) is a single-stranded, alternating copolymer—that is, the sequence of bases along the chain is ATATATA . . . . What kind of structure would you expect this to assume in solution?

1–2. Poly(dA-dT) is an alternating copolymer having the base sequence ATATAT . . . as in problem 1–1. If [^{14}N]poly(dA-dT) is mixed with [^{15}N]poly(dA-dT) and subjected to renaturing conditions, it might be expected that the double-stranded DNA that results would have the following composition: 25% ^{14}N·^{15}N, 25% ^{15}N·^{15}N, and 50% ^{14}N·^{15}N. However, this is not the case; what is observed is: 46% ^{14}N·^{14}N, 48% ^{15}N·^{15}N, and 4% ^{14}N·^{15}N. Explain this observation.

1–3. A protein contains six cysteines. If all were engaged in disulfide bonds and if all possible pairs of cysteines could be joined, how many different protein structures would be possible?

1–4. Hershey circles are formed by heating a linear DNA that has single-stranded cohesive ends to a temperature that favors end joining. At very low DNA concentration, only Hershey circles are formed. What structures might be expected at very high DNA concentrations?

1–5. Which polypeptide would you expect to be more compact at neutral pH, polylysine or polyvaline? Which would show the greatest dependence of shape on pH? Why?

1–6. Adenine, guanine, cytosine, and thymine have low solubility in water, whereas deoxynucleotides are much more soluble. The solubility of the bases increases when solvents less polar than water are added to the water—for example, methanol or ethylene glycol. What would you expect to happen to DNA if it were placed in 100% ethylene glycol?

1–7. Suppose that you hybridize two proteins (as in Figure 1-17B), each consisting of three identical subunits. How many different hybrid types are possible if the subunits are arranged linearly? If arranged in a triangle?

1–8. Suppose that the proteins in Problem 1-7 are hexamers, with the subunits producing a hexagonal array. How many hybrid types are possible?

1–9. If a protein contains four identical subunits, what kinds of forces might be holding them together?

1–10. If a single strand of DNA, whose base sequence is AGCTAACGCGA, were mixed with another strand whose sequence is TCGATTAGTCATGCGCT under conditions that favor hybridization, what type of molecule would result? Suppose that the second strand had the sequence TCGATTAGTA-CTGCGCT instead. Draw the resulting structure.

1–11. Calculate the fraction of (hybrid) [$^{14}N \cdot ^{15}N$]DNA produced by denaturation and renaturation of a mixture of one part [^{14}N]DNA and three parts [^{15}N]DNA. What fractions of the total remain (a) heavy and (b) light?

1–12. Write down the structure of the hexapeptide H_2N-glycine-tryptophan-proline-isoleucine-valine-methionine-COOH. How many peptide bonds are there?

1–13. A particular polypeptide chain folds in such a way that there is a cluster of 4 leucines, 3 isoleucines, 6 valines, and 3 phenylalanines at one small region on the surface of the protein. What would you guess to be the state of this polypeptide chain in aqueous solution?

1–14. The structure of most proteins is determined by the sum of many weak interactions. Covalent disulfide bonds between cysteines are also common. The biological activity of such proteins, for example, an enzyme, is usually lost if the disulfide bonds are broken or if reagents are added that eliminate hydrophobic interactions. Thus even in such a protein, hydrophobic interactions are important. What is the likely role of the disulfide bond?

1–15. For every 100 molecules in a protein sample, there are 53 molecules having a molecular weight of 42,000, 38 whose molecular weight is 38,000, and 9 whose molecular weight is 15,000. Calculate M_n and M_w.

1–16. A protein sample contains monomers and dimers. Thirty percent of the total weight of protein consists of monomers. What is the relative number of monomers and dimers?

1–17. What are the values of M_n and M_w for a mixture of molecules consisting of 6% (by weight) monomers, 90% dimers, and 4% trimers, if the molecular weight of the monomer is 25,000?

1–18. A sample of DNA molecules, whose molecular weight is 22×10^6, contains some molecules whose molecular weight is 10^6; these smaller molecules amount to 2% by weight of the sample. What are the values of M_n and M_w for the sample?

1–19. A protein molecule has a molecular weight of 24,340 and $\bar{v} = 0.72 \text{ cm}^3/\text{g}$. Sedimentation studies suggest that the end-to-end distance is 36 Å. What can you say about the shape of the molecule?

1–20. A protein molecule has a molecular weight of 37,650 and $\bar{v} = 0.70 \text{ cm}^3/\text{g}$. The observed radius of gyration is 64.5 Å. What can you say about the shape of the molecule?

1–21. What is the radius of gyration of four small spheres placed at the corners of a square 10 cm on a side? Two spheres weigh 5 g and are at opposite corners of the square. The other two weigh 1 g each.

1–22. How would the end-to-end distance of (a) a random coil (b) a rodlike molecule vary with increasing temperature?

1–23. A sample of DNA is heated to a temperature sufficiently high for the two strands to separate. It is then cooled but the strands remain separate. Compare the end-to-end distance and the radius of gyration of the DNA before and after the heating-cooling cycle by stating which value is larger. Do this for two conditions, namely, 0.01 M NaCl and 1 M NaCl. In 0.01 M NaCl, the negatively charged phosphates in the sugar-phosphate chain repel one another. In 1 M NaCl, each phosphate is neutralized by a Na^+ ion.

1–24. Consider a molecule whose structure is maintained by two non-cooperative interactions. What determines the temperature at which the molecule undergoes a transition to the random-coil configuration?

1–25. Two hypothetical protein molecules consist of 96 glycines, 2 lysines, and 2 aspartates. In molecule A, the lysines are adjacent and the aspartates are adjacent; each lysine is hydrogen-bonded to one aspartate. In molecule B, the lysines are in positions 16 and 80 in the polypeptide chain and the aspartates are in positions 35 and 65. Hydrogen bonds exist between lysine-16 and aspartate-35 and between lysine-80 and aspartate-65. As the temperature is increased, both A and B undergo a helix-coil transition. For which molecule is the transition cooperative? Explain.

1–26. Some DNA molecules contain mirror-image sequences called palindromes—for example

. . . ———————————————— . . .
A B C D D′ C′ B′ A′
A′ B′ C′ D′ D C B A
. . . ———————————————— . . .

in which a prime denotes a complementary base pair. It has been proposed that such a molecule might also have the "looped-out" structure.

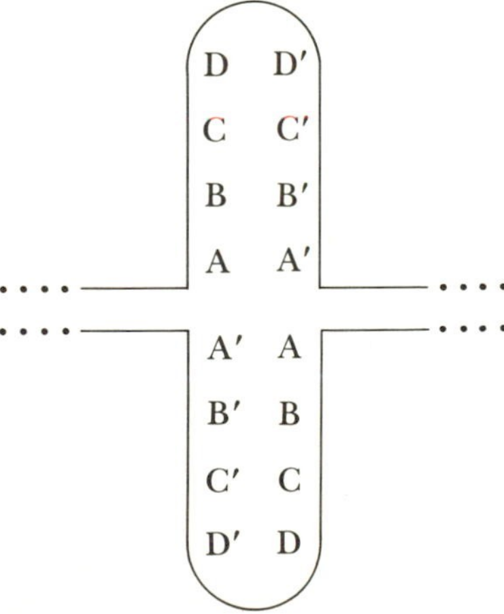

The sequences are typically 10–30 bases long. Which structure do you think is more stable and why?

One

DIRECT OBSERVATION

2

Light Microscopy

Too often the biochemist or molecular biologist forgets that the chemical and physical systems and the molecules and macromolecules under study are located in cells. Not only is it worthwhile and satisfying to see the cells themselves, but also microscopic observation can yield quantitative information, especially with the important techniques of polarization, interference, and fluorescence microscopy. However, the microscope with its array of lenses, apertures, and screws can seem baffling. This chapter attempts to make it less so.

Simple Theory of Microscopy

The theory of image formation by a lens can be presented in terms of either geometric or physical optics. Geometric optics easily explains focus and aberrations; howevers, physical optics is necessary to understand why images are not perfectly sharp and how contrast is obtained.

The theory of geometric optics is presented in numerous texts (refer to the Selected References near the end of this chapter) and is given here only briefly. There are two rules of geometric optics from which all else follows: (1) light travels in a straight path and (2) the path bends (refracts) at an interface between two transparent media (Figure 2-1).

An ideal simple lens with two convex spherical surfaces, not necessarily having the same radii of curvature, has two focal points (Figure 2-2). If two lines representing light rays coming from a given point on an object

(O) are drawn such that one ray is parallel to the lens axis and the other passes through the focus (F), they will emerge from the other side of the lens in a direction such that the parallel one will pass through the second focus (F′) and the other will be parallel to the optic axis, as shown in Figure 2-2. The intersection of these rays defines the image at O′. This simple construction defines an object plane and an image plane. The distances indicated in the figure obey the relation $aa' = ff'$ and the magnification of the image is $-(f/a)$, in which the minus sign indicates that the image is inverted.

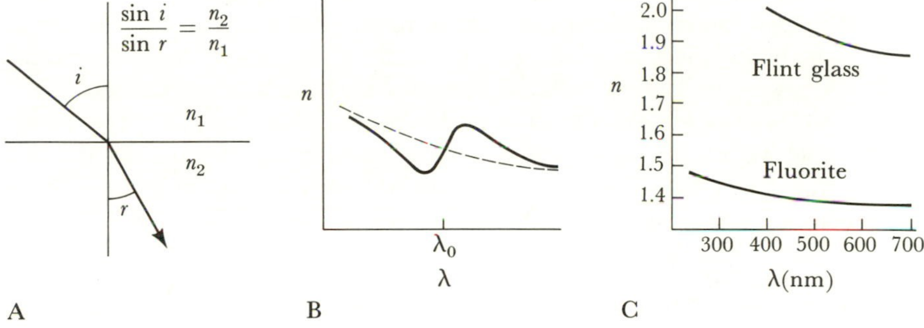

Figure 2-1
A. Refraction at a surface: n_1 and n_2 are the refractive indices of the medium on either side of the surface; *i* and *r* are the angles of incidence and refraction, respectively. The equation (Snell's Law) describes the relationship. As drawn, medium 1 has a higher index of refraction than medium 2; for example, media 1 and 2 might be air and glass respectively. B. Dependence of *n* on λ: the dashed line is the kind of curve found far from an absorption maximum; the solid curve is the relation near an absorption maximum, λ_0. C. *n* versus λ for two materials used in the construction of microscope lenses.

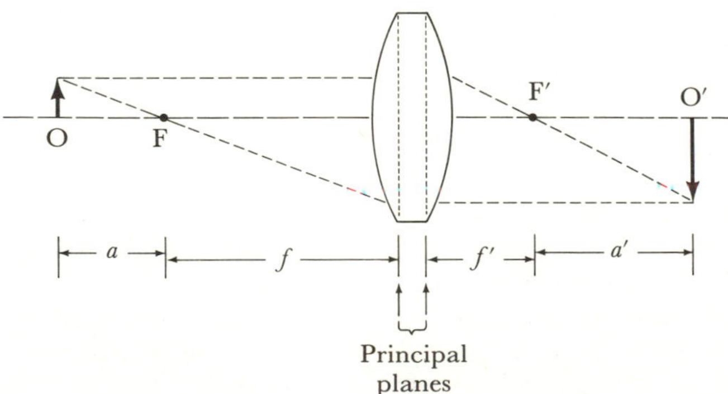

Figure 2-2
Image formation by a simple lens.

The simple construction just described is for an ideal lens. However, real lenses fail to bring all rays from a given point on an object to a unique focus; that is called *aberration*. The most easily understood aberration is *chromatic aberration* (Figure 2-3A), which results from the fact that, because the index of refraction of any substance depends on wavelength (Figure 2-1), the position of the focal points is wavelength-dependent. Hence, in white light, which consists of light of many wavelengths, images will be fuzzy, being the superposition of a large number of images, not all of which can simultaneously be in focus. Chromatic abberation is simple to correct. One way is by using a lens system consisting of several types of glass for which the relations between the index of refraction (n) and wavelength, (λ) balance to make n independent of λ. Such a lens system is called *anachromatic*. Another way is to use monochromatic light or, more practically, a filter to reduce the range of wavelengths in the light. It is good practice to use a filter even when a color-corrected lens system is in use because the color correction is not perfect.

Other aberrations exist even for monochromatic light. The major aberrations are called point-imaging because they result from the fact that all rays from a single point do not pass through the same image point. The principal point-imaging aberration is *spherical aberration*

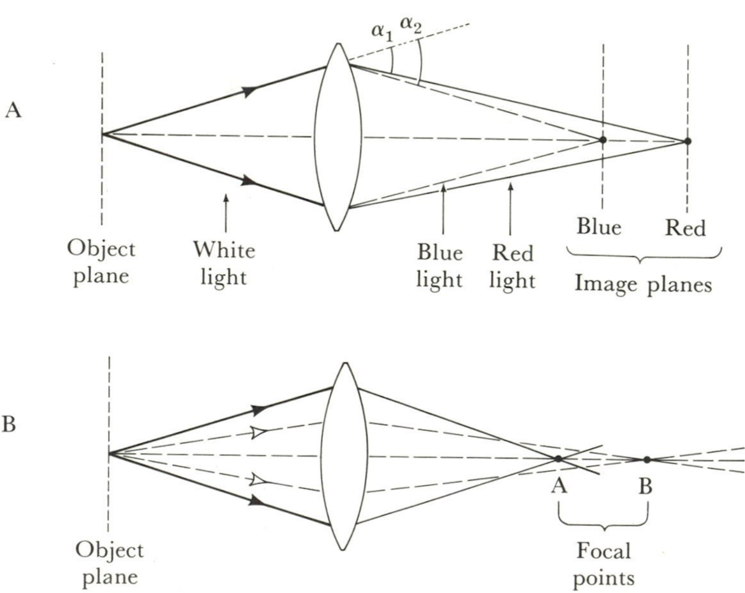

Figure 2-3
Two common aberrations of simple lenses: (A) chromatic aberration in which focal point depends on wavelength; (B) spherical aberration in which focal point depends on which part of the lens surface is producing the refraction.

(Figure. 2-3B), in which rays from a single point on the optical axis of the lens are refracted by different parts of the lens and therefore do not come to focus at the same point on the axis. A second important point-imaging aberration is called *coma* because it gives points a comet shape so that symmetry in the object is always lost in the image. By appropriate lens construction, these aberrations are easily corrected. A lens so corrected is an *aplanatic* lens. Another aberration is *astigmatism.* Except for observation at the highest resolution (as in electron microscopy where it is extremely important), this aberration is frequently left uncorrected but, if it is corrected, the lens is called an *anastigmat.* All commerically available microscopes are corrected for aberrations before being sold, but it is rare to find a lens that is corrected for both astigmatism and for chromatic aberration.

The theory of physical optics explains image formation and resolution, which will be important in a later discussion of phase-contrast microscopy. Within the framework of physical optics, light is not thought of as rays traveling in straight lines but as electromagnetic radiation that is diffracted (bent) at edges and apertures and can interfere constructively or destructively. Because a lens is in a sense an aperture, all light passing through the lens is diffracted. What this means is that an illuminated point in the object plane appears in the image plane as a circle of light surrounded by a series of bright concentric rings resulting from constructive interference. In microscopy this pattern is called an *Airy disc* (Figure 2-4A). It can be shown that the radius of the first dark ring surrounding the central disc is $0.61\ \lambda/n \sin U$, in which λ is the wavelength of the light, n is the index of refraction on the object side of the lens, and U is the angle made by the lens axis and a line drawn from an axial object point to the edge of the aperture (Figure 2-4B). Two object points that are close together will therefore appear as two tiny discs in the image plane and the resolution of these points (i.e., the ability to state unequivocally that there are two points) is determined not only by the separation of the image points defined by geometric optics, but by the size of the tiny discs. The precise point at which resolution is lost can-can be determined (Figure 2-4C); however, the convention that the resolution limit is $0.61\ \lambda/n \sin U$ was adopted long ago. Resolution is increased by decreasing the wavelength, increasing n, and increasing U; increasing U can be accomplished either by decreasing the distance from object to lens or by increasing the lens diameter. (We will return to these points later.) The quantity $n \sin U$ is called the *numerical aperture*, or NA, and is always printed on microscope objectives. That resolution increases as angle U increases can be explained in the following way. As illuminating light falls on an object, it is diffracted by the object itself. The different orders of diffraction will leave the points of the object at various angles, which increase with the order of diffraction. If each order of diffraction is thought of as part of the information supplied by the object for forming an image, then to maximize the information all the diffracted light must be gathered. This clearly means having the maximum collection angle possible between an object and the lens aperture—that is, a large value of U.

Parts of the Microscope

A basic microscope consists of only three components: a source of light of uniform brightness, an object holder, and a magnifying lens corrected for various aberrations. However, to have very high magnification, the focal length must be very small and the eye would have to be virtually on the opposite surface of the lens. To avoid the practical problems of making a lens of very short focal length with adequate corrections while

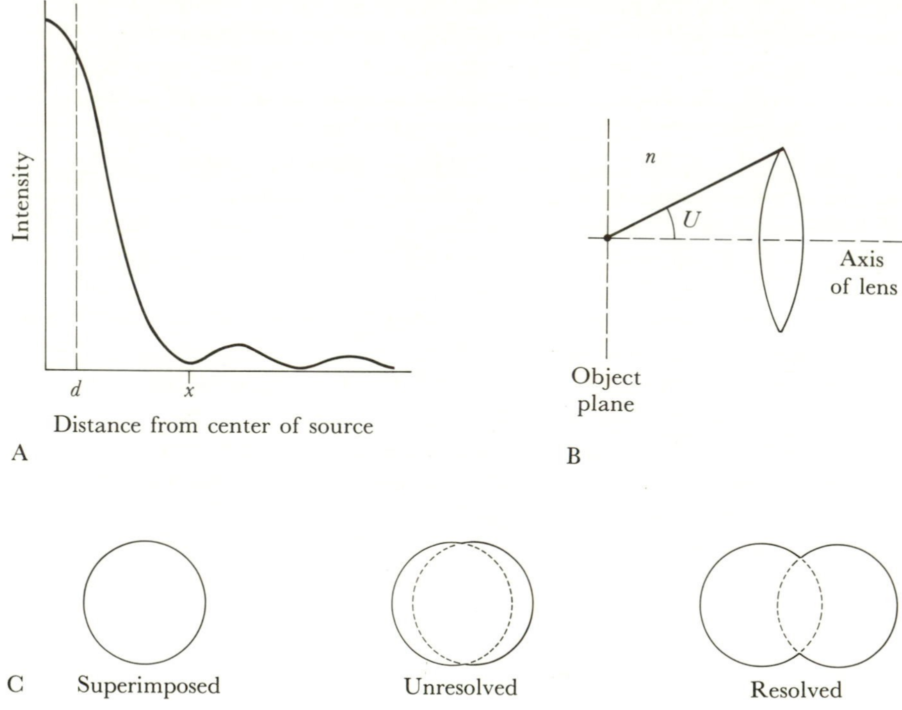

Figure 2-4
A. Plot of the intensity of light surrounding a light source of diameter d. The image is a bright central spot surrounded by bright rings and its configuration is called an Airy disc. As d becomes smaller, x increases. The size of the central region is principally responsible for the poor resolution between closely spaced point sources. As λ decreases, x decreases so that resolution is improved by decreasing the wavelength. B. Illustration of the angle U. The refractive index to the right of the object plane is n. C. Resolution of two point objects.

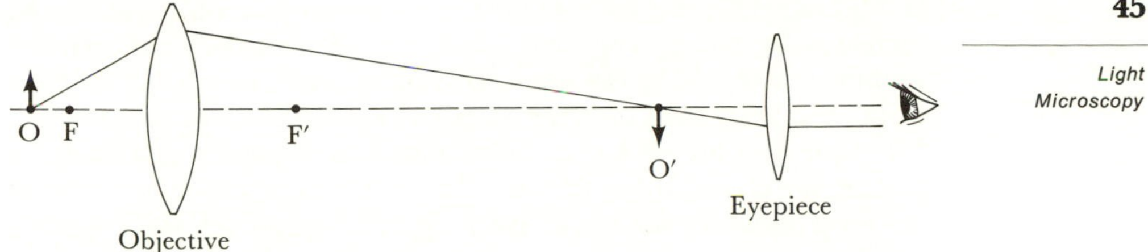

Figure 2-5
Optics of a compound microscope: F and F′ are the focal points of the objective.

providing a reasonable distance from lens to eye, a secondary lens system called an *eyepiece* is introduced. The basic lens system (the objective lens) plus the eyepiece constitute a *compound microscope*. The basic optics of a compound microscope are shown in Figure 2-5. The object is placed just past the focal point on the object side of the lens. Thus a magnified (10–100 ×) and inverted image is formed far from the focal point on the image side. The eyepiece then magnifies this image somewhat (5–20 ×) and focuses it onto the retina of the eye, which can be at a convenient working distance.

Objective Lenses

Microscope objectives are fairly well standardized with respect to magnification and NA. Table 2-1 gives some of the parameters of typical objectives and Figure 2-6 shows a diagram of two typical objectives, a 40 × "high dry" and a 100 × oil-immersion lens. Note that in general NA increases with magnification so that magnification is not "empty"

Table 2-1
Properties of Objective Lenses.

Magnification	NA	Focal length (mm)	Working distance (mm)*	Diameter of field (mm)
10	0.25	16	5.50	2.00
20	0.54	8	1.40	1.00
40	0.65	4	0.60	0.50
40	0.95	4	0.25	0.20
95†	1.32	2	0.10	0.05

* Distance from front surface of lens to sample.
† Oil-immersion lens.

(Empty magnification refers to the fact that magnification can always be increased by adding more lenses but, if resolution does not increase, nothing is gained.) The NA is normally increased by decreasing the focal length because the lens diameter decreases with increasing magnification. With a lens of high NA (e.g., 1.30), this is accomplished essentially by placing the object within the lens. This is the significance of the oil in so-called oil-immersion lenses. By using an oil whose refractive index is the same as that of the glass cover-slip above the object and that of the front lens element, all surfaces between object and lens are eliminated (Figure 2-6), except, of course, where the object touches the cover-slip. The front lens element of an oil-immersion objective is usually greater than a hemisphere so that with the oil the object can be thought of as being within a spherical lens. This results in gathering all light diffracted in the forward direction.

Many oil-immersion lenses have a residual chromatic aberration so that the magnification of the blue image is about twice that of the red. This is objectionable for certain critical work and can be eliminated by using a color-correcting *compensating eyepiece* (see next section) with a

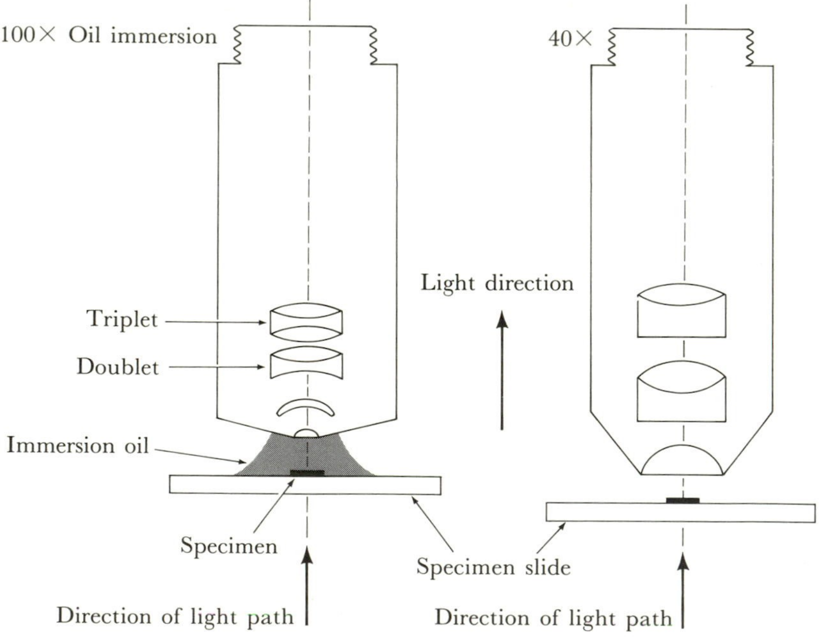

Figure 2-6
Typical objective lenses containing several lens elements for correcting various aberrations.

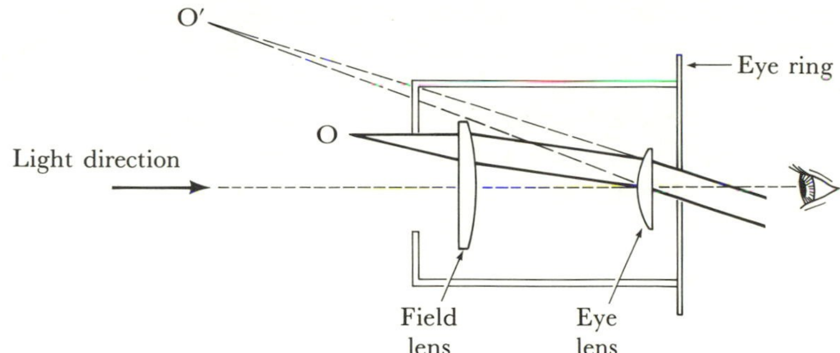

Figure 2-7
A Ramsden eyepiece. The image at O produced by the objective is magnified to produce the final image at O′.

partially corrected objective called an *apochromat*.* The best solution for all chromatic problems is to use monochromatic illumination.

A minor aberration called *curvature of field* (long straight lines appear curved) is usually left uncorrected because normally only a small part of the field is being observed. However, curvature of field can be objectionable in photomicrography because a large area is usually examined. *Flat-field* objectives are available to correct this aberration but they are expensive and their purchase is warranted only if a microscope is used frequently for photomicrography.

Eyepieces

The principal function of the eyepiece is to deliver the image of the objective to the eye. There are various kinds of eyepieces, including Ramsden (Figure 2-7), Huygens, Kellner, and compensating. The first three are interchangeable and differ only in the manner of inserting grids, pointers, and other reference points. The compensating eyepiece is designed as a further correction of chromatic aberration. Ramsden and Huygens eyepieces each consist of two lenses at opposite ends of a tube. In the compensating and Kellner eyepieces, one of the lens elements consists of a pair of lenses cemented together (a doublet). Eyepieces come in magnifications ranging from 5 to 20. However, it should always be remembered that increasing magnification without increasing resolution is of little

* Apochromats use CaF_2 (fluorite) instead of glass for some of the lens elements because it has greater n and can be made with a slightly higher NA (1.35). However, such lenses cost more, they have some fluorescence (so they cannot be used in fluorescence microscopes), and the fluorite elements scatter light so that special types of illumination may be required.

value and the range of magnification of eyepieces is mostly for convenience in viewing. Hence a 40× objective with a 5× eyepiece gives greater resolution of objects at high magnification than a 10× objective with a 20× eyepiece, because the 40× objective has a higher NA (the resolving power is determined solely by the objective); this is true even though both combinations give 200× magnification.

Condenser Lenses

For ordinary (bright-field) microscopy, the illumination of the object must satisfy two criteria: First, it must provide a beam of light whose divergence when leaving the object plane is at least as great as the angle U, to make use of the resolution capability of the objective (in jargon used by microscopists, the incident light must "fill the numerical aperture" of the objective). Second, the illumination should be uniform across the specimen and, for convenience, it should be possible to control the intensity and to eliminate stray light. To accomplish these ends, a source of light of small dimensions is employed, which is focussed down to a very small area by using a set of lenses mounted in a single unit called a *condenser*; this is simply an objective operated with the light path reversed. The aberrations of a condenser are not usually corrected as stringently as are those of an objective lens because the resolving power of an objective is not significantly affected by small aberrations of the condenser. There are three basic kinds of condensers. The Abbé and aplanatic (which are for all practical purposes interchangeable) and the achromatic (which improves the color correction).

Probably the most misunderstood point in the operation of a microscope is condenser adjustment. The condenser is used to provide either *critical* or *Köhler illumination* (Figure 2-8A). In critical illumination the light source is focused by the condenser lens so that the object is illuminated by a solid cone of light. Operationally, this means that the position of the condenser with respect to the object is adjusted so that the spot of illuminating light is as small as possible if seen by the naked eye and as bright as possible if viewed through the eyepiece and objective *when focused on the object*. Critical illumination is very commonly used; it requires the use of a diffusing screen in front of the light source to avoid imaging the structure of the light filament on the object. This illumination method is somewhat wasteful of light but is simple to obtain. The condenser lens must be chosen so that its NA is at least as great as that of the objective in order to fill the objective aperture.

Köhler illumination is more intense and precisely controlled. An accessory lens, called a field lens, (Figure 2-8B) is placed between the light source and the condenser so that it forms an image of the source in the focal plane of the condenser. All light rays leave the condenser as parallel beams called pencils of light. They pass through the object at various angles, which increase with the distance of the source points from the optic axis. Hence the object is illuminated by a set of pencils

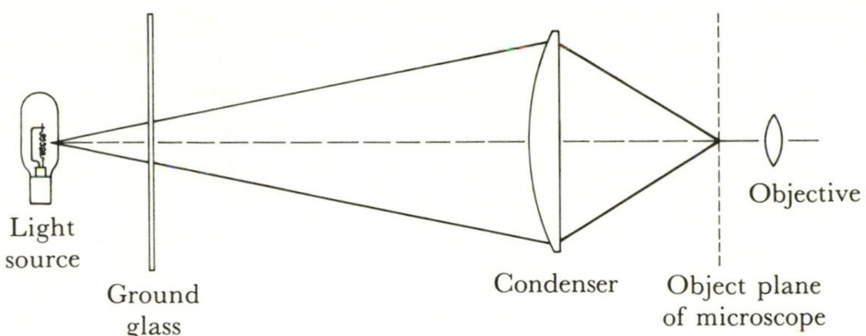

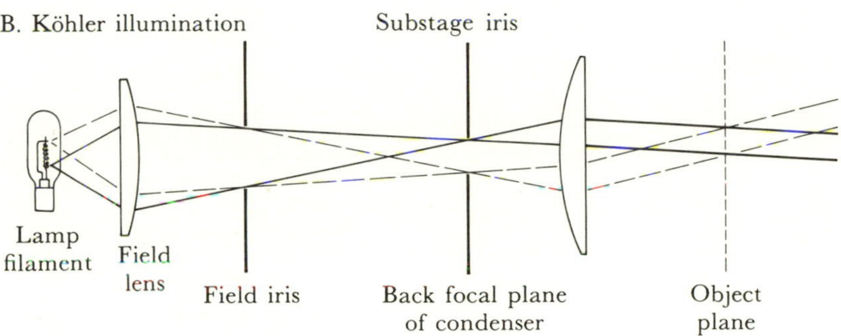

Figure 2-8
Types of illumination used in microscopy.

that constitute a cone of light.* The condenser can be adjusted so that the field lens is imaged in the object plane and the lamp is imaged at the back focal plane of the objective. This is done by decreasing the field iris (Figure 2-8B) and focusing it on the object with the condenser. This provides illumination as even as that of the field lens. A condenser or substage iris is usually added to control the NA of the illuminating beam.

A complete plan of a typical microscope is shown in Figure 2-9.

Adjusting a Microscope

Three adjustments must be made in setting up a microscope for use: (1) the light source and all components must be centered on the optic axis of the instrument; (2) the objective must be focused; and (3) illumina-

* It is not actually a cone because the filament of the light source is not round. However, because of the size and placement of the field iris, only a small circular part of the luminous filament is used as a source.

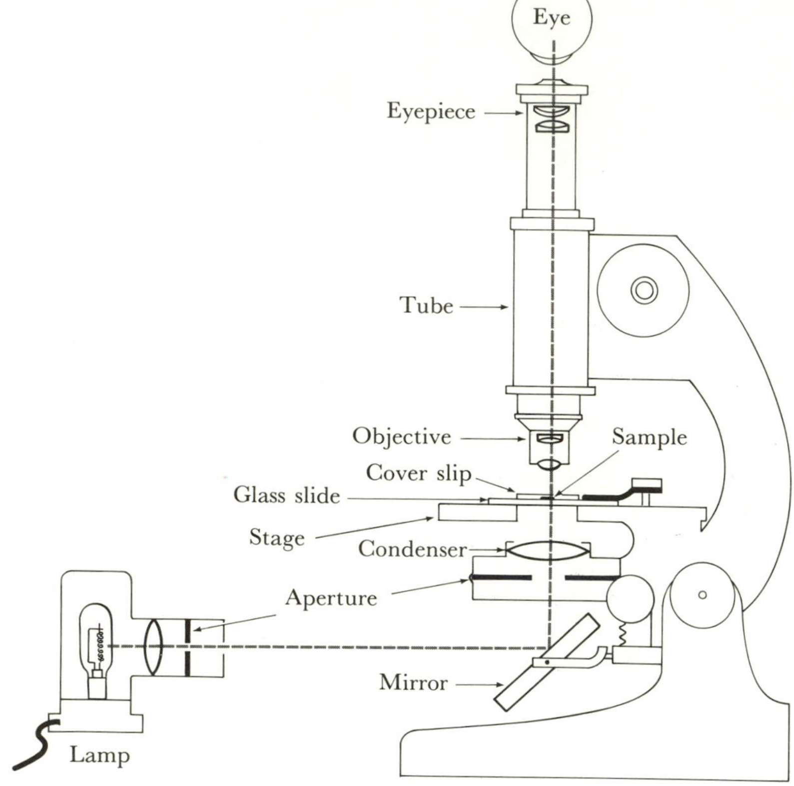

Figure 2-9
A complete microscope. Both tube and condenser can be separately moved along the optic axis for focusing.

tion must be adjusted. In most bright-field (i.e., standard) microscopes the condenser, objectives, and eyepieces are coaxial so that only the light source must be centered. This is done by focusing on a microscope slide, removing the eyepiece, looking down the microscope tube, and moving the light source (which has adjusting screws) until the light is in the center of the objective lens. If the centering of the condenser is also adjustable, the condenser is first removed, the light source is centered as described above, the condenser is returned to the system, and the adjusting screws on the condenser are moved to center the source. The condenser is then focused on the object for critical illumination or, as discussed above, for Köhler illumination. The substage iris (Figure 2-8B) should always be adjusted so that the NA of the lens is filled, which can be determined by looking down the tube without an eyepiece and noting that the lens is fully illuminated. To eliminate stray light and glare, the field iris can be reduced so that only that part of the field containing the object is illuminated. Sometimes the field is too bright for comfortable viewing. The intensity should never be reduced by changing apertures; instead, either a neutral density filter should be inserted in front of the source or the voltage applied to the light source should be reduced.

Contrast

To be seen, the image of an object must differ in light intensity from that of the surrounding medium. The difference in intensity between object and medium is called *contrast*. Unfortunately, most biological specimens (i.e., cells and their components) are transparent so that contrast is near zero. In the past, the solution to this problem was to stain the specimen—that is, to apply colored compounds that react with particular components of the cells. Hundreds of such substances are known, most of which have a preference (though rarely absolute) for proteins, lipids, or nucleic acids. Whether dyes exist that are really specific for particular chemical groups is a matter of some controversy. The technology of staining is described in many texts on microscopy and histochemistry (see the Selected References near the end of the chapter). Five special methods have been developed to create contrast and to obtain quantitative information from microscopy: dark-field, phase-contrast, interference, polorization, and fluorescence microscopy. We will examine them separately and give examples of the uses of each.

Dark-Field Microscopy

Let us consider critical illumination of an object, remembering that the light coming from the condenser forms a solid cone (Figure 2-8B). If there is no object in the object plane, the field will appear to be uniformly illuminated (bright field). If an opaque disc containing a transparent annulus is inserted just below or within the condenser (Figure 2-10), the object plane will be illuminated by a *hollow* cone of light that passes through the object plane and emerges as a second hollow cone. If the size of the annulus is sufficiently large, the hollow cone will surround the objective and no light will enter the lens; in this case the field appears black. However, if an object is present in the object plane, it will diffract light and some of the orders of diffraction will be at an angle such that they enter the objective lens. Even though the intensity of the diffracted light is low, a bright image will be produced against a black background. Such an image does not have much detail (because information is lost by discarding the zero order of light), but the contrast is exceptional and governed mostly by the intensity of the light source and the degree to which internal reflections in the system have been eliminated. Hence dark-field microscopy is useful for counting small particles that are difficult to see with a bright field. Otherwise it is not often used because of its poor resolution. If dark-field accessories are not available for a particular microscope, a dark field can be produced by gross misalignment of the light source and condenser so that little or no light enters the objective. Figure 2-11 shows a comparison of unstained bacteria seen by bright-field, dark-field, and phase microscopy.

Dark-field condensers are currently in use in high-quality fluorescence microscopes; this will be explained in a later section.

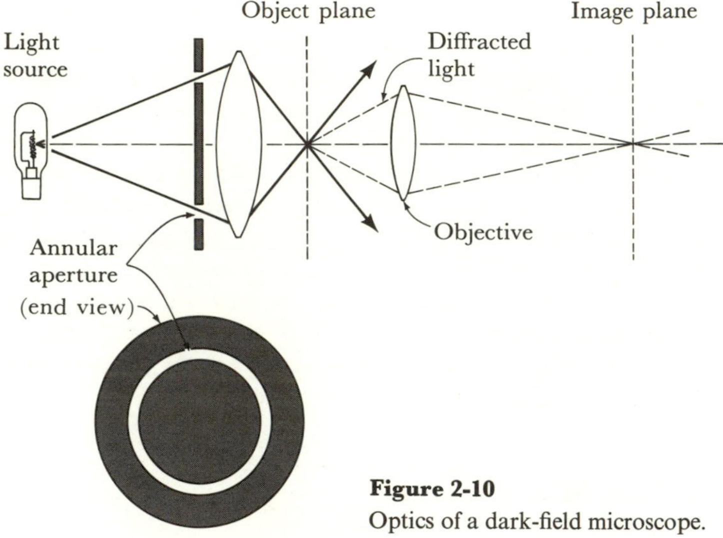

Figure 2-10
Optics of a dark-field microscope.

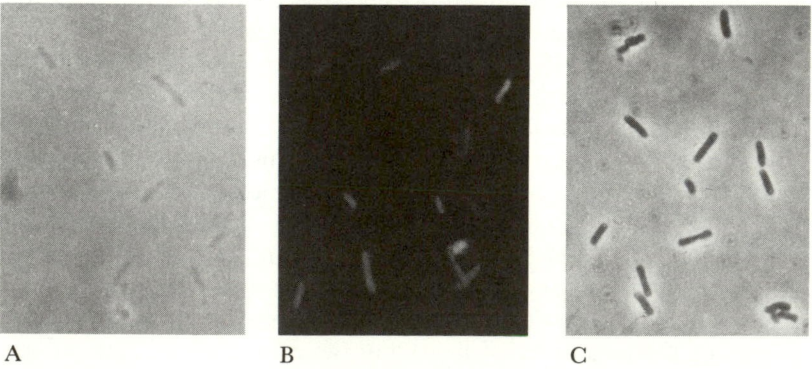

Figure 2-11
Bacteria (*E. coli*) seen by (A) bright-field, (B) dark-field, and (C) phase microscopy. Note that they are nearly invisible in part A.

Phase-Contrast Microscopy

Consider a transparent disc in a transparent medium illuminated by parallel light falling perpendicularly on the disc. For all practical purposes, the disc is invisible because the intensity of the light passing through disc and medium is the same. However, the light transmitted by disc and medium differs in phase if the refractive indices of disc and medium are different, because a light wave is retarded when it passes through transparent matter (see Chapter 16).* Neither our eyes nor photographic film can distinguish light of different phases—hence the disc is invisible.

With small objects, the formation of an image cannot be adequately described by geometric optics; it is actually the result of two diffraction processes. The incident light is first diffracted by the object and then by the aperture of the objective. This means that every point of the object and every point of the objective contributes to the formation of every point of the image by interference of all of these diffracted waves.

Let us now consider the light that falls on the objective after passing through a point of the transparent disc. This light consists principally of zero-order diffracted light—that is, a direct beam, which might be called *undeviated light*. The diffracted light forms maxima at various angles from the zero-order wave and may be called *deviated* light; it differs in phase by one-quarter wavelength from the undeviated. The light passing through the medium is also broken down into deviated and undeviated. However, when all of these waves are recombined by the objective and projected onto the image plane, the result is an image consisting only of phase differences.

In phase microscopy the phase difference is converted into an intensity difference by the following simple trick (Figure 2-12). An annular aperture (such as in dark-field microscopy) is placed in the focal plane below the condenser so that the object is illuminated by a *hollow* cone of light. If no object is present, the hollow cone passes through the object plane and falls on the objective, forming a tiny ring of light. In the back focal plane of the objective (i.e., on the eyepiece side of the objective), an image of the condenser annulus forms. At this point a *phase plate* is introduced, consisting of a disc either with an annular groove one-quarter wavelength in depth or with an additional layer of one-quarter wavelength. For the latter arrangement, a transparent film is usually deposited directly on one of the lens surfaces in the objective. The size of this phase plate coincides exactly with that of the image of the condenser annulus. Therefore all light passing through the condenser annulus will be advanced or re-

* It is important to understand the meaning of phase. If two sine waves moving along a single axis in the same direction have maxima and minima at the same points, they are said to be in phase. If the maxima of one and the minima of the other occur at the same point, they are one-half wavelength, or 180°, out of phase and the two waves, if superimposed, would cancel one another out.

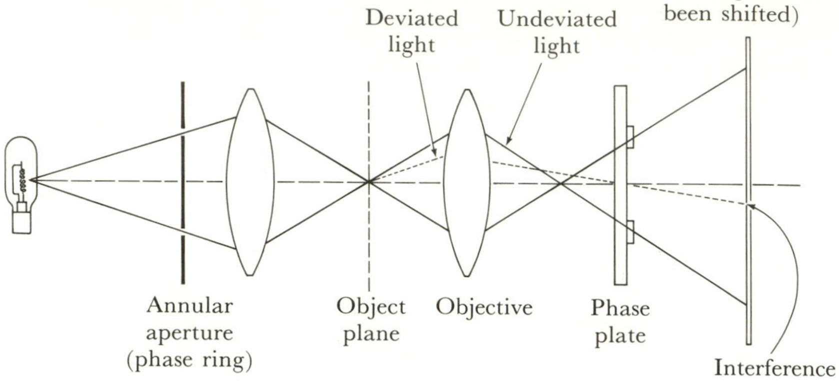

Figure 2-12
Optics of a phase-contrast microscope.

tarded by one-quarter wavelength, depending on whether the phase plate is a groove or a layer. If we now think of an object that diffracts the incident light, it is clear that the zero-order light (undeviated) will pass through the phase plate, whereas the diffracted light (which goes off at an angle, or is deviated) will pass through the remainder of the objective aperture. Furthermore, to make the intensity of deviated and undeviated light more nearly the same (zero-order light is always more intense than the higher orders), a thin film of aluminum or silver (neither of which alters phase but merely removes light) may be deposited on the phase plate.

When the deviated and undeviated light are recombined by the objective, they differ in phase by one-half wavelength—the condition for destructive interference. Hence, the background will appear dark, the actual intensity depending on the relative intensities of the deviated and undeviated light (determined by the thickness of the aluminum in the phase plate). With an object present, the light has two components—that which passes through the object and that which passes through the medium—and the light deviated by the object will interfere destructively only with undeviated light that is at the same position in the image plane. Remember that the object itself introduces a phase difference because its index of refraction differs from that of the medium. This phase difference, when superimposed on the difference created by the phase plate, results in a loss of destructive interference. Hence the object will appear brighter than the background. (If the undeviated light were advanced by one-quarter wavelength, the object would be dark against

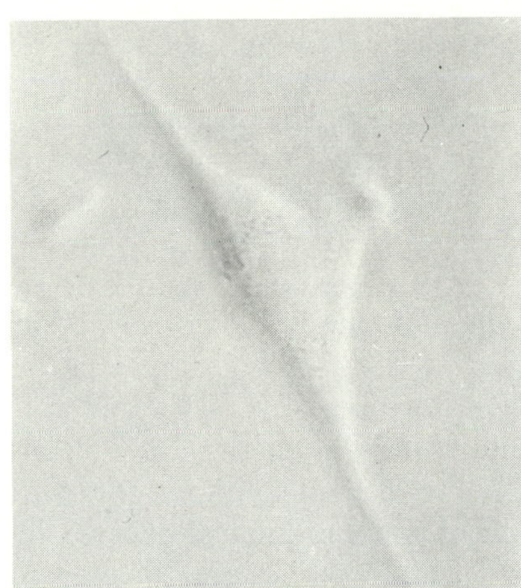

A

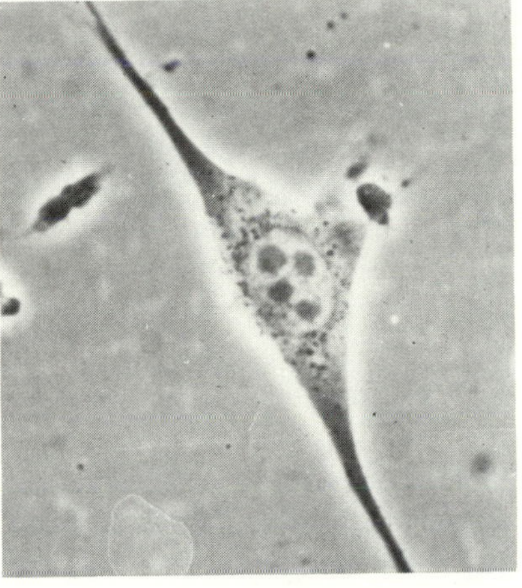

B

Figure 2-13
A mouse fibroblast seen by (A) bright-field and (B) phase-contrast microscopy. Note the vastly improved contrast in part B.

a bright background.) Clearly, the darkness of each objects depends on its particular refractive index and thickness.*

In this simple way, "invisible" objects can be made visible. The only disadvantage of phase microscopy compared with interference microscopy (described in the next section) is that objects are surrounded by unavoidable halos caused by diffraction by the phase plate. Figure 2-13 (as well as Figure 2-11) shows a comparison of the same cell viewed by bright-field and phase-contrast microscopy.

Alignment of a phase microscope is critical not only because the criteria for standard microscopy must be satisfied, but also because the condenser annulus must be precisely centered and focused on the phase plate. The method of alignment is described in the appendix at the end of this chapter.

Uses of the Phase Microscope

The only requirement that a sample must meet for phase microscopy is that it is nonabsorbing because the microscope is based on phase differences only. For absorbing objects (e.g., stained cells), intensity

* Manufacturers of lenses make lenses with various amounts of advancement and retardation. These are called dark phase, light phase, medium-dark phase, and so forth. The choice of the appropriate type is strictly empirical—that is, the lens that gives the greatest detail for a particular object is the one to be used. Medium-dark phase is especially good for viewing individual animal cells in liquid.

differences are added to the phase effects, often resulting in reduced contrast and poor image quality.

The phase microscope is of great value in visualizing the organelles of living cells (Figure 2-13). Nuclei contrast strongly with the cytopalsm and cytoplasmic components such as mitochondria, vacuoles, and fat droplets are easily seen. Nucleoli and chromosomes are especially visible. A membrane can be seen clearly as a dark line with a bright halo on either side although, because of the halo, unit membranes are not easily distinguished from double membranes. It is difficult to see detail in small objects such as bacteria because of the halo effect on the cell wall. However, by placing bacteria in a medium whose refractive index equals that of the cell wall or cytoplasm (e.g., concentrated solutions of bovine serum albumin), internal details, such as nuclear bodies, become relatively easy to see.

Interference Microscopy

Like the phase microscope, the interference microscope is capable of converting phase differences into intensity differences, with the advantages that there is no halo effect and that certain quantitative measurements are possible. The most effective form of the instrument (the Dyson instrument) employs a complex system of silvered surfaces and reflectors that divides the light emerging from an object so that a fraction of it passes through the surrounding medium and a phase-shifting plate. When this fraction recombines with the other fraction of the light, that is, that which has not undergone the phase shift, interference occurs. In this way phase shifts introduced by regions in the object having different indices of refraction and/or thickness are converted into intensity differences. If white light is used, these regions have different colors, whereas with monochromatic light only intensity differences are seen.

This brief treatment is not meant to explain fully how this relatively complex instrument works. For a detailed explanation the references given near the end of the chapter should be consulted. The main point is that the student of biology should be aware of the existence of this rather uncommon instrument because it has the following capabilities.

1. Without the halo produced by the phase microscope, contrast of objects is enhanced. Small detail not easily discernible by the phase microscope can thus be seen.

2. Quantitative measurements not possible with the phase microscope can be made. The difference in optical path between a particle and the surrounding medium can be measured. Because optical path is the product of index of refraction and thickness, one can be measured if the other is known. Furthermore, the concentration of a known material can be determined if the index of refraction and the *specific*

refractive increment (change in index of refraction per unit amount of solute) is known. If n_p and n_s are the indices of refraction of a protein and the solvent, respectively,

$$n_p - n_s = \alpha c$$

in which α is the specific refractive increment and c is the concentration in grams per 100 milliliters. Because n_p and α do not vary enormously from one protein to another, a rough estimate of the protein concentration within a cell can be easily made.

Polarization Microscopy

Note: Before reading this section, Chapter 16 should be consulted for a discussion of plane-polarized light.

Structures consisting of elongated particles in a parallel array or stacked discs embedded in a medium whose refractive index differs from that of the structure particles exhibit *form birefringence* (Figure 2-14). This means that the structures will pass plane-polarized light only if the plane of polarization is parallel to the long axis of the particles. This will be true even if the particles themselves are not intrinsically birefringent—that is, if the particles transmit polarized light with equal probability for all angles of incidence. (Form birefingence is the same phenomenon encountered when fibrous molecules are subjected to flow through a tube or to the shear gradient between coaxial cylinders, as discussed in Chapter 19.)

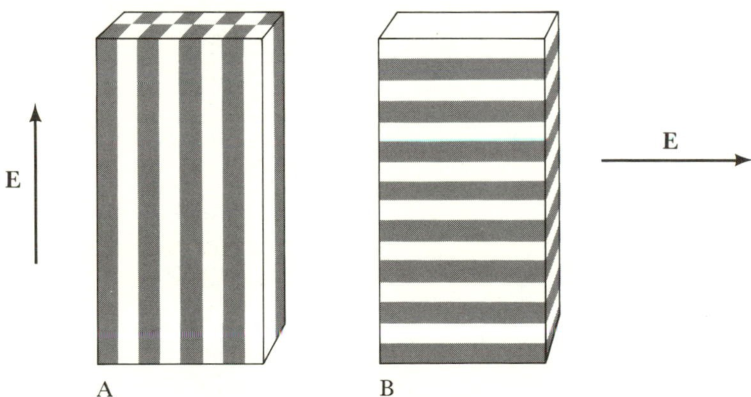

Figure 2-14
Structures showing form birefringence: (A) positive and (B) negative. The plane of the **E** vector passes by each is shown.

Form birefringence is easily observed in cellular material using a polarization microscope. It is of great importance not only because it can be used analytically to determine orientation, but also because in some cases it is the only property that can be used to make a structure visible (e.g., cases in which the particles cannot be stained or in which their concentration or specific refractive increment is too low to generate a phase difference large enough to make it visible by phase-contrast or interference microscopy).

The essential components of a polarization microscope are a polarizer located between the light source and the condenser; a rotating stage or sample holder; an analyzer, situated between the objective and eyepiece, that can be set so that its axis is perpendicular to the polarizer axis (in which case the polarizer and analyzer are said to be *crossed*); and a compensator (Figure 2-15). Because polarized light is partly depolarized by reflection and refraction, the condenser is adjusted so that the object is illuminated by nearly parallel light.

When the polarizer and analyzer are crossed and either no object or an *isotropic* object (having no preferred axis for the refraction of polarized light) is in the object plane, the field appears uniformly dark. If a birefringent object is present and lying so that its axis is at an angle other than 0° or 90° with respect to the plane of polarization, it will resolve the polarized light into two components, one parallel and one perpendicular to the plane of the analyzer (Figure 2-16). Hence some light will pass through the analyzer so that the object will appear bright against a dark background. As the object is rotated, it will be invisible when lined up with either the polarizer or the analyzer and will have maximum brightness at an angle of 45°. (This is the reason for the rotating stage.)

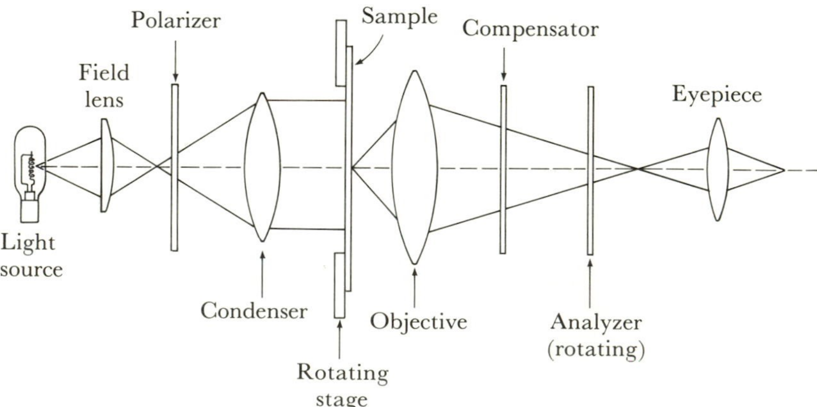

Figure 2-15
Optics of a polarization microscope.

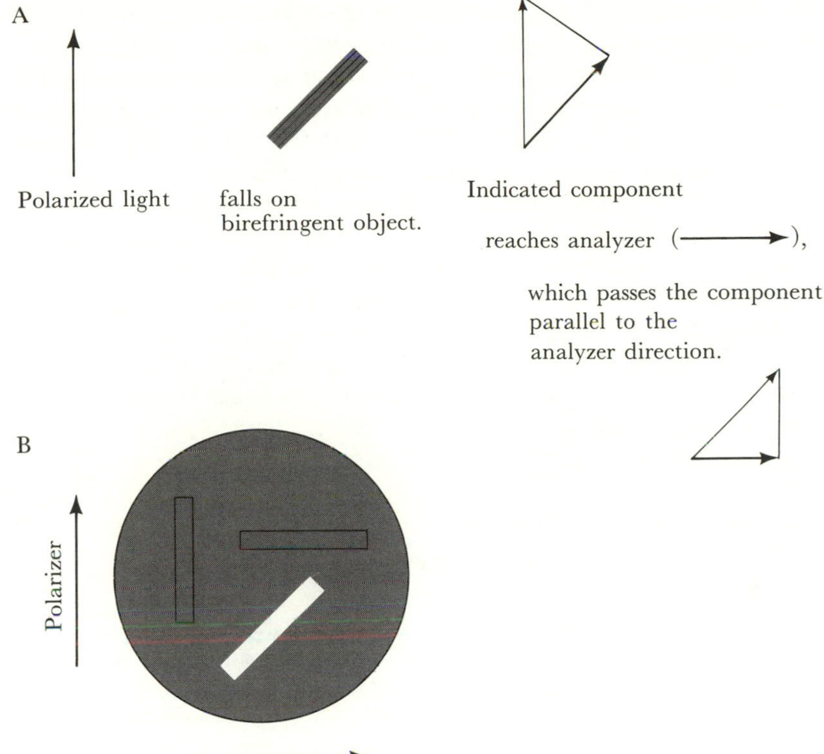
A
Polarized light
falls on birefringent object.
Indicated component
reaches analyzer (⟶),
which passes the component parallel to the analyzer direction.
B
Polarizer
Analyzer

Figure 2-16
A. Resolution of polarized light by a birefringent object. B. The appearance of a birefringent object aligned with the polarizer, aligned with the analyzer, and at 45° to both.

Let us consider a rodlike structure with form birefringence oriented so that its long axis is at 45° with respect to the crossed polarizer and analyzer. Brightness will be maximal if the structure consists of either parallel elongated molecules or stacked discs, because both the axis and the plane of the discs, being perpendicular, would be at a 45° angle. The question is how to distinguish the parallel ("positively" birefringent) from the perpendicular ("negatively" birefringent) structure. This is done with a *compensator*. To understand how the compensator works, it is necessary to remember that an object is birefringent if the index of refraction of light polarized parallel to the axis differs from that for light polarized in the perpendicular direction. For positive birefringence, the parallel direction has the lower index of refraction and therefore allows the light to travel faster than in the perpendicular direction. If the velocities of the parallel and perpendicular components were equalized, the net birefringence would be zero and the object would become invisible. Consider the effect of interposing between the object and the analyzer a thin layer of a birefringent crystal (e.g., mica, gypsum) whose slow and fast direction are known and indicated. If the slow direction of the crystal is parallel to the fast direction of the birefringent sample, the velocities of the parallel and perpendicular components approach one another and the brightness of the object decreases. Therefore, if such a unit (which is the compensator) is rotated, the sign of the birefringence can be identified by means of the position of the compensator—that is, that which produces minimum or maximum brightness. The value of this capability will be described in some of the examples that follow.

Use of the Polarizing Microscope in Biology and Biochemistry

Often, the polarizing microscope can give detailed information about molecular architecture in a relatively short time. In some studies of living cells, this is the only applicable method because the more precise and sophisticated techniques described in later chapters require either a dried (i.e., dead) sample or a large volume.

Example 2-A ☐ Orientation of molecules.

Birefringence seen in a cell indicates that there is a structure containing oriented molecules. The sign of the birefringence tells how the components are oriented. For example, studies of muscle cells and sperm tails indicated that they contain molecules arranged parallel to the fiber direction long before such structures were seen in the electron microscope.

Similarly, the birefringence of chloroplasts and the rod cells of the retina indicate that they have a lamellar (stacked-disc) structure. Oriented fibers of DNA also show the disc structure due to the purine·pyrimidine base pairs.

Birefringence measurements can also give some indication of a mixture of two oriented components. For instance, the myelin sheath of nerve cells is positively birefringent and consists of proteins and lipids. When the lipids are extracted by solvents, the positive birefringence increases so much that one may conclude that the lipid molecules must be nearly perpendicular to the highly oriented proteins.

Example 2-B ☐ Visualization of structures not otherwise visible.

There are cell structures that are not visible by either bright-field or phase-contrast microscopy but are easily seen in a polarization micro-

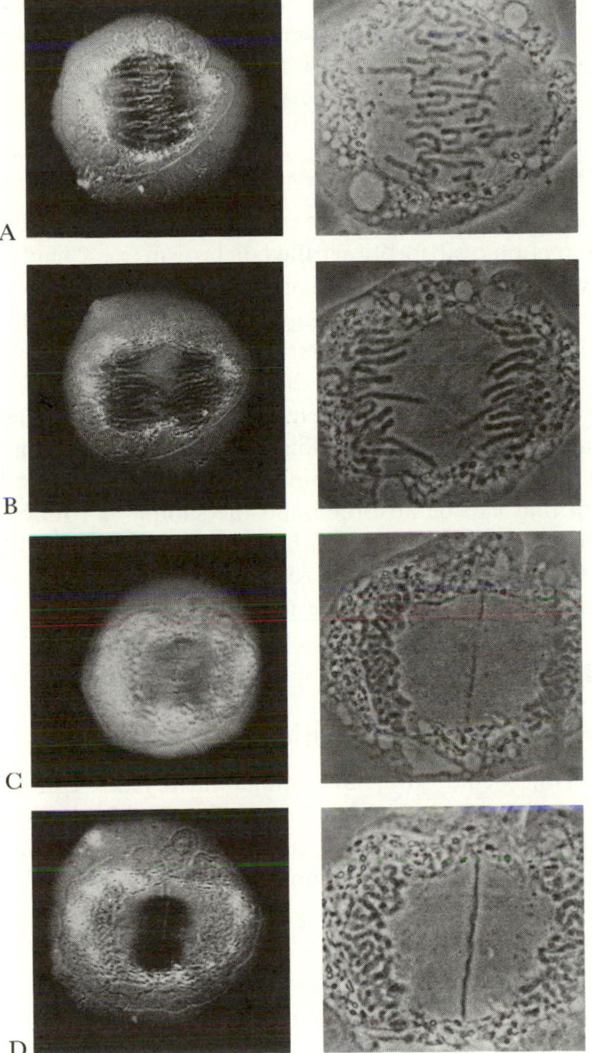

Figure 2-17
Cell division of *Haemanthus* endosperm seen by (left) polarization and (right) phase-contrast microscopy. The same cell has been followed through various stages of mitosis: (A) metaphase—the mitotic spindle is visible in the polarization micrograph by virtue of its birefringence and the chromosomes are visible by phase contrast; (B) anaphase—the mitotic spindle has begun to disintegrate; (C) telophase—the spindle is nearly gone and the septum separating the cell is forming; (D) the phragmoplast (an oriented array of molecules on the cell surface) is clear by polarization microscopy, whereas only the well-formed septum is seen by phase contrast. The magnitude on the right is 1.6 times that of the left. The cells on the left do not fill the illuminated area, which has been reduced by an aperture to eliminate stray light, which lowers contrast. [Photomicrographs courtesy of Robert Haynes and Raymond Zirkle.]

scope. An example is the mitotic spindle (Figure 2-17) and the orientation of molecules in the phragmoplast (boundary region) between plant cells undergoing cell division. By time-lapse photography and polarization microscopy, it is even possible to follow the development of these structures in living cells.

Example 2-C

☐ Identification of helical arrays.

If positively birefringent fibers are observed looking down the fiber axis, they look dark for all rotations. However, if the fiber is helical, the fiber appears alternately dark and light, the dark corresponding to the regions where the rung of the helix points directly at the observer. In this way helical structures can be identified. For example, the inclusion bodies (orderly aggregates) of tobacco mosaic virus in tobacco leaf cells appear helical in cross-section.

Use of the Polarizing Microscope to Measure Dichroism

All the objects discussed in the examples are transparent. Many objects possess a preferred direction of absorption called dichroism (see Chapter 14), which again may be intrinsic or by virtue of form. Intrinsic dichroism is most common in chemicals containing conjugated rings in which absorption is maximal when the plane of polarization is in the plane of the rings and parallel to a particular axis of the ring system.

A polarizing microscope can be used to detect dichroism by removing the analyzer and compensator because the object itself will behave as an analyzer. Hence, if a dichroic sample is illuminated with plane-polarized light of a wavelength that can be absorbed and the sample is rotated, there will be an angle at which it will appear darkest. If the dichroism of the molecules known to be in a given structure is known, the orientation of the molecules in that structure can be determined. This technique has not had widespread use, but the following examples should indicate the great power of the method.

Example 2-D □ Orientation of absorbing groups in crystals.

The orientation of the heme group in hemoglobin crystals can be determined by viewing the crystals with blue light. Because absorption along the *b*-axis of the crystal is much greater than in the perpendicular direction, the flat surface of the heme group must lie along the *b*-axis.

Example 2-E □ Orientation of macromolecules in large structures.

The absorption of polarized ultraviolet light by stretched chromosomes indicates the net orientation of DNA in the chromosome because, for DNA, ultraviolet absorption is maximum perpendicular to the helical axis (i.e., parallel to the plane of the base pairs).

Fluorescence Microscopy

Fluorescence is an extraordinarily sensitive method for detecting minute quantities of material because, for any fluorescent material, *the total intensity of the fluorescence is proportional to the intensity of the incident light* (see Chapter 15). The principal problem, however, in detecting fluorescence is that of separating the fluorescence from the incident light. With a fluorescence spectrophotometer, this is accomplished by viewing the sample at right angles to the incident beam. With a fluorescence microscope, this is not easily done and three other methods are used. The most common procedure uses optical filters (Figure 2-18).

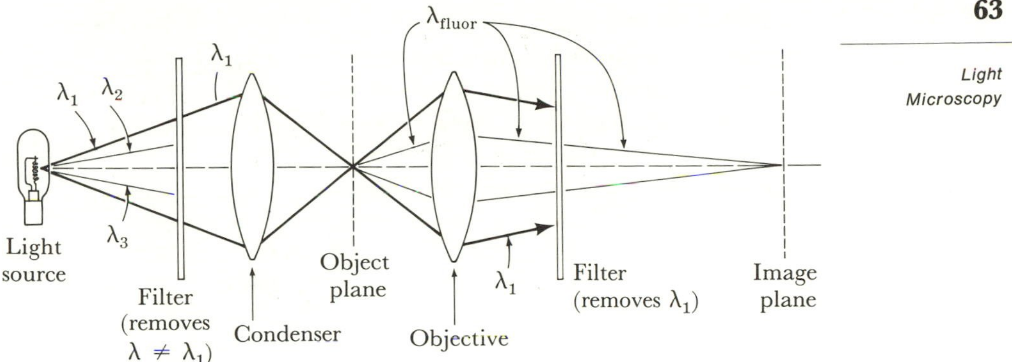

Figure 2-18
Optics of a fluorescence microscope.

A filter that allows the exciting wavelengths but not any wavelengths in the fluorescence spectrum to be transmitted is placed between the light source and the condenser. Another filter, called a barrier filter, which transmits the fluorescence wavelengths but not the exciting light, is placed somewhere between the objective and the eye. Hence, in the absence of a fluorescent object, no light reaches the eye and the field is black. If an object that fluoresces is present, it will contrast strongly with the background. In the second procedure, the condenser and the first filter are replaced by a dark-field condenser. The annulus of this condenser is designed so that no exciting light enters the objective lens. This eliminates the need for the barrier filter. The third method uses a complex optical system that enables the sample to be illuminated from above; that is, the incident light moves away from the objective. Again no filters are needed. This is the best system but is also the most expensive.

To analyze weak fluorescence, it is necessary to use microscope slides, cover-slips, and lens elements made of nonfluorescing glass because any of these can lighten the background significantly. Apochromatic lenses must never be used because the CaF_2 elements in them are strongly fluorescent.

Because few cell components are very fluorescent (other than molecules like tryptophan, tyrosine, and riboflavin, which are found throughout the cell) and even fewer can be excited by short-wavelength visible or near ultraviolet light (because short-wavelength ultraviolet radiation requires special lenses and illumination systems), extrinsic fluors are always used—that is, a fluor that binds to particular cell components is added. Common fluors are acridine orange (for nucleic acids), fluorescein, and quinacrine.

Fluorescence microscopy is used for three purposes: to visualize components difficult to see, to localize substances by means of specific binding, and to determine orientation by means of fluorescence polarization. How this is done can be seen in the following examples.

Example 2-F ☐ Visualization of nucleic acids in animal and plant cells.

Acridine orange binds to both DNA and RNA but gives a green fluorescence with DNA and, if the dye concentration is high, an orange fluorescence with RNA (Figure 2-19). Hence, at low dye concentration, eukaryotic cells show a bright green nucleus and pale green cytoplasm; at high concentrations, the cytoplasm becomes orange. During mitosis, chromosomes glow a bright green and their morphological characteristics are easily seen *in the living cell.* In a cell infected with a DNA virus, viral inclusions can be detected and counted as bright green spots in the cytoplasm.

Example 2-G ☐ Visualization of small organelles.

The yeast nucleus is very difficult to see because of its small size. If acridine orange is added at low concentration, a bright green polar object is seen outside of the vacuole (Figure 2-19). At high concentrations, the vacuole remains colorless, the nucleus becomes a brighter green, and the cytoplasm glows orange. This was actually the first demonstration that yeast contained DNA, because yeast has such a large RNA/DNA ratio that, in the early 1950s, chemical tests failed to show its presence.

Example 2-H ☐ Fluorescent antibody technique.

Fluorescein can be covalently linked to antibodies prepared against various cell fractions and against proteins (Figure 2-20). Addition of

A

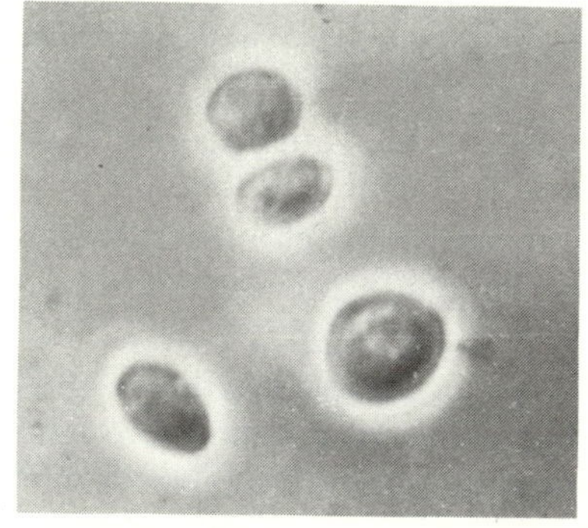

B

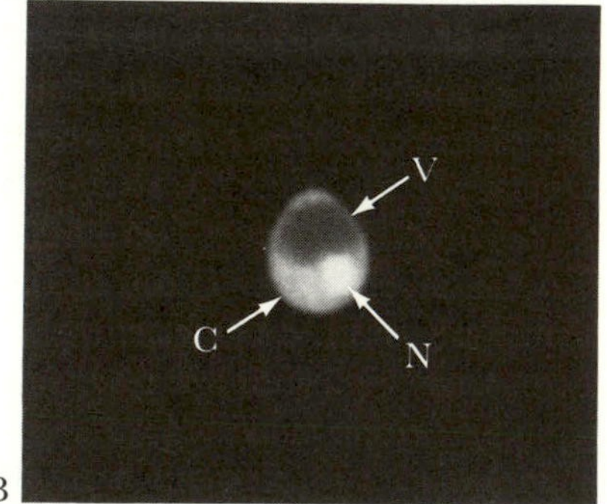

Figure 2-19

Photomicrographs of resting cells of baker's yeast: (A) as seen by phase-contrast microscopy; (B) as seen by fluorescence microscopy. Note the lack of detail and the bright halo around the cells in part A. In part B, the fluorochrome—acridine orange—has been added, causing the nucleus (N) to become bright green and clearly distinguishable from the cytoplasm (C) containing RNA, which becomes orange, and the vacuole (V), which contains no nucleic acid and is therefore non-fluorescent.

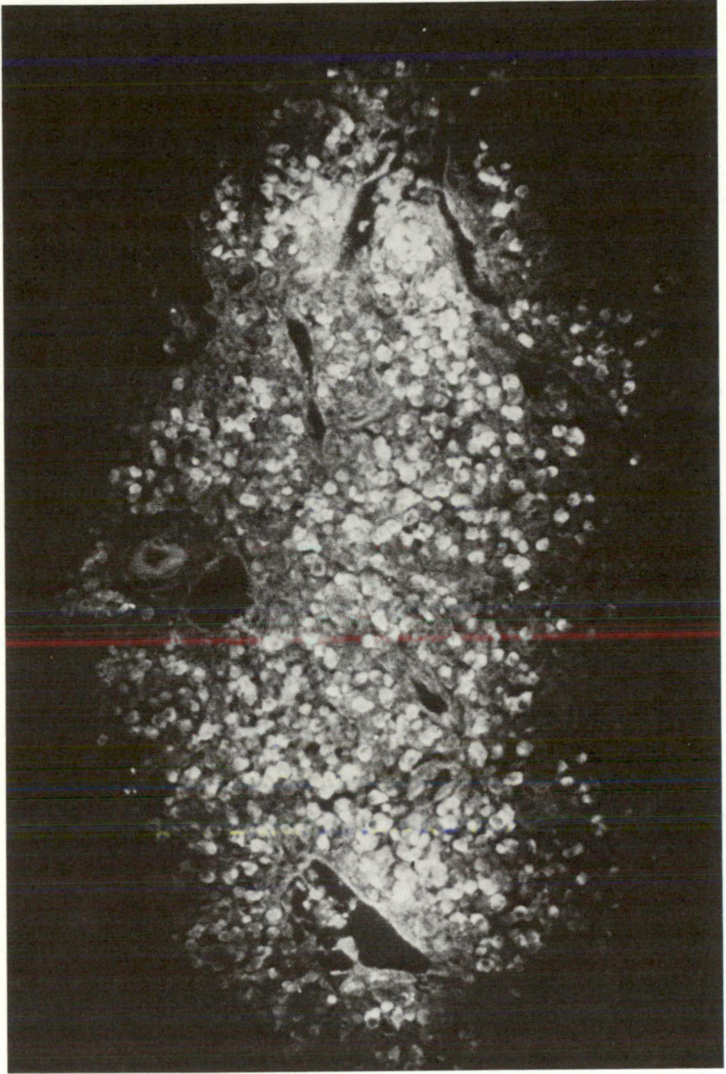

Figure 2-20
An example of the use of the fluorescence microscope. This cross-section of a rabbit popliteal lymph node shows cells containing antibody against bovine serum albumin (BSA) on the fourth day after a booster injection. Many cells show the yellow green fluorescence of fluorescein (white) against the bluish background of the other cells (grey). The antibody is in the cytoplasm. The tissue was fixed in 95% ethanol at 4°C overnight, then dehydrated in cold 100% ethanol, cleared in cold xylene, and finally embedded in paraffin below 60°C. Sections were then made, deparaffinized, and hydrated. Then the section was reacted with BSA (0.5 mg/ml in saline) and washed; thus, BSA was bound only in regions containing antibody, and unbound BSA was washed off. Anti-BSA labeled with fluorescein (two molecules per molecule of protein) was layered over the section, to react with whatever BSA had been bound. This revealed the antibody that was originally in the cells. In this way, it was shown that cells in the lymph node had synthesized anti-BSA. [Courtesy of Albert Coons.]

the fluorescent antibody to cell sections (thin slices) or to cells made permeable to protein by treatment with acid or acetone allows localization of these substances. For example, viral antigens, cell-membrane components, histones, and many other substances have been localized in individual cells or in tissues by this method. The proteins of actin and myosin in muscle fibers were distinguished by fluorescein-labeled anti-actin and anti-myosin.

Example 2-I ☐ Polarization-fluorescence microscopy.

It is known that acridine orange intercalates between the DNA base pairs. The plane of polarization of the fluorescence of acridine orange is also known and, hence, the angle of polarization of the fluorescence with respect to the DNA helix axis. Acridine orange binds tightly to chromosomes, presumably by DNA binding. Therefore, from the polarization of the fluorescence of the chromosomes, the orientation of the DNA molecule with respect to the chromosome can be determined.

Example 2-J ☐ Identification of chromosomes by quinacrine fluorescence.

Quinacrine binds to DNA, either in solution or in chromosomes. Quinacrine mustard binds even more efficiently by reacting with free amino groups of nucleotides in DNA. When it is added to cells, chromosomes are not only fluorescent but show characteristic patterns of bright and dark bands. The cause of the pattern of bands is not yet clear; probably there is enhanced binding of quinacrine to regions high in adenine·thymine base pairs and to regions to which certain histones are preferentially bound. The value of this method is that chromosomes that are very difficult to distinguish morphologically can be easily distinguished by their patterns of bands. Furthermore, chromosomal abnormalities not otherwise recognized can be observed; therefore, this method is attaining great clinical importance. Figure 2-21 shows chromosomes stained with quinacrine mustard.

APPENDIX

Alignment of a Phase-Contrast Microscope

To obtain phase contrast, it is important that the image of the condenser annulus (sometimes called the phase ring) is precisely focused on the phase plate. This is accomplished in the following way. The condenser and eyepiece are removed and the light source is adjusted so that it is roughly centered in the objective when viewed down the tube. An object is then put on the stage, the eyepiece is reinserted, and the objective with desired magnification is swung into position. The objective is then focused on the object. The condenser is replaced and focused so that an image of the lamp iris is in focus in the object plane (i.e., Köhler illumination). A

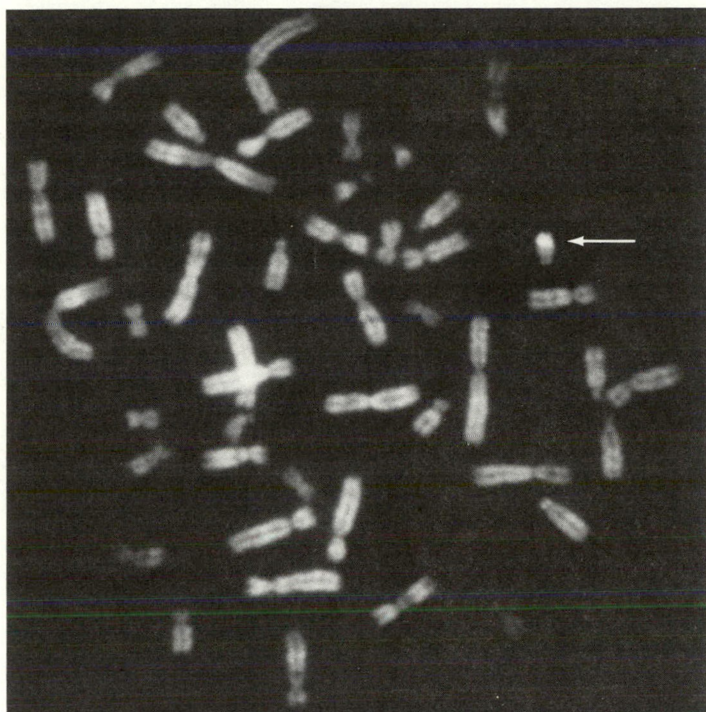

Figure 2-21
Chromosomes visualized by fluorescence microscopy. These are metaphase chromosomes from peripheral (blood circulation) leukocytes of a normal human male that have been stained with quinacrine mustard. Note that the chromosomes do not fluoresce with equal intensity but are banded. Chromosomes that are morphologically indistinguishable by phase-contrast microscopy or by bright-field microscopy using standard histological stains can be easily distinguished by their banding patterns. Note in particular the brightly fluorescent Y chromosome (indicated by the arrow). The background is dark because only the chromosomes are fluorescent. Details of the utility of the method can be found in T. Casperson, G. Lomatka, and L. Zech, *Hereditas* 67(1971):89. [Photomicrograph courtesy of Edward Modest.]

special lens, called a Bertrand lens, is slid into place just below the eyepiece. The Bertrand lens, which is focused on the back focal plane of the objective, allows the position of the phase plate, which is in this plane, to be viewed.

The appropriate condenser annulus is swung into position (there is a separate annulus for each objective because the size of the annulus and the phase plate must be matched). The condenser annulus is then moved by means of adjusting knobs until it is concentric with the phase plate. A slight refocusing of the condenser is usually necessary at this point so that the image of the condenser annulus is precisely superimposed on the phase plate. The Bertrand lens is then removed and the system is ready for use.

Gurr, E. 1965. *The Rational Use of Dyes in Biology and General Staining Methods.* Williams & Wilkins.

Martin, L. C., and W. T. Welford. 1971. "The Light Microscope," in *Physical Techniques in Biological Research,* vol. 1A, edited by G. Oster, pp. 2–70. Academic Press.

Oster, G. 1955. "Birefringence and Dichroism," in *Physical Techniques in Biological Research,* vol. 1, edited by G. Oster and A. W. Pollister, pp. 439–459. Academic Press.

Osterberg, H. 1955. "Phase and Interference Microscopy," in *Physical Techniques in Biological Research,* vol. 1, edited by G. Oster and A. W. Pollister, pp. 378–437. Academic Press.

Slayter, E. M. 1970. *Optical Methods in Biology.* Wiley. This excellent book explains just about everything.

Zernike, F. 1942. "Phase Contrast, a New Method for the Microscopic Observation of Transparent Objects," parts 1 and 2. *Physica* 9:686–693; 974–985. The theory of phase contrast was developed in these two papers.

Several manufacturers of optical equipment (e.g., Leitz, Olympus, and Zeiss) have excellent brochures that describe the operation of various types of microscopes. Their booklets on fluorescence microscopy are especially informative.

PROBLEMS

2–1. A microscope with an objective whose numerical aperture is 1.32 is used to look at an opaque disc containing regularly spaced holes 0.01 micron (μm) in diameter separated by 20 μm. Light having a wavelength of 540 nm is used and the holes are seen as points of light using a 10 × eyepiece. Explain why the fact that the number of holes per square millimeter can be counted does *not* conflict with the value of the limit of resolution of the system.

2–2. What is the limit of resolution of a system using an objective having NA = 1.28 and using light of wavelength 453.7 nm?

2–3. Explain why the brightness of an image decreases as the magnification increases.

2–4. If you had a bacterial culture and you wanted to count the number of bacteria per milliliter, a reasonable way would be to count the number of particles in a measured volume using a phase-contrast microscope. But, if you wanted to count viruses, this method would not work because they are too small to be seen. Explain how such a count might be obtained using a fluorescence microscope and acridine orange. What property should the virus have to be optimally counted? Would it be easier to visualize a large DNA or RNA virus?

2–5. Would there by any difference in appearance between a stacked-disc and a stacked-ring structure in a polarizing microscope?

2–6. Explain why the image quality will be poor if an absorbing object is used in phase microscopy.

2–7. Explain how you could use a polarizing microscope to show that an object is dichroic. Would there be any significant difference between eliminating the polarizer and eliminating the analyzer?

2–8. A particular type of cell structure can be stained a light pink. However, it is barely visible with an ordinary bright-field microscope. What simple modification to the microscope could you make to increase the contrast between the structure and the surrounding medium?

2–9. If an object becomes positively birefringent when stretched, what can you guess about its molecular structure?

2–10. It is usually possible to vary the intensity of the illuminating source for all microscopes; for ease of viewing and to prevent eye strain, the intensity is kept relatively low. For what type of microscopy would the maximum illuminating intensity be desirable?

2–11. Give several reasons for using cover-slips in preparing samples for microscopy. What would happen to the image if the sample were in liquid but no cover-slip were used?

2–12. If a cell were suspended in a medium whose index of refraction is the same as that of the cell wall, how would the cell appear?

2–13. A microscopic sac contains protein molecules lined up as follows.

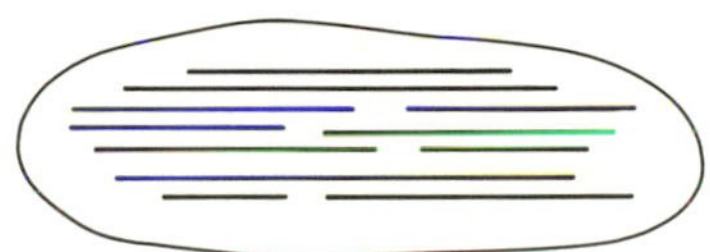

The index of refraction n of the contents of the sac is 1.03. If the sacs are suspended in water ($n = 1.33$), they are visible by phase microscopy and appear dark against a light background.

a. How would they look in a phase microscope if they were suspended in a fluid having $n = 1.003$?
b. Repeat part (a) for $n = 1.03$.
c. Suppose you put them in a polarization microscope with crossed polarizer and analyzer. If you rotate the sample through 360°, the sample would be alternately bright and dark. How many times will it be bright in one complete rotation?

2–14. A rod cell in the eye has the following structure.

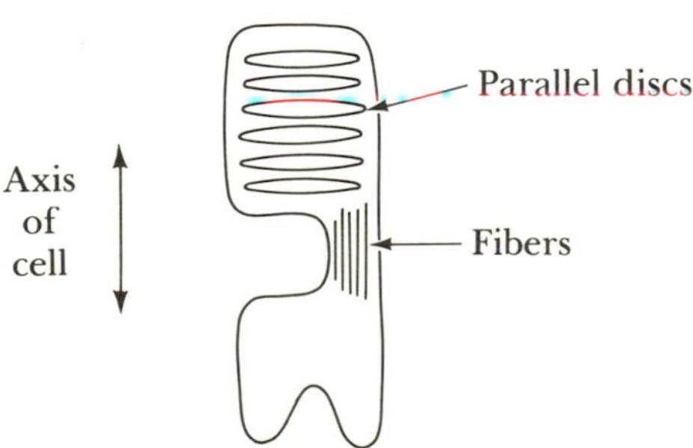

a. Using a polarizing microscope, at what angle with respect to the analyzer would the cell be most visible? Draw a picture showing the brightest regions when the cell is at this angle.
b. What simple modification could you make to the microscope to determine the orientation of the parallel discs? That is, how could you show that they are perpendicular and not parallel to the long axis of the cell?

2–15. Bacteria in water appear in a phase microscope as dark, homogenous rods against a light background. If put in 20% albumin solution, they appear light gray against a light background. In 30% albumin, they are not visible. Explain.

2–16. Sperm are being observed with a polarizing microscope. Four cells each oriented 45° with respect to the axes of the polarizer and analyzer are observed. They appear as follows.

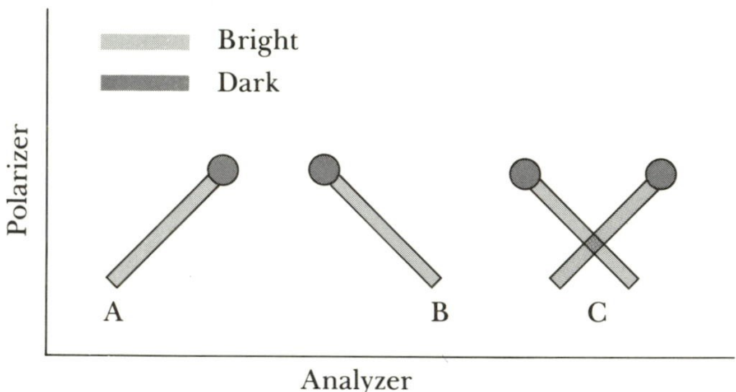

Explain
a. Why the tail is light and the head is not.
b. Why A and B both have the same brightness.
c. Why there is a discontinuity in the brightness at the junction point in C.

2–17. A round, flat object is observed with a polarization microscope with polarizer and analyzer perpendicular to one another. The object appears to contain a bright cross (whose arms intersect to form 90° angles), which is oriented at 45° with respect to the polarizer axis. The pattern does not change when the object is rotated. What is the probable structure of the object? What would be observed if the analyzer were removed from the microscope and the object rotated?

2–18. The fluorescent dye acridine orange binds to both single-stranded and double-stranded polynucleotides. At low concentration it binds much more efficiently to double-stranded molecules. When bound to single-stranded molecules, the fluorescence is green or orange at low or high concentrations, respectively. Draw a picture of a cell, showing the nucleus and the cytoplasm, as seen in a fluorescence microscope if the acridine orange is added (a) at a very low concentration, (b) at a low concentration but higher than in part (a), and (c) at a very high concentration.

2–19. If acridine orange is added to animal cells in mitosis (i.e., when the chromosomes have condensed), the chromosomes appear bright green when viewed with a fluorescence microscope using light that excites acridine orange fluorescence. We will use this technique to interpret the following experiment. An instrument exists (the ultraviolet microbeam) that allows a small part of a cell to be irradiated with an intense beam of ultraviolet light. If a a part of a chromosome is irradiated with a spot whose diameter is larger than the cross-section of a chromosome, five observations can be made: (1) the chromosome remains intact; (2) if observed with a phase microscope, the irradiated part of the chromosome appears pale compared with the remainder of the chromosome (which is normally black); (3) if observed with an interference microscope, using white light, the irradiated part is a different color from that of the remainder of the chromosome; (4) if acridine orange irradiated part is not fluorescent; and (5) if the cell is stained with dye that produces a red color wherever deoxyribose is present, the irradiated part is unstained, whereas the remainder is red.

What is the probable effect of the ultraviolet irradiation? What can you say about the probable composition of chromosomes? What can you say about the physical structure of chromosomes?

3

Electron Microscopy

The limit of resolution of the light microscope is roughly 2000 Å (see Chapter 2), which is insufficient to visualize cell organelles, viruses, and macromolecules of current interest. This is possible, however, with the electron microscope for which the limit is less than the diameter of a uranium atom (approximately 5 Å) under special conditions.

There are few instruments that present such a bewildering array of knobs and meters as does the electron microscope; furthermore, few techniques require greater skill and attentiveness to detail. Nonetheless, for most applications the instrument is simple to use and sample preparation is not complicated. There is no doubt that every laboratory should have access to an electron microscope and that every biochemist should be proficient in its use. This chapter will not supply specific procedures or technical details but will indicate how electron microscopy (EM) is done and what its potential is.

Simple Theory of Operation

An ordinary light microscope consists of a light source, a condenser for focusing the light on or near the object, an object holder (i.e., the slide and the stage), an objective lens for focusing the image, and an eyepiece for projecting the image formed by the objective onto the eye or photographic film (see Chapter 2). This is also true of an electron microscope, except that the light is replaced by an electron beam, the sample holder is a wire screen called a *grid*, and the lenses are electromagnets rather than glass.

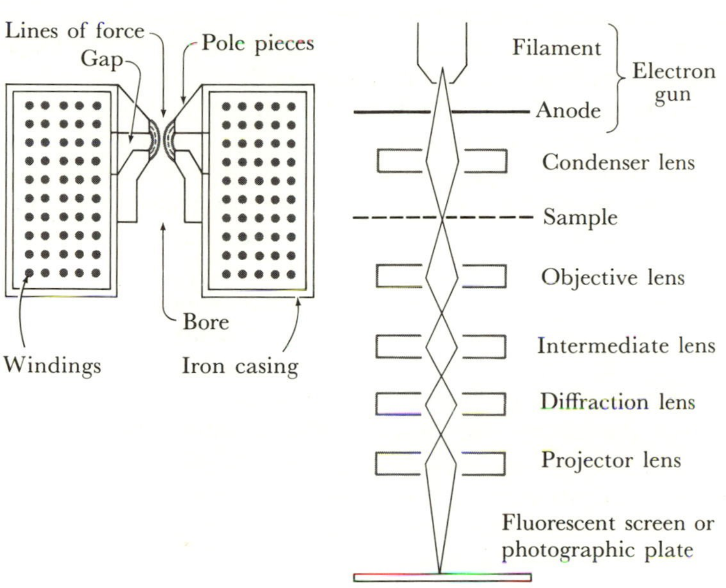

Figure 3-1
A. An electromagnetic lens. Although there is tremendous variation in details of design, every lens has an iron casing containing copper windings through which an electric current passes and pole pieces to concentrate the magnetic lines of force. The unit shown here is not drawn to scale; a typical lens is thirteen or more centimeters in diameter with a bore having a diameter measurable in micrometers. B. The path of electrons through the lens system of an electron microscope. The electrons are moving downward from the filament. The components and the beam are not drawn to scale. Some microscopes have two condensers; some lack the diffraction lens.

A schematic representation of an electromagnetic lens, which basically consists of an axially* symmetric electromagnet through which the electron beam passes, is shown in Figure 3-1A. (The focusing action of a magnetic field is complex and will not be explained here. For further information see the Selected References near the end of the chapter.) A magnetic field is generated by windings of copper wire. For high magnification, lenses of short focal length are needed and it can be shown that focal length decreases as the magnetic field increases. The magnetic field could be made larger by increasing the current in the windings, but this would generate a considerable amount of heat. Instead the

* It is frequently stated that electromagnetic lenses are *cylindrically* symmetric, meaning that they have the shape of a cylinder. Axially symmetric is a more precise term because the symmetry is with respect to the axis.

field is concentrated by enclosing the wires in a soft iron casing and by inserting two conical pieces of soft iron called *pole pieces*, each of which contains a small orifice through which the beam passes. The magnetic field lines are then as indicated in Figure 3-1B.

As mentioned earlier, this brief description is not intended to explain how an electromagnetic lens works, but mainly to describe some of the parts and to introduce the terminology commonly encountered in electron microscopy. Suffice it to say that a divergent beam of electrons can be brought to focus within certain limits (i.e., there is substantial spherical aberration) at a point on the axis of the lens.

The optical system of an electron microscope is diagrammed in Figure 3-1B. The illumination source consists of a white-hot tungsten filament, which emits electrons. The potential of the anode, to which electrons are drawn, is normally from 40 to 100 kilovolts greater than that of the filament. The filament and the anode together constitute the *electron gun*. The anode contains a small orifice through which some of the fastest electrons pass. This hole plus a small aperture just below it collimate the electrons to form a beam. The beam is slightly divergent because the electrons are deflected toward the edge of the orifice of the anode owing to its positive potential. The divergent beam is then made to converge onto the specimen by an electromagnetic condenser lens. The beam is rarely focused sharply on the sample because an intense beam could destroy it.

The image is formed by what is often called the subtractive action of the sample. That is, some of the electrons are scattered from the atoms of the object. The pattern of this loss of electrons generates the image pattern (in much the same way that the light intensity is reduced by an absorbing object in the light microscope). The *objective lens*, which is adjusted so that the sample is precisely at its focal point, then refocuses the beam to produce an image. This image is then magnified in several stages by three electromagnetic lenses called the *diffraction*, *intermediate*, and *projector lenses*. The final projector lens forms the image on either a fluorescent screen or a photographic plate.

The electron microscope differs in three respects from the light microscope. First, because electrons do not travel very far in air, the entire microscope column must be in a high vacuum; hence, the object must always be dry. If it were not dry, the water in the sample would boil in the vacuum of the column. This requirement for dryness means that an object cannot be alive. The technique for drying with a minimum amount of destruction will be discussed in the section below on sample preparation. Second, because the magnification of an electromagnetic lens is proportional to the magnetic field, which in turn is proportional to the current in the windings, the magnification can be varied continuously by varying the current through the windings of the lens. In light optics, magnification is fixed by the set shape of the glass lens; hence, many objectives are needed to cover a range of magnification. Third, none of the primary aberrations (i.e., the spherical and chromatic aberrations discussed in Chapter 2) can be corrected in standard electromagnetic lenses because the magnetic lenses are always convergent. (The Crewe microscope described in a later section partly corrects spherical aberration.) To reduce spherical aberration and thereby improve image quality, the lenses are operated at very small numerical apertures. (Numerical aperture (NA) is discussed in Chapter 2.) This has the effect of severely limiting the resolution allowable by the Compton wavelength of the electron because the limit of resolution is $0.61\ \lambda/\text{NA}$. For an electron, the Compton wavelength, λ, is hc/E, in which h is Planck's constant, c is the velocity of light, and E is the energy of the electron. For a 60-kV electron this is 0.03 Å. However, because the numerical aperture of a magnetic lens is normally about 0.0005, the practical limit is more like 4 Å. (As will be discussed later, the nature of the sample actually allows this limit to be reached only rarely.) Nonetheless, the resolution is still approximately 500 times as great as that obtained with a light microscope.

Methods for Preparing Samples and Producing Contrast

Preparation of Specimen Supports

The great capability of all matter for scattering electrons requires that a sample be very thin—otherwise no beam will get through to form an image. In practice, the maximum thickness is approximately 0.1 micron (0.1 μm; 1000 Å) for 100 Å resolution and from approximately 50 Å to 100 Å for 10 Å resolution. This poses no real problem in observing viruses, fibrils, or macromolecules, but for most cells, which range from 1 to 50 μm in thickness, it is necessary to make thin sections (see Embedding, Sectioning, and Staining). This requirement clearly means that the sample support (i.e., the equivalent of a microscope slide) must also be very thin, uniform in thickness, and without obvious structure at high magnification.

The specimen support, or *grid* (Figure 3-2) for all samples consists of a disc cut from a rigid copper (in some cases, platinum) mesh with

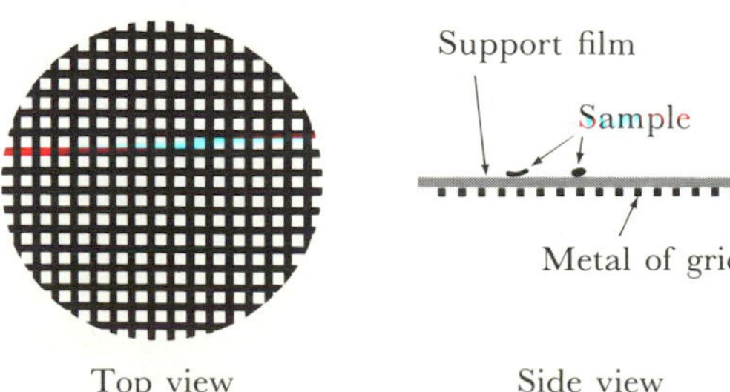

Figure 3-2
A specimen support or grid, which usually consists of a fine copper mesh (the grid itself) overlaid with a thin film of plastic or carbon (the support film). The sample rests on the support film.

openings approximately 75 μm per side, overlaid with a thin "electron-transparent" film called the support film.* Electron transparency indicates that its electron-scattering power is both low and uniform. Commonly used films consist of layers from 100 Å to 200 Å thick of either carbon or various plastics (Parlodion, Formvar). Unfortunately, no support film is truly structureless and for high-resolution work with macromolecules whose dimensions are comparable to film thickness, variations in the intensity of the background produce a "granularity," the grains ranging from 5 Å to 10 Å. This seriously limits the attainable resolution. Films are prepared in one of three ways (Figure 3-3). Parlodion films are prepared by placing a drop of a solution of Parlodion in amyl acetate on a water surface. (A liquid surface is used because it is very smooth.) The droplet spreads, the solvent evaporates, and a thin film of the plastic forms. Formvar films are prepared by dipping a smooth glass microscope slide into a solution of the plastic and then removing it. When dry, the thin film on the glass will slip onto a water surface if the slide is slowly lowered into the water. (This dipping method is preferred by some microscopists for Parlodion.) Carbon films are prepared by evaporating onto freshly cleaved mica (which is a molecularly smooth surface, being a single plane of a crystal), and floating the film onto a water surface as is done in preparing Formvar films. In all cases, the film is mounted on grids in either of two ways: (1) it can be lowered onto the grids—which have been placed on the bottom of the container beforehand—by draining the water off, or (2) the grids can be placed on the film from above (as shown in Figure 3-3) and the entire support picked up by touching the surface with a sheet of plastic or absorbent paper. It should be notice that there is an inherent difference between methods 1 and 2 for attaching the film to the grid. In method 1 the sample is ultimately placed on the side of the film that faces the air; in method 2 the sample is on the water side of the film.

The support films just described are hydrophobic. Since most biological particles and macromolecules are hydrophilic, it is often difficult to transfer the sample from the solution to the support film. Various means of sample preparation are used that avoid this difficulty. A useful alternative is to alter the support film so that it becomes hydrophilic. There are numerous ways to do this but the two most important are the glow discharge and the polylysine procedures. In the glow-discharge method, the grid holding the support film is placed in a chamber having a partial vacuum and a high voltage is applied between two metal plates. The air molecules become ionized and produce a glow discharge as in fluorescent lamps. The support film is exposed to the glowing plasma and gradually acquires a charge that is retained for about an hour. The charged film is able to draw proteins from solution and bind the protein

* The word "grid" is used by microscopists to mean both a disc cut from the mesh but not coated with a support film and the coated mesh.

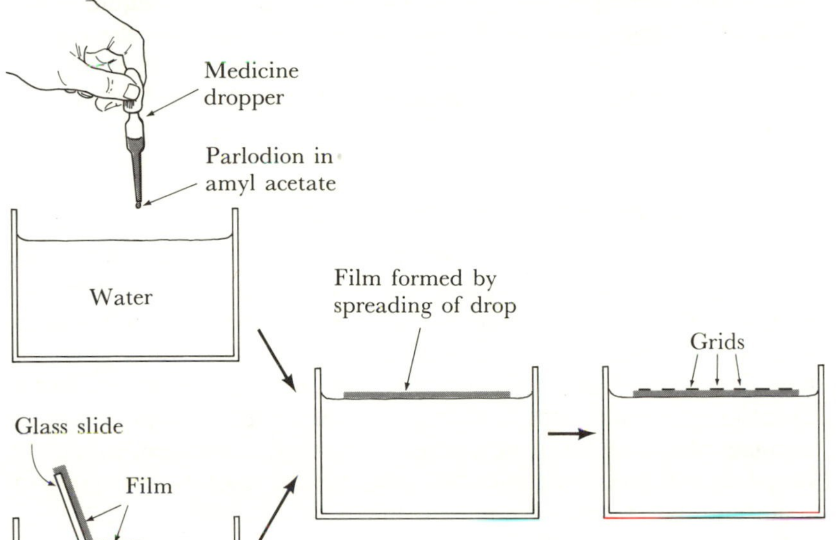

Figure 3-3
Preparation of a plastic support film. In one method, a droplet of Parlodion dissolved in isoamyl acetate is placed on a clean surface of water. The droplet spreads and the solvent evaporates, leaving a thin film of the plastic. In the other method, a glass microscope slide is coated with a thin film of either Parlodion or Formvar by dipping the slide into a solution of the plastic considerably less concentrated than the solution used with the droplet method. In some cases, the film is a thin layer of carbon that has been evaporated onto the glass or onto a mica sheet. When the slide is lowered into the water, the film comes off the glass and floats on the water. Several grids are then placed on the surface of the film and a piece of absorbent paper is placed on the grids. When the paper is lifted up, the grids adhere to it. An alternative method is to use a vessel with a bottom drain. The grids are placed on a screen platform under the water surface before forming the film. The film is then formed and the vessel is drained. As the water level drops, the film comes in contact with the grids.

molecules tightly. The charged film also binds nucleic acids but rather inefficiently. It may be the case that the charge predominately has a negative sign (nucleic acids are negatively charged) because negatively charged proteins are also poorly bound. In the polylysine procedure, the film is glow discharged and then a droplet of the synthetic polypeptide polylysine is placed on each support film for about thirty seconds. The positively charged polylysine molecules bind tightly. The liquid is then removed leaving a monomolecular film of polylysine on the support film. In this way the support film aquires a permanent positive charged (from the ε-amino group of lysine) and can adsorb nucleic acids very strongly.

Sample Preparation and Contrast Enhancement

The intrinsic contrast of biological material is poor because the scattering of the carbon atoms in the support film is of roughly the same magnitude as that of all the principal atoms (C, N, O, P, S) of the material. The usual method for correcting this situation is to deposit heavy metals of very high scattering power on the structure in such a way that the pattern of metal somehow indicates the features of the sample. Useful metals are osmium, platinum, lead, and uranium, although chromium, palladium, tungsten, and gold are sometimes used. Several standard methods for sample preparation and contrast enhancement follow.

Embedding, sectioning, and staining

If the material under observation is too thick for the passage of electrons, a thin slice or section must be made. To prepare a thin section, the sample must be made rigid so that it can be cleanly cut. This process, called *embedding*, consists of the gradual replacement of the aqueous material of the sample with an organic monomer (e.g., methyl methacrylate) that can be hardened by polymerization. The usual procedure is to place the sample in a solution of a *fixative*, which is usually a dilute formaldehyde solution or a mixture of acetic acid and ethanol. The fixative denatures and often cross-links proteins and other structures and presumably fixes all structures in place so that they will not be moved or disrupted by further handling. The "fixed" sample is then transferred to ethanol–water mixtures with gradually increasing concentrations of ethanol until the sample is in 100% ethanol. This process is called *dehydration*. The sample is then transferred to an alcoholic solution of the monomer, which is then stimulated to polymerize.

After the sample has become solid, the plastic containing the supposedly undisrupted sample is sliced with an ultramicrotome (a kind of knife) into layers from 500 Å to 1000 Å thick. The sections are then stained (although staining is sometimes done before embedding) by exposure to solutions of salts of molybdenum, tungsten, lead, or uranium, or to the vapor of osmium tetroxide. (The word *staining* refers to the deposition of a metal by a chemical reaction or the formation of a complex with certain components of the sample, to increase the electron density.) These stains react with proteins and other macromolecules and aggregates and thereby put electron-dense material in the sample. Stained preparations are beautiful to look at (Figure 3-4) and appear to contain considerable detail, yet it must be realized that what is being observed is the distribution of metal atoms and therefore of the chemical groups that can react with a particular stain. An example of a type of artifact that can arise is the deposition of osmium on opposite sides of a thick membrane, producing two black lines separated by an unstained space, which can be mistaken for a double membrane. The embedding and sectioning procedures themselves can induce distortion because of uneven permeation of the sample by the organic monomer and because of the cutting itself.

Only a single layer through a sample is observed when looking at a thin section and this may not always be adequate. To get a picture of the entire sample, a large number of sections are normally examined. An elegant though tedious method is *serial sectioning* in which successive sections are collected in sequence and examined.

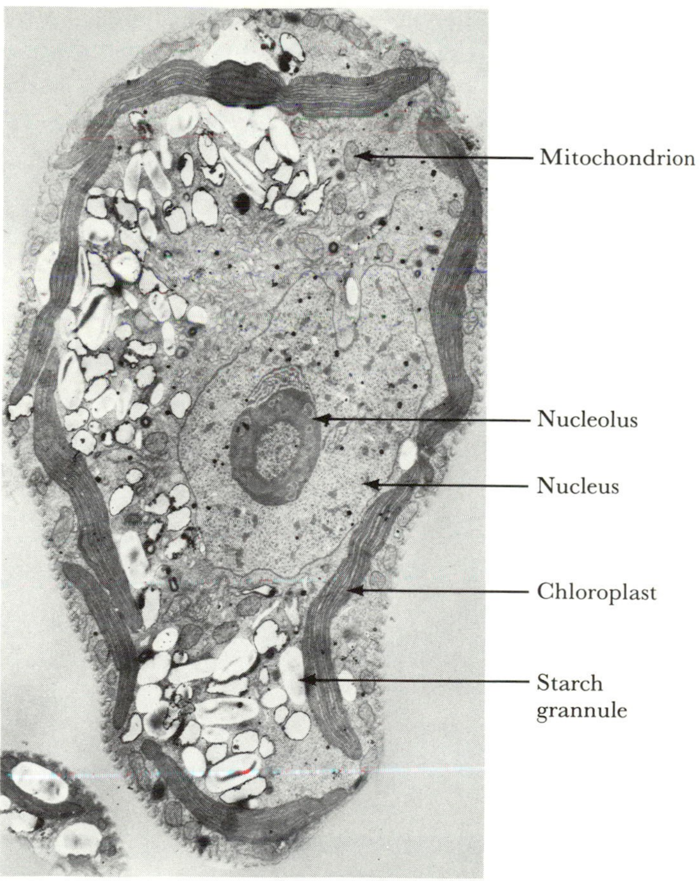

Figure 3-4
Electron micrograph of an ultrathin slice of *Euglena gracilis* stained with osmium tetroxide. [Courtesy of Jerome Schiff and Nancy O'Donohue.]

Replica formation

The method used for observing the surface of an electron-opaque or easily destroyed specimen is called replica formation. The specimen is coated first with a thin layer of platinum and then with a supporting layer of carbon (for strength), both deposited by vacuum evaporation or shadow-casting (see Figure 3-8 and the accompanying discussion). This bilayer is then floated off onto water and picked up on a grid (Figure 3-5). The replica is thus a facsimile of the surface of the object—that is, the contours are the same as those of the sample. This method has been used to study the surfaces of viruses, membranes, and certain protein crystals that are immediately destroyed by the electron beam.

Freeze-etching and the critical-point technique

In replica formation, the water in the sample must be removed before preparing the replica because the production of a film by shadow-casting must be in vacuum. This presents a problem in that structures usually

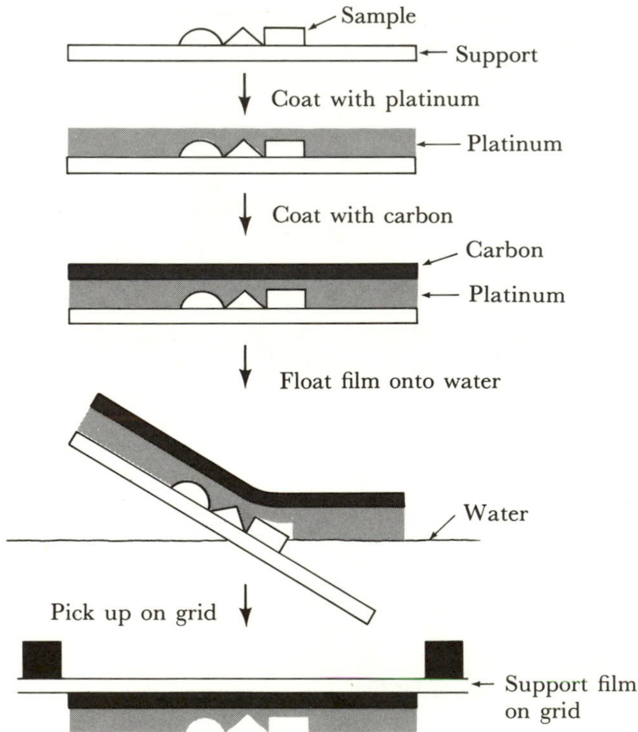

Figure 3-5
Formation of a carbon-platinum replica. In some cases, the support film is applied to the carbon layer before floating on water; a bare grid is then used to pick up the film.

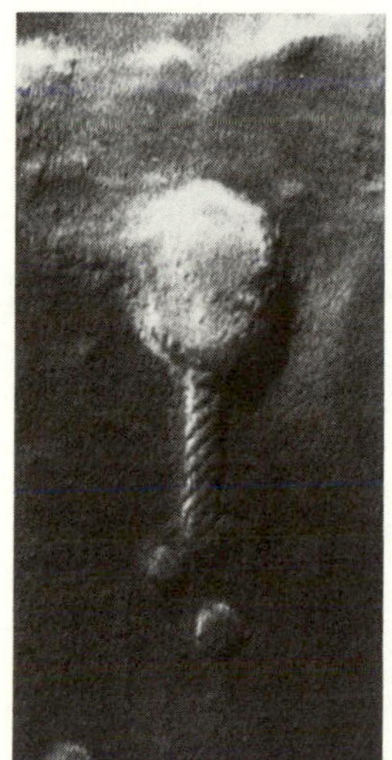

Figure 3-6
An *E. coli* T2 phage prepared by freeze-etching, a slight modification of the critical-point method in which a replica is made. The surface details of the phage tail are particularly clear. This should be compared with the similar T4 phages shown in Figures 3-10 and 3-12. [Courtesy of Manfred Bayer.]

collapse during air drying as a result of surface tension effects accompanying the phase changes that occur during evaporation of the solvent. Freeze-etching and the critical-point method avoid the production of artifacts due to drying. In freeze-etching, the sample is rapidly frozen, sectioned or fractured, and placed in a vacuum with conditions of pressure and temperature such that the water sublimes from the surface of the sample. A replica of this surface is then prepared by evaporating platinum or carbon while it is still in the vacuum. An example of a replica prepared by the freeze-etching method is shown in Figure 3-6.

The critical-point method makes use of the fact that no liquid phase can exist above a "critical temperature" characteristic of each substance. The procedure follows. First, a wet sample is soaked in ethanol. The ethanol is then exchanged with liquid CO_2 under pressure at 15°C. The temperature of the specimen is then raised above 31°C (the critical temperature) and the liquid CO_2 becomes a gas. Presumably, all three-dimensional relations are preserved. A replica can then be prepared or the sample can be observed directly in the microscope, if it has been stained beforehand. This method is especially useful in preserving macromolecular structures if molecules are deposited on a film from a solution.

Freeze-fracture

The replica method and freeze-etching have been combined in a technique that allows visualization of the *internal* structure of extended objects consisting of two or more layers. This method, *freeze-fracture* electron microscopy, is the best procedure currently available for the study of biological membranes.

A sample containing membranes is frozen and then fractured by the impact of a microtome blade (Figure 3-7A). Often the cleavage plane of the membrane, which consists of two layers, as shown in the figure, lies along the middle of the bilayer. The ice is then sublimed away and a replica is made by successive coating with platinum and carbon. A replica of the membrane of a red cell prepared in this way is shown in Figure 3-7B. Micrographs of this kind obtained by Jonathan Singer

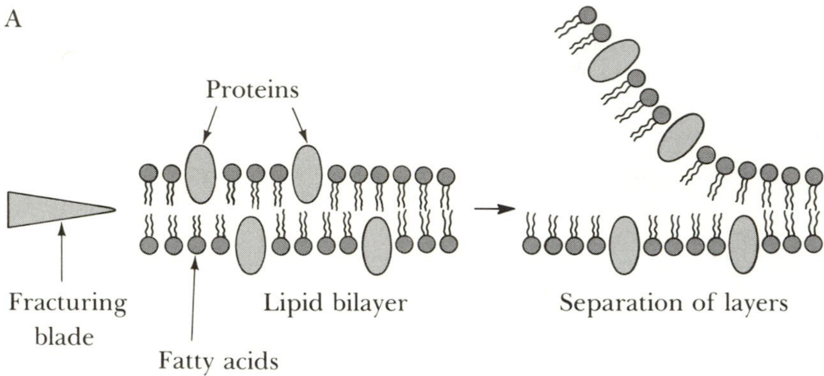

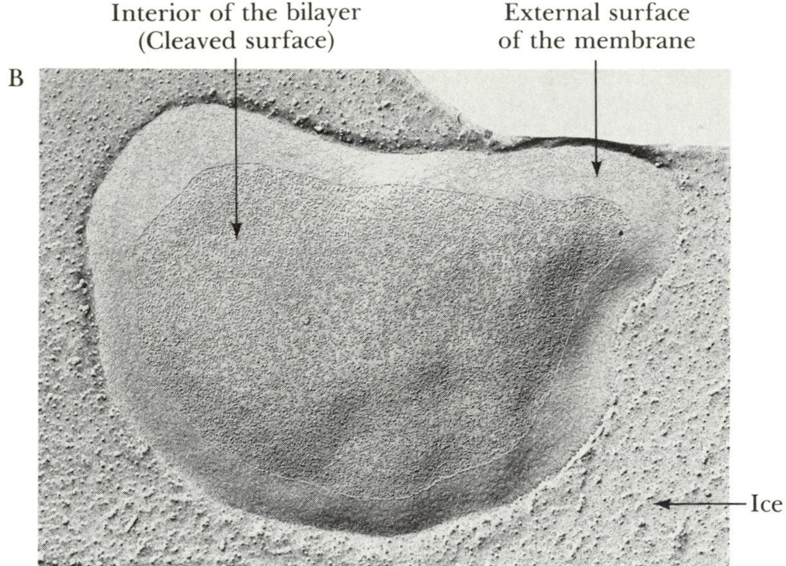

Figure 3-7

A. Schematic drawing of a lipid bilayer membrane being fractured. The tailed circles represent the fatty acids and the ellipses represent the so-called integral proteins postulated by Singer and Nicholson. B. Electron micrograph of a replica of the interior of the plasma membrane of a red blood cell prepared by freeze-fracture. The 75-Å globular particles strewn over the surface are thought to be the integral proteins. A portion of the external surface of the membrane is also visible. [Courtesy of Dr. Vincent Marchesi.]

and Garth Nicholson provided the first evidence for the presence of large protein molecules embedded in many biological membranes, as shown schematically in part (A) of the figure.

Shadow-casting

A great deal of electron microscopy is concerned with the structures of particles—such as viruses, phages, and ribosomes—and of macromolecules. The sizes of such objects as well as limited information about their structures can be obtained by shadow-casting. The particles (in solution or suspension) are applied by spraying a suspension of the particles onto a grid overlaid by a support film. The liquid quickly evaporates, the sample is placed in vacuum, and a heavy metal is applied by evaporation. This requires boiling a metal and is typically done as follows. A thin metal wire is wrapped around a tungsten wire (Figure 3-8) or small lumps of metal or a metal oxide are placed in a tungsten wire basket. An electric current is passed through the tungsten wire until

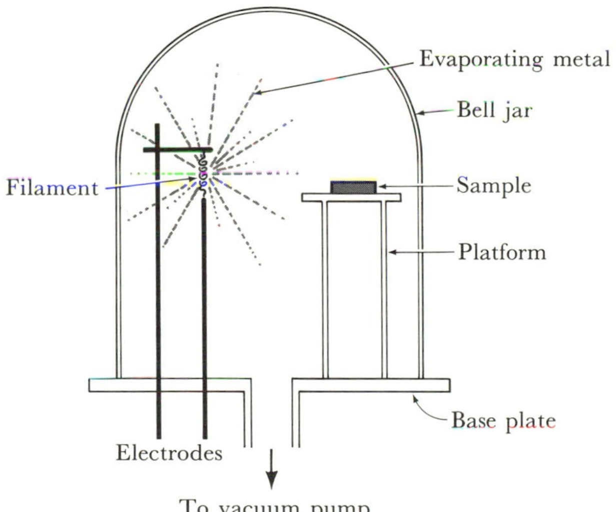

Figure 3-8
Apparatus for vacuum evaporation or shadow-casting. The bell jar is evacuated by vacuum and diffusion pumps. The tungsten filament, around which the metal wire to be evaporated is wrapped, is heated to the boiling point of the metal. In 10 or 15 seconds, the metal is boiled away and forms a film on the sample. The thickness of the film is proportional to the amount of metal put on the filament and inversely proportional to the square of the distance from filament to sample.

it becomes white hot. At this temperature the metal to be evaporated boils away. The metal atoms are projected in all directions and, if the vacuum is good, in straight lines. If evaporation is from an acute angle (Figure 3-9), metal will pile up on only one side of the sample and will cover the grid except in the shadow of the particle. If the vertical (H) and horizontal (L) distances from the evaporation source to the specimen are known, the height (h) of the particle above the surface of the grid can be calculated from the length (d) of the shadow cast by the specimen because $h/d = H/L$. Hence, the dimensions of the particle can be determined. Figure 3-10 shows a shadowed preparation of a phage.

The amount of metal deposited on the sample affects contrast. If there is too little, the sample is not visible and, if there is a great excess, the sample may be totally buried. The correct amount of metal is determined empirically by using metal wire of a particular diameter and counting the number of turns of this wire around the tungsten that, when used, results in optimal contrast. When metal oxide grains are used in a tungsten wire basket, the grains are usually weighed.

Some metals, for instance, 100% platinum, may yield variable contrast. For platinum this is a result of formation of a platinum-tunsten alloy. The platinum in the alloy does not evaporate so that less metal is deposited on the sample. If wire consisting of 80% platinum and 20% palladium is used, this problem does not occur because palladium prevents formation of the platinum-tungsten alloy.

The magnitude of the vacuum during shadow casting also affects the picture quality. If the pressure is too high, a significant fraction of

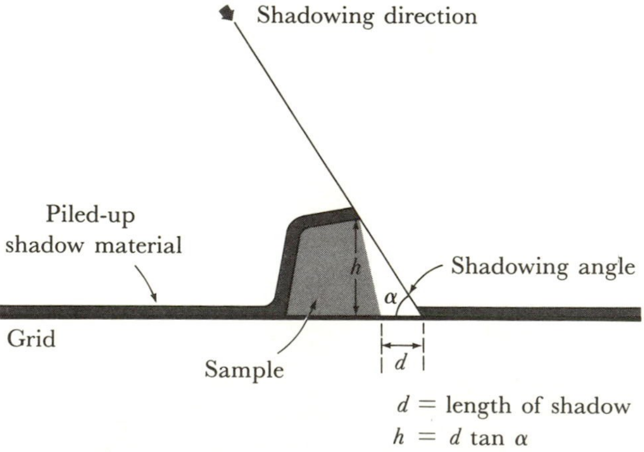

Figure 3-9
Determination of the height of an object from the length of its shadow. Because evaporation is not done from directly above the sample, there is a region on the support film on which there is no metal. This region is called the shadow.

Figure 3-10
Electron micrograph of *E. coli* bacteriophage T4 prepared by shadowing. The shadows are indicated by arrows. [Courtesy of Jonathan King.]

the metal atoms collide with air molecules and fail to reach the sample. One can, of course, compensate somewhat for this effect by increasing the amount of metal wire. This is not a particularly effective technique for the following reason. Most of the metal atoms evaporate as single atoms but some of these are in large clusters. These clusters occur infrequently but they are less likely to be deflected if they collide with an air molecule. At higher pressures the clusters have a higher probability of reaching the sample than do individual atoms. Thus, by increasing the amount of metal evaporated when the pressure is too high, one merely enhances the fraction of molecules that arrive as clusters. The sample is then very coarsely shadowed and most of the detail will be lost.

Shadow-casting has not been too successful for smaller macromolecules because of the small size of the shadow and the granularity of the support film. Many of the problems can be avoided by using the *negative-contrast procedure* described next.

A special use of shadow-casting, the Kleinschmidt technique for observing nucleic acid molecules, will be discussed later.

Negative-contrast technique

The negative-contrast (incorrectly and commonly called negative staining) procedure of Brenner and Horne consists of embedding small particles or macromolecules in a continuous stain or electron-opaque film (Figure 3-11). The stain penetrates the interstices of the particle but not the particle itself. The image is a result of the relative intensity of the beam at every point, which is proportional to the thickness of the opaque material at that point. Hence, contrast is achieved by virtue of the particle reducing the effective thickness of the opaque film—that is, the particles are seen in outline. Figure 3-12 shows a negative-contrast picture of a phage. In this procedure the sample is either mixed with the stain and sprayed on the grid or sprayed on the grid first and then sprayed with the stain. The interpretation of negative stained samples is sometimes difficult because various patterns can be observed depending on (1) the thickness of the stain, (2) whether it has penetrated the interstices of the particle, (3) whether it lies above and below the particle, and (4) whether any of it has adsorbed specifically to the sample (positive staining). For this reason it is usually necessary to look at a large number of preparations and particles. Nonetheless, the negative-contrast method has been used successfully for a wide variety of phages

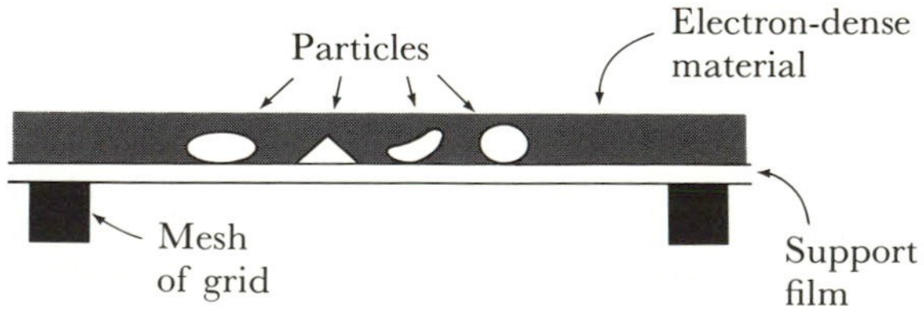

Figure 3-11
The negative-contrast method. Four particles are embedded in an electron-dense material. As the beam passes through the sample, the attenuation will depend on the total thickness of the stain; therefore, more electrons will pass through the regions containing each particle.

and viruses. The most useful stains for negative contrast are phosphotungstic acid salts and uranyl acetate, nitrate, or formate. It should be realized, of course, that resolution in the negative-contrast method depends on the size of the opaque atoms (i.e., approximately 5 Å).

In recent years, the negative-contrast procedure, when coupled with glow-discharged support films, has become one of the most important techniques for studying the structure of individual protein molecules and the arrangement of subunits in complex proteins. A droplet containing the molecule of interest is placed on a glow-discharged carbon

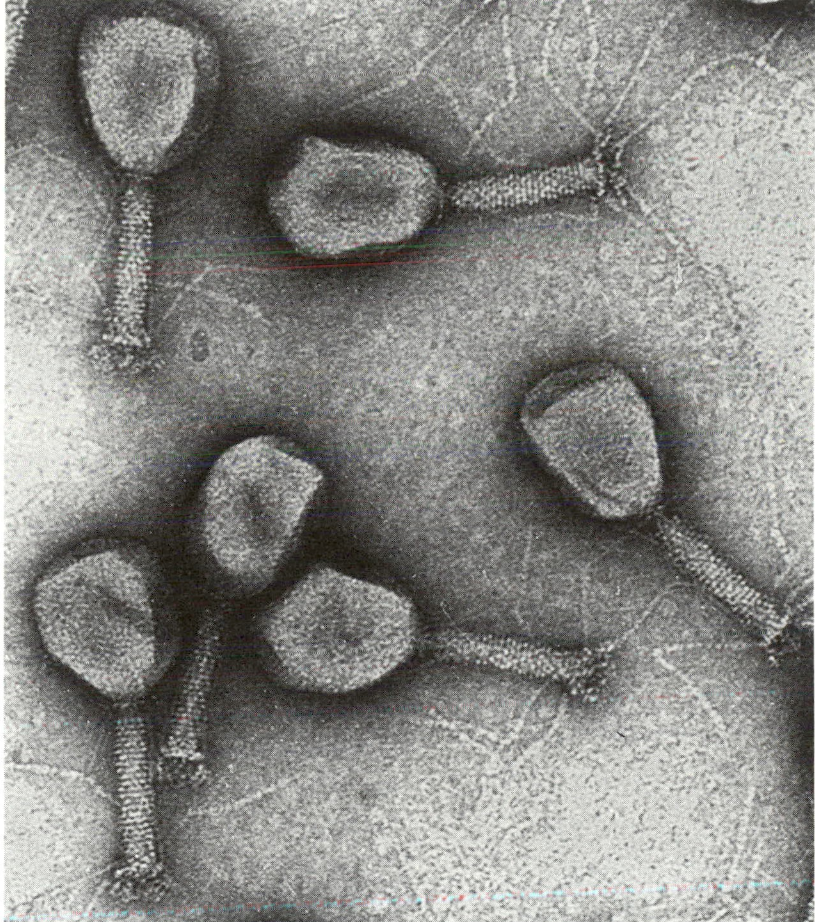

Figure 3-12
Electron micrograph of *E. coli* phage T4 prepared by the negative-contrast procedure. The stain is phosphotungstic acid. Compare with Figure 3-10. The individual black dots are tungsten atoms in the phosphotungstic acid. [Courtesy of Jonathan King.]

film. After about thirty seconds, at which time most of the molecules have adsorbed to the film, the liquid is removed and the grid is washed. For best results uranyl formate, which provides an extremely fine-grained and uniform background, is used. Figure 3-13 shows electron micrographs of the proteins aspartyl transcarbamylase and fibrinogen. The protein molecules are adsorbed to the film in various orientations, allowing numerous views of the molecule so that the arrangement of different parts of the molecule can be seen.

Proteins have also been visualized after binding to polylysine-coated carbon films, followed by negative staining. However, the simple glow-discharged films usual yield better micrographs.

Positive staining

Positive staining has not had widespread use for most macromolecules because it is not usually possible to attach a sufficiently large number of heavy atoms to obtain good contrast, although it has been possible with large molecules and structures such as ribosomes, DNA, RNA polymerase, and collagen. The collagen results are especially beautiful and deserve description because they indicate the great analytical power of this technique. When tropocollagen molecules are allowed to aggregate side-by-side to produce collagen and the resulting structure is stained with phosphotungstic acid at pH 4.2 (binding to positive groups), a pattern of bands is observed (Figure 3-14). If uranyl acetate, which binds negative groups, is used, the banding pattern is exactly the same. Therefore the sideways aggregation probably includes the conjunction of positive and negatively charged groups. Analysis of the banding pattern of several different types of collagen aggregates shows that the tropocollagen molecule can form head-to-head or head-to-tail fibers

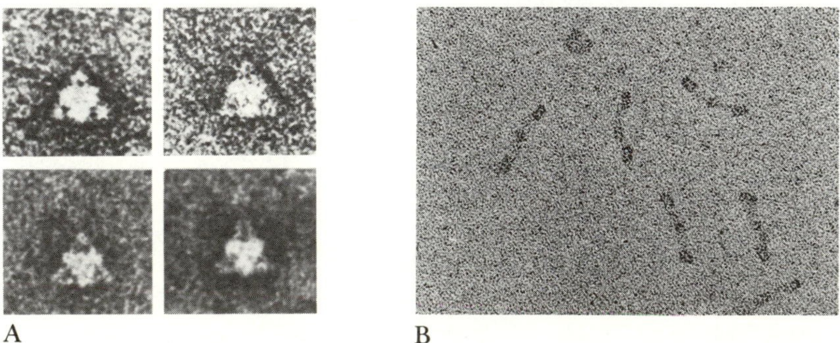

Figure 3-13
Electron micrographs of several molecules prepared by the negative contrast method. (A) Four molecules of aspartyl transcarbamylase. (B) A field of fibrinogen molecules. [Courtesy of Robley Williams.]

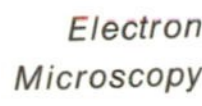

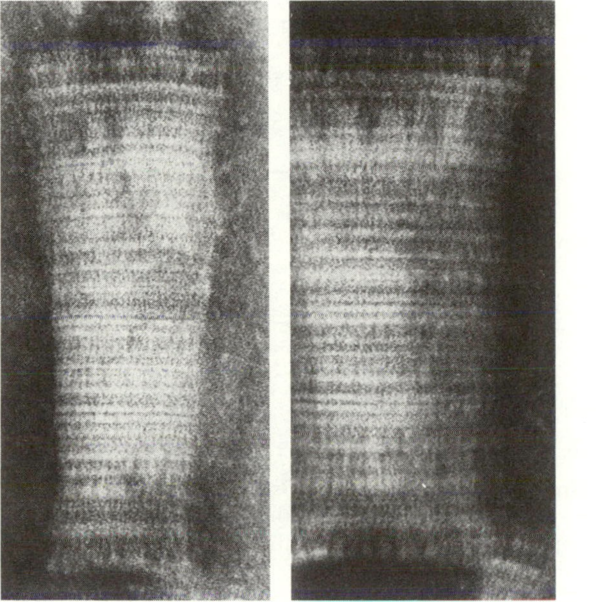

Figure 3-14

A. Electron micrographs of two segment-long-spacing aggregates of tropocollagen molecules positively stained by phosphotungstic acid. The length of the aggregate is approximately 0.28 μm. The bands are the parts of tropocollagen that bind the stain. [Courtesy of Peter F. Davison.] B. A. schematic representation of the collagen fiber. Each arrow represents a tropocollagen molecule; the molecules are attached end-to-end and aggregated sideways but in a quarter-staggered array. This model was derived from an analysis of the pattern of bands of this and other forms of collagen. [From A. J. Hodge, J. Highberger, G. Deffner, and F. O. Schmitt, *Proc. Natl. Acad. Sci. U.S.A.* 46(1960):186.]

and that in the standard structure the sideways aggregation involves a displacement of one-quarter of the molecular length from one tropocollagen to the next.

Kleinschmidt spreading with positive staining and rotary shadowing

Probably the most spectacular method of sample preparation of the past decade has been the Kleinschmidt procedure for visualizing DNA. In a single step, artifacts due to drying are eliminated and extraordinary contrast is obtained. This technique is now used in almost all biochemical laboratories and can be learned in an afternoon. A drop of a DNA solution in 0.5 M to 1.0 M ammonium acetate containing 0.1 mg/ml of cytochrome *c* is allowed to flow down a glass slide onto the surface of 0.15 M to 0.25 M ammonium acetate (Figure 3-15). As the drop touches

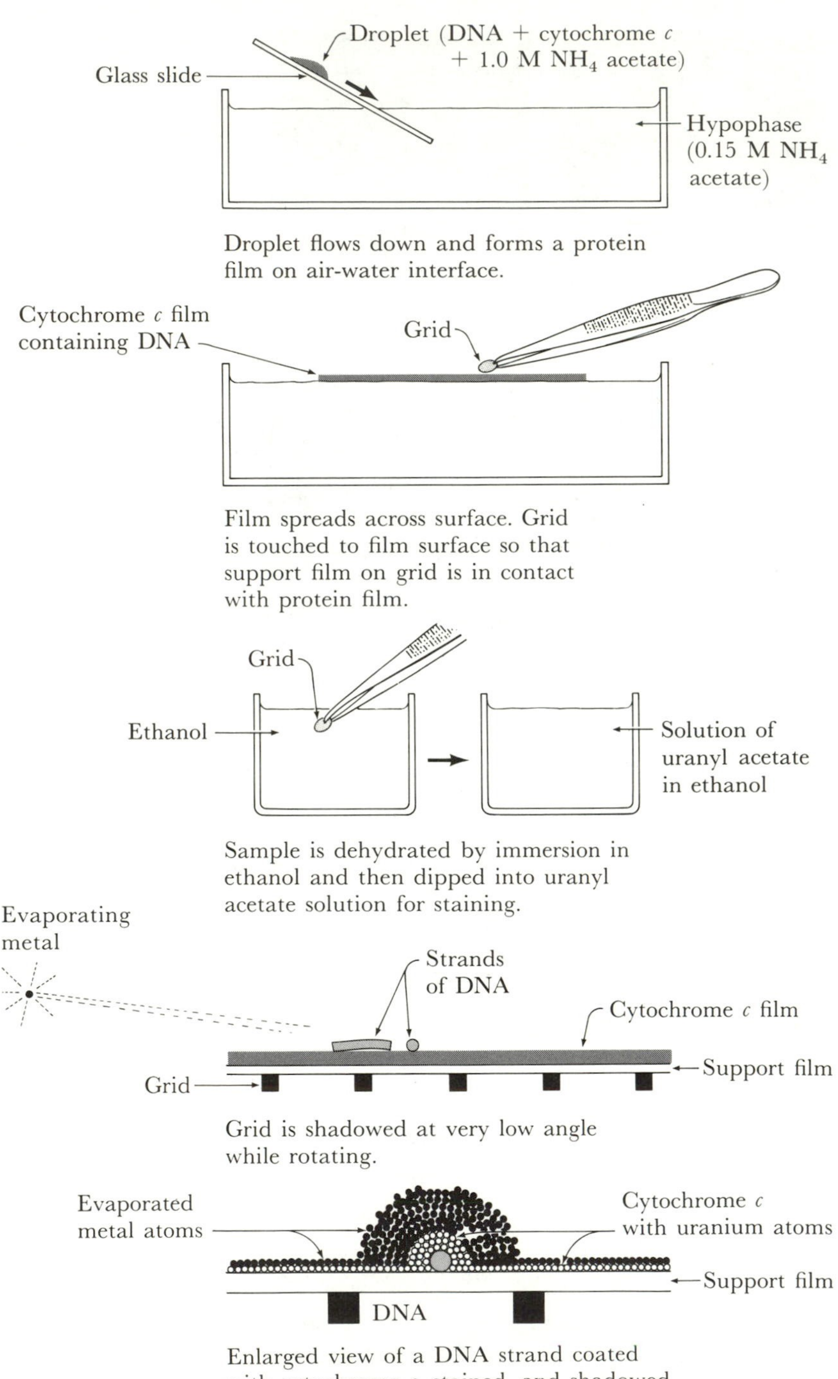
Droplet (DNA + cytochrome c + 1.0 M NH4 acetate)
Glass slide
Hypophase (0.15 M NH4 acetate)
Droplet flows down and forms a protein film on air-water interface.
Cytochrome c film containing DNA
Grid
Film spreads across surface. Grid is touched to film surface so that support film on grid is in contact with protein film.
Grid
Ethanol
Solution of uranyl acetate in ethanol
Sample is dehydrated by immersion in ethanol and then dipped into uranyl acetate solution for staining.
Evaporating metal
Strands of DNA
Cytochrome c film
Support film
Grid
Grid is shadowed at very low angle while rotating.
Evaporated metal atoms
Cytochrome c with uranium atoms
Support film
DNA
Enlarged view of a DNA strand coated with cytochrome c, stained, and shadowed.

Figure 3-15
Preparation of DNA for electron microscopy, using the Kleinschmidt method.

the surface, a film of denatured cytochrome *c* spreads across the surface. This film contains somewhat extended DNA molecules to which a thick (100–200 Å) layer of denatured cytochrome *c* binds. If a grid is touched to the denatured protein film, a drop containing a part of the film is transferred to it. When the grid with the adhering drop is immersed in alcohol, the aqueous phase is removed and the film adheres tightly to the support film on the grid. As used at present, the technique includes a preliminary positive staining with uranyl acetate; the protein absorbed to the DNA becomes stained as well as the background film but, owing to the excess protein bound to the DNA, good contrast is achieved. Contrast is enhanced (or created, if staining is not used) by shadow-casting a metal (usually platinum) at a very small angle while the sample is rotating. Because the DNA coated with protein projects above the protein film, metal piles up against the DNA-protein complex like snow drifting against a fence—but on both sides and on all molecules, regardless of orientation, because of the rotation. The contrast is extraordinary, as shown in Figure 3-15. This method can be used to determine the length of DNA and whether it is circular or supercoiled. Under certain conditions (usually by incorporation of the denaturant formamide into all solutions and keeping the salt concentration low), single-stranded polynucleotides become extended and are easily visualized. Single-stranded DNA and RNA are distinguishable from native DNA by their relative thinness and kinkiness (Figure 3-16). In laboratory jargon the

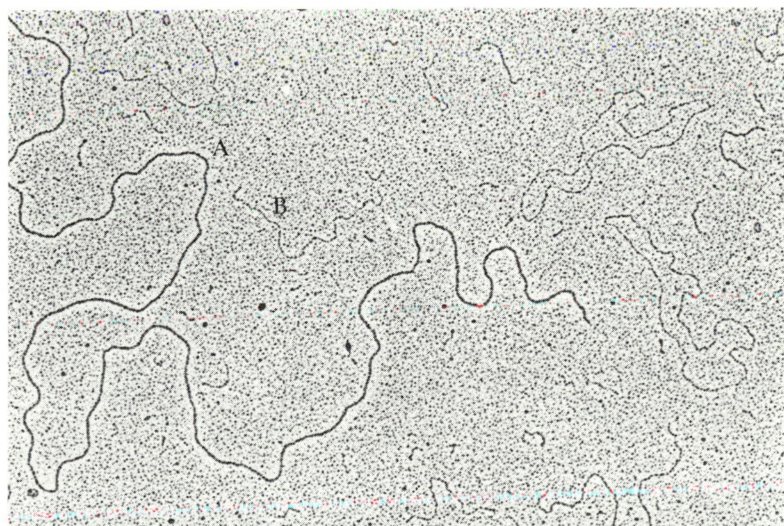

Figure 3-16
Electron micrograph of (A) double-stranded DNA and (B) single-stranded DNA prepared by the Kleinschmidt method using the formamide to prevent collapse of the single strands. Note that single-stranded DNA is kinkier than double-stranded DNA.

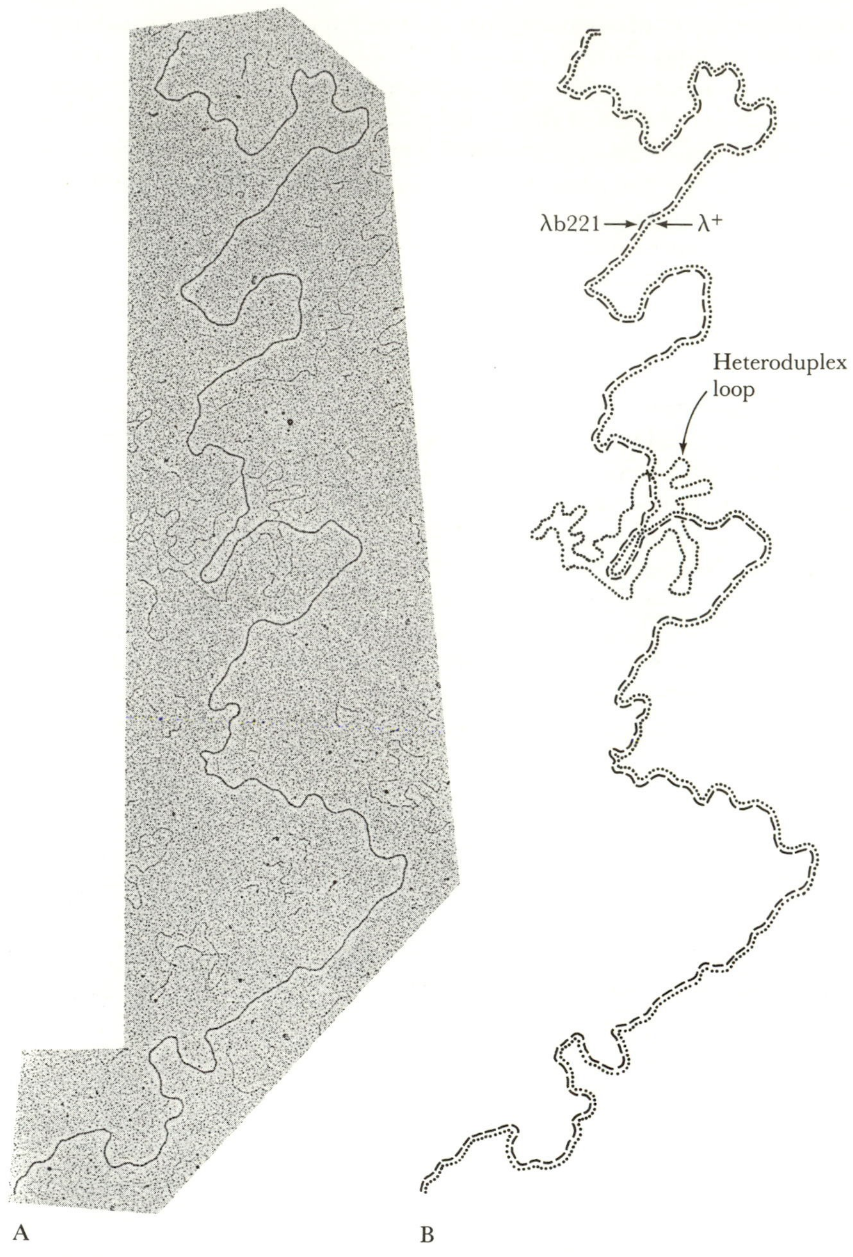

Figure 3-17

A. Electron micrograph of a heteroduplex of λ^+ and $\lambda b221$ (a deletion phage); the DNAs were denatured and renatured. B. The dashed line in the drawing is a single strand of $\lambda b221$ DNA. The dotted line is single-stranded λ^+ DNA. The heteroduplex loop is the section of λ^+ corresponding to the deletion of $\lambda b221$. This is a formamide spreading. [Courtesy of Manuel Valenzuela.]

technique using formamide is called *formamide spreading*; the usual procedure for preparing double-stranded DNA is called an *aqueous spreading*.

In a variation of this method, the diffusion method, a cytochrome *c* film is formed on a DNA solution and DNA molecules diffuse upward and adhere to the film. This is a slow process but allows much smaller DNA concentrations to be used.

One important application of the Kleinschmidt technique is seen in the *heteroduplex method*. Here single strands from two different DNA molecules are allowed to hybridize (see Chapter 19). Homologous regions (i.e., regions having complementary base pairs) show up as double-stranded DNA but nonhomologous regions remain as single-strand loops (Figure 3-17).

Special procedures are often needed when attempting to localize protein molecules bound to DNA molecules. This is because often the solutions needed for the spreading contain salts, detergents, or other substances that are not compatible with a stable DNA-protein interaction.

It has been found that preincubation of the sample with the dialdehyde, glutaraldehyde, stabilizes the interaction. Glutaraldehyde produces protein-protein, protein-nucleic acid, and nucleic-acid–nucleic-acid covalent cross-links that prevent dissociation of the components. The chemistry of the reaction is not known. Glutaraldehyde is always worth trying when a sample of any kind is found to dissociate during preparation. It has also had wide use in preserving thin sections of cells and tissues of plants and animals; it is thought by some microscopists to be the best fixative available.

A recently developed variant of the Kleinschmidt technique replaces the cytochrome *c* with benzalkonium chloride, a surfactant that also forms a film to which DNA adheres. The DNA is shadowed with platinum but, because the DNA is not coated with protein, it appears much narrower than when cytochrome *c* is used. Positive staining is still used under conditions that allow the uranium atoms to bind directly to the DNA. This procedure gives better resolution of macromolecules (e.g., RNA polymerase) bound to the DNA (Figure 3-18), however it is not as reliable as the cytochrome *c* procedure. Probably the best method for viewing proteins bound to nucleic acids uses adsorption of the complexes to polylysine-coated support films followed by negative staining. This will be shown shortly (page 101).

Special procedures used with the Kleinschmidt technique

There are several procedures that can be used to localize specific regions of a DNA molecule. One of the most important is *denaturation mapping*.

When a solution of DNA molecules is heated, the increased kinetic energy causes disruption of the hydrogen bonds that are responsible for

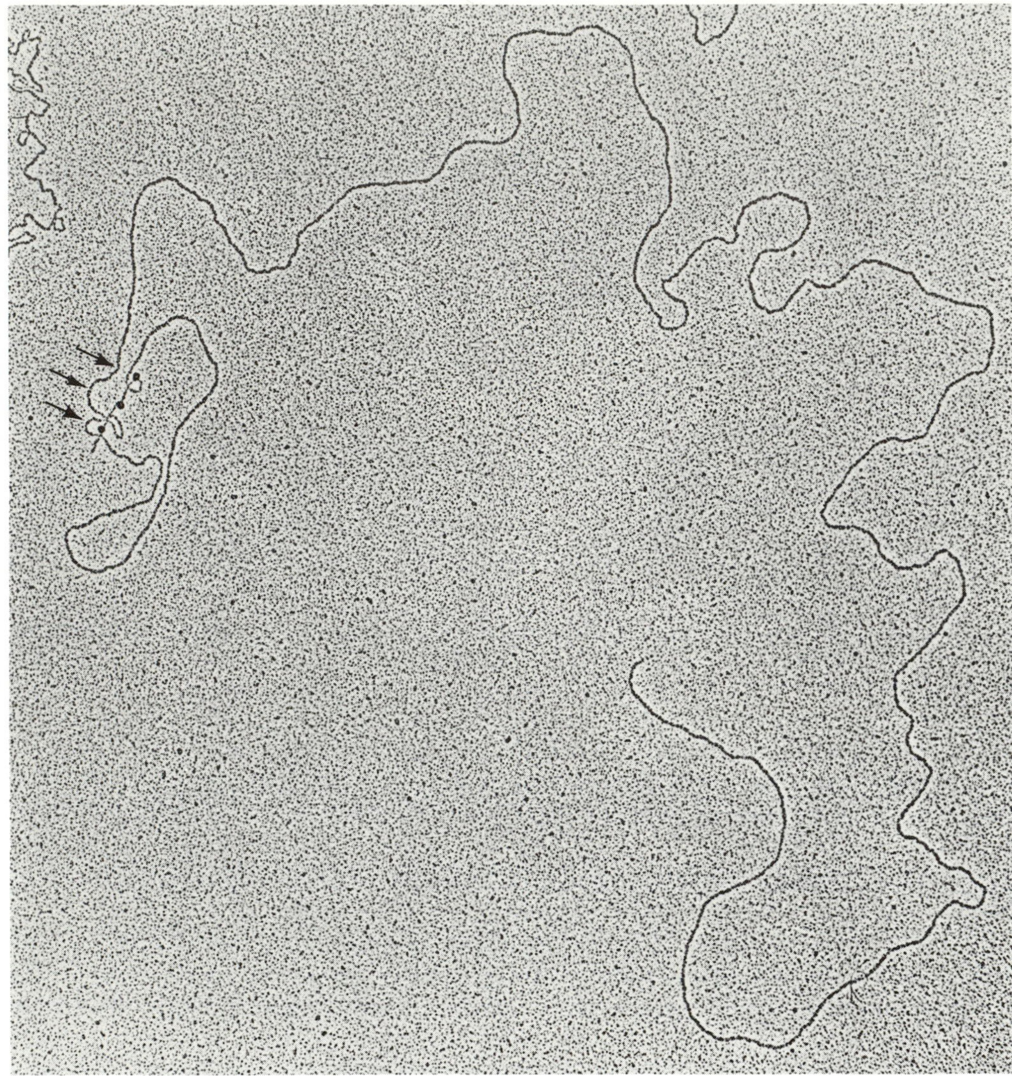

Figure 3-18

DNA visualized by the benzalkonium chloride method described in the *Proceedings of the National Academy of Sciences* [72(1975): 83]. Molecules prepared for electron microscopy are narrower than if prepared using the Kleinschmidt method. Hence, bound protein molecules are easily seen. This micrograph shows three spherical molecules (arrows) of *E. coli* RNA polymerase bound to phage T7 DNA. The width of the photograph is approximately 3 μm. [Courtesy of T. Koller.]

base pairing. The adenine·thymine (A·T) pairs are joined by two hydrogen bonds whereas there are three hydrogen bonds in a guanine·cytosine (G·C) base pair (Chapter 1). Thus the A·T pairs are broken at a lower temperature than G·C pairs. It is possible to choose a temperature at which a few particularly (A + T)-rich sequence are totally single-stranded and the remainder of the molecule is double-stranded. If formaldehyde is added, it reacts nearly irreversibly with the amino groups on the bases that are not hydrogen-bonded; then the sample can be cooled without reformation of any base pairs. Such a molecule is said to be *partially denatured.* As an alternative to elevated temperature, high pH (plus formaldehyde) can be used to create partially denatured DNA–in fact, this procedure is used more often than high temperature despite its requirement for controlling the pH to ± 0.05 pH unit.

Partially denatured DNA can be prepared for electron microscopy by a formamide spreading. An electron micrograph of such DNA shows each denatured region as a section of DNA consisting of two separated single strands (Figure 3-19A); the single-stranded regions are usually called *denaturation loops* or *bubbles.* It is found that the positions of the bubbles are nearly identical for all molecules in a sample of molecules of the same type. A histogram showing the positions obtained by measuring a large number of molecules is called a *denaturation map* (Figure 3-19B). The important aspect of this method is that the denaturation maps of two different types of DNA molecules (for example from two different phages) are *never* the same (Figure 3-19C). Thus a *denaturation map is a unique way to identify a particular type of DNA molecule.* Following are several samples of the use of denaturation mapping.

Example 3-A

☐ Localization of the position of the ends of a linear DNA molecule in the circular form.

The bubbles in the linear DNA molecule isolated from *E. coli* bacteriophage λ particles are in the positions shown schematically in Figure 3-20. Circular molecules are isolated from bacteria infected with phage λ. A partially denatured circular molecule is also shown in the figure. Comparison of the patterns of the bubbles indicates that the ends of the linear molecule must be joined at the position of the arrow.

Example 3-B

☐ Identification of very long molecules, isolated from *E. coli* infected with phage λ, as polymers (concatemers) of λ DNA.

When isolating DNA from cells infected with phages, one usually recovers a mixture of phage and bacterial DNA. Anne Skalka and her coworkers partially denatured a portion of the mixture and observed that there were two classes of linear molecules. Class 1 consisted of molecules of variable length having many patterns of bubbles that were quite different from that of λ DNA; presumably this class consisted of broken fragments of *E. coli* DNA. Class 2 consisted of molecules that were also of variable length and longer than a individual λ DNA

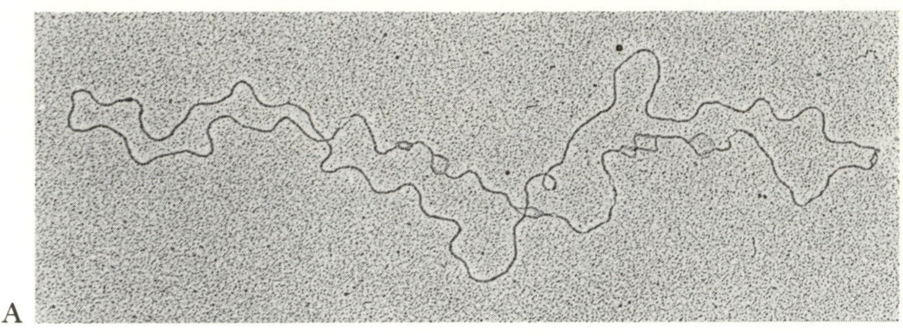
A

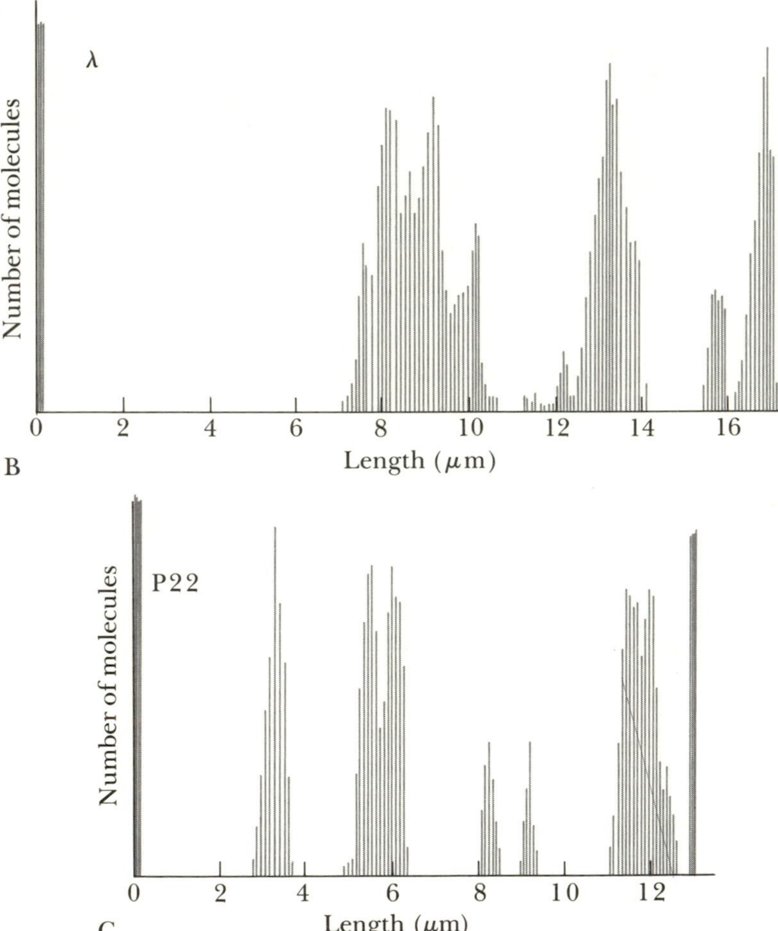

Figure 3-19
A. An electron micrograph of a partially denatured DNA molecule showing bubbles. [Courtesy of Manuel Valenzuela.] B. A denaturation map of phage λ DNA showing the number of molecules having separated strands at particular distances from one end of the molecule. The heavy vertical lines indicate the ends of the DNA molecule. The molecules are oriented left to right so that bubbles are nearly in the same positions. C. Same as B but for phage P22 DNA.

molecules and had a denaturation map that consisted of one or more λ maps arranged in tandem, as shown in Figure 3-21. These molecules are tandem units of λ DNA known as *concatemers* (see figure). The denaturation map also shows that all of the concatemers do not have the same termini. Later work showed that these concatemers arise by a special mode of DNA replication called rolling-circle replication.

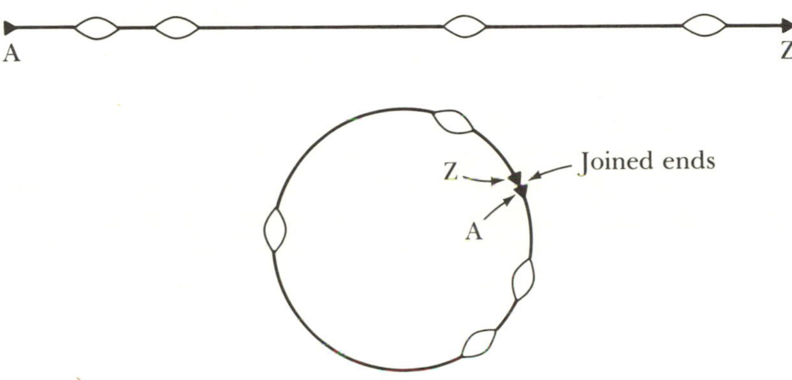

Figure 3-20
Partially denatured linear and circular forms of the same molecule.

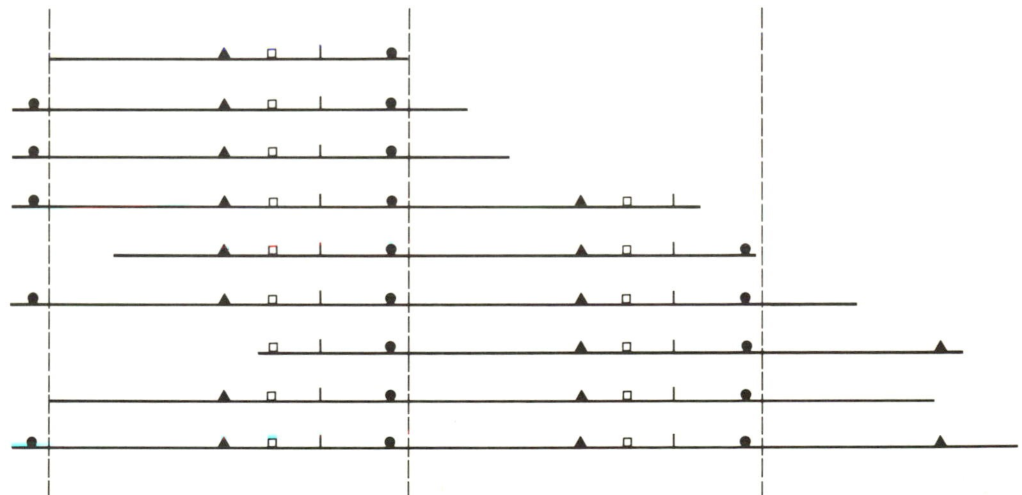

Figure 3-21
Concatemers, isolated from *E. coli* infected with phage λ, showing the positions of denaturation bubbles. The four bubbles are denoted by a triangle, open square, vertical line, and solid circle to aid in aligning the concatemers. The vertical dashed lines indicate the positions of the ends of λ DNA molecules of unit length.

Example 3-C □ Determination of the direction of replication of λ DNA.

In a series of clever experiments, Ross Inman, who originated the technique of denaturation mapping, and his colleague, Maria Schnös, isolated replicating circular DNA molecules from *E. coli* infected with phage λ and obtained denaturation maps of these molecules. A few representative molecules are shown in Figure 3-22. If only a single replication fork were moving clockwise, the stationary fork (the left one in the figure) would remain at a fixed position with respect to the bubbles and the moving fork would shift as the replication loop enlarged. The fixed position is called the replication origin; that is, the site at which replication begins. However, this is not what is observed. Instead, as the replicating loop enlarges, both forks are displaced by increasing distances from the three upper bubbles. Thus, both forks are moving or replication is bidirectional, that is, both clockwise and counterclockwise. Furthermore the position midway between the two replication forks is always at the same position with respect to the bubbles. Hence the clockwise and counterclockwise forks must move

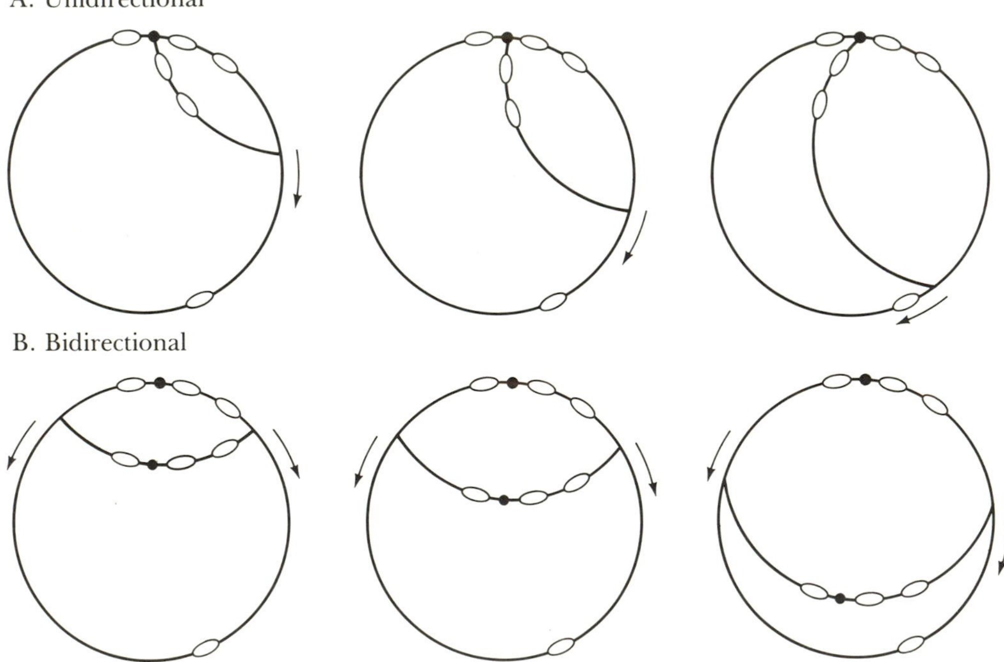

Figure 3-22
Examples of partially denatured circular DNA molecules in three different stages of replication. (A) The types expected if replication were unidirectional and clockwise. (B) The types expected if replication were bidirectional and both replication forks moved at the same rate. The arrows indicate the direction of movement of the fork and the dot represents the site of initiation of replication.

at the same rate and the midpoint is the replication origin. It should be noticed that each of the conclusions just stated depend on the measurement of positions with respect to the denaturation bubbles.

There are several valuable procedures used to determine the regions of a DNA molecule from which particular RNA molecules are transcribed. Each of these involves formation of an RNA·DNA hybrid molecule using conventional techniques of renaturation. These procedures, which are shown schematically in Figure 3-23, are the following.

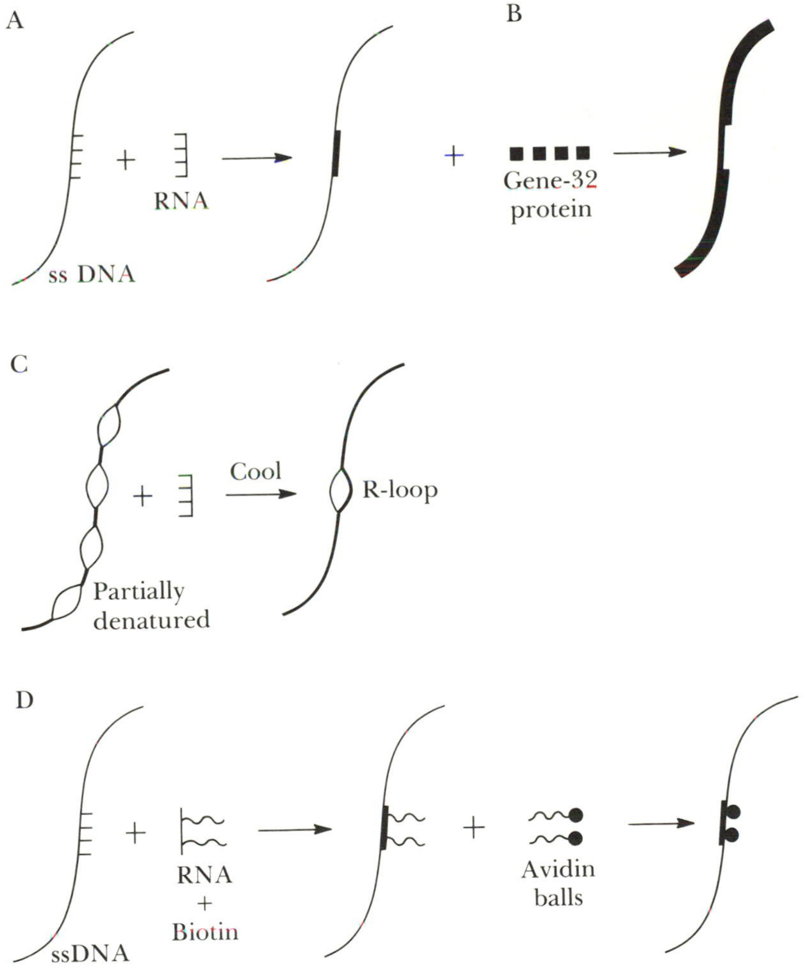

Figure 3-23
Four methods of visualizing DNA·RNA hybrids. (A) Direct annealing of RNA to single-stranded (ss) DNA; (B) addition of T4 gene-32 protein after treatment in (A); (C) annealing of RNA with partially denatured DNA; (D) annealing of biotin-coupled RNA to ssDNA followed by addition of avidin-coated polystyrene spheres.

1. *Annealing of RNA to a single strand of DNA.* In this procedure, a DNA sample is denatured and renatured with excess RNA molecules. Two types of molecules result—the original double-stranded DNA and a single strand of DNA to which the RNA molecule has annealed. The RNA molecule is generally of much lower molecular weight than the DNA molecule so that, using the formamide spreading method, one observes a thin molecule with a single thickened region, which is the double-stranded DNA·RNA hybrid. By measuring the lengths of the single-stranded regions, one can identify the segment of the DNA molecule from which the RNA is transcribed. Even more information can be obtained if the individual strands of DNA are separately isolated. (This can be done by the CsCl–poly(I-G) technique described in Chapter 11, Example 11-V.) Using these strands separately, it is found that the DNA·RNA hybrid region is observed only when one of the strands is used; this must be the strand that is copied and the other is the complementary strand.

2. *Annealing of RNA to single-stranded DNA followed by binding of T4 gene-32 protein.* Precise localization with method 1 is difficult because it is usually observed that the lengths of the single-stranded regions flanking the DNA·RNA hybrid vary by as much as 20%. This is a result of local stretching and collapsing of the single strands. This problem can be eliminated by use of single-stranded DNA-binding proteins. The T4 gene-32 protein binds very tightly to single-stranded DNA but not to double-stranded DNA. Single-stranded DNA becomes so heavily coated with gene-32 protein that it appears even thicker than the DNA·RNA hybrid, as shown in Figure 3-22B. The coated strands have a fixed and regular mass per unit length; hence a measurement of the relative lengths of the three regions shown in the figure allows the transcribed region to be localized.

3. *Annealing of RNA to partially denatured DNA: R-loops.* Under certain conditions a DNA·RNA hybrid is more stable than double-stranded DNA. Thus, if partially denatured DNA and RNA are held in these conditions, DNA·RNA hybrid molecules can form at the expense of double-stranded DNA. When such DNA is returned to standard conditions at which double-stranded DNA is stable, there results a molecule having what is known as an R-loop, as shown in Figure 3-23C. The R-loop is a large internal loop, one branch of which is single-stranded DNA; the other branch is the DNA·RNA hybrid where position can be determined accurately by measuring the lengths of the regions flanking the R-loop.

4. *The avidin-biotin procedure* (*Figure 3-23D*). The preceding methods fail if the RNA molecule is so short that the thickening due to a DNA·RNA hybrid is undetectable against the normal granularity of DNA. This is always the case with transfer RNA (tRNA) molecules, which consist of only 78–90 nucleotides and have an extended length

of about 0.025 μm. In order to solve this problem, an elegant procedure has been developed that makes use of the extraordinarily tight binding between the vitamin biotin and the biotin-inhibitor avidin. Biotin is covalently linked at one or a few sites to a tRNA molecule. This chemical coupling does not interfere with hydrogen-bond formation so that the tRNA molecule can hybridize with the DNA molecule. Thus there are biotin groups, which cannot be seen, at the site on the DNA molecule at which the tRNA molecule has hydrogen-bonded. Then, either very small polystyrene balls or phages, to which many avidin molecules are bound, are added to the DNA·tRNA hybrid molecules and the sample is prepared for electron microscopy using the formamide technique. Thin strands of single-stranded DNA are seen on each of which is superimposed either a polystyrene ball or a phage *at the position of the tRNA molecule.* In this way, the tRNA molecule is located on the DNA molecule.

Several variants of the basic Kleinschmidt method are important when the amount of a valuable DNA is very small. In the usual Kleinschmidt procedure 0.05 ml of a solution of DNA at 0.5 μg/ml, is spread on about 100 ml of hypophase (0.05 + 0.5 = 0.025 μg of DNA is used). Ross Inman introduced a microprocedure in which 0.001 ml of a DNA solution at 0.5 μg/ml is spread on a 0.1-ml drop. In this procedure only 0.0005 μg of DNA is needed but the concentration must still be 0.5 μg/ml. Dimitrij Lang introduced a technique, which he called *microversion*, in which a lesser amount of DNA at a lower concentration can be used. A 0.1-ml sample is prepared at a DNA concentration of 0.0001 μg/ml (for a total of 10^{-5} μg of DNA) and containing a small amount of formaldehyde and an amount of cytochrome *c* sufficient to form a film on the surface of the drop. The formaldehyde denatures the cytochrome *c*, which then moves to the surface forming a film. The droplet, which is protected from evaporation by keeping it in a humid atmosphere, is allowed to rest for up to 16 hours. During this time, most of the DNA molecules in the droplet diffuse to the surface and adsorb there. A grid is touched to the surface of the droplet to pick up the film and this is followed by the usual staining and shadowing protocol. Microversion is a valuable procedure but often the DNA molecules are not spread well enough to see all parts of the molecule clearly.

Visualization of nucleic-acid–protein complexes on polylysine films

Of the various procedures used to visualize individual protein molecules bound to DNA, none is superior to the polylysine procedure because of its simplicity and reproducibility. A droplet of an aqueous solution containing a nucleic-acid–protein complex is placed on a polylysine-coated carbon film. After removal of the liquid, a negative stain is added. Figure 3-24 shows an electron micrograph of RNA polymerase bound to DNA, prepared in this way.

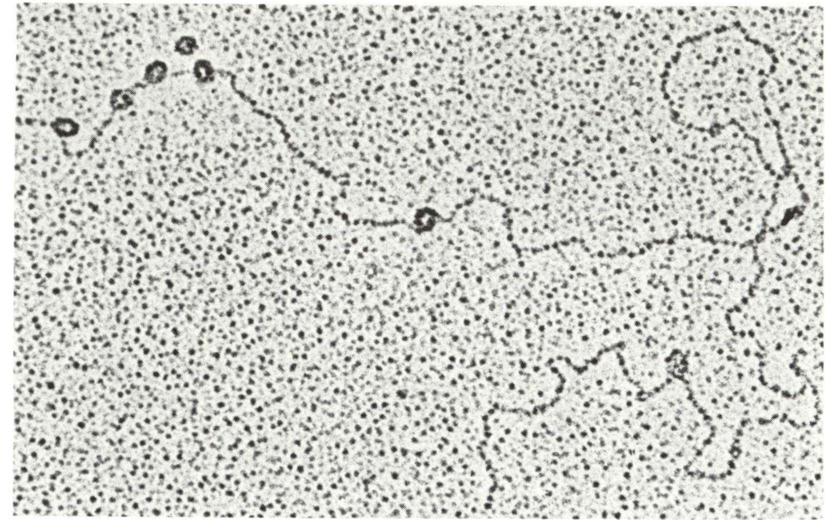

Figure 3-24
Electron micrograph of RNA polymerase bound to DNA, prepared by the polylysine procedure. [Courtesy of Robley Williams.]

Improving the Quality of the Image and Image Reconstruction

Minimal Beam Exposure

The kinetic energy of 60-kV electrons is very great; in colliding with the atoms of a sample, the sample temperature can rapidly rise many hundreds of degrees. This disrupts covalent bonds and can even cause vaporization of parts of the sample. For example, if a sample of a radioactive (^{35}S-labeled) protein is applied to a grid, the amount of radioactivity on the grid decreases considerably after only a few seconds of exposure to the electron beam. In studies of the alteration of proteins by electron beams produced by a linear accelerator, it was shown that disruption is detectable at an average dose of 1 electron Å^2; this is far smaller than the dose received by a sample during routine operation of an electron microscope. If the sample is stained or shadowed with a heavy metal, the apparent effect is much less; the biological molecule is probably mostly destroyed after a few seconds but the pattern of heavy metal atoms remains and it is this pattern that is observed. However, the loss of the sample certainly causes an alteration of the array of the metal atoms and considerably reduces the information obtained from the micrograph.

For high resolution in examining biological structures, a procedure developed by Robley Williams can be used. This method, *minimal-beam-*

exposure microscopy, consists of localizing the electron beam on a small portion of the sample to allow focusing and then moving either the beam or the sample so that an unexposed region of the sample can be photographed. Figure 3-25 shows a pair of negative-contrast electron micrographs of tobacco mosaic virus taken without prior exposure and after exposure to the beam for a few seconds. The "before" micrograph shows the helical array of the protein subunits of the virus coat. It is possible to measure the pitch of the helix and to count the number of helical turns per unit length of the virus. The hollow core of the virus is also clearly seen. The "after" micrograph, whereas still quite good, is fuzzy in comparison, lacking the detail of the former. The high-quality micrographs already shown in Figure 3-13 were taken by the minimal-exposure method.

Three-Dimensional Reconstruction of Molecular Structure from Unstained Samples

Accessories are available by which a sample can be tilted with respect to the electron beam. This makes it possible to photograph a single particle or structure at several angles. A great deal of information can be obtained

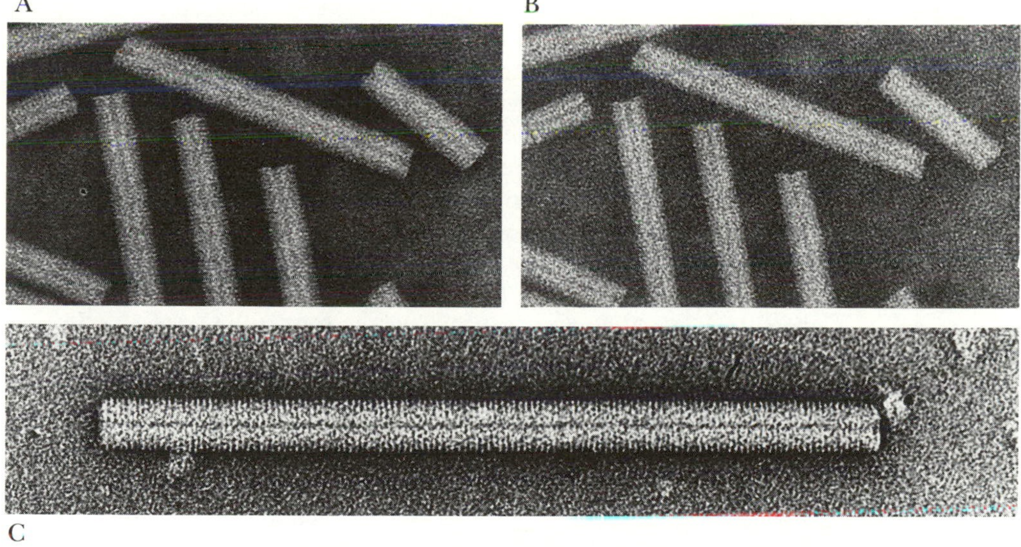

Figure 3-25
Electron micrographs of tobacco mosaic virus. (A) Particles photographed by the minimal beam exposure method. (B) The same field after a thirty-second exposure to the electron beam. Note that the details of the apparent stacked disc structure are lost after exposure to the beam. (C) A higher-magnification micrograph of a single particle photographed by the minimal-beam-exposure method. [From R. C. Williams and H. W. Fisher, *J. Mol. Biol.* 52(1970):121–123. By permission of the publisher and the author.]

by this technique. For example, a rectangular prism might always be on a grid with its largest face perpendicular to the electron beam. By viewing at an angle, the thickness of the prism can be determined. Of course, the object of study does not usually have such a simple shape. David DeRosier and Aaron Klug developed a computer program that allows reconstruction of the three-dimensional structure from micrographs taken at many angles. This has been used to determine the fine points of the structure of viruses and, more recently, of membranes.

The principal limitations of the image-reconstruction technique (which has generally provided satisfactory information) are that the sample is progressively damaged by repeated beam exposure, as explained in the previous section, and that the sample is either negatively or positively stained; this limits the resolution of the technique. A variation of the method that is applicable to extended orderly arrays, such as membranes and crystalline arrays, has been developed by R. Henderson and P. Unwin. This procedure uses unstained specimens that are photographed by a very brief exposure to the beam. The number of electrons passing through each region of the sample is so small that there is virtually no molecular damage. However, the micrograph obtained with such low doses of electrons appears featureless because of the statistical fluctuation of the number of electrons passing through the sample without scattering. This background is superimposed on the very weak image of the sample because only a small number of electrons are scattered by the sample. The statistical problem can be solved if the sample is an orderly array of identical molecules because, in theory, if the weak images of each molecule could be superimposed, these images would reinforce one another and produce a pattern that reflects the structure of the molecules, whereas the background would tend to cancel out and become uniform. In order to get a reliable and informative pattern, it is necessary to scan a region of the film containing about 3×10^3 to 10^4 molecules; the precise number depends on the degree of regularity of the sample. Actually the best procedure is to use the electron microscope as an electron diffraction instrument and obtain an electron diffraction pattern (a discussion of which is beyond the scope of this book). Henderson and Unwin obtained such a pattern on film using the electron microscope and, by measuring the blackening of the film and processing the data by means of an appropriate computer program, a two-dimensional image, or more precisely, a two-dimensional pattern of electron density, was constructed. Such patterns can be obtained from micrographs taken at several angles and these patterns can be recombined to form a three-dimensional image by a modification of the DeRosier-Klug procedure. Figure 3-26 shows the first result of application of this procedure; this is a model of a single protein molecule in the purple membrane of *Halobacterium halobium*, a marine photosynthetic bacterium. At the current state of this technique, the limit of resolution is about 7 Å; the limitation is imposed primarily by optical aberrations in the microscope. If these are eliminated, which seems to be a likely possibility in the near future,

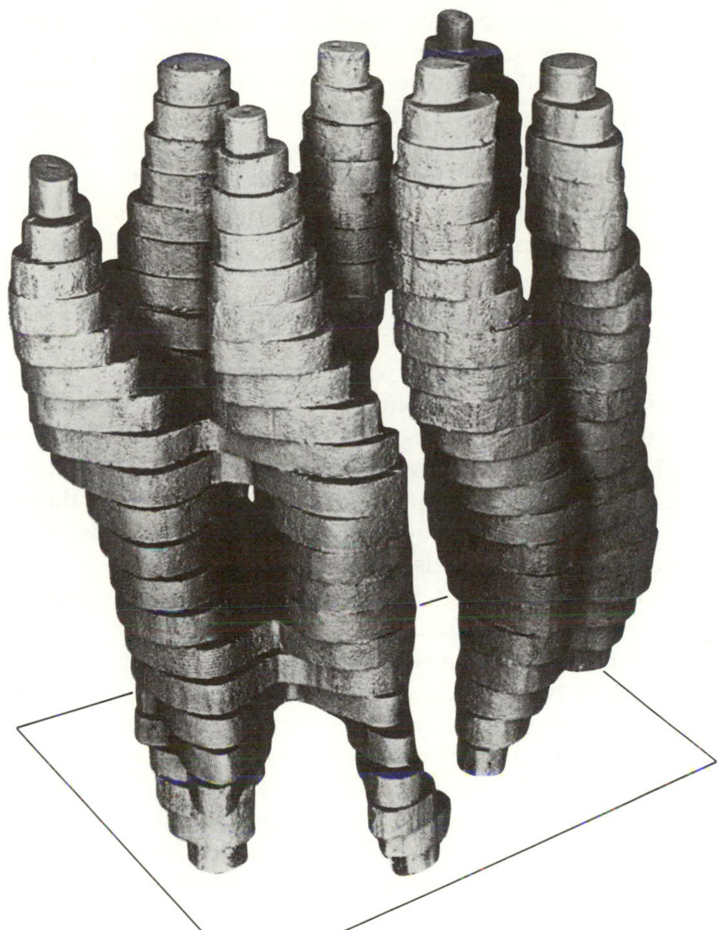

Figure 3-26
A model of a single protein molecule in the purple membrane, obtained by the reconstruction method of Henderson and Unwin. [From R. Henderson and N. Unwin. Reprinted by permission from *Nature* (*Lond.*) 257:28. Copyright © 1975 Macmillan Journals Limited.]

the limit of resolution should be about 3 Å; this is imposed by beam damage and alterations of the structure that occur when the sample is placed in the vacuum of the microscope column.

Image Rotation and Rotational Filtering

The fine details of organized arrays of the molecules in virus particles are lost in the negative-contrast method because of the minor disarray of the heavy metal atoms on the surface of the particle. When the particle has rotational symmetry, the random variation (noise) of the surface pattern can be smoothed out by image rotation. The principle is like that of the Henderson-Unwin technique; that is, if a large number of micrographs of nearly identical objects are superimposed, the noise will cancel whereas the details of the structure (the signal) will be enhanced. With a rotationally symmetric sample, Roy Markham showed that it is not necessary to superimpose many micrographs but only to rotate the image about its axis of rotational symmetry and combine the views obtained at various angles of rotation. This is because each subunit is presumably identical and images of the subunits should enhance one another when superimposed. If a particle has sixfold symmetry. then six 60° rotations should be equivalent to combining six images. The procedure used is to project an image of a particle from the electron micrographic photographic film onto a turntable, on which is placed a sheet of unexposed photographic paper. The image is centered on the axis of rotation of the turntable. The paper is then rotated through $360/n$ degrees and after each turn the paper is exposed. If n is the degree of rotational symmetry of the object, fine detail will appear in the developed picture, If n is not the degree of rotational symmetry, all detail is lost. Sometimes a particle might have two kinds of rotational symmetry. For example, it might consist of two concentric circles of proteins, the outer circle having sixteen molecules and the inner one having nine molecules. This would become evident by preparing photographs with $n = 16$ and $n = 9$; each photograph would illustrate one of the kinds of symmetry.

The rotational motion can also be accomplished by computer processing using rotational filtration. Here the blackening of a micrograph is measured. A computer then performs mathematical rotation and provides a printout of a processed image. Figure 3-27 shows an example of this procedure for an extended and contracted baseplate of *E. coli* phage T4; panel (A) shows the original electron micrographs and panel (B) shows the sixfold rotationally filtered images.

Special Mechanisms of Image Formation

Dark-Field Electron Microscopy

Dark-field electron microscopy allows for substantially increased contrast, as is the case with the light microscope. A dark field can be obtained either by using hollow-cone illumination, as in light microscopy (Chapter 2), or by making the beam fall on the sample at an angle such that it does not enter the imaging system. The image is formed by the scattered and diffracted electrons alone. Dark-field electron microscopy has not been used very much, but for observation of macromolecules it deserves greater attention. It is especially useful if the contrast of a sample is poor

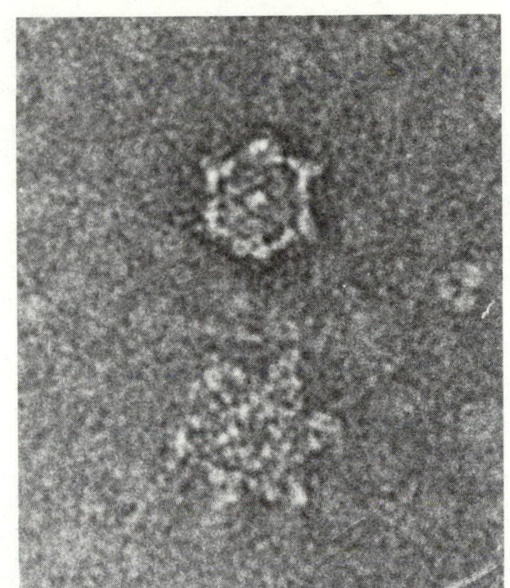

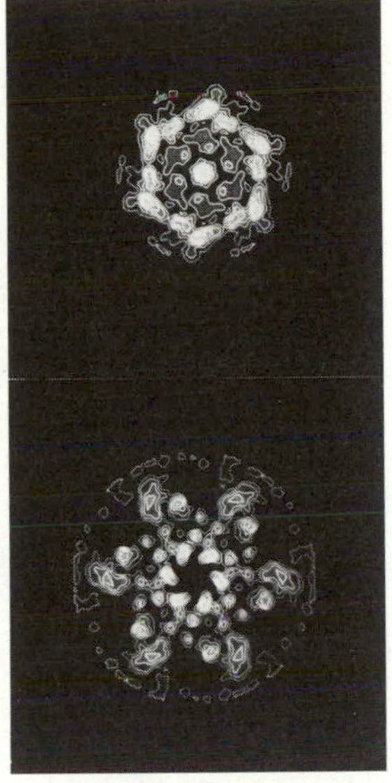

A B

Figure 3-27
Electron microscopy of the baseplate of phage T4. An extended baseplate is shown at the top and a contracted one below it. (A) Original micrograph. (B) Sixfold rotationally filtered image. [From R. A. Crowther, E. V. Lenk, Y. Kikuchi, and J. King, *J. Mol. Biol.* 116(1977):489. By permission of the publisher and the authors.]

because contrast is enhanced considerably by dark-field microscopy. A micrograph of a DNA molecule visualized in this way is shown in Figure 3-28.

The Crewe Microscope

In the ordinary transmission electron microscope, the incident beam covers the entire sample. As the electrons interact with the sample, they do so in several different ways: (1) essentially no interaction at all (i.e., traveling through the interatomic spaces), the most abundant class of electrons; (2) inelastically scattered (i.e., with loss of energy) by the orbital electrons of the atoms of the sample, the second most abundant; and (3) elastically scattered (i.e., without loss of energy) from atomic nuclei, the least abundant. The ratio of the last two classes is a characteristic of each element, because the size of the nuclear target and hence the cross-section for inelastic events increases greatly for the larger atoms.

A special new microscope has been designed to take advantage of these facts. To do this, the beam is focused into a very small (approximately

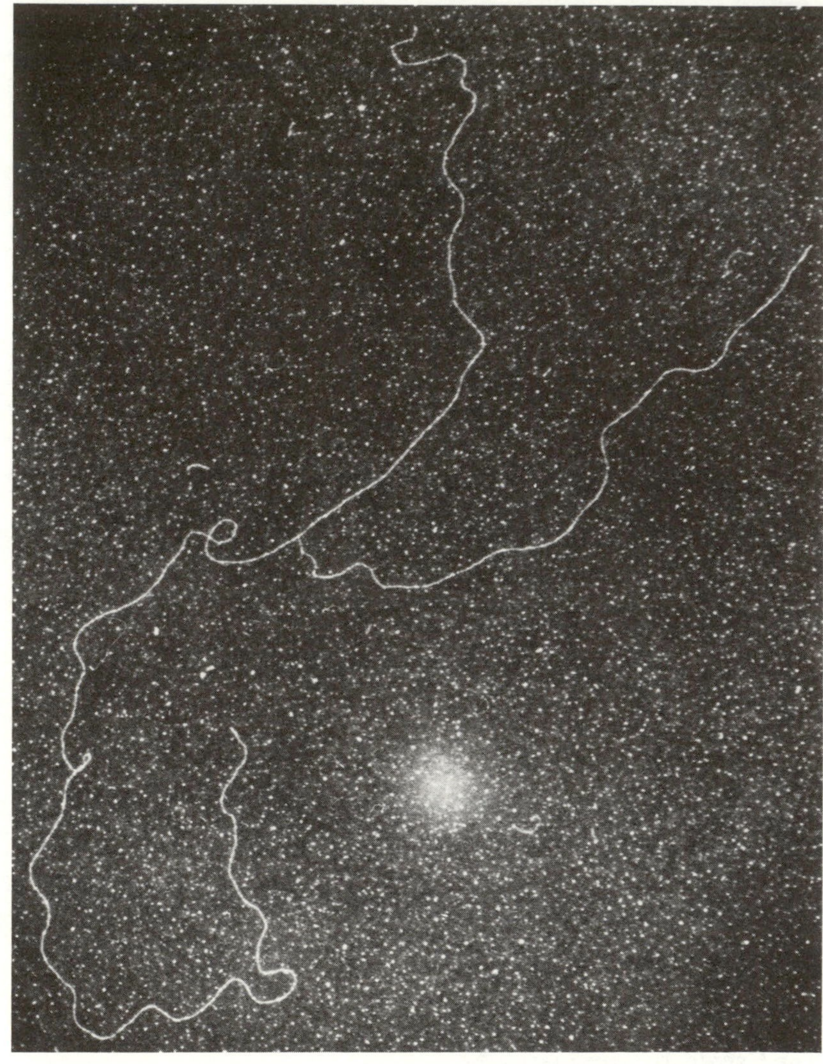

Figure 3-28
Dark-field electron micrograph of replicating *E. coli* T7 DNA. [From D. Dressler, in *Control Processes in Virus Multiplication*, edited by D. C. Burke and W. C. Russel, Cambridge University Press, 1975.]

5 Å) spot. The spot is swept across the sample as in a television set. As the beam moves, the ratio of the latter two classes of electrons is measured at each point by an electron-energy spectrometer. This ratio is converted into an image on a television screen by suitable electronic circuits; the information attained can also be processed by computer analysis. This new and important step in electron microscopy gives a new element of

analysis to electron microscopy because individual atoms can be identified. In some cases, the limit of resolution is improved and pictures at 2 Å resolution have been obtained. Figure 3-29 shows several examples of molecules visualized by this procedure.

The Backscatter Scanning Microscope

The scanning electron microscope (SEM) is a device that has produced the many beautiful photographs of cell surfaces seen in the past few years. This microscope is limited to about 200 Å in resolution and operates on a very different principle from that of the transmission electron micro-

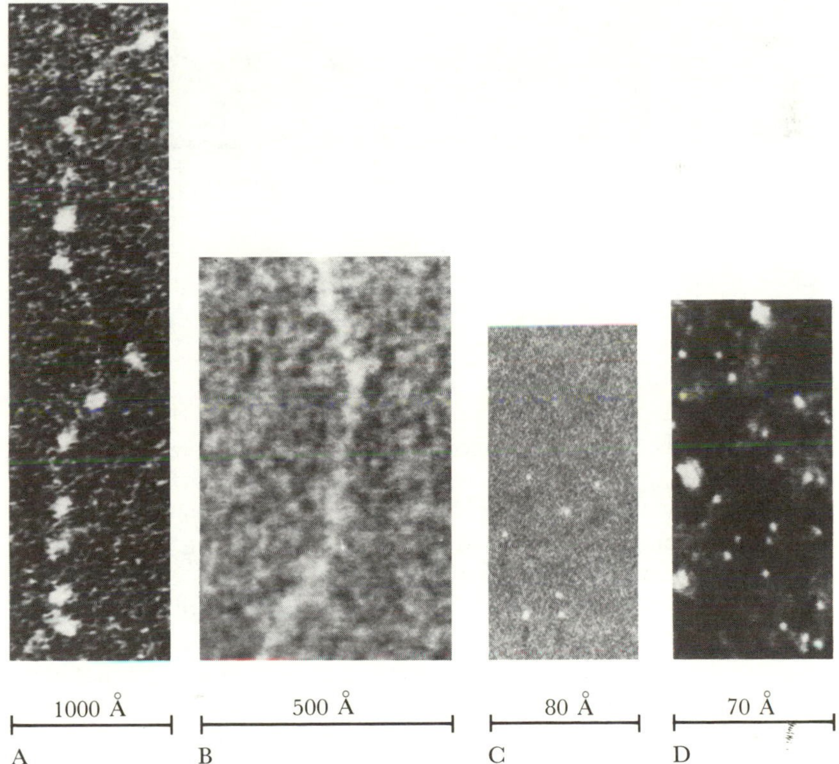

Figure 3-29

A. Electron micrograph of calf thymus chromatin depleted of very-lysine-rich histone (H1). A fiber of DNA, 20 Å or 30 Å wide, is seen coated with protein particles, whose average diameter is 135 Å, spaced approximately 260 Å apart. The bar is 1000 Å. [From J. P. Langmore and J. Wooley, *Proc. Natl. Acad. Sci. U.S.A.* 72(1975):2691–2695.] B. Part of an unstained T7 DNA molecule showing a fiber from 20 Å to 30 Å thick. C. Mercury atoms. D. Uranium atoms. [All micrographs were obtained with the Crewe scanning transmission electron microscope and were kindly provided by John Langmore and Albert Crewe.]

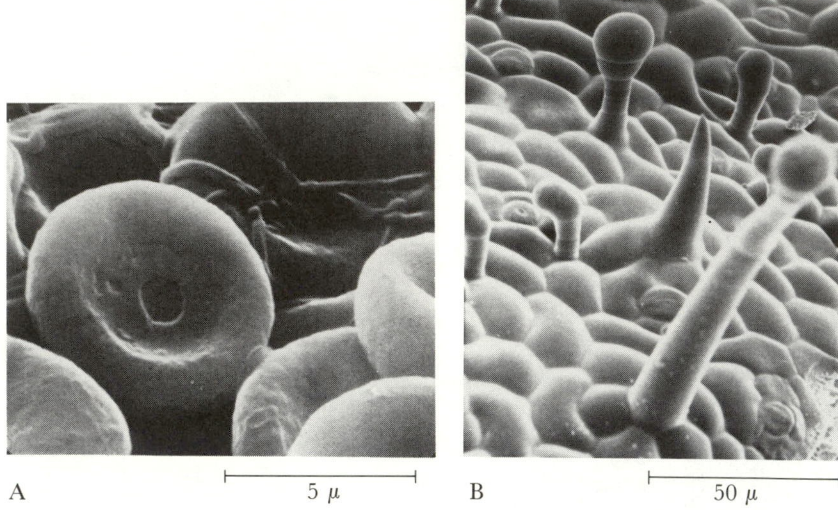

Figure 3-30
Scanning electron micrograph of (A) a human red blood cell and (B) the surface of a geranium leaf. [Courtesy of Thomas Hayes.]

scope. Like the Crewe microscope, the beam is collimated into a small (100 Å) spot, and the spot is swept across the sample surface, which has been coated with a thick (200 Å) layer of gold or other heavy metal. As the beam impinges on the metal and penetrates into it a short distance, electrons are emitted from the gold either as secondary emissions or as directly backscattered electrons from the beam. Because of an unexpected angular relationship between the number of electrons emitted and the angle of the surface to the incident beam, which is close to, but not identical with, the way that light reflects from the surface of an object, the image formed by the collected electrons gives dramatic images of the surface being examined. An example of a scanning micrograph is shown in Figure 3-30.

SELECTED REFERENCES

Crewe, A. V. 1971. "A High-Resolution Scanning Electron Microscope." *Sci. Am.* 224 :26–35.

Davis, R. W., M. Simon, and N. G. Davidson. 1971. "Electron Microscope Heteroduplex Methods for Mapping Base Sequence Homology in Nucleic Acid," in *Methods in Enzymology*, vol. 21, edited by L. Grossman and K. Moldave, pp. 413–428. Academic Press.

Finch, J. T. 1975. "Electron Microscopy of Proteins," in *The Proteins*, 3rd edition, edited by H. Neurath and R. L. Hill, pp. 413–497. Academic Press.

Fisher, H. W., and R. C. Williams. 1979. "Electron Microscopic Visualization of Nucleic Acids and of Their Complexes with Proteins." *Annu. Rev. Biochem.* 48:649–680.

Greenstone, A. 1968. *The Electron Microscope in Biology*. St. Martin's Press.

Haggis, G. H. 1967. *The Electron Microscope in Molecular Biology*. Wiley.

Hall, C. E. 1966. *Introduction to Electron Microscopy*. McGraw-Hill. A classic.

Henderson, R., and P. N. T. Unwin. 1975. "Three-dimensional Model of Purple Membrane Obtained by Electron Microscopy." *Nature* (*Lond.*) 257:28–32.

Kleinschmidt, A. K. 1968. "Monolayer Techniques in Electron Microscopy of Nucleic Acid Molecules," in *Methods in Enzymology*, vol. 12B, edited by L. Grossman and K. Moldave, pp. 361–376. Academic Press.

Markham, R., S. Frey, and G. J. Hills. 1963. "Methods for the Enhancement of Image Detail and Accentuation of Structure in Electron Microscopy." *Virology* 20:88–102.

Meek, G. A. 1976. *Practical Electron Microscopy for Biologists*. Wiley.

Oliver, R. M. 1973. "Negative Stain Electron Microscopy of Protein Macromolecules," in *Methods in Enzymology*, vol. 27, edited by C. H. W. Hirs and S. N. Timasheff, pp. 616–672. Academic Press.

Slayter, E. M. 1970. *Optical Methods in Biology*. Wiley.

Williams, R. C. 1977. "Use of Polylysine for Adsorption of Nucleic Acids and Enzymes to Electron Microscope Specimen Films." *Proc. Natl Acad. Sci. U.S.A.* 74:2311–2315.

Williams, R. C., and H. W. Fisher. 1970. "Electron Microscopy of Tobacco Mosaic Virus under Conditions of Minimal Beam Exposure." *J. Mol. Biol.* 52:121–123.

Wu, M., and N. Davidson. 1978. "An Electron Microscopic Method for the Mapping of Proteins Attached to Nucleic Acids." *Nucleic Acid Res.* 5:2825–2846.

Younghusband, H. B., and R. B. Inman. 1974. "The Electron Microscopy of DNA." *Annu. Rev. Biochem.* 43:605–619.

PROBLEMS

3–1. A sample of RNA-containing viruses is observed by the negative-contrast method. Two types of particles are observed—those that appear uniformly light against a dark background and those that have large dark centers. What is the structure of each?

3–2. A spherical virus is mixed with polystyrene spheres, 750 Å in diameter. After shadowing with gold, the length of shadow of the polystyrene spheres is 1250 Å and that of the virus 820 Å. What is the diameter of the virus? Some of the viruses have shadows ranging from 150μm to 200 Å. What are these?

3–3. DNA no. 1 has a length of 10 μm; no. 2 is 9.5 μm. For most of their length, their base sequences are identical. A region from 4.0 μm to 4.5 μm (measured from what is arbitrarily called the left end) of no. 1 is deleted in no. 2.

The region from 7.1 μm to 7.8 μm of no. 1 is replaced in no. 2 by sequences not found in no. 1. The two DNAs are mixed, denatured, and renatured according to the heteroduplex procedure. How will the heteroduplexes appear?

3–4. A sample of linear double-stranded DNA (all molecules identical) is digested briefly with an exonuclease attacking only the 5′-*P* ends of DNA strands until approximately 3% of the DNA is removed. It is known that these molecules are terminally redundant—that is, the gene order is ABCD . . . XYZABC in which the length of segment ABC is 1% that of the total DNA. When these treated molecules are exposed to renaturing conditions, a new type of structure appears. Describe this structure, including the lengths of the various regions in terms of the percentage of original length.

3–5. A protein structure contains sixty spherical subunits arranged so that six are in a hexagonal array and ten hexagons are stacked one above the other. Draw the types of structures that would be observed by (a) the negative-contrast method, (b) shadowing, and (c) the replica technique.

3–6. Support films are always made of plastic or pure carbon. They are very fragile and frequently break. Stronger films could be made of metals such as chromium. Would such metal films be useful? Explain.

3–7. In measuring the length of DNA molecules, a large enough number of molecules must be measured to get a statistically significant value. Clearly, as larger molecules are studied, there are fewer molecules per grid hole. One solution to this problem is to increase the DNA concentration. However, why is this not reasonable if the molecules are longer than, say, 20 μm?

3–8. A virus-infected cell appears to be approximately 50 μm in diameter. A thin section, 100 Å thick, contains twenty-two viruses approximately 400 Å in diameter. Roughly how many viruses are there in this cell? What assumptions have you made to perform this calculation?

3–9. A particular cell is roughly cylindrical, 10 μm long and 1 μm in diameter. It contains a set of stacked discs (stacked on the cylinder axis), 1 μm in diameter and 400 Å long. A collection of cells is embedded in plastic, sliced into sections 200 Å thick, and stained with a heavy metal stain. What types of structures will be seen?

3–10. Answer the following questions.

a. Does the limit of resolution increase or decrease with increasing accelerating voltage?
b. What is the Compton wavelength for 200-kV electrons?
c. What are several advantages and disadvantages of using 10^4-kV electrons?

3–11. What factors determine the choice of thickness of the support film?

3–12. What is the purpose of a fixative? What criteria might you use for good and bad fixation?

3–13. What methods of sample preparation might you use to study each of the following: ribosomes, a large protein molecule, the arrangement of protein subunits of a virus, the surface of a membrane, the inner regions of a membrane, the inner structure of a bacterium, the number of empty virus particles in a suspension of viruses.

3–14. A grid coated with particles is shadowed at an angle of 30 degrees. The length of the shadow is 0.022 μm. What is the height of the particle?

3–15. The horizontal distance between a sample and an evaporating metal source is 8.4 cm. The vertical distance is 2.3 cm. Particles are seen with a shadow length of 0.069 μm. What is the height of the particles?

3–16. Answer the following questions about shadowing.

a. How would the DNA appear if shadowing were done twice, the second time at right angles to the direction of the first, rather than while rotating the sample?

b. How would a DNA sample appear if too little and too much metal were deposited?

c. Silver is much less expensive than platinum. Why is silver not used?

3–17. A sample of circular DNA from the virus SV40 (Simian virus 40) is mixed with circular ϕX174 DNA, which is a standard of useful length because the number of base pairs per ϕX174 DNA molecule is precisely known. Several hundred molecules of both types are measured. The observed lengths of the ϕX174 DNA molecules vary by 7% from the mean but the variation of the SV40 DNA is 30%. What can you conclude about the actual length of the DNA isolated for SV40 virus?

3–18. When examining a field of view in the electron microscope, it is usually observed that when the center of the field is sharp and clear, the edge of the field seems always to be slightly out of focus. This almost never occurs with a good light microscope. What is the problem with the electron microscope?

3–19. You wish to get a picture of a phage attached to a bacterium so you allow 10 phages to adsorb to each of a very large number of bacteria. The adsorbed complexes are shadowed using conditions known to make phages visible. A large number of bacteria are seen in the microscope but you never see one attached to a bacterium. At 500 phages per bacterium you now see attached phages. What is the simple problem at the low multiplicity of infection?

3–20. A sample consisting of linear DNA molecules is denatured to produce single-stranded DNA. It is then observed by electron microscopy using the Kleinschmidt procedure. No time has been allowed for renaturation yet the molecules clearly have double-stranded termini. Furthermore, they are only 90% of the length of the original undenatured molecule.

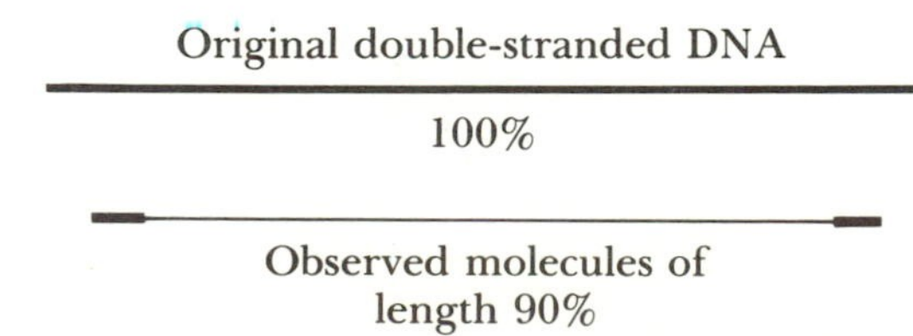

What feature of the molecule is responsible for this observation?

3–21. You have isolated what you think is a unique DNA molecule from mitochondria—that is, a catenane or two nicked circles linked as in a chain. Indeed in the electron microscope, you see molecules like this. (If this is traced, it can be seen that it consists of two overlapping circles.) You

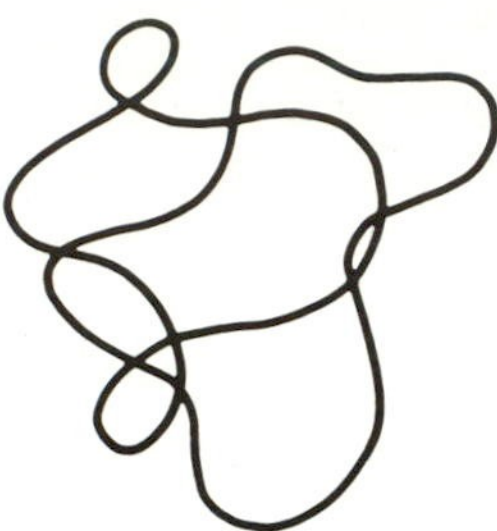

want to know what fraction of the molecules are catenanes. The simplest way is to count them. To get statistical accuracy, you do many spreadings and count the molecules in many different grid holes. The data obtained are the following:

	Grid hole no.	Total molecules	No. molecules that look like catenanes
	1	125	16
	2	123	18
Spreading 1	3	95	10
	4	90	8
	5	85	8
	1	62	5
	2	58	4
	3	57	5
Spreading 2	4	55	5
	5	40	3
	6	39	3
	7	38	3
	1	38	3
	2	38	3
	3	36	2
	4	35	4
Spreading 3	5	29	2
	6	28	3
	7	28	4
	8	27	3
	9	20	2

a. Why do you think that the fraction scored as catenanes is not the same in the three spreadings?

b. What fraction of the molecules are catenanes?

3–22. You have a collection of phage DNA molecules all of the same molecular weight (20×10^6). You prepare electron micrographs, photograph, and measure the length of the molecules. It is known that the mass per unit length is $2 \times 10^6/\mu m$ when measured by Watson and Crick using X-ray diffraction. You measure the length of 100 molecules and indeed the mean value is 10 μm. However, they are not all the same length having a range of 9.3–10.7 μm. What experimental factors might produce this variability?

3–23. You are examining phages by the negative-contrast method using phosphotungstic acid. Most of the phages appear white against a dark background. Occasionally phage particles are seen that appear rather dark gray against the same black background. *Some* of these particles have a small rupture in the phage head. What are the dark particles (including the ones that are not obviously ruptured)?

3–24. You are trying to measure the molecular weight of a small particle by electron microscopy. Since the particle density is known accurately to be 1.32 g/cm^3, you can measure the molecular weight from the particle volume. You measure the diameter of several particles and obtain a value of 530 $\pm$ 25 Å. Is this a useful measurement if the error in molecular weight is to be less than 8%?

Two

GENERAL LABORATORY METHODS

4

Measurement of pH

Because essentially all biochemical reactions depend strongly on pH (which is defined as $-\log[H^+]$, in which $[H^+]$ is the molar hydrogen ion concentration*) it is important to be able to measure pH accurately. This is accomplished with a commerical pH meter by simply immersing two electrodes into a solution and reading the pH value on a dial. However, it is important to know how the instrument measures pH, because several factors can cause the observed value to differ from the actual pH.

A pH meter measures the voltage between two electrodes placed in the solution. The heart of the system is an electrode whose potential is pH-dependent. The most commonly used pH-dependent unit is the *glass electrode.* The action of this electrode is based upon the fact that certain types of borosilicate glass are permeable to H^+ ions but not to other cations or anions. Therefore, if a thin layer of such glass is interposed between two solutions of different H^+ ion concentrations, H^+ ions will move across the glass from the solution of high to that of low H^+ concentration. Because passage of a H^+ ion through the glass adds a positive ion to the solution of low H^+ concentration and leaves behind a negative ion, an

* Strictly speaking, pH is defined in terms of the thermodynamic activity of the H^+ ion and not the molar concentration. This definition is important because the activity depends on the concentration of other electrolytes present in the solution. However, in biological systems the total ionic strength is always sufficiently low that the effect of the electrolyte concentration on the pH is small enough to be ignored. In fact, it is a general "rule" of biochemistry that "activity, for all practical purposes, equals concentration." Whereas this is not always the case, the "rule" is almost universally applied.

electric potential develops across the glass. The magnitude of this potential is given by the equation

$$E = 2.303 \frac{RT}{F} \log \frac{[H^+]_1}{[H^+]_2} \qquad (1)$$

in which E is the potential, R the gas constant, T the absolute temperature, F the Faraday constant, and $[H^+]_1$ and $[H^+]_2$ the molar H^+ concentrations on the inside and outside of the glass, respectively. Clearly, if the H^+ concentration of one of the solutions is fixed, the potential will be proportional to the pH of the other solution.

A diagram of a glass electrode is shown in Figure 4-1. The glass electrode contains 0.1 M HCl in contact with the H^+-permeable glass. Connection to the voltmeter is by means of a silver wire coated with silver chloride, which is immersed in the HCl.

The circuit is completed by immersing into the solution a reference electrode that has been selected to be pH-*in*dependent (see Figure 4-1). The type most commonly used contains a Hg-Hg_2Cl_2 paste in saturated KCl; this is called a *calomel* electrode. If a high-temperature operation is required, Ag-AgCl is used instead of Hg-Hg_2Cl_2. In both cases, the KCl serves to make contact between the Hg-Hg_2Cl_2 or Ag-AgCl unit

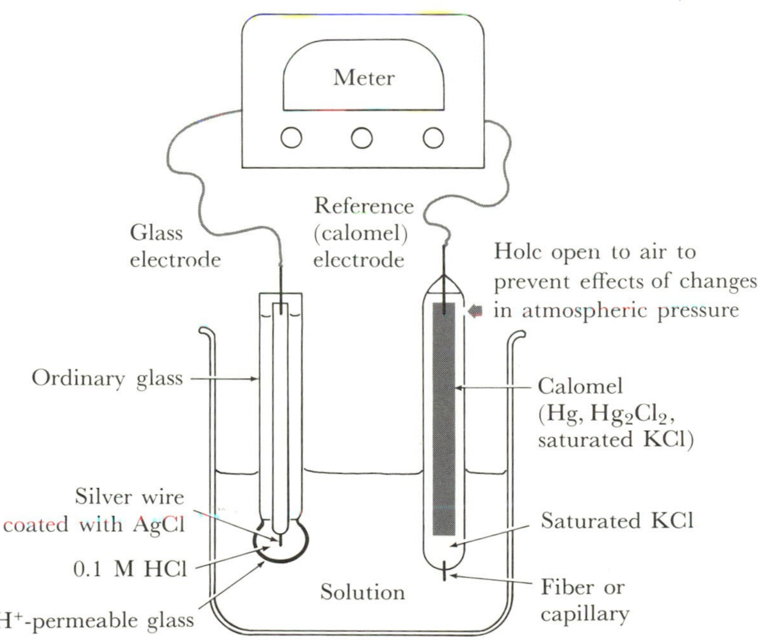

Figure 4-1
Glass and reference electrodes of a pH meter.

and the solution being measured. This unit is encased in a tube made of glass that is impermeable to H^+ ions (so that its potential is pH-independent). Electrical contact between the KCl within the electrode and the solution is by means of a fine fiber or capillary in the glass casing. (The KCl slowly flows into the sample. In cases in which the Cl^- ion is undesirable, a Hg-$HgSO_4$ reference electrode can be used.)

The voltage measured by such a system is primarily the difference between that of the glass and the reference electrodes. However, there are three other potentials present in the circuit: (1) the so-called liquid-junction potential of the reference electrode resulting from the fact that K^+ and Cl^- do not diffuse at the same rate so that a charge is generated at the interface between the KCl solution in the reference electrode and the sample; (2) a poorly understood potential called the asymmetry potential, which develops across glass even when the pH on both sides is the same; and (3) the potential of the Ag-AgCl in the glass-electrode unit, which is itself an electrode because of its contact with the Cl^- of the HCl. These three potentials and that of the reference electrode itself are relatively independent of pH and of ionic strength (in the range normally encountered) and therefore can be considered to be constant. Hence, the voltage, V, measured with the total system may be expressed as the difference between the fixed potentials and that of the glass electrode:

$$V = E_{\text{fixed}} - \frac{2.303RT}{F}\log\frac{[H^+]_1}{[H^+]_2} \qquad (2)$$

Because the glass electrode is normally filled with 0.1 M HCl,

$$[H^+]_1 = 10^{-1}$$

and because pH $= -\log H^+$,

$$V = E_{\text{fixed}} + \frac{2.303RT}{F} - \frac{2.303RT}{F}\cdot \text{pH}$$

or

$$V = \text{constant} - \frac{2.303RT}{F}\cdot \text{pH} \qquad (3)$$

Therefore, the voltage generated is linearly related to the pH of the solution.

To avoid determining the constant in equation (3), and *because the concentration of the* HCl *in the glass electrode changes by repeated use*, a pH meter is normally standardized against a solution of known pH. (The KCl concentration of the reference electrode does not change because the solution is saturated and contains undissolved crystals.) That is, the electrodes are placed in a standard buffer (typical standards are pH 4, 7, and 10) and the meter is adjusted to read the pH of the standard. Such standardization is generally sufficient. However, because of minor perturbations producing slight nonlinearity, for precise determination of pH, it is advisable to use a standard whose pH is within one or two pH units of the unknown.

Note that equation (3) states that the relation between the measured voltage and the actual pH is temperature-dependent.* Hence, to determine pH it is necessary to adjust the pH meter (by means of a knob usually labeled "Temp" or "Temperature Compensation Control") to the temperature of the solution being measured. This adjustment introduces or removes a resistance in the electrical circuit so that the voltage change per pH unit increment is always the same. Clearly, because of this temperature dependence as well as that due to the effect of temperature on ionization, it is important that the measurement itself does not induce temperature changes. This point is mentioned here because, in fact, the temperature might be expected to rise during a measurement as a result of current flow through the solution. However, because the glass electrode has very high resistance, only a very small current is drawn by the voltmeter. Furthermore, the solution being measured usually has very low resistance so that the change in temperature (proportional to the resistance times the square of the current) over a short interval is very small.

Complications of pH Measurement

Dependence of pH on Concentration of Ions

The pH of a buffer depends not on the concentration of the buffer ions but on a thermodynamic quantity called the activity. This parameter is strongly affected by the total concentration of ions in the solution (actually the ionic strength) so that the pH of a buffer will vary both with its own concentration and with the concentration of other salts in the solution. This is especially important in the use of commercially available pH standards because they are usually concentrated solutions to be diluted 25-fold—a dilution factor that must be rigorously adhered to. Similarly, if a buffered solution is being prepared as a concentrate (i.e., a stock solution), the pH of this stock solution should not be adjusted to the value desired for the diluted solution. Instead, the concentrated solution must be prepared so that, when the pH of a dilution (i.e., to the concentration desired for a particular experiment) is measured, it will have the required value.

* It is important to realize that this temperature dependence is not the same as that which makes the pH of a buffer vary with T—that effect is the result of the temperature dependence of the dissociation constant of an acid, whereas this dependence is the effect of T on the potential.

Electrode Contamination or Alteration

Any substance that can be adsorbed by the H^+-permeable glass of the glass electrode can affect the pH reading by affecting the permeability to H^+ ions. This frequently happens with protein solutions because a thin protein film can form on the glass. Fortunately, such a film can be removed by treatment with detergents or acid.

The commonly used buffer Tris[tris-(hydroxymethyl)aminomethane] has been found to react with the components of several commercially available electrodes. Errors of as much as one pH unit have been observed. The manufacturers of electrodes usually indicate in their catalogs which electrodes are suitable for use with this buffer.

The Sodium Error

General-purpose glass electrodes are almost always somewhat permeable to sodium ions. Therefore, a potential related to the Na^+ concentration can be produced in the same way as with H^+ ions. If Na^+ is present in a solution whose pH is to be determined, the measured pH decreases as the Na^+ concentration increases, because the electrode detects the sum of the H^+ and Na^+ concentration. Hence the H^+ can *appear* to be greater, which means that the observed pH (i.e., $-\log[H^+$ *and* $Na^+])$ is lower than the actual pH ($-\log[H^+]$). This effect is most noticeable at high pH (when NaOH is used) and can be as high as one or two pH units in 1 M Na^+. It is important to remember this because the Na^+ ion is so ubiquitous.

Special Na^+-impermeable glass electrodes are commercially available and can be used to prevent the sodium error if the presence of the Na^+ ion is necessary. Alternatively, because the permeability of the glass to other alkali metal cations (e.g., K^+) is very low, if present at all, the Na^+ error can be prevented by use of potassium salts. If work at high pH is necessary, it is best to use KOH rather than NaOH, if possible.

It should be noted that the frequently encountered statement that general-purpose electrodes read low at high pH is not strictly true, because it is not the OH^- but the Na^+ that causes the difficulty. The so-called high-pH electrodes are simply glass electrodes made with Na^+-impermeable glass.

Buffers

Buffers are used to maintain constant pH. The first criterion for choice of a particular buffer is its pK value, which determines that the range in which buffering occurs is appropriate to the problem at hand. However, each buffer has other chemical characteristics, which make it more or less appropriate for a particular experiment. A few of the more important buffers are discussed in this section.

Phosphate Buffers

Phosphate buffers have been widely used because of their very high buffering capacity. Because of the great solubility of both the Na and K salts, any ratio of Na^+ to K^+ ions can be selected. Furthermore, because the ions are strongly charged, high ionic strength is obtained without the need for excessive molarity. On the other hand, it is not possible to prepare a phosphate buffer that simultaneously has a low ionic strength and a high buffering capacity.

Phosphate buffers have two major disadvantages: (1) they are somewhat toxic to mammalian cells and (2) Ca^{2+} and, to a lesser extent, Mg^{2+} ions are bound by phosphate ions.

Buffers based on K_3PO_4 are the most useful buffers for the pH range of 12.0–12.5.

Formate and Acetate Buffers

These are useful buffers in the range of pH 3–4.5 (formate) and 4–5.5 (acetate). The principal use is in chromatography and electrophoresis. Ammonium formate–formic acid is a completely volatile buffer system and may be employed when a sample is to be concentrated by drying in a vacuum.

Cacodylate, EDTA, and Barbiturate Buffers

Each of these systems is useful from pH 6 to 7. Cacodylate, $H_2[(CH_3)_2As]_2O_2$, was commonly used in spectral studies of nucleic acid because it has no absorption in the range of wavelengths used to study nucleic acids. However, since about 1965 it has rarely been used because of its toxicity.

Ethylenediaminetetraacetate (EDTA) is primarily a chelating agent and is capable of reducing the concentration of divalent cations, such as Mg^{2+} and Ca^{2+}, to vanishingly small values. It is not often used for buffering and is usually added to other buffers when divalent cations are to be avoided. Use of EDTA is an effective way to inactivate nucleases, which almost invariably require the Mg^{2+} ion as a cofactor. In optical studies of nucleic acids its concentration must be kept below 0.001 M because EDTA has significant absorption in the ultraviolet range.

Barbiturate (also called Veronal) was in common use at one time because it is an effective buffer. Its single disadvantage is its strong ultraviolet absorption. It is not often used today since it is a controlled drug (a sleeping potion).

Glycylglycine and Triethanolamine Buffers

These two buffers are commonly used in enzymological studies in which buffering at pH 7.5–8.0 is needed. Both Mg^{2+} and Ca^{2+} ions can be freely added to these buffers. Neither has significant ultraviolet absorption and both are tolerated by most enzymes. Glycylglycine can be cleaved by some proteases so it is not often used with crude protein extracts that may contain proteins. Triethanolamine is somewhat volatile and is sometimes used when molecules are to be purified and the buffer subsequently removed.

Boric Acid and Glycine Buffers

These are the buffers of choice for the pH ranges of 8.7–9.7 (borate) and 9.5–10.1 (glycine). Some bacteriophages are stabilized (in an unknown way) by borate. The use of borate requires some care, though, because of weak toxicity. Both buffers have low ultraviolet absorbance and tolerate divalent cations.

Carbonate Buffer

This is the principal buffer used in the pH range of 10–10.8. It has many limitations, however, due to the insolubility of most metal carbonates and a large change of pH with increasing temperatures (because of loss of CO_2).

Tris Buffer

Tris buffer is probably the most widely used buffer in biochemistry because of its high buffering capacity, its low toxicity, its low interference with most biochemical reactions, and its availability in very pure form. The p*K* of Tris buffer is 8 so that its buffering capacity is high between pH 7.5 and 8.5. Unfortunately it is commonly used at pH 7 at low concentration (0.01 M); in this state, its buffering capacity is poor. Tris also has the following four drawbacks: (1) it reacts with certain electrodes, as mentioned earlier; (2) its pH varies more with temperature than does that of most buffers—pH increases approximately 0.03 pH unit per degree Celsius from 25°C to 5°C; (3) its concentration dependence is also greater than that of most buffers—the relative pH of a buffer at 0.01, 0.05, and 0.1 M is x, $x + 0.05$, and $x + 0.1$, respectively; (4) it reacts with some metal cations (e.g., Cu^{2+}, Nl^{2+}, Ag^{+}, Ca^{2+}).

Good's Buffers

Until the development by Norman Good of the buffers described in this section, most of the buffers that were available were not designed for biochemical studies. The most common problems are the following: lack of buffering between pH 7.5 and 8.0 (for example, phosphate and glycylglycine) or between 7.0 and 7.5 (Tris), inhibition of some enzymes (phosphate and Tris), precipitation of polyvalent cations (phosphate and Tris), strong effects of concentration and temperature on pH (Tris), absorption of ultraviolet light (barbiturate), and toxicity (cacodylate).

Table 4-1
Good's Buffers.

pK_a (20°C)	Buffer	Useful pH range
6.15	MES	5.8– 6.5
6.62	ADA	6.2– 7.2
6.80	PIPES	6.4– 7.2
6.88	ACES	6.4– 7.4
7.15	BES	6.6– 7.6
7.20	MOPS	6.5– 7.9
7.50	TES	7.0– 8.0
7.55	HEPES	7.0– 8.0
8.00	EPPS	7.6– 8.6
8.15	Tricine	7.6– 8.8
8.35	Bicine	7.8– 8.8
9.55	CHES	9.0–10.1
10.40	CAPS	9.7–11.1

SOURCE: N. E. Good, G. D. Winget, W. Winter, T. Connelly, S. Izawa, R. M. M. Singh. 1966. *Biochemistry*, 5: 467–477.

In 1966, Good examined a large number of zwitterionic buffers and found a set that spanned the range of pH from 6 to 10.5. These buffers, known as Good's buffers, lack the difficulties just mentioned and in addition are resistant to hydrolysis and to the action of most enzymes. Because of their rather lengthy chemical names, they are known by abbreviations. These buffers are listed in Table 4.1.

For some reason the Good buffers have not become as prevalent as might be expected. This may be a result of their high cost compared to other buffers. Three of the buffers, HEPES, Tricine, and TES, have been shown to be highly effective in animal-cell culture medium. Surprisingly they have not been widely used, as researchers have conservatively continued with a cumbersome carbonate buffer whose pH is maintained by continually flowing a 5% CO_2–95% air mixture through the culture chamber.

Types of Electrodes

A manufacturer's catalog of pH-meter electrodes can be bewildering. However, careful perusal indicates that there are actually only a few types and combinations thereof.

Glass ("H^+ permeable") electrodes are of three types: general purpose, high-temperature, and low Na^+ error ("high pH" electrode).

Reference ("H^+ impermeable") electrodes are of three types: general-purpose (usually calomel), high-temperature (usually Ag-AgCl), and chloride-free (uses Hg-$HgSO_4$ instead of chloride). Some are constructed with a hole in the tip covered with a ground glass sleeve to allow rapid equilibration with samples of very high viscosity or with slurries or emulsions.

Combination electrodes consist of a glass and a reference electrode in a single unit and are used almost exclusively in biochemistry. Their particular advantage is that, with a single unit, smaller volumes of solution can be measured. Their disadvantages are higher cost and the fact that they must be discarded if one of the elements fails.

There also exist electrodes for measurements other than of pH. For example, metallic electrodes can be used to measure redox potential and the concentration of specific ions. Platinum is usually used for oxidation-reduction measurements because it is resistant to chemical attack. Silver electrodes are used for direct potentiometric determination of Cl^-, Br^-, I^-, S^{2-} and SH^- (i.e., any ion whose silver salt has low solubility). These measurements are performed with the millivoltmeter attachment of most commercial pH meters. For example, in measuring Cl^-, solutions having different chloride concentrations are measured and the reading on the millivoltmeter is plotted against $[Cl^-]$. The concentration of a sample can then be determined from this standard curve. Redox potentials are similarly measured by using standard solutions of known redox potential.

Ion-selective electrodes are of particular interest, especially in neurobiology and membrane biology, in which changes of ion concentration are measured repeatedly. The best type of ion-selective electrode currently available consists of a glass that (like the H^+ ion electrode) is permeable to only one ion. For example, such glasses are prepared by fusing silicon with other metal oxides (not necessarily related in any obvious way to the particular ion. For example, the H^+ ion glass electrode consists of fused SiO_2, Na_2O, and CaO in appropriate ratios. A glass prepared from SiO_2, Al_2O_3, and Li_2O is permeable to the Na^+ ion; its behavior is described by an equation similar to equation (2). Glasses having other compositions allow the selective measurement of cations Li^+, K^+, Rb^+, Cs^+, Ag^+, Cu^+, Tl^+, and NH_4^+. Other electrodes, using thin crystals, have also been manufactured for measuring ion concentrations. These electrodes use an impermeable glass having an opening across which is cemented a thinly sliced crystal. For example, a LaF_3 slice can be used to measure the F^- ion and crystals of Ag_2S and AgX (in which X is a halide) are used to measure the S^{2-} and X^- ions respectively. Liquid membranes have also been used; these consist of an inorganic salt in an organic solvent held in the pores of a porous glass or plastic. For example, a liquid membrane of $Ca(C_{10}H_{21}O)_2PO_2$, calcium decyl phosphate, is sensitive to the Ca^{2+} ion.

The Use of pH Paper

A variety of color indicator papers exist for pH determination, some supposedly sensitive to 0.2 pH unit. They are somewhat useful for rough testing, but it is important to know that errors of several pH units can be introduced by high salt concentration, protein, and certain chromatographic materials such as DEAE-cellulose. In some cases, the dyes in the paper can react with organic substances in the solution and this can result in color changes that are unrelated to pH. Finally, in weakly buffered solutions, the dyes in the paper can act as a buffer and change the pH or give a false reading.

SELECTED REFERENCES

Barrow, G. M. 1979. *Physical Chemistry*, chapter 18. McGraw-Hill. This chapter has a good discussion of the measurement of pH and of ion-selective electrodes.

Bates, R. G. 1973. *Determination of pH: Theory and Practice*. Wiley.

Buck, R. P. 1971. "Potentiometry: pH Measurements and Ion-Selective Electrodes," in *Techniques of Chemistry*, vol. I, edited by A. Weissberger and B. W. Rossiter, part IIA, chapter 2. Wiley-Interscience.

Dole, M. 1941. *The Glass Electrode*. Wiley.

Durst, R. A., ed. 1973. "Ion-Selective Electrodes in Science, Medicine, and Technology." *Am. Sci.* 59:353–361.

Eisenman, G., R. Bates, G. Matlock, and S. M. Friedman. 1965. *The Glass Electrode*. Interscience.

Fischer, R. B. 1974. "Ion-Selective Electrodes." *J. Chem. Educ.* 51:387–390.

Levine, I. N. 1978. *Physical Chemistry*, chapter 14. McGraw-Hill. This chapter has an excellent discussion of the glass electrode and ion-selective electrodes.

Williams, V. R., W. L. Mattice, and H. B. Williams. 1978. *Basic Physical Chemistry for the Life Sciences*, chapter 5. W. H. Freeman and Company. This chapter has a good discussion of the measurement of pH and of buffer systems.

Brochures supplied by the manufacturers of Beckman, Corning, and Radiometer pH meters are good sources of information.

PROBLEMS

4–1. What is the H^+ ion concentration at pH 2 and pH 10?

4–2. A pH meter is to be standardized with a pH 7 buffer. By accident, a pH 8 standard buffer is used but the meter is adjusted to read pH 7. A sample is then tested and the meter indicates that its pH is 6.2. What is the actual pH of the sample?

4–3. After operating for several years, electrodes frequently become useless in that it is no longer possible to obtain an on-scale meter reading with a standard buffer. There are many possible causes of this. Sometimes the electrode can be regenerated by soaking it for a few days in 1 M HCl.

a. When this technique works, what do you think had been the cause of failure?

b. Why is 1 M HCl used rather than 0.1 M HCl?

4–4. Name several factors that should be considered when selecting a buffer.

4–5. How would you go about preparing a buffer that is to be used at 80°C?

4–6. You are studying the killing of bacteria by ultraviolet light. You observe that a larger dose is required to kill 99% of the cells when they are suspended in barbiturate buffer than when they are in phosphate buffer, although both buffers are at the same pH. Propose an explanation.

4–7. An enzyme is found to have very little activity in a phosphate buffer at pH 6.8 compared to the activity in MOPS buffer at pH 6.8. Propose an explanation.

5

Radioactive Labeling and Counting

Many biochemical analyses require the detection of minute (10^{-14}–10^{-6} mole) quantities of material. However, chemical tests are rarely responsive to less than 10^{-7} mole. This limitation has been alleviated by the development of radiotracer technology through which extraordinarily sensitive detection of radioisotopically labeled material has allowed studies of many substances in quantities of 10^{-12} mole to become routine. In addition, the use of radioactivity has permitted the development of powerful experimental approaches to various types of problems. Such approaches employ the *double-labeling* technique for following two substances simultaneously or for distinguishing two identical substances synthesized at different times; the *pulse-chase* method for following a substance at a time after its synthesis without the interference of material concurrently synthesized; and *exchange analysis* for measuring participation in reactions.

In this chapter, these and other techniques will be described in some detail. The methods for detecting and measuring radioisotopes will also be presented herein. Autoradiography has been excluded from this chapter because its technology and applications are rather different from the subject of this chapter. It is presented in Chapter 6.

Types of Radiation Used in Biochemistry

Nuclear radiation is a result of the spontaneous disintegration of atomic nuclei. Of the several kinds of emitted radiation those of importance in isotopic labeling are β particles (emitted electrons) and γ rays (photons).*

* Alpha particles are rarely used for two reasons: (1) they are difficult to detect because they are strongly absorbed by the samples themselves; and (2) there are few α-emitting isotopic labels that can be satisfactorily used for biological materials.

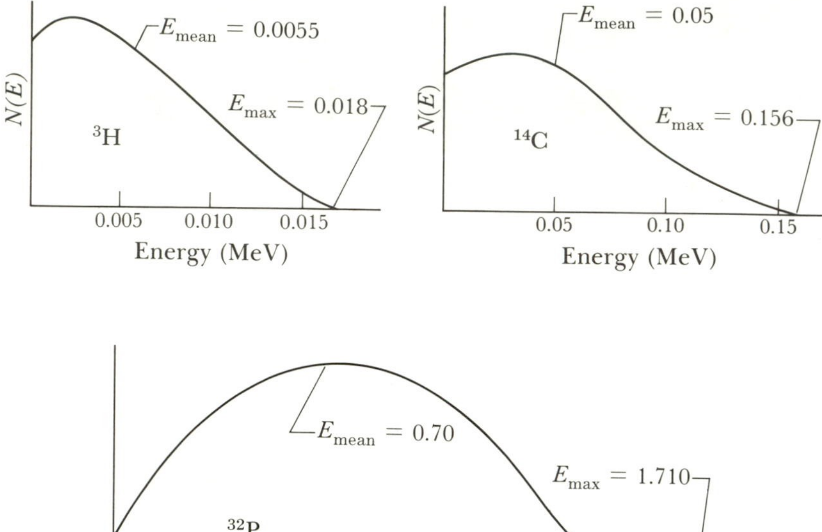

Figure 5-1
Beta spectra for ^3H, ^{14}C, and ^{32}P: E refers to the energy of the particles and $N(E)$ is the number of particles emitted with energy E. This is, in fact, a measure of the probability of emission of a particle at that energy.

Beta Particles

For a particular β-emitting nucleus, β particles are emitted having a continuous range of energy from zero to a maximum value (E_{max}) characteristic of the particular isotope. A plot of the relative probability of emission of a β particle as a function of energy is called a β spectrum. The spectra for several commonly used radioisotopes are given in Figure 5-1. The energy of a given β emitter is traditionally described by stating E_{max}, even though the fraction of particles with energy near E_{max} is very small. A better description is probably given by E_{mean}, the mean energy, because a large fraction of the particles have energy near this value. E_{mean} is roughly $\frac{1}{3}E_{max}$. These spectra are important in discriminating between different isotopes in the same sample and will be discussed again in the section on liquid scintillation counting.

When β particles pass through matter, their energy is dissipated mostly by ionization and/or excitation of the atoms with which they collide. These interactions are detected by Geiger-Müller (ionization) and scintillation (excitation) counters, which will be described in detail in a later section.

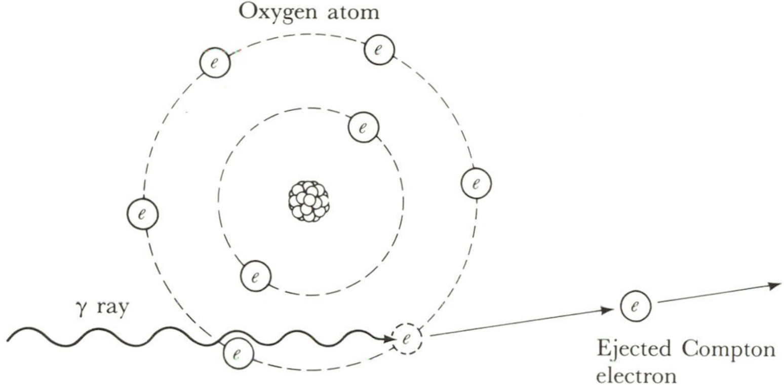

Figure 5-2
Ejection of an electron by the interaction of a γ photon with an oxygen atom.

Gamma Rays

A γ ray is a form of electromagnetic radiation and for a given radioisotope is emitted with one or more discrete energy values rather than over a continuous range, as with β particles. A γ ray is uncharged and therefore does not directly ionize atoms in its path. However, it can interact with an orbital electron of an atom and eject it from the orbit, or with a nucleus to produce an electron-positron pair (Figure 5-2). In both types of interaction, the secondary electrons produced by the absorption of a γ photon are like β particles and can ionize and excite other atoms. Hence, the detection of a γ ray is ultimately accomplished in the same way as is a β particle. In practice, as will be seen in a later section, because of the low probability of interaction (often spoken of as the high penetrability of γ rays), special detectors are needed for γ counting.

Properties of the Radioactive Decay of Chemical Compounds

Radioactive decay is a random process in the sense that, in a collection of radioactive atoms, each nucleus has the same probability of decay as any other nucleus; furthermore the decay of a particular nucleus in no way affects the time of decay of any other nucleus. This has the effect that at any particular time, the number of atoms of a radioactive material decaying per unit time is proportional to the number of atoms present at that time. That is, if N is the number of atoms present at time t, and dN the number of atoms disintegrating in the interval dt, then

$$-\frac{dN}{dt} = \lambda N. \tag{1}$$

Table 5-1
Characteristics of Commonly Used Isotopes.

Isotope	Particle emitted	E_{max} (MeV)	Half-life
^{3}H	β	0.018	12.3 years
^{14}C	β	0.155	5568 years
^{24}Na	β	1.39	14.97 hours
	γ	1.7, 2.75	
^{32}P	β	1.71	14.2 days
^{35}S	β	0.167	87 days
^{40}K	β	1.33, 1.46	1.25×10^{9} years
^{45}Ca	β	0.254	164 days
^{131}I	β	0.335, 0.608	8.1 days
	γ	0.284, 0.364, 0.637	

in which λ is the decay constant.* Thus, if N_0 is the number of atoms at $t = 0$,

$$N = N_0 \, e^{-\lambda t} \tag{2}$$

which is sometimes called the law of radioactive decay. An exponential decay equation states that, *in any given time interval, the radioactivity (disintegration rate) will decrease by the same fraction*; hence, it is convenient to express the decay constant as the half-life, $\tau_{1/2}$, the time required for the activity (decay rate) to decrease by one-half. Hence, $\tau_{1/2} = -(\log_e \frac{1}{2})/\lambda = +0.693/\lambda$. This simply means that after one half-life, one-half of the initial activity remains, after a second half-life, one-quarter of the activity, and so forth. Half-lives for some of the isotopes commonly used in biological studies are given in Table 5-1. (Note the enormous range in half-lives.) The principal reason for knowing the half-life of an isotope being used in an experiment is that, if the half-life is short compared with the time it takes to do the experiment, the amount detected will depend on the time at which the measurement is made; hence, if radioactivity is being used to determine the amount of material present at a given time, a correction for the changing count rate must be made. The half-life also determines the maximum specific activity (see page 133) that is obtainable and is therefore a factor in selecting an isotope for a particular experiment.

* λ is the conventional notation for the decay constant. It is totally unrelated to the wavelength of electromagnetic radiation, which is also denoted by λ.

Radioactivity is expressed in units of *curies*. One curie (Ci) is defined as the number of disintegrations per second per gram of radium and equals 3.70×10^{10} disintegrations per second. For most biological applications, quantities much less than one curie are normally used and the milli- (mCi) or microcurie (μCi) is employed. Furthermore, in practice, a minute is the standard time unit—hence, 1 μCi = 2.22×10^6 disintegrations per minute. For reasons that will be discussed later, it is fairly difficult to determine the absolute number of disintegrations because radiation counters detect only a fraction of them; hence activity is usually stated as detected counts per minute (cpm). For highly quantitative experiments, in which the amount of radioactivity is used to calculate the absolute amount of material present, the value of cpm is is converted into disintegrations per minute (dpm) by dividing by the efficiency of counting.

Radioisotopes and isotopically labeled compounds are rarely isotopically pure—that is, the radioisotope is usually diluted by the presence of chemically identical nonradioactive isotopes. (A few isotopes—e.g., ^{32}P—can be prepared in a pure, or *carrier-free*, state, but they are rarely used as such.) The relative abundance of a radioisotope is described by the *specific activity*—that is, the disintegration rate per unit mass. This is normally expressed as activity (Ci, mCi, μCi) per millimole or micromole (mmol, μmol). Unfortunately, specific activity can be expressed in different ways: for example, the specific activity of [^{14}C]glycine (which contains two carbon atoms) may be stated as 50 μCi/μmol or 25 μCi/μmol *C atom*. Hence, it is important to note the units of specific activity.

Radiochemical purity is a complication often difficult to deal with. For example, when an isotopically labeled biochemical is synthesized, the reaction mixture usually contains a variety of labeled products, which are separated to produce radiochemically pure compounds. Manufacturers of radiochemicals usually supply an assay of the radiochemicals present in the sample and thereby indicate the degree of purity. However, radiochemical purity can change in time for two reasons. First, radiation emitted by one molecule can alter an identical labeled molecule either directly by ionization followed by chemical rearrangement or indirectly by ionizing the solvent (radiolysis) and producing reactive species that can attack the originally pure compound. This is especially true of radiochemicals stored in aqueous solution. A common event with ^{3}H- and ^{32}P-labeled compounds stored at concentrations greater than 100 μCi/ml is the production of H_2O_2 following ionization of water. This substance can attack a large number of chemical compounds. In some laboratories radiochemicals are stored with a small amount of the enzyme catalase (1 μg/ml), which catalyzes breadkdown of H_2O_2.*

* The H_2O_2 concentration can be so high in some stocks of carrier-free H_3PO_4 and in very concentrated [^{3}H]thymidine that additional of these substances to a bacterial culture results in rapid death of the bacteria. Treatment of the stocks with catalase prior to adding it to the cell culture avoids this problem.

Second, if a substance contains two isotopically labeled atoms (which is uncommon except in the so-called uniformly labeled compounds), the decay of one to produce a new atom will result in a molecular rearrangement. For example, consider a substance with two ^3H atoms in each molecule. Beta decay of one ^3H nucleus converts the hydrogen into helium. Because helium does not participate in chemical-bond formation, it will fall off and a chemical rearrangement must occur, thus forming a new radioactive compound containing the second ^3H atom. Similarly, ^{14}C decays to form ^{13}N and, because the valences of carbon and nitrogen differ, a chemical rearrangement must occur. For the most part these problems can be minimized by following the manufacturer's directions for storage, by using freshly made compounds, and, when necessary, by purifying the compounds by chromatography or electrophoresis before use.

Measurement of Beta Activity by Methods Employing Gas Ionization

A β particle that passes through a gas may dislodge an orbital electron from one of the atoms of the gas, which results in the production of an ion pair—the dislodged electron plus the remaining positively charged ion. The β particle often has sufficient energy to ionize several atoms successively. If the gas is contained in a chamber in which there are two charged electrodes, the secondary electrons and the positive ions will be attracted to the anode and cathode, respectively, and this can be recorded as a tiny pulse of charge or current. A medium-energy, single β particle will produce from 10^2 to 10^3 ion pairs per centimeter, but not all of these ions can be collected by the electrodes because the ion pairs rapidly recombine in the wake of the particle. The efficiency of collection of the ions depends on the voltage between the electrodes. For example, with very low voltage most ions recombine before reaching the electrodes but, as the voltage is increased, a greater fraction of the ions is captured before they have recombined. Ultimately, a voltage is reached at which all ions are collected and *saturation* is achieved. This property of ion chambers is shown in Figure 5-3. At a much higher voltage (well above that required for saturation), the dislodged orbital electrons are accelerated toward the anode at such high velocities that they also cause ionization of the gas atoms, producing what is called an *avalanche of ions or gas amplification.* At even higher voltage, the number of secondary ions collected becomes proportional to the number formed in the original ionization (the proportional region). This is followed by a second region of saturation (the Geiger-Müller region) in which all possible ion pairs, both primary and secondary, are collected; ultimately, the voltage is so high that the gas is ionized by the applied voltage even when no β particles are present (the continual discharge region). The complete dependence of the system on voltage is shown in Figure 5-3.

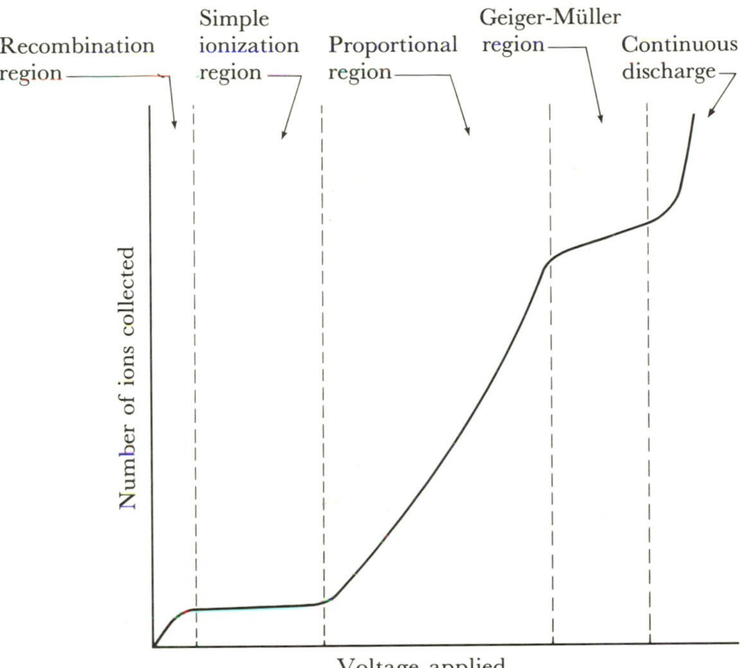

Figure 5-3
Output (number of ion pairs) of a Geiger-Müller tube as a function of the voltage between anode and cathode.

Ionization detectors can be operated in either the proportional or the Geiger-Müller region. Proportional counting has the advantage that particles of different energies can be distinguished by pulse-height analyzers (see page 142), because the size of the current pulse received by the electrode is proportional to the energy of the original charged particle entering the chamber. In this way, two different radioisotopes could be present in the same sample and the relative amounts of each could be determined from their β spectra by the procedure described on page 143 for scintillation counting. The disadvantage of the proportional counter is that extraordinarily stable, high-voltage power supplies are necessary because small fluctuations in voltage can produce large fluctuations in current. Proportional counting has been replaced by liquid scintillation counting (see next section) and will not be discussed further.

In the Geiger-Müller (G-M) region, maximal gas amplification is achieved and voltage fluctuations have little or no effect. This results in a reliable, sensitive, relatively inexpensive, and stable counting device—the Geiger-Müller counter.

A diagram of the most commonly used type of Geiger-Müller counter is shown in Figure 5-4—the so-called *end-window counter*. The detector

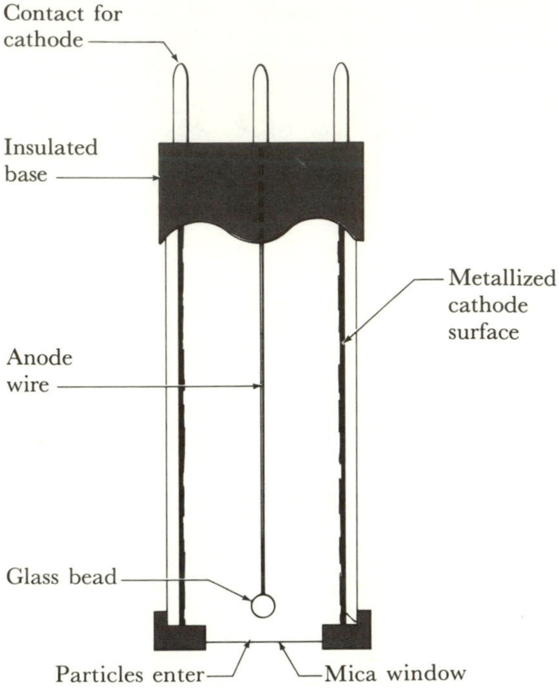

Figure 5-4
An end-window Geiger-Müller tube.

consists of a cylinder, the inner wall of which is metallized and serves as the cathode, an axial wire anode, and an end window (typically made of mica or a plastic called mylar) through which β particles enter the gas-filled chamber. An electronic system measures the current produced by the capture of electrons by the anode. The tube is usually filled with an inert gas such as helium, neon, or argon to which has been added a small amount of a *quenching agent*—usually butane, propane, ethanol, chlorine, or bromine. The quenching agent prevents continuous ionization, which would cause the detector to fail to respond to any other than the first incoming particle. When a positive ion nears the cathode, it sometimes pulls off an electron and becomes a neutral atom again. This recombination results in the production of both X rays and short-wavelength ultraviolet radiation, either of which can ionize the gas molecules and form secondary electrons, producing the subsequent avalanche of electrons. It is this that causes a self-generating, continuous ionization. However, if a quenching gas (e.g., butane) is present that has a lower ionization potential than the major gas (e.g., helium), the positive ions of the helium will collide with the butane and acquire an electron from the butane to form a neutral helium atom and a positively charged butane ion. These butane ions move to the cathode and pick up an electron to become neutral again; however, because of the physical and chemical properties of the butane, the excess energy of recombination is not converted into electromagnetic radiation but instead breaks the chemical bonds of the butane resulting in the destruction of the molecule. After a certain number of discharges no quenching gas will be left. Actually, for most commercial G-M tubes, from 10^8 to 10^{10} pulses are possible before the tube becomes useless. This limitation can be eliminated by adding an entrance and an exit port and passing the gas continuously through the tube. This is called a *flow tube.* The sensitivity of the tube can also be improved by eliminating the window; many of the β particle either fail to pass through the window or lose sufficient energy during the passage that they are no longer able to ionize the gas. To do this, the tube is mounted so that it is in close contact with the sample holder. Because the gas would be lost whenever the sample was inserted or removed, the gas is continually flushed through the tube. Hence, this modification is known as a *windowless flow counter.*

Measurement of Radioactivity by Liquid Scintillation Counting

The ideal instrument for measuring radioactivity would detect all decays but no such instrument exists. Geiger-Müller windowless counters are very efficient detectors of high-energy β particles such as those from ^{32}P but are very inefficient for low-energy particles such as those from ^3H. The two major factors that limit efficiency are that not all emitted particles reach the detector and, of those that do, not all are counted. The main reason for failing to reach the detector is that the geometry is usually such that some particles are emitted in a direction that misses the detector. For example, even if a sample were flush with the front surface of a G-M detector, one-half of the particles would still be emitted in a direction away from the tube. The principal reason for failing to detect a particle that does enter the G-M tube is that the particle may have insufficient energy to cause ionization of the gas. By looking at the β spectra in Figure 5-1, it can be seen that the particle energy can be as low as zero (there is no E_{min}), so that some fraction will always be less than the energy required to ionize the gas. It can also be seen that the shape of the spectra is such that, with decreasing E_{mean}, there is an ever-increasing fraction of particles having energy in this very low range. This is in fact the main problem in detecting low-energy β particles with all counters.

The geometric problem could be solved if the sample were contained *within* the detector. In this way, losses due to failure to reach the detector would be limited to those particles whose range is so short that they fail to leave the sample itself ("self absorption"). Such a solution to the geometric problem is provided by the technique of *liquid scintillation counting.*

In liquid scintillation counting, the sample is dissolved or suspended in a solvent containing one or more substances that are fluorescent (Table

Table 5-2

Solvents and Fluors Commonly Used in Liquid Scintillation Counting.

Solvents

Toluene

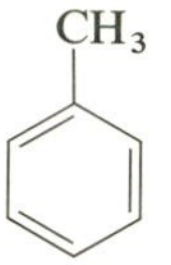

CH_3

1,4-Dioxane

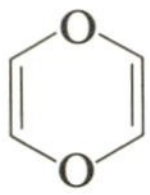

O O

p-Xylene

CH_3

CH_3

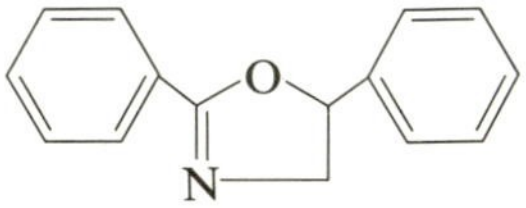

Fluors

PPO

O N

POPOP

N O N O

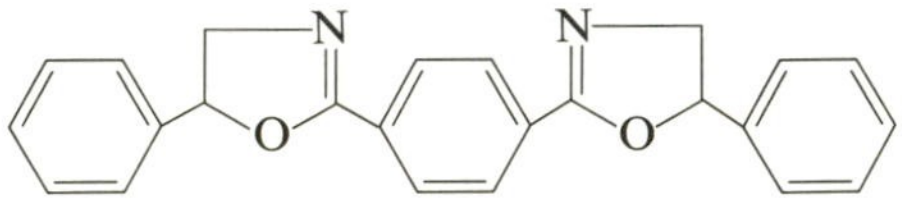

5-2). In brief, the emitted particle causes a pulse of light, which is detected by an optical device (a photomultiplier tube) that converts it into an electrical pulse that can be counted.

The Scintillation Process

Let us first understand how the decay produces detectable fluorescence. Consider a β particle that leaves the sample and enters the solvent in which the sample has been placed. In most solvents, the energy of the particle either would be dissipated as heat or would cause a chemical alteration (e.g., ionization or dissociation). However, in certain solvents, the energy is absorbed by the solvent molecules, which then are raised to an excited state. The excited molecule then returns to the ground state and gives up its energy by emission of a photon of light (see Chapter 15 for a discussion of the fluorescence process). The wavelengths of these emitted photons are very short and are not detectable by most photodetectors and so a fluorescent substance (a *fluor*) at a fairly low concentration must be added; this substance (called the *primary fluor*) efficiently absorbs the photons emitted by the excited solvent molecules and then reemits photons at a longer wavelength. The photons emitted by the fluor must be detected by a photomultiplier tube. However, if the wavelengths of the photons emitted by the fluor are not in the region of highest sensitivity of the photomultiplier, an electrical pulse will not be generated. Therefore, a second fluor is almost always added. The secondary fluor absorbs the photons emitted by the primary fluor and reemits them as fluorescence at a longer wavelength, which the photomultiplier can detect with high efficiency. It should be noted that neither primary nor secondary fluors absorb much of the original energy of the β particles (because they are present at exceedingly low molarity) and are simply wavelength shifters. Secondary fluors are unnecessary with most scintillation counters made after 1973.

This sequence of events converts the energy of the emitted particle into a flash of light—that is, a collection of photons emitted in an extremely short interval ($\sim 10^{-9}$ second). Note that the light need not come from a single point because of the multiple transfer of energy. The efficiency of detection of the original decay depends on (1) the properties of the solvent (i.e., the fraction of the absorbed energy that is converted into excitation and the fraction of the excitation energy that is transferred to the primary fluor rather than being dissipated as heat); (2) the number of photons produced by the primary and secondary fluors; (3) the geometry of the photomultiplier tubes (i.e., the efficiency by which the photons are gathered); (4) the signal-to-noise ratio of the photomultiplier (*noise* refers to pulses produced in the absence of light); and (5) the circuitry that converts the charge of the photomultiplier into a voltage.

The Scintillation Cocktail

A solution of fluors is called a scintillation cocktail. These solutions are of two types: those which are miscible with aqueous solutions and those which are not. The solvent most commonly used for the water-accepting cocktail is dioxane. This solvent contains peroxides, which can destroy many types of samples, so that dioxane containing peroxide inhibitors must be used. One must also be aware of the mild toxicity of dioxane. The most common water-immiscible cocktail employs toluene as a solvent. However, for most purposes the less frequently used xylene-based cocktails are preferable because xylene has a lower vapor pressure, lower flashpoint, a 7% greater efficiency of detection of decays than toluene (pure *p*-xylene is 12% more efficient but about four times as expensive), and lower toxicity than toluene so that it is safer. Toluene is still in more common use because of its lower cost. There are about fifteen different fluors in use. There is little reason to choose one instead of another as they are nearly equivalent in effectiveness. One, known as butyl-PBD, is slightly more efficient but has numerous problems and should probably be avoided.

The properties of the cocktails that accept water will be discussed in greater detail in a later section.

Background Noise

Noise is a significant problem in the design of liquid scintillation counters—a problem that can be readily understood by observing the magnitude of the electrical pulse resulting from a single β decay. To give a rough idea of what happens, a 50-keV β particle (e.g., from ^{14}C) will yield a few hundred photons in the commonly used solvent-fluor systems. These photons result in roughly 50 photoelectrons (10%–20%) being produced by the photocathode of the photomultiplier. This is amplified approximately 10^6-fold by the multiplier to yield 5×10^7 electrons, or a charge of 8×10^{-12} coulomb, which can be converted into an electrical pulse of 0.16 V. For ^3H, E_{mean} is 5.5 keV and the pulse corresponding to this energy is about 0.018 V. The low voltage of these pulses is the source of the problem because the thermal noise of photomultipliers at room temperature produces pulses on the order of 0.005 V, which corresponds to a 1.5-keV particle. Referring to Figure 5-1, it can be seen that a substantial fraction of emitted β particles (a fraction that increases with decreasing E_{mean}) produces pulses of lower voltage than does thermionic noise.

This problem has been attacked in two ways. The first was to place the photomultiplier and the samples in a freezer at a temperature ranging from 0° to 5°C, which reduced the noise level by a factor of about 4; now with the advent of improved low-noise photomultipliers, room temperature ("ambient") operation is possible. Thermionic noise results in a single pulse, whereas β particles usually result in the production of several photons. Hence, the second way of attacking the problem requires the use of two photomultipliers for viewing the sample; the outputs of the photomultipliers are fed into a *coincidence circuit*, which registers a count only when two pulses are *simultaneously* (i.e., within microseconds) received by the photomultipliers. In this way, the single noise pulses are discarded. The timing of the coincidence circuit is such that it will only very rarely be the case that one noise pulse will follow another so closely in time that it will be recorded; only when two photons have been produced as a result of a *single* β particle will the coincidence circuit register a pulse. Coincidence circuitry reduces the background count rate from about 10^5 cpm to 15 cpm (i.e., the number of pulses that pass the coincidence circuit).

The necessity of using coincidence circuitry eliminates the possibility of 100% efficiency of detection of low-energy β particles because a minimum of two photons (one for each photomultiplier) is required and a certain number of the decays do in fact yield only one. However, this disadvantage is more than offset by the tremendous efficiency resulting from the huge reduction in background.

Several other sources of background counts exist although they are of a lesser order of magnitude. They are, however, the ones encountered by the user of counter rather than the designer. One source is the radioisotope ^{40}K found in glass. If the sample is a solution, it must be in a container, which is usually a glass vial fabricated from "low-potassium" glass. This is necessary because of the large amount of the radioisotope ^{40}K found in ordinary glass. However, even low-potassium glass contains significant amount of ^{40}K and contributes a background of about 15 cpm—the same amount contributed by thermionic noise. For some operations requiring very low backgrounds, polyethylene vials can be used to eliminate the ^{40}K background, but they are not resistant to all solvents used in scintillation counting and are therefore not in general use. Another source of background is Ĉerenkov radiation, the wavelength of which is in the region of efficient response of the photomultipliers. Ĉerenkov radiation is produced by the responses of the scintillation solvent (without added fluor) and the vial itself to cosmic rays. It produces pulses in the low-to-mid-energy range and contributes another 10 cpm. Yet another source is environmental activity. When the sample vial is filled with solvent and fluors, environmental radioactivity adds about 40 cpm to the background. Taken together these effects, including thermionic noise, produce a background ranging from about 80 cpm to 90 cpm. However, because emitted β particles always have a maximum energy, it is always possible to require (by means of pulse-height analysis and by the use of discriminators—see page 142) that pulses with energies above that value be rejected. This results in a reduction of background to about 35 cpm with no loss of counting efficiency.

Quenching

The ability to detect radioactivity depends on the signal-to-noise ratio. So far, the discussion of background has been concerned with maximizing this ratio by reducing noise. Now the problems associated with *maximizing the efficiency of production of photons* (i.e., maximizing the signal) will be considered.

Quenching refers to a reduction in the efficiency of transferring energy from the β particles to the photomultiplier (by any means). Quenching results in a decreased (perhaps to zero) number of photons per β particle and therefore the production of a pulse of reduced voltage. The three most common mechanisms are *chemical*, *color*, and *dilution* quenching.

Chemical quenching

When a sample is added to a scintillation solution, the sample itself may contain substances that either absorb some of the energy of the β particles without emitting any photons or absorb the photons emitted by the excited solvent molecules without fluorescing. The most common chemical quenchers are water, acids, salts, dissolved oxygen, peroxides, peroxides, and chlorinated hydrocarbons. In many cases, the quencher competes with either the primary or secondary fluor for the excitation energy. Thus, quenching can often be reduced by increasing the concentrations of the fluors.

Color quenching

Colored samples can absorb some of the photons emitted by the secondary fluor so that they never leave the counting vial. It is important to realize that a lack of visible color in the sample is no criterion for the absence of color quenching because the human eye cannot determine whether a substance absorbs in the near ultraviolet, which is the range of wavelengths emitted by the fluor. It is, though, generally true that substances that appear yellow, red, or brown probably introduce severe color quenching. If it is not too great, color quenching can also be reduced significantly by increasing the fluor concentration.

Dilution quenching

The dilution of the solvent and fluor by the sample reduces the probability of a scintillation event. If the sample is a liquid, there is no way to avoid this; however, it can be corrected for in the analysis of data and will be described in a later section.

Proportional Counting and Pulse-Height Analysis

The greatest value of scintillation counting is the ability to determine the ratio of two radioisotopes present in a mixture. This is possible because the voltage pulse produced as a result of a decay is proportional to the energy of the emitted particles. The resolution of the two isotopes is accomplished with a *pulse-height analyzer* equipped with *discriminators.* A pulse-height analyzer is an electronic instrument that can sort fluctuations (pulses) in current or voltage. As each pulse is detected, its magnitude (pulse-height) is also registered. The instrument is used in circuitry that is designed to count pulses in different voltage *intervals.* For example, it can be adjusted to detect pulses either greater than zero or some other value, less than a certain value, or between any pair of values. The controls that determine the voltage levels defining the voltage range are called *discriminators.* Two discriminators are normally employed to define an interval that excludes low-energy photomultiplier noise and high-energy environmental noise. With more than two discriminators, it is possible to count two intervals simultaneously. Counting in two ranges simultaneously requires two counting circuits, called *scalers.* Each scaler is said to be counting in a *channel* defined by the particular voltage *levels,* frequently designated L1, L2, L3, and so forth. Decay rate can be plotted against energy or pulse height by counting pulses between a variety of pairs of voltage levels. Figure 5-5 shows a plot for ^{3}H and ^{14}C. Using this ability to determine the decay rate in a given pulse-height interval, the ratio of two isotopes can be determined. How this is done can best be seen by example.

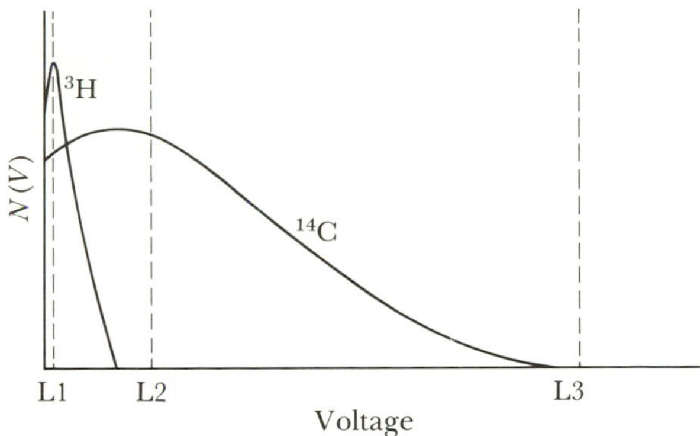

Figure 5-5
Simultaneous counting of ^{3}H and ^{14}C. The ^{3}H and ^{14}C spectra are plotted on a single scale: L1 is the lower limit for most counters, below which thermal noise becomes severe; L2 and L3 are the voltage levels used in Example 5-A. The y-axis is an arbitrary scale giving the count rate in a small interval centered on a particular voltage, V.

Example 5-A

☐ Ratio of ^{3}H to ^{14}C activity in a single sample.

Consider a radioactive sample containing both ^{3}H and ^{14}C labels and counted in the two voltage ranges defined by L1, L2, and L3 of Figure 5-5. Data obtained for the sample are given in Table 5-3. By counting ^{3}H and ^{14}C standard samples using the same instrument settings, the amount of ^{3}H and ^{14}C present in each channel can be calculated as shown in Table 5-3.

What is done follows. The distribution of counts from the ^{3}H and ^{14}C standard samples in channels A and B tells how the ^{3}H and ^{14}C radioactivities are distributed in the experimental sample assuming that there is no quenching of the sample with respect to the standard. Line 1 shows that only 0.0015 of all ^{3}H counts are in channel B, which is approximately zero for the sample. Hence, all counts in channel B are due to ^{14}C, as indicated in line 2. The data for the ^{14}C standard sample shows that the count rate in channel A is 0.146 times the count rate of ^{14}C in channel B. Because line 2 shows that channel B contains only ^{14}C counts, the calculation of line 3 can be carried out. Hence, of the 500 cpm in channel A, 333 are from ^{14}C and the remainder are from ^{3}H, as shown in line 4. Because there is virtually no ^{3}H in channel B, the amount in channel A is all of the ^{3}H, as stated in line 5. The total ^{14}C is then the sum of the ^{14}C in channels A and B, as shown in line 6. The ^{3}H-to-^{14}C ratio can then be calculated as in line 7. Note that the simplicity of the calculation

Table 5-3

Calculation of $^3H/^{14}C$ Ratio in a Doubly Labeled Sample Counted in Two Channels of a Scintillation Counter.

Sample	Radioactivity (cpm) recorded: L1–L2 (channel A)	Radioactivity (cpm) recorded: L2–L3 (channel B)	B/A	A/B
Experimental	500	800		
^3H standard	16,840	25	0.0015	–
^{14}C standard	4250	10,200	–	0.416

Calculations for sample

1. ^3H in channel B = (0.0015)(500) = 0.75 cpm
2. ^{14}C in channel B = 800 − 0.75 ≅ 800 cpm
3. ^{14}C in channel A = (0.416)800 = 333 cpm
4. ^3H in channel A = 500 − 333 = 167 cpm
5. Total ^3H = 167 cpm
6. Total ^{14}C = 333 + 800 = 1133 cpm
7. $^3H/^{14}C$ = 167/1133 = 0.147

comes about by selecting L2 so that there are, for all practical purposes, no ^3H counts in channel B. It should also be noted that, if there were a great deal more ^{14}C, the ^3H determination would be inaccurate. For example, suppose that there were twenty times as much ^{14}C as ^3H. Then the ^{14}C in channel A would be $20 \times 333 = 6660$ cpm and the count rate registered in channel A would be $6660 + 167 = 6827$ cpm. The ^3H value would then be obtained by subtracting 6660 from 6827.

To understand why subtracting large numbers is a problem, consider the statistics of counting. According to the theory of statistics, the standard error of a measurement is $N^{\frac{1}{2}}$ in which N is the total counts (not cpm). This means that there is a 68% probability that the "true" value is in the range $N \pm N^{\frac{1}{2}}$. If the samples in the example with $20 \times$ ^{14}C were counted for one minute, the ^3H would be obtained from the difference between 6827 ± 82 and 6660 ± 81—which could range from 4 to 330 cpm. Clearly, the error in the difference would be huge. If the samples were counted for 10 minutes, the difference would be $(68{,}270 \pm 261) - (66{,}600 \pm 259)$, which could range from 115 to 219 cpm, an improvement but still a large error. With the $^3H/^{14}C$ ratio used in the Example, the range is from 164 to 180 for 10-minute counting periods.

To indicate the value of double-label analysis, consider an extension of this example. Suppose that the ^3H is in leucine at a specific activity of 5 Ci/mmol and the ^{14}C is in tyrosine at a specific activity of 4 Ci/mmol. Let us assume that the counting efficiency for ^3H and ^{14}C is 30% and 90%, respectively. Therefore the actual ^3H and ^{14}C count rates are (measured ^3H count rate/^3H efficiency) and (measured ^{14}C count rate/^{14}C efficiency) or $167/0.30 = 556$ cpm and $1133/0.90 = 1259$ cpm, respectively. The amounts of leucine and tyrosine are therefore (amount ^3H/specific activity of leucine) and (amount ^{14}C/specific activity of tyrosine). The molar ratio of leucine to tyrosine is therefore ($^3H/^{14}C$) × (specific activity of tyrosine/specific activity of leucine) $= (556/1259) \times (4/5) = 0.353$.

Example 5-A assumes that the distribution of counts between channels A and B for ^3H is the same in the experimental sample as in the ^3H standard sample and that this is also true for ^{14}C. This means either that there is no quenching or that the quenching in the sample is the same as in the standards. If this is not the case, the relative degrees of quenching would have to be determined. As will be seen in the section on sample preparation, this can be a formidable problem in doubly labelled samples. Hence, it cannot be overemphasized that the samples used as standards *must* be prepared in an identical manner to the samples containing both labels. A common error made by novices is to use the standards provided by the manufacturer of the counter for determining the ratio of count rates in the channels. These standards are typically radioactive toluene in a toluene-based solvent in an argon atmosphere—an ideal unquenched situation almost never encountered in practice.

Sample Preparation

The method of sample preparation can have a great effect on counting efficiency.

In G-M counting, the sample is usually deposited on an aluminum disc called a *planchet* and dried, if it is in liquid. Frequently, insoluble samples are collected on the surface of a thin membrane filter (Chapter 7) and cemented onto the planchet. The principal problem in sample preparation for G-M counting is self-absorption—that is, if the sample is too thick, many of the β particles fail to reach the counter because they are absorbed by the sample itself, or they lose sufficient energy that they cannot ionize the gas. For β particles as energetic as those from ^{32}P, self-absorption is rarely a problem; for ^{14}C, it can be a problem if the mass of the sample exceeds 1 mg; for ^3H, it is so severe that G-M counting is rarely possible. The problem of self-absorption is circumvented by using samples of various thickness (i.e., various volumes), determining the count rate per unit volume for each, and determining whether the count rate is proportional to volume. If this is not the case,

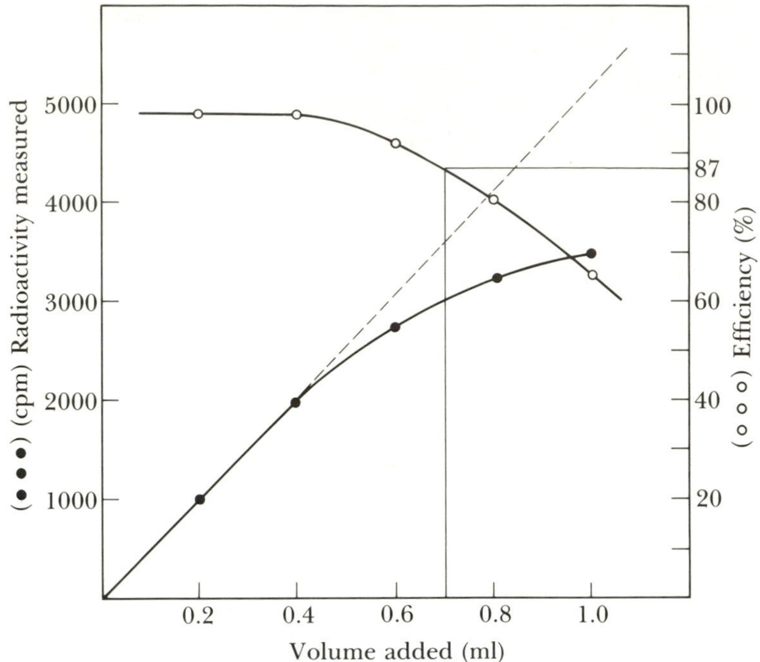

Figure 5-6
Evidence of self-absorption: the measured radioactivity is not proportional to the volume added; the dashed line indicates the count rate that would be expected if there were no self-absorption; the ratio of the solid curve to the dashed line gives the efficiency (open circles). Hence, if a similar 0.7-ml sample yielded 4500 cpm, the corrected value would be 4500/0.87 = 5184 cpm.

a graph is made of count rate versus volume and is used to correct the data. Figure 5-6 illustrates how this is done. It is often necessary to add a separate compound of high activity to an identical sample in order to have a count rate that is high enough to obtain a statistically significant correction curve.

For scintillation counting, the ideal sample is one that is dissolved in an unquenched counting solution because then every radioactive atom is in intimate contact with the solvent. Such a sample is called a homogeneous counting system. If the sample is insoluble (that is, if it is a heterogeneous system), it will exist in the form of small particles or aggregates in the solvent and the energy of some of the β electrons will be reduced by passage through the aggregate—that is, there will be self-absorption. Again, this is not a significant problem for high-energy particles but must be considered for ^{3}H. In general, sample solubility is a problem in biological experimentation because the fluors used are usually soluble only in nonpolar or weakly polar solvents, in which most biological molecules are insoluble.

A homogeneous system can be produced by various methods, some of which are discussed later in this section. It is usually assumed that the efficiency of counting of a homogeneous system will also be greater than that of a heterogeneous system containing the same radioactive substance. In general, this will be true but the possibility that this is not the case must always be considered. For instance, a sample might contain a substance that quenches when it is soluble yet fails to quench when the substance is insoluble. Thus, solubilization of a sample to produce homogeneity would introduce a quenching agent. The means of testing this possibility and the procedures for creating a homogeneous system are described below.

There are three basic methods for preparing biological samples for scintillation counting: (1) the scintillation solvent is altered in a way that permits the addition of aqueous solutions so that soluble material remains in solution; (2) insoluble substances are chemically or physically converted into a soluble form; (3) insoluble substances are collected on nitrocellulose, glass fiber, or paper filters and then dried. The first two methods introduce quenching and this must be dealt with by methods described next; the third method is rapid and convenient but also presents several problems.

Alteration of Counting Fluid to Permit Water Uptake

The usual counting fluids are toluene-based and cannot take up water without severe quenching. Dioxane-based solutions can accept water (as much as 20%), but dioxane is less efficient in energy transfer than toluene. Dioxane-based fluids can be improved by the addition of naphthalene to increase efficiency plus a variety of alcohols or ethers to maintain naphthalene solubility when water is added. A particularly common dioxane-based fluid is called Bray's solution. In each system the additives produce quenching, but this can usually be ignored if relative amounts of radioactivity are being measured because it will be the same for all samples. However, water added to the sample and the salts dissolved in the water produce quenching, the degree of which increases as more water is added. The usual recourse is to add water or a solution to each sample so that the total amount of added material is the same for all samples in a set. An alternative is to determine the degree of quenching in each sample, which will be described shortly.

The mean range of ^{3}H β particles in water is approximately 0.5 micron (μm). Hence, if a precipitate or insoluble material could be kept dispersed so that the particle size were much less than 0.5 μm, self-absorption losses would be small. This can be done in several ways. If a surfactant such as Triton-X100 or an emulsifier such as Cab-O-Sil (silica gel) is added to toluene, water can be added up to 10% by volume

Table 5-4
Several Scintillation Cocktails.

Cocktail	Composition	Application
Toluene-based	PPO, POPOP, toluene	Non-aqueous sample
Triton–toluene	PPO, POPOP, toluene Triton-X100	Aqueous sample
Dioxane-based	Bis-MSB, PPO, naphthalene in dioxane	Aqueous sample
Bray's solution	PPO, POPOP, ethylene glycol, methanol, naphthalene, in dioxane	Aqueous sample
Toluene for thin-layer chromatograms	PPO, non-ionic surfactant, water, and toluene	Sample on a solid support as in thin-layer chromatography
Radioimmunoassay	Aromatic solvents, non-ionic surfactant, and quench-resistant fluors	For radioimmunoassay, aqueous solutions containing proteins and salts

and the water droplets will be dispersed in the emulsion as particles <0.1 μm in diameter. Other emulsifying mixtures are PCS and Aquasol (New England Nuclear Corporation). An alternative is to filter the sample onto finely divided diatomaceous earth (Celite, Johns-Manville Corporation) and count in any emulsifying mixture. It is not clear, though, why the latter method works. When using samples in suspension, it is usually necessary to use a gelling agent to prevent the finely dispersed sample from settling to the bottom of the counting vial. Thus this procedure is often called *suspension counting* or *gel counting*. This procedure is especially valuable in counting regions of thin-layer chromatograms (see Chapter 8), in which case the chromatographic material, which is a powder, can be scraped off of the supporting glass plate and placed directly in the scintillation cocktail.

The characteristics of several types of cocktails are summarized in Table 5-4.

Solubilization of Sample

Acidic substances, such as proteins and polypeptides, acidic polysaccharides, and nucleic acids, can be solubilized by reaction with organic bases—primary examples are Hyamine 10-X-hydroxide and primene 81-R (high-molecular-weight quaternary and primary amines, respectively). A technique that is sometimes used with ^{14}C-labeled material is to incinerate the sample and collect the radioactive CO_2 by bubbling through Hyamine, 2-phenylethylamine, or 2-aminoethanol. Insoluble bases, such as metallic ions, are solubilized by 2-ethylhexanoic acid or dialkyl hydrogen phosphates. Most of these methods are technically complex and introduce significant quenching.

Collection on Membrane Filters

Insoluble samples are commonly filtered or dried onto paper filters—a convenient way to remove water. But this popular method has two disadvantages: (1) the efficiency of counting depends on the orientation of the paper in the counting vial (because the paper is opaque in the solvents and intercepts some of the photons) and (2) very small molecules (e.g., amino acids and purines) penetrate the paper and become inaccessible to the solvent.

A great improvement is to collect fine precipitates on nitrocellulose membrane filters (e.g., Millipore, Schleicher and Schuell, and Gelman) or, even better, on fiberglass papers, as described in Chapter 7. This is probably the most common method of sample preparation in modern biochemistry because of speed and convenience, although self-absorption can be a problem, especially if the concentration of material is >1 mg/cm^2. This will be discussed in more detail in the next section. Nitrocellulose filters present a special problem that is not commonly recognized. After a sample has been collected, the filter is usually dried in a 100°C oven to remove water from the pores of the filter, in order to prevent chemical quenching. Sometimes there is a slight yellowing (or charring) caused by the heating. The yellow material leaches out into the scintillation cocktail and absorbs the blue-violet photons emitted by the fluor. This yellowing often goes unnoticed and is a frequent cause of the lack of reproducibility in duplicate samples. This problem is best avoided either by washing the sample in an organic solvent (for example, ethanol) to aid in rapid drying or by reducing the drying temperature to 80°C. The replacement of nitrocellulose filters with fiberglass filters eliminates the problem of yellowing of the support. Fiberglass is superior because, also for unknown reasons, it gives a higher efficiency in toluene-based solutions and, as indicated earlier, lacks the problem of variable color quenching resulting from the yellowing of nitrocellulose filters in the course of drying at a high temperature.

A filter supporting a precipitate is a heterogeneous system consisting of three components, namely, the precipitated sample, the filter, and the cocktail. In general, the efficiency of counting of such a sample is not optimal. At first glance, it may seem that the lower efficiency is a result of the attenuation of the β particles passing through the filter material itself. However, this is surprisingly unimportant because the efficiency of counting an insoluble sample is little affected by its being on a filter. Presumably, the particles are seated so high on the filter that they are nearly surrounded by the solvent. In most experiments low efficiency is of no concern because the filtration method is used most often when comparing the *relative* amount of radioactivity in a large number of nearly *identical* samples (for example, in the assay of an enzyme or in measuring DNA synthesis in a bacterial culture at various times). However, with some types of samples, the efficiency of counting is not the same in all of the individual samples of a set of supposedly identical samples. It has been shown that as long as all water is removed from the filter, the lower counting efficiency and the irreproducibility is a consequence of self-absorption by the sample rather than the filter. Self-absorption losses will increase with particle size and total mass of the precipitate and reduced spatial distribution of the precipitate on the filter so that variability in any of these factors can cause irreproducibility in counting efficiency. The means of measuring and correcting for self-absorption is discussed in the section that follows. The point made here is that it is commonly thought that the filter is the problem so that methods are used to solubilize the filter. Two such methods are to use a dioxane-based cocktail or to add ethylene glycol monoethyl ether (Cellusolve) to a toluene- or xylene-based cocktail. However, most biological precipitates are not soluble in these solvents so that the three-component heterogeneous system has only been converted to a two-component heterogeneous system. The point cannot be made too emphatic that solubilizing the filter but not the precipitate is of no value.

There is one solubilization procedure that is effective if done properly. In the preceding section, solubilization of most biological substances by alkali was described. Filters made of cellulose acetate or nitrocellulose are also soluble in concentrated NH_4OH, NaOH, and KOH. Some types of filters, for example, MF-Millipore, produce a brown solution with NaOH and KOH but not with NH_4OH. The brown material introduces serious color quenching so that NH_4OH is usually the preferred substance. A final problem is that some samples are phosphorescent at alkaline pH so that it is good practice to add acid to neutralize the sample. Once solubilization and neutralization are complete, it is of course necessary to use a cocktail that can accept water. The dioxane-based cocktails are invariably preferable.

The use of fiberglass filters avoids many of the problems just mentioned. This is because any acceptable procedure can be used to solubilize the sample, and the filter, which is of course inert, can then just be washed to extract material from its interstices and then removed.

Methods for Precipitating Macromolecules

In a great many experiments in biochemistry and molecular biology, proteins or nucleic acids are labeled. Generally these are precipitated and collected on membrane filters. The most common protocol for precipitation is to add concentrated trichloroacetic acid (often referred to as TCA) or perchloric acid (PCA) to make a final concentration of

5% acid. The mechanism of precipitation of nucleic acids by trichloroacetic and perchloric acids differs from that of proteins. Nucleic acids having more than about 20 bases are quite insoluble in acid and, if enough material is present, they form a sizeable precipitate. Proteins, on the other hand, are not always insoluble at low pH nor do they necessarily form precipitated particles that are retained by a filter. However, both trichloroacetic and perchloric acids form tight multimolecular complexes with protein molecules, resulting in significant aggregation. Both trichloroacetic and perchloric acids precipitate protein and nucleic acids with equal efficiency. However, if 1 M HCl is used instead, both DNA and RNA form collectible precipitates whereas many proteins remain soluble in 1 M HCl. This distribution allows one to precipitate DNA and RNA from a protein–nucleic-acid mixture without precipitating protein. This is especially valuable in nucleic acid samples having sufficient protein to cause problematic self-absorption.

It is sometimes necessary to separate DNA from RNA, for example, if a compound such as [^{3}H]uridine, which can label both molecules, is used. This separation can be accomplished in two ways. If the DNA is to be retained, the sample can be incubated in 1 M NaOH at 37°C for 1 hour or at 20°C for 16 hours prior to acidification; in alkali, RNA is degraded to acid-soluble mononucleotides. If the RNA is to be retained, the sample can be pretreated at neutral pH with a DNase (an enzyme that depolymerizes DNA).

One final technique should be mentioned. Bacterial DNA is often labeled by growth of the bacteria in the presence of [^{3}H]thymidine. Acid precipitation of the bacterium usually results in trapping of unincorporated [^{3}H]thymidine inside the precipitated bacteria. To avoid this, the bacteria are incubated for several minutes in 1 M NaOH, which breaks open the bacteria. When acidified, no free [^{3}H]thymidine is retained in the precipitate. This method is also useful when labeling proteins with a radioactive amino acid.

Determination of the Efficiency of Counting by Pulse-Height Analysis

An evaluation of the efficiency of counting is clearly necessary if the work being done requires that the *absolute* number of disintegrations be known, although in most experimental situations this is not the case. In the more common situation, in which radioactivity in several samples is compared, it is necessary to know only that the efficiency of counting is the same in all samples. This is not always so, because the samples may have different chemical compositions, which might lead to different degrees of chemical quenching, different colors, or variable amounts of dilution quenching. Furthermore, it is often necessary to compare a soluble sample with an insoluble one, or two insoluble samples with different physical properties.

Chemical quenching is the type encountered most frequently. The *channels ratio*, the *external standard ratio*, and the *internal standard ratio* methods are the most common procedures used to evaluate the degree of quenching. The first two are based on the fact that quenching reduces counting efficiency by shifting the spectrum to a lower energy range (Figure 5-7A) so that counts are apparently lost because a larger fraction have an energy that the scintillation counter fails to detect–that is, they are in the one-photon range that the coincidence circuit rejects or they are below the L1 level needed to reduce photomultiplier noise. However, it should be noted that, if the quenched and unquenched spectra are examined between two pairs of levels, L1 → L2 and L2 → L3, quenching causes a *relative increase* in the number of counts in channel A compared with B (i.e., the channels ratio changes), although the total count rate decreases. Hence, in the channels ratio method, a set of standards is prepared with various amounts of quenching (e.g., by the addition of such quenching agents as water or chloroform) and a curve is prepared relating the observed count rate to the ratio of count rates in channels A and B (Figure 5-7B). Hence, for a sample with suspected quenching, the sample is counted in the same pair of channels used for the standard, the ratio is calculated, and the percentage of quenching is determined from the curve shown in Figure 5-7B. For example, if a particular sample had 500 cpm and the channel ratio showed 20% efficiency, the count rate in an unquenched sample would be 500/0.20 = 2500 cpm.

An important assumption is made in the channels ratio method—namely, that all quenching agents produce the same value of A/B for a particular degree of quenching. In other words, the shape of the energy spectrum for a sample quenched by one reagent is assumed to be the same as that of another sample quenched by a different reagent, if the degree of quenching is the same in both cases. In fact, this assumption

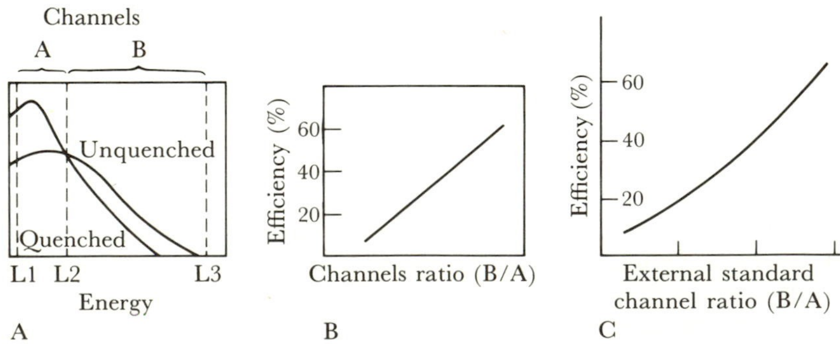

Figure 5-7
A. Shift in spectrum caused by quenching. The area remains constant. B. Correction curve for channels ratio method of quench correction. C. Correction curve for external standards method of quench correction.

is correct as long as the mechanism of quenching is the same, that is, if the quenching is solely chemical quenching causing reduced efficiency of energy transfer from a solvent molecule to a primary fluor. Note that the validity of the assumption does *not* mean that all quenching agents are equally effective; that is, it is not the case that the degree of quenching is proportional to molarity and unrelated to the identity of a quenching agent.

Example 5-B

☐ Calculation of the correct amount of radioactivity in a quenched sample using the channels ratio correction.

A sample containing ^3H is counted and found to have a count rate of 6300 cpm. Quenching is suspected so a channels ratio test is performed. Counting in channels A and B yields 1350 cpm and 4950 cpm respectively. A set of quenched standards is prepared using a toluene-based cocktail and increasing amounts of chloroform as a quenching agent. Each sample contains an amount of [^3H]toluene, whose disintegration rate is 10,000 dpm. These are counted in channels A and B and the following data are obtained.

A (cpm)	B (cpm)	A + B	Efficiency, (A + B)/10,000	B/A
1300	8700	10,000	1.00	6.7
1250	6250	7500	0.75	5.0
1200	4800	60,000	0.60	4.0
1125	3375	4500	0.45	3.0
1000	2000	3000	0.30	2.0
750	750	1500	0.15	1.0

The sample in question has a value of B/A = 4950/1350 = 3.67. If the data in the table are used to prepare a graph of efficiency versus B/A, it is found that a value of B/A = 3.67 corresponds to an efficiency of 0.55. Thus, the radioactivity in the sample is (1/0.55)(6300) = 11,454 dpm.

The *external standard ratio* procedure requires the use of a radiation source (typically ^{137}Cs) external to the counting vial but contained within the instrument and bombarding the sample with γ rays. When the γ rays pass through the solvent, some electrons are produced, which in turn are detected in the same manner as β particles. In a quenched sample, the energy spectrum of these electrons is also shifted and a similar plot of counting efficiency versus the channel ratio can be made (see Figure 5-7C). The advantage of the external standard ratio method over the channels ratio method is that the γ source provides a very high count rate ($\sim 10^5$ cpm) and therefore great statistical reliability (remember the $N^{\frac{1}{2}}$ rule of Example 5-A). This method is especially valuable if the activity of the sample in question is too low to use the channels ratio method.

Example 5-C ☐ Use of the external standards method for quench correction.

The quenched sample of Example 5-B is γ-irradiated with a source producing 2.35×10^5 decays per minute. The sample indicates 1.293×10^5 dpm. The ratio 1.293/2.35 is 0.55 so that the sample contains $6300/0.55 = 11{,}454$ cpm. A second sample that is known to be highly quenched has an observed count rate of 1582 cpm. When γ-irradiated, the decay rate is 3.05×10^4 dpm of which $3.05 \times 10^4 - 1582 = 2.89 \times 10^4$ dpm is a result of the γ-irradiation. Thus the radioactivity of this sample is $[(2.35 \times 10^5/2.89 \times 10^4)](1582) = 12{,}864$ cpm.

Modern instruments with associated computers are available that utilize the external standard method and automatically correct for quenching before printing out the data.

If a sample is quenched and contains two labels, special techniques are required for quench correction (when it is possible). Descriptions of such techniques are usually given in instrument manuals or in the specialized texts listed near the end of this chapter.

The third method of quench correction—the *internal standard ratio* method—is carried out in the following way. After counting a sample, a known amount of a nonquenching, labeled compound is added and the sample is recounted. From the ratio of the observed activity to the amount known to have been added, the efficiency of counting can be calculated. Although this method seems perfect, it has two problems: (1) the volume of added material is usually so small that there is a significant error (3%–5%) in volume measurement and (2) the sample is destroyed and cannot be recounted if desired. This method is not often used because it is considered tedious by most workers.

Both the internal and external standard methods can be used to correct for color and dilution quenching. However, colored samples are more easily dealt with by oxidizing them with H_2O_2, which usually eliminates the color. Dilution quenching is best dealt with by successive dilutions of the sample with the fluor solution. The count rate/unit volume of counting solution can then be plotted against sample concentration and extrapolated to zero concentration. This method is time-consuming but satisfactory.

It is important not to confuse self-absorption with quenching. The quench corrections discussed above correct only for changes in the β spectrum associated with the responsiveness of the scintillation fluid and give no information about the extent of self-absorption. With insoluble samples or particles collected on a filter, losses are caused by the physical properties of the sample itself, as explained on page 149. The problem of self-absorption is difficult to solve and often the principal cause of low counting efficiency. This is especially true in studies of the synthesis of macromolecules, such as proteins and nucleic acids, because the most common means of sample preparation is to precipitate the macromolecule with acid (hydrochloric, perchloric, or trichloroacetic) and collect the precipitate on a filter. If self-absorption is taking place, an increase in efficiency can sometimes be achieved by reducing the sample size, if sample is very large and on a filter, because the precipitate may have formed a thick layer. However, dilution does not usually eliminate self-absorption because the particles of the precipitate of most substances have a minimum size that is not reduced by dilution. Unfortunately, in low-energy β particles such as those from ^{3}H, self-absorption due to the small particles of the precipitate is already severe. In general, self-absorption cannot be eliminated in any way other than by the use of solubilizing agents.

The best way to *count relative amounts* of radioactivity in different insoluble samples is to make all samples as nearly identical as possible. One method of accomplishing this is to add an excess and fixed amount of a precipitable material (e.g., DNA or bovine serum albumin) to each sample so that *variation* in the amount of precipitate from one sample to the next is reduced. If a soluble sample is to be compared with an insoluble one (e.g., to relate the specific activity of a protein to that of an amino acid used in the synthesis), the only way to obtain high precision is to solubilize the samples and count them in identical mixtures.

Gamma-Ray Detection

Gamma emitters have recently become of great value in biochemistry, especially in radioimmunoassay (see Chapter 10) in which antibody or other proteins are labeled with the γ emitter ^{126}I. This radiation is easily detected and counted by means of commercially available γ counters, using an external sample-scintillation detector, which work in the following way. The sample is placed in a glass, or plastic, tube and inserted into a large thallium-activated NaI crystal [NaI(Tl)]; this is called "well" counting (Figure 5-8). Because γ rays have very high energy, they leave the sample and the container virtually without loss of energy. With an efficiency of a few percent, they produce electrons while tranversing the crystal. These electrons excite adjacent parts of the fluor crystal and thereby produce fluorescence, which is detected by a photomultiplier tube. As in liquid scintillation counting, the appropriate circuitry converts the light into pulses, which are counted. Coincidence circuitry is not necessary in γ counting because the γ energy is so high compared with the background noise of the photomultipliers that the voltage defining the lowest level of the pulse-height analyzer (i.e., L1 of liquid scintillation counting) can be set above that of the background pulses. In fact, the system usually uses many photomultipliers *acting independently* to increase the efficiency of the counter.

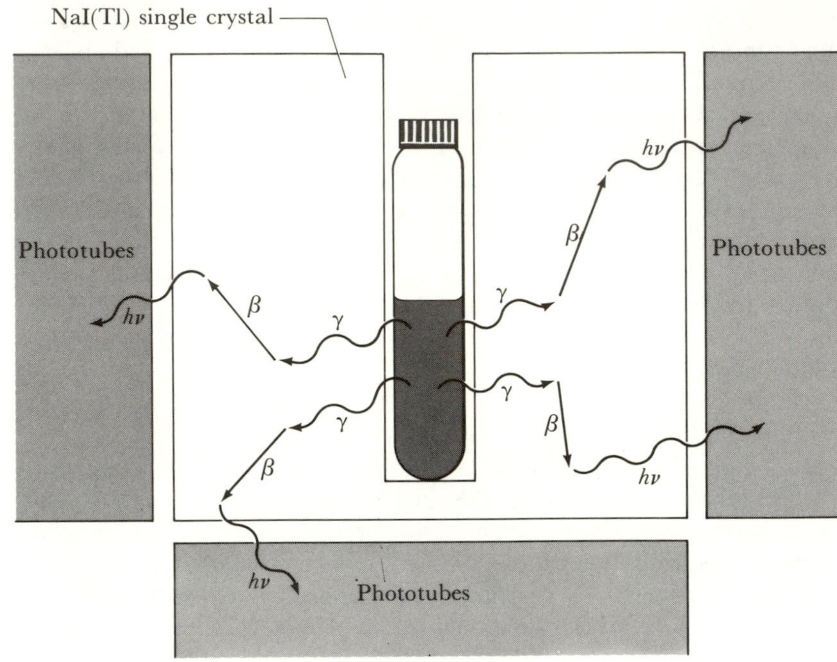

Figure 5-8
Well counting of γ rays. The γ rays interact with the crystal and produce β particles that excite the crystal, which then emits photons ($h\nu$) detectable by the phototubes.

As in liquid scintillation counting, some double-label experiments are performed using different γ-emitting sources. Again, because the pulses produced are related to the energy of the γ ray, pulse-height analysis can be carried out to distinguish different isotopes—for example, ^{126}I and ^{131}I. Detailed discussions of γ spectrometry can be found in the references given near the end of the chapter.

Examples of the Use of Radioactive Materials

This section includes many examples of the use of radioactivity in biochemistry and molecular biology. Radioisotopes can be used (1) to test material in quantities that are too small for direct chemical testing, (2) to distinguish molecules that are identical but are in different chemical locations, (3) to analyze mixtures that are too complex for traditional chemical analysis, and (4) to demonstrate participation in a reaction in which the products are chemically indistinguishable from the reactants. The first situation is illustrated in Examples 5-E and 5-G, the second in 5-E and 5-K, the third in 5-H and 5-I; and the fourth in 5-L.

☐ Tracer analysis. **Example 5-D**

One of the earliest uses of radiochemicals in biology was as tracers to work out biochemical pathways. For instance, the following reaction sequence, F → F → G → I → J, might be proposed with the reaction F → G as an irreversible step. Let us assume that each of the substances contains five carbon atoms and that each reaction consists of a substitution onto the carbon chain—that is, the carbon chain remains intact throughout the reaction sequence. If this were the case, then addition of [^{14}C]E to the reaction mixture would result in appearance first of [^{14}C]F, then [^{14}C]G, [^{14}C]H, and [^{14}C]I in sequence. If E → F is reversible, then addition of [^{14}C]F would yield both [^{14}C]E and [^{14}C]G; since reaction F → G is irreversible, [^{14}C]E would gradually disappear as [^{14}C]G, [^{14}C]I, and [^{14}C]J appeared. Of course, if [^{14}C]G were added, it would be converted to [^{14}C]I and [^{14}C]J but no [^{14}C]E and [^{14}C]F would be found, which would indicate that the F → G reaction is irreversible.

Additional information can be obtained by use of a ^{3}H label. Suppose in the reaction G → I, the hydrogen on the central (C-3) carbon atom is removed. This could be detected by using E labeled with a C-3 tritium. Thus addition of [^{3}H]E would result in the appearance of [^{3}H]F and [^{3}H]G but not [^{3}H]I or [^{3}H]J.

Knowledge of these pathways from tracer studies permits specific labeling of macromolecules. For instance, radioactive DNA can be prepared by growing cells in the presence of radioactive thymine or thymidine. If the incorporation of radioactive thymidine, which is soluble in acid, into acid-insoluble material is to be used as a measure of DNA synthesis, it is essential that no radioactivity gets into RNA or protein since both of these macromolecules are insoluble in acid. In bacteria, thymidylic acid (thymidine 5′-monophosphate) is made from deoxyuridylic acid, which is connected by a variety of pathways to uridine and cytosine, components of RNA. Thus if thymidylic acid, labeled in a C or H atom in the pyrimidine ring, is added to bacteria, some radioactive RNA is made. However, thymidylic acid is 5-methyldeoxyuridylic acid. Thus if the thymidylic acid contains a ^{3}H atom on the 5-methyl group, then occasional conversion to deoxyuridylic acid would entail loss of the methyl group and hence of the radioactivity. Therefore, there is no way for radioactive RNA to be made from thymidine labeled with ^{3}H or ^{14}C in the 5-methyl group.

☐ Reactions of *Escherichia coli* DNA polymerase I. **Example 5-E**

If purified DNA polymerase I is added to a reaction mixture containing a buffer, Mg^{2+}, DNA, and the four deoxyribonucleotide 5′-triphosphates (of adenine, thymine, guanine, and cytosine) labeled with ^{32}P in the α position (nearest to the deoxyribose) and incubated at 37°C, the ^{32}P becomes insoluble in 10% trichloroacetic acid and can be collected on a membrane filter—an indication of polymerization to DNA because DNA, but not the nucleotide triphosphates, is insoluble in trichloroacetic acid. Note that newly synthesized DNA is distinguishable from the template DNA by virtue of the radioactivity of the new DNA. If only three deoxynucleotide 5′-triphosphates are present, or if Mg^{2+} or DNA is not present, or if the nucleotides are ribo- instead of deoxyribo-, the ^{32}P remains soluble. Hence, polymerization requires all four *deoxy*ribonucleotides, Mg^{2+}, and DNA. If any of the triphosphates is replaced by mono- or diphosphate, ^{32}P remains soluble (i.e., there is no polymerization); hence, the triphosphate group is necessary in the substrate. If the deoxyribonucleotide 5′-triphosphates carry the ^{32}P in the β or γ position, and ^{14}C in the deoxyribose moiety, the ^{14}C, but not ^{32}P, becomes acid insoluble. (The ^{14}C and ^{32}P can be distinguished by liquid scintillation counting.) Hence, the β and γ phosphate groups are removed in the reaction; only the α phosphate remains in the polymer. Thus, by the use of radioactivity, the different phosphate atoms can be distinguished.

Example 5-F ☐ Molecular weight of a DNA molecule by end-group labeling.

The enzyme *E. coli* polynucleotide kinase transfers a γ phosphate from adenosine 5′-triphosphate (ATP) to a 5′-hydroxyl terminus of DNA. DNA normally contains two 5′-phosphoryl termini (one for each polynucleotide strand), which can be converted into 5′-hydroxyl groups by the enzyme alkaline phosphatase. In the laboratory at Brandeis University, 4.7 μg of purified, homogeneous (meaning that all molecules are the same size) phage λ DNA with 5′-hydroxyl termini was reacted with [γ-^{32}P]ATP at a specific activity of 3 mCi/μmol, using polynucleotide kinase. When acid-precipitated and counted, the sample showed 1870 cpm, using a counter that detected ^{32}P at 85% efficiency. Hence, 1870/0.85 or 2200 dpm of ^{32}P were precipitated. In units of microcuries, this is $2200/(2.2 \times 10^6) = 10^{-3}$ μCi. From the specific activity, 10^{-3} μCi is equivalent to 3.3×10^{-7} μmol of ATP, or 1.98×10^{11} molecules. Because 4.7 μg of λ DNA was added, the weight of a single λ DNA molecule is $4.7/9.9 \times 10^{10} = 4.7 \times 10^{-11}$ μg or 31.0×10^6 atomic weight units. Here, the radioactivity served to measure phosphorylation of 5′-hydroxyl groups with very high sensitivity. New terminal phosphoryls could also be distinguished from old 5′-phosphoryl groups.

Example 5-G ☐ Identification of "buried" tyrosines in a hypothetical protein.

Tyrosine is the only amino acid that can be iodinated efficiently. A protein is known from amino acid analysis to contain six tyrosine residues. When 1 μg of purified protein reacts with ^{131}I, 2550 cpm can be precipitated with trichloroacetic acid and collected on a fiberglass filter. If the reaction takes place in the presence of a denaturing agent—so that the protein is totally unfolded—3820 cpm are acid insoluble.

Hence, (2550/3820) $\times$ 6 $\simeq$ 4 tyrosines are available for iodination when the protein is in the native configuration. Hence, 6 − 4 = 2 are unavailable and therefore "buried" in the three-dimensional structure. In this case, radioactivity allows the detection of iodine in the protein when the amount of iodine is too small for chemical detection.

Example 5-H

□ Permeability of a bacterium to adenylic acid (AMP).

Not all radioactive chemicals added to a bacterial culture can penetrate the bacteria; some are excluded. If a radiochemical enters the bacteria and the bacteria are collected by filtration, radioactivity will appear on the filter. If it is excluded, the radioactivity will pass through the filter and any residue can be washed away. For example, if either [^{32}P]AMP or [^{3}H]AMP (with ^{3}H in the purine ring) is added to a culture of the bacterium *E. coli* and after a period of growth the bacteria are collected on a membrane filter and washed thoroughly with a buffer, neither ^{3}H nor ^{32}P is found on the filter. Hence, AMP is not taken up by the bacterium. In this case, radioactivity allows the added AMP to be distinguished from the AMP already present in the cell. If there were an uptake of an amount equal to 0.1% of that already within the bacterium, the increase by that amount would be undetectable by chemical procedures; using radioactive material, the 0.1% could be detected against a zero background because the internal material is unlabeled.

Because both [^{32}P]AMP and [^{3}H]AMP are being tested in this example, the lack of uptake of either label also shows that the cells do not cleave external AMP to form adenosine and phosphate since it is known that these two substances can enter the cell. This is an important point because there are examples, such as thymidine, in which cleavage (in this case to thymine) occurs before uptake.

Example 5-I

□ Identification of a substance by precipitation with antibody.

If the bacterium *E. coli*, growing in a medium containing [^{14}C]glucose as the sole carbon source, is infected with phage T4 and the infected cells are allowed to grow, phage-mediated proteins are synthesized. If the infected cells are broken open and antibody specific to purified T4 tail fibers is added, ^{14}C is precipitated. If the cells are broken at various times, the rate of synthesis of tail fibers can be determined from the amount of precipitated ^{14}C as a function of time. A similar analysis can be made with cells infected with phage T5 and antibody specific to T5 tail fibers. However, if the cells are simultaneously infected with T4 and T5 and are later disrupted, ^{14}C can be precipitated by anti-(T4 tail fibers) but not by anti-(T5 tail fibers). Hence, T5 tail fibers are not synthesized in such a dual infection. Such an experiment would be virtually impossible without radioactivity because of the small quantities and the difficulty in identifying tail fibers in a mixture as complex as a cell lysate plus antiserum.

Example 5-J ☐ Purification of a protein for which there is no biological or chemical assay.

Proteins are usually detected by virtue of having a measurable biological activity (e.g., enzymatic or inhibitory to some reaction) or a distinguishable physical property (such as the spectral property of hemoglobin). But detection by these means is not always possible, in which case the double-label method described herein is used.

If a phage-infected bacterium capable of making a particular protein is grown in a medium containing [^{3}H]leucine, all proteins will be ^{3}H-labeled. If a mutant bacterium that makes only a fragment of the protein is grown in the presence of [^{14}C]leucine, all proteins will be labeled, but there will be no ^{14}C-labeled protein corresponding to the intact protein of interest. If two cultures thus labeled are mixed and the total protein is isolated and fractionated by chromatography, most fractions will have a fixed ratio of ^{3}H to ^{14}C. The protein of interest will be in a fraction that has a high ratio of ^{3}H to ^{14}C (Figure 5-9). Hence, purification schemes can be based on maximizing this ratio until a fraction is obtained that contains ^{3}H but no ^{14}C. If all proteins are labeled, the protein will be chemically pure (which is, of course, not the case in a phage-infected cell because most bacterial proteins are nonradioactive).

Example 5-K ☐ Sedimentation analysis of bacteriophage DNA structures in a superinfected lysogenic bacterium.

If a stable lysogen of the *E. coli* bacteriophage λ is infected with another λ phage, no phage are produced. The fate of the incoming DNA can be investigated by isolating all intracellular DNA and analyzing it by zonal centrifugation through a preformed density gradient (see Chapter 11). The analysis could be based on the fact that the λ phage DNA has a different molecular weight (3×10^7) from that of *E. coli* DNA (2.6×10^9) were it not for the fact that *E. coli* DNA usually breaks into pieces of many sizes during isolation. The question then is how to distinguish the λ and the *E. coli* DNA. This is easily done if the λ DNA is radioactive because, if radioactivity is used as an assay, all radioactivity will be in λ DNA. Therefore, [^{3}H]λ can be used to infect an *E. coli* lysogen. The total DNA is isolated and analyzed by centrifugation. After sedimentation, the centrifuge tube is fractionated and the radioactivity in each fraction determined. The data obtained in an experiment of this type are shown in Figure 5-10A and indicate that the major fraction of the DNA is converted into a nonreplicating twisted circle. Figure 5-10B shows the result that would have been obtained had radioactive phage not been used. Because the ratio of bacteria to phage DNA is about 150, it would have been difficult to recognize any of the peaks seen in Figure 5-10A.

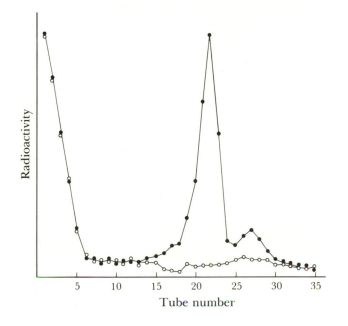

Figure 5-9
The use of double labeling in protein purification. A noninducible lysogenic strain of the bacterium *E. coli* was ultraviolet-irradiated to suppress protein synthesis. The cells were then infected either with phage λ wild-type in the presence of [^{3}H] leucine or with λcIsus10 (which makes a short fragment of the λ repressor) in the presence of [^{14}C] leucine. Proteins were isolated from each culture, mixed, and chromatographed on DEAE-cellulose (see Chapter 8). The elution pattern of the chromatographic column is shown as the amount of ^{3}H and ^{14}C in each fraction eluting from the column. A peak of ^{3}H activity appears for which there is no corresponding ^{14}C peak. This represents a protein produced in the λ wild-type infection but not in the λcIsus10 infection—that is, the λ repressor. This was the first means of assaying the λ repressor. Solid circles, ^{3}H; open circles, ^{14}C. [Courtesy of Mark Ptashne.]

Example 5-L

☐ Studies of exchange reactions.

Because exchange reactions result in the production of a substance chemically identical with that of the starting material, traditional chemical analysis fails to give any evidence for a reaction. However, exchange—or, more general, participation—is easily studied with radioisotopes, as the following two examples will show.

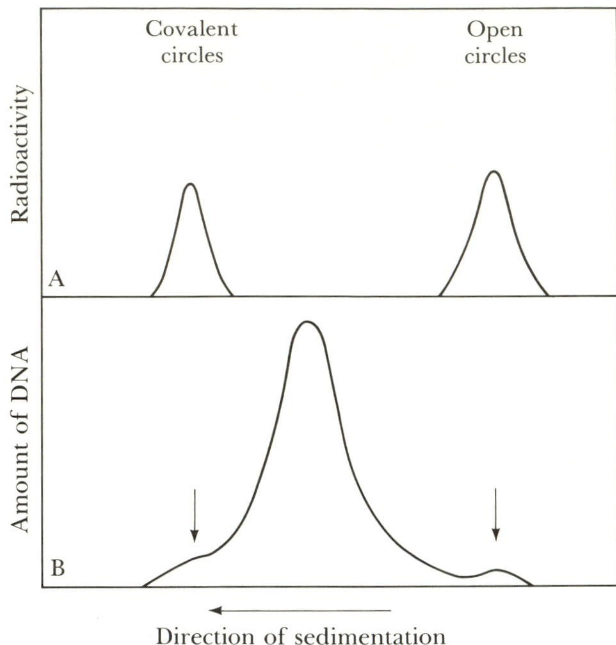

Figure 5-10
Comparison of experiments done with and without the use of radioactivity. A strain of the bacterium *E. coli*, lysogenic for phage λ, is infected with λ phage. It is known that the injected DNA is converted into a mixture of open and covalent circles that are easily distinguished by sedimentation in an alkaline sucrose gradient (Chapter 11). The graphs illustrate two such sedimentation runs. In graph A, the phage are labeled with [^{3}H]thymidine and the sedimentation pattern shows radioactive DNA only. The two forms are easily distinguished. In graph B, the phage is nonradioactive and the DNA concentration throughout the gradient is plotted. Because the bacterial DNA is in great excess, the phage DNA appears as barely visible blips (see arrows).

Pyridoxal phosphate is a cofactor for many enzymatic reactions. If [^{3}H]pyridoxal phosphate is used, if is found that if the ^{3}H is in certain positions it will appear in water as ^3H_2O. The labeled enzyme-bound pyridoxal phosphate can be removed from the water by gel-exclusion chromatography (see Chapter 8) and the ^{3}H in the water can be counted by the direct addition of the water to a scintillation mixture that accepts water. If ^{3}H is found in the water, direct evidence is provided that pyridoxal phosphate participated in the reaction by means of a proton transfer.

DNA is a double-stranded polynucleotide held together by hydrogen bonds. If DNA is added to ^3H_2O and then removed at various times by gel-exclusion chromatography, ^3H appears in the DNA. From the kinetics of exchange and the effects of various denaturing agents, it can be shown that hydrogen bonds are continually breaking and reforming. This phenomenon is called "breathing" and has also been used to determine the rate of protein folding and unfolding (see Chapter 19).

Example 5-M

☐ Identification of the active sites of enzymes.

Radioisotopes play an important role in identifying the active sites of enzymes because enzymatic reactions require such low enzyme concentrations that huge volumes would be necessary (if, in fact, sufficient enzyme were available) to obtain enough material for traditional chemical analysis. The following is a simple example of the determination of the number of binding sites.

That 1-amino-2-bromoethane covalently binds to the active site of bovine plasma amine oxidase can be easily demonstrated if the second carbon is labeled with ^{14}C. Measuring the ratio of moles of ^{14}C bound mole of enzyme tells the number of binding sites. By hydrolyzing the protein to amino acids and by identifying the type of chemical linkage, the site of binding (i.e., the amino acid) can be identified. If the amino acid sequence of the enzyme being studied is known, hydrolysis to peptides identifies the location of the amino acid in the protein.

SELECTED REFERENCES

Birks, J. B. 1964. *Theory and Practice of Scintillations Counting.* Pergamon.

Bransome, E. D., ed. 1970. *The Current Status of Liquid Scintillation Counting.* Grune and Stratton.

Dyer, A., ed. 1972. *Liquid Scintillation Counting.* Heyden.

Fox, B. W. 1976. *Techniques of Sample Preparation for Liquid Scintillation Counting.* American Elsevier.

Horrocks, D. L., ed. 1974. *Application of Liquid Scintillation Counting.* Academic Press.

Kobayashi, Y., and D. V. Maudsley. 1974. *Biological Application of Liquid Scintillation Counting.* Academic Press.

Millipore Corporation. *Multiple Sample Filtration and Scintillation Counting.*

Neame, K. D., and C. A. Homewood. 1974. *Liquid Scintillation Counting.* Halstead Press.

Noujaim, A. A., C. Ediss, and L. I. Wiebe. 1976. *Liquid Scintillation: Science and Technology*. Academic Press.

Stanley, P E., and B. A. Scoggins. 1974. *Liquid Scintillation Counting: Recent Developments*. Academic Press.

Some of the best information available can be obtained from the manuals for various scintillation counters and the brochures supplied by manufacturers of radiochemicals.

PROBLEMS

5–1. The specific activity of [methyl-^{3}H]thymidine is 6 Ci/mmol. What fraction of the thymidine molecules is radioactive? If DNA having a molecular weight of 25×10^6 is labeled with this thymidine and if 50% of its base pairs are A·T, how many ^{3}H nuclei are there per DNA molecule? If a counter has 52% efficiency, what weight of DNA will give 1000 cpm?

5–2. The specific activity of a sample of [^{32}P]ATP is 5.3 Ci/mmol on January 23. What will the specific activity be on February 18 of the same year?

5–3. A good way to dispose of ^{32}P-contaminated glassware is to store it in a lead container. How long should glassware contaminated with 2×10^6 cpm be stored so that the radioactivity is no more than 100 cpm? Could storage be used to decontaminate glassware containing ^{3}H?

5–4. If ^{3}H, ^{14}C, and ^{32}P could each be used with equal simplicity in a double-label experiment, which pair of isotopes would you choose? Why?

5–5. In an experiment to measure DNA synthesis, would [^{3}H]thymidine or ^{32}P be a better choice? Why?

5–6. A scintillation counter is adjusted such that the ratio of counts in channels A and B is 1000 for ^{3}H and 0.2 for ^{14}C. Using the same levels, a sample is counted and has 1450 cpm in channel A and 1620 cpm in channel B. What is the ratio of ^{3}H to ^{14}C in this sample?

5–7. Referring to Problem 5–6, if the sample contains [^{3}H]thymidine at 6 Ci/mmol and [^{14}C]uracil at 0.5 Ci/mmol and the counting efficiencies are 25% for ^{3}H and 82% for ^{14}C, what is the molar ratio of thymidine and uridine in the sample? Suppose that the thymidine is in DNA that is 43% G + C and the uridine is in RNA that is 28% U. What weights of DNA and RNA are in the sample?

5–8. Because ^{32}P causes strand breakage in DNA, to study only unbroken DNA it is advisable to incorporate only one ^{32}P per DNA. In that way, after a ^{32}P has decayed, the DNA is no longer radioactive and hence not detectable. If the minimum concentration of phosphate in a growth medium that supports the growth of a DNA-containing bacteriophage is 10^{-3} M, how much pure ^{32}P must be added per milliliter of solution to obtain one ^{32}P per phage if the DNA has a molecular weights of 20×10^6?

5–9. A sample counted for one minute shows a count rate of 752 cpm. For how many minutes should it be counted to have a 1% probable error?

5–10. In a scintillation counter, the sample is observed by two photomultiplier tubes on either side of the sample vial. The sample vials can hold as much as 20 ml of solvent. Do you think the count rate will be affected by the volume of the solvent if the sample is collected on a filter? What about the background?

5–11. A set of quenched standard samples is counted in two channels. The following count rates are observed even though all samples have the same amount of added radioactivity (i.e., 1000 cpm).

Sample	Channel A (cpm)	Channel B (cpm)
1	502	500
2	478	425
3	445	354
4	410	290
5	379	227
6	332	170

An unknown sample is counted. It has 1822 cpm in channel A and 1211 cpm in channel B. What is the actual count rate in this sample?

5–12. A substance lowers the efficiency of counting of a ^{3}H-containing sample tenfold.

a. What mechanism(s) of quenching would cause a tenfold reduction of a ^{14}C-containing sample?

b. What mechanism(s) would cause a very different degree of quenching of a ^{14}C-containing sample and would the quenching be greater or less?

5–13. A particular kind of cell must be grown in a complex growth medium containing a huge number of nutrients. The medium has never been analyzed. You wish to label the protein with [^{14}C]leucine. Add 100 μl of [^{14}C]leucine at a specific activity of 50 μCi/μg and a concentration of 25 μg/ml to 10 ml of growth medium. After allowing the cells to grow in this medium for 1 hour, collect 0.1 ml, acid precipitate, and count in a counter with 70% efficiency for ^{14}C. Suppose that the counter shows shows 1251 cpm. The 0.1-ml sample is known to contain roughly 25 μg of protein and the protein is 4% leucine. How much [^{14}C]leucine must be added per milliliter of medium to increase the amount of collected radioactivity to approximately 5000 cpm?

5–14. Suppose that you have in your laboratory a Geiger counter with an efficiency of 22% for ^{14}C and a background of 6 cpm and a scintillation counter with an efficiency of 72% for ^{14}C and a background of 38 cpm. In an experiment with ^{14}C, you expect your sample to have very low activity—that is, from 75 to 100 cpm. Which counter should you use? Why?

5–15. A population of bacteria has been labeled for a period of time with [^{14}C]leucine. Various aliquots are taken and collected on a filter and counted. Appropriate precautions have been taken so that the cells are broken and no [^{14}C]leucine, other than that in protein, is being counted. The counter is set to count in two channels and the following data are obtained.

Sample	Channel A (cpm)	Channel B (cpm)
^{14}C standard	16,075	64,300
10^9 cells	10,160	15,005
8×10^8 cells	8,032	15,948
6×10^8 cells	6,001	14,982
4×10^8 cells	3,989	12,021
2×10^8 cells	1,555	5,444
10^8 cells	998	4,001
8×10^7 cells	801	3,199

If 10^9 cells contain 1 mg of protein, what is the specific activity of the labeled protein, expressed in μCi of ^{14}C/mg of protein? Assume that the counting efficiency of unquenched ^{14}C is 100% in this counter.

5–16. You are trying to measure the relation between the rate of protein and DNA synthesis in a bacterium using growth conditions that are rather difficult to achieve. Because of this difficulty you decide that the best thing to do is to add both [^{3}H]thymidine and [^{14}C]leucine simultaneously (leucine and thymidine are not interconvertible so there is no problem). Samples are taken at various times. One-third of each sample is precipitated with trichloroacetic acid, collected on a filter, and washed adequately to remove all ^{3}H and ^{14}C radioactivity that is not in DNA and protein. (Note that you have used only part of each sample so you will have some left if you need it.) The data are the following (background count rates are subtracted already).

Sample	Channel A (cpm)	Channel B (cpm)
^{3}H standard	9874	0
^{14}C standard	998	4,010
t = 0	0	0
t = 1	2046	8,021
t = 2	4102	15,994
t = 3	6052	23,998
t = 4	8196	32,008

a. After going over the data, you realize that it is not worth much. Why not?
b. Since you have two-thirds of the sample left, you realize that there is something you can do to improve the results. What might it be and why would it be better?

5–17. You are studying membrane synthesis in bacteria by labeling lipids with [^{3}H]oleic acid. You have a bottle of the oleic acid that you have just purchased and use it in a particular growth medium. You find that the count rate in the labeled bacteria is 52,000 cpm per 10^9 bacteria. (There is no quenching or other counting problems.) To be sure that the radioactivity is in lipid, you check for solubility in an organic solvent—you find that if 10^8 cells are shaken with $CHCl_3$, 5150 cpm are found in the $CHCl_3$. Treatment of broken cells with the protease trypsin causes no loss of radioactivity measured. Two months later you repeat the experiment (using the same [^{3}H]oleic acid) as part of a larger experiment and discover to your surprise that 145,000 cpm are contained in 10^9 bacteria and that only 25% of the radioactivity is extractable by $CHCl_3$. Treatment of broken cells with trypsin results in a 50% loss of radioactivity. What is going on?

5–18. You are doing an experiment in which you wish to measure RNA and DNA synthesis simultaneously. You are using cells to which thymidine and uridine must be added because the cell cannot make its own. You label with [^{3}H]thymidine and [^{14}C]uridine (assume ^{14}C only gets into RNA). The data obtained are the following.

Sample	Channel A (cpm)	Channel B (cpm)
^{3}H standard (i.e., pure ^{3}H DNA collected on a filter)	52,480	35
^{14}C standard (i.e., pure ^{14}C RNA collected on a filter)	1350	4620
5-min sample	5500	1000
10-min sample	11,000	2000
15-min sample	16,500	3000

On calculating the amount of ^{3}H and ^{14}C you know that there is something wrong with the experiment because the specific activities of the uridine and thymidine are the same and in general more RNA than DNA is made per unit time.

a. What shows that there is a problem?
b. What do you think is the source of the problem?
c. How would you solve the problem?

5–19. Polynucleotide kinase puts one ^{32}P from [γ-^{32}P]ATP onto the 5′ ends of double-stranded DNA. You can measure the molecular wieght of a DNA by end-group labeling. If you have a DNA whose molecular weight is in the 50–100 million range and your reaction mixture cannot contain more than 1 μg of DNA, what specific activity of [^{32}P]ATP (one ^{32}P per ATP) would you need if you thought that the count rate would have to be more than 500 cpm in order to be reliable?

5–20. You are studying protein synthesis and choose [^{14}C]glycine as a label because it is inexpensive. In a control experiment, you add chloramphenicol, which is known to reduce protein synthesis to 1% of its normal level; to your surprise the amount of ^{14}C that is acid-precipitable is 80%

of the value seen without the drug. Assuming that the chloramphenicol solution is satisfactory, what is the explanation of this value?

5–21. You are studying a reaction in which [^{3}H]$CHCl_3$ is the product. This is measured by adding your sample to a toluene-based scintillation solvent. Most of the material in the reaction precipitates (including the original radioactive reactant) and is easily removed. The $CHCl_3$ remains because it is miscible with toluene. You discover that 0.1 ml of sample has a count rate of 5000 cpm. However, 0.05 ml yields 4000 cpm and 0.02 ml yields 3600 cpm. This sounds like self-absorption but self-absorption has never been observed when the sample is a liquid miscible with the solvent. You then test for quenching in the following way. You extract the original sample with benzene to remove the $CHCl_3$, mix different volumes (i.e., 0.1, 0.05, and 0.02 ml) of the remainder of the sample with the scintillation mixture, and then add 10,000 cpm of [^{3}H]toluene to each. You observe 10,000 cpm for each starting volume. What is going on? How would you solve the problem so that the count rate will indicate how much material is present?

6

Autoradiography

Autoradiography is a method by which radioactive material can be localized—for example, within a particular tissue, cell, cell part, or even molecule. In this technique, a sample containing a radioactive substance is put in direct contact with a thick layer of a photographic emulsion specially designed for autoradiography. Radioactive atoms decay in the sample and the emitted radiation activates individual silver halide grains in the emulsion and renders them susceptible to conversion into metallic silver by a photographic developer.* On chemical development, the resulting pattern of grains shows the distribution of radioactive material within the specimen (Figure 6-1). Observation is by microscopy. The image gives two specific bits of information: the location of the radioactive material with respect to the object or its parts, and its intensity, which is related to the amount of radioactive material present. This chapter describes how autoradiography is performed, how the appropriate isotope and preparative technique are selected, the problems arising in their use, and how the kinds of information obtained can be used to supply answers to particular questions.

Nuclear Emulsions Used in Biological Studies

Nuclear emulsion differs from standard photographic film principally in the high ratio of silver halide to gelatin—roughly equal volumes in nuclear emulsions—and in the small size of the grains (0.02–0.3 μm).

* All photographic film consists of a suspension of silver halide crystals in gelatin. The crystals have the unusual property that, on exposure to light or radiation, they are activated in the sense that various reducing agents (developers) become capable of chemically converting the silver halide into metallic silver. In the absence of such activation, they are resistant to chemical reduction.

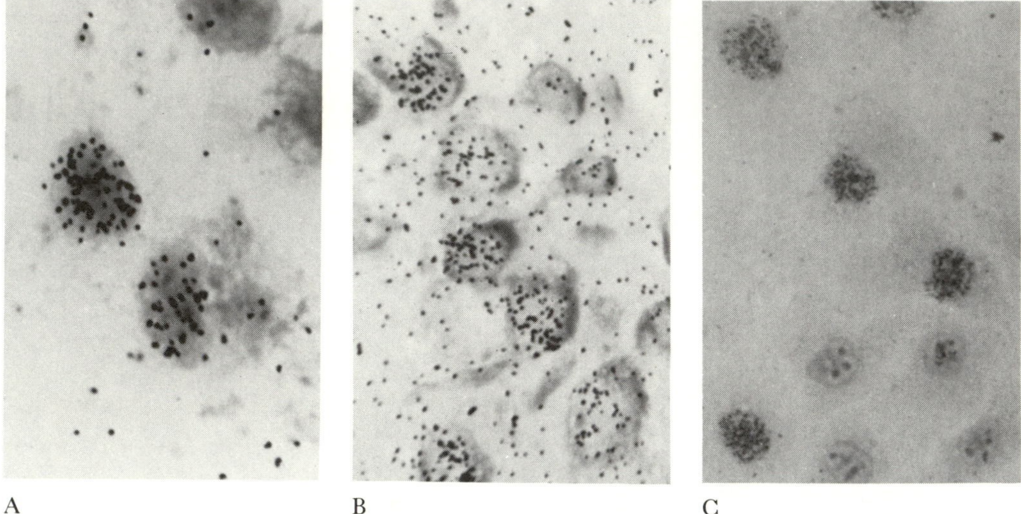

A B C

Figure 6-1

A. Autoradiogram showing concentration of radioactivity in nuclei isolated from a guinea-pig uterus that had been exposed for 15 minutes to 1,2,6,7-[^{3}H] progesterone. The nuclei were placed on a slide and overlaid with nuclear emulsion. After eleven days the autoradiogram was developed to make the silver grains visible and then stained with methyl green-pyronin to make the nuclei visible. This shows that progesterone or a compound derived from progesterone is rapidly localized in the nuclei of uterine cells. Magnification 1100 ×. [From W. Stumpf, in *Methods in Cell Biology*, vol. 3, edited by D. M. Prescott, Academic Press, 1976.] B. Autoradiogram of a section of a rat hippocampus showing nuclear concentration of [^{3}H]corticosterone in neurons one hour after intravenous injection. After the thin section was mounted on a slide, nuclear emulsion was applied and the autoradiogram was exposed for ninety-five days, developed, and then stained with methyl green-pyronin to make the nuclei visible. The high concentration of grains over the nucleus shows that a great deal of corticosterone is in the nucleus. There are grains between the nuclei also: to determine whether these are due to ^{3}H in the cytoplasm or to background, it is necessary to determine the number of grains per unit area of emulsion in a region far from the tissue section. [From W. E. Stumpf and M. Sar, in *Methods in Enzymology*, vol. 36, edited by B. W. O'Malley and J. G. Hardman, Academic Press, 1975.] C. Autoradiographic detection of DNA synthesis in animal cells. African green monkey kidney cells were grown for six hours in a growth medium containing [^{3}H] thymidine. The cells were washed, dehydrated, and covered with autoradiographic film. After exposure and development of the film, the nuclei (gray regions) were stained. Such an experiment allows the identification of the cells that replicated their DNA during the six-hour period (i.e., those whose nuclei are covered with grains). Hence, the fraction of replicating cells can be measured. [Courtesy of James A. Robb.]

The emulsions in common use are of three types: *premounted*, *liquid*, and *stripping*. A premounted emulsion is a relatively thick (50–1200 μm) layer of emulsion that has been mounted on a glass microscope slide. A liquid emulsion is supplied as a shredded gel, which must be melted; the sample, which is usually mounted on a glass microscope slide, is dipped into the molten gel and withdrawn, and the emulsion hardens and forms a film whose thickness depends on the concentration of gelatin in the liquid. Stripping film is supplied as a thin (about 5 μm) film mounted on glass. It is removed from the glass with a razor blade and then placed on a water surface. The sample, which has been premounted on a glass microscope slide, is placed under the floating film and lifted up into the film, thus transferring the film to the microscope slide. This is allowed to dry and the thin emulsion adheres tightly to the slide. These processes are shown diagrammatically in Figure 6-2.

Isotopes Commonly Used in Biological and Biochemical Studies

The radioisotopes most commonly used are of three energy types—high (e.g., ^{32}P), medium, (e.g., ^{14}C and ^{35}S) and low (e.g., ^{3}H); almost all are β emitters. On occasion, α emitters such as polonium and thorium are used. The autoradiographic properties of these isotopes are given in the next section.

Track Length of Various Emitted Particles

As a particle emitted by a radioactive source passes through a nuclear emulsion, it continually loses energy by collisions with nuclei and orbital electrons. Some of this energy produces defects in the silver halide crystals and thereby renders them developable (i.e., they are exposed). The pattern of grains in the emulsion is called a *track*, which has three parameters—length, grain density (either grains per unit length or grains per total track length), and shape (e.g., linear, curved, angled, etc.). These parameters are determined by the mass of the particle, the particle energy, the emulsion, and the development of the emulsion. The effects of mass and energy are described next.

Alpha Particles

Alpha particles are heavy, have two positive charges, and usually have an energy between 4 and 8 MeV.* These massive particles are relatively

* The energy of emitted particles is measured in electron volts, the energy acquired by an electron falling through a potential difference of 1 volt, or 1.602×10^{-19} joule. The usual units are keV and MeV: 10^3 and 10^6 electron volts respectively.

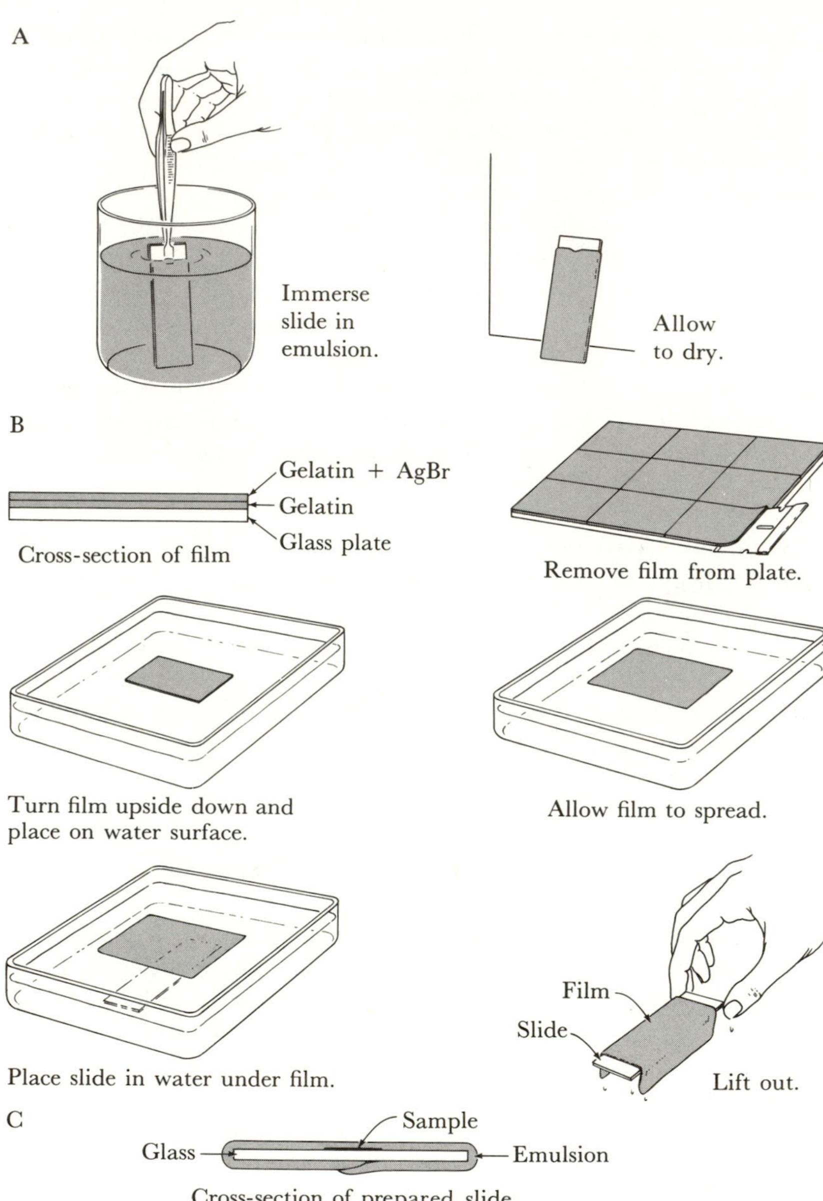
A
Immerse
slide in
emulsion.
Allow
to dry.
B
Gelatin + AgBr
Gelatin
Glass plate
Cross-section of film
Remove film from plate.
Turn film upside down and
place on water surface.
Allow film to spread.
Place slide in water under film.
Film
Slide
Lift out.
C
Sample
Glass
Emulsion
Cross-section of prepared slide

Figure 6-2

Methods of putting emulsions onto a sample. (A) Dipping method. A glass microscope slide to which the sample is affixed is dipped into molten emulsion. A thin layer adheres to the glass and solidifies at room temperature. (B) Stripping method. A sheet of commercially available stripping film formed on a glass plate is scored with a razor blade to make rectangular pieces approximately 1 × 3 inches (2.5 × 7.5 cm) The film is stripped from the plate by forcing a razor blade under the edge of the film. The film is inverted, floated on water, and picked up as shown. (C) Cross-section of a finished slide with sample and emulsion prepared by the stripping method.

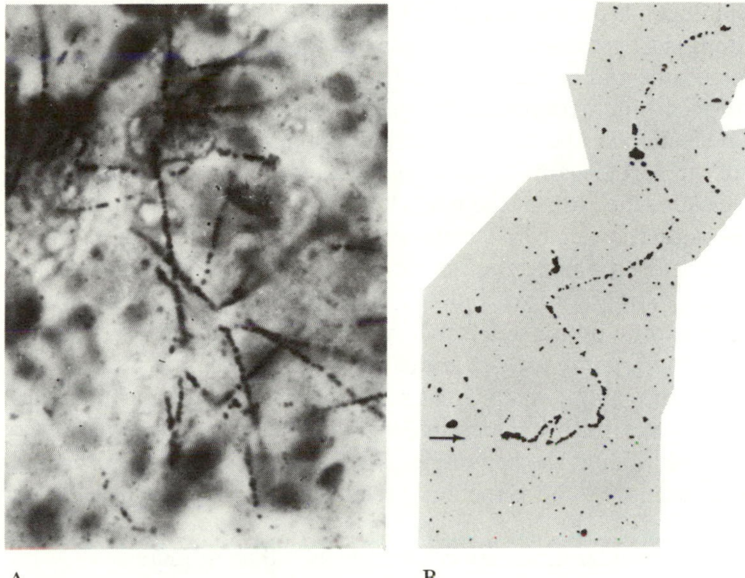

Figure 6-3
A. Alpha particle tracks emitted from thorotrast (thorium dioxide) particles in a thin section of rabbit spleen. The thin section was overlaid with liquid emulsion, which was then allowed to harden. After exposure, the emulsion was developed and the nuclei stained. Note the high grain density and the straightness of the tracks and compare them with part B. The tracks are from 40 μm to 45μm long. [Autoradiogram supplied by Hilde Levi.] B. A. track produced by a 300-keV (medium energy) β particle showing that grain density increases at the end of the track (arrow). The track is approximately 250 μm long. [From R. H. Herz, *Photographic Action of Ionizing Radiations*, Wiley-Interscience, 1969.]

unaffected by collision with electrons, tend to maintain a straight path following such a collision, and have a tremendous disrupting effect on orbital electrons as they pass through an emulsion. This results in excitation of just about every silver halide crystal that they traverse and therefore produces a very high grain density. Because an α particle interacts with a very large number of electrons per unit distance, it loses energy rapidly and has a relatively short track length (usually between 15 μm and 40 μm). Figure 6-3A is a photograph of alpha tracks in a thick emulsion.

Beta Particles

Beta particles are electrons and are therefore easily scattered by orbital electrons. As β particles collide with other electrons, they rapidly lose energy and are sharply deflected in each collision. The magnitude of the deflection depends on the energy of the particle; at very high energy, the momentum is so great that the particle has a greater tendency to move in a straight line and be minimally deflected. This means that, as energy is lost in each interaction with an orbital electron, the probability of greater deflection in the next interaction increases. Hence, because the energy of the particle continuously decreases, the encounters with other electrons cause the path of the particle to become more and more tortuous. Because the electron density of matter is very great, these sharp deflections tend to balance out so that over a short distance the track remains fairly straight (occasionally a β particle will pass close enough to a nucleus to be both accelerated by the positive charge and sharply deflected). The grain density (a measure of the number of interactions per unit distance along an *apparent* path) increases as the particle loses energy, which means that the grain density will always be greater at the end of a track than at the beginning (Figure 6-3B); this is the principal way of determining the direction of movement of a particle in an emulsion.

The same considerations apply to α tracks—that is, increasing grain density toward the end of a track. The principal difference between α and β tracks is that, because of the charge and great mass of the α particle, it interacts with more electrons per unit distance than does a β particle; it therefore loses energy at a greater rate per unit distance and for a given energy has a shorter track length than a β.

As described in Chapter 5, Figure 5-1, each isotope has a wide range of energies. Hence, the track lengths for a particular isotope will also show a great range of values.

Physical Arrangements Between Emitting Source and Emulsion

In this discussion, *emitter* means a nucleus that is decaying and *source* refers to a collection of potential emitters.

There are three basic source-emulsion relations in common use (Figure 6-4): (1) the source is embedded in the emulsion and the emulsion

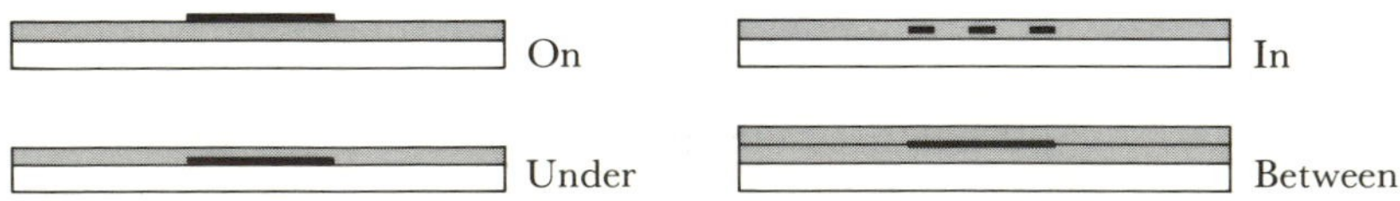

Figure 6-4
Four source–emulsion relations. The one labeled "under" is the arrangement for stripping film or for dipping. The "between" arrangement is used if a very thick emulsion is required.

is thicker than the maximum track length; (2) the source is on a surface—usually a glass microscope slide—and is covered with an emulsion whose thickness is greater than the maximum track length; and (3) the source is on a surface and the emulsion thickness is much less than the maximum track length. With types 1 and 2, the entire track length can be seen—although with type 2 only one-half of the number of tracks are seen (the other half never entering the emulsion). With type 3, only the very beginning of the track is seen. We will see later how these three types have different applications.

Figure 6-5 shows schematic diagrams of tracks of low-, medium-, and high-energy isotopes prepared in each of the three ways. Carefully note the enormous differences.

Factors Governing Choice of Isotope and Choice of Emitter-Emulsion Relation

In any experiment, high resolution, high efficiency, and low background are requisite and how they can be attained is discussed in this section.

Resolution

Resolution can mean the ability to determine the position of the emitting source, the ability to separate the individual grains to get an accurate grain count, and the ability to separate two emitting sources. This discussion is concerned primarily with the first two points.

The isotope itself affects resolution. As the energy of the emitter is increased, the tracks will be longer and have fewer grains near the point of emission (see Figure 6-5). This in general decreases the ability to localize this point because the grains in the low-density region near the source cannot be readily distinguished from the background grains. Hence, isotopes such as ^{14}C, which give curved tracks, cannot be used for high-resolution localization of a source. If the energy is great enough that the tracks are long and straight, as is the case with ^{32}P, resolution is improved if the specific activity of the source is great enough to produce many tracks. In this case, the tracks can be easily extrapolated to a well-defined origin. This will be described later in Example 6-A.

If either the source itself is very thick or the source is far from the emulsion (instead of being in direct contact with it), resolution will be lost because grains will again be further from the source.

The size of the grains also affects resolution because a single grain can be exposed by an interaction anywhere within the crystal. Therefore, large grains can have their centers considerably off the path of the particle, thus giving lower resolution in localizing the source. Furthermore, if a source consists of a set of emitters that are very near one another and are an isotope, such as ^{3}H, for which a decay can expose

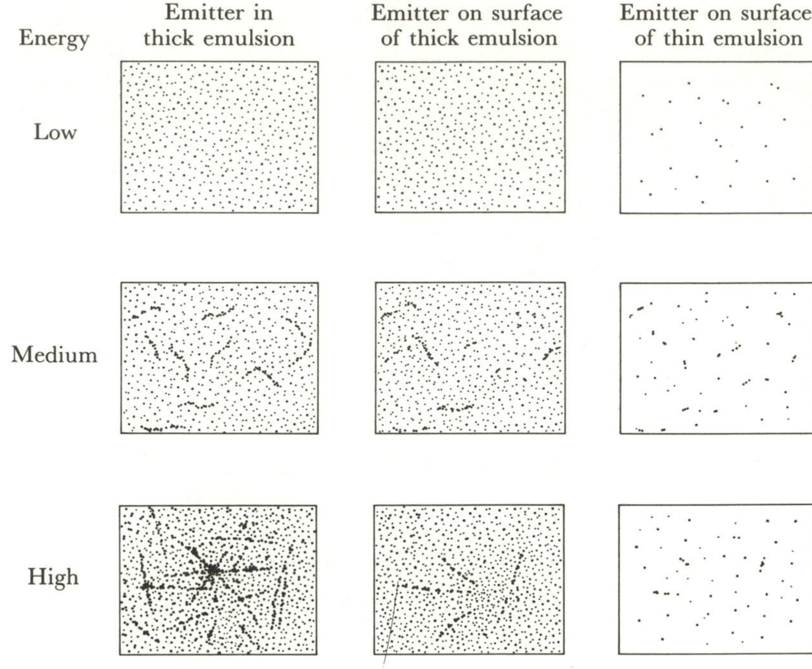

Figure 6-5
Drawings of low (^3H), medium (^{14}C), and high (^{32}P) energy tracks *in* thick emulsions and *on* thick and thin emulsions. Note that high-energy tracks become dots on a thin emulsion and low-energy tracks are totally obscured by background with thick emulsions.

only one grain, to count decays (e.g., to determine the number of emitters) each grain must be sufficiently small that exposed grains will be separated by unexposed ones. Another problem encountered in grain counting with low-energy isotopes is that a single grain can be exposed only once. Hence, if the grain size is too large (larger than the space between the emitters), there will no longer be correspondence between the number of decays and the number of grains.

The sensitivity of the emulsion—that is, the energy required to activate a crystal—also affects resolution. To understand this, one must remember that the grain density is always greater at the end of the track than at the beginning because more energy is lost per unit distance when the particle energy is low. Hence, if the emulsion is insensitive and requires a large energy loss to expose a grain, then those particles emitted at high energy will not expose a grain until near the end of the track. This results in a loss of grains near the source and a comparable increase far from it—which decreases resolution.

Efficiency

Ideally, every decay should produce a track. However, because a particle must reach the emulsion, this does not always happen. There are two basic problems: first, if the sample is on the surface of the emulsion, only one-half of the decays will enter the emulsion; and, second, even if the sample is embedded in the emulsion, there will be self-absorption of the energy by the sample owing to its finite thickness. For high-energy isotopes such as ^{32}P, self-absorption is not a severe problem—for example, a 5-μm-thick sample (about the maximum thickness for most cells or tissue sections) absorbs less than 1% of the energy. For the less energetic isotopes such as ^{14}C and ^{35}S, there is 82% and 70% transmission at a thickness of 5 μm and 10 μm, respectively. For low-energy isotopes such as ^{3}H, self-absorption is severe; at 0.5 μm the efficiency is only 16% and at 5 μm, 4%. In small cells such as a bacterium, ^{3}H can be detected at an efficiency of 27%.

Because most grains are produced near the end of a track, emulsion thickness also affects efficiency. Indeed, with ^{32}P, which has long tracks, the efficiency is proportional to emulsion thickness up to a thickness of 50 μm to 100 μm. For ^{14}C, maximum efficiency is reached at an emulsion thickness between 3 μm and 5 μm. Because the ^{3}H β particle has a range of only 1 μm, nothing is gained by using thicker emulsions (in fact, matters will be worse because of increased background).

Background

Developed emulsions that have not been exposed to a radioactive sample contain dark grains called *background*. How, then, can a track be identified? For high-energy isotopes, this is no problem because tracks are dense with grains. For ^{14}C, the grain density of tracks ranges from high to low. The convention adopted by autoradiographers is that a minimum of four grains in a straight line is required to define a track because such a configuration has a low probability of occurring in an unexposed emulsion. However, for ^{3}H, a track is usually just one large grain, and so it is necessary to reduce background to a very low level if this isotope is used.

Background has many causes: accidental exposure to light, the presence of chemicals and metal ions (especially Cu) in the sample, mechanical pressure (slight pressure on undeveloped emulsion produces heavy blackening on development), certain conditions of development, electric sparks, and background radioactivity (e.g., ^{40}K in glass, ^{14}C in gelatin, and cosmic rays). There are two types of background—long tracks and individual grains. Long tracks are seen mostly in thick emulsions. For individual background grains, the grain density (i.e., grains per unit volume) is constant and the number of grains per unit area increases with emulsion thickness. Therefore, there is never any reason to use an emulsion thicker than is necessary to achieve maximum efficiency; furthermore, it is often advisable to sacrifice efficiency and use very thin emulsions to improve the ability to recognize tracks—this is usually the case with ^3H.

Background can form before the sample is applied or during exposure (i.e., the time elapsed between placing sample and emulsion in contact and development) and development. A prior background of tracks is easily dealt with if liquid emulsions (see page 182) are used because the tracks (but not the individual grains) are destroyed when the emulsion is melted. The prior background of stripping film (see next section) is reduced because the film contains a latent image fader (a chemical that reverses the effect of exposure of a grain), which is removed at the time of applying the sample because the fader dissolves in the water used to float the film. Background subsequent to sample application is normally minimized by scrupulous cleanliness and by appropriate storage and development.

Exposure Time and Latent Image Fading

An autoradiogram must be exposed for a sufficiently long time that the pattern of exposed grains is visible. The required time is dependent on the sensitivity of the emulsion and the specific activity of the radioactive material—the exposure time decreases as the specific activity increases. Two other factors, which are less easily controlled, are important; these are the rate of accumulation of background grains and latent image fading. Background increases with exposure time; it is of greatest importance with the weakly energetic isotopes in which only single grain is exposed by one decay event. With the very thick emulsions used with the more energetic isotopes, the many individual grains, which collectively are called *fog*, are not usually a serious problem. However, long heavy tracks produced by highly energetic cosmic rays may be significant; these usually have a much higher grain density than any isotope being used so that they are easily recognized.

Latent image fading refers to the fact that exposed grains gradually revert to the unexposed, and hence undevelopable, state. This is of great concern when exposures of several weeks or more are required. For example, it may be the case that after a time the increase in the number of grains exposed is exactly counterbalanced by the loss of exposed grains. If this is occurring, it may not be possible to reach an acceptable grain density. Latent image fading is a natural process of restoration of the normal state of silver halide crystals but is considerably accelerated by high humidity and temperature. Hence, in practice emulsions are exposed below 5°C in a box containing a drying agent such as anhydrous calcium chloride. (They are of course protected from all light and external radioactivity.) These conditions, plus the use of the highest possible specific activity to reduce the exposure time, minimize the problem of both latent image fading and background.

Techniques for Placing Emulsion and Sample in Contact

Temporary Contact Method

The sample is placed in contact with the emulsion, held in place by pressure during exposure, and then removed before development. A thin protective sheet may be placed between the sample and the emulsion to prevent exposure due to chemicals in the sample. This method is used to identify spots in chromatograms and electrophoresis gels and to localize radiochemicals in large biological samples such as leaves, bone sections, and large tissue sections. The emulsion used for this technique is usually X-ray film. Semiquantitative information can be obtained using this method if certain conditions are met. However, because a grain cannot be exposed twice, it is important that the exposure be set so that blackening of the film is proportional to the amount of radioactivity; if exposure is extended to the point where areas of very low activity produce blackening of the film, the high-activity regions may have passed the point of linear response of the emulsion. Two examples of the use of this method are shown in Figure 6-6. Note that, with samples of this type, observation is not made by light microscopy because neither grains nor tracks are being counted.

Permanent Contact Methods

The three permanent contact methods in use are: mounting on a preformed emulsion, coating with or dipping into melted emulsion, and applying stripping film.

Mounting on a preformed emulsion

Either a drop of the sample is dried onto the emulsion or a tissue section is floated on water and the emulsion (which comes premounted on a glass slide) is brought up under it so that the tissue lies on the film. This is a very simple method but has three disadvantages: development may be nonuniform because the developer must penetrate the sample, the grains and tracks must be viewed through the sample, and background with preformed emulsions is high. On the other hand, sensitivity is high. If the sample consists of radioactive particles that can be diluted so that resolution is no problem and the purpose of autoradiography is to count the particles, this method is excellent. For such purposes, sensitivity can be doubled by pouring liquid emulsion on top of the deposited sample. A classic experiment using this method follows.

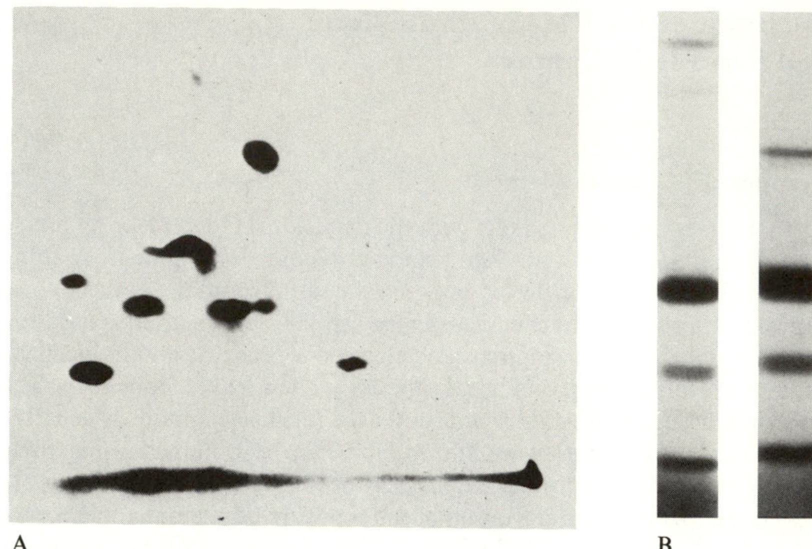

Figure 6-6
A. Autoradiogram of a two-dimensional chromatogram-electrophoregram. The major protein subunit of polyoma virus was iodinated with ^{125}I (which reacts principally with tyrosine) and digested with trypsin, spotted on paper, chromatographed, and then electrophoresed in the direction perpendicular to the chromatography. The paper was dried and placed in contact with X-ray film. The blackening of the film corresponds to the positions of the tyrosine-containing peptides. [Courtesy of William Murakami.] B. Autoradiographic detection of the bands of a gel electrophoregram of proteins labeled with ^{35}S. There are two samples and movement is vertical. Each band represents a single protein type. Each sample consists of proteins made by a particular phage mutant. Note that the proteins made by each mutant differ. [Courtesy of H. Echols.]

Example 6-A ☐ Determination of the number of DNA molecules in a bacteriophage—the "star" experiment.

The DNA of *Escherichia coli* bacteriophage T2 was labeled with ^{32}P to very high activity—approximately 100 ^{32}P atoms per phage. DNA was then extracted from some of the phage and two samples were applied to the emulsion: (1) an aliquot containing a certain number of phages and (2) an aliquot containing *all* of the DNA from the same number of phages. A liquid emulsion was then poured on top so that the sample was embedded. After exposure and development, heavy tracks coming from a single point were observed (see Figure 6-7). These configurations represented ^{32}P β particles emitted by a single phage or a DNA molecule and were called "stars." It was found that

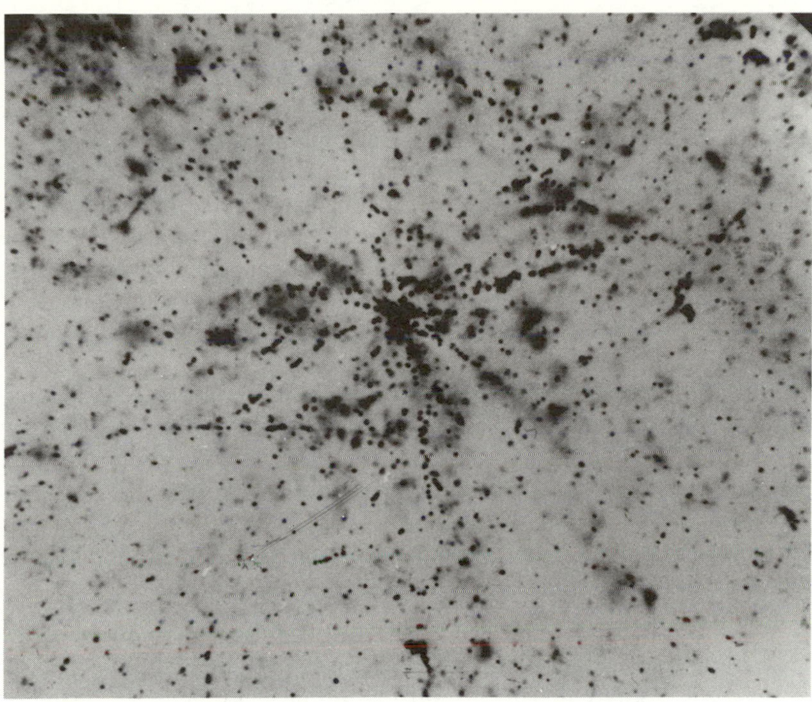

Figure 6-7
A star. A bacteriophage labeled with ^{32}P is embedded in emulsion. All ^{32}P β tracks originate from a single point, which allows the number of tracks (rays) per star to be counted. To count all rays it is necessary to focus the microscope up and down in order to recognize those that are not in the plane of the source. [Courtesy of Charles A. Thomas.]

the average number of stars per droplet was the same for both DNA and phage, suggesting that there was only one DNA molecule per phage. If there had been two DNA molecules of equal size per phage, the number of stars in the DNA sample would have been twice that of the phage sample. However, if the phage had consisted of, for example, one piece of DNA containing 80% of the DNA of the phage and ten 2% pieces, the 2% pieces would not have given stars (they would not have been recognized because they would rarely have had more than one track or ray) and the number of stars in phage and DNA would have been equal. However, the number of rays per star was also counted and found to be the same for both DNA and phage. Hence, both DNA and phage contained the same number of ^{32}P atoms, indicating that there was only one DNA molecule per phage. It was, of course, necessary to determine the number of rays per star with high statistical precision so that an 80% piece could be distinguished from a 100% piece.

Dipping method

The sample is mounted on a microscope slide and the slide is dipped into a melted emulsion (see Figure 6-2). This method gives the most intimate contact between sample and emulsion, thus maximizing efficiency for low-energy emitters. Dipping has the advantages that very thin emulsions can be obtained by appropriate dilution and choice of temperature, sample preparation is very simple and rapid, and liquid emulsion can be prepared with silver halide grains of minimum size, thus improving resolution of single grains. Its disadvantage is nonuniform thickness, but this is no problem if ^{3}H is used; as indicated earlier, ^{3}H β particles rarely penetrate past the first micrometer of crystals.

The ability of the dipping method to provide resolution of individual ^{3}H β particles is excellent. Hence, the dipping method has been used to determine the relative amount of DNA (as [^{3}H]thymidine) in a cell grown in various ways, as shown in the following example.

Example 6-B ☐ Demonstration that amino acid starvation prevents reinitiation of a round of DNA synthesis in bacteria.

If a strain of the bacterium *E. coli* requiring thymine and leucine for growth is grown in a medium containing [^{3}H]thymidine ([^{3}H]dT), which is a DNA precursor, and the amino acid leucine, [^{3}H]dT is incorporated into DNA as replication proceeds. If the leucine is removed from the medium, incorporation of [^{3}H]dT continues for about one generation and then stops. The residual amount of DNA synthesized during the leucine-starvation period is exactly what is calculated for a *population* of cells randomly distributed in all stages of DNA synthesis in which all cells engaged in replication complete synthesis of only that molecule whose replication was in progress at the start and do not reinitiate synthesis. However, to confirm this interpretation it is necessary to show that the distribution of incorporation *per cell* does agree with the proposed distribution of ages of the individual cells because it is possible that these distributions are quite different—for example, most cells might make no DNA and a few might make more than one copy and the agreement might be fortuitous.

To test this, a culture of bacteria was divided into two parts. One was grown in medium containing [^{3}H]dT and leucine for many generations; the other was grown in nonradioactive thymidine and leucine for many generations and then was shifted to a medium containing [^{3}H]dT but no leucine and grown until [^{3}H]dT incorporation stopped. Dilutions of each culture were applied to microscope slides and dried and both cultures were prepared for autoradiography using the dipping method. After exposure for several days and development, the entire slide with the cells and emulsion was dipped into a cellular stain that made the bacteria visible with

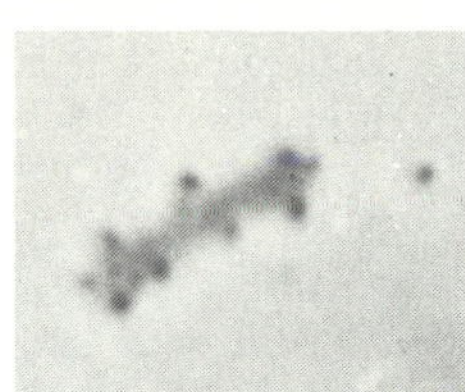

Figure 6-8
Autoradiogram of a bacterium labeled with [^{3}H]thymidine. Each grain represents the decay of a single ^{3}H nucleus. The bacterium and the grains are at different levels in the sample making it difficult to get both cells and grains in sharp focus. In this picture, the grains are in focus so that the bacterium appears somewhat fuzzy. [Courtesy of William Howe.]

the light microscope. Figure 6-8 shows an example of stained cells with ^{3}H grains. The number of cells with 0, 1, 2, 3, . . . grains was tabulated. The first culture showed the distribution of grains per cell, representing a full complement of DNA. The second culture showed the amount of DNA synthesized in each cell during the leucine-starvation period. A statistical analysis of the distribution obtained for this culture agreed with the interpretation. This experiment confirmed the important hypothesis that protein synthesis is necessary for initiating DNA synthesis.

Stripping-film method

Stripping film consists of a 5-μm layer of nuclear emulsion on a 10-μm gelatin layer with the gelatin in contact with a sheet of glass. If a cut is made through the gelatin with a scalpel, the film can be pulled or stripped from the glass. In the stripping method the film is placed on a water surface, the emulsion side on the water, where it swells for 2 or 3 minutes to about 140% of its initial area. A slide containing the sample is dipped in the water under the film and lifted up with the film draped on it. As it dries, it shrinks and adheres tightly to the slide (see Figure 6-2). Both emulsion and gelatin remain in contact with the sample during exposure, development, and microscopic observation. The sample is usually stained after development to make it visible.

Stripping film is very easy to work with and gives high resolution in the range of 0.5 μm to 5.0 μm. The emulsion thickness is more uniform than in any other method, so that valid comparisons of the radioactivity of different structures or parts of cells can be based on grain counts. However, it does have a few disadvantages: it has lower sensitivity than other emulsions; if stripping is done at low humidity, there are flashes of static charge that increase background; and the procedure is fairly slow compared with dipping. It is certainly the most widely used autoradiographic method, although some workers argue that the dipping method has all of the advantages of stripping film plus that of speed; probably the choice between stripping and dipping is a matter of preference.

One of the classic experiments of molecular biology was done with stripping film.

Example 6-C ☐ Demonstration of the number of conserved subunits of chromosomes.

In 1957, Matthew Meselson and Franklin Stahl showed that DNA replicated semiconservatively (see Chapter 11, Example 11-Q); that is, each strand served as a template for the replication of the other so that the two parental strands segregated into separate daughter, double-stranded helices. The chromosomes of root cells of the bean *Vicia faba* can be labeled with [^{3}H]thymidine by growth for a long period in a medium containing [^{3}H]dT. Autoradiograms of these cells in mitosis show that all of the grains are localized above the chromosomes (see Figure 6-9). If, after extensive labeling, the cells are grown in a nonradioactive medium for a period, autoradiograms of samples taken at various times show that, after one generation, in the cells that are in mitosis both daughter chromosomes contain

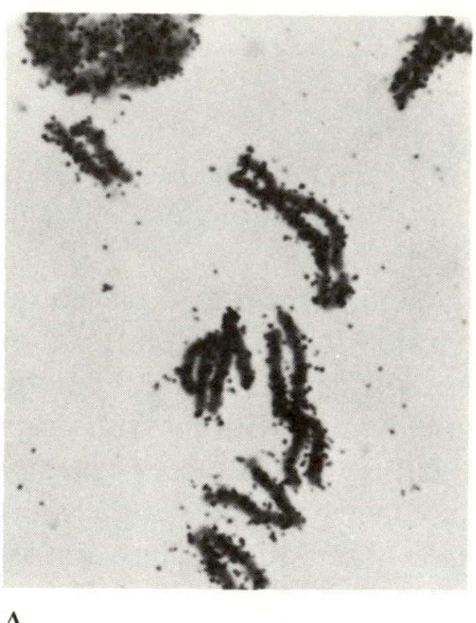

A

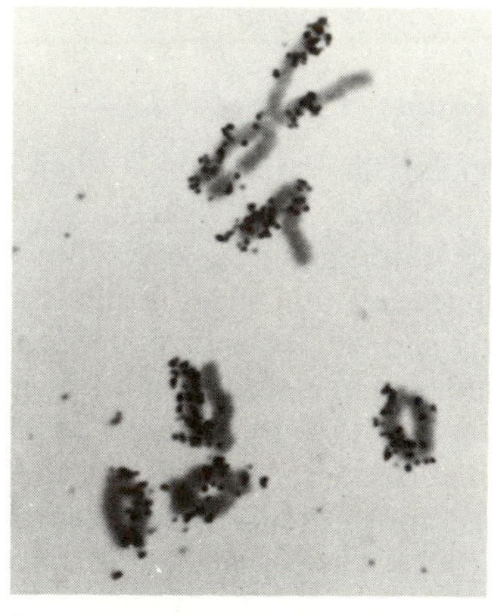

B

Figure 6-9
Autoradiograms showing chromosomes of *Vicia faba* at the (A) first and (B) second divisions after labeling with [^{3}H]thymidine during one synthetic phase. In part A, both sister chromatids are labeled. In B, only one of each pair is labeled. Note that in the uppermost X-shaped configuration there is label in both upper arms but distal and proximal to the point of intersection (the centromere). This is an example of sister chromatid exchange. [From J. Herbert Taylor, *Molecular Genetics*, part 1, Academic Press, 1963. Autoradiograms courtesy of J. Herbert Taylor.]

radioactivity but the grain count per chromosome is half that of the original chromosome. After many cell divisions, the grain count per chromosome remains constant at this value of one-half and there are never two labeled sister chromosomes separating during anaphase. In this way, J. Herbert Taylor showed that a chromosome contains two subunits that segregate at the first division after synthesis and are henceforth conserved. This result agrees with the Meselson-Stahl experiment if the subunits are the individual polynucleotide strands of the DNA.

High-Sensitivity, Indirect Autoradiography

There are few procedures in biochemistry or molecular biology that take as long as autoradiography; this is because the required exposure time can range from several days to several months. However, in 1974 several new techniques were developed that combine the technology of scintillation counting and autoradiography; these techniques reduce the exposure time about 20- to 50-fold for certain types of autoradiography.

Sensitization by Addition of Fluors to the Emulsion

As discussed on page 177 in the section concerning the efficiency of detection of a decay, weakly energetic particles, such as tritium β particles, are detected with much less than 100% efficiency. Exposure time could be shortened if the efficiency could be increased. There are several types of autoradiography in which one does not count grains, but in which proportionality between the number of grains and the number of decays is needed. Some examples are the localization of DNA in a cell using labeling with [^{3}H]thymidine or localizing a band in an electrophoregram or a chromatogram. In these cases, one looks for a blackened area on the film, which may be a collection of either individual grains or tracks. The efficiency of detection in these cases can be enhanced by increasing the number of individual grains or individual tracks or by increasing the number of grains per track. These effects are all produced if a fluor such as PPO or POPOP is added to the emulsion. Many decays of weakly energetic isotopes never reach a AgBr crystal. However, just as in scintillation counting, each decay can produce ultraviolet photons merely by the charged particle passing through the gelatin in which the crystals are embedded. The wavelength of these photons is too low to expose a AgBr crystal. However, the fluors convert the energy of this photon to visible light, which can expose the crystals. This has the effect of increasing both the number of grains exposed by a collection of ^{3}H atoms and the number of grains per track, which decreases the exposure time required to produce either a particular number of individual grains or a specific amount of blackening in the film. The two methods in most common use are to add the fluor to liquefied emulsion and use the dipping method or to soak stripping film in a solution of the fluor. Very recently, film containing the fluor has become available.

Sensitization by Preflashing

In the techniques to be described radioactive decays are converted to light and the sensitization results from the fact that the emulsion is able to respond to the light with greater blackening than from the radioactive decay. However, the response of any photographic emulsion to light is not completely linear, as shown in Figure 6-10. This has two effects—(1) the maximum amount of blackening per exposure compared to the background is not achieved and (2) the relative blackening of two regions cannot be used to measure the amount of radioactivity in the two regions. Both problems are eliminated by *preflashing*—that is, exposing the film to a millisecond flash of light prior to placing it in contact with the sample. The duration of the pre-exposure is chosen so that the point on the curve indicated by an arrow is reached. This means (1) that *all* of the time of exposure to the radioactivity-generated light produces blackening and (2) that the degree of blackening above the background is proportional to the amount of radioactivity.

Sensitization by Addition of Fluors to the Sample

An important technique in use today is electrophoresis in gels (see Chapter 9). In this procedure, individual macromolecules are separated by electrophoresis through a gel. Each type of molecule forms a band and the pattern and positions of the bands are used to identify and to determine the molecular weight of the molecules. In the analysis of protein mixtures, it is common to use ^{3}H- or ^{35}S-labeled molecules and to localize the bands by autoradiography. In order to see all bands, an

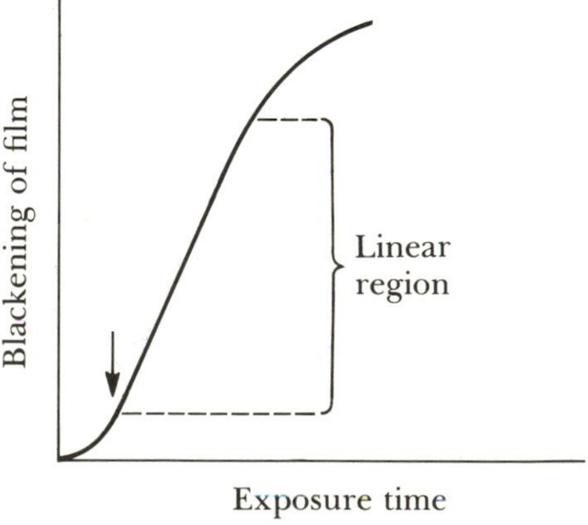

Figure 6-10
The nonlinear response of photographic film to light. The arrow indicates the point on the curve that one attempts to reach by preflashing, as described in the text.

exposure of many days (or weeks) or a very high specific activity is required. The sensitivity can be increased considerably by soaking the gel in a solution containing PPO or other fluors. A commercially available material is ^{3}H-Enhance (New England Nuclear). Following soaking of the gel and a step to form a microprecipitate of the fluor, the gel is dried and placed in contact with preflashed X-ray film. By this procedure, the exposure time or the specific activity can be reduced up to fiftyfold.

Intensifying Screens

High-energy emitters such as ^{32}P (used to label nucleic acids) and the γ-emitting ^{125}I (used to prepare radioactive proteins), are detected very poorly in most nuclear emulsions. However, they are extremely useful labels in biochemical autoradiography, especially in gel electrophoresis. Addition of a fluor to the emulsion, as described in the preceding section, is not of much value because most of the radioactivity passes through the film without significant loss of energy; the same is true of the use of fluors in the sample. However, high efficiency of detection is achieved by the use of intensifying screens (which have been in use in the medical profession for many years). An intensifying screen consists of a plastic sheet that is impregnated with a solid phosphor that emits light after absorbing high-energy radiation. Typical phosphors are $CaWO_4$ (Dupont Cronex Lightning Plus, Ilford Fast Tungstate), $BaFCl \cdot Eu^{2+}$ (Dupont Quanta II), and rare-earth oxysulfides (Kodak Lanex); the heavy metals in the fluors are especially efficient at absorbing high-energy radiation. The procedure is to prepare a sandwich consisting of the radioactive sample (e.g., a dried gel), X-ray film, and the screen. Some emitted radiation passes through the film, producing a direct autoradiographic image; that which passes through the sample is absorbed by the screen. The light produced by the screen superimposes a photographic image over the autoradiographic image. The film is selected to be sensitive to both the radiation and the wavelength produced by the screen. The film is also preflashed in order to achieve maximum sensitivity. Using a screen and preflashed film increases the efficiency of detection of ^{32}P and ^{125}I by factors of 10.5 and 16 respectively.

For details of the characteristics of different screens and films, the reader should consult the paper by Laskey and Mills, listed in the Selected References.

Molecular Autoradiography

The principal use of autoradiography has been to *localize* radioactivity in cells or tissues. However, it is also possible to use autoradiography to *visualize* or determine physical parameters of molecules and this is

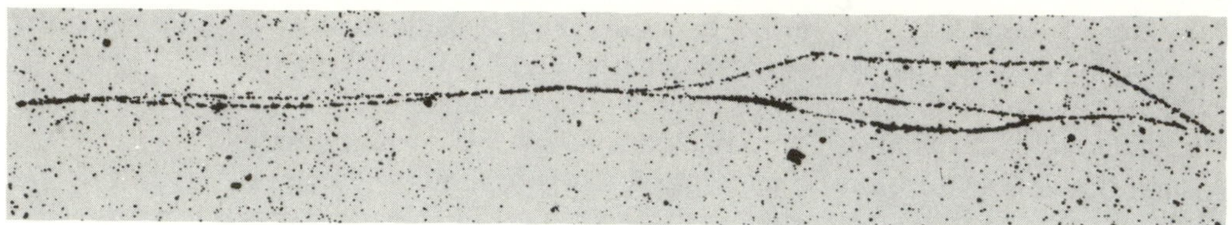

Figure 6-11
Autoradiogram of a replicating *E. coli* chromosome. [Courtesy of John Cairns.]

called molecular autoradiography. (The first example of this use was given in Example 6-A, in which the molecular weight of the DNA molecule of a phage was determined.) This powerful method, which has not had widespread use, has been used in several studies of DNA structure to obtain information not available by any other method. A well-known example is given below.

Example 6-D ☐ Visualization of a replicating *E. coli* DNA molecule.

John Cairns grew the bacterium *E. coli* for many generations in a medium containing [^{3}H]dT; he isolated unbroken DNA, extended it by adsorption to nitrocellulose filters (see Chapter 7), and prepared stripping-film autoradiograms. The beautiful pictures that he obtained are exemplified in Figure 6-11. The autoradiograms showed that the *E. coli* DNA molecule is a circle and gave the length of the molecule. (Even though DNA is easily visualized by electron microscopy [see Chapter 3], electron micrographs of such large molecules [$M = 2.6 \times 10^9$] in extended forms have never been obtained because their preparation for electron microscopy causes breakage of the DNA. If conditions producing breakage are avoided, the molecules are usually tangled.)

In a variant of this experiment it was possible to demonstrate which part of the molecule had replicated (information necessary to formulate a model of replication). This was done by growing the cells for a little *more* than one generation in a medium containing [^{3}H]dT. After one generation only a single strand of the DNA is labeled; however, continued replication for an additional short period produces a short section of DNA, *both strands of which contain* [^{3}H]*dT*. Hence, the DNA replicated in the second generation will have *twice the grain density* (grains per unit length) of the remainder. In this way, doubly replicated DNA was identified autoradiographically and as expected it was always connected to the replication forks. A remarkable finding was that there were two regions of double grain density—one attached to each fork—a result that showed that both forks were growing parts and that DNA replicates bidirectionally (Figure 6-12).

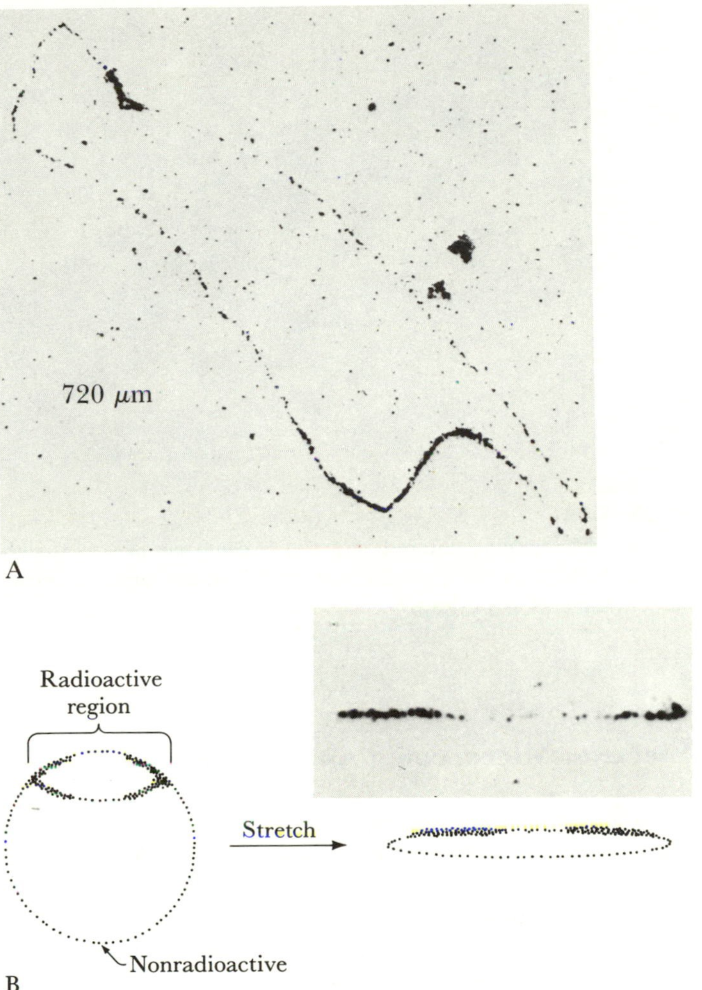

Figure 6-12
A. An autoradiogram of an entire *E. coli* chromosome obtained by labeling, with tritiated thymidine, a synchronized culture of a temperature-sensitive initiation-defective strain while it was engaged in a synchronous round of replication. The intensity of labeling was increased near the end of this round of replication and kept at this high level while the culture was allowed to reinitiate the next round of replication. This strategy of labeling will heavily label those parts of the chromosome replicated at the beginning and at the end of the replication cycle. If replication is bidirectional, then these two regions of labeling should be symmetrically disposed on the chromosome 180° apart, as is the case. [From R. L. Rodriguez, M. S. Dalbey, and C. I. Davern, *J. Mol. Biol.* 74(1973):599–604.] B. Another way of looking at the replicating region of a DNA molecule. Unlabeled *E. coli* were grown for 10 minutes in a medium containing [^{3}H]thymidine with low specific activity and then for 5 minutes with very high specific activity. The DNA was isolated and prepared for autoradiography in a way that produces stretching. Because of this stretching, the pattern observed is the result of a superposition of two DNA strands. An interpretation of the pattern is shown in the diagram under the autoradiogram. The fact that the region of low grain density is flanked by two high-density regions also demonstrates bidirectional replication. [Courtesy of R. Rodriguez and C. I. Davern.]

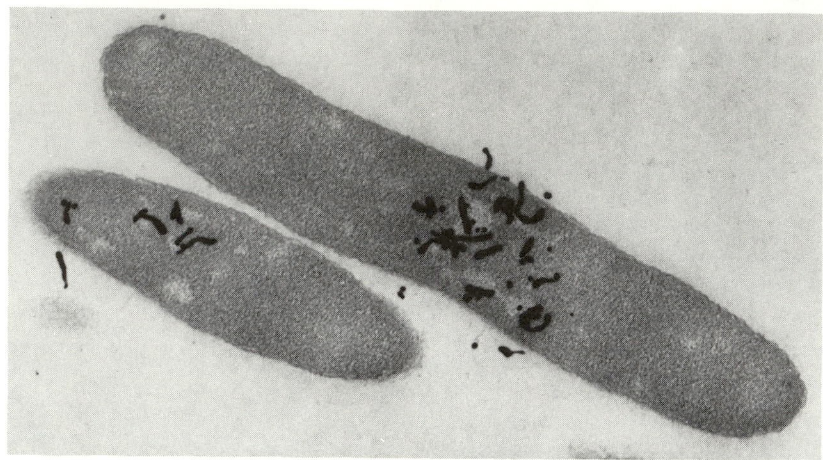

Figure 6-13
Electron-micrographic autoradiogram of *E. coli* bacteria labeled with [^{3}H]thymidine. Note that by electron microscopy the grains appear as threads of silver rather than dots. [Courtesy of Lucien Caro.]

Electron-Microscopic Autoradiography

Autoradiograms are normally observed with the light microscope. Lucien Caro and Robert van Tubergen developed a high-resolution method using the electron microscope. This necessitated the use of very thin emulsions that would not interfere with observations of the specimen with the electron microscope. A thin section of tissue or cell embedded in plastic and adsorbed to an electron-microscope grid is dipped in an emulsion so dilute that only a single layer of silver halide crystals is deposited. The emulsion used has very tiny (<0.1 μm) crystals. After exposure and development, the tissue section is stained with uranyl acetate and examined with the electron microscope (Figure 6-13). A resolution of 0.1 μm is obtainable. This method is not often used but deserves more attention.

The technique has also been applied to electron microscopy of individual molecules. If a DNA molecule is labeled with [^{3}H]thymidine, prepared for electron microscopy by the Kleinschmidt method, and for autoradiography, a molecule can be seen overlaid with individual grains. Hideyuki Ogawa has used this technique to identify products of genetic recombination between two bacteriophages. *E. coli* bacteria were infected with a mixture consisting of radioactive phages and nonradioactive, but density-labeled, phages. DNA was isolated and fractionated according to density by centrifugation in CsCl (see Chapter 11). Fractions were taken whose density indicated that the DNA contained portions of both types of molecules. These were prepared, as just described, for combined electron microscopy and autoradiography. Many molecules having an unusual structure were observed by electron microscopy; these had grains over only discrete portions of the molecules, indicating that they were indeed recombinant DNA molecules.

SELECTED REFERENCES

Cairns, J. 1962. "The Bacterial Chromosome and its Manner of Replication as seen by Autoradiography." *J. Mol. Biol.* 6:208–213. This is the first and classic example of an entire molecule visualized by autoradiography.

Caro, L., and Van Tubergen, R. P. 1962. "High Resolution Autoradiography." *J. Cell. Biol.* 15:179–188. The method of electron-microscopic autoradiography.

Davison, P. F., D. Freifelder, R. Hede, and C. Levinthal. 1961. "The Structural Unity of the DNA of the T2 Bacteriophage." *Proc. Natl. Acad. Sci. U.S.A.* 47:1123–1129. In this paper, autoradiography was used to show that the DNA content of one bacteriophage is one DNA molecule; the paper is the corrected version of the paper by Levinthal and Thomas listed below.

Gabran, P. B. 1972. *Autoradiography for Biologists.* Academic Press.

Gude, W. D. 1968. *Autoradiographic Techniques: Localization of Radioisotopes in Biological Material.* Prentice-Hall.

Laskey, R. A., and A. D. Mills. 1977. "Enhanced Autoradiographic Detection of ^{32}P and ^{125}I Using Intensifying Screens and Hypersensitized Film." *FEBS Lett.* 82:314–316.

Levinthal, C., and C. A. Thomas. 1957. "Molecular Autoradiography: The β-ray Counting from Single Virus Particles and DNA Molecules in Nuclear Emulsion." *Biochim. Biophys. Acta* 22:453–465.

Rogers, A. 1979. *Techniques of Autoradiography*, American Elsevier.

Brochures are available from Eastman Kodak and Ilford that give valuable technical information.

PROBLEMS

6–1. When bacteriophages adsorb to bacteria, the phages inject all of their DNA into the bacterium. If the phages are X-irradiated, they lose viability. It is known that the X rays produce double-strand breaks in the DNA. Design an autoradiographic procedure to determine whether an X-irradiated phage injects all of its DNA or only a fragment.

6–2. A phage is labeled with forty-two atoms of ^{32}P (on the average) per phage. The efficiency of the detection of ^{32}P by autoradiography in a thick film is 94%. How many days should the film be exposed to get an average of twelve rays per star?

6–3. Suppose that you suspect that a large cell (50 μm in diameter) incorporates a particular compound only into its cell wall and that other workers have argued that is is incorporated entirely into the cytoplasm. Could these alternatives be distinguished by autoradiography? How?

6–4. A strain of *E. coli* is claimed to have four complete chromosomes if grown in a particular very rich growth medium. How could this be proved by autoradiography? (Hint: It is possible to allow a cell to divide and form a microcolony on agar and then autoradiograph the microcolony when only a few hundred cells are present.)

6–5. You wish to identify radioactive spots on a paper chromatogram by contact with X-ray film. You could use ^3H or ^{14}C. Which would you use? Why?

6–6. A bacterial mutant has the property of synthesizing protein at 42°C at $\frac{1}{10}$ the rate that it does at 37°C. Design an autoradiographic experiment to determine whether all of the bacteria in a population share this property or whether there is a normal and an abnormal fraction of the population. Indicate which biochemical, isotope, and emulsion you would use. Why would it be necessary to use different exposure times for the control and the mutant populations?

6–7. A curious mutant of *E. coli* has the property of dividing asymmetrically and producing a normal cell and one having approximately $\frac{1}{10}$ the volume at a fairly high frequency. The small cells never divide again. How could you determine whether they have the normal amount of DNA? of RNA?

6–8. Nucleoli are dense bodies in the nuclei of many eucaryotic cells; in animal cells, they are usually about $\frac{1}{10}$ the diameter of the nucleus. Can whether they contain DNA be determined by autoradiography? Explain.

6–9. Cells of male strains of *E. coli* form pairs with cells of female strains and transfer DNA from male to female to produce genetic recombinants (i.e., females that have lost some of their own genes and replaced them with male genes.) Genetic tests indicate that no recombinants are obtained if males are mixed with males or females with females. This could be due either to lack of homosexual pairing or to lack of transfer. Design an autoradiographic experiment to determine whether homosexual pairing occurs.

7

Membrane Filtration and Dialysis

A common operation in chemistry and biochemistry is the separation of one substance from another. Before the era of centrifugation, electrophoresis, chromatography, and so on, this was accomplished by filtration, when possible. Originally, filters were of fine cloth (e.g., cheesecloth was used to separate curds from whey) and in fact cheesecloth is sometimes still used for preliminary clarification of tissue extracts. Later, porous paper replaced cloth and ultimately, to control the size of the particles retained, papers were developed having different pore sizes. To retain particles smaller than those retained using the finest papers, filters were developed consisting of a cellulose acetate, nitrocellulose, or fiberglass matrix. An even finer membrane is dialysis tubing, which passes small molecules and ions but retains macromolecules and macromolecular aggregates. A variant of dialysis tubing is the so-called molecular filter, which separates small macromolecules from large macromolecules.

In the course of using nitrocellulose membranes, it was found that they also have useful adsorptive properties (unrelated to their ability to filter), which enable them to bind particular macromolecules (e.g., single-stranded polynucleotides and some proteins) that are actually smaller than the pores of the filter; this property has contributed significantly to modern analytical techniques, and will also be described in this chapter.

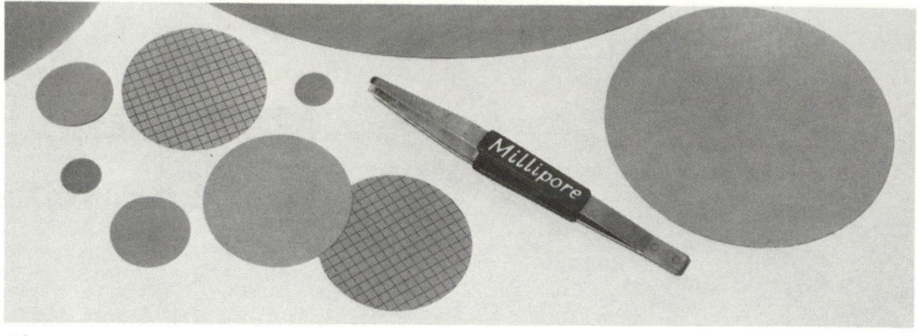

A

B

C

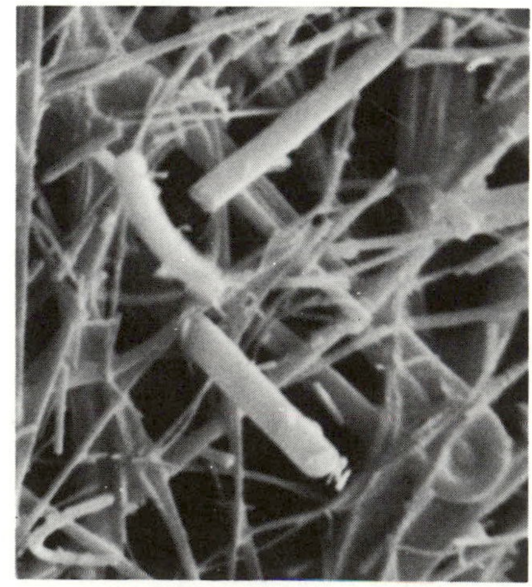

D

Figure 7-1
A. Photograph of Millipore membrane filters, which are available in a variety of sizes and are sometimes ruled. B. Scanning electron micrograph of a Millipore filter, type HA, whose average pore size is 0.45 μm. C. Pore-size distribution of the filter shown in part B. D. Scanning electron micrograph of a Millipore Microfiber Glass filter (picture width, 20 μm). Note the greater variety in the size of the pores than is found in the membrane filter shown in B. [Photographs courtesy of Millipore Corporation.]

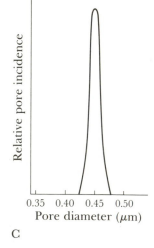

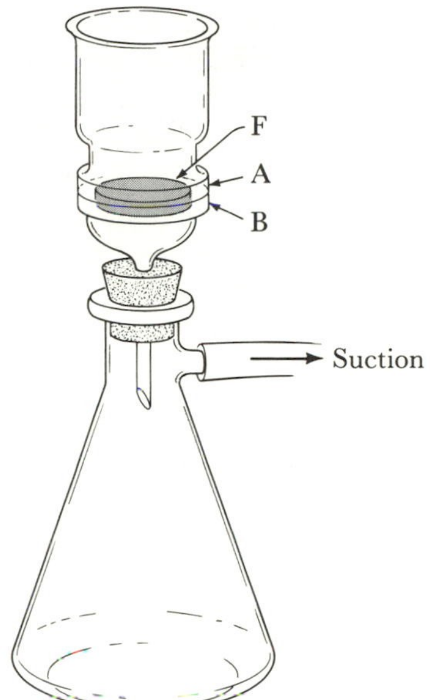

Figure 7-2
Typical apparatus for suction filtration. Rims A and B are held together by a spring clamp; F is a fritted glass support on which the filter is placed.

Nitrocellulose Filters and Membrane Filtration

A nitrocellulose filter consists of a close network of nitrocellulose fibers.* The method of manufacture allows control of the maximum size of particles passing through the filter. The pores of the filter are not circular but irregularly shaped and account for roughly 80% of the surface area (Figure 7-1A–C). The filters are sufficiently thin that retained particles do not penetrate the filter but tend to remain on the surface; hence the particles can be easily washed off the surface. The best-known manufacturers are Millipore, Schleicher and Schuell, and Gelman. The range of maximum particle size passed is from 0.01 μm to 14 μm. The most commonly used filter has a pore size of approximately 0.45 μm.

Because of both the small pore size and the surface tension, liquids do not easily pass through these filters with gravity as the driving force so that pressure or suction is usually employed. A typical suction apparatus is shown in Figure 7-2. With the vacuum developed by a water aspirator, the flow rate through a 0.45-μm filter is approximately 65 ml min^{-1} cm^{-2}. A second consequence of the small pore size is that the filters can be used only for small quantities of material because they easily clog; the amount of material retained before there is a significant decrease in flow rates is approximately 250 $\mu g/cm^2$ of area. To avoid the clogging problem, the filters are sometimes used with a thick paper or fiberglass "prefilter" to

* Although commonly called nitrocellulose filters, some are mixtures of cellulose nitrate and cellulose acetate (e.g., the common MF filters made by Millipore Corporation).

remove large particles that might easily produce clogging. These filters are, however, excellently suited for the collection of microprecipitates.

Two little-known facts about nitrocellulose filters are important. First, nitrocellulose is hydrophobic so that, in order for the filters to be wettable, most, if not all, filters contain a detergent or surfactant, the identity of which is not disclosed by the manufacturers. Second, the filters are very brittle when dry and many types contain small amounts of glycerol to increase pliability. If the filters are not prewashed, these materials may contaminate the filtrate—a considerable disadvantage if it is the filtrate that is wanted.

Nitrocellulose filters come in white, black, or green circles or sheets, with or without ruled grids to aid in particle counting. The colors are used for the microscopic identification of some microorganisms. Because nitrocellulose is soluble in a variety of solvents and attacked by certain chemicals, the manufacturers supply similar membrane filters of cellulose acetate, nylon, teflon, and polyvinyl chloride for use in certain solvent systems.

Fiberglass Filters

Fiberglass filters consist of a network of fine glass fibers (Figure 7-1D). This network cannot be made as fine as that of nitrocellulose filters so that to have strength and the ability to retain particles of small size, fiberglass filters are relatively thick—approximately 0.25 mm. Also with some brands (e.g. the Millipore Microfiber Glass filter) there is an acrylic binder for strengthening the filter.

The fiberglass filters operate on a different principle from membrane filters. The latter are a class of filter called a *screen filter*—that is, they are very thin and retain particles solely on their surface. Fiberglass filters and ordinary paper filters are examples of *depth filters*. That is, they are thick compressed fibers in which there is a maze of passages. The porosity of the filter is determined by the degree of compaction of the fibers. Particles are trapped within the matrix of the filter rather then being retained on the surface so that the capacity (i.e., mass of retained solid per unit area of the filter) is much higher than that of a screen filter.

The advantages of fiberglass filters are very high flow rate (120 ml min^{-1} cm^{-2}), great capacity before clogging, resistance to almost all solvents, ability to be heated to high temperatures for rapid drying, and low cost (about one-third the cost of membrane filters per unit area).

There are several disadvantages to fiberglass filters, and the principal one is inherent in all depth filters. That is, because the orientation of the component fibers is random, there is not an absolute pore size as there is with a membrane filter. Furthermore there is greater variability in the pore size, as shown in Figure 7-1D. However, the variability is very often within acceptable limits; for example, a 5-μm fiberglass filter will retain 98% of particles in the 5–6-μm range—that is, only 2% of the particles in this range can pass through the filter. Other disadvantages are that sometimes tiny pieces of glass fiber are found in the filtrate, only a small number of porosities are available, and, because fine particles penetrate the relatively thick filter, quantitative removal is not always possible.

In the following section, the uses of nitrocellulose and fiberglass filters are described and, when possible, which of the two types to use is indicated (in some cases it is a matter of preference).

Filtration with Nitrocellulose and Fiberglass Filters

Clarification of Solutions

For a variety of chemical and biological applications, it is necessary to remove particulate matter from a liquid, such as bacteria, dust, and fine precipitates. If the particles are very fine and if there is not too much material, either a nitrocellulose or a fiberglass filter can be used unless the particle is too small to be retained by the finest fiberglass filter. If there is a great deal of coarse material, prefiltration with a large-pore fiberglass filter is necessary to avoid clogging (most prefilters are made of fiberglass because of the high capacity).

Some proteins, viruses, and bacteriophages adsorb to nitrocellulose so that the filter must be pretreated in some way if larger particles are to be removed without losing any of these materials. Passing 0.1% bovine serum albumin in H_2O through the filter followed by extensive washing with buffers of low ionic strength usually saturates the adsorption sites and eliminates the binding of other proteins. Single-stranded polynucleotides bind to both nitrocellulose and fiberglass filters in moderate ionic strengths and this is not prevented by the bovine serum albumin wash. However, at low ionic strength this is not a problem.

Most adsorption problems can be eliminated by use of pure cellulose acetate filters (for example, Millipore Celotate filters), a fact that is not commonly known. These filters are made with the widely used pore size of 0.5 μm.

If the solution is to be clarified for special studies such as measurement of absorbance, fluorescence, optical activity, or nuclear magnetic resonance spectroscopy, either type of filter is satisfactory. However, if light scattering or viscometry is to be done (in which dustfree solutions are absolutely necessary), fiberglass filters are probably unsatisfactory because tiny glass particles sometimes enter the filtrate.

Collection of Precipitates for Counting Radioactivity

As described in Chapter 5, for counting macromolecular radioactivity, it is almost always necessary to separate radioactive macromolecules from small radioactive molecules and to concentrate the sample to a small volume to avoid diluting the scintillation cocktail or adding water. This is usually accomplished by acid precipitation, using either perchloric or trichloroacetic acid (hydrochloric acid can be used for polynucleotides but it is ineffective with protein). Following acidification, the precipitates can be collected rapidly on either a nitrocellulose or a fiberglass filter and washed, using the apparatus shown in Figure 7-2. For liquid scintillation counting, fiberglass should be used because it produces slightly higher counting efficiency (for unknown reasons) and it can be dried at high temperatures without the charring that sometimes produces color quenching (see Chapter 5). For Geiger counting, the two filter types are equivalent. (See Chapter 5, pages 149–151, for a more detailed description of the collection of precipitated macromolecules.)

Media Transfer for Growing Bacteria

In some experiments with growing bacteria, it is necessary to change the growth medium very rapidly—for example, if a radioactive compound or essential nutrient is to be removed. For this purpose, nothing is superior to membrane filtration. Only two problems must be considered: (1) nitrocellulose filters contain both wetting agents and glycerol, which must be washed out before use (otherwise they are added to the second medium when the cells are eluted); and (2) many bacteria adsorb to nitrocellulose (approximately 2×10^6 cells/cm^2), obviating quantitative transfer at low cell density (this can usually be prevented by passing 0.1% bovine serum albumin—approximately 0.25 ml/cm^2—through the filter before collecting the cells). Following filtration, cells are removed from the filter by pipetting growth medium onto the filter or by immersing the filter in the medium and agitating gently. This technique has been used extensively for bacteria, using nitrocellulose filters and has recently been used successfully with animal cells, using fiberglass filters.

Nitrocellulose Filters in Binding Assays

Nitrocellulose filters bind proteins and single-stranded DNA under certain conditions. This binding can be used as a basis for a large number of enzymatic and physicochemical assays and as a means of purification of various materials. The binding properties of two different brands of membrane filters are shown in Table 7-1. It should be noted that the binding of proteins is independent of ionic strength, but that single-stranded DNA adheres to these filters only at relatively high ionic strength. The poor binding of single-stranded DNA to the Millipore brand is not understood, although it could be because these filters contain other esters of cellulose in addition to nitrocellulose. RNA does not bind to any of these filters.

The use of these binding properties is best seen by the following examples.

Table 7-1
Binding Various Substances to Nitrocellulose Filters.

Substance	Schleicher and Schuell	Millipore	Ionic strength (M)*
Proteins	+	+	0.01–1.0
Single-stranded DNA	+	poorly	0.15
Single-stranded DNA	–	–	0.01
Double-stranded DNA	–	–	>0.05
Double-stranded DNA	+[†]	+[†]	0.001–0.005
Single-stranded RNA	–	–	0.01–1.0
Complex of protein and double-stranded DNA	+	+	0.01–1.0

NOTE: Some of the data have been provided by Dr. Andrew Braun.
* Ionic strength has been tested only in the range indicated.
[†] Binding is inefficient.

Example 7-A

☐ Purification of covalently closed, circular DNA.

Some strains of bacteria and some animal cells contain covalently closed, circular, double-stranded DNA molecules whose polynucleotide strands contain no interruptions. They frequently account for only a small fraction of the total DNA. If DNA is treated with alkali, hydrogen bonding is destroyed and the single strands separate; however, if the DNA is a covalently closed circle, the strands remain physically entangled. When the DNA is returned to neutrality and incubated at a slightly elevated temperature, the covalently closed circles rapidly reform at a rate that is independent of DNA concentration; the single strands derived from linear DNA will ultimately rejoin, but this is a very slow process and strongly dependent on DNA concentration. A circular DNA molecule containing one interruption in one of the strands will separate to form a linear and a circular strand; reformation of a double-stranded circle has the same requirements as reformation of linear DNA. If alkali-treated and neutralized DNA is passed through a Schleicher and Schuell filter at high ionic strength (Table 7-1), only double-stranded covalent circles will pass through (because all other components are single strands that bind to the filter) and will thereby be purified (Figure 7-3). For large-scale preparations, columns of powdered nitrocellulose are used.

Example 7-B

☐ Assay of messenger RNA.

Filters containing bound, single-stranded DNA are prepared by passing a solution of single-stranded DNA through the filter and then

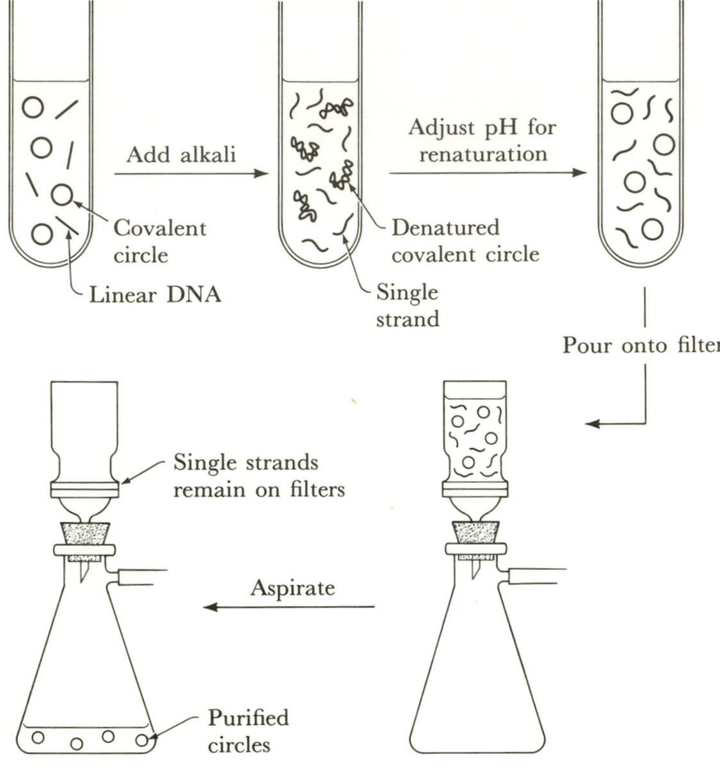

Figure 7-3
Plan for the purification of covalent circles with nitrocellulose filters. A DNA sample is denatured by treatment with alkali. The pH is then adjusted to a value at which covalent circles rapidly reform but separated strands do not. The mixture is filtered. The double-strand circles pass through the filter; the single strands adhere to the filter surface. The circles are thereby purified.

drying it in vacuum. This DNA remains bound even if the filter is immersed in water. If radioactive RNA and the filter are placed in a small vial under renaturing conditions leading to the formation of a hydrogen-bonded complementary DNA·RNA hybrid (see Chapter 1) and then the filter is placed under suction and washed, most of the unbound RNA will be washed through the filter (Figure 7-4). If treated with the enzyme pancreatic ribonuclease (which hydrolyzes single-stranded RNA but fails to attack RNA in a double-stranded structure), the remaining unpaired RNA is digested. The filter is then extensively washed and only specifically bound (hybridized) RNA remains, which can be detected by its radioactivity.

If the washed filter is then heated to above the melting temperature for the DNA·RNA hybrid (Chapter 1), the specifically bound RNA is released; this is the best way to purify specific messenger RNAs.

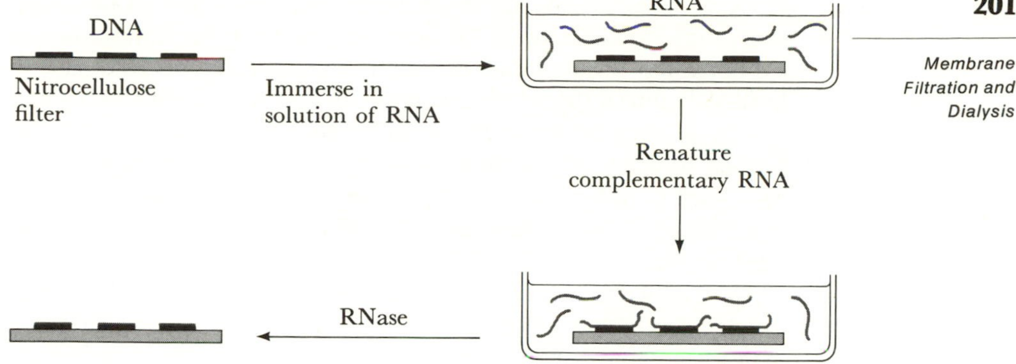

Figure 7-4
Method of hybridizing RNA to nitrocellulose filters containing bound, single-stranded DNA.

☐ Purification of mRNA from mammalian cells. **Example 7-C**

In mammalian cells the mRNA, but not other types of RNA, contains long terminal regions in which the only base is adenine (the so-called polyadenylic acid or poly(A) tail). For unknown reasons, in certain ionic conditions poly(A) binds to Millipore filters although RNA molecules do not. This is the basis of a procedure for purification of mRNA developed by G. Brawerman and his colleagues [G. Brawerman, J. Mendecki, and S. Y. Lee, *Biochemistry*, 11(1972): 637–641]. In this method, polysomes (complexes consisting of one mRNA molecule and several ribosomes) are isolated by standard centrifugation techniques. The polysomes are dissociated to yield a mixture of proteins, mRNA, and very large quantities of ribosomal RNA (rRNA). In the original experiments, 9 mg of RNA was filtered and 0.04 mg, which was mostly mRNA, was retained on the filter. After washing the filter to remove all unbound rRNA, the mRNA was eluted from the filter with sodium dodecyl sulfate. It should be emphasized that this procedure only works with mammalian, poly(A)-containing mRNA.

☐ Assay of complementary single-stranded DNA. **Example 7-D**

Filters containing nonradioactive, bound, single-stranded DNA, prepared as in Example 7-B, can be used to assay complementary single-stranded DNA. However, if radioactive single-stranded DNA is added to such filters, it will, of course, also bind, whether or not there is prebound DNA (see Table 7-1). However, by washing a filter already containing nonradioactive DNA with bovine serum albumin and polyvinylpyrollidone, the remaining single-stranded DNA binding sites become saturated so that no radioactive single-stranded

DNA can be bound. By immersing such a pretreated filter in a solution containing radioactive DNA and subjecting it to conditions of hybridization as in Example 7-B, the complementary radioactive DNA will anneal to the previously bound DNA. Unbound DNA can then be washed away and the bound radioactivity counted. This can be used to identify a particular species of DNA. For example, if the bacterium *E. coli* is infected with phage λ and incubated in a medium containing [^{3}H]thymidine, both *E. coli* and λ DNA will be synthesized. If the radioactive DNA is converted into single-stranded DNA (by thermal or alkaline denaturation) and annealed to filters to which is bound either *E. coli* or λ DNA, the relative amounts of radioactivity bound to each filter will indicate the fraction of each in the original radioactive mixture.

Example 7-E ☐ Assay of proteins that bind to double-stranded DNA.

If a mixture of proteins is incubated under appropriate conditions with radioactive double-stranded DNA, some DNA-protein complexes will form from special DNA-binding proteins (Figure 7-5). Because protein binds to the filter, any radioactive DNA complexed with protein will adhere by means of the protein. These assays are best done with Millipore filters because the background due to unbound DNA is very low (see Table 7-1). This simple assay has been used in the purification of repressor proteins because they are detectable only by virtue of their ability to bind to specific sites on double-stranded DNA. It has also been used to detect DNA-enzyme

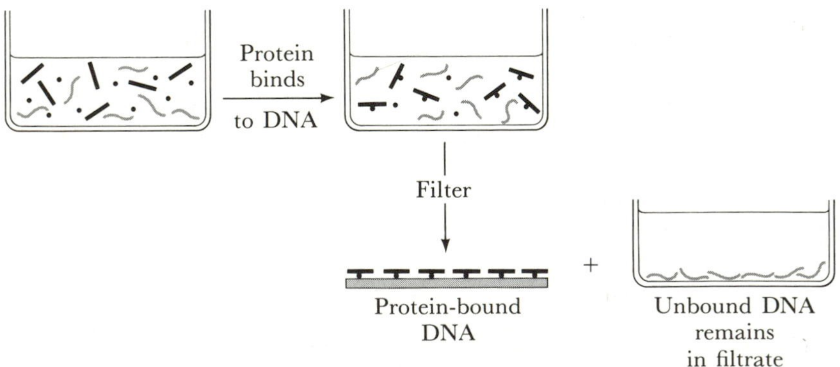

Figure 7-5
Filter-binding assay of protein-bound DNA. The sample contains protein (solid circles) and two types of DNA. The protein binds to the DNA shown as a heavy line but not to that indicated by the light line. When filtered through nitrocellulose, all protein binds. DNA molecules bound to the protein remain on the filter, whereas all other DNA molecules pass through. The amount of bound radioactive DNA can be a measure of the amount of binding protein.

intermediates in certain reactions and to measure the number of binding sites for particular proteins (e.g., RNA polymerase) on a DNA molecule.

☐ Assay for a specific aminoacyl-tRNA synthetase. **Example 7-F**

An aminoacyl-tRNA synthetase is an enzyme that attaches a particular amino acid to the appropriate tRNA. Purification of these enzymes is facilitated if their activity is assayed by binding to a nitrocellulose filter. The synthetase is a protein and hence binds to the filter; tRNA does not bind. Thus, if various protein fractions are added to a sample containing a particular radioactive tRNA (for example, alanyl-tRNA) and the mixture is filtered, there will be radioactivity bound to the filter only when an aliquot is added that contains alanyl-tRNA synthetase.

☐ Elucidation of the genetic code by membrane binding. **Example 7-G**

The protein-synthesizing complex contains ribosomes, mRNA, tRNA, and a variety of protein factors. Since ribosomes contain many proteins, they will be bound to a nitrocellulose filter; radioactive tRNA is not retained unless it is bound to the ribosome. If an mRNA that contains many different codons is added to a mixture of ribosomes, protein factors and alanyl-tRNA charged with [^{14}C]alanine, ^{14}C will be retained on the filter. The tRNA is bound to the complex by an interaction between six nucleotide bases—the three bases of the anticodon of the tRNA and the three bases of the codon in the mRNA. If a trinucleotide, whose base sequence is that of the alanine anticodon, is added in excess prior to addition of [^{14}C]alanyl-tRNA, binding of the [^{14}C]alanyl-tRNA to the ribosome complex is inhibited. This inhibition will only occur if an alanyl anticodon triplet is added. Thus, by testing all 64 trinucleotides, the anticodon sequence of alanine can be identified. This can be repeated with [^{14}C]leucine to identify the leucine anticodons, and so forth for each amino acid.

☐ Detection of antigen-antibody interactions by membrane binding. **Example 7-H**

Antibodies are proteins and therefore bind to membrane filters. Many antigens are small organic molecules that cannot bind to the filter unless they are complexed with an antigen. This is an important means of detecting the interaction in the radioimmunoassay (see Chapter 10).

Polycarbonate Filters

Polycarbonate filters differ in several ways from both the nitrocellulose and the fiberglass filter. First, they are made of a continuous sheet of very thin polycarbonate, in which cylindrical holes have been made by bombardment with nuclear particles. This contrasts with the fibrous network of the other filters. Second, they are thin, very flexible, transparent, and wettable and have no need for detergents and softening agents. Thus, they do not require preliminary washing. Third, they have a very small surface area. The fibrous filters are mostly open space, as already stated, but the surface of each of the fibers provides a large area for potential adsorption. The available surface area of a polycarbonate membrane is about 3% of the area of a nitrocellulose filter having the same diameter. This has the effect that adsorption to the filters is very poor, which makes them very useful as a means of sterilizing protein solutions and virus suspensions. These filters cannot of course be used for the binding assays already described.

Polycarbonate membranes, which are marketed as Unipore membrane filters (manufactured by Nucleopore Corp. but sold by Bio-Rad), have not yet gained the popularity of other filters as general filters. However, they have a special use in cell biology and clinical testing based on the flexibility and transparency of the filter. For example, several animal cell types are able to grow well on the surface of a filter that is placed in cell-culture medium. Because of the transparency of the filter the cells are visible by light microscopy while still on the filter. Furthermore the cells can be dehydrated by standard techniques and prepared for scanning electron microscopy or sectioned with a microtome and prepared for standard transmission microscopy. Finally the membranes are extremely useful for collecting and concentrating cancer cells from body fluids such as cerebrospinal fluid and urine. The cells are easily examined by light microscopy.

Miscellaneous Uses of Membrane Filters

A membrane filter can also serve as an immobilizing agent. For example, in matings between chromosome-transferring male (Hfr) strains of the bacterium *E. coli* and female recipient strains, Brownian motion disrupts the mating pair. In the construction of bacterial strains with special properties, it is often necessary to transfer a gene that is normally passed from male to female only very late in the mating event. However, the transfer of "late" genes is rare owing to the disruption described above. If the females are collected on a membrane filter and the males are filtered on top of them, mating proceeds normally if the filter is placed on the surface of a nutrient agar plate. Because the cells are immobilized there is little disruption and the frequency at which late genes are transferred is increased considerably. At the end of the mating period, the cells can be washed from the filter and plated onto appropriate agar plates.

Membrane filters have also been useful in quantitative assay of cellular proteins and nucleic acids. These assays, which are mainly colorimetric tests, are difficult with animal cells since these cells are usually in complex media containing large quantities of proteins and of substances that interfere with color formation and because a significant fraction of the cells may either burst or be lost if they are collected by centrifugation. Membrane filtration eliminates these problems. A typical assay can be carried out as follows. A few milliliters of a cell suspension is passed through a filter and the sediment is washed by filtering numerous small volumes of a buffer. The filter is then placed in NaOH in which both filter and cells dissolve. This is then filtered through an alkali-resistant filter to remove debris and appropriate tests are carried out. For example, addition of $CuSO_4$ produces a color indicative of protein, acidification and determining the optical absorbance at 269 nm gives the total amount of RNA plus DNA, and acidification following by addition of diphenylamine yields the DNA concentration.

Membrane filters can also be used to purify radiochemicals. For instance, solutions of many radioactive compounds such as $H_3{}^{32}PO_4$ and [^{14}C]adenosine triphosphate accumulate significant quantities of polymers as a result of radioactive decay. The polymers often produce excessively high background in labeling experiments in which acid-insoluble and acid-soluble radioactivity are to be separated. Often, filtration of the stock radiochemical removes the polymer.

Membrane filters are valuable for sterilizing chemicals that cannot tolerate the high temperature of steam sterilization. Some examples are thiamine, biotin, tryptophan, and streptomycin. Such solutions are easily sterilized if the apparatus shown in Figure 7-2 is first steam sterilized and the nonsterile solution is filtered into the sterile suction flask.

Several important techniques are based upon the ability to collect samples very rapidly in membrane filters. One example is in transport studies, in which one wishes to measure the rate at which a substance is taken up by cells or organelles. Since uptake is often complete in less than one minute, a collection technique such as centrifugation, which frequently takes five minutes or more, has very limited value. If a radiochemical is added to a suspension of cells or organelles, samples can be taken, collected, and washed in less than ten seconds, which vastly improves the resolution of such an analysis.

Dialysis and Molecular Filtration

Conventional filtration separates *particulate* matter from fluids by passing suspensions through porous material. However, with so-called semipermeable membranes it is possible to separate *dissolved* molecules by virtue of their molecular dimensions.

The best-known method is *dialysis*, in which an aqueous solution containing both macromolecules and very small molecules is placed in a collodion bag, which is in turn placed in a large reservoir of a given buffer (Figure 7-6). Small solute molecules (except those that are highly charged) freely pass through the membrane until equilibrium is reached.*

* Equilibrium in this case means that the concentrations inside and outside the bag are the same.

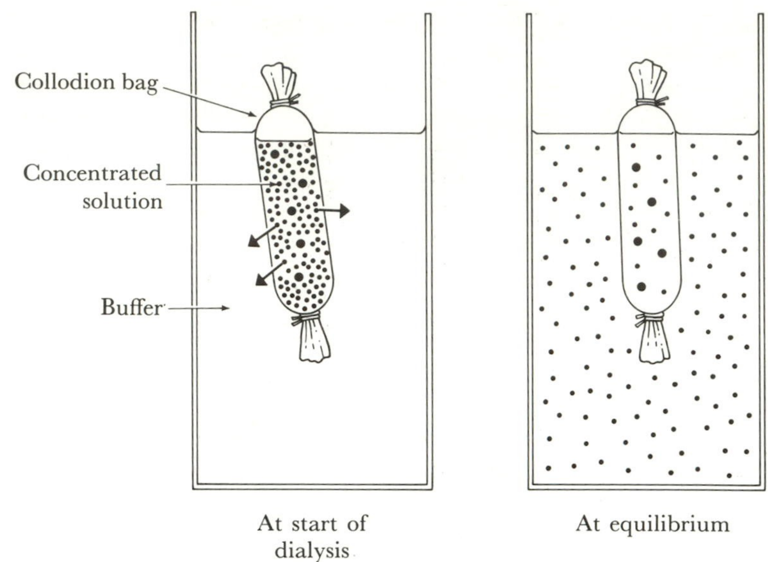

Figure 7-6
Dialysis. Only small molecules (dots) diffuse through the collodion membrane. At equilibrium, the concentration of small molecules is the same inside and outside the membrane. Macromolecules remain in the bag.

To convert the solute composition within the bag into the composition wanted, the external fluid must be repeatedly changed to maintain the required final composition. If the volume within the bag is 10 ml and the external volume is 1 l, the maximum dilution of dialyzable molecules initially inside the bag is 1/101. For example, if 10 ml of 0.1 M NaCl is dialyzed against 1 l of distilled water, the concentration at equilibrium is (0.1)(1/101) = 0.00099 M. Similarly, if the external fluid is 0.1 M $MgCl_2$, the final composition of both the internal and external fluid is 0.00099 M NaCl and (100/101)(0.1) = 0.099 M $MgCl_2$. A 10^6-fold reduction of the initial internal concentration could be accomplished by dialysis against a 10^6-fold excess external volume (1000 l). A more reasonable procedure would be to dialyze against a 10^3-fold excess twice.

Water also passes freely through the bag, thus causing concentration or dilution of non-dialyzable material, depending on whether the internal solution is less or more concentrated than the external solution (osmotic effect).

Several factors affect the rate of dialysis. The driving force for movement of a particular molecule through a dialysis membrane is the ratio of the higher to the lower concentration of that molecule on the two sides of the membrane. As molecules move through the membrane, there is a depletion of molecules on the high-concentration side and an accumulation on the low-concentration side. Mixing on both sides occurs by diffusion and convection so that a small concentration difference is maintained. However, to maximize the rate of movement, it is necessary to agitate the liquids. Usually a stirrer of some kind is placed in the external fluid. A significant increase in rate can also be obtained by stirring the fluid inside the bag. This is accomplished by placing a glass ball in the bag and rocking the entire system. (There are several mechanical devices for doing this.) As time goes on, the internal and external concentrations approach one another and the rate of dialysis decreases. Figure 7-7 shows a typical plot of the internal concentration of a dialyzable salt as a function of time. The main point to be noted is that there is a rapid initial drop followed by a slow approach to equilibrium. The time required for the first tenfold decrease is much less than that for the second tenfold decrease. Thus, a thousandfold decrease is accomplished most rapidly by adding fresh external fluid each time a tenfold decrease has occurred, as indicated in the figure. The composition and pore size of the membrane and the size and charge of the dialyz-

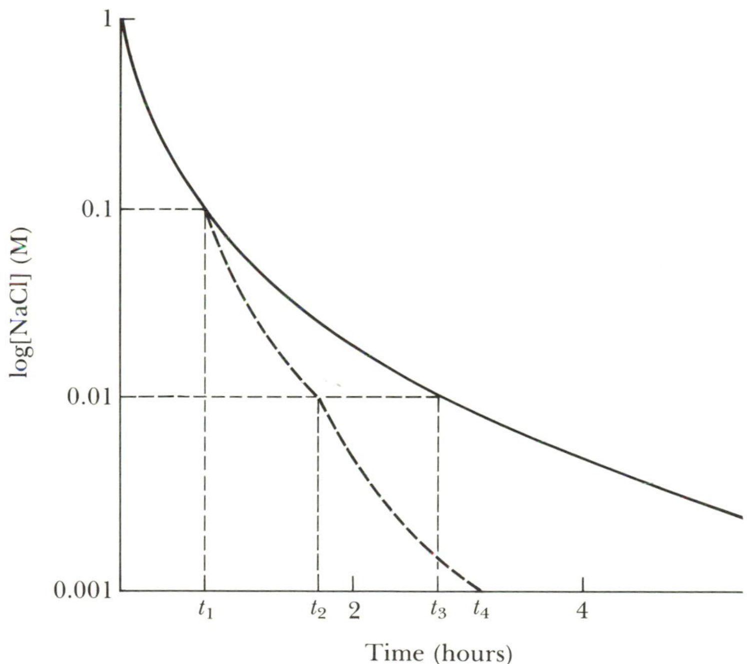

Figure 7-7
Concentration of NaCl in a dialysis bag initially containing 1 ml 1 M NaCl immersed in water, as a function of time (solid line). At the times t_1 and t_3 the concentration has dropped tenfold and 100-fold, respectively. The dashed line shows the result of transferring the bag to fresh water at time t_1, after which the 100-fold decrease is reached at time t_2, which is less than t_3. At t_2 there is a second transfer to fresh water resulting in completion of a 1000-fold total dilution by t_4; without these two transfers it would have taken infinite time to achieve this dilution.

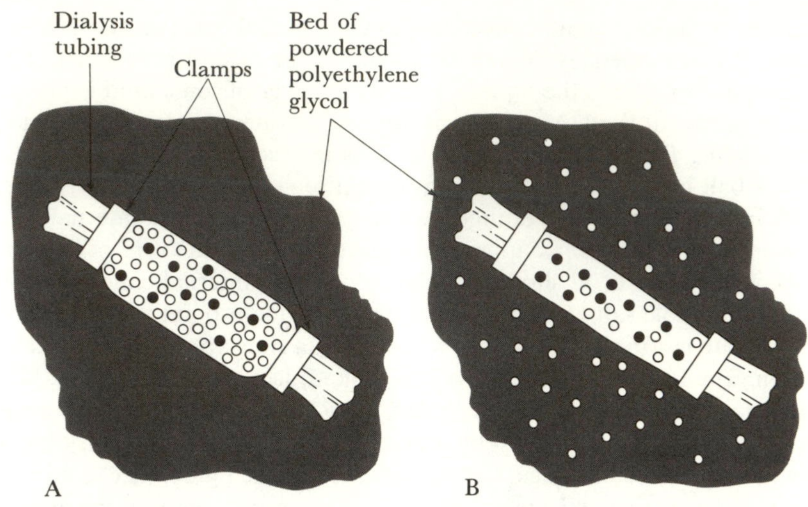

Figure 7-8
Reverse dialysis: (A) a solution of macromolecules (solid circles) and solvent molecules (open circles) is placed in a dialysis bag that is packed in dry polyethylene glycol powder; (B) the solvent molecules leave the bag and enter the polyethylene glycol phase. Neither the polyethylene glycol nor the macromolecules can pass through the membrane and so the solution is concentrated.

able molecule are also important factors that affect the dialysis rate; the effect of these factors is usually determined empirically. The dialysis of highly charged molecules is often very slow; the rate can be improved considerably if the ionic strength of the solution is increased. An interesting effect of concentration occurs with detergents and other amphipathic compounds (long-chain substances having a polar terminus and a long nonpolar segment). Above a critical concentration, these substances form large aggregates called *micelles*. The concentration, called the *critical micelle concentration* (cmc), is specific for each type of molecule. Micelles are so large that there are very few membranes through which they can pass. However, if the concentration is lowered by dilution to less than the critical micelle concentration, dialysis is usually possible because micelle formation is a reversible process.

A variation called *reverse dialysis* (Figure 7-8) can be used to concentrate the material in the bag. The filled bag is packed in a dry, water-soluble polymer (which cannot enter the membrane) such as polyethylene glycol. Water then leaves the bag to equilibrate with the dry external phase. Sucrose can also be used but, because it is a dialyzable substance, it will enter the bag as water is removed. The sucrose can be removed by ordinary dialysis; it is necessary to move the clamps shown in Figure 7-8 very near one another—otherwise the concentrated sample will be diluted again. One must be careful when using reverse dialysis. This is because equilibrium is never reached and water and salts are continually removed until the sample is totally dry. If this occurs, most macromolecules become irreversibly bound to the dialysis tubing and hence, for all practical purposes, they are lost.

Semipermeable membranes that allow small but not large macromolecules to pass through are also commercially available. The type most commonly used is Spectropor tubing (Spectrum Medical Industries, Inc.), which is used in much the same way as dialysis tubing. It is available in several grades, with a molecular weight cutoff (i.e., maximum size passed) ranging from 6000 to 14,000.

An important modification of dialysis tubing is the Diaflo (Amicon Corporation) or the Pellicon (Millipore Corporation) membrane. This is a very thin (0.1–1.0 μm) polymer membrane having a pore size that ranges from 2 Å to 100 Å mounted on a thicker (50–250 μm) supportive layer of an open-celled "sponge;" it is sometimes called a *skinned* membrane. It comes with a wide variety of molecular weight cutoffs, ranging from 500 to 100,000. Such membranes can be used for concentration (using a membrane through which only water passes), desalting in preparation for chromatography, and fractionation by molecular size. The flow rate through these membranes is so low that they are operated under pressure. Special holders are usually required for their use (Figure 7-9). Modifications of these instruments are available for handling especially large or small volumes or for multiple samples.

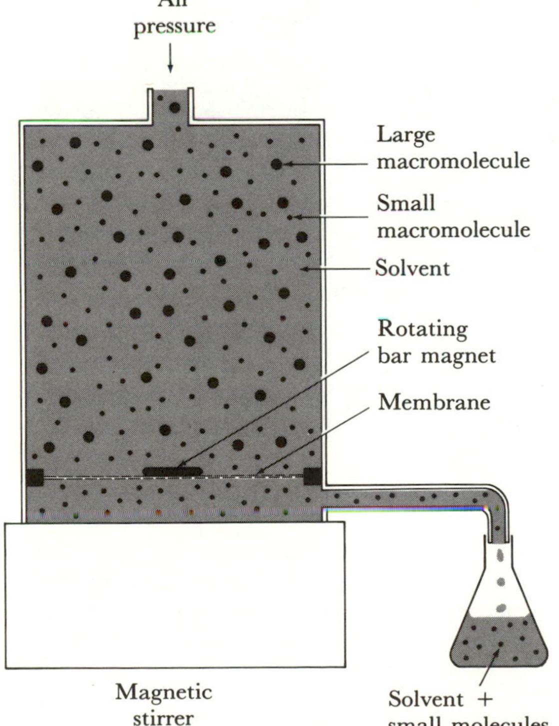

Figure 7-9
Apparatus for molecular filtration. A solution containing small and large macromolecules is forced against the molecular membrane. Molecules larger than the cutoff size fail to pass through the membrane. Smaller molecules and the solvent do pass through. A rotating bar magnet is used to prevent the membrane from clogging. Either small or large molecules can be purified in this way. The large molecules are also concentrated.

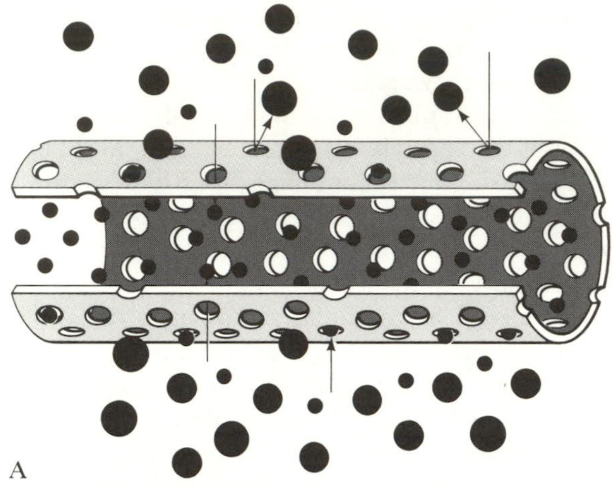
A

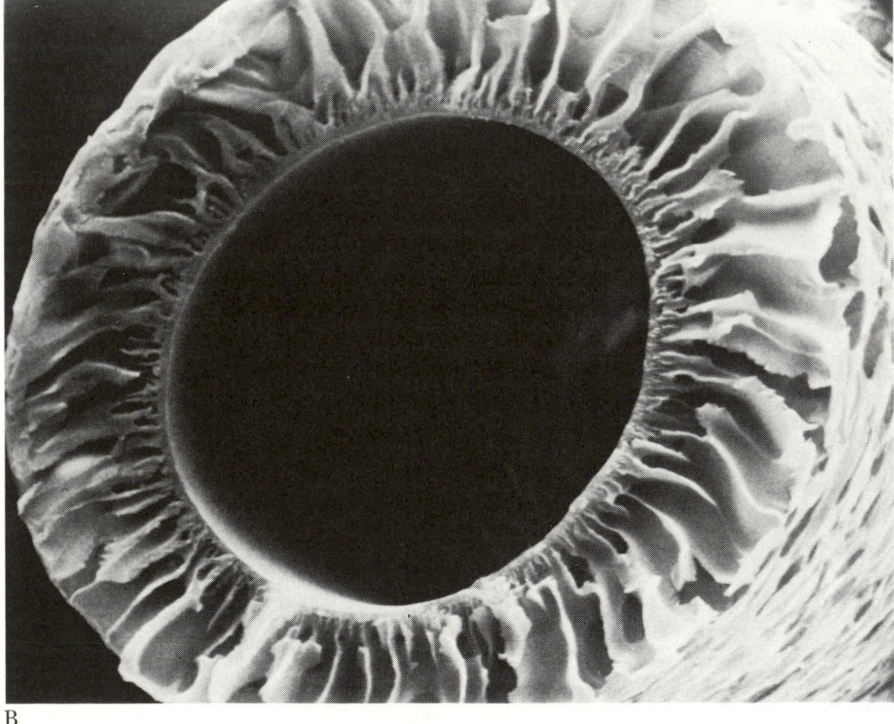
B

Figure 7-10

A. The operation of a hollow fiber. The fiber is initially immersed in a solution of large and small molecules. The small molecules pass through the pores of the fiber to the inner channel and are swept away by flowing water. The large molecules cannot pass through the pores. [Courtesy of Bio-Rad Laboratories.] B. Electron micrograph of a hollow fiber. [Courtesy of Amicon Corporation.]

The principal problem with skinned membrane filters is a phenomenon called concentration polarization. This is nothing more than a concentrated layer of macromolecules that accumulates on the surface of the skin and both reduces the flow rate and the molecular weight cutoff. It is minimized but not completely eliminated by continuous agitation of the solution, as shown in Figure 7-9.

Glass Fiber Dialysis

Semipermeable glass fibers are valuable devices for both dialysis and concentration. They are hollow-bore fibers whose glass walls contain pores of controlled size. Molecules smaller than the pores pass freely through the wall of the fiber (Figure 7-10A). These fibers are usually used in bundles, thus providing a very large surface area. They are normally used in a unit of the type shown in Figure 7-11. To change the buffer in a sample containing a macromolecule to a second buffer, the sample is placed in the vessel and a large volume of the second buffer is allowed to flow through the fibers (Figure 7-11A). The small molecules

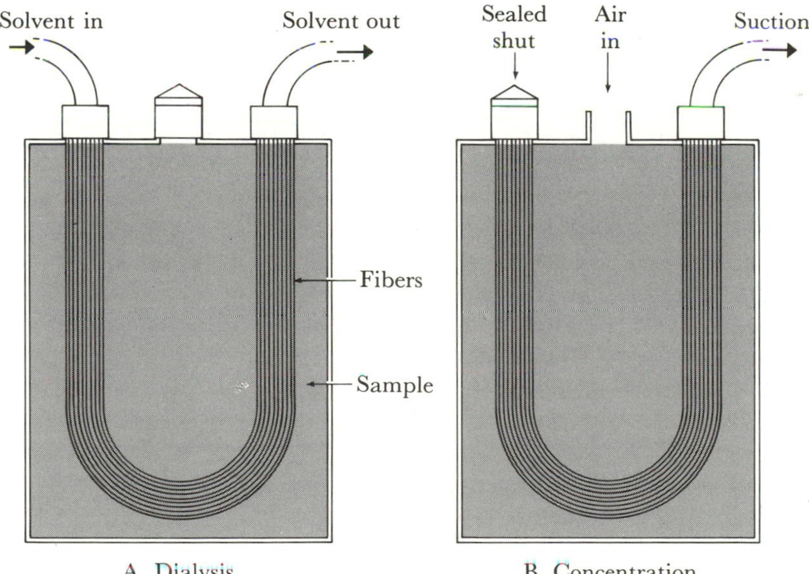

Figure 7-11
Two ways of using hollow fibers: (A) dialysis, in which a solvent flows through the fibers and small molecules enter the fibers, thus reducing the concentration of small molecules in the sample; and (B) vacuum is applied to the fiber bundle and the solvent and small molecules enter the fiber, thus concentrating any macromolecules in the sample.

of the two buffers rapidly exchange through the pores of the fibers; because the buffer within the fibers is in excess, the first buffer is replaced by the second. The macromolecules fail to penetrate the pores of the fibers and remain outside. If desalting is required, water is passes through the fibers. For concentrating samples, the arrangement shown in Figure 7-11B is used. The sample is again in the vessel and suction is applied to the fibers. The pressure differential then forces the solvent and small solute molecules into the fibers, thus concentrating macromolecules in the solution. Other physical arrangements that allow the sample to pass through the fibers are possible but are usually not done because the fibers easily clog.

SELECTED REFERENCES

Bøvre, K., and W. Szybalski. 1971. "Multi-step DNA-RNA Hybridization Techniques," in *Methods in Enzymology*, vol. 21, edited by L. Grossman and K. Moldave, pp. 350–382. Academic Press.

Gilman, A. G. 1970. "A Protein-Binding Assay for Adenosine 3′:5′-Cyclic Monophosphate." *Proc. Natl. Acad. Sci. U.S.A.* 67:305–311.

Kennell, D. 1971. "Use of Filters to Separate Radioactivity in RNA, DNA, and Protein," in *Methods in Enzymology*, vol. 12A, edited by L. Grossman and K. Moldave, pp. 686–692. Academic Press.

Millipore Corporation. *Molecular Filtration.*

Nirenberg, M., and P. Leder. 1964. "RNA Codewords and Protein Synthesis." *Science* (Wash., DC) 145:1399–1407. This paper describes how the genetic code was worked out using a membrane-filter-binding assay.

Parks, J. S., M. Gottesman, K. Shimada, R. L. Pearlman, and I. Pastan. 1971. Isolation of the Gal Repressor." *Proc. Natl. Acad. Sci. U.S.A.* 68:1891–1895.

Riggs, A. D., G. Reiness, and G. Zubay. 1971. "Purification and DNA-Binding Properties of the Catabolite Gene Activator Protein." *Proc. Natl. Acad. Sci. U.S.A.* 68:1222–1225.

Riggs, A. D., H. Suzuki, and S. Bourgeois. 1970. "Lac Repressor-Operator Interaction." *J. Mol. Biol.* 48:67–83. In this paper binding constants were measured using a membrane-filter-binding assay.

Excellent brochures are supplied by H. Reeve Angel, Inc. (*properties of paper and fiberglass filters*), *Amicon Corporation* (*Diaflo membranes*), *Millipore Corporation* (*nitrocellulose filters and Pellicon membranes*), *and Schleicher and Schuell* (*nitrocellulose filters*).

PROBLEMS

7–1. Suppose that you are collecting a precipitate from an acetone solution. Would a nitrocellulose or fiberglass filter be better? Why?

7–2. If you were collecting a precipitate from an aqueous solution, what criteria would you use in deciding between filtration and centrifugation?

7–3. Particles presumably smaller than the pores of a particular nitrocellulose filter are efficiently collected on the filter. They might adsorb or they might aggregate and form large clusters. How could you distinguish these alternatives?

7–4. Linear and circular DNA molecules having the same molecular weight can separated on nitrocellulose filters if the flow rate is very high. At low flow rate there is no separation. Which would remain on the filter? Why?

7–5. Growth medium for biological material must always be sterile. Sterilization is normally done by raising the temperature of the solution to 135°C. However, this is not always possible because some of the components may be destroyed by heating. Explain how filtration could be used for sterilization. Describe what components must be sterilized and how you would do it.

7–6. In preparing protein or nuclei acid polymerization reaction mixtures for scintillation counting, a common procedure is to add trichloroacetic acid to the mixture. The polymer is insoluble in acid, whereas the monomer remains soluble. When this mixture is filtered, the polymer remains on the filter and the monomer passes through. However, a small fraction of the monomer often adsorbs tightly. If 10^6 cpm of monomer is in the mixture, 0.02% is polymerized, and 0.01% adsorbs, then the background radioactivity will be too high for reliable counting. Given that proteins and nucleic acids are soluble in alkali, design a procedure for reducing the background. State the type of filter that will be used and estimate the background resulting from the improved procedure.

7–7. Suppose that you are collecting bacteria on a filter in order to transfer them to a new growth medium. Transferring is done by placing the filter in the new medium and agitating violently to wash off the bacteria. In a series of experiments, you have checked the recovery of bacteria from the filter as a function of the total number of bacteria collected and have obtained the following results:

Number of bacteria collected	Percentage washed off filter
5×10^8	98
3×10^8	100
1×10^8	89
5×10^7	50
3×10^7	35

What would the recovery be (approximately) if 10^7, 10^6, and 10^5 bacteria were collected? Explain.

7–8. That dialysis tubing adsorbs both proteins and nucleic acids is well established. Explain how you might determine whether this is occurring for a given protein or nucleic acid. Enzymes sometimes lose biological activity on dialysis. How could you distinguish adsorption from inactivation for a given enzyme?

Three

SEPARATION AND IDENTIFICATION OF MATERIALS

8

Chromatography

A goal of biochemistry is to separate and identify chemical compounds. Chromatography is one of the most effective techniques for accomplishing this. Although it is generally acknowledged that the method was developed in 1906 by the Russian botanist Mikhail Tswett, who separated plant *pigments* (hence the name), it was described in 1855 by Karl Runge, a German chemist who separated inorganic materials by paper chromatography; in fact Pliny the Elder reported the separation of dyes on papyrus and devised a papyrus chromatographic test for iron. However, chromatography did not become a serious technique until the work of Archer Martin and R. L. M. Synge in 1944, who later received a Nobel Prize for developing the methodology of partition chromatography.

Today there are many kinds of chromatography—adsorption, partition, ion-exchange, and molecular-sieve—and many specialized techniques for using them—column, paper, thin-layer, and gas chromatography. Technical modifications can be introduced if chromatography is to be used for large-scale work (i.e., for producing large quantities of a relatively pure material) rather than as an analytical procedure. This chapter describes each of these procedures and many of the modifications.

Simple Theory of Partition Chromatography

Grossly dissimilar substances like iron filings and glass particles can be easily separated with a magnet; sand and sugar can be separated by dissolving the sugar in water. However, if the substances are similar in physical and chemical properties, separation procedures become more complex and subtle. In chromatography, substances are placed in a system consisting of two physically distinguishable components—a *mobile phase* and a *stationary phase*—and molecular species separate because they differ (many of them only slightly) in their distribution between these two phases. The relative movement of each molecule is a result of a balance between a driving force (i.e., the movement of the mobile phase) and retarding effects. The retarding effects that we will be considering first are partition and adsorption.

In discussing chromatography, the following standard terminology is used. The stationary phase is the *sorbent*. If the sorbent is a liquid held stationary by a solid, the solid is called the *support* or *matrix*. The mobile phase is the *solvent* or *developer* and the components in the mixture to be separated constitute the *solute*.

The theory of *partition chromatography* is that, in general, if two phases are in contact with one another and if one or both phases contain a solute, the solute will distribute itself between the two phases. This is called partitioning and is described by the *partition coefficient*, the ratio of the concentrations of the solute in the two phases.

Partition chromatography will be described in terms of the operation of a *column*—that is, a tube filled with a sorbent and a solvent. A solution containing the solute is layered on top of the sorbent and allowed to enter the sorbent. The solvent is then allowed to pass continually through the column. Although the sorbent and solvent within the column are certainly continuous from the top of the column to the bottom, the column can be thought of as consisting of a large number of individual layers ("theoretical plates"), each containing the two phases. Consider 256 *identical* molecules that distribute themselves equally between the two phases, one stationary and one mobile, and a column with eighteen plates (Figure 8-1). In the uppermost plate (the *origin*) the 256 molecules are distributed so that 128 are in each phase. When the mobile 128 molecules from plate 1 enter the second plate, they redistribute 62 and 64 in that plate; the 128 remaining in plate 1 also redistribute 64 and 64, as shown in Figure 8-1. If the mobile phases each advance again by one plate, redistribution again occurs. After twenty successive transfers, the situation shown in the graph is achieved. Suppose at the origin there also were 256 identical molecules *of a different type* (shown in italics), which distribute with three times as many in the stationary phase. Figure 8-1 shows the distribution of these molecules after twenty transfers also. The distributions of these two kinds of molecules are different and a

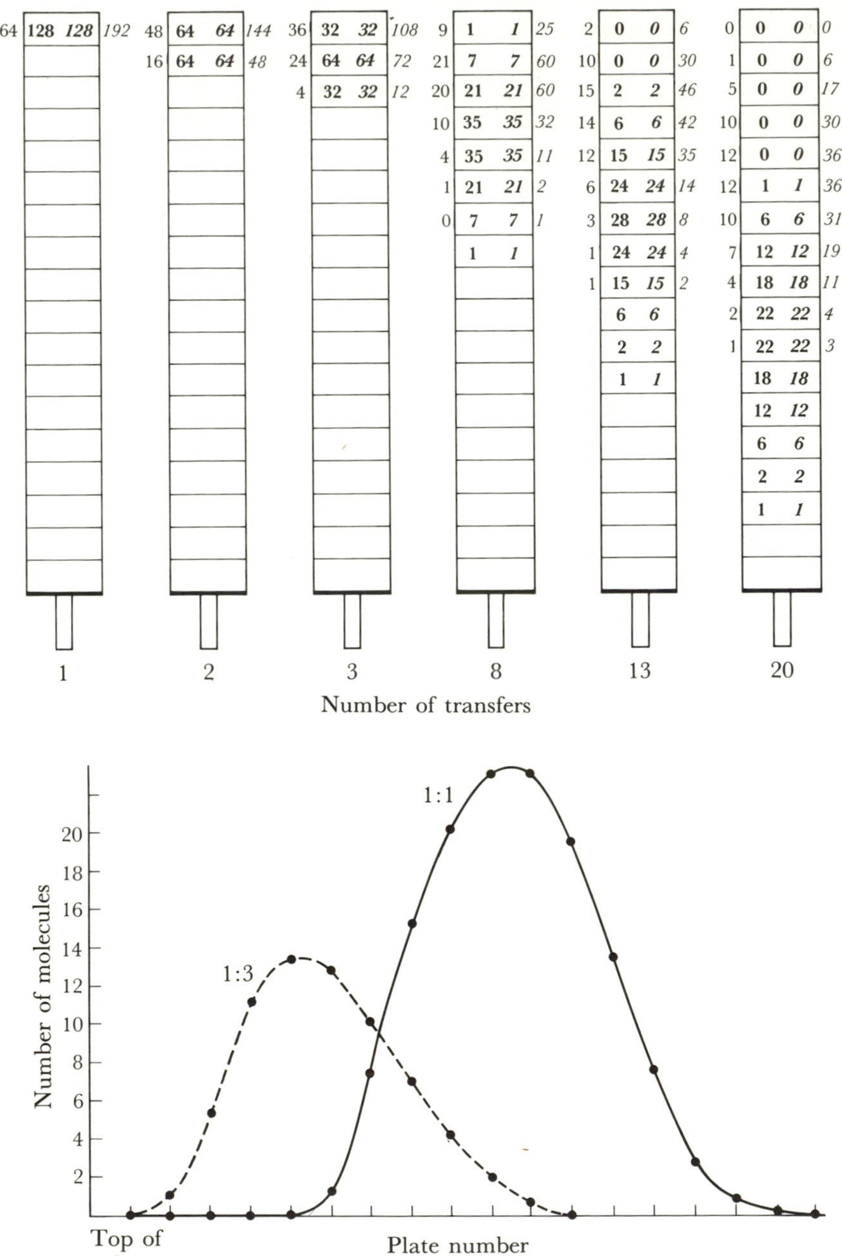

Figure 8-1

Principle of operation of a column in which separation is based on partition. The column has arbitrarily been divided into eighteen theoretical plates. Five hundred twelve molecules are loaded onto the column. Two hundred fifty-six of these (bold type) distribute equally between mobile phase (roman type) and stationary phase (italic); they are the 1:1 class. The 1:3 class (light type) distribute so that 25% are in the mobile phase and 75% in the stationary phase. A transfer means that all molecules in the mobile phase advance to the next plate. Following each transfer the number of molecules of each class redistributes according to the 1:1 or 1:3 rule. The graph shows the distribution of each after twenty transfers. Note that the 1:3 class moves more slowly through the column.

substantial fraction of the molecules have separated. As the number of theoretical plates is increased (i.e., if the column length is increased), greater separation will result. It should be noted that, as the number of plates increases, material is spread throughout a greater part of the column. In a real situation, in which the total number of molecules is huge (i.e., $>10^{16}$), the number in plate 1 would not reach zero. On the other hand, the degree of spreading would decrease in the sense that a *greater percentage* of the material is in a small fraction of plates. For example, after eight transfers, 40% of the molecules illustrated in Figure 8-1 that distribute 1:1 are contained in $\frac{4}{8}$ of the plates, whereas, after twenty transfers, 40% are in $\frac{6}{16}$, or $\frac{3}{8}$, of the plates. These considerations should make it clear that in the ideal case, in which the distribution is determined only by partition, the *resolution of two substances will improve as the length of the stationary phase increases.*

Examples of Partition Chromatography

The two most common types of partition chromatography are *paper* and *thin-layer chromatography*. In both cases, the matrix contains a bound liquid: water molecules are bound to cellulose in paper chromatography, and the solvent used to form the thin layer is bound to the support in thin-layer chromatography (see pages 229–232). (These techniques are sometimes thought of as types of adsorption chromatography, because adsorptive effects do enter into the degree of separation; however, the principal mode of action is by partitioning.) Other examples of partition chromatography are *gas-liquid* and *gel chromatography*, which are described in detail on pages 232 and 238.

Partition chromatography may also be carried out in columns by using a matrix that does not adsorb the solutes. Common supporting materials are diatomaceous earth (e.g., Celite), silica gel, cellulose powder, and certain cross-linked dextrans (e.g., Sephadex LH20). The stationary phase is created by suspending the support or washing the column with the appropriate sorbent. In this way the sorbent either coats the particles of the support and is retained by adsorption or simply penetrates the interstices of the particles and is held there by capillarity. It is important to realize that the stationary phase does not fill the spaces between the particles—that space is to be occupied by the mobile phase. Typical stationary-phase materials are hydrophobic solvents, such as benzene, for the separation of nonpolar materials or hydrophilic solvents, such as an alcohol, for polar materials. Typical mobile phases are alcohols or amides for the nonpolar material or water for polar substances. Note that the stationary-phase materials are liquids.

Partition chromatography is used primarily for molecules of small molecular weight. To reduce diffusional spreading (i.e., the broadening of peaks), very small starting zones and rapid separation is necessary. How this is done will be described in the sections on paper and thin-layer chromatography.

Simple Theory of Adsorption Chromatography

Consider a solid surface containing a wide variety of binding sites—for example, regions that are electron-rich (negatively charged), electron-poor (positively charged), nonpolar, and so forth—and a liquid containing solute in contact with the surface. If binding is reversible, the number of molecules bound to the surface will depend on the solute concentration. This dependence (of which there are three types) is shown in Figure 8-2. Curves of this sort are called *adsorption isotherms*. The most common is the convex curve—that is, binding sites with high affinity are filled first so that additional amounts of solute are bound less tightly. The binding isotherm is a characteristic of a particular molecule and sorbent. If a given concentration of a molecule is applied to the surface (which in practice is usually a collection of particles in a column or on a solid support) and solvent is allowed to flow across the surface, a fixed amount will bind and the remainder will move along. The advancing material is retarded by its adsorption to the column material. Also the fraction bound is not a constant amount but decreases with decreasing concentration. (This differs from partition chromatography in which the fraction bound is constant.) The rate at which the substance moves is related to the strength of binding—that is, the tighter the binding, the slower the movement. Clearly then, molecules can be separated if they have different adsorption isotherms because they will be retarded to different extents.

Types of Adsorption Chromatography

Adsorption chromatography uses a mobile liquid phase and a solid stationary phase with the one exception of gas-liquid chromatography (page 232). Separation is either in columns or on thin layers. Table 8-1 gives the common adsorbents and their uses.

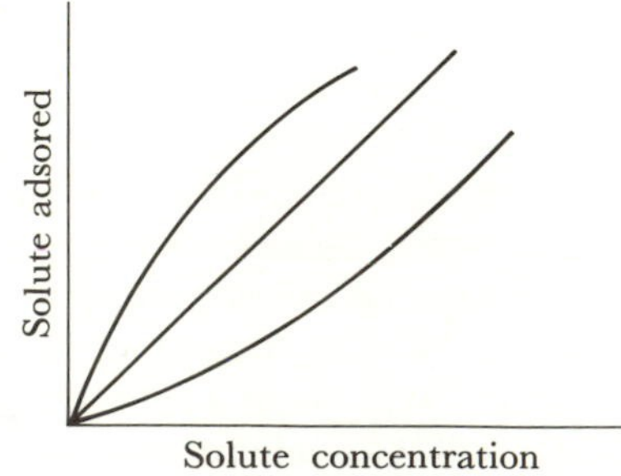

Figure 8-2
Three common distributions of solute between adsorbent and solvent as a function of solute concentration. These curves are called adsorption isotherms. Because of the concentration dependence of the fraction bound, the distribution as a function of the number of transfers is more complicated than that of Figure 8-1.

Table 8-1
Common Materials Used in Adsorption Chromatography.

Material	Substances separated
Alumina	Small organic molecules, proteins
Silica gel	Sterols, amino acids
Activated carbon	Peptides, amino acids, carbohydrates
Calcium phosphate gel	Proteins, polynucleotides
Hydroxyapatite	Nucleic acids

An important variation of adsorption chromatography is ion-exchange chromatography (page 248). This differs mainly in that the composition of the mobile phase is such that, as the material is being applied to the adsorbent, the solute becomes immobilized. Migration does not begin until a new mobile phase is added. This is not different in principle but is a special case in which adsorption is very strong.

It is important to realize that partition and adsorption chromatography are rarely exclusive in that adsorptive effects may be present in paper, thin-layer, and sometimes gel chromatography.

Operation of Columns

Probably the most common way to hold the stationary phase or support is in a column. In column chromatography, a tube is filled with the material constituting the stationary phase, plus a solvent. Then a small volume (i.e., a thin lamella) of sample is placed on the stationary phase and allowed to enter the column (this is called *loading* the column). The chromatogram is then developed by flowing a solvent (the mobile phase) through the column (Figure 8-3). The latter process is called *eluting* the column. As different substances move through the column, they separate and appear in the effluent when particular volumes of liquid have passed through the column. The total volume of material, both solid and liquid, in the column is called the *bed volume*. The volume of the mobile phase is the *void*, *retention*, or *hold-up* volume. The amount of liquid that must be added to produce a peak of a particular solute in the effluent is the *elution volume*.

The manner in which the *bed* is formed in a column is called *packing*. It is important that the bed be homogeneous and free of bubbles, cracks, or spaces between the walls. The uneven flow resulting from inhomogeneity is called *channeling*. The usual effect of channeling is to distort the elution pattern so that single substances appear in multiple peaks.

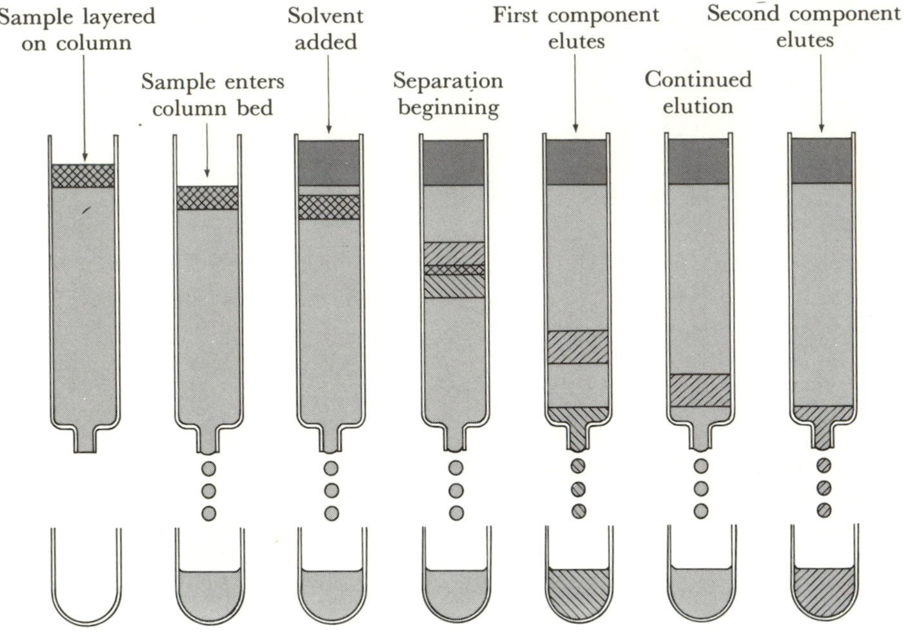

Figure 8-3
Operation of a column showing the loading of the column and various stages of elution.

This is a result of the rapid movement of the mobile phase down through the column, reestablishing partitioning or adsorption at a new point.

The liquid leaving the column (the eluent) is usually collected as discrete fractions, using an automatic collector such as that shown in Figure 8-4. The separated components are then found and identified by testing aliquots of each fraction—for example, spectral measurements, chemical tests, radioactivity, and so forth. In cases in which analysis is by the absorption of light, an automatic, continuously recording spectrophotometer is used. The sample passes through a tube before fractionation and the optical density at an appropriate wavelength is measured continuously and plotted on a chart recorder.

Elution

Columns are eluted in one of three ways. In the simplest method a single solvent flows *continually* through the column. This is a common method in ion-exchange chromatography (see page 248) and in gel chromatography.

Stepwise or batch elution is commonly used if the column is being operated for preparative purposes. The column is eluted with one solvent

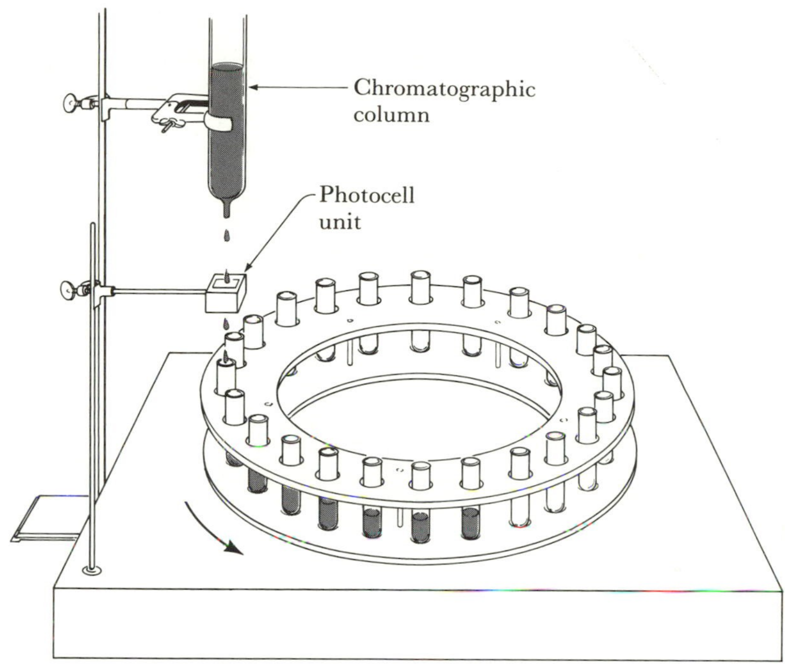

Figure 8-4

A simple fraction collector. A measured number of drops or a measured volume falls into one of the tubes; the tubes then advance until the next tube is centered under the liquid outflow. The drops are detected by a photocell unit containing a small light and a photocell; each drop interrupts the light beam and is counted.

until a predetermined volume has been applied. Then a second solvent is added. This method has the advantage that conditions can be arranged so that a particular material can be eluted in a rather small volume. For example, consider a mixture of substances, one of which (X) is strongly retarded and the others weakly retarded when solvent A is used; X is weakly retarded by solvent B. Hence, if the column is extensively washed with solvent A, most of the material is removed from the column but most of X remains bound. If the column is then washed with solvent B, X is rapidly eluted in a relatively small volume.

The third method is *gradient elution*, which consists either of changing the ratio of two solvents or of increasing the concentration of one or more of the components in the solvent (e.g., the salt concentration). The latter is the most common way of eluting adsorption and ion-exchange columns. Gradients are prepared by means of apparatus of the type shown in Figure 8-5. A gradient maker consists of two vessels, a *reservoir* and a *mixing chamber*, which are connected at their bases. The reservoir contains the more concentrated solvent. Liquid leaves the mixing cham-

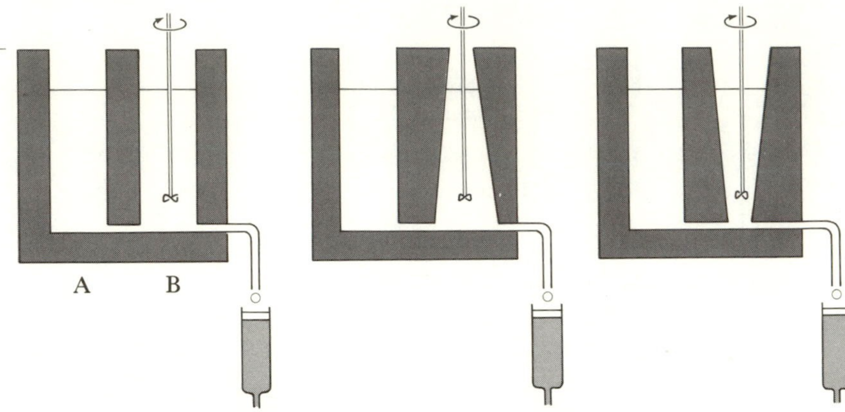

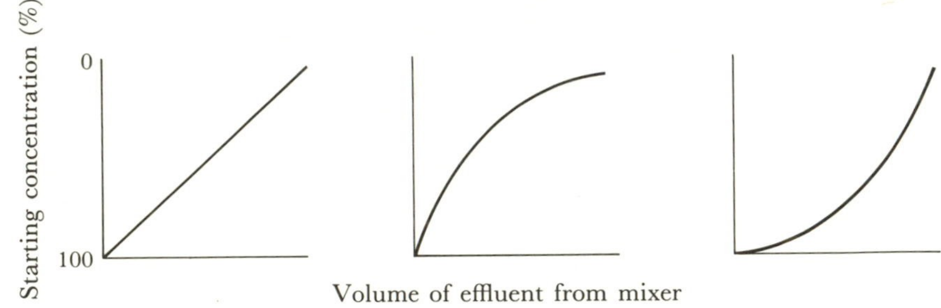

Figure 8-5
Three types of gradient-making apparatus and the resulting gradients. As liquid drains from the mixing chamber, liquid from the reservoir flows in to maintain constant hydrostatic pressure. If the densities of the two liquids are the same, the heights in the two chambers will be identical. If the fluid in B is denser than that in A, the level in B will be lower than that of A. A linear gradient results only if the volumes of the reservoir and mixing chamber are equal.

ber and enters the column. Because the hydrostatic heads (not necessarily the heights, because the densities of the liquid in the two vessels may differ) must be equal, liquid simultaneously flows from the reservoir to the mixing vessel. If the chambers have the same shape, the gradient is linear; concave and convex gradients can also be prepared, as shown in Figure 8-5.*

* In Figure 11-24A an apparatus is described for preparing concentration gradients in centrifuge tubes. In forming these gradients the reservoir contains the solution having the *lower* concentration. This should not be confused with the gradients shown in Figure 8-5.

Supports

The support of a column consists of particles, which come in various sizes and shapes. The form of the particle is an important parameter in the operation of a column because it determines both the flow rate of the mobile phase and the available surface area per unit volume of the support. If the particles are very large, the flow rate will be very high and there may not be enough time for molecules in the mobile phase to come to equilibrium with the stationary phase. This will have the effect of reducing separation. On the other hand, if the flow rate is too low, diffusional spreading can reduce the separation also. In the extreme case, the particles can be so small that no liquid can flow through the column. On the other hand, the smaller the particle the greater is the surface-to-volume ratio; this is equivalent to a greater number of theoretical plates, which improves resolution. Clearly some balance has to be made between these factors. The shape that can be packed most closely and still have a mobile phase is the sphere. Thus, manufacturers have devoted great effort to prepare spherical supports—the term used to describe this shape is *beaded*. The size is described as *mesh size*. This refers to the size of a particle that can just pass through a standard sieve; for example, a 200-mesh screen has 200 openings per running inch (80 per cm) so that the larger the mesh number, the smaller is the particle. The best mesh size must be determined empirically. Fortunately this has been done to some extent by manufacturers who have made the following judgement:

50–100 mesh: High flow rate for preparative applications.

100–200 mesh: Standard material for analytical work.

200–400 mesh: For high resolution analytical work.

>400 mesh: For extremely high resolution in high-performance liquid chromatography (HPLC, see later section).

Figure 8-6 shows a photomicrograph of a beaded support.

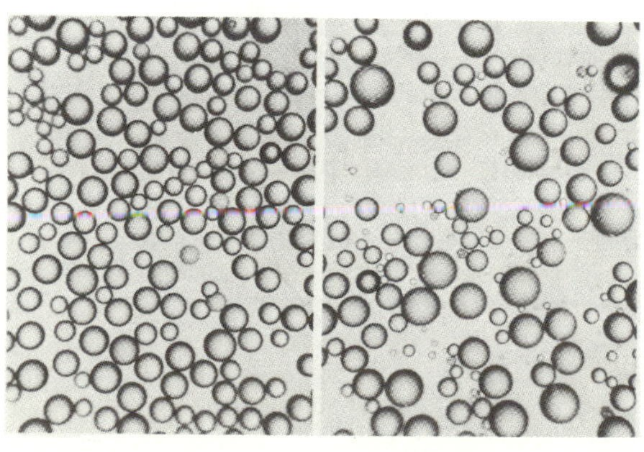

Figure 8-6
Photomicrograph of a 100–200 mesh ion-exchange resin based on a cross-linked polystyrene lattice with ionic functional groups. [Courtesy of Bio-Rad Laboratories.]

Partition chromatography can be performed on columns of cellulose. Paper chromatography is a variant of this procedure in which the cellulose support is in the form of a sheet of paper. Cellulose contains a large amount of bound water even when extensively dried. Partitioning occurs between the bound water and the developing solvent. Frequently, the solvent used is water itself so that it is reasonable to ask whether the mode of action in this case is adsorption. Certainly, some adsorptive effects are present but, because the physical structure of bound water is very different from that of "free" water, partitioning can still occur.

Experimental Procedure for Paper Chromatography

Accompanying the change in the form of the support (from column to paper) is a change in methodology. In paper chromatography there is no effluent and *substances are distinguished by their relative positions in the paper after the solvent has moved a given distance.*

A tiny volume (approximately 10–20 μl) of a solution of the mixture to be separated is placed at a marked spot on a strip or sheet of paper (Figure 8-7) and allowed to dry. This spot defines the *origin.* The paper is then placed in a closed chamber and one end is immersed in a suitable solvent (the mobile phase). Capillarity draws the solvent through the

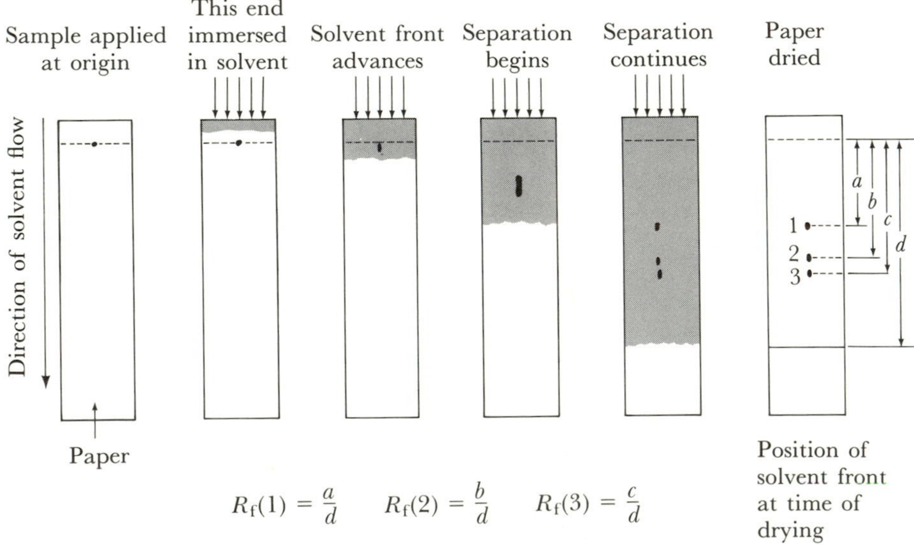

Figure 8-7
Spotting and developing a paper chromatogram with a sample containing three components.

paper, dissolves the sample as it passes the origin, and moves the components in the direction of flow. (Note that, because the sample must be dissolved before it can migrate, a factor in determining the separation of the components is the rate of solubilization into the mobile phase.) After the *solvent front* (see Figure 8-7) has reached a point near the other end of the paper, the sheet is removed and dried. The spots, which may or may not be visible, are then detected and their positions marked. The ratio of the distances traveled by a spot and by the solvent is called the R_f; its value is always less than one. Values of R_f depend on the substance, the paper, and the solvent.

Paper chromatograms can be developed by either ascending or descending solvent flow. The experimental apparatus for each is shown in Figure 8-8. There is little difference in the quality of the chromatograms and the choice is usually a matter of personal preference. Descending chromatography has two advantages: (1) it is faster because gravity aids the flow and (2) for quantitative separations of materials with very small R_f values, which therefore require long runs, the solvent can run off the paper. Its only disadvantage is the care with which the apparatus must be assembled because dirt or poor contact where the paper passes over the support bar can result in inhomogeneous flow and consequent streaking. In a typical experiment, the paper is 10–40 cm long and the development time ranges from about two to twelve hours.

A particularly useful variant is two-dimensional paper chromatography. In this method, after chromatography has been carried out in a single direction, the paper is dried and then rechromatographed at right

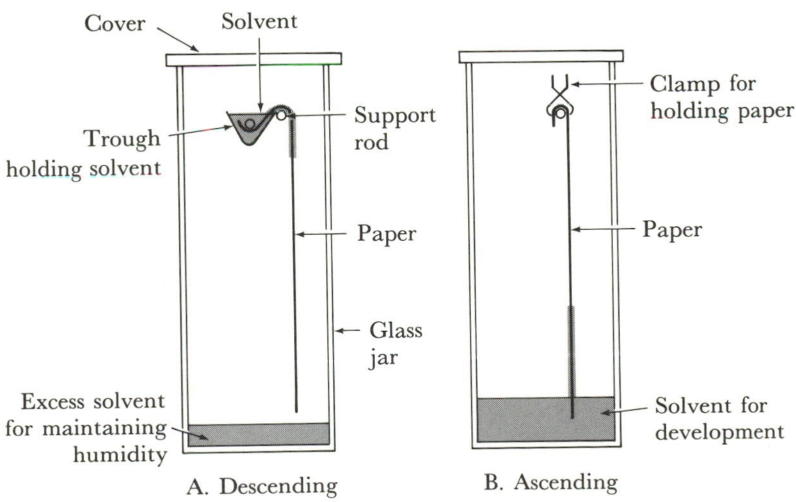

Figure 8-8
Experimental arrangement for descending and ascending paper chromatography.

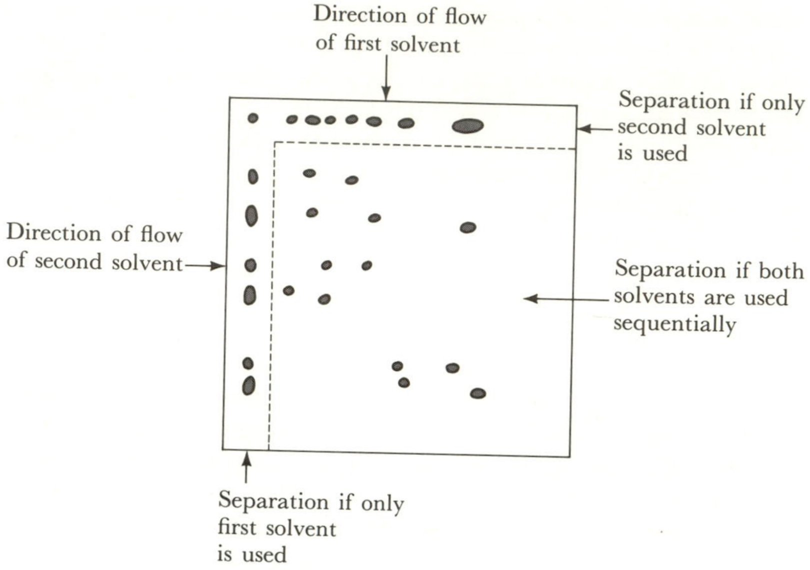

Figure 8-9
Two-dimensional paper chromatography.

angles to the original direction of flow, using a different solvent system (Figure 8-9). In this way, substances that fail to separate in the first solvent can often be separated in the second. If the same solvent system were used in both directions, all spots would lie on the same diagonal and little, if anything, would be gained.

What two-dimensional chromatography actually accomplishes can be seen in the following simple example. Consider a mixture of four substances A, B, C, and D. In solvent 1,A and B separate well but C and D form a single spot; in solvent 2,C and D separate whereas A and B form a single spot. In order to separate the four substances, one might try to find a solvent system in which all four separate. However, if the R_f values in solvents 1 and 2 are known, it is much simpler to do two-dimensional chromatography—use solvent 1 in one direction and solvent 2 in the other direction.

Detection and Identification of Spots

Spots in paper chromatograms can be detected by their color, by their fluorescence, by the chemical reactions that take place after the paper has been sprayed with various reagents, or by radioactivity (Figure 8-10). The autoradiographic detection of spots with X-ray film is described in Chapter 6. Identification is usually based on comparison with standards

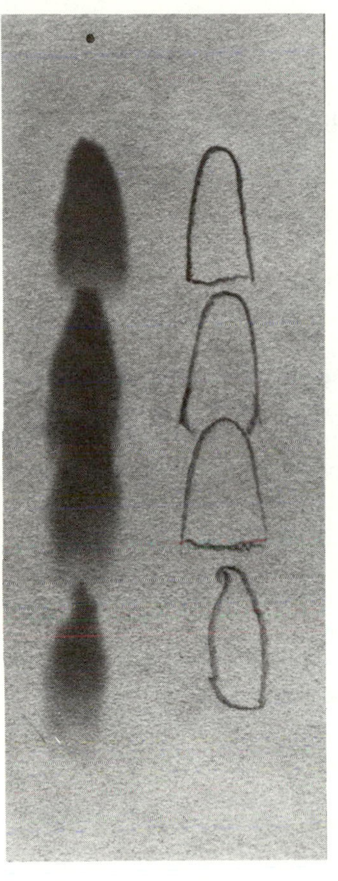

Figure 8-10
A paper chromatogram in which four amino acids have been separated. After the paper had dried, it was sprayed with ninhydrin, which produces a colored product with amino acids. The dot shows the origin. The column on the right shows a drawn outline of each spot.

of known R_f or by elution. Elution is accomplished by cutting out the spot and soaking the paper in the appropriate solvent; this can often be done in a quantitative way.

Thin-Layer Chromatography

Thin-layer chromatography (TLC) originally developed from a need to separate lipids. Although paper chromatography was faster than column chromatography, papers could be prepared only from cellulose products, which were not of great value for nonpolar materials. Thin-layer chromatography has the advantages of paper but allows the use of any substance that can be finely divided and formed into a uniform layer. This includes inorganic substances—such as silica gel, aluminum oxide, diatomaceous earth, and magnesium silicate—and organic substances—such as cellulose, polyamide, and polyethylene powder.

Thin-layer chromatography is used almost exclusively for the separation of small molecules. For macromolecules, thin-layer gel chromatography, described in a later section, is used.

In thin-layer chromatography, the stationary phase is a layer (0.25–0.5 mm) of sorbent spread uniformly over the surface of a glass or plastic plate. The plates are prepared in the following way (although prepared plates of many different sorbents are now commercially available and in common use).

A slurry of the sorbent is made in a solvent specified for that particular sorbent. For very small chromatograms, a microscope slide is coated with the sorbent by dipping it into the slurry. For larger chromatograms, several layers of narrow tape are put along the two edges of a plate of glass that will be vertical during the chromatography. The number of layers of tape determines the thickness of the final sorbent layer. The slurry is poured onto the glass at an untaped edge and is spread evenly by sliding a glass rod, the ends of which rest on the tapes, along the plate. The plate is then dried and the tapes are removed. (More elegant apparatus for coating plates are commercially available.) Henceforth, the plates are treated as in paper chromatography. The sample is then applied with a micropipette and dried. The thin-layer chromatography plate is placed in a chamber containing the solvent and developed by ascending chromatography (Figure 8-11). After the solvent front has almost reached the top, the plate is removed from the chamber and dried. If desired, the dried plate can be rechromatographed at right angles with a second solvent for two-dimensional work. Spots are usually located as in paper chromatography by natural color, by fluorescence, or by spraying various reagents that react with the substances in the spots to produce color. Commonly used sprays are ninhydrin for amino acids; rhodamine B for lipids; antimony chloride for steroids and terpenoids; sulfuric acid plus heating for almost any organic substance (produces charring); potassium permanganate in sulfuric acid for hydrocarbons; anisaldehyde

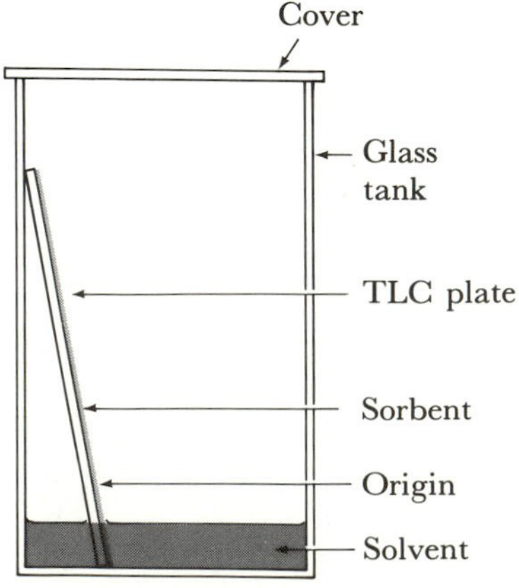

Figure 8-11
Typical arrangement for ascending development of a thin-layer chromatographic plate.

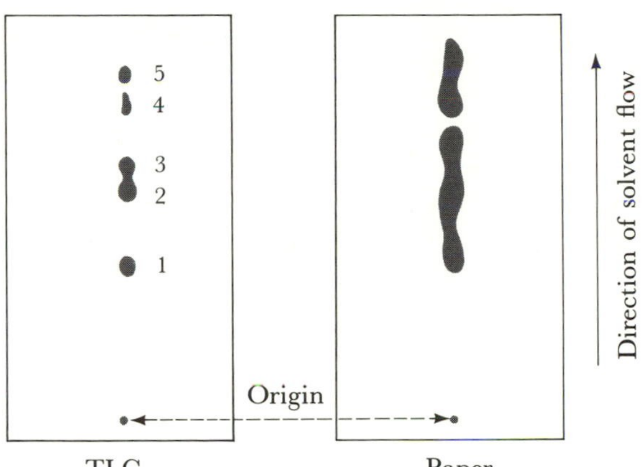

Figure 8-12
Comparison of separation by thin-layer and paper chromatography of adenine (1), adenosine (2), hypoxanthine (3), inosine (4), and uridine (5), both on cellulose and developed by water.

in sulfuric acid for carbohydrates; bromine vapor for olefins; and so forth. Material can be eluted from the chromatogram by scraping off the sorbent and eluting the powder with a suitable solvent. A typical thin-layer chromatogram is shown in Figure 8-12.

Thin-layer chromatography is widely used because, compared with paper or column chromatography, it offers the following advantages: greater resolving power because spots are smaller, greater speed of separation, a wider choice of materials as sorbents, easy detection of spots, and easy isolation of substances from the chromatogram.

Two factors are responsible for the great resolving power of thin-layer chromatography: First, the weight ratio of sorbent to solute is from 10^3 to 10^4:1, whereas in column chromatography the ratio is normally about 50:1. Second, because the particles can be made very fine (<0.1 mm), the surface-to-volume ratio is very high, yielding a large active area for a given amount of sorbent. Particles of such small size cannot be used in columns because the weight of the material causes compacting, clogging, and extremely slow flow rates. This is detrimental because, with very slow flow rate, diffusional spreading increases.

Paper chromatography is limited to cellulose products and those few materials that can be made into paper. Thin-layer chromatography is not limited in this way. However, even for chromatography with materials that can be made into paper, thin-layer chromatography is advantageous because it is faster and has higher resolution. The problem with papers is that the fibrous structure and associated capillarity of the fibers tends to increase spot size. The materials used in thin-layer chromatography can be pulverized to eliminate the fibrous structure. With smaller spots, running time can be decreased because it takes less time for the spots to separate. Often, thin-layer chromatography can be completed in a matter of minutes. This is a particular advantage if the solvent required for optimal separation is not known, because a large number of solvent systems can be tested in one day.

The smaller size of the spots also means that the material in a spot is more concentrated; thus, a smaller amount of material is detectable. In fact, for many substances, thin-layer chromatography is from fifty to one hundred times as sensitive as paper chromatography and samples as small as one nanomole can be detected. For amino acids, the detectable amount of material is about one-tenth that required for paper chromatography, which can be of great value if only small samples of purified proteins are available for amino acid or peptide analysis. For nucleotides, about one-hundredth of the amount required for paper is needed.

Gas-Liquid Chromatography

In the types of partition chromatography described so far, the sample is carried in a liquid mobile phase past a liquid stationary phase, the liquid being immobilized by adsorption to or absorption by a supporting solid. In gas-liquid chromatography (GLC), the mobile phase is a gas; the stationary phase is again a liquid adsorbed either to the inner surface of a tube or column (*open-tubular*, or *capillary operation*) or to a solid support (*packed-column operation*), such as diatomaceous earth, Teflon powder, or fine glass beads. The liquid is usually applied as a solid dissolved in a volatile solvent such as ether. For example, beads are dipped into a solution of polyethylene glycol in ether. When the ether evaporates, each bead is coated with polyethylene glycol. At the temperatures used for gas-liquid chromatography, the polyethylene glycol melts and remains on the bead as a liquid film. The sample, which may be any compound capable of being volatilized without decomposition, is introduced as a liquid with an inert gas—such as helium, argon, or nitrogen—and then heated. This gaseous mixture passes through the tubing, which is arranged as shown in Figure 8-13. For packed-column operation, the tubing is approximately 0.5 cm in diameter and from 1 m to 20 m long. For the capillary method, the length is from 30 m to 100 m. For very high resolution, capillary systems with 2 km of tubing are used. The vaporized compounds continually redistribute themselves between the gaseous mobile phase and the liquid stationary phase, according to their partition coefficients, and are thereby chromatographed. At the end of the column a suitable detector is used.

Detection of Substances in the Effluent

Three types of detectors are commonly used in analytical work: the thermal conductivity cell (sensitivity, approximately 10 μg), the argon ionization detector (sensitivity, 10^{-5} μg), and the flame ionization detector (sensitivity, approximately 10^{-5} μg). The action of the thermal conductivity cell is based on the fact that the electrical resistance of a wire is temperature dependent. If a gas flows at a constant rate past a hot wire, the wire will be cooled to a temperature determined by the flow rate and

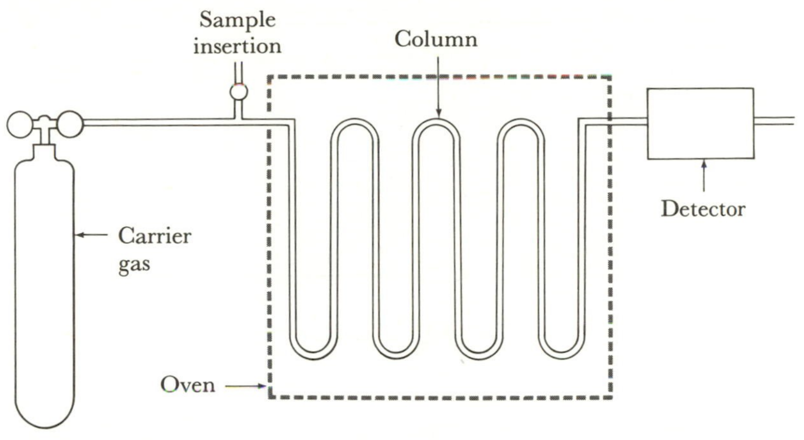

Figure 8-13
Apparatus for gas-liquid chromatography.

thermal conductivity of the gas. Hence, at constant flow rate, the temperature, and therefore the resistance, is a characteristic of a particular gas. If the composition of the gas changes, there will be a change in the thermal conductivity of the gas and therefore of the resistance; this is usually plotted on a chart recorder as a function of time of flow of the gas.

With an argon ionization detector, the carrier gas, argon, passes through a detection chamber (at the exit port of the column) similar to a Geiger-Müller counter and is ionized by being bombarded with β particles. As explained in Chapter 5, when the positive argon ion nears the cathode, it pulls off an electron and is neutralized. X-rays are produced by this recombination and they ionize more argon atoms. This causes a self-generating, continuous ionization so that a constant current is produced by the Geiger-Müller tube. If a compound is present whose ionization potential is less than argon, it can collide with and transfer an electron to an argon atom. The compound is then positively charged. When it approaches the cathode, it picks up an electron and is also neutralized. However, with most organic compounds, the excess energy of recombination does not lead to the production of electromagnetic radiation but breaks chemical bonds. Hence, whenever such a compound is present, the current between the electrodes decreases. This current is measured and plotted as a function of time. A calibration is also necessary, as with the thermal detector. The sensitivity of this type of detector is 0.1 μg.

The flame ionization detector is shown schematically in Figure 8-14. Hydrogen gas is burned with air in the presence of the column effluent. Any carbon in the sample is burned and converted into carbon dioxide. For reasons that are not clear, electrons and negative ions are produced and detected as a current, which is then converted into voltage differences.

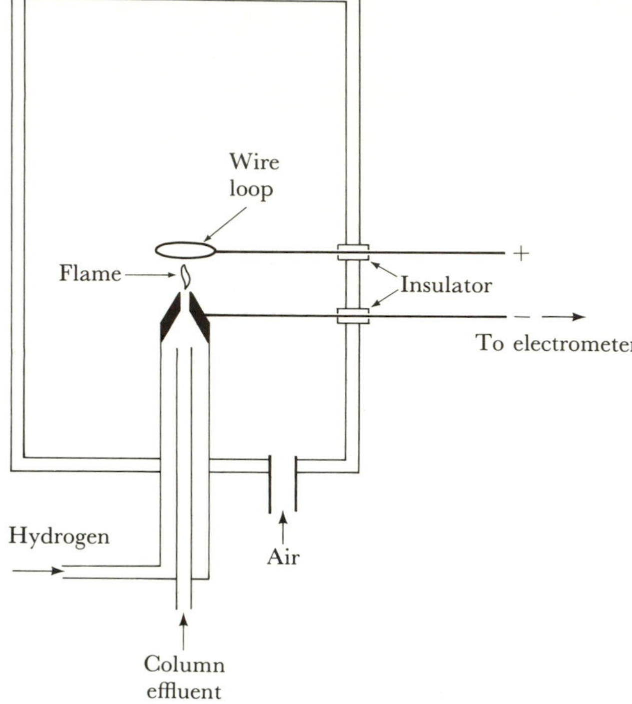

Figure 8-14
A flame ionization detector.

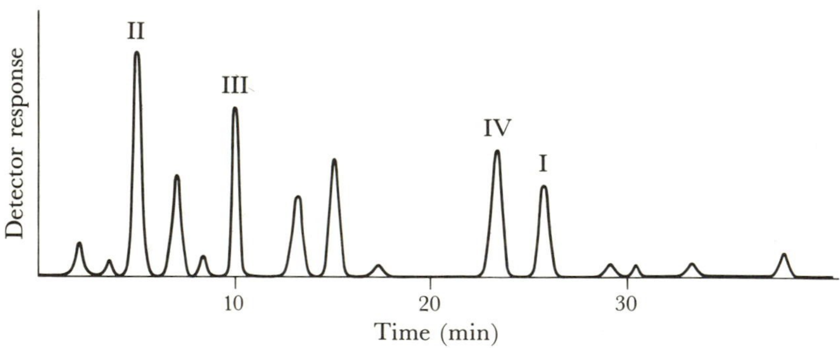

Figure 8-15
A gas chromatogram of methyl esters of fatty acids prepared from a rat liver that had been perfused with 25 mM glucose and 2 mM arachidonate (I). Pentadecanoate (II) was added before homogenizing and extracting the liver to obtain a measure of the recovery of fatty acids from the original sample; methyl heptadecanoate (III) was added to the mixture before injection into the gas chromatographic column to quantitate the amount injected. A flame ionization detector was used. The peaks represent normal components of rat liver; peak IV appears only if animals are fed diets free of essential fatty acids. [Courtesy of John Lowenstein.]

This detector operates linearly from 0.01 μg to 5 mg and essentially counts carbon atoms. A typical gas chromatogram is shown in Figure 8-15.

Identification of Components from the Detector Output

Each of the detectors indicates the amount of material emerging from the column as a function of time. However, there is no direct indication of the identity of the material producing a particular peak nor of the amount in the peak. There are various means of identifying peaks. For example, if the substances present in a mixture are known in advance (and the purpose of the chromatography is to determine the amount of each), peaks may be identified by preparing a duplicate sample containing a small amount of an added known substance to the mixture and rechromatographing this mixture. If the known substance is the same as one of the components in the mixture, the size of that peak will increase (Figure 8-16). If the peaks are part of a homologous series (e.g., ethyl esters of fatty acids), the logarithm of the retention time (i.e., the time between injection of the sample and its appearance in the detector) can be plotted as a function of the number of carbon atoms (Figure 8-17) and usually a straight line will result. The addition of one or two of the standards of the homologous series then provides a calibration for the chromatogram. Of course, this method can be used only if it is known that all substances are part of a homologous series.

The identification of unknown substances is difficult and there is no general method. Sometimes the effluent gas is bubbled through test solu-

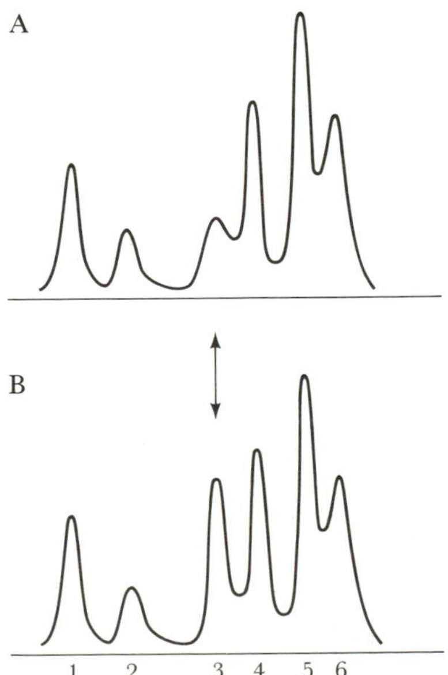

Figure 8-16
Identification of a peak in a gas chromatogram by the addition of a known compound. The mixture contained (1) phenol, (2) *o*-cresol, (3) *p*-cresol, (4) *m*-cresol, (5) 2,4-xylenol, and (6) 2,5-xylenol. The sample was run (A) and then *p*-cresol was added and the sample was rerun (B). Note that peak 3 has increased.

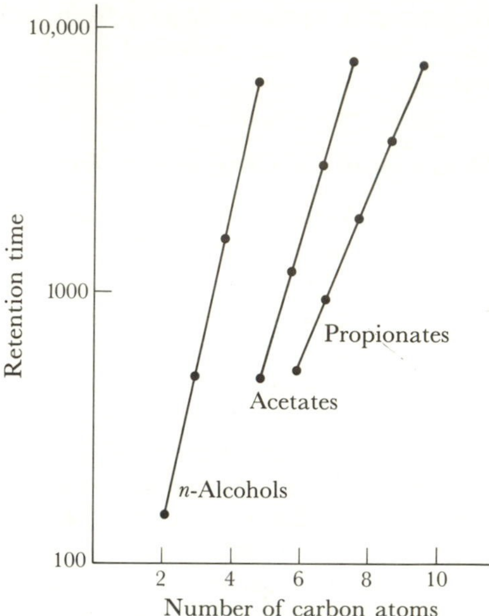

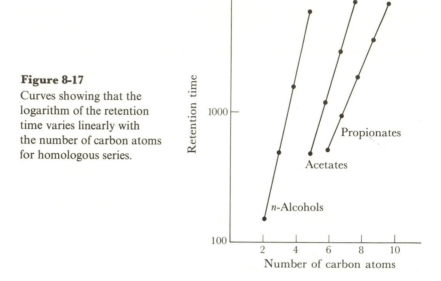

Figure 8-17
Curves showing that the logarithm of the retention time varies linearly with the number of carbon atoms for homologous series.

tions that give particular colors with specific functional groups (e.g., aldehyde, alcohol, etc.). Another method consists of chemical alteration and rechromatographing against various standards or analysis by other physical techniques. This is often done if what the substances are can be surmised. Of course, when substances are to be identified, it is always necessary to use a nondestructive detector (e.g., thermal conductivity cell).

The most useful method of identification is mass spectroscopy. This technique is based on the fact that a charged particle in motion is deflected by a magnetic field. The amount of deflection depends on the momentum (mass × velocity) of the particle; a particle with a larger momentum will be deflected less than one having lower momentum. If a beam consisting of particles having different masses but the same velocity is placed in a magnetic field, the deflection will depend on mass. This method can be used to measure the molecular weight of atoms and molecules simply by ob erving the deflection of each particle; commercial mass spectrometers are available that can accomplish this quite simply. The sensitivity is so great that individual isotopes are easily distinguished. For molecules, the procedure has certain complications because molecules in the gas phase are uncharged. Since only a charged particle is deflected in a magnetic field, it is necessary to introduce a charge, and this is accomplished by irradiating the molecules with an electron beam. However, the irradiation often causes fragmentation of some of the molecules so that a sample initially consisting of a single molecule is converted to a collection of ions having different masses. This is actually no real problem because the spectrometer is designed to measure the relative amounts of each ion—that is, a spectrum of masses is obtained. Mass spectra have been obtained for thousands of organic components so that a substance can often be identified merely by comparing the spectrum obtained with known spectra. This process is aided by the fact that the non-fragmented ion is usually present. In sophisticated gas chromatographic systems, the effluent is passed directly into a mass spectrometer so that identification occurs during development of the chromatogram. In exploratory experiments, spectra may be found that do not correlate with known spectra. Such substances can also be identified using standard but tedious methods of mass spectroscopy. The single disadvantage of the on-line use of the mass spectrometer is its expense.

The quantitative determination of the amount of material in a peak is not a straightforward procedure. For any given compound, the amount of material is proportional to the area of the peak. However, the proportionality constant varies with each substance. Hence, it is first of all essential to be able to identify the material in the peak. Quantitation then requires that a standard curve be prepared; that is, various quantities of the particular substance are chromatographed and a graph is drawn to relate peak area to amount of material.

Advantages of Gas-Liquid Chromatography

The separation of small molecules in gas-liquid chromatography is excellent because of the great length (and hence the large number of theoretical plates) of the column. Sensitivity and speed are extraordinary, with 10^{-12} gram being detectable for many substances. Because the rapidity of development of chromatograms depends on the rate of diffusion between the mobile and stationary phases and because the diffusion rate of gases is much greater than that of liquids, the gas chromatogram can be run approximately one thousand times as fast as that produced by liquid column chromatography. Hence, separations are frequently achieved in less than a minute. Furthermore, by using a nondestructive detector and condensing the samples at the collection end, it is also possible to use gas-liquid chromatography preparatively. Large preparative instruments can purify gram quantities of material.

Uses of Gas-Liquid Chromatography

Gas-liquid chromatography can be used with any substance that can be volatilized. This includes thousands of organic compounds. Nonvolatile substances can also be examined if converted into volatile ones by oxidation, acylation, alkylation, and so forth.

The principal use in biological samples has been the separation of alcohols, esters, fatty acids, and amines. Hence, it is especially valuable in the study of intermediary metabolism and in working out enzyme reaction mechanisms. It has been extensively used to identify the components of flavorings and wines and for detecting pesticides in biological material.

Gas-liquid chromatography plays an important role in organic analysis also. For instance carbon and hydrogen can be determined to 0.5% and 0.1% accuracy, respectively, by burning the sample in a dry stream of oxygen free of CO_2 and H_2O and determining the amount of CO_2 and H_2O. Functional groups can also be identified—for example, alkoxy groups are detected by iodination to form an alkyl iodide, which is easily identified. The position of a double bond can be determined also by cleaving the bond by oxidation or ozonolysis and chromatographing the products.

Gel Chromatography

Gel chromatography (or molecular-sieve chromatography, as it is sometimes called) is a special type of partition chromatography in which separation is based on molecular size.

Simple Theory of Gel Chromatography

The basis of gel chromatography is quite simple. A column is prepared of tiny particles of an inert substance that contains small pores. If a solution containing molecules of various dimensions is passed through the column (Figure 8-18), molecules larger than the pores move only in the

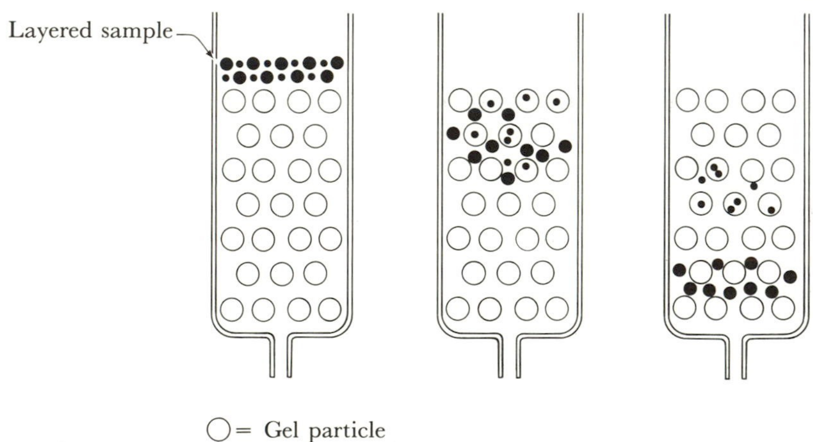

Figure 8-18
Separation of two molecules by passage through a column containing particles of a porous gel. The molecules larger than the pores move more rapidly than the smaller ones because the smaller ones move in and out of the pores.

space between the particles and hence are not retarded by the column material. However, molecules smaller than the pores diffuse in and out of particles with a probability that increases with decreasing molecular size; in this way, they are slowed down in their movement down the column. As long as the material of which the particles are made (i.e., the gel) does not adsorb the molecules, the probability of penetration is the principal factor determining the rate of movement through the column. Hence, molecules are eluted from the column in order of decreasing size or, if the shape is relatively constant (e.g., globular or rodlike), decreasing molecular weight.

In detailed analysis of the mechanism of gel chromatography, it is clear that this steric effect, although the principal factor, does not alone explain the chromatographic behavior of all molecules. Another important factor is the charge of the molecule, although this is only manifested at very low ionic strength when highly charged small molecules seem to be excluded from the pores even though the size is sufficient. This is probably due to electrostatic repulsion between the molecules, thus limiting the number of molecules in a pore at any given time. At very low ionic strength, there are also apparently adsorptive effects with some types of gels.

Materials in Gel Chromatography

A gel is a three-dimensional network whose structure is uaually random. The gels used as molecular sieves consist of cross-linked polymers that are generally inert, do not bind or react with the material being analyzed, and are uncharged. The space within the gel is filled with liquid and this liquid occupies most of the gel volume.

The gels currently in use are of three types: dextran, agarose, and polyacrylamide. They are used for aqueous solutions.

Dextran is a polysaccharide composed of glucose residues and produced by the fermentation of sucrose by the microorganism *Leuconostoc mesenteroides*. It is prepared with various degrees of cross-linking to control pore size and is supplied in the form of dry beads of various degrees of fineness that swell when water is added. Swelling is the process by which the pores become filled with the liquid to be used as eluant. It is commercially available under the trade name Sephadex (Pharmacia Fine Chemicals, Inc.). Sephadex cannot be prepared in a form that can accept molecules whose molecular weight is greater than 6×10^5. This is because the degree of cross-linking would have to be so small that the gel would have no strength and in a column the beads near the bottom would be crushed. This problem is solved by increasing the length of the cross-linking molecule. That is, alkyl dextran is cross-linked with *N,N′*-methylenebisacrylamide, which enables strong beads to be made. The product is called Sephacryl S-300 and it fractionates in the range 10,000 to 1.5×10^6.

Agarose (which is obtained from certain seaweeds) is a linear polymer of D-galactose and 3,6-anhydro-1-galactose and forms a gel that is held together without cross-links by hydrogen bonds. It is dissolved in boiling water and forms a gel when cooled. Agarose is not used in chromatography as a continuous gel but is formed into porous beads having various degrees of fineness. The concentration of the material in the gel determines the size of the pores—which are much larger than those of Sephadex. This makes it useful for the analysis or separation of large globular proteins or long, linear molecules such as DNA. Agarose is supplied as wet beads called Sepharose (Pharmacia Fine Chemicals, Inc.) and Bio-gel A (Bio-Rad Laboratories). A variant of agarose has been made in which the agarose chains are cross-linked to provide greater strength and stability. This substance is called Sepharose CL; it can be used with concentrated denaturing solutions (6 M guanidinium chloride, 8 M urea) and at temperatures up to 70°C.

Polyacrylamide gels are prepared by cross-linking acrylamide with *N*,*N'*-methylenebisacrylamide. Again, the pore size is determined by the degree of cross-linking. These gels differ from dextran and agarose gels in that they contain a polar, carboxylamide group on alternate carbon atoms, but their separation properties are much the same as those of the dextrans. Polyacrylamide gels, which are marketed as Bio-gel P (Bio-Rad Laboratories), seem to be as useful as the dextrans, although they have been used less frequently. They do have an advantage over the dextrans in that they are commercially available in a wider range of pore sizes.

Mixed gels of polyacrylamide and agarose are also available from LKB (Stockholm) and are called Ultragel. This material is claimed to give higher resolution than the pure gels but at present there are not enough examples of its use to substantiate the claim.

The gels heretofore described swell in water and a few organic solvents: glycol, formamide, and dimethylsulfoxide. They fail to swell in pure alcohols, hydrocarbons, and most polar and nonpolar organic solvents. However, substances such as lipids, steroids, and certain vitamins are more easily handled in such solvents. Several gels have been developed for this purpose. For gel chromatography in nonpolar organic solvents, a cross-linked polystyrene gel (Styragel, Dow Chemical Co.; Bio-beads S, Bio-Rad Laboratories) has been used successfully. For polar organic solvents, there is a methylated Sephadex. A hydroxypropyl derivative of cross-linked dextran (Sephadex LH) can be used for both polar and nonpolar solvents and is excellent for separating lipids, fatty acids, and steroids. However, Sephadex LH also adsorbs many molecules because of its hydrophobic groups so that separation is based on a combination of gel, adsorption, and partition chromatography.

The pore size determines the range of molecular weights in which fractionation occurs. Table 8-2 gives representative values for some of the materials. There is usually no difficulty in selecting the appropriate material. However, the gel beads come in various sizes: coarse, medium,

Table 8-2
Materials Commonly Used in Gel Chromatography.

Material and trade name	Fractionation range* (molecular weight)
Dextran	
Sephadex G-10	0–700
Sephadex G-25	1000–5000
Sephadex G-50	1500–30,000
Sephadex G-75	3000–70,000
Sephadex G-100	4000–150,000
Sephadex G-150	5000–300,000
Sephadex G-200	5000–800,000
Polyacrylamide	
Bio-gel P-2	100–1800
Bio-gel P-6	1000–6000
Bio-gel P-60	3000–60,000
Bio-gel P-150	15,000–150,000
Bio-gel P-300	60,000–400,000
Agarose	
Sepharose 2B	2×10^6–25×10^6
Sepharose 4B	3×10^5–3×10^6
Sepharose 6B	10^4–20×10^6
Bio-gel A-0.5 M	10,000–500,000
Bio-gel A-15 M	40,000–15×10^6
Bio-gel A-150 M	1×10^6–150×10^6

* The molecular weights listed are for globular proteins when the molecular weight is less than 10^6 and for viruses and large protein aggregates for higher molecular weights. The upper limits actually depend on the excluded volume rather than on the molecular weight so that for fibrous proteins the upper limits are smaller. For double-stranded DNA, the upper limit is about $\frac{1}{4}$ the value indicated; for single-stranded nucleic acids, the limits are approximately those given.

fine, and superfine. The rule is the coarser the bead, the more rapid the flow rate and the poorer the resolution. This is because as the flow rate increases, the time available for the molecules to equilibrate between the mobile phase and the pore space decreases. Hence, superfine is used if maximum resolution is required—for example, for analytical work—but it is very slow. Fine, which is standard, is recommended for most preparative work in which columns are fairly large and flow rate is of some concern. The coarser grades are for very large preparations in which resolution is less important than time.

The *capacity* of a gel is an important parameter. It is usually expressed as the maximum number of grams of a particular molecule (e.g., a protein or a nucleotide) that can penetrate the pores of a particular weight of gel; the stated weight may be either the wet or the dry weight. This means that any molecules exceeding this stated limit will pass through in the void volumes.

Advantages of Gel Chromatography

For the separation of molecules whose molecular weights differ, gel chromatography is unsurpassed for the following reasons.

1. Because the chromatographic behavior of almost all substances in gels is independent of temperature, pH, ionic strength, and buffer composition, separations can be carried out under virtually all conditions. For very labile materials (e.g., enzymes), this means that the conditions for maximum stability can be maintained.
2. Because there is virtually no adsorption, very labile substances are not affected by the chromatography. For example, some enzymes are inactivated or altered by binding to adsorbent surfaces or ion-exchange resins.
3. There is less zone spreading than with other chromatographic techniques (for reasons that are unclear).
4. The elution volume is related in a simple manner to molecular weight.

Estimation of molecular weight

It has been observed for a variety of gel types that a plot of a parameter, K, versus log M (molecular weight) yields a straight line except for very small and very large molecules (see Figure 8-19). To date, the only deviations are small aromatics, highly charged molecules in very low

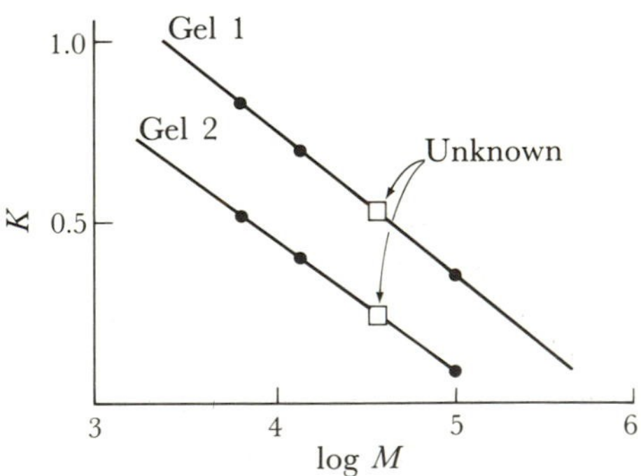

Figure 8-19
Typical data obtained for determining the approximate molecular weight of proteins by gel chromatography, using three known proteins (solid circles) and two different types of gels (1 and 2). The open squares represent the sample (K is defined in the text). Using two different gel types is valuable because usually, if a protein violates the K/log M proportionality (because of its shape), the measured values of M will differ for the two gels.

($\ll$0.01 M) ionic strength, and a few rodlike molecules such as collagen and fibrinogen. The parameter K is defined as $(V_e - V_0)/V_s$, in which V_e is the volume of solvent required to elute the molecule of interest, V_0 is the void volume or the volume required to elute a molecule that never entered the stationary phase, and V_s is the volume of the stationary phase. In gel chromatography, the volume of the gel (which is mostly the volume of its pores) is V_s, and V_0 is the volume for a substance completely excluded by the gel; V_0 is usually measured by passing the excluded high-molecular-weight material, dextran blue, through the column, and V_s is measured as $V_t - V_0$ in which V_t is the total volume of the column.

For globular proteins and for many carbohydrates, this technique is very reliable when various types of Sephadex and Bio-gel P are used. Most important, the estimation of M can be made from unfractionated cell extracts as long as the protein can be assayed; almost every other method requires highly purified samples. Hence, to determine M simply requires the inclusion of several molecules of known M to define the straight line and M of the sample can be calculated by interpolation. The precision of these determinations is about 10%. The error is primarily a result of the breadth of the band during elution. A somewhat more precise variation of this technique will be described in the section on thin-layer gel chromatography.

The determination of a molecular-weight distribution is frequently important in the characterization of natural and synthetic polymers. Again, with physical methods such as centrifugation, relatively pure material is needed. However, with gel chromatography, it is possible to determine the distribution by merely measuring the amount of the substance as a function of elution volume. This is done by preparing a calibration curve and using a relatively simple mathematical correction for zone spreading.

The polyacrylamide gels are quite valuable in the separation and identification of carbohydrates. For example, glucose oligomers having from one to eleven monomers are well-separated. This technique is thus useful in following the stages of polymerization of a simple sugar and has been applied to a study of assembly of the polysaccharides of the cell wall of bacteria.

Application of Gel Chromatography

Gel chromatography is basically a separation procedure and has its most widespread use in purifying enzymes and other proteins and in fractionating nucleic acids. The dextran gels are especially valuable for unstable proteins as explained earlier. In preparative work, both dextrans and polyacrylamides have been used in volumes ranging from a few milliliters to several liters. For industrial purposes, 1000-liter columns have even been employed. Various species of RNA and viruses have been successfully fractionated and purified, using agarose gels.

In large-scale purifications of macromolecules in which various types of fractionation procedures are needed, it is often necessary to remove salts, change buffers, concentrate, and remove substances such as phenol and the detergents used in the isolation and purification of nucleic acids. This is often time-consuming (e.g., by precipitation or dialysis) and for this reason can result in the loss of unstable samples. However, gel chromatography provides a very rapid way to accomplish this. For example, salts and small molecules can be rapidly removed because they are retarded by all of the gels. This process is known as *desalting*. Buffer exchange can be accomplished merely by passing a solution of macromolecules through a column previously equilibrated with the desired buffer. Because the macromolecules move more easily through the gel than the components of the original buffer, they will be eluted in the buffer with which the gel was equilibrated. Concentrating macromolecules is easily accomplished by the addition of dry gel particles, using a type whose pore size is smaller than the molecules being examined. As the beads swell, they imbibe water but not the macromolecules, thus concentrating the macromolecules. With most procedures for concentrating macromolecules, there is a problem in that the salts are also concentrated and excessive salt concentrations can alter (sometimes dissociate or denature) some macromolecules. This is no problem if gels are used because they imbibe the salts as well as the water.

Specialized examples of gel chromatography in preparative work (e.g., purifications) follow.

1. In the chemical synthesis of various reagents, it is usually necessary to separate the product from the reactants. For example, in preparing fluorescent antibodies (see Chapters 2 and 10) by reacting antibody with fluorescein isothiocyanate, the conjugated protein must be separated from unreacted dye. This can be done with Sephadex, using gels that pass large proteins in the void volume. Of course, the unreacted protein is not separated from the conjugated protein but, for the fluorescent antibody technique, this is usually unnecessary.

2. In the assay of enzymes or the determination of cofactor requirements, the enzyme preparation sometimes contains inhibitors of small molecular size or the cofactors themselves. Also, in physical studies of some molecules (e.g. in fluorescence spectroscopy), interfering substances may be present. Such small molecules are easily removed with the dextran or polyacrylamide gels.

3. Similarly, there are frequently contaminants of large molecular size in mixtures being assayed for small molecules. The small-pore dextrans are useful in such cases. Also, proteins must often be freed of nucleic acids; this can sometimes be done by using an agarose gel, which impedes all proteins and passes nucleic acids in the void volume.

4. For most physical analyses of nucleic acids, the sample must be free of protein. This is also easily done. An interesting example from my own laboratory is the preparation of DNA from crude cell extracts for electron microscopic analysis, using the Kleinschmidt cytochrome *c* method (see Chapter 3). Small (20–50 μl) samples of cell extracts containing DNA are passed through tiny Sepharose columns that have been equilibrated with the solution needed for electron microscopy. The DNA comes through the column in the void volume in the appropriate buffer, whereas all other material is retained. Following the addition of cytochrome *c*, the sample is ready to prepare for microscopy.

5. The most common use of gel chromatography is in the purification of proteins (see Appendix, step 7). To purify a protein from a cell extract, it is usually necessary to use a sequence of separation procedures based on such parameters as solubility in certain solutions, charge, molecular weight, and so forth. The step in which size separation takes place almost invariably uses gel chromatography.

Gel chromatography is also a valuable analytical tool. The determination of molecular weight mentioned previously is an important example of this. Other examples are listed below.

1. In studying RNA metabolism, various fractions of RNA are usually distinguished by zone centrifugation or even better by polyacrylamide-gel electrophoresis. Gel chromatography with agarose is also of great use. An example is shown in Figure 8-20.

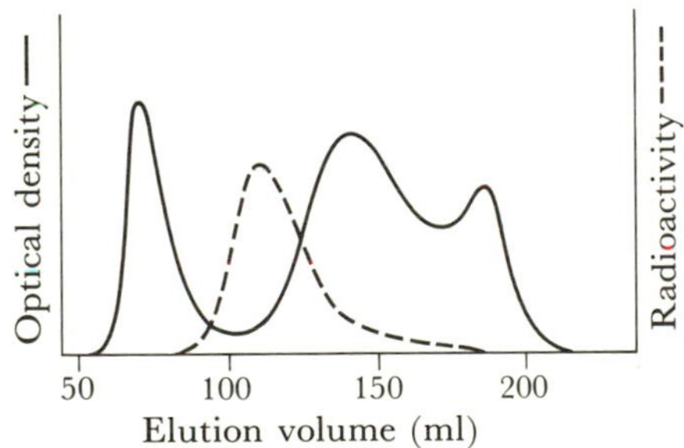

Figure 8-20
Separation of various species of nucleic acids by chromatography on Sepharose. KB cells were infected with ^{32}P-labeled poliovirus. The cells were lysed shortly after infection, and nucleic acids were isolated and chromatographed. The total cellular nucleic acid was detected by measuring the optical density at 260 nm. The polio RNA was revealed by its radioactivity. The peaks in order from left to right are KB cell DNA, polio RNA, KB ribosomal RNA, and KB transfer RNA. [Courtesy of Pharmacia Fine Chemicals, Inc.]

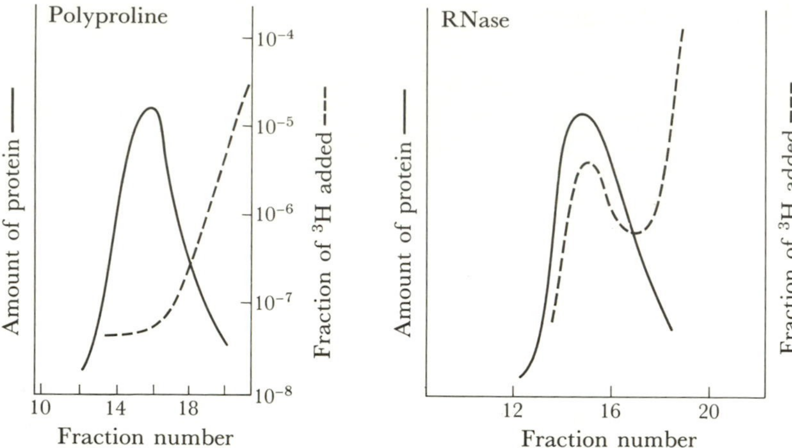

Figure 8-21
One of the earliest ^{3}H-exchange experiments. Polyproline and RNase were incubated in ^3H_2O for several hours to allow maximum exchange. The samples were then diluted in H_2O and at various times applied to a Sephadex G-25 column to separate the protein and ^3H_2O. In the graph at the left, there is no ^{3}H associated with polyproline, from which it can be concluded that either it never became tritiated or exchange is very rapid. In the graph at the right, it is clear that the RNase has ^{3}H bound to it. [Redrawn with permission from W. Englander, *Biochemistry* 2(1965): 798–807. Copyright by the American Chemical Society.]

2. Plasma protein fractions must often be determined quantitatively in the diagnosis of certain human diseases. This can be done directly with dextran gels and has been developed as a reliable test for macro- and hyperglobulinemia.

3. The tritium-exchange (see Chapter 19) method for examining protein or DNA structure requires that the macromolecule be very rapidly separated from ^3H_2O. This can be done in about 10 seconds using charged gels because the ^3H_2O is strongly retarded in all gels. If the smallest pore size is used, the macromolecule comes through rapidly in the void volume or shortly thereafter. An example is shown in Figure 8-21.

4. Gel chromatography can be used to study binding between proteins and small molecules either by separating the product and the reactants or by passing protein through a column equilibrated with the small molecule. A simple calculation allows the determination of binding constants. The great value of gel chromatography in studies of chemical equilibrium is that a gel column can be operated over a wide range of concentrations, pH, ionic strength, and temperature because the pore size of the gel is unaffected by these factors.

Thin-Layer Gel Chromatography

As indicated earlier, thin-layer chromatography has the advantages of quick separation, high sensitivity, simple equipment, and ready elution and is rapidly replacing paper chromatography as an analytical method. Thin-layer gel chromatography (TLG) offers these same advantages plus the opportunities afforded by the gels, especially the ability to use molecules of high molecular weight.

Thin-layer gel chromatography is similar to thin-layer chromatography in that a thin layer of material is spread on a glass plate, the sample is spotted, and a mobile phase traverses the layer. However, there are important differences.

1. In thin-layer chromatography, the support is dried before applying the sample. However, because gels cannot be dried and easily rehydrated, in thin-layer gel chromatography, the sample is spotted onto a wet layer equilibrated with the appropriate solvent.
2. Thin-layer chromatography is done with an ascending mobile phase (see Figure 8-8), whereas in thin-layer gel chromatography the descending method is used. (This is because capillarity moves the eluant in dry material but gravity is needed for the already wet gel.) The plate is put into an airtight chamber to avoid drying of the gel and connected to a reservoir at both ends by filter-paper bridges (Figure 8-22). Liquid flows through the layer at a rate determined by the angle (usually 20°). As in thin-layer chromatography, the run terminates before the material of interest enters the lower reservoir. However, there is a continuous flow of liquid so that, unlike thin-layer chromatography, there is no solvent front. Therefore, there is no measurement of R_f values and positions must be measured with respect to added standards.

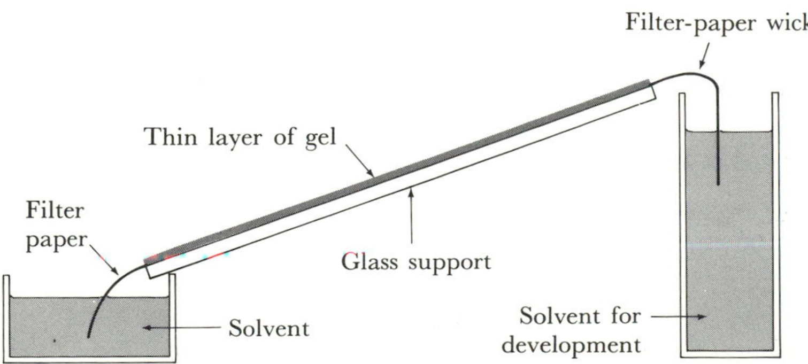

Figure 8-22
Experimental arrangement for thin-layer gel chromatography. The gel and the filter paper are usually enclosed in various ways to prevent evaporation.

Applications of Thin-Layer Gel Chromatography

Whereas thin-layer chromatography is primarily used for amino acids, sugars, oligosaccharides, lipids, steroids, and other small molecules, thin-layer gel chromatography is applied to proteins, peptides, nucleic acids, nucleotides, and other large hydrophilic substances.

The most important use of thin-layer gel chromatography is in determining molecular weights of proteins and peptides. Plots similar to those in Figure 8-19 can be obtained in which the *y*-axis is the *relative distance* migrated. This gives fairly accurate values, especially if several determinations are made.

Thin-layer gel chromatography is being used in clinical diagnosis to detect pathological proteins from blood serum, spinal fluid, and urine. It is, of course, of great value is assaying the purity of proteins, enzymes, and so forth. It has also been developed as a modification of immunodiffusion, immunoelectrophoresis, and isoelectric focusing.

Ion-Exchange Chromatography

An *ion exchanger* is a solid that has chemically bound charged groups to which ions are electrostatically bound; it can exchange these ions for ions in aqueous solution. Ion exchangers can be used in column chromatography to separate molecules according to charge; actually other features of the molecule are usually important so that the chromatographic behavior is sensitive to the charge density, charge distribution, and the size of the molecule.

The principle of ion-exchange chromatography is that charged molecules *adsorb* to ion exchangers reversibly so that molecules can be bound or eluted by changing the ionic environment. Separation on ion exchangers is usually accomplished in two stages: first, the substances to be separated are bound to the exchanger, using conditions that give stable and tight binding; then the column is eluted with buffers of different pH, ionic strength, or composition and the components of the buffer compete with the bound material for the binding sites.

Properties of Ion Exchangers

An ion exchanger is usually a three-dimensional network or matrix that contains covalently linked charged groups. If a group is negatively charged, it will exchange positive ions and is a *cation exchanger*. A typical group used in cation exchangers is the sulfonic group, SO_3^-. If an H^+ is bound to the group, the exchanger is said to be in the acid form; it can, for example, exchange one H^+ for one Na^+ or two H^+ for one Ca^{2+}. The sulfonic acid group is called a *strongly* acidic *cation* exchanger. Other commonly used groups are phenolic hydroxyl and carboxyl, both *weakly* acidic cation exchangers. If the charged group is positive—for example, a quaternary amino group—it is a strongly basic *anion exchanger*. The most common weakly basic anion exchangers are aromatic or aliphatic amino groups.

The matrix can be made of various materials. Commonly used materials are dextran, cellulose, agarose and copolymers of styrene and vinylbenzene in which the divinylbenzene both cross-links the polystyrene strands and contains the charged groups. Table 8-3 gives the composition of many common ion exchangers.

The *total capacity* of an ion exchanger measures its ability to take up exchangeable ions and is usually expressed as milliequivalents of exchangeable groups per milligram of dry weight. This number is supplied by the manufacturer and is important because, if the capacity is exceeded, ions will pass through the column without binding.

Table 8-3
Properties of Various Ion Exchangers.

Matrix	Exchanger*	Functional group	Trade name
Dextran	SC	Sulfopropyl	SP-Sephadex
	WC	Carboxymethyl	CM-Sephadex
	SA	Diethyl-(2-hydroxypropyl)-aminoethyl	QAE-Sephadex
	WA	Diethylaminoethyl	DEAE-Sephadex
Cellulose	C	Carboxymethyl	CM-cellulose
	C	Phospho	P-cel
	A	Diethylaminoethyl	DEAE-cellulose
	A	Polyethyleneimine	PEI-cellulose
	A	Benzoylated-naphthoylated, diethylaminoethyl	DEAE(BND)-cellulose
	A	*p*-Aminobenzyl	PAB-cellulose
Styrene-divinyl-benzene	SC	Sulfonic acid	AG 50
	SA	Tetramethylammonium	AG 1
	SC + SA†	Both of the above	AG 501
Acrylic	WC	Carboxylic	Bio-Rex 70
Phenolic	SC	Sulfonic acid	Bio-Rex 40
Epoxyamine	WA	Tertiary amino	AG-3

* C, cationic; A, anionic; S, strong; W, weak.
† Mixed-bed resin.

The *available capacity* is the capacity under particular experimental conditions (i.e., pH, ionic strength). For example, the extent to which an ion exchanger is charged depends on the pH (the effect of pH is smaller with strong ion exchangers). Another factor is ionic strength because small ions near the charged groups compete with the sample molecule for these groups. This competition is quite effective if the sample is a macromolecule because the higher diffusion coefficient of the small ion means a greater number of encounters. Clearly, as buffer concentration increases, competition becomes keener.

The *porosity* of the matrix is an important feature because the charged groups are both inside and outside the matrix and because the matrix also acts as a molecular sieve. Large molecules may be unable to penetrate the pores; so the capacity will decrease with increasing molecular dimensions. The porosity of the polystyrene-based resins is determined by the amount of cross-linking by the divinylbenzene (porosity decreases with increasing amounts of divinylbenzene). With the Dowex and AG series, the percentage of divinylbenzene is indicated by a number after an X—hence, Dowex-50-X8 is 8% divinylbenzene.

Ion exchangers come in a variety of particle sizes, called *mesh size.* Finer mesh means an increased surface-to-volume ratio and therefore increased capacity and decreased time for exchange to occur for a given volume of the exchanger. On the other hand, fine mesh means a slow flow rate, which can increase diffusional spreading. (However, see the section on high-performance liquid chromatography.)

Such a collection of exchangers having such different properties—charge, capacity, porosity, mesh—makes the selection of the appropriate one for accomplishing a particular separation difficult. How to decide on the type of column material and the conditions for binding and elution is described in the following sections.

Choice of Ion Exchanger

The first choice to be made is whether the exchanger is to be anionic or cationic. If the materials to be bound to the column have a single charge (i.e., either plus or minus), the choice is clear. However, many substances (e.g., proteins), carry both negative and positive charges and the net charge depends on the pH. In such cases, the primary factor is the stability of the substance at various pH values. Most proteins have a pH range of stability (i.e., in which they do not denature) in which they are either positively or negatively charged. Hence, if a protein is stable at pH values above the isoelectric point, an anion exchanger should be used; if stable at values below the isoelectric point, a cation exchanger is required. These considerations are shown graphically in Figure 8-23.

The choice between strong and weak exchangers is also based on the effect of pH on charge and stability. For example, if a weakly ionized substance that requires very low or high pH for ionization is chromatographed, a strong ion exchanger is called for because it functions over the entire pH range. However, if the substance is labile, weak ion exchangers

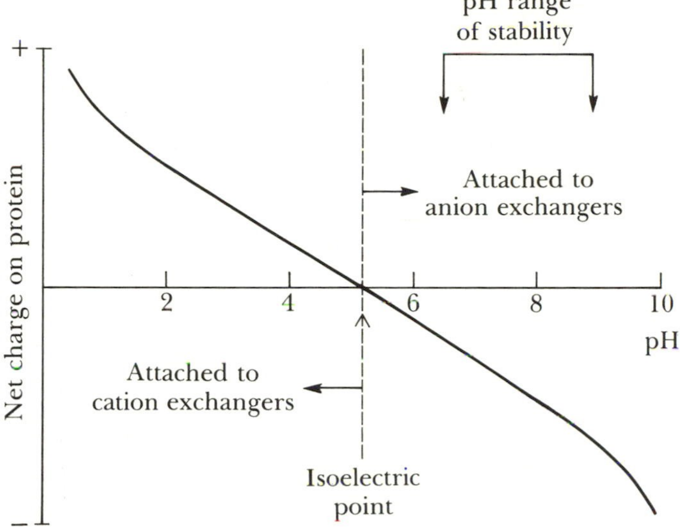

Figure 8-23
The net charge of a protein as a function of pH, showing the pH ranges for stability and for binding to cation and to anion exchangers. This protein would be chromatographed on an anion exchanger using a buffer adjusted to pH 6.3–8.4.

are preferable because strong exchangers are often capable of distorting a molecule so much that the molecule denatures. The pH at which the substance is stable must, of course, be matched to the narrow range of pH in which a particular weak exchanger is charged. Weak ion exchangers are also excellent for the separation of molecules with a high charge from those with a small charge, because the weakly charged ions usually fail to bind. Weak exchangers also show greater resolution of substances if charge differences are very small. If a macromolecule has a very strong charge, it may be also impossible to elute from a strong exchanger and a weak exchanger again may be preferable. In general, weak exchangers are more useful than strong exchangers.

The Sephadex and Bio-gel exchangers offer a particular advantage for macromolecules that are unstable in low ionic strength. Because the cross-links in these materials maintain the insolubility of the matrix even if the matrix is highly polar, the density of ionizable groups can be made several times greater than is possible with cellulose ion exchangers. The increased charge density means increased affinity so that adsorption can be carried out at higher ionic strengths. On the other hand, these exchangers retain some of their molecular sieving properties so that sometimes molecular weight differences annul the distribution caused by the charge differences; the molecular sieving effect may of course enhance the separation also.

Choice of Porosity and Mesh Size

Small molecules are best separated on matrices with small pore size (high degree of cross-linking) because the available capacity is large, whereas macromolecules need large pore size. However, except for the Sephadex type, most ion exchangers do not afford the opportunity for matching the porosity with the molecular weight.

The cellulose ion exchangers have proved to be the best for purifying large molecules such as proteins and polynucleotides. This is because the matrix is fibrous, and hence all functional groups are on the surface and available to even the largest molecules. In many cases however, beaded forms such as DEAE-Sephacel and DEAE-Biogel P are more useful because there is a better flow rate and the molecular sieving effect aids in separation.

Selecting a mesh size is always difficult. Small mesh size improves resolution but decreases flow rate, which increases zone spreading and decreases resolution. Hence, the appropriate mesh size is usually determined empirically.

Choice of pH, Buffer, and Ionic Conditions

Because buffers themselves consist of ions, they can also exchange, and the pH equilibrium can be affected. To avoid these problems, the *rule of buffers* is adopted: use *cationic buffers with anion exchangers* and *anionic buffers with cation exchangers.* Because ionic strength is a factor in binding, a buffer should be chosen that has a high buffering capacity so that its ionic strength need not be too high. Furthermore, for best resolution, it has been generally found that the ionic conditions used to apply the sample to the column (the so-called *starting conditions*) should be near those used for eluting the column.

Techniques of Ion-Exchange Chromatography

The basic principle of ion-exchange chromatography is that the affinity of a substance for the exchanger depends on both the electrical properties of the material and the relative affinity of other charged substances in the solvent. Hence, bound material can be eluted by changing the pH, thus altering the charge of the material, or by adding competing materials, of which salts are but one example. Because different substances have different electrical properties, the conditions for release vary with each bound molecular species. In general, to get good separation, the methods of choice are either continuous ionic strength gradient elution or stepwise elution. (A gradient of pH alone is not used because it is difficult to set up a pH gradient without simultaneously increasing ionic strength.) For an anion exchanger, either pH and ionic strength are gradually increased or ionic strength alone is increased. For a cation exchanger, both pH and ionic strength are increased. The actual choice

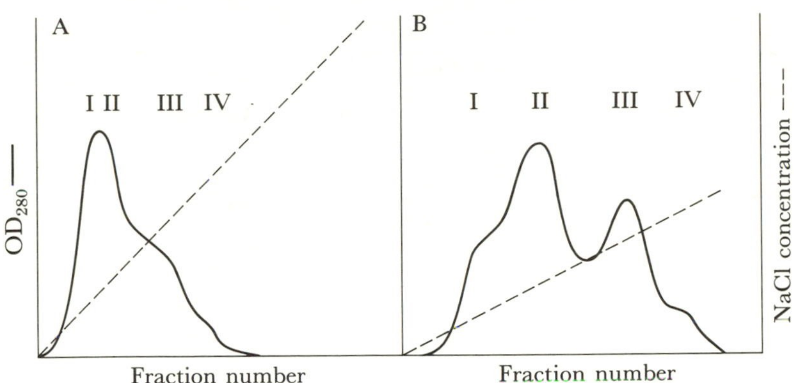

Figure 8-24
Separation of four proteins by ion-exchange chromatography on DEAE-cellulose. Note the strong effect of the steepness of the NaCl gradient. In part A, the steeper gradient allows only partial separation of two fractions. With the shallower gradient of part B, proteins II and III clearly separate and the existence of I and IV becomes obvious.

of the elution procedure is usually a result of trial and error and of considerations of stability. For example, for unstable materials, it is best to maintain fairly constant pH. Figure 8-24 shows a typical ion-exchange chromatogram, using gradient elution.

For resolution of very similar materials, elution can be carried out without varying pH or ionic strength. For example, suppose that the ionic conditions are selected so that the binding of a molecule with many binding sites is so weak that there is a finite probability that simultaneous dissociation of all bonds will occur in a given time. Hence, the molecule is in equilibrium with the exchanger and will move down the column at a finite rate. The migration rate will then depend precisely on this probability so that molecules having slightly different affinities for the exchanger will migrate at different rates and will therefore separate. (This is, of course, formally equivalent to adsorption chromatography.) A simple variation of this method is called the *starting-condition procedure* in which the sample is adsorbed and eluted in the same solution. In general, resolution is improved if the flow rate is very slow; however, with decreasing flow rate, diffusional spreading becomes more of a problem and resolution can be reduced. The proper flow rate is usually determined empirically. Figure 8-25 shows a typical chromatogram developed by the starting-condition procedure.

In deciding on the elution conditions, it is important to consider how the eluting fluid will affect the assay for the material. For example, if spectral analysis is to be used, the buffer should not absorb in the required wavelength range. If the sample is to be assayed by its radioactivity using

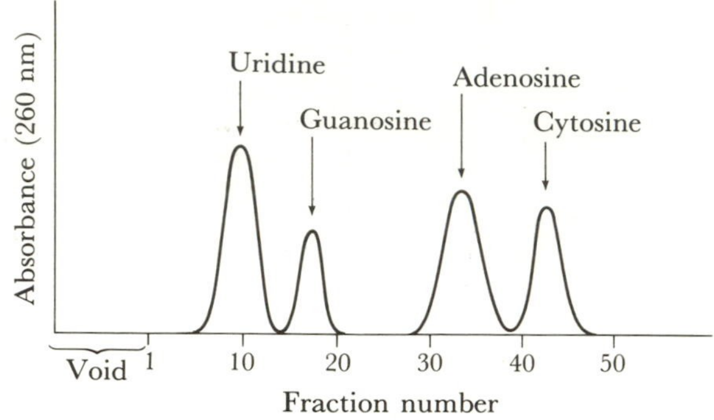

Figure 8-25
Separation of RNA nucleosides by chromatography on Dowex 50W4X. The sample, which is an RNA hydrolysate treated with alkaline phophatase to remove terminal phosphate, was applied and eluted with 0.4 M ammonium formate, pH 4. The absorbance can be used to calculate the amount of each substance because the relation between absorbance at 260 nm and concentration is known.

Geiger counting, volatile buffers help to prevent encasing the radioactivity in crystals resulting from drying down the buffer. If the radioactivity is to be detected by scintillation counting, it is important that the eluting liquid does not contain quenchers.

Applications of Ion-Exchange Chromatography

In principle, any substance that is charged can be chromatographed on an ion exchanger. Resin exchangers are most useful for small organic molecules and can even be used to separate metallic ions (e.g., Ca^{2+} from Mg^{2+}). Proteins and polysaccharides are best used with the cellulose, dextran, and polyacrylamide exchangers. The dextran and polyacrylamide exchangers have also been widely used for the separation of nucleotides, amino acids, and other biologically important small molecules.

One important use of ion-exchange chromatography is in desalting. Gel chromatography is an effective means of removing ions from solutions of macromolecules; a matrix is used that can be penetrated by the ions but not by the macromolecule, which passes through the column in the void volume. This method might fail if the molecule of interest is also small although it is likely that some separation would occur. An ion-exchanger called an *ion-retardation resin* (for example, AG11-A8, Bio-Rad) solves the problem. These resins contain both positively charged (quaternary ammonium) and negatively charged (carboxyl) groups and therefore can remove both anions and cations. An uncharged molecule passes through the column bed without retardation. These resins can also be used to desalt weak organic acids and bases (for example, amino acids) from inorganic ions. Usually, organic ions have a lower charge density than inorganic ions and hence bind less tightly to the column. Thus, an organic ion is retarded less than an inorganic ion.

A related resin called a *mixed-bed resin* is used to prepare deionized water. These resins contain equivalent amounts of groups in the H^+ and OH^- forms. As ions are taken up, H^+ and OH^- ions, which are just the components of water, are released. This is sometimes used as an alternative to distillation.

Chelex (Bio-Rad) is an ion-exchanger that is particularly valuable. It consists of cross-linked styrene-divinylbenzene with linked iminodiacetate ions. This ion is a powerful chelating agent and can remove divalent cations from a solution containing an excess of univalent cations. Consider a protein solution containing Ca^{2+} ions, which are to be removed and replaced by Mg^{2+} ions. Without an exchanger the procedure would be to dialyze the solution exhaustively against a solution containing Mg^{2+} ions, which might take up to two days. Alternatively a chelating agent could be added, removed by dialysis and then when the concentration of the chelating agent is sufficiently low, Mg^{2+} ions could be added. This again would take a day or so. This can be contrasted with passage through Chelex followed by addition of Mg^{2+} ions, a procedure that could be completed in much less than an hour.

Ion-exchangers have been found to be an effective way to separate weakly polar substances by using the exchanger as a matrix for partition chromatography. For example, a mixture of sugars in a mixed solvent containing a polar solvent and a non-polar or weakly polar solvent is applied to the column. The polar solvent binds to the matrix forming a polar stationary phase and the sugars partition between this phase and the mobile weakly polar or non-polar phase. This is called *reverse-phase chromatography*.

By use of extremely fine particles, a high-resolution system has been developed for performing ion-exchange chromatography. This is described in the following section.

High-Performance Liquid Chromatography

Ideally the resolving power of a column increases with the length of the column and the number of theoretical plates per unit length. The number of plates increases as the available surface area per unit length of the column becomes greater—in other words, as the matrix particles become smaller. However, the flow rate of a column drops as the particle size decreases and this allows time for significant diffusional spreading to occur. Accompanying this spreading is, of course, a loss of resolution. Diffusional spreading can be reduced if the transit time of the mobile phase in the column is made quite small; this can be accomplished by establishing a pressure difference across the top and bottom of the column to force the liquid through the bed. The use of very fine particles, approximately 10 μm in diameter and high pressure to maintain an adequate flow rate is called *high-performance* or *high-pressure liquid chromatography* or simply HPLC.

High-performance liquid chromatography is characterized by very rapid separation with extraordinary resolution of peaks. For instance, a separation may often be accomplished in a matter of minutes or at most an hour. Furthermore only a very small volume of the sample is needed because the particles are so small and close-packed that the void volume is a very small fraction of the bed volume. Also, the concentration of the sample need not be very great because the bands are so narrow that there is very little dilution of the sample.

With a typical column having a diameter of 3–6 mm and a length of 10–20 cm operating at pressures of 50–200 pounds per square inch (5–20 kg/cm^2), a sample of 0.01–0.1 ml can be used.

At the present time, this procedure has been used principally with ion-exchange and adsorption chromatography of small molecules, peptides, small carbohydrates, and tRNA. The restriction to small molecules is a result of the need to have a high degree of cross-linking of the matrix. Without this cross-linking the particles lack mechanical strength and are crushed by the pressure used.

Figure 8-26 shows a typical difference in the separation of four com-

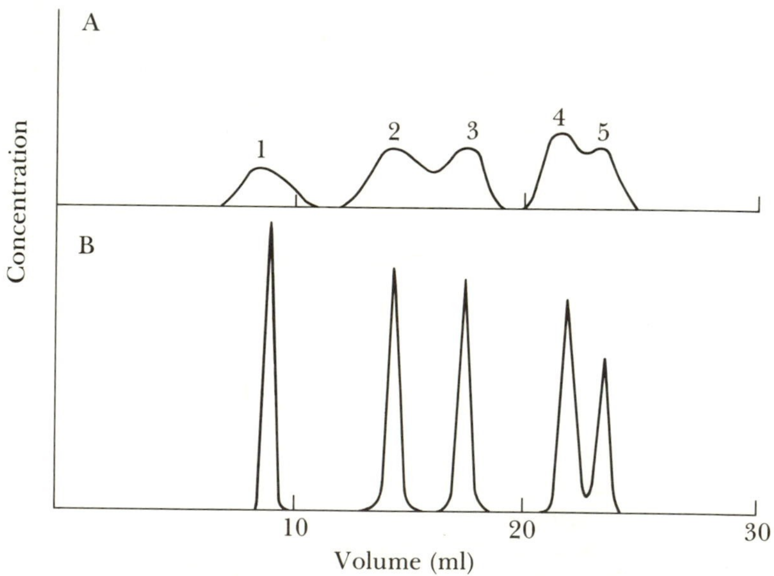

Figure 8-26
Comparison of normal ion-exchange chromatography (upper panel) and high-performance liquid chromatography (lower panel) in the separation of several components of the citric acid cycle; (1) α-ketoglutaric acid; (2) citric acid; (3) malic acid; (4) fumaric acid; and (5) succinic acid. The times to obtain the separations shown are 180 and 20 minutes for the upper and lower panels, respectively.

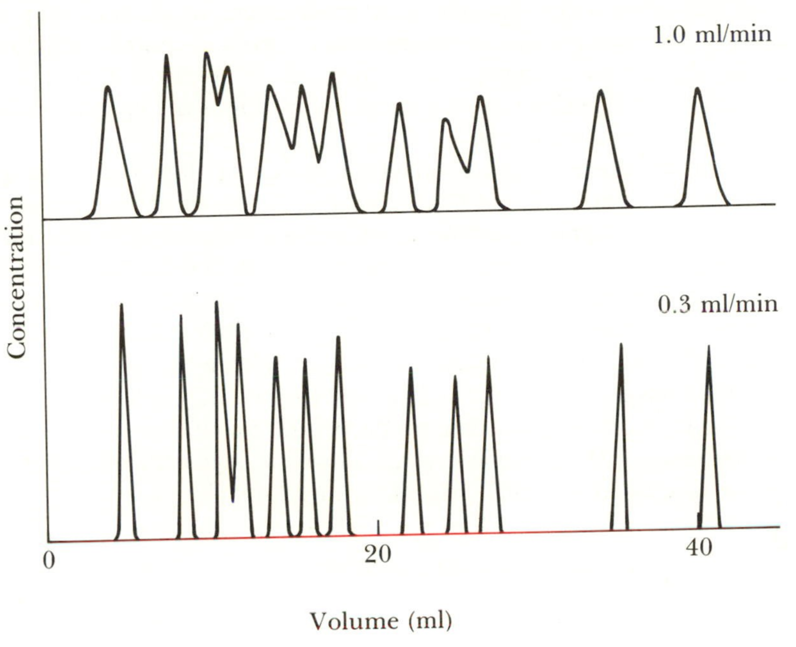

Figure 8-27
Effect of flow rate on the separation of the twelve ribonucleoside mono-, di-, and triphosphates using the same high-performance liquid chromatography resin. Resolution is diminished at the higher flow rate.

ponents by high-performance liquid chromatography and ordinary chromatography with larger particles.

The pressure used affects the resolution of the bands. As we have already mentioned, at atmospheric pressure diffusional spreading causes the bands to overlap. However, if the pressure (and hence the flow rate) is too high, there may be insufficient time for molecules in the mobile phase to equilibrate and the bands also broaden. Thus, if an experimenter is too eager to complete a separation, resolution can be lost. An example of this is shown in Figure 8-27.

Affinity Chromatography

Affinity chromatography is a type of chromatography that makes use of a specific affinity between a substance to be isolated and a molecule that it can specifically bind (a *ligand*). The column material is synthesized by covalently coupling a binding molecule (which may be a macromolecule or a small molecule) to an insoluble matrix. The column material is then specifically able to adsorb from the solution the substance to be isolated. Elution is accomplished by changing the conditions to those in which binding does not occur.

Affinity chromatography has to a great extent revolutionized preparative chromatography in biochemistry. Often a complex mixture can be applied to a column containing an immobilized ligand and purification of the substance of interest, which binds the ligand, can be accomplished in a single step.

Several requirements must be met for success in affinity chromatography: (1) the matrix should be a substance that does not itself adsorb molecules to any significant extent and that has a broad range of physical, chemical, and thermal stability; (2) the ligand must be coupled without altering its binding properties; (3) a ligand should be chosen whose binding is relatively tight; (4) it should be possible to elute the substance of interest without destroying the sample. The most useful matrix materials are agarose and polyacrylamide because they exhibit minimal adsorption, maintain good flow properties after coupling, and tolerate the extremes of pH and ionic strength as well as high concentrations of denaturants, which are often needed for successful elution.

Methods of Ligand Immobilization

Since 1970, a large number of methods have been developed for coupling ligands to matrix materials. The most common procedure is to link a coupling agent to the matrix material and then add the ligand. Many of these matrix-coupler units are commercially available; these represent the coupling systems in most common use and will be the only ones described. Details for their use are provided in brochures supplied by the manufacturers. (Others may be found in some of the Selected References at the end of the chapter.) These systems are the following.

1. *Cyanogen-bromide-activated agarose.* Cyanogen bromide (CNBr) reacts strongly with the amino group and is extremely useful in coupling enzymes, coenzymes, inhibitors, antigens, antibodies, nucleic acids, and most proteins to agarose. It is perhaps the most general coupler available and is usually tried first. It is marketed principally under the trade name, CNBr-activated Sepharose (Pharmacia Fine Chemicals).

2. *6-aminohexanoic-acid and* 1,6-*diaminohexane–agarose.* In the case of some small ligands, steric interference occurs because the ligand is too near the matrix surface. When this occurs, the problem is usually eliminated by inserting a six-carbon atom spacer between the matrix and the ligand. This can be done using agarose to which 6-aminohexanoic acid (CH-Sepharose) or 1,6-diaminohexane (AH-Sepharose) are coupled, the former having a free carboxyl group and the latter a free amino group. Ligands having a free amino or carboxyl group can be coupled to CH-Sepharose and AH-Sepharose respectively using carbodiimide. A variant of this material is called activated CH-Sepharose. The carboxyl group is esterified by *N*-hydroxysuccinimide. If any substance containing a primary amino group

(for example, a protein) is added to this material, a reaction in which a stable amide bond is formed occurs spontaneously.

3. *Epoxy-activated agarose.* Agarose to which is coupled 1,4-bis-(2,3-epoxypropoxy)butane, which contains free oxirane groups, allows linkage of sugars, carbohydrates, or any ligand containing a hydroxyl, amino, or thiol group. No additional reagents are required.

4. *Thiopropyl-agarose.* This material has 2-thiopyridyl groups linked to agarose by a spacer of a 2-hydroxypropyl residue. The 2-thiopyridyl group can react with terminal thiol groups, such as in a low-molecular-weight thiol compound or a sulfur-containing protein. It is of special value in that valuable proteins and enzymes can be decoupled by simple chemical reaction (for example, exposure to cysteine.)

5. *Carbonyldiimidazole-activated agarose.* Coupling of *N*-nucleophiles with cyanogen bromide (CNBr) occasionally produces material in which the chromatographic specificity is not as great as expected. This is because when an *N*-nucleophile reacts with the cyanate ester of CNBr, the result is an isourea linkage that carries a potential charge and thus can act as an ion exchanger; this is shown in Figure 8-28A. This problem is avoided by use of carbonyldiimidazole, in which case the coupling is an uncharged *N*-alkylcarbamate, also shown in Figure 8-28B. This recently developed material is marked as Reacti-gel (6X) (Pierce Chemical Co.) and references to a more detailed description of its properties is given at the end of this chapter.

6. *Aminoethyl- and hydrazide-activated polyacrylamide.* Polyacrylamide is not as generally useful as agarose because it fails to work properly for molecules or structures having a high molecular weight. On the other hand, usually more ligands can be coupled for a particular volume of matrix so that the capacity of a polyacrylamide support column is greater than that of agarose. Bio-Rad Laboratories markets a variety of polyacrylamide supports called Bio-Gel. The two most commonly used materials contain the linkers —CONH—CH_2—CH_2—NH_2 and —CONH—NH_2, to which can be coupled carboxyl

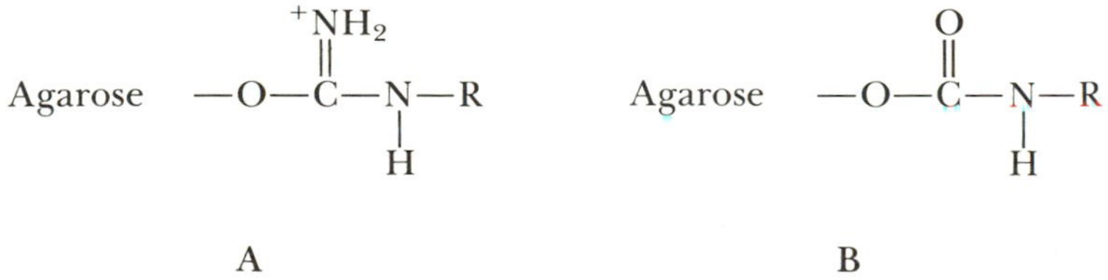

$$\text{Agarose}\quad \text{—O—}\overset{\overset{+}{N}H_2\;\|}{C}\text{—}\underset{\underset{H}{|}}{N}\text{—R} \qquad\qquad \text{Agarose}\quad \text{—O—}\overset{O\;\|}{C}\text{—}\underset{\underset{H}{|}}{N}\text{—R}$$

A B

Figure 8-28
Comparison of two types of coupling agents. A carries an isourea linkage, which is charged. B carries an uncharged substance, carbonyldiimidazole.

and amino ligands, respectively, by standard techniques. Many other ligands can be coupled using more complex procedures; these are described in detail in the book by W. B. Jakoby listed in the Selected References.

Uses of Affinity Chromatography

The ligands that are used in affinity chromatography are either group-specific or substance-specific. By group-specific is meant a ligand that binds a class of substances such as thiol groups, RNA molecules, and so forth. By substance-specific is meant a ligand that binds a unique substance—for example, a particular DNA molecule as a ligand for binding a particular repressor protein molecule.

There are a few general principles that have evolved for isolation of certain kinds of substances. For example, in the purification of enzymes either the substrate, a tight-binding inhibitor, or a cofactor can be coupled to the matrix. If a mixture of proteins or even a crude cell extract is passed through the column, only materials that bind remain on the column. In some cases, enzymes can be substantially purified directly from very complex mistures. However, because there are often many enzymes that can bind to the substrate or cofactor, only partial purification usually results.

In the purification of antibodies, the ligands used are various antigens such as proteins, viruses, or bovine serum albumin coupled with haptens; in fact, this is the method of choice for antibody purification. Transport proteins can be purified by using as a ligand the substance that is being transported. For example, if thyroxin, estradiol, or various hormones and drugs are used as ligands, proteins that transport these substances and various binding proteins can be purified.

A large number of matrix-ligand systems have become standard materials for separations of certain types. Many are commercially available. Some of these materials are the following:

1. *Lectin-Sepharose, using various lectins.* Lectins are a class of substances (usually isolated from plant seeds but occasionally from other sources) that bind to a variety of polysaccharides and glycoproteins found in the outer membrane of many blood cells and other animal cells. These are usually coupled by cyanogen bromide to agarose. ConA-Sepharose, which contains the lectin concanavalin A, was the first material of this sort used and has been widely used in the isolation of polysaccharides and glycoproteins; other lectins that have been used are lentil lectin (which is useful in purifying histocompatibility antigens), wheat germ agglutininin (which selectively binds *N*-acetylglucosaminyl residues and will separate different classes of intact lymphocytes), and *Helix pomatia* lectin (which is a hemaglutinin obtained from a snail and can fractionate human peripheral lym-

phocytes to yield pure T-cells). In recent years, these lectins have aroused great interest because of their ability to separate virus-induced tumor cells from normal animal cells. Lectins themselves are usually purified by affinity chromatography using carbohydrate ligands. The following table shows several such ligands and the lectin that they easily purify, the carbohydrate being coupled to agarose.

Carbohydrate	Lectin
Lactose	Castor bean, peanut
Maltose	Lentil, jack bean
N-Acetyl-D-galactosamine	Soybean
N-Acetyl-D-glucosamine	Wheat germ
D-Galactosamine	Clam
L-Fucose	Lotus

2. *5′-adenylic-acid(AMP)–and 2′,5′-adenosine-diphosphate(ADP)–Sepharose (agarose).* NAD^+ and ATP are cofactors for a great many dehydrogenases and kinases; a large number of these will bind to 5′-AMP–Sepharose. Enzymes that use $NADP^+$ as a cofactor show affinity for 2′,5′-ADP.

3. *Protein-A–agarose.* Protein A is derived from the cell wall of the bacterium *Staphylococcus aureus.* It has high affinity for the F_c region of human immunoglobulin G (IgG) and can be used to isolate these molecules. The F_c region is not part of the antigen binding site so that once IgG molecules are bound, the column can then be used to purify antigens from serum or cell extracts.

4. *Cibracron-blue–agarose.* Cibracron blue is a dye that binds a variety of proteins such as kinases, dehydrogenases, DNA polymerase, nitrate reductase, succinyl-CoA transferase albumin and blood coagulation factors. It is an example of a group-specific ligand, probably binding to any enzyme having an affinity for nucleotides since it is structurally similar to nucleotides (although it is a very complex compound).

5. *Polynucleotide- and lysine-agarose.* The use of polyuridylic acid [poly(U)] as a ligand allows purification of mRNA molecules containing a polyadenylic acid [poly(A)] sequence, as in the case with many mammalian mRNA molecules. Some nucleic-acid-binding proteins have also been purified. Poly(A)-agarose binds nuclear mRNA molecules having a poly(U) sequence and has been used to fractionate ribonucleoproteins, RNA polymerases and many viral mRNA molecules. Lysine-agarose binds ribosomal RNA and the protein plasminogen.

6. *Boronate-polyacrylamide.* This material has very high capacity for binding low-molecular-weight compounds with *cis*-diol groups such as ribonucleotides, various sugars, catecholamines and many co-enzymes.

7. *Heparin-agarose.* Heparin is a substance used in medicine to prevent blood clotting; it reacts with several components in blood. It has been found to be useful in affinity columns through its ability to bind a large number of proteins. However, in a complex mixture of proteins such as in a tissue or cell extract, only a small number of proteins usually bind. Some of the proteins and other substances that have been purified by immobilized heparin are the following: bone collagenase, hepatitis B surface antigen, plasma antithrombin III, uterine estradiol receptor, murine myeloma DNA polymerase, ribosomes, adrenal tyrosine hydroxylase, and several androgen receptors.

8. *Acriflavine-agarose.* Acriflavine and other aminoacridines are well-known for their ability to intercalate reversibly between the base pairs of DNA. However, they also bind weakly to phosphorylated heterocycles, for example, nucleotides. The binding is primarily electrostatic, between the positive charge on the acridine ring and the negatively charged phosphate group. Immobilized acriflavine successfully separates nucleotides, oligonucleotides having different numbers of monomers, and AMP from cyclic AMP (adenosine 3′5′-phosphate).

9. *Iminodiacetic-acid–agarose.* In this case the ligand is a chelating agent that binds the Zn^{2+} and Cu^{2+} ions quite tightly. However, the ion is still free to form stable coordination complexes with available thiol and imidazole groups in proteins. The column is charged by passing solutions of either $ZnCl_2$ or $CuSO_4$ through the column. A protein solution can then be loaded on the charged column. Bound proteins can be eluted by lowering the pH and increasing the ionic strength, which together break down the coordination bonds. This method, which is sometimes called *metal chelate chromatography*, has been used to purify many proteins and is especially effective for gamma globulins, ceruloplasmin, transferrin, β-lipoproteins, lactoferrin, and haptoglobins. The Zn^{2+} column is currently the method of choice for isolation of human interferon.

Hydrophobic Interaction Chromatography

Most proteins have hydrophobic regions on their surfaces. In strongly polar solutions, hydrophobic regions tend to interact in an effort to minimize contact with water (or, more precisely, to reduce the decrease in the entropy of water near a hydrophobic site). This principle is used in hydrophobic interaction chromatography; a column is prepared of material having a hydrophobic stationary phase and proteins are loaded on the column in solutions of high ionic strength. Elution is accomplished by reducing the polarity of the mobile phase.

The most common materials in current use are octyl-agarose and phenyl-agarose (both marketed by Pharmacia Fine Chemicals). Proteins are loaded on the column in 0.01 M sodium phosphate, pH 6.8 containing 1 M ammonium sulfate, a solution having a high ionic strength but lacking the disruptive effect of concentrated solutions such as 1 M NaCl or 1 M KCl. This solution has the additional advantage that often an early step in protein purification is concentration of the protein in 2–4 M ammonium sulfate so that media changes prior to column loading are unnecessary. The strength of the interaction of the protein depends upon the hydrophobicity of the protein and the matrix. For elution, the polarity of the solution is lowered either by lowering the ionic strength, adding ethanol or ethylene glycol, raising the pH, or adding a detergent; alternately, chaotropic substances such as the CNS^-, ClO_4^-, CF_3COO^- ions or urea, which tend to disrupt water structure and thereby decrease hydrophobic interactions, may be added.

Hydrophobic interaction chromatography has been used successfully in the purification of many enzymes, globulins, albumins, and membrane proteins.

Covalent Chromatography

So far, all of the techniques of chromatography that have been discussed are based upon noncovalent interactions. In covalent chromatography, substances in the mobile phase form covalent bonds with the stationary phase and elution is accomplished by breakage of these bonds. At present, the technique is only used for the purification of substances that can form disulfide bonds with the stationary phase. Two materials are commercially available. Activated thiol-Sepharose contains 2,2′-dipyridyl disulfide, which under mild conditions, forms mixed disulfides with the thiol groups of proteins. Thiopropyl-Sepharose is activated by dithiothreitol or 2-mercaptoethanol, after which bonds can be formed with thiol groups in proteins. With both materials elution is accomplished by the addition of L-cysteine. The affinity of a substance is determined by the number of available thiol groups and the ease with which L-cysteine can break the disulfide bonds. The technique is extremely useful in the purification of proteins, such as papain and urease, which have a large number of free thiol groups.

Elution Procedures in Affinity Chromatography

The interactions that maintain the complex of the ligand and the molecules of interest are the usual ones—electrostatic interactions, hydrophobic effects, hydrogen bonding, and van der Waals forces. However, owing to the specificity of the interaction, binding is sometimes quite strong and special techniques of desorption are required. The procedures are of two types—those that act directly at the binding site and those that alter the structure of the molecule of interest and thereby indirectly affect the binding site.

If the binding is primarily electrostatic, desorption can be accomplished by a gradient of increasing ionic strength. When it is a hydrophobic interaction, solvents with reduced polarity are effective—for instance, 10% dioxane–water or 50% ethylene glycol–water. Changing pH can alter the degree of ionization of charged groups either on the ligand or in the binding site. However, with proteins there may be an indirect effect resulting from a conformational change. Desorption by this procedure usually requires lowering of the pH; the limitation of this method is the stability of the matrix, ligand, and molecule of interest at low pH.

Chaotropic substances, such as the ClO_4^-, CF_3COO^-, CNS^-, and CCl_3COO^- ions, are able to break down very strong interactions because of their general disruptive effect on macromolecules such as nucleic acids and proteins. Denaturants such as urea and guanidinium chloride have the same effect. These substances are very effective in elution but must be used with caution because they may irreversibly alter the structure of the molecule of interest. However, they provide the method of choice for elution of antibodies because the structure of these molecules is usually altered reversibly and normal structure is restored as long as the molecules are not totally denatured.

Electrophoresis (Chapter 9) is another means of removing very tightly bound molecules. That is, the molecules are forced to move in an electric field; the matrix is immobilized so that the molecules of interest are actively torn from the matrix.

An interesting means of desorption is called *affinity elution*. In this procedure molecules are added that compete with the ligand for the binding site. For example, the $NADP^+$-dependent dehydrogenases, such as lactic dehydrogenase, bind the cofactor $NADP^+$. These enzymes bind to 2′,5′-ADP–Sepharose using the $NADP^+$ binding site, as long as no $NADP^+$ is present. NADH competes with $NADP^+$ so that NADH is an effective eluting agent. A concentration gradient of NADH has been used successfully to separate isozymes of lactic dehydrogenase because each of the isozymes has a different strength of binding of NADH. An alternate mode of affinity elution is to add a substance that binds more tightly to the ligand than does the molecule of interest. An example of this is the elution of glycoproteins from ConA-Sepharose, which binds sugars (see page 260), by the addition of various sugars. The enzyme horseradish peroxidase is a glycoprotein that is used in certain immunoassays (Chapter 10). It is necessary to couple the enzyme to immunoglobulins and then free the complex from free enzyme and free immunoglobulin. This has been done in two steps. First, free peroxidase, which is smaller than the complex, is removed by gel chromatography. Second, the mixture of the complex and free immunoglobulin is adsorbed to ConA-Sepharose; only the complex binds. Elution is accomplished with methyl-α-D-mannoside, which displaces the complex by binding to the concanavalin A.

Special Techniques in the Chromatography of Nucleic Acids and of Proteins That Bind Nucleic Acids

Complex molecules such as proteins and nucleic acids have such a variety of functional groups, binding sites, and stabilizing forces that their behavior on standard chromatographic materials is difficult to predict. Instead, ingenious chromatographic procedures based on specific molecular interactions have been developed.

DNA-Cellulose Chromatography

DNA-cellulose binds many proteins that bind to DNA and can therefore be used to separate these proteins from nonbinding proteins or other substances. The preparation of DNA-cellulose consists of mixing cellulose powder with a solution of either single- or double-stranded DNA and then drying the mixture in a vacuum. The powder is then washed to remove unbound DNA and redried for storage. The mechanism of binding the DNA to the cellulose is unknown. With prolonged use, columns made of this material tend to lose DNA; to avoid this, the dried DNA-cellulose powder is sometimes resuspended in ethanol and irradiated strongly with ultraviolet light. This chemically couples the DNA to the cellulose in an unknown way.

DNA-cellulose chromatography is used primarily to purify DNA-binding proteins. A mixture of proteins in a buffer having low ionic strength (in which DNA-binding proteins adsorb to the DNA) is passed through the column. The column is then washed to remove unbound proteins and eluted with a gradient of increasing ionic strength. The proteins are thereby removed, with those that bind more weakly eluting first.

This method was first used by Bruce Alberts to detect DNA-binding proteins in *E. coli* infected with phage T4. An example of this work is shown in Figure 8-29.

The technique has now become a standard step in the purification of polymerases, nucleases, repressors, and so forth. Because both single- and double-stranded DNA can be used, proteins can also be separated according to the relative binding to each.

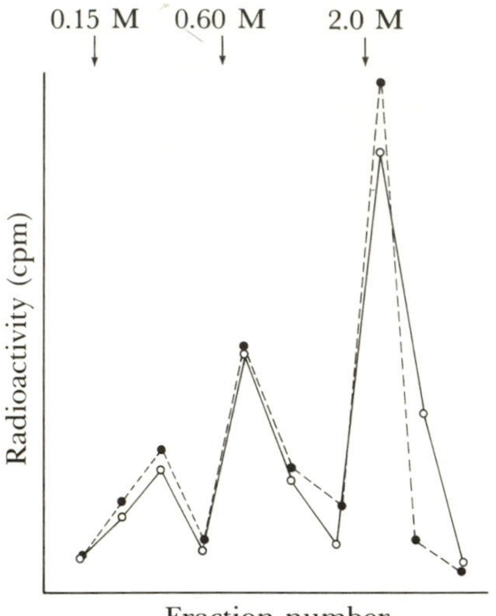

Figure 8-29
An example of DNA-cellulose chromatography. *E. coli* cells were infected with phage T4 in medium containing [^{14}C]leucine; another culture was infected with a T4 mutant in gene 32 in medium containing [^{3}H]leucine. Proteins were isolated, mixed, and loaded onto a column of single-stranded DNA cellulose. The column was eluted stepwise with 0.15 M. 0.60 M, and 2.0 M NaCl. In the third step, there was a deficiency of [^{3}H]leucine, indicating that this fraction contains ^{14}C-labeled gene-32 protein. This was the first use of DNA-cellulose chromatography to purify a DNA-binding protein. Open circles, ^{14}C; solid circles, ^{3}H. [Redrawn from B. M. Alberts, *Fed: Proc.* 29(1970): 1154–1175.]

Methylated Albumin-Kieselgur (MAK) Columns

Methylated serum albumin adsorbs to kieselgur (diatomaceous earth) and binds tightly. Joseph Mandell and Alfred Hershey showed that columns of this material bind single- and double-stranded DNA and RNA and can fractionate them if they are eluted with a gradient of increasing ionic strength. Using various conditions, this method has been used to separate native DNAs according to molecular weights, base composition, and degree of glucosylation; to separate single- from double-stranded DNA; to separate double-stranded DNA with single-strand ends from those without such termini; to separate transfer RNA from ribosomal RNA; and to separate different transfer RNAs from one another. An example of the use of this column material is shown in Figure 8-30.

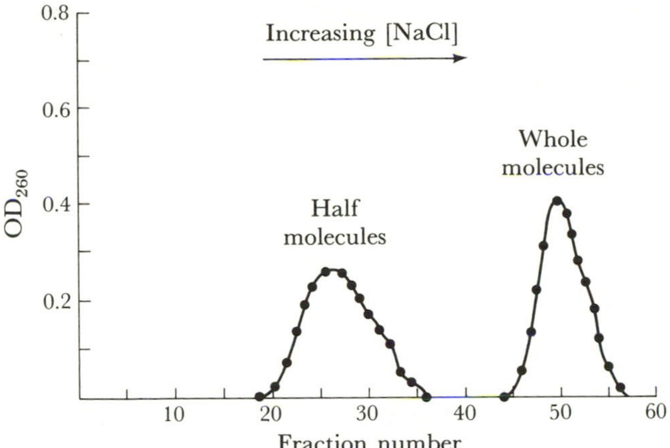

Figure 8-30
Separation of equal amounts of whole and half T2 DNA molecules by methylated albumin-kieselgur chromatography. The half molecules were prepared by hydrodynamic shearing (see Chapter 19). Elution is by an increasing NaCl concentration gradient. Note that the larger molecules are bound more tightly and need a highter NaCl concentration for elution. Because all "halves" are not the same size, the zone for half molecules is broader than for whole ones.

Separation of DNA Fractions Having Different Base Composition

Two affinity systems have been developed for separating DNA according to base composition. Both use a bonded polyacrylamide having a large pore size so that molecular sieving is minimized. One system uses an immobilized dye, malachite green, which binds preferentially to A · T pairs. The DNA is applied in dilute phosphate buffer; in this solvent all DNA molecules bind tightly. Elution is accomplished with increasing concentrations of sodium perchlorate ($NaClO_4$). DNA having a low A · T content is eluted at lower $NaClO_4$ concentrations than high-A · T DNA. The $NaClO_4$ is then removed either by dialysis or by passage through a desalting gel. The malachite green gel has been used successfully for large-scale separation of DNA fragments generated by restriction endonucleases. This system can also be used to concentrate DNA; that is, a large volume of a dilute DNA solution is applied to the column after which the DNA is eluted in a small volume of concentrated $NaClO_4$.

The dye, phenyl neutral red, which interacts preferentially with G·C pairs, constitutes the second system. Phenyl neutral red gel also fractionates DNA by base composition but is actually more useful in separation of supercoiled molecules from other forms of DNA. The dye binds more tightly to DNA that has single-stranded regions. This is a property of supercoiled DNA, which therefore is eluted at a higher salt concentration than are non-supercoiled forms since these are less strongly bound.

Hydroxyapatite Chromatography

Hydroxyapatite (HA) is a crystalline form of $Ca_{10}(PO_4)_6(OH)_2$ prepared from $CaHPO_4 \cdot 2H_2O$ crystals. Columns of hydroxyapatite bind substances that interact with calcium, such as DNA, RNA, chromatin, and phosphoproteins. Small polynucleotides and oligonucleotides bind weakly and can easily be separated from samples of high-molecular-weight nucleic acids. For phosphate-containing compounds binding is apparently by means of the phosphate groups in these compounds; binding requires a low phosphate concentration and elution is accomplished by increasing the phosphate concentration in the eluting buffer. However, there are other theories about the mechanism of binding. It is clear that binding is not simply an electrostatic effect because for polynucleotides, whose molecular weight is greater than a few million, there is virtually no dependence of the strength of binding on molecular weight; this contrasts with methylated-albumin-kieselguhr chromatography of nucleic acids.

Hydroxyapatite also binds most proteins, even though they contain no phosphate groups. The ionic strength of the solution must be low and elution is accomplished by raising the phosphate concentration or the ionic strength. For unknown reasons hydroxyapatite chromatography is unusually good for resolving complex mixtures of proteins and there are many instances in which a protein is purified with hydroxyapatite when all other chromatographic techniques fail. The mechanism of binding of proteins is not known. Binding to hydroxyapatite is the basis of a very effective concentrating procedure in which a dilute solution is adsorbed to hydroxyapatite and then eluted in a very small volume. Gradient elution of complex mixtures also gives very good separation of the proteins in the mixture. The resolution is so extraordinary that presumably binding is sensitive to very slight changes in the configuration of the protein surface.

Hydroxyapatite chromatography has been very useful in nucleic acid research, principally in the separation of single-stranded from double-stranded DNA. At low phosphate concentrations, both forms of DNA adsorb to hydroxyapatite. A small increase in the phosphate concentration of the eluant causes selective desorption of single-stranded DNA and a further increase results in release of double-stranded DNA. An example of this is shown in Figure 8-31.

The ability to separate single- and double-stranded DNA is the basis of a fractionation procedure called *thermal chromatography*. Both forms

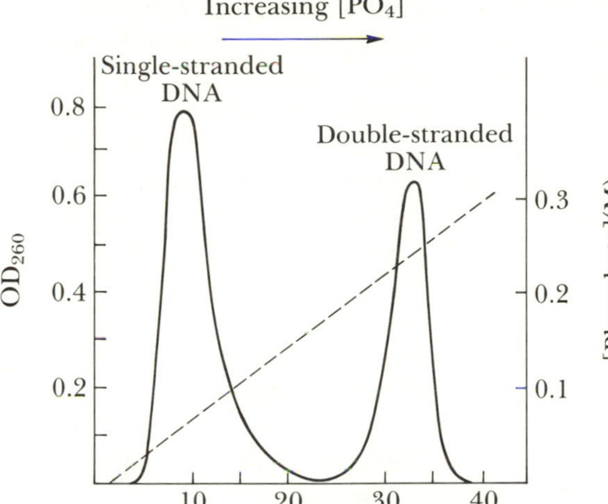

Figure 8-31
Separation of single- and double-stranded DNA by chromatography on hydroxyapatite. Elution is by a gradient of increasing phosphate concentration (dashed line).

of DNA bind well to hydroxyapatite at high temperatures but the elution pattern changes. As the temperature is increased, double-stranded DNA molecules are denatured and converted to single-stranded DNA molecules, which are eluted at a low phosphate concentration. Thus, by raising the temperature and eluting at a phosphate concentration that removes only single-stranded DNA, a mixture of molecules can be fractionated according to melting temperature. Since the melting temperature decreases with increasing content of A·T base pairs, thermal chromatography fractionates DNA samples according to base composition. Single-stranded DNA is recovered, however.

The affinity of hydroxyapatite for double-stranded DNA is so great compared to most other molecules that it is possible to choose a phosphate concentration at which few molecules bind other than double-stranded DNA. This has been used as the basis for rapid isolation and purification of DNA from cell extracts. The procedure is outlined in the following. Cells are broken open by addition of detergents to a concentrated cell suspension. The extract is applied to a hydroxyapatite column. Using the appropriate phosphate concentration, RNA and most proteins fail to bind. The column is then washed with buffer and then the DNA molecules are eluted with a small volume of concentrated phosphate buffer. This procedure has worked successfully with a variety of animal cells and bacteria.

The dye ethidium bromide intercalates between the base pairs of double-stranded DNA. This causes the base pairs to move apart slightly and the helix unwinds. Ultimately a sufficient number of ethidium bromide molecules are bound that the configuration of the DNA molecule is altered to the point where no additional dye molecules can bind. However, if the DNA is a covalent circle, there is another constraint, namely, that the untwisting of the helix causes positive supercoiling. A naturally occurring supercoil is negatively supercoiled so that addition of ethidium bromide initially removes supercoiling and then causes the molecule to become a positive supercoil. Ultimately the DNA molecule is so supercoiled that no additional dye molecules can bind. This limit is reached when a number of ethidium bromide molecules per unit molecular weight of the DNA is less than the limit for a linear molecule. If the circle is a nicked circle, there is free rotation and no supercoiling so that the binding limit of a nicked circle is the same as that of a linear molecule. Of all forms, the covalent circle binds less ethidium bromide then the other forms at saturating concentrations of the dye. The binding of ethidium bromide affects the affinity of DNA for hydroxyapatite and allows rapid and efficient separation of covalent circles from linear molecules and nicked circles.

One of the most important uses of hydroxyapatite is the separation of single- from double-stranded DNA in the technique known as c_0t analysis. This is described in Chapter 19.

APPENDIX

Purification of an enzyme: an example of the use of many chromatographic procedures

Justice can hardly be done to the subject of enzyme purification in a few paragraphs; nonetheless, it is instructive to examine a typical purification scheme to see how different chromatographic procedures are made use of. The example is the purification of *E. coli* DNA polymerase I. The method is that of Paul Englund, modified from the original procedure from Arthur Kornberg's laboratory.

Step 1. Breakage of bacteria

The first step is almost always to break open the bacteria. This can be done by sonication (see Chapter 19) or by alternate freezing and thawing to weaken the cell wall, followed by stirring in a food blender with fine glass beads. After most of the cells have been broken, the glass beads, unbroken cells and cell walls are removed by centrifugation.

Step 2. Partial removal of the nucleic acids

Nucleic acids are precipitated by the addition of streptomycin sulfate. This step reduces the viscosity of the material, thus allowing substantial concentration, and eliminates one of the major cell constituents. It is done at this point because the nucleic acids would bind to some of the chromatographic columns used later.

Step 3. Final removal of nucleic acids and preliminary removal of some proteins: autolysis

Magnesium chloride is added to the mixture to activate nucleases. After prolonged incubation, almost all nucleic acids are degraded to nucleotides and small oligonucleotides, which remain soluble when the proteins are precipitated in step 4.

Step 4. Ammonium sulfate fractionation

Proteins differ in their solubility in concentrated salt solutions and hence can be separated from one another by precipitation at high ionic strength. Many salts are possible, but ammonium sulfate is preferred because it does not significantly affect pH; it is inexpensive, very soluble, and does not destabilize proteins (see Chapter 19). In fact, many proteins are stabilized by the NH_4^+ ion.

Hence, the extract is treated with $(NH_4)_2SO_4$ at a concentration that precipitates proteins other than the polymerase. This precipitate is removed by centrifugation and more $(NH_4)_2SO_4$ is added to the supernatant to reach a concentration that precipitates the polymerase. This precipitate is collected and redissolved in an appropriate buffer in preparation for step 5.

Step 5. DEAE-cellulose ion-exchange chromatography

A DEAE-cellulose column is loaded with the material from step 5, using conditions of pH and ionic strength such that the column adsorbs all nucleic acids without binding the polymerase. This is a necessary step because small amounts of nucleic acids that remain after step 4 would interfere with adsorption to phosphocellulose in the next step. The void volume is then collected for step 6.

Step 6. Phosphocellulose chromatography

Chromatography on a phosphocellulose column separates most of the proteins from polymerase. The few remaining proteins are eliminated in the next step.

Step 7. Gel chromatography on Sephadex

Fractions from phosphocellulose chromatography containing polymerase are chromatographed on Sephadex and the separation is based on molecular weight. This is a useful step because the polymerase has a very high molecular weight compared with that of the remaining proteins. All proteins in the effluent from step 6 separate on the Sephadex. The fractions containing the polymerase are therefore pure. They are then treated with $(NH_4)_2SO_4$ to precipitate the polymerase in order to concentrate the purified enzyme. The precipitate is then dissolved in a small volume of a suitable buffer.

Bernardi, G. 1971. "Chromatography of Proteins on Hydroxyapatite," in *Methods in Enzymology*, edited by W. B. Jakoby, pp. 325–339. Academic Press.

Bobbit, J. M., A. E. Schwarting, and R. J. Gutter. 1968. *Introduction to Chromatography.* Reinhold.

Chovin, P. 1962. "Gas Chromatography," in *Comprehensive Biochemistry*, vol. 4, edited by M. Florkin and E. H. Stotz. Elsevier.

Cuatrecasas, P., and C. B. Anfinsen. 1971. "Affinity Chromatography," in *Methods in Enzymology*, vol. 22, edited by W. B. Jakoby, pp. 345–378. Academic Press.

Fisher, L. 1969. "An Introduction to Gel Chromatography," *Laboratory Techniques in Biochemistry and Molecular Biology*, edited by T. S. Work and E. Work. American Elsevier.

Heftman, E. 1975. *Chromatography.* Van Nostrand-Reinhold.

Helferich, F. 1962. *Ion Exchange.* McGraw-Hill.

Himmelhoch, S. R. 1971. "Chromatography of Proteins on Ion-exchange Adsorbents," in *Methods in Enzymology*, vol. 22, edited by W. B. Jakoby, pp. 287–321. Academic Press.

Hoffman-Ostenhof, O. 1978. *Affinity Chromatography.* Pergamon.

Ingram, V. M. 1959. "Abnormal Human Haemoglobins. 1. The Comparison of Normal Human and Sickle-Cell Haemoglobins by 'Fingerprinting.'" *Biochim. Biophys. Acta* 28:539–545. The development of fingerprinting.

Jakoby, W. B., ed. 1974. "Affinity Chromatography." *Methods in Enzymology*, vol. 34B. Academic Press. This contains numerous pertinent articles.

Mandell, J. D., and A. D. Hershey. 1960. "A Fractionating Column for Analysis of Nucleic Acids." *Anal. Biochem.* 1:66–77. MAK chromatography was first described here.

Martin A. J. P., and R. L. M. Synge. 1941. "A New Form of Chromatography Employing Two Liquid Phases." *Biochem. J.* 35:1358–1368. Partition chromatography was first described in this paper.

Morris, C., and P. Morris. 1976. *Separation Methods in Biochemistry.* Wiley.

Peterson, E. A. 1970. "Cellulose Ion Exchangers," in *Laboratory Techniques in Biochemistry and Molecular Biology*, vol. 2, edited by T. S. Work and E. Work, pp. 228–400. North-Holland.

Purnell, H. 1962. *Gas Chromatography.* Wiley.

Randerath, K. 1966. *Thin-layer Chromatography.* Academic Press.

The best information by far can be found in brochures available free of charge from Pharmacia Fine Chemicals, Inc.; for example:

Affinity Chromatography-Principles and Methods

Sephadex, Gel Giltration in Theory and Practice

Sephadex Ion Exchangers

Thin-layer Gel Filtration

CNBr-activated Sepharose 4B

and from Bio-Rad Laboratories; for example, Materials, Equipment and Systems for Chromatography, Electrophoresis and Membrane Techniques.
The latest information can be found in such journals as Journal of Chromatography and Journal of Gas Chromatography.

Chromatography

8–1. An ion exchanger is used to separate single-stranded polynucleotides. A gradient of increasing salt concentration at pH 2 is used for elution. Which elutes first, triadenylic acid or pentaadenylic acid?

8–2. An interesting enzyme, whose molecular weight is 56,000, is used in catalytic amounts to cleave a membrane protein, whose molecular weight is 65,000, into two fragments having molecular weights 43,000 and 22,000. Gel chromatography is used to assess the extent of reaction at various times. Three major fractions of proteins are eluted at each time. What are they and in what order are they eluted?

8–3. A globular protein aggregates to form either a tetrahedral tetramer or a linear tetramer. If a mixture is chromatographed with a Sephadex gel, which form will elute first?

8–4. The R_f values of substances A and B are 0.34 and 0.68 when chromatographed on paper using ethanol as a solvent. What is the ratio of the distance moved after three hours, assuming that neither substance has run off the paper?

8–5. Sepharose can be melted and will harden to form a continuous gel. If a glass rod is inserted a few millimeters into the liquid gel and removed after the gel has hardened, the gel will contain a small well.

a. Suppose a solution containing high-molecular-weight DNA, which cannot penetrate the gel, and 1 M NaCl is placed in the well. Assuming that the gel is prepared with 0.01 M NaCl, how will the composition of the solution in the well change with time?

b. The well is filled with a solution containing 0.01 M NaCl and two small RNA molecules, A and B, whose molecular weights are 1×10^6 and 3×10^6. Assuming that both A and B can penetrate the gel, what is the geometric distribution of A and B several hours later and which molecule will have moved further?

8–6. Suppose that an enzyme is dissociated into four identical subunits and that you want to test for the enzymatic activity of the individual subunits, but you must be sure that there are no tetramers remaining in the sample. What chromatographic system would you choose to free the monomers from the tetramer?

8–7. In attempting to purify a protein, it is found that the protein is tightly bound to DNA. When DNase is added to destroy DNA, it is found that the DNase binds to the complex without digesting the DNA. Therefore, DNase treatment cannot be used to eliminate the DNA. To dissociate the DNA-protein complex, 2 M NaCl is needed; at this concentration neither DNA nor protein is retarded by ion exchangers. How might the protein be purified?

8–8. In preparing hybrid DNA by renaturation of denatured DNA, there is usually some unrenatured DNA. How would you remove this DNA?

8–9. In comparing a molecule of DNA that is circular with one that is linear, both of molecular weight 6×10^6, which form would elute first from an agarose column? What about native and denatured ribosomal RNA? (Ribosomal RNA has considerable secondary structure.)

8–10. A procedure has been devised for purifying a particular enzyme using gel chromatography. In an effort to increase the amount of material to be handled, you use a sample whose concentration of protein is ten times that normally used with this procedure. The enzyme activity now elutes principally in the void volume. Explain what happened.

In general, to increase the amount of material to be handled, should the diameter of the column be increased? Should the length be increased? or should both?

8–11. If chromatographed on paper, the substances A, B, C, and D have the following R_f values in the solvent systems indicated.

Substance	Butanol–H_2O	Isopropanol–HCl	Ethanol	Acetic acid
A	0.23	0.31	0.42	0.09
B	0.24	0.20	0.51	0.62
C	0.38	0.58	0.40	0.64
D	0.41	0.56	0.53	0.10

What would be the best way to separate A, B, C, and D?

8–12. Twenty milligrams of a protein mixture is applied to a DEAE-cellulose column. Thirty percent of the protein is known to be enzyme X. After eluting the column, the total amount of protein in all samples is 18.9 mg. There is no detectable enzyme X activity. Give a possible explanation for the lack of enzymatic activity.

8–13. A DNA molecule consisting of a 6×10^6 dalton double-strand piece and a single-strand extension equal in length to the double-strand segment is adsorbed to hydroxyapatite. Given the fact that there is very little dependence of elution behavior on molecular weight in this range, if the column were eluted with an increasing gradient of $[PO_4]^{3-}$, where would you expect this molecule to elute—in the single-strand or double-strand region or elsewhere?

8–14. An enzyme is known to require a high concentration of Mg^{2+} for activity. If the Mg^{2+} is removed, the protein is irreversibly denatured. Suppose that, in establishing a purification scheme, you try both ion-exchange and gel chromatography and that, in both cases, the enzyme loses activity. Explain why this might happen. In view of your explanation, what modifications could you make to improve the situation?

8–15. One hundred units of a protein contained in a crude cell extract is chromatographed on an ion exchanger. Twenty 5-ml fractions are taken. Fraction 3 has 1.5 units. No other fraction has any enzymatic activity and no enzyme activity can be found by extensive washing of the column. Fractions 8 and 12 are mixed together and yield a solution having 250 units of activity.

a. What was in fractions 8 and 12?
b. Explain how the activity in the 8–12 mixture could be greater than 100 units.
c. What would happen if the 8–12 mixture were passed over the column again? That is, would there be activity in any of the fractions?
d. What would happen if the 8–12 mixture were passed through a gel column? Do you think that any of the fractions from this column would have enzymatic activity?
e. Suppose you mix together all 20 fractions. Would you expect to get 100 units, 250 units, or something quite different?

8–16. A pure protein is chromatographed on a gel. It is applied to the gel at a concentration of 1 mg/ml. The resulting elution pattern is shown in A. It

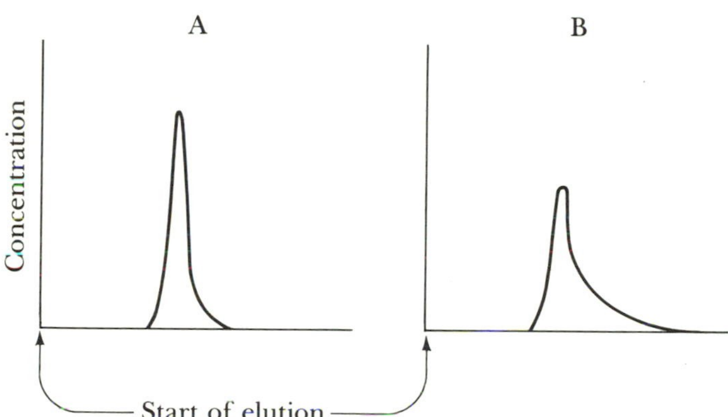

is chromatographed by someone else using the same gel but applying the protein at a concentration of 0.1 mg/ml. The pattern the second person gets is shown in B. The important thing to note is that only one side of the peak has been stretched out. Explain the difference.

8–17. Certain drugs, such as morphine, bind to specific receptors in neural tissue. Design a procedure for the partial purification of the receptors.

8–18. Two proteins have the same molecular weights. At pH 5.5 both have considerable secondary structure and are at least 75% α helix. At pH 8.5, one of them loses all structure and is a random coil; when returned to pH 5.5, its structure is restored. Devise a procedure for separating these proteins.

9

Electrophoresis

Most biological polymers are electrically charged and will therefore move in an electric field. The transport of particles through a solvent by an electric field is called electrophoresis.

A useful way to characterize macromolecules is by their rate of movement in an electric field. This property can be used to determine protein molecular weights, to distinguish molecules by virtue of their net charge or their shape, to detect amino acid changes from charged to uncharged residues or vice versa, and to separate different molecular species quantitatively. How this is done is described in this chapter.

Theory of Electrophoresis

The detailed theory of electrophoresis is highly complicated and at present incomplete; a simple description of the electrophoretic principle is sufficient for an understanding of how the technique is used for most purposes.

In many ways, electrophoresis is like sedimentation (Chapter 11): a force is applied and countered by viscous drag. If a particle with charge q, suspended in an insulating medium, is in an electric field, E, the particle will move at a constant velocity, v, determined by the balance between the electrical force, Eq, and the viscous drag, fv, in which f is the frictional coefficient, that is,

$$Eq = fv \quad (1)$$

It is important to note that the velocity is proportional to the voltage and that the electrical current does not appear in equation (1). This point will be discussed in some detail later in this chapter.

The mobility, u, is defined as the velocity per unit field, or

$$u = v/E = q/f \tag{2}$$

Because mobility depends on the frictional coefficient, which in turn is a function of some of the physical parameters of the molecules, in principle the value of u should give information about the size and shape of the molecule. However, the supporting medium is normally not an insulator but an electrolyte consisting of charged ions, and this introduces great complexity. The principal complication is that a charged particle suspended in an electrolyte attracts the ions and hence is surrounded by an ion atmosphere that shields the particle from the applied field. However, this ion atmosphere is partly disrupted both by the field and by the motion of the particle through the medium. To date, the theory of electrophoresis has failed to account adequately for these complications, as well as several others, so that electrophoresis has not turned out to be very useful in supplying *detailed* information about macromolecular structure. It is, however, enormously useful as both an analytical and a preparative tool* because of its ability to separate different molecules. Although the theory is correct in providing the working rule of electrophoresis—that is, that mobility increases with q, decreases with f, and is zero for uncharged molecules—in practice, the optimal conditions for separation are almost always determined empirically.

Types of Electrophoresis

There are two types of electrophoresis: *moving-boundary* and *zone.*

Moving-Boundary Electrophoresis

In moving-boundary electrophoresis, macromolecules are present throughout a solution and the position of the molecules (actually, the boundary separating the solution from the solvent) as a function of time is determined by schlieren optics (Chapter 11). This method, which is in many ways equivalent to boundary sedimentation, is an analytical method that has been used primarily for the determination of mobilities

* Reference will often be made to preparative and analytical procedures. By a preparative procedure is meant a technique designed to provide relatively large quantities of pure (or nearly pure) material to be used in later experiments. An analytical procedure is used in determining purity, evaluating the number of components in a mixture and possibly the proportions of each, detecting changes in charge, and so forth—that is, obtaining information to find answers to particular questions.

and isoelectric points of proteins. However, because there is not usually much to be gained from the quantitative determination of mobility, moving-boundary electrophoresis is rarely used. Further information about moving-boundary electrophoresis can be found in the Selected References near the end of the chapter.

Zone Electrophoresis

In zone electrophoresis, a solution is applied as a spot or a band, and particles migrate through a solvent that is almost always supported by a chemically inert and homogeneous medium such as paper or in a gel. It is used to analyze mixtures, to determine purity, to assay for changes in mobility and/or conformation, and for purification. The quantitative determination of mobility is never done because of the virtual impossibility of accounting theoretically for various effects of the support material.

Zone electrophoresis is the most common type of electrophoresis in use at present; a spot or thin layer of solution is placed in contact with a semisolid or gelatinous medium, an electric field is applied, and molecules migrate on or through the supporting material. Because the molecules are applied in a zone, small samples are used and the separation of solutes can be complete. In some types of electrophoresis (for example, on paper and cellulose acetate), the major function of the supporting medium is to prevent mechanical disturbances and convection that arise both from temperature changes and from the high density of concentrated macromolecular solutions. However, the supporting medium sometimes adsorbs various molecular species or acts as a molecular sieve and therefore has a chromatographic effect (see Chapter 8) that can either aid or detract from the separation. In starch gel electrophoresis, there is a molecular-sieve effect in addition to adsorption. In pure gel electrophoresis, using non-adsorbing gels such as agarose or polyacrylmide, the molecular-sieve effect predominates.

Methods for using various supporting substances are described in the following section.

Paper electrophoresis

Low-voltage paper electrophoresis is rarely used today, having been superceded by gel electrophoresis; however, it is useful to begin a discussion of zone electrophoresis by examining this method because it is very simple and illustrates the basic technique.

Figure 9-1 shows a typical low-voltage (20 V/cm) paper-electrophoresis arrangement. The paper strip is first dipped in the buffer solution and then placed in the tank as indicated. The sample is then applied either as a spot or a line. The paper is enclosed in a tank to prevent evaporation and voltage is applied. When separation is completed, the paper is removed and dried. If the sample contains sufficient material,

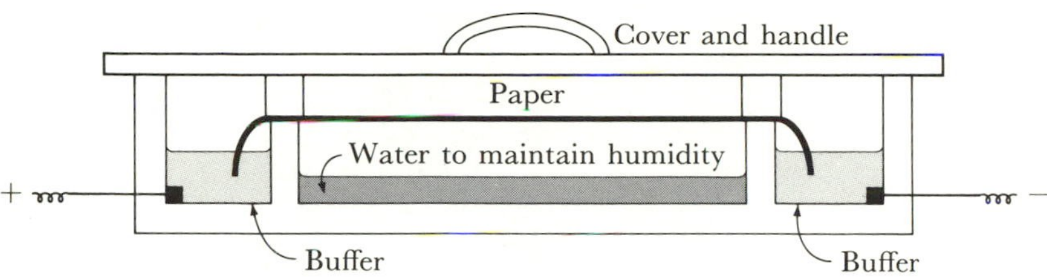

Figure 9-1
Experimental arrangement for low-voltage paper electrophoresis. A sample is spotted onto paper as in paper chromatography (see Figure 8-6) but usually near the middle. The paper is saturated with a buffer and both ends are placed in a buffer reservoir to make electrical contact with the power supply. The system is enclosed to prevent drying of the paper. At the end of a predetermined time, the paper is removed and dried.

each component can be located either by its color or fluorescence or by staining with various dyes. If quantitative measurement is required, the dye can be eluted and determined spectrophotometrically. Because dye uptake is rarely quantitative, the accuracy is about 20%. If the sample is radioactive, spots can be located by cutting up the paper and counting the radioactivity (see Chapter 5) or by autoradiography of the whole sheet using X-ray film (see Chapter 6). A particularly useful technique is to combine stains with radioactivity to determine which substances are radioactive. Low-voltage paper electrophoresis was most useful for the analysis of protein mixtures that are poorly separated by chromatography.

Sometimes material must be eluted from the paper for further analysis. Elution is accomplished in three ways, each of which begins with cutting out the region of the paper containing the material to be eluted. (1) The paper is soaked in a small volume of eluting buffer. (2) The spot is cut out as a V-shaped piece of paper. A base of the triangle is folded over the edge of a beaker containing the elution buffer. The buffer flows through the paper and drops containing the substance to be eluted are collected from the lower vertex of the triangle opposite the base. (3) The material is electrophoresed from the spot into a small volume.

Low-voltage electrophoresis is inefficient for small molecules (e.g., amino acids and nucleotides) because their small charge results in low mobility (i.e., slow separation) and their small size allows considerable diffusional spreading. In *high-voltage electrophoresis*, the speed of separation is increased by the use of a potential gradient of 200 V/cm. The high voltage produces a high current, which heats the paper so that cooling is necessary. This can be accomplished by immersing the paper in a large volume of an immiscible and nonconducting liquid or by pressing it against cool paper or glass. The immersed-strip technique

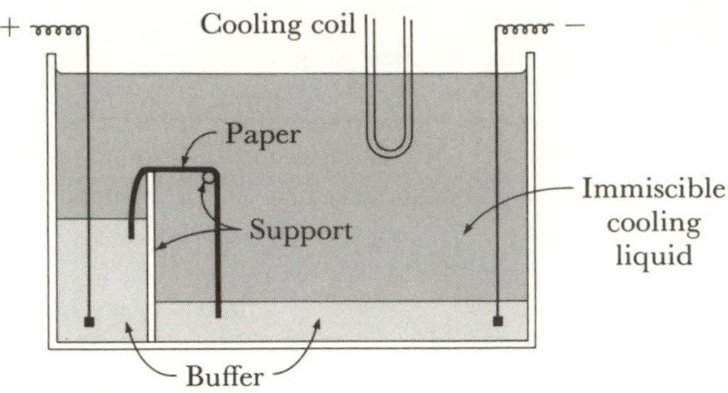

A. Immersed-strip method

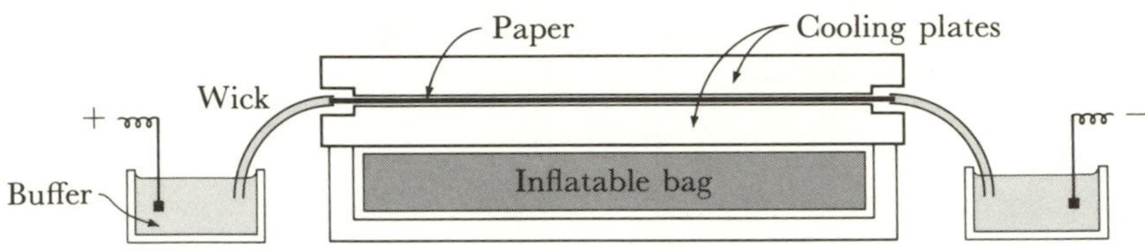

B. Enclosed-strip method

Figure 9-2
Two experimental arrangements for high-voltage electrophoresis. Because the higher current produces sufficient heat to destroy the sample and evaporate the solvent, the paper is either immersed in a cooling liquid or placed between two cooling plates. The inflatable bag supports the cooling plates, preventing bowing.

(see Figure 9-2 for the arrangement) is not often used because the required coolants (e.g., toluene, carbon tetrachloride, or various oils) are toxic and sometimes inflammable. The *enclosed-strip method*, shown in Figure 9-2, is far more common. Note that the paper is not immersed in the buffer. After electrophoresis is complete, the paper is dried and spots are identified as they are in the low-voltage method.

High-voltage paper electrophoresis is of great value for resolving amino acids and peptides. Because complete resolution of mixtures is not always possible in a single high-voltage electrophoresis operation, it is frequently coupled with chromatography in the *two-dimensional separation* technique. The sample is first electrophoresed and then chromatographed at right angles, or vice versa. This technique is discussed in a later section (page 312).

Cellulose acetate strip electrophoresis

Many biological macromolecules adsorb to cellulose (i.e., paper) by means of the cellulose hydroxyl groups. Adsorption impedes movement (see Chapter 8) and therefore causes the tailing of spots or bands, which reduces resolution. This can be avoided by using a cellulose acetate membrane instead of paper where most of the hydroxyls have been converted into acetate groups, which are generally nonadsorbing. The elimination of this skewing improves resolution so that separations are more rapid at low voltage. The fact that spots are smaller also means that the material in the spot is more concentrated and easier to detect. The low adsorption of cellulose acetate also reduces background staining, thus improving the sensitivity of detection.

Two other advantages in using cellulose acetate are that the material is transparent, which aids in the spectrophotometric determination of material, and is easily dissolved in various solvents, thus allowing simple elution of material.

For simplicity of operation coupled with high resolution, cellulose acetate cannot be surpassed. However, for maximum resolution, gel electrophoresis is the method of choice.

Gel electrophoresis

The use of gels such as starch, polyacrylamide, agarose, and agarose-acrylamide as supporting media provides enhanced resolution, particularly for proteins and nucleic acids. The reasons are not clear but certainly include a combination of reduced diffusion by the gel network and the separating action of gel chromatography ("molecular sieving").

The earliest work in gel electrophoresis of proteins was done with *starch gels.* Figure 9-3 shows a typical arrangement for starch gel electrophoresis. The gel consists of a paste of potato starch whose grains have been burst by heating in the buffer. When the gel is mounted horizontally, as shown, the sample is applied to a slot cut with a razor blade either as a single solution or as a slurry with starch grains. The slot is sealed with wax or grease and the voltage is applied. After electrophoresis, the semi-rigid gel is removed and frequently sliced into two or three layers, each to be differentially stained. The various components appear as a series of bands in the gel (Figure 9-4).

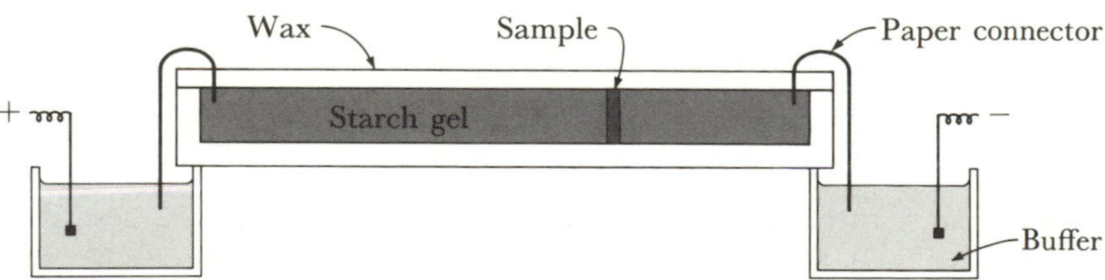

Figure 9-3
Arrangement for starch gel electrophoresis. The gel is formed in place by allowing the starch grains to swell in a buffer. The gel is cut and the sample is placed in the cut. The system is covered with wax to prevent drying.

Figure 9-4
A starch gel electrophoregram of separated proteins. When electrophoresis is finished, the gel is stained to make the proteins visible. Note that proteins have migrated both in positive and in negative directions, depending on their charge.

Starch gel electrophoresis was very useful at one time because of ease of visualization of bands of proteins by chemical stains, and efficient elution of enzymes, which facilitated assay of enzymatic activity. Sometimes amounts of enzymes too small to be seen by stains could be assayed merely by placing a slurry of starch grains containing the enzyme into the standard assay medium. Starch gel electrophoresis provided the first evidence for the existence of isozymes, that is, different forms of the same enzyme or different molecules (or arrangement of subunits) having the same enzymatic activity. It is still the method of choice for analyzing many mixtures of isozymes; however, for general use with proteins, starch gel has been superceded by polyacrylamide gels.

Polyacrylamide gels have replaced starch gels because the amount of molecular sieving can be controlled by the concentration of the gel and the adsorption of proteins is negligible. Polyacrylamide is currently the most effective support medium in use for proteins, small RNA molecules, and very small fragments of DNA. (For nucleic acids that are too large for the polyacrylamide pores, agarose and agarose-acrylamide are superior.) Polyacrylamide gel is prepared by cross-linking acrylamide with *N*,*N'*-methylenebisacrylamide in the container in which the electrophoresis is to be carried out. It is, of course, a continuous gel and is not in the beaded form commonly used in column chromatography.

There are basically two experimental arrangements used for polyacrylamide gel electrophoresis: column gels and slab gels. The slab gels can be run either vertically or horizontally; the horizontal mode is essential when the gel concentration is low because a soft gel can be crushed under its own weight in the vertical configuration. Two arrangements that have commonly been used are shown in Figure 9-5 and 9-6. Several important variations of the horizontal mode will be described later (page 293). Slab gels have, for the most part, replaced column gels because a large number of samples can be run simultaneously.

Polyacrylamide gel electrophoresis is probably the most versatile and useful electrophoretic system for the analysis and separation of proteins and RNA molecules. Two variations that have been developed for special purposes are described next.

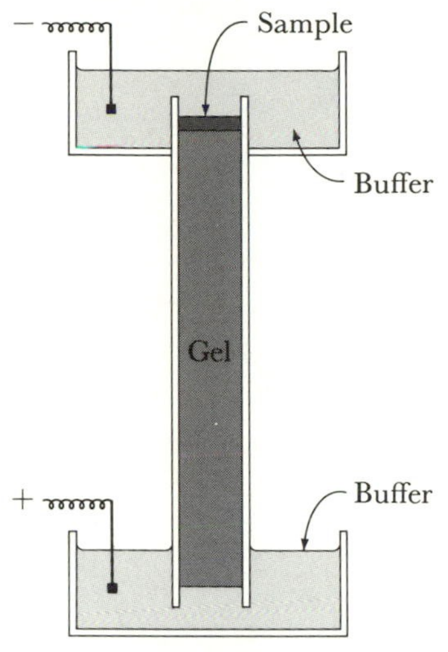

Figure 9-5
Experimental arrangement for gel electrophoresis in a column. The gel is polymerized in the column and then placed in the buffer reservoirs. The sample is suspended in a concentrated sucrose solution and layered on the gel with a fine pipette. After electrophoresis, the gel is pushed out of the column and then either stained with the dye Coomassie blue to make the bands visible or, if a radioactive sample is used, sliced and the radioactivity in each slice measured.

*Sodium dodecyl sulfate–polyacrylamide gel electrophoresis**

In 1967 A. Shapiro, Edward Vinuela, and Jacob Maizel showed that the molecular weights of most proteins could be determined by measuring the mobility in polyacrylamide gels containing the detergent sodium dodecyl sulfate (SDS). This technique was improved two years later by Klaus Weber and Mary Osborn and became a standard technique for the determination of the molecular weights of single polypeptide chains. At neutral pH, in 1% SDS and 0.1 M mercaptoethanol, most multichain proteins bind SDS and dissociate; disulfide linkages are broken by the mercaptoethanol, secondary structure is lost, and the complexes, consisting of protein subunits and SDS assume a random-coil configuration. Proteins treated in this way behave as though they have uniform shape and an identical charge-to-mass ratio. This is because the amount of SDS bound per unit weight of protein is constant—1.4 g of SDS/g of protein. The charge then is determined by the bound SDS rather than the intrinsic charge of the amino acids. Since the SDS-clad random coils all have the same ratio of charge to mass, one might expect the driving force on each protein molecule to be proportional to the number of peptide bonds per molecule; that is, the mobility would be proportional to the molecular weight. However, the gel is a molecular sieve through which each molecule must pass. The smaller the molecule the more easily it can find its way through the tortuous passages of the gel; thus the mobility increases with decreasing molecular weight. (Note how this differs from gel chromatography in which the smaller molecules

* The acronym PAGE-SDS is often used for *p*oly*a*crylamide *g*el *e*lectrophoresis when sodium dodecyl sulfate (SDS) is used.

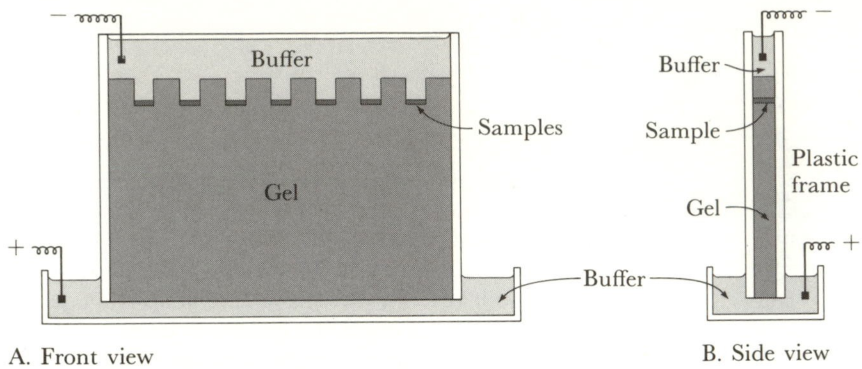

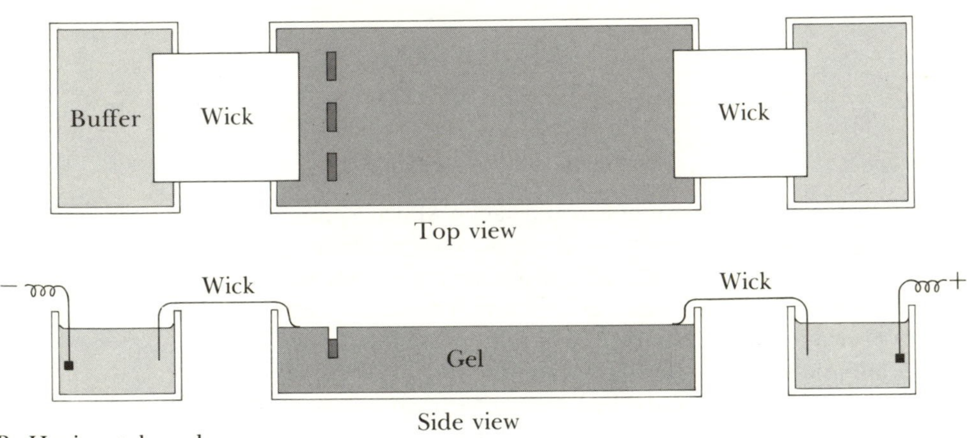

Figure 9-6
Two types of apparatus for gel electrophoresis. In both the vertical and horizontal modes, the gel is allowed to polymerize or harden in place. An appropriately shaped mold called a comb is suspended on the top of a frame (vertical mode) or in the horizontal gel during polymerization or hardening in order to make notches to serve as sample wells. After electrophoresis the slab is removed from the frame. The gel is then stained by immersing it in a staining solution. Excess stain is removed by washing or by adding an agent that couples the stain covalently to the macromolecule, after which free stain is removed by electrophoresis.

move more slowly through a column because they are more often side-tracked by penetrating the gel beads.) Weber and Osborn showed that if a series of proteins of known molecular weight are electrophoresed in a column gel, they will separate into a series of bands (Figure 9-7) and a plot of the distance migrated versus log M gives a straight line (Figure 9-8). Hence, if a protein of unknown molecular weight is electrophoresed

Figure 9-7
Electrophoresis of several proteins in a polyacrylamide gel containing 10% sodium dodecyl sulfate. Each band represents 2.5 μg of protein. [Courtesy of Bio-Rad Laboratories.]

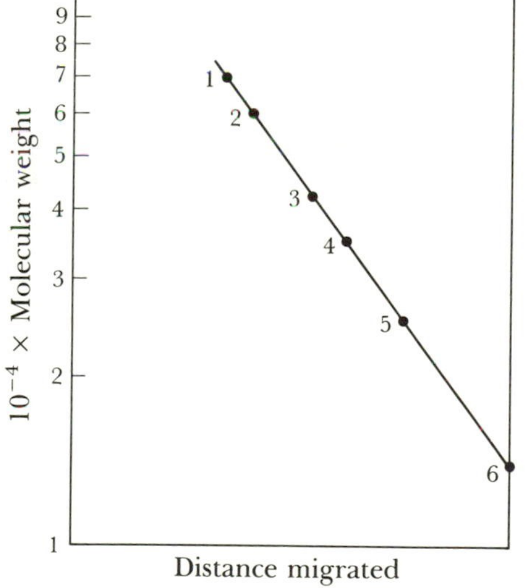

Figure 9-8
Typical semilogarithmic plot of M versus distance migrated for determining molecular weight by SDS–polyacrylamide gel electrophoresis. Proteins were reduced to eliminate disulfide bonds. The proteins are: (1) bovine serum albumin; (2) catalase; (3) ovalbumin; (4) carboxypeptidase A; (5) chymotrypsinogen; (6) lysozyme.

with two or more of known molecular weight, then the unknown can be calculated to an accuracy ranging between 5% and 10%. The size of the error is determined by the accuracy of the slope of the plot; this is increased as the number of standards used is increased. Greatest accuracy is obtained if the known values for M bracket the unknown. This is certainly the most common way of estimating the molecular weight of protein subunits in use today (although thin-layer gel chromatography—see page 247—will probably prove to be of equal value). Note that the sodium dodecyl sulfate gel technique can also be used to determine whether S—S groups are present because, if they are, the mobility will depend on whether the protein is treated with the S—S cleaving agent, mercaptoethanol. This is because the effective volume and hence the friction factor always changes if S—S bonds are broken. This complements the viscosity measurements described in Chapter 11. It should also be noted that some glycoproteins do not bind SDS as well as do simple proteins and tend to move slowly; thus molecular weights of glycoproteins determined by this method are usually overestimated.

Polyacrylamide gel and polyacrylamide-agarose gel electrophoresis of single-stranded nucleic acids.

The phosphate group of each nucleotide carries a single strong negative charge that is much greater than any of the charges on the bases. Thus the charge-to-mass ratio of all polynucleotides is independent of the base composition. Using free electrophoresis (that is, without a supporting gel) Norman Davidson showed that there is no molecular weight dependence on the mobility of nucleotides ranging in size from a single nucleotide to polynucleotides 1.7×10^5 nucleotides long. Thus, in electrophoresis of nucleic acids, the molecular-sieving effect is also the principal factor in separation because the charge-to-mass ratio is nearly the same for all polynucleotides. Hence, small molecules move faster than large ones. Because naturally occurring nucleic acid molecules are very large, the pore size of the gel must be large, that is, the gel must be dilute. To strengthen the gels, especially for large molecules ($M > 5 \times 10^6$), agarose (a highly porous polysaccharide) is added or sometimes agarose alone is used. Electrophoresis is done in slab gels or column gels. For RNA molecules (e.g., messenger and ribosomal RNA), resolution is far better than in zonal centrifugation (Figure 9-9) and electrophoresis is certainly the method of choice. Furthermore the distance D migrated is related to the molecular weight M by the equation

$$D = a - b \log M \tag{3}$$

in which a and b are constants, so that M is measurable if two samples of known molecular weight are included. Separation in polyacrylamide and agarose is also very good for single-stranded DNA up to $M = 50 \times 10^6$. For reasons that are unclear, in certain buffer systems single-stranded DNA molecules separate not only by size, but apparently on some other basis. For example, the two polynucleotide strands obtained by denaturing several *E. coli* phage DNAs are separable in 0.6% agarose using a Tris-phosphate buffer. We will see that this is an essential step in the determination of the base sequence of DNA by the Maxam-Gilbert procedure to be described.

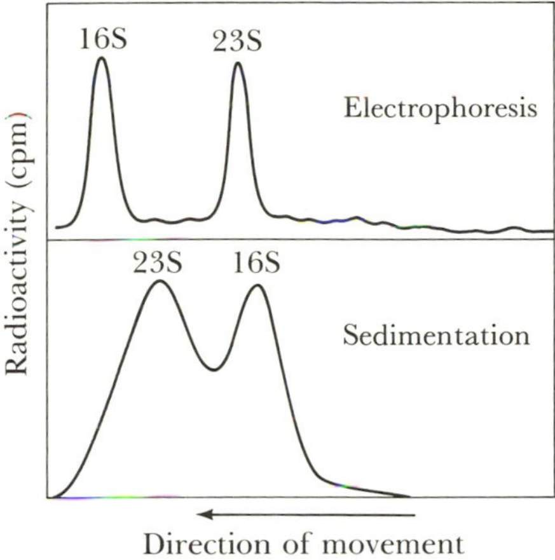

Figure 9-9
Separation of ribosomal RNA species (16S and 23S) by polyacrylamide gel electrophoresis compared with sedimentation through sucrose gradients. Note that the electrophoretic bands are narrower than the zones in sedimentation; also the relative positions of the two classes of molecules are reversed.

Detection of nucleic acids in gels

Nucleic acid bands can be detected in gels in a variety of ways. Radioactivity is usually measured by slicing the gel, solubilizing it in 0.5 M NaOH or H_2O_2, adding a suitable scintillation cocktail, and counting the sample with a scintillation counter. With material of high molecular weight, slicing can be difficult because the gel is necessarily soft to insure large pore size. An alternative is to detect radioactivity by autoradiography; in this procedure, the gel is dried and pressed firmly against autoradiographic film. Another procedure uses the dye Stains-All, which has the property of staining DNA, RNA, and proteins, and yielding a different color for each; however, the use of Stains-All requires a high concentration of macromolecules.

Electrophoresis of double-stranded DNA has been developed to such a high art and has so many variants that it will be described in a later section.

Disc electrophoresis in polyacrylamide gels

For maximum resolution, disc electrophoresis—an important refinement of zone electrophoresis—is used. The term "disc" does not refer to the narrow bands that the technique yields, but to the fact that the system uses a pH *dis*continuity. The physical arrangement is shown in Figure 9-10. The gel system is prepared in a vertical column or in a slab and consists of three separate regions: the uppermost or sample gel, the middle or spacer gel, and the lower separation gel. The sample and spacer gels are less concentrated (larger pore size) than the separation gel and are prepared in a buffer of lower ionic strength and different pH. The larger pore size in the upper gels means that molecules will be impeded less and will move faster than in the separation gel. Similarly, the lower ionic strength means higher electrical resistance so that the electric field (V/cm) is greater in the upper region and hence makes the

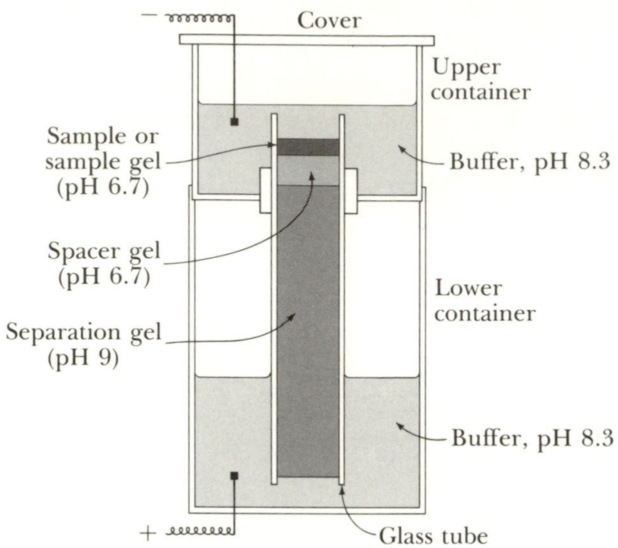

Figure 9-10
Apparatus for disc electrophoresis. The gels are separately polymerized in the column and then placed in the buffer reservoir. Sometimes a sample gel is not used and the sample is in a dense sucrose solution, which is layered with a fine pipette on the spacer gel but under the buffer. The buffer in the reservoir is usually Tris-glycine. The other buffers are Tris-chloride. After electrophoresis, the gel is pushed out. To make the band visible, the sample is either stained or, if the sample is radioactive, sliced transversely, each slice being counted in a scintillation counter. In some cases, the gel is sliced vertically and placed against autoradiographic film. The slab-gel arrangement (Figure 9-6) can also be used for disc electrophoresis.

molecules move faster than in the lower gel. This rapid movement through the upper gels results in an accumulation of material at the boundary between the spacer and separation gels. However, for reasons that are beyond the scope of this book (see Selected References for the physics of stacking) the individual components, though very close, are also *arranged in stacks* in order of mobility. Hence the spacer gel is also called the *stacking gel.* As the molecules proceed downward through the separation gel, the various zones separate according to mobility. At the end of the operation, the gel is removed from the glass tube. The zones can be identified in many ways. First, the gel can be stained by immersion in dye followed by extensive washing to remove unbound dye. (In some cases, destaining can be done electrophoretically. This is possible because, in the staining process, the dye becomes covalently coupled to the protein and the protein is coupled to the gel. Unbound dye is free to move out of the gel when the field is applied again.) The gel can be photographed if necessary or, if quantitative information is necessary, dye uptake can be measured by a recording densitometer. Second, the gel can be sliced transversely into many layers, each assayed for radioactivity, enzymatic activity, optical density at particular wavelengths, or dye uptake. Third, if radioactive, the gel can be sliced vertically and assayed by autoradiography.

The principal uses of disc electrophoresis are to determine the purity of a presumably pure protein and to analyze the components of a mixture with very high resolution (i.e., if there is a very large number of components).

Purity is, of course, a relative term and a single band may result because the impurities are at too low a concentration to be detectable. Therefore, in purity tests a large amount of protein must be used. Because the limit of detection by staining is about 1 μg, to see if something is 99% pure at least 100 μg should be used—and that quantity is sufficient only if the impurity is a single protein! Even at very high concentrations (0.5–1.0 mg) a single band is not a criterion of purity because the impurities might not have been resolved by the particular pH and/or buffer. Hence, to have some confidence about homogeneity, it is usually necessary to use several buffers and a variety of pH values.

The following example illustrates the use of disc electrophoresis to resolve a complex mixture.

Example 9-A

☐ Analysis of proteins made during infection of the bacterium *E. coli* by phage λ.

If *E. coli* is infected with phage λ a large number of phage-mediated proteins are synthesized. The relative proportions of each and the timing of their synthesis have been studied by disc electrophoresis, using a radioactive label to detect the protein. If a mixture of all the proteins in an infected cell were electrophoresed and then stained, the gel would be nearly continuously stained because there would be thousands of proteins present. However, if the bacteria are irradiated with ultraviolet light, bacterial protein synthesis is prevented because of damage to the cellular DNA. Thus, if irradiated bacteria are infected with λ in the presence of [^{3}H]leucine, phage but not bacterial proteins will be radioactive. Hence, to study the sequence of protein synthesis during infection with λ such an infection was performed, [^{3}H]leucine was added at various times after infection, and the cells were harvested several minutes after each addition of [^{3}H]leucine. The proteins were isolated and then separated by disc electrophoresis, and the bands were detected autoradiographically. It was observed that different bands were formed at different times.

When a phage mutant that makes only a fragment of a particular protein was studied, one band was missing and a new one (the fragment) appeared. Hence, the missing band corresponds to the gene that has the mutation. By the use of many such mutations, each band could be identified; in this way the sequence of synthesis of different proteins was elucidated. It is worth emphasizing that this analysis was done in the presence of all the bacterial proteins but, because they were not radioactive, they were not seen.

Electrical Parameters in Electrophoresis

The source of electrical power in electrophoretic apparatus is a power supply. These instruments are designed so that they can deliver either a constant voltage or a constant current to the electrophoresis setup. Often confusion exists concerning whether the voltage or the current is the factor that determines how far a molecule moves in a particular time interval; this is important if electrophoresis experiments are to be reproducible. The confusion arises because equation (1) states that the velocity is proportional to the applied voltage yet laboratory workers have observed that the velocity is proportional to the current and not to the voltage. The explanation, we will see, is that the velocity is in fact proportional to the voltage *seen by the molecules* and not to the applied voltage and that the value of the voltage across the gel is determined by the current. This statement will be made clear by examining the complete electrical circuit of an electrophoresis set-up, as shown in Figure 9-11.

An electrophoresis circuit consists of a voltage supply with voltage V and several resistors in series. The resistors are the wire leads from the voltage supply, the buffer reservoirs, the buffer-saturated paper wicks

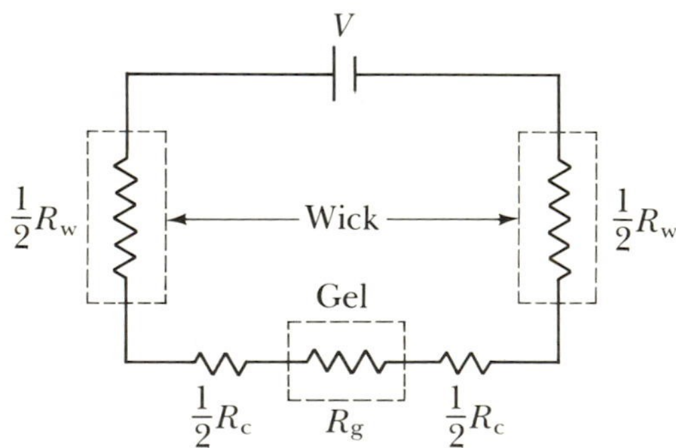

Figure 9-11
Schematic diagram for the electrical circuit of a gel electrophoresis set-up showing the five major resistances—$\frac{1}{2}R_w$ is the resistance of one wick, $\frac{1}{2}R_c$ is the resistance of the contact between a wick and the gel, and R_g is the resistance of the gel. V is the power supply voltage.

(when they are used), the contacts between the wicks and the gel, and the gel itself. The total resistance R of the circuit equals the sum of the resistances of each component because all elements of the circuit are in series. The current I in the circuit is V/R, according to Ohm's Law. The resistances of the wicks R_w, the contacts R_c, and the gel R_g have the largest values and are most important. In any circuit, the voltage across each component equals the product of the current and the resistance so that these voltages are $V_w = IR_w$, $V_c = IR_c$ and $V_g = IR_g$ respectively. From equation (1) the velocity of a particle is proportional to E, the electric field seen by the particle. This field is the potential difference per unit length within the gel or V_g/l, in which l is the length of the gel. Thus the velocity of the particle is proportional to V_g, which is *not equal* to V but *proportional* to V. The value of $R_w + R_c$ is usually much greater than R_g so that in practice the voltage across the gel is about 10% of the voltage produced by the power supply. As electrophoresis is normally performed, a gel of fixed thickness and composition is used so that R_g is the same from one day to the next. However, the values of R_w and R_c are difficult to reproduce. Thus, because $I = V/(R_w + R_c + R_g)$, I would also be variable if V were set to have the same value in all experiments. In order to have reproducible velocity, V_g, and not V, must be reproduced; because of the invariability of R_g (if l is constant), this condition is met by maintaining the *current* and not V at a constant value from one experiment to the next. In summary, the voltage V_g across the gel is the driving force that moves the molecules; this voltage is maintained constant by adjusting the circuit to have constant current. Often in the scientific literature the conditions for a particular electrophoretic experiment are described in terms of volts per centimeter. Unfortunately the value stated is invariably the voltage of the power supply divided by the length of the gel, which is certainly uninformative. In order to state the value of the electric field, one must measure V_g with a high-resistance voltmeter. The following should also be noted.

1. If a gel is increased in length from l_1 to l_2, the resistance increases from R_g to $R_g(l_2/l_1)$; however, if the current is kept constant the electric field will be the same in both gels.

2. If a gel is increased in thickness from d_1 to d_2, the resistance decreases from R_g to $R_g(d_1/d_2)$ and to maintain a constant electric field, the current must be *increased* from I to $I(d_2/d_1)$.

3. The current must not be made too high in an effort to increase mobility because the temperature of the gel may increase to a value that can alter the properties of the gel (for instance, melt it or change the electrical resistance in a nonuniform way) or even denature a protein. According to the laws of electrical circuitry, the electrical power or wattage produced in the gel is I^2R_g so that increasing I can cause a very rapid rise in temperature.

Agarose-gel Electrophoresis of Double-Stranded DNA

Naturally occurring DNA molecules usually have molecular weights that are so high that the molecules cannot penetrate even a weakly cross-linked polyacrylamide gel. Agarose solves this problem. A 0.8% agarose gel can accept DNA molecules whose molecular weight is as high as 50×10^6 and this gel is still sufficiently rigid that it can be used in the vertical configuration shown in Figure 9-6. Concentrations as low as 0.2% agarose have been used in the horizontal mode and DNA molecules having a molecular weight of more than 150×10^6 can be electrophoresed in such a gel. The introduction of agarose and the technology associated with DNA electrophoresis have revolutionized the study of DNA since its introduction in the early 1970s. In the following, its most important applications are described.

Separation of DNA by Molecular Weight

The resolution of DNA molecules by their molecular weight using agarose gel electrophoresis is truly extraordinary; good separation is achieved with molecules whose molecular weights differ by as little as 1%. Furthermore the range of applicability is from molecules containing less than ten base pairs to about 3×10^5 base pairs. For the different size ranges, gels of different concentrations must be used: 0.8–1.5% agarose for molecules up to about 5×10^4 base pairs and 0.2–0.4% for the very large molecules. This is necessary because very large molecules cannot penetrate the more concentrated gels and very small molecules would pass through the dilute gels so easily that their velocity becomes only weakly dependent on molecular weight.

Both the vertical and horizontal modes shown in Figure 9-6 are used. However, in recent years two new methods of performing horizontal gels have become popular. These procedures eliminate the wicks and make running times more reproducible. The most useful set-up uses agarose legs and is depicted in Figure 9-12A. In some experiments, electrophoresis must be carried out for a very long time and the gels tend to dry and crack. To avoid this, the submarine system shown in Figure 9-12B is used. In both cases the technique is very simple. First, the well is filled with buffer. Then a sample of about 25 microliters of DNA in buffer containing 50% glycerol is placed in each well. The glycerol raises the density of the sample so that it can be layered beneath the buffer. The DNA concentration of each sample is adjusted so that there is 0.001 to 0.01 microgram of DNA for each band that is expected. The electric field is then applied. When electrophoresis is complete, the DNA is visualized by soaking the gel in a dilute solution of the fluorescent dye ethidium bromide, whose quantum yield (see Chapter 15) is considerably enhanced when bound to native DNA. The gel is washed to remove

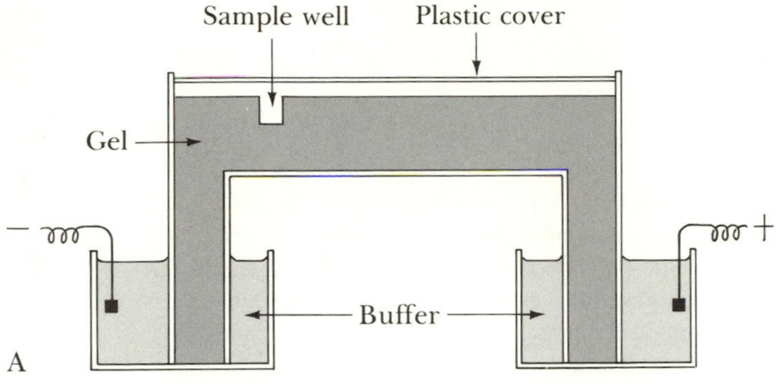

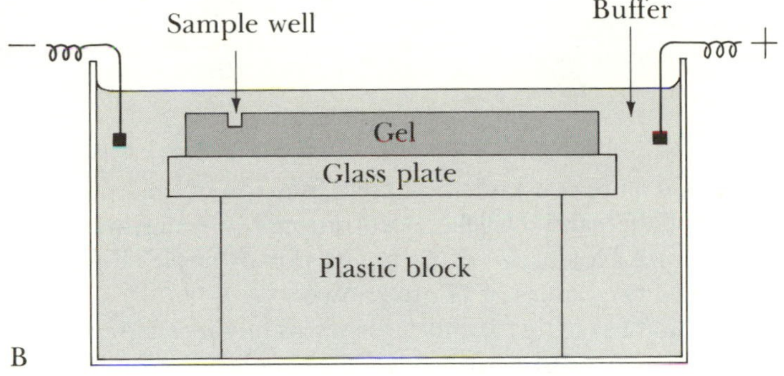

Figure 9-12
Two modes of performing horizontal gel electrophoresis: (A) agarose legs; (B) submarine gels.

unbound dye from the agarose and then it is illuminated with ultraviolet light to excite the fluorescence of the ethidium bromide. (The fluor acridine orange has also been used in the author's laboratory.) The gel with its fluorescent bands is usually photographed. Figure 9-13 shows the result of an electrophoretic separation of several DNA molecules.

Sometimes it is valuable to be able to visualize the bands before electrophoresis is complete. The submarine gel system is particularly useful when this is required. Both gel and the buffers are prepared containing ethidium bromide and, when viewing is desired, the glass plate that supports the gel is lifted out and the gel is illuminated with ultraviolet light. If the bands have not migrated far enough to achieve the desired resolution, the plate can be put back into the buffer.

The molecular weights of the DNA of each band are determined from the distance of the band from the origin. The equation that describes the relation between the molecular weight (M) and the distance (D) is again

$$D = a - b \log M$$

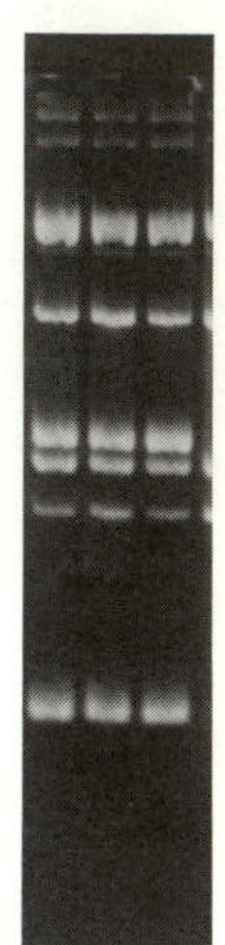

Figure 9-13
Slab-gel electrophoresis of DNA fragments in 1% agarose. The three lanes each contain the same sample, thus indicating the reproducibility of the technique. The gel was 1.4% agarose. After electrophoresis, the gel was soaked in a solution of ethidium bromide, illuminated with exciting light, and photographed to reveal the fluorescence produced by the dye bound to DNA. Because the fluorescence is strongly enhanced when bound, the gel itself does not show significant fluorescence. Each horizontal band represents a distinct DNA molecule. [Courtesy of Elizabeth Levine.]

in which a and b are constants that are a function of the time of electrophoresis, the buffer, and the gel concentration. Molecular weights can be obtained from a plot of D versus log M (for example, Figure 9-8), as long as at least two values of M are known.

An important use of this technique is restriction mapping, a procedure that can be used to distinguish one DNA molecule from other types. This is described in the following example.

Example 9-B ☐ Distinguishing two phage DNA molecules by restriction mapping.

Many bacteria synthesize enzymes called restriction endonucleases. These enzymes produce double-strand breaks in DNA at particular base sequences at least four base pairs long. Each kind of restriction endonuclease (several hundred different enzymes have been isolated from numerous organisms) acts at one and only one base sequence. These sequences occur sufficiently rarely that a small DNA molecule, for instance, that of a phage or virus, will be cleaved in no more than five positions and perhaps not at all. A large DNA molecule, such as that of a bacterium, may receive several hundred cuts. Since the susceptible sequences are rare, it is highly unlikely that two different DNA molecules will be cut in the same pattern. Therefore the number and size of the fragments produced by a particular enzyme are a fingerprint of the DNA molecule. For instance, consider two molecules whose molecular weights are the same and that are cut at the positions indicated in Figure 9-14. Molecule A will yield fragments whose molecular weights in terms of fraction of the whole molecule are 0.06, 0.10, 0.13, 0.17, 0.25 and 0.29 whereas the fragments of molecule B will have molecular weights 0.08, 0.15, 0.19, 0.23, and 0.27. The electrophoretic patterns of each are also shown in the Figure and are easily distinguishable.

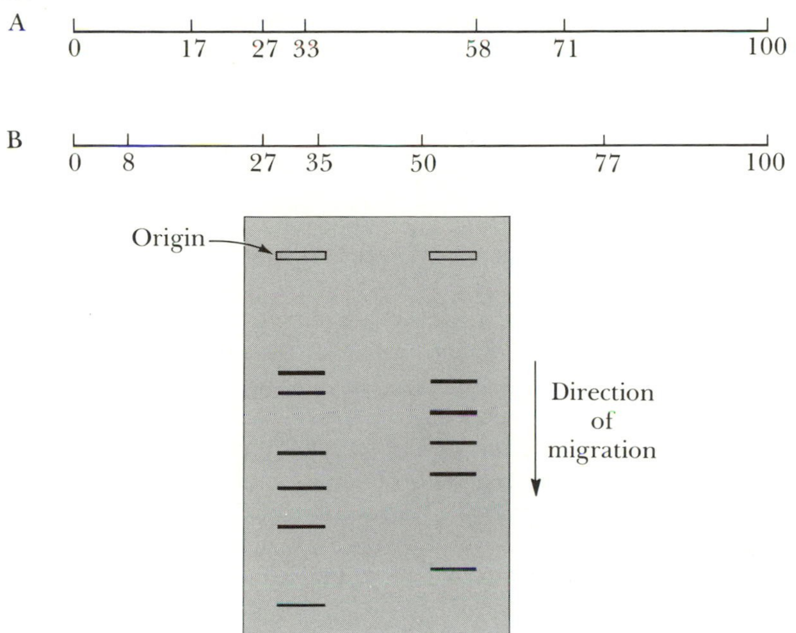

Figure 9-14
Two phage DNA molecules A and B that have the same molecular weights but different sites at which a particular restriction endonuclease makes cuts. The lower panel shows the result of electrophoresis of the fragments of each molecule digested by the restriction enzyme.

Restriction mapping enables one to identify a deletion or deletion-substitution of a known DNA molecule, as shown in the following example.

Example 9-C

☐ Location of a deletion and a deletion-substitution of phage DNA molecule A of Example 9-B.

Suppose two variants of molecule A are isolated, each having a molecular weight 95% that of the wild-type molecule. After treating variant 1 with the restriction enzyme used in Example 9-B, it is found that the 0.25 fragment is not present but a new fragment, whose size is 0.20, is observed. This means that the 5% deletion must be in the 0.25 fragment. Variant 2 has a different pattern; it lacks both the 0.13 and the 0.25 fragments but has a 0.33 fragment. Thus the deletion has removed the restriction site at position 0.58; the 0.33 fragment is the piece between sites 0.33 and 0.71 minus the 0.05 deletion. A third variant is known to have unit size but contain several bacterial genes.

The variant lacks the 0.06 and 0.25 fragments but now has 0.04 and 0.27 fragments. Thus the substitution is in the fragment between sites 0.33 and 0.58 and has introduced a new site at position 0.30.

Restriction mapping can also identify another form of a DNA molecule, as shown in the following example.

Example 9-D ☐ Detection of a circular form of a DNA molecule.

When phage A of example 9B infects its host bacterium, two types of phage DNA molecules can be isolated. By centrifiguration, it is guessed that type I is a linear DNA molecule and type II is circular. Restriction mapping of type I gives the same pattern as is found with DNA isolated from the phage. However with type II, the 0.17 and the 0.29 molecules are absent and a 0.46 (= 0.17 + 0.29) fragment is found. This is consistent with the circular form. If type II is heated to 70°C and then analyzed, the 0.17 and 0.29 but not the 0.46 fragments are found. Thus type II is a Hershey circle (Chapter 1).

Often a particular region of a DNA molecule is to be isolated in order to study the base sequence of a particular section such as an operator or a promoter. This can be done using several restriction enzymes that act at different sites. After treatment with the first enzyme, a fragment containing the region of interest can be isolated from the gel. This fragment can then be broken down with a second, third, etc. enzyme. How this is done is shown in the following example.

Example 9-E ☐ Isolation of the region 0.505–0.508 from DNA molecule A of example 9-B.

DNA molecule A of example 9-B is treated separately with restriction enzymes I, II, and III. The restriction maps shown in Figure 9-15

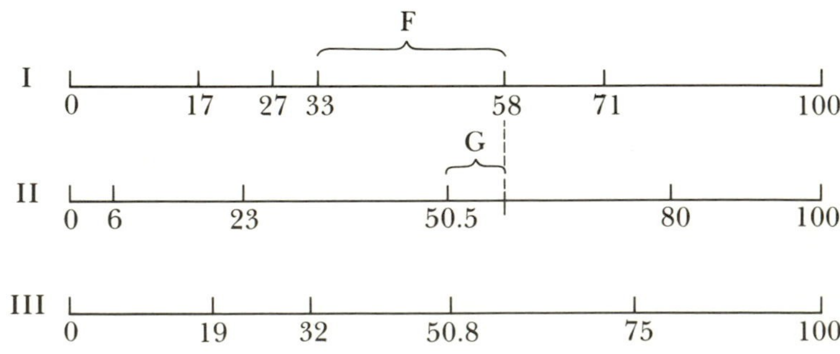

Figure 9-15
Positions of cuts of a particular DNA molecule that is cleaved by three restriction endonucleases I, II, and III.

are obtained. After treatment with enzyme I the piece of agarose containing fragment F is cut out of the gel. Fragment F DNA is isolated, treated with enzyme II, and electrophoresed. Two bands of size 0.505 − 0.33 = 0.175 and 0.58 − 0.505 = 0.075 (fragment G) are observed. Fragment G is then isolated, treated with enzyme III, and electrophoresed. Two bands of size 0.508 − 0.505 = 0.003 and 0.580 − 0.508 = 0.072 are seen. The 0.003 band (the band of interest) is isolated.

Separation of DNA Molecules According to Shape

The ability of a molecule to pass through a gel depends upon both the size and the shape of a molecule. For example, a long, rigid, rodlike molecule will encounter greater resistance than a molecule having the same size and collapsed into a dense sphere. Thus the linear, circular, and supercoiled forms of a DNA molecule do not have the same mobility.

It is difficult to predict the relative mobility of these forms of DNA because in a gel not only is the value of the frictional coefficient important (see equation 2) but also the flexibility of the molecule affects its ability to make its way through the tortuous passages of the gel. In fact, a linear molecule has a higher mobility than a nicked circle even though the circle is more compact because apparently the free ends of the linear form can snake their way through the gel with greater ease than the circular form. However, a supercoil has higher mobility than a nicked circle because the supercoil is much more compact. Curiously the compact supercoil is not as rapid as the highly flexible linear form so that, of the three, the linear form has the greatest mobility. The relative mobilities of the three forms, also depend somewhat on molecular weight. However, roughly speaking, if the linear form has a mobility of one unit, a naturally occurring supercoil would have a mobility of about 0.95 but the mobility of nicked circle ranges from 0.1 to 0.5.

A nicked circle and a non-supercoiled covalent circle have the same shape and flexibility. If a single twist is introduced into a covalent circle, the mobility of the DNA increases. A second twist increases the mobility gain. In general, as twists are introduced, the mobility of a circle increases stepwise, one increment for each twist. These increases are used as the basis of an assay for supercoiling enzymes, as shown in the example that follows.

Example 9-F

☐ Assay of the enzyme DNA gyrase.

DNA gyrase is one of a class of enzymes called DNA topoisomerases. It is found in the bacterium *E. coli* and is able to introduce negative superhelical turns into a covalent circle. It is responsible for the production of supercoiled DNA in *E. coli.* DNA gyrase has been assayed by mixing non-supercoiled covalent circles with the enzyme and, after an incubation period, electrophoresing the mixture. The

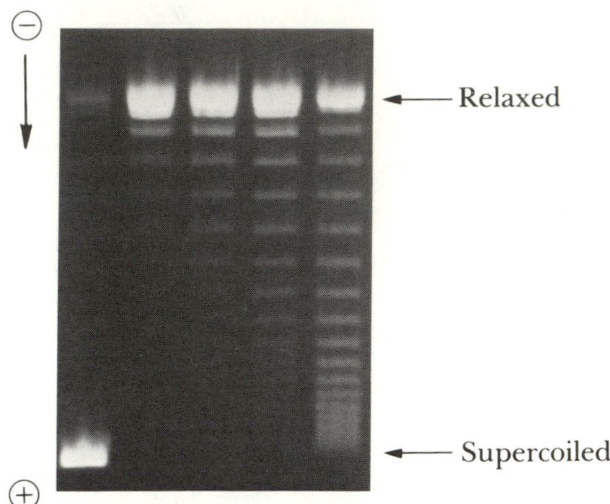

Figure 9-16
Agarose gel electrophoretic pattern of circular DNA exposed to the enzyme DNA gyrase for various periods of time. In the lane to the far left, the upper band contains DNA that has not yet been exposed to the enzyme; the lower band is supercoiled DNA added to mark position. [Courtesy of J. E. Germond.]

result of an assay is shown in Figure 9-16. Lane 1 shows the bands of non-supercoiled and naturally occurring supercoiled DNA. Lanes 2, 3, 4, and 5 show the reaction mixture as a function of time. Many intermediate bands are present, each one representing molecules with one more turn than the immediately adjacent slower band.

Example 9-G □ Detection of plasmids by gel electrophoresis.

Many bacteria contain small, independently replicating, circular DNA molecules known as plasmids. They are extraordinarily useful in the new recombinant DNA technology. An important technique is the introduction of a gene or gene cluster from one organisms into a plasmid. When this is done, the genes of interest replicate as part of the plasmid. If a plasmid is used for which there are numerous (50–3000) copies per cell, the gene cluster is amplified; that is, more gene products are synthesized by the cell and a large amount of the DNA that constitutes the genes can be isolated merely by isolating the plasmid DNA.

There are many methods to create a plasmid that contains additional DNA. These are described in the more recent textbooks on molecular biology and in some of the Selected References given at the end of this chapter. Each of these methods results in a plasmid DNA having a higher molecular weight than the original plasmid. However, following any of these procedures, only a few percent of the bacteria in a population have the new plasmid. The procedure that follows [W. M. Barnes *Science* (*Wash., DC*), 195(1977):393–394] enables the bacteria containing the new plasmid to be isolated.

Several hundred bacteria, a few of which contain the new plasmid, are placed on a nutrient agar surface and allowed to grow to produce a bacterial colony. A colony is picked up with a toothpick, most of it is suspended in a drop of a solution containing sodium dodecyl sulfate, and the cells break open. (A small fraction of the colony is saved.) A small aliquot is placed in a well of a 1% agarose gel. The process is repeated with many colonies—usually a gel having sixteen wells is used. The samples are then electrophoresed and the gel is stained with ethidium bromide. The DNA of the bacterial chromosome is so large that it cannot penetrate the gel but the plasmid DNA migrates with little or no interference from the chromosomal DNA. Most of the bands that are observed migrate to the same position—each of the colonies producing these bands contain the original plasmid—but an occasional band is found that has not moved as far as the others. Since a larger DNA molecule moves more slowly than a smaller molecule, such a band consists of the new plasmid DNA that contains the additional bacterial DNA. A portion of each colony was saved so that this can be used to grow more bacteria containing the desired genes.

The Southern Transfer Procedure

When the large DNA of bacteria or eukaryotic cells is exposed to a restriction endonuclease, a great number of fragments are produced. Usually there are so many fragments that none can be resolved by electrophoresis in an agarose gel. Instead, DNA appears to be present throughout the gel. The transfer-and-hybridization method developed by Edwin Southern and known as the Southern transfer enables the fragment containing a particular gene to be identified. This procedure has many uses in the molecular biology of eukaryotic cells but only two will be described.

The principle underlying the procedure is that a specific DNA can be localized by hybridization with radioactive DNA or RNA molecules. The complications are that hybridization does not occur easily within a gel and that the DNA that has been electrophoresed is usually double-stranded. The solution is to denature the DNA and then remove the single-stranded DNA from the agarose after electrophoresis. DNA (or any other molecule) in a gel can be removed simply by squeezing the gel because the DNA is not bound to the gel matrix and the gel is mostly liquid. Thus the following procedure is used (Figure 9-17).

Following electrophoresis the gel is soaked in a denaturing solution (usually NaOH) so that all DNA in the gel is converted to single-stranded DNA. A large sheet of nitrocellulose membrane filter (Chapter 7) is then placed on top of several sheets of ordinary filter paper. The gel is then placed on the nitrocellulose filter and covered with a glass plate. A weight is placed on the top of the stack and the liquid is squeezed out of the gel. The liquid passes downward through the nitrocellulose filter, which adsorbs the DNA molecules very tightly (see Table 7-1, Chapter 7). The remaining liquid passes through the nitrocellulose and is absorbed by the filter paper. DNA molecules do not diffuse very much so that if the gel and the nitrocellulose filter are in firm contact, the

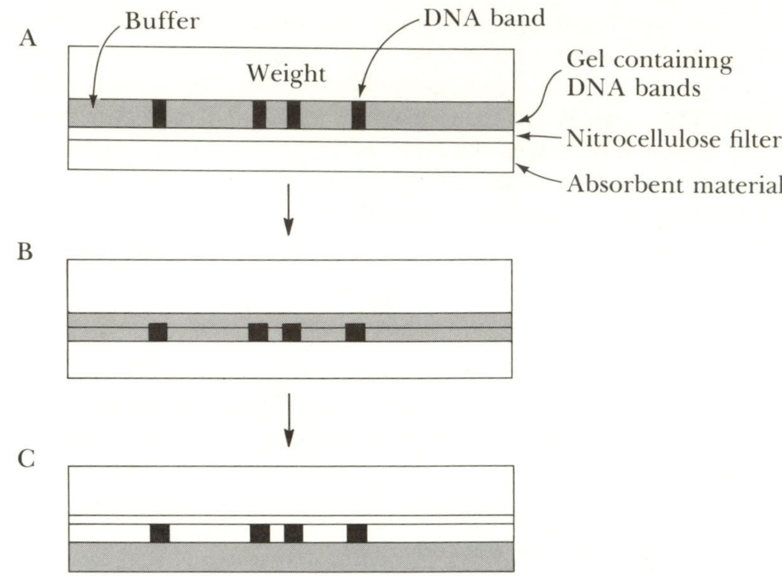

Figure 9-17
The Southern transfer technique. Panel A shows the stack at the time the weight is applied. The weight forces the buffer (shaded area), which carries the DNA, into the nitrocellulose, as shown in panel B. The lowest layer absorbs the buffer but not the DNA, which is bound to the nitrocellulose, as shown in panel C.

positions of the DNA molecules on the filter are identical to the positions in the gel. The nitrocellulose filter is then dried in vacuum, which insures that the DNA remains on the filter during the hybridization step. The dried filter is then moistened with a very small volume of a solution of ^{32}P-labeled DNA or RNA, placed in a tight-fitting plastic bag to prevent drying, and held at a temperature suitable for renaturation usually for 16–24 hours. The filter is then removed, washed to remove unbound radioactive molecules, dried and autoradiographed with X-ray film (Chapter 5). The positions of the blackening on the film indicate the location of the DNA molecules whose DNA base sequences are complementary to the sequences of the added radioactive molecules. The Southern transfer procedure has many uses. One common use is to determine whether a particular sequence is present in a collection of molecules. This is shown in Example 9-H. Another use is to determine the size of the molecule containing a particular sequence, as shown in Example 9-I.

Example 9-H ☐ Identification of a plasmid containing hemoglobin DNA.

Recombinant DNA techniques are often used to prepare bacterial plasmids containing a particular mammalian gene. Sometimes the so-called "shotgun method" is used in which a particular plasmid is allowed to pick up random fragments of mammalian DNA. Then many plasmids containing mammalian DNA are isolated and from this collection is selected the plasmid containing the gene of interest. In the case of hemoglobin DNA, one would use the Barnes method described in example 9-G to obtain a gel containing plasmids from 16 to 20 different bacterial colonies. This plasmid DNA could be transferred by the Southern technique and hybridized with hemoglobin RNA of specific activity. If a plasmid band becomes radioactive, it would have to contain hemoglobin DNA.

Example 9-I

☐ Location of integrated viral DNA molecules in virus-induced cancer cells.

Certain mammalian cell lines become cancer cells (transformed cells) following infection with oncogenic viruses, such as simian virus 40 (SV40). It is believed that the viral DNA, which is circular, is broken to form a linear molecule and then inserted into the mammalian chromosomal DNA. This idea is easily tested by a Southern transfer. DNA is isolated from a transformed cell. In the course of isolation, the DNA is broken by hydrodynamic shear into thousands of fragments having about a hundredfold range of molecular weights. The viral DNA is very small (molecular weight $= 3.6 \times 10^6$) and is not broken. The DNA sample is electrophoresed. By ethidium bromide fluorescence only a broad smear of DNA is observed because of the breakage of the DNA. The SV40 DNA can be localized by a Southern transfer using ^{32}P-labeled SV40 DNA as a probe. If the SV40 DNA is not integrated, radioactivity is found only in a single band in the position expected for intact SV40 DNA. Possibly three discrete bands would be found if the normally supercoiled SV40 DNA were converted in part to nicked circles and linear molecules during isolation. However, if the DNA were integrated, radioactive DNA would be found at all positions in the gene; this is because the DNA is fragmented to random sizes so that because the DNA is isolated from a huge number of cells, the SV40 sequences will be found in fragments of all sizes. The latter was observed indicating that the SV40 DNA is integrated.

In an elegant variation of this experiment, it was shown that the position of integration in the chromosome is not always the same. The DNA was treated with a restriction endonuclease that makes one cut in SV40 DNA but hundreds of cuts in mammalian DNA. After electrophoresis and a Southern transfer, there were two bands of radioactive SV40 DNA. The explanation for the two bands is as follows. The integrated DNA is necessarily flanked by two restriction cuts made in the chromosomal DNA; this would generate a fragment having a unique cut but, because the SV40 suffers a single cut, two fragments are produced. It is unlikely that the three cuts would be located such that the two fragments were equal in length; thus, two fragments of different size are formed and these account for the two bands. A collection of transformed cell lines was used for such an analysis by Gary Ketner and Thomas Kelly [*Proc. Natl. Acad. Sci. U.S.A.* 73 (1976):1102–1106]. If the SV40 molecule were always in the same position, the band pattern after a Southern transfer would be the same for all samples. However, if the positions were not the same, it would be expected that the flanking cuts in the chromosome plus the cut in the viral DNA would generate a pair of fragment whose size differed for each cell line. A pair of bands was found for each cell line but at different positions on the nitrocellulose filter used in the transfer. Thus, integration occurs at different positions in different cell lines.

Northern and Western Transfers

Variations of the Southern transfer procedure have been developed in which macromolecules other than DNA can be transferred from a gel to paper. The transfer of RNA and protein has been (humorously) named a "Northern" and "Western" transfer, respectively. Since neither RNA nor proteins bind efficienctly to paper, a pretreatment of the paper is required. Either a nitrobenzyloxymethyl group or an aminobenzyloxymethyl group is covalently coupled to paper via the methyl group. (Paper prepared in this way is commerically available from Schleicher and Schuell, Inc. under the trade name Transa-Bind.) The nitro group or the amino group is then diazotized, which produces an active group that can be covalently coupled to RNA and proteins. The transfer is then carried out using the same protocol as in a Southern transfer. The result is covalent binding of RNA or protein to the paper as shown below.

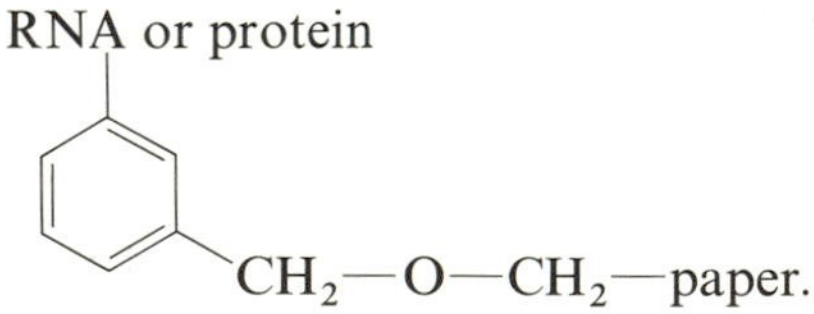

RNA or protein

$$\text{(RNA or protein)}-C_6H_4-CH_2-O-CH_2-\text{paper.}$$

The bound RNA can be hybridized either with radioactive RNA or single-stranded DNA; proteins are usually identified by reaction with radioactive antibodies. As in the Southern transfer procedure, radioactivity is detected by autoradiography of the paper. For further information see J. Alwine, D. J. Kemp, and G. Stark, *Proc. Natl. Acad. Sci. U.S.A.* 74(1977):5350–5354 (Northern); and J. Renart, J. Reiser, and G. Stark, *Proc. Natl. Acad. Sci. U.S.A.* 76(1979):3116–3120 (Western).

A very recently developed technique, known as *protein blotting*, has been used to detect proteins that bind double-stranded DNA [B. Bowen, J. Steinberg, U. K. Laemmli, and H. Weintraub, *Nucleic Acids Res.* 8 (1980):1–20]. In this technique proteins in a gel are transferred to a nitrocellulose filter; as indicated in Table 7-1, double-stranded DNA does not bind to nitrocellulose. If the filter is incubated with [^{32}P]DNA, the DNA will stick to DNA-binding proteins but not anywhere else. Thus by autoradiography, the bands corresponding to DNA-binding proteins can be identified.

All of the transfer procedures described so far utilize diffusion as a means of moving molecules from a gel to paper or nitrocellulose. This is a slow process in which there is both diffusional spreading of the bands and some loss of material. A novel technique, known as *Blitz blotting*, uses electrophoresis between the gel and either diazotized paper or nitrocellulose to reduce the transfer time about tenfold. [M. Bittner, P. Kupferer, and C. F. Mouis, *Anal. Biochem.* 102(1980):459–471]. Transfer occurs with little or no loss and there is no significant increase in band width.

Sequencing of DNA Molecules

Several new methods for determining the base sequence of DNA were developed in 1977. In each method double-stranded DNA is subjected to several treatments that generate a family of short single-stranded molecules. The treatments chosen cause base-specific cleavage of the single-stranded DNA; for instance, the polymer may break only at the 5′ side of deoxyadenosine residues. This means that once the length of a fragment has been determined by gel electrophoresis to be n nucleotides long, one can conclude that adenine was at position $n + 1$. Although each of the sequencing methods is important, we shall describe only the procedure of Allan Maxam and Walter Gilbert [*Proc. Natl. Acad. Sci. U.S.A.* 74(1977):560–564]. The other procedure, developed by Frederick Sanger, is described in papers listed in the Selected References at the end of this chapter. Gilbert and Sanger shared the 1980 Nobel Prize in Chemistry for this work.

The Maxam-Gilbert procedure (Figure 9-18) is the following. A large DNA molecule is fragmented by digestion with one or more restriction endonucleases until fragments are obtained that have no more than 100 base pairs. The fragments are electrophoresed in a gel and one fragment whose sequence is to be determined is isolated. Each of the single strands comprising the fragment has one 3′-hydroxyl and one 5′-phosphate terminus. The two 5′-phosphate groups are at opposite ends of the fragment because the two strands are chemically antiparallel (Chapter 1). The 5′-phosphate groups are enzymatically removed and replaced with [^{32}P]phosphate, which is radioactive. The two strands must be separated prior to sequencing each strand. This is done by denaturing the radioactive fragment with alkali and then electrophoresing it in a polyacrylamide gel at neutral pH. The two separated strands, which are somewhat collapsed and partially hydrogen bonded at neutral pH, usually have slightly different shapes and hence different mobilities. Thus, the strands form two bands in the gel and they can be

Production of **G + A** Fragments

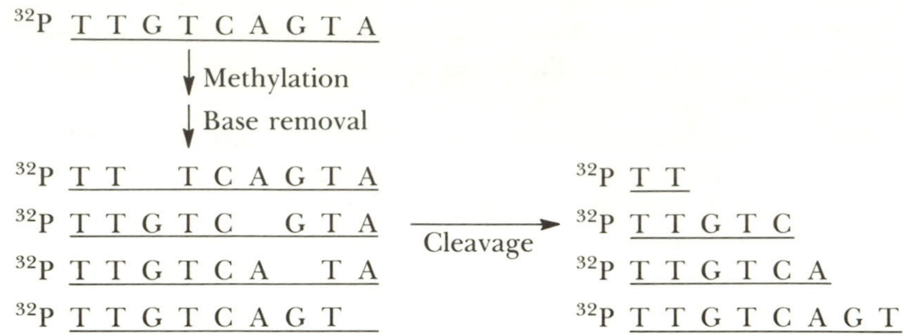

Production of **C + T** Fragments

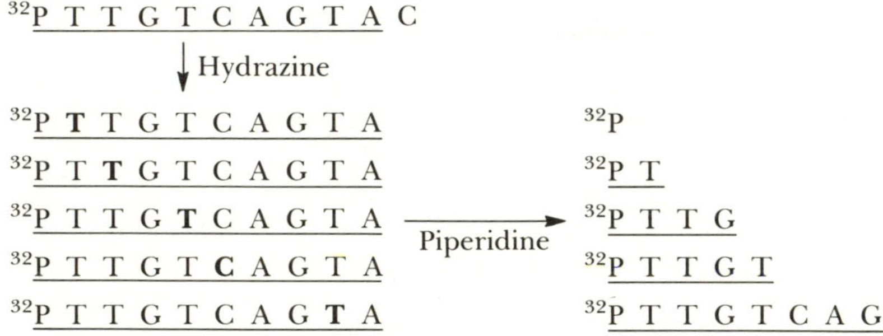

Figure 9-18
A diagram showing the production of the **G + A** and **C + T** fragments by the Maxam-Gilbert sequencing procedure. In the hydrazine-treated molecules the bold-faced letters indicate the affected bases. The generation of the **G-only** and **C-only** fragments is not shown but is explained in the text.

eluted and thereby purified. A sample containing only one of the two strands is divided into two portions. To one (I) dimethyl sulfate is added. This methylates both adenine (A) and guanine (G), but the reaction with G is five times as fast as that with A. The reaction is not carried to completion but only to the extent that one A or G base in fifty to one hundred bases is methylated. Since the DNA strand contains no more than 100 bases and roughly half are G or A, then, on the average, no more than one base per strand is methylated; of course, the particular A or G that is methylated differs in each strand. The methylated DNA is also divided into two aliquots, Ia and Ib. Sample Ia is heated to remove all methylated bases and then treated with alkali, which cleaves the sugar-phosphate chain at the site of the base that has been removed. This generates a set of fragments of varying size and the number of nucleotides in each fragment is determined by the positions of G and A.

Sample Ia is said to contain the **G-only** fragments because **G** was primarily methylated. Sample Ib is treated with diluted acid, which removes mainly methylated A, and then with alkali to cleave the sugar-phosphate chain at the site of an A. Thus fragments are produced whose size is determined mainly by the position of the A—these are called the **G + A** fragments. Note that every **G-only** fragment is also present in the **G + A** collection and that because A is methylated at one-fifth the rate of G, there are fewer A fragments than G fragments in the **G + A** collection. This can be made clearer by considering a 100-base molecule that contains only two G's, one each at positions 18 and 32 and two A's at positions 14 and 50. All remaining bases are either cytosine or thymine. In a collection of 10^5 molecules there are 2×10^5 G's and 2×10^5 A's that are available for reaction. If reaction conditions are such that 1 G in 100 and 1 A in 500 are methylated, then 2000 G's and 400 A's react. All the G's are presumably equally reactive as are all the A's so that the following radioactive fragments are produced:

A + G fragments		**G-only** fragments	
no. of fragments	nucleosides/fragment	no. of fragments	nucleosides/fragment
200	49 (ended with A)	0	49
1000	31 (ended with G)	1000	31
1000	17 (ended with G)	1000	17
200	13 (ended with A)	0	13

Two important points should be noted: (1) When a 5′-^{32}P-labeled molecule is cleaved, only one of the two fragments produced contains ^{32}P. (2) If the cleavage is at position $n + 1$ (counting from the ^{32}P-labeled terminus), the number of bases in the fragment is n. This means that there is no fragment that identifies the terminal nucleotide so that this identification requires a different technique.

The two samples Ia (**G + A**) and Ib (**G-only**) are electrophoresed at an elevated temperature in a 20% polyacrylamide gel containing 8 M urea, a denaturant that prevents hydrogen bonding. The urea and the high temperature keep the fragments single stranded so that the molecules are fractionated exclusively by their size. The pore size of the gel is very small because of the high concentration of polyacrylamide, and fragments differing in length by only one nucleotide are resolved. The gels are very long (about 200 cm) in order to allow sufficient length for about fifty bands to be resolved. For technical reasons, the mononucleotide that would identify the penultimate base cannot be detected unambiguously so that this assignment is carried out by a different procedure; this will be discussed shortly. The bands are located by autoradiography on X-ray film. Since each fragment contains only

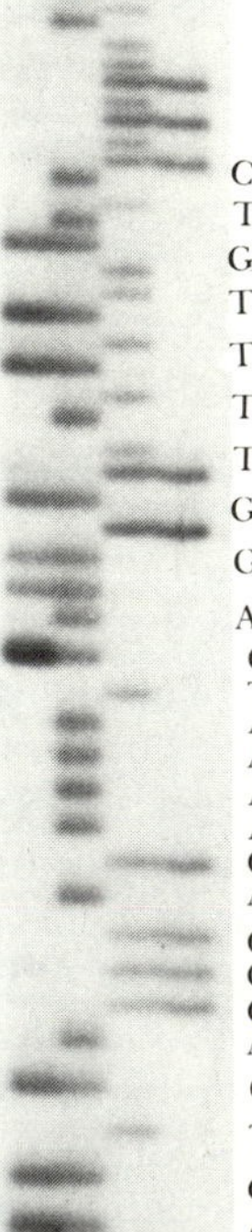

Figure 9-19
A Maxam-Gilbert sequencing gel. The sequence is read from the bottom. The first thirty-five bases of the readable sequence are identified in the figure. The remainder are not labeled, for the sake of clarity. The next seventy bases, which are easily read from the gel, are TCTCTTATATA AAACACCCGCCTTCCATAGAGTGTGTAATAGTG TCAGTTGAGTATGTACATGCGTATAG. [Courtesy of Allan Maxam.]

one ^{32}P atom, each fragment contributes equally to the blackening of the film so that the total blackening is proportional to the number of the fragments and not to the size. Thus, in the **A** + **G** lane, there will be four bands—two intense ones corresponding to polymers containing 17 and 31 nucleotides and two light ones corresponding to polymers containing 13 and 49 nucleotides. In the **G-only** lane, there will be light bands at exactly the position of the light bands in the **A** + **G** lane. In practice, five times as much DNA is put in the **G** + **A** band to make the bands sufficiently visible that there is no question about their presence.

Sample II is used to identify the positions of cytosine (C) and thymine (T). This sample is also divided into two portions IIa and IIb, which are reacted with hydrazine in either dilute buffer (IIa) or 2 M NaCl (IIb). Each sample is then treated with piperidine, which breaks the sugar-phosphate chain at sites of bases that have reacted with hydrazine. In sample IIa (**C** + **T**) the reaction occurs equally with C and T; in sample IIb the reaction is with **C-only**. The length of these fragments is, as in the case just explained, determined by the positions of C and T. Both **T** and **C** fragments produce bands in one lane and these are of equal intensity; the other band contains **C** fragments only.

All four samples Ia, Ib, IIa, and IIb are usually electrophoresed simultaneously so that all bands are seen in a single gel. This enables the sequence to be read directly from the gel. Figure 9-19 shows a photograph of a sequencing gel. The shortest fragments are those that move the fastest. Each fragment contains the original [5′-^{32}P]phosphate group so that the sequence is read from the bottom to the top of the gel. Note that *when a band in the figure is labeled with base X, this means that X was removed from the* 3′ *end of the molecule to generate the fragment.*

It was mentioned above that neither the first nor the second base, measured from the 5′ end, can be identified by this procedure. This information can be obtained by sequencing the complementary single strand, in which case the two unidentified bases from the first strand are complementary to the 3′-terminal bases of the second strand. Sequencing both strands also had the advantage that the sequence determined from one strand confirms the sequence of the other strand.

Although the gels can resolve two fragments differing in length by one nucleotide, they are not capable of distinguishing one hundred different fragments simultaneously, unless the gel is exceedingly long. To avoid this problem, one usually divides all samples into three aliquots and electrophoreses each set of four samples for different times. For example, run 1 might give the sequence of bases 3 through 40, run 2 bases 35–75 and run 3 bases 65–100. By observing the common sequences (for example, bases 35–40 of runs 1 and 2), the complete sequence can be worked out, as indicated in the figure.

This method has been used to determine the base sequence of a DNA molecule having more than 37,000 base pairs. These larger molecules are broken down with restriction endonucleases to a set of fragments each of which is separately sequenced.

Sequencing a Protein-Binding Site: DNase Footprinting

Determination of the base sequence with which a DNA-binding protein interacts is essential to understanding the mechanism of action of these proteins. If the binding is sequence-specific, a variation of the sequencing procedure just described enables the bases in that sequence to be identified. The technique, which is called DNase footprinting or simply footprinting, is based upon the assumption that if a protein binds to a particular sequence of bases, the phosphodiester bonds in that sequence are shielded so that they are resistant to cleavage by a DNase. The procedure is shown in Figure 9–20.

Consider a sequence in a DNA molecule that contains a binding site in a region of the DNA that is bounded by two known sites of cleavage by two *different* restriction enzymes (e.g., *Hae*III and *Bam* enzymes). The DNA is cut first by the *Hae*III enzyme and the 5′ terminus is labeled with ^{32}P by polynucleotide kinase. Then the DNA is cut again with *Bam* enzyme and the fragment is isolated. This fragment has a ^{32}P at *only one end*. Let us assume that this fragment consists of 39 nucleotides. If the fragment is digested with a DNase whose cleavage is independent of sequence, and the number of cuts is, on the average, one per fragment, and the DNA is then denatured and electrophoresed in the denaturing gel used in the Maxam-Gilbert sequencing procedure, there will result a family of ^{32}P-labeled single-stranded fragments ranging in length from 1 to 38 nucleotides. However, suppose, as shown in the figure, that the DNA-binding protein is present and phosphodiester bonds 15 through 26 are protected, then the family of ^{32}P fragments will range in size from 1 to 14 nucleotides and from 27 to 39 nucleotides. The sizes of the missing fragments indicate the region of the fragment to which the protein is bound. An independent determination of the sequence of the fragment by the Maxam-Gilbert procedure shows which bases are in the protected sequence.

If the labeled fragment is prepared by making the *Bam* cut before the *Hae*III cut, the other 5′ terminus is labeled and binding to the complementary DNA strand can be examined.

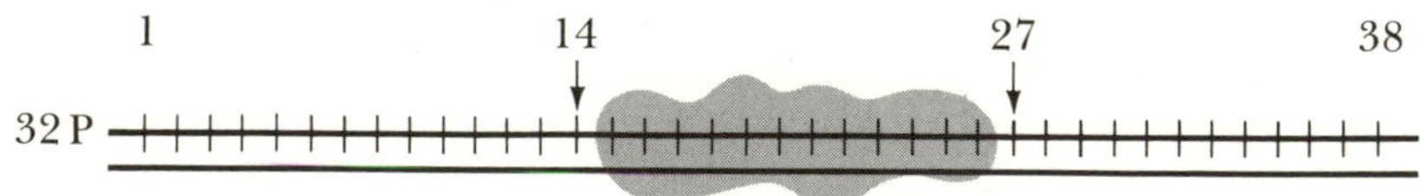

Figure 9-20
A DNA fragment containing thirty-nine nucleotides and labeled with ^{32}P at one 5′ terminus. The thirty-eight phosphodiester bonds of the labeled strand are numbered. A protein is bound to the DNA and protects bonds 15 through 26.

Gradient Electrophoresis— A High-Resolution Method in Separating Macromolecules

A pore-gradient gel is a gel in which the size of the pores decreases in the direction in which the molecules migrate in an electric field. The gradient is formed by an increase in the concentration of polyacrylamide in the direction of migration. When the electric field is applied, molecules move through the gel according to their mobility. As the pore size decreases, larger molecules slow down and ultimately stop moving at a point in the gel at which the pore size is so small that no further forward motion is possible. Electrophoresis is continued until the system is at equilibrium and all of the molecules are at rest. The result is a set of very narrow bands. An important point is that molecules separate only by size and not by mobility. Thus given sufficient time, a small molecule that is weakly charged will pass a larger strongly charged molecule even though the larger one may have a much higher mobility. This method is useful for separating native proteins—it is not necessary to add sodium dodecyl sulfate, which dissociates subunits, as in the SDS-PAGE technique.

In many cases it is not necessary to wait until equilibrium is reached because good separation may be achieved much sooner; in this case, separation *is* based on the relative mobilities, which may be enhanced in a pore-gradient gel.

Isoelectric Focusing

Proteins are ampholytes—that is, they contain both positively and negatively charged groups. All ampholytes have the property that their charge depends on pH; they are positively charged at low pH and negatively charged at high pH. For every ampholyte, there exists a pH at which it is uncharged, and this is called the *isoelectric point*. At the isoelectric point, the ampholyte will not move in an electric field. If a protein solution is placed in a pH gradient, the molecules will move until they reach a point in the gradient at which they are uncharged; then they will cease to move. With a mixture of different proteins, each type of molecule will come to rest at a point in the pH gradient corresponding to its own isoelectric point. This method of separating proteins according to their isoelectric points in a pH gradient is called isoelectric focusing and it is the electrophoretic analog of centrifugation to equilibrium in a density gradient (see Chapter 11). The process of the migration of two different proteins with different isoelectric points is shown schematically in Figure 9-21.

The pH gradient is established in an unusual way. If it were established by simply allowing two buffers at different pH to diffuse into one another

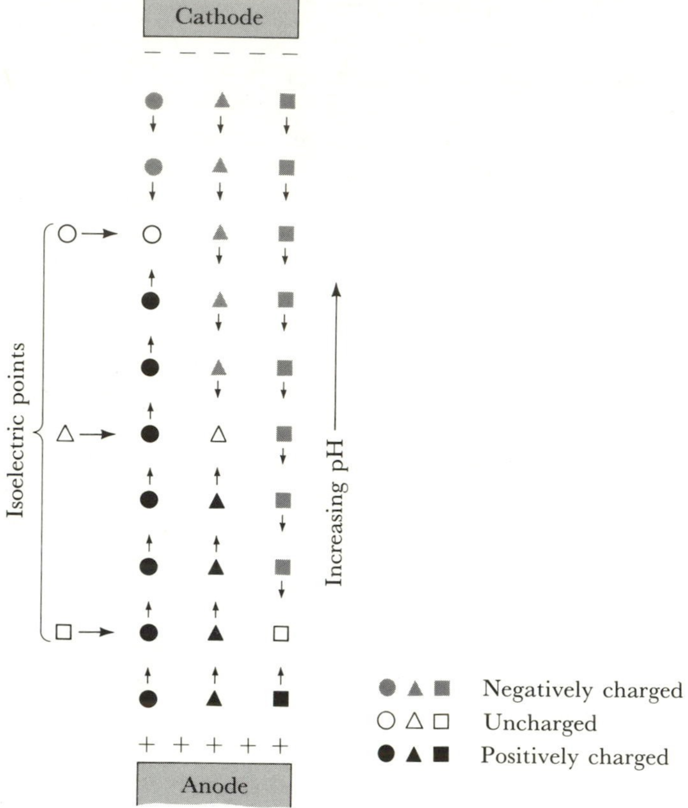

Figure 9-21
The process of migration in isoelectric focusing. Proteins move until they reach a position in the pH gradient at which they are uncharged.

or by mixing two buffers in the way that is standard for preparing concentration gradients (Chapter 8), the resulting gradient would not be stable in an electric field because the buffer ions would migrate in the field; fractionation of the macromolecules could not occur because the macromolecules would migrate much more slowly than the pH gradient is disrupted. The method used to produce a stable pH gradient consists of distributing a mixture of synthetic, low-molecular-weight (300–600) polyampholytes (multicharged structures) that cover a wide range of isoelectric points (up to one thousand different isoelectric points per interval of one pH unit). These polyampholytes are usually mixed polymers of aliphatic amino and either carboxylic or sulfonic acids. They are commercially available as Ampholine, Pharmalyte, and Bio-Lyte. A pH gradient is established by starting with a mixture in distilled water of polyampholytes having isoelectric points covering a range of either two or seven pH units (depending on the resolution required). Before the application of an electric field, the pH throughout the system is constant and is averaged from all the polyampholytes in the solution. When the field is applied, the polyampholytes start to migrate. Because of their own buffering capacities, a pH gradient is gradually established. Soon each particle will come to rest in this self-established gradient at the point corresponding to its own isoelectric point. If the mixture contains proteins of different isoelectric points, they will migrate (but much more slowly) to the positions corresponding to their isoelectric points as long as the concentration and buffering capacity of the protein (which is also a polyampholyte) is not so high that the pH gradient is disrupted.

When the goal of an experiment is to purify a particular protein, an arrangement of the type shown in Figure 9-22 is used. The pH gradient is formed in a water-cooled glass column containing a cathode tube and a cathode. The tube is filled with a uniform concentration of the polyampholyte prepared in a sucrose concentration gradient to eliminate convection. The sample material is also contained in the polyampholyte suspension. The cathode tube is filled with a strong base (typically triethanolamine) and the main column is overlaid with phosphoric acid; the anode is in this acid layer. The valve at the bottom of the cathode tube

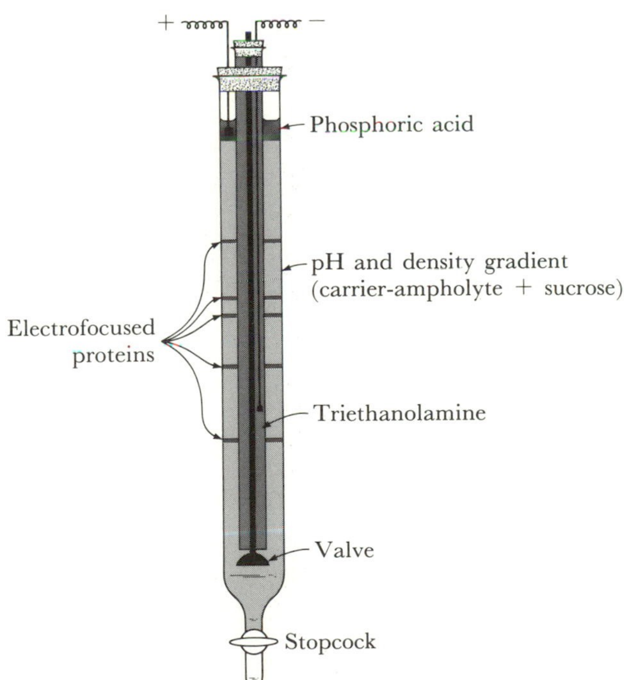

Figure 9-22
Apparatus for isoelectric focusing. A description of its operation is given in the text.

is opened, followed by the application of a few hundred volts between the electrodes. The polyampholyte near the cathode will have a negative charge and will move to the anode. From one to three days later, the system will be at equilibrium and the proteins distributed throughout the pH gradient according to their own isoelectric points. The tube is then drained and fractionated through the stopcock at the bottom. The various proteins can be detected by spectrophotometry, enzyme activity, or radioactivity.

A substantial improvement in the resolution of individual molecules is obtained by using as a supporting medium a continuous agarose or polyacrylamide gel rather than a sucrose concentration gradient. When this is done, proteins whose isoelectric points differ by 0.01 pH unit are readily separated. The single requirement of the gel is that the pore size must be so large that molecular sieving is unimportant and the isoelectric point is the single factor that determines the position of the band. This means using 2% polyacrylamide or, even better, 1% agarose. Tubular gels are sometimes used but, because of their greater capacity, horizontal slab gels are preferable. The gel initially contains the polyampholyte. The proteins may be present throughout the gel initially but they are usually applied in a single zone as in ordinary gel electrophoresis. Following focusing, the bands are visualized by staining with Coomassie brilliant blue and Crocein scarlet. This mode of isoelectric focusing is used to test the purity or composition of a sample. Two examples of isoelectric focusing in a gel are shown in Figure 9-23. Panel A shows a standard isoelectric focusing gel; panel B shows the high resolution obtainable when polyampholytes are used that produce a smaller pH gradient—this is called *narrow-range amplification.*

A variant of this procedure can be used as a preparative method. In the technique just described, the gel is continuous. If the gel is in beaded form instead (such as Sepharose, see Chapter 8), isoelectric focusing can also be carried out, exactly as above. However, after equilibrium is achieved, the beads containing the protein of interest can be removed with a spatula and eluted as in chromatography and the protein is thereby purified.

Combined Electrophoresis and Chromatography

The separation of molecules is often vastly improved by combining electrophoresis and chromatography. The usual method is to use a two-dimensional separation; that is, a sample is applied to a planar support and chromatographed in a particular dimension. The support, which is usually a paper of some kind, is then dried and electrophoresis is carried out in the perpendicular direction. Separation, such as that shown in Figure 9-24, can easily be obtained. It is of course irrelevant whether the chromatography or the electrophoresis is performed first.

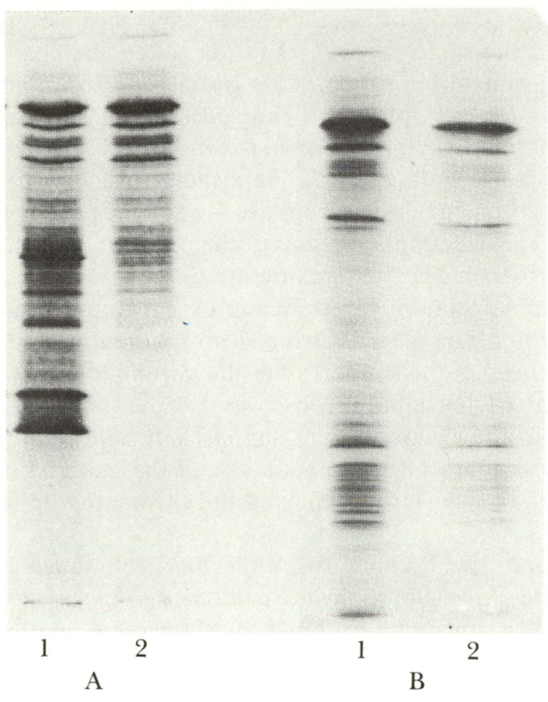

Figure 9-23
Two protein mixtures (1 and 2) separated by isoelectric focusing. (A) Normal system; (B) high-resolution system. [Courtesy of Bio-Rad Laboratories.]

Chromatography alone

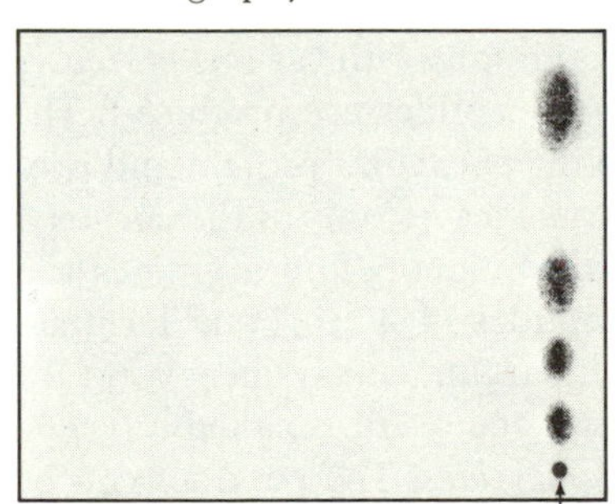

Chromatography followed by electrophoresis

1 2 3 4 5 6 7 8 9 10 11 12 13 14 15 16 17

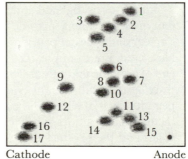

Cathode Anode

Figure 9-24
Two-dimensional chromatography combined with electrophoresis to separate amino acids: chromatography in lutidine and water in a ratio of 2 to 1; electrophoresis at pH 2.25. The numbered spots are (1) tryptophan; (2) tyrosine; (3) leucine; (4) phenylalanine; (5) methionine; (6) threonine; (7) hydroxyproline; (8) proline; (9) alanine; (10) serine; (11) glutamine; (12) glycine; (13) glutamic acid; (14) asparagine; (15) aspartic acid; (16) arginine; (17) lysine.

In the procedure just described, both electrophoresis and chromatography are performed on the same support. However, often the chromatography is best performed on a support that is different from that used in electrophoresis—for example, on a charged support such as DEAE-cellulose (diethylaminoethyl) paper or in a thin layer. Frederick Sanger developed a simple technique for accomplishing this. Electrophoresis is first performed on the desired support—usually a paper or cellulose acetate film. The electrophoregram is then placed in firm contact with the support desired for chromatography. For example, a sheet of paper or cellulose acetate is layed on top of a prepared thin layer. A buffer or some liquid is placed on the electrophoregram and allowed to pass slowly through the support onto the chromatographic support. The composition of this liquid is chosen so that the substances of interest do not bind to the electrophoresis support but bind tightly to the chromatographic support. Usually about 80% of the material is transferred. The electrophoregram is then removed and chromatography is performed.

In the following sections, we examine two important techniques in which electrophoresis and chromatography are combined—*fingerprinting*, which is performed on a single support, and *homochromatography*, in which a transfer is performed.

Fingerprinting: A Special Application of Electrophoresis and of Paper Chromatography to the Study of Proteins

An important problem in molecular biology is to identify an isolated protein or to identify the site of an amino acid change in a mutant protein. The fingerprinting technique developed by Vernon Ingram allows this to be done relatively simply.

If a protein is digested under defined conditions with various *proteases* (enzymes that break peptide bonds), small peptides are produced. The number and types of peptides depend on the particular protein and protease used. For example, the enzyme trypsin cleaves only at the carboxyl end of arginine and lysine. Thus, a protein containing four arginines and three lysines will usually yield eight peptides (4 + 3 + one terminal peptide), each of which will end either with arginine or lysine, except for the terminal peptide. When trypsin is used, the result is called a *tryptic digest* and the peptides are called *tryptic peptides*. The peptides can be separated by two-dimensional paper chromatography or by chromatography in one direction and electrophoresis in the other direction (which was Ingram's original procedure). The result of the separation is a distribution of spots that is rarely, if ever, duplicated by a second protein. Such a peptide map is called a *fingerprint*—an example is shown in Figure 9-25. Note that it is not the two-dimensionality that makes the pattern unique but only the amino acid sequence and the specificity of the protease cleavage. The two-dimensional system merely improves the resolution, enabling the experimenter to see more spots clearly. The term fingerprint is, however, only used for the two-dimensional pattern.

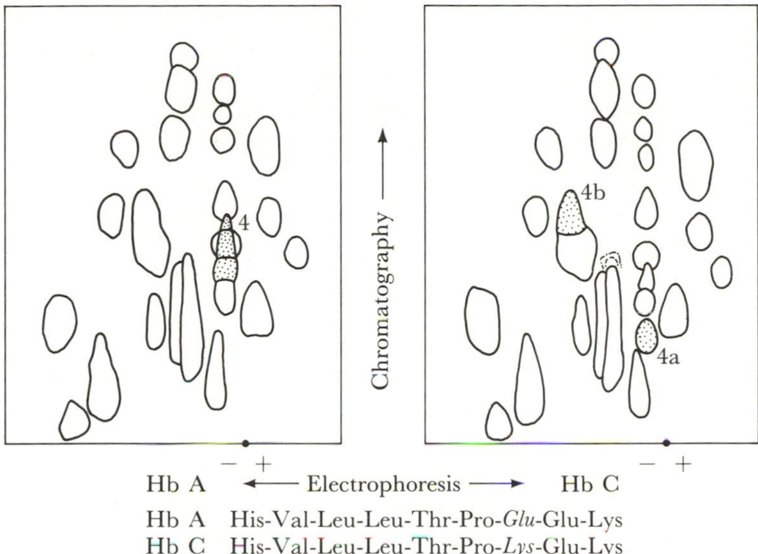

Figure 9-25
Tracings of fingerprints of trypsin digests of two different human hemoglobins. The peptides in the digests were first chromatographed and then electrophoresed. Note that the shaded peptide in normal hemoglobin A is missing in the variant, hemoglobin C, in which two new peptides, 4a and 4b, appear. The amino acid sequences of peptides 4, 4a and 4b are also shown. The conversion of glutamic acid (Glu) into lysine (Lys) in the variant creates two new peptides because trypsin cannot cleave between two molecules of glutamic acid but can cleave the peptide bond between lysine and glutamic acid. [From J. A. Hunt and V. M. Ingram, *Nature* (*Lond.*) 181(1958):1062–1063.]

The fingerprint of a mutant protein with a single amino acid change differs from that of a wild-type protein (Figure 9-25). If the amino acid change does not affect the site of cleavage by the protease, a single spot in the fingerprint will disappear and one new spot will appear. If it does affect the site of cleavage, several spots will be altered. By eluting the original and mutant spots and determining the amino acid composition of each (this can be done by complete enzymatic hydrolysis to amino acids followed by paper chromatography), the amino acids exchange can be determined.

The fingerprinting technique has had several important applications in molecular biology, among which are the following.

Proof that a protein (*A*) *is a product of cleavage of a larger protein* (*B*). If all spots in a fingerprint of protein A are found in protein B, it is likely that protein A is a part of the amino acid sequence of protein B. Hence, either protein B is made from protein A by the addition of amino acids or protein A is derived from protein B by hydrolysis. Usually, kinetic studies of the synthesis of A and B can distinguish these possibilities. One way to do this is to add a radioactive amino acid for a very short period of time (a "pulse" label) to a culture of cells capable of making both A and B. If radioactive B is made before radioactive A, then A probably results from hydrolysis of B. If A is synthesized before B, then B is probably made by addition of amino acids to A.

Identification of the amino acids inserted by various suppressor transfer RNAs. Certain mutations cause the premature termination of amino acid sequences in proteins. This termination is reversed by the presence of suppressor tRNA molecules, which insert an amino acid at the site of premature termination and thereby allow continued synthesis to the natural terminus. A protein results that has a single amino acid replacement. Fingerprints of such proteins allow the identification of the amino acid inserted by each of the known suppressor tRNAs. To determine this, it is necessary that the amino acid substitution causes a change in the fingerprints; this is not always the case as, for example, when one amino acid is replaced by another one that is chemically similar (e.g., isoleucine for leucine). If the substitution creates or eliminates a tryptic scission, the number of peptides in the wild-type and mutant protein will not be the same. If the substitution alters the chromatographic or electrophoretic properties, both mutant and wild-type proteins will lack a peptide that is present in the tryptic digest of the other. In both cases just described, a determination of the amino acid sequence of the varying peptides yields the required information.

Identification of base changes produced by particular mutagens. Because the genetic code is well known, it is possible to identify base changes that produce particular mutations. For example, suppose that a mutagen is found to produce frequent changes from phenylalanine to isoleucine (which can be determined by fingerprinting, using the technique of the preceding paragraph). The DNA triplets corresponding to phenylalanine are AAA and AAG and to isoleucine are TAA, TAG, and TAT. Because most mutations change only a single base, the particular mutagen would have to lead to frequent replacement of A with T.

Separation of Large Oligonucleotides by Homochromatography

Determination of the nucleotide sequence of large RNA molecules (e.g., ribosomal RNA and phage RNA) is important in modern molecular biology. It is impractical to determine directly the sequence of RNA molecules having several thousand bases so the RNA is broken down to fragments having fifty bases or less. Enzymes are available (e.g., T1 RNAse) that can cleave RNA molecules at specific sites and yield a collection of unique fragments. If these fragments can be purified, the sequence of each can be determined by straightforward sequencing methods. A two-dimensional procedure in which fragments are first separated electrophoretically by size and then chromatographically by base composition allows their purification. This is done in the following way.

A radioactive RNA sample is first digested with an enzyme to produce fragments. The sample is applied at one corner of a cellulose acetate sheet and then electrophoresed in a buffer containing 7 M urea. The high urea concentration prevents all intrastrand base pairing and allows separation based primarily on the number of nucleotides per fragment. The separation depends slightly on the base composition so that the fragments do not separate as cleanly as they might in a polyacrylamide gel. The cellulose acetate film is then placed onto a thin-layer plate containing a mixture of cellulose and DEAE-cellulose. The cellulose acetate is then covered with several layers of filter paper and a weighted glass sheet to maintain firm contact. Water is then allowed to flow onto the filter paper and the fragments transfer to the thin layer. There is no simple elution buffer that will allow good chromatographic separation of the fragments because the chromatographic behavior depends on size, base composition, and base sequence. Thus, Sanger chose to elute with a large quantity of transfer RNA (tRNA) that had seen partially digested with different RNAses. This elution procedure, called homochromatography, is based on the idea that in the tRNA digest there will be oligonucleotides of many sizes and sequences and these will compete with the radioactive fragments for binding to the thin-layer support. Thus, the nonradioactive fragments will tend to displace the desired radioactive fragments from the thin-layer material. This method is in fact very effective. After the thin-layer chromatogram is developed, the layer is autoradiographed to locate the radioactive spots. The spots are then scraped off the glass supporting the thin layer and the purified fragments are eluted.

Large-Scale Preparative Electrophoresis

Block Electrophoresis

A slurry of starch grains (or glass powder, silica gel, cellulose, polyvinyl chloride, or agar) in a buffer is poured into a large rectangular trough, which is mounted so that there is an electrode at either end. The sample is applied by removing starch from a large groove near one electrode, mixing the protein solution with the starch, and refilling the groove. After electrophoresis, the block is cut up and each section eluted.

Column Electrophoresis with Polyacrylamide

This procedure employs the apparatus shown in Figure 9-26. Material continually flows into the elution buffer, which flows through as indicated. The elution buffer is collected by a fraction collector.

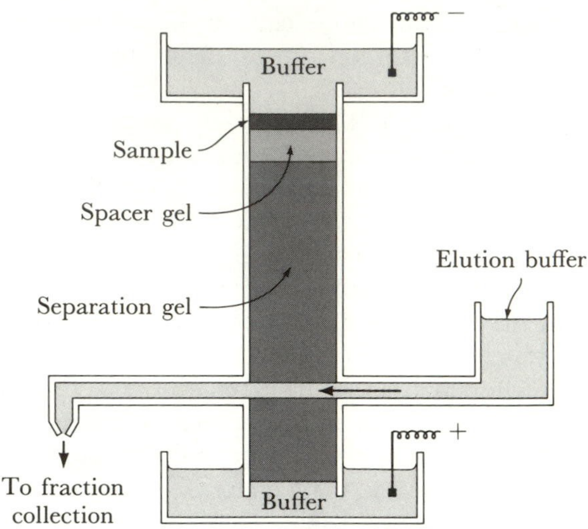

Figure 9-26 Apparatus for large-scale disc electrophoresis.

Immunoelectrophoresis

A variety of special techniques that combine electrophoresis and immunology has been developed for identifying molecules and for determining the amount of a particular molecule present in a mixture. These procedures, immunoelectrophoresis and rocket electrophoresis, are described in Chapter 10.

SELECTED REFERENCES

Ames, G. F., and H. Nikaido. 1976. "Two-dimensional Gel Electrophoresis of Membrane Proteins." *Biochemistry*, 15:616–623.

Brownlee, G. G., and F. Sanger. 1969. "Chromatography of ^{32}P-Labeled Oligonucleotides on Thin Layers of DEAE-Cellulose." *Eur. J. Biochem.* 11:395–399. This contains the details of homochromatography.

Catsimpoolas, N. 1976. *Isoelectric Focusing.* Academic Press.

DeWachter, R., and W. Fiers. 1971. "Fractionation of RNA by Electrophoresis on Polyacrylamide Gel Slabs," in *Methods in Enzymology*, vol. 21, edited by L. Grossman and K. Moldave, pp. 167–178. Academic Press.

Efron, M. 1960. "High Voltage Paper Electrophoresis," in *Chromatographic and Electrophoretic Techniques*, vol. 2, edited by I. Smith, pp. 158–189. Interscience.

Galas, D. J., and A. Schwartz. 1978. "DNAase Footprinting: A Simple Method for the Detection of Protein-DNA Binding Specificity." *Nucleic Acids Res.* 5:3157–3170.

Gordon, A. H. 1968. "Electrophoresis of Proteins in Polyacrylamide and Starch Gels," in *Laboratory Techniques in Biochemistry and Molecular Biology*, vol. 1, edited by T. S. Work and E. Work, pp. 1–149. North-Holland.

Heftman, E. 1975. *Chromatography.* Van Nostrand-Reinhold. This explains many electrophoretic techniques.

Karol, P. J., and M. H. Karol. 1978. "Isotachophoresis." *J. Chem. Educ.* 55:626–631. This explains the physics of stacking in disc electrophoresis.

Lewis, L. A. 1980. *CRC Handbook of Electrophoresis.* CRC Press.
Maurer, H. R. 1971. *Disc Electrophoresis.* deGruyter.
Maxam, A., and W. Gilbert. 1977. "A New Method for Sequencing DNA." *Proc. Natl. Acad. Sci. U.S.A.* 74:560–564.
Moody, G. J., and J. D. R. Thomas. 1975. *Practical Electrophoresis.* Merrow.
Ornstein, L. 1964. "Disc Electrophoresis." *Ann. N.Y. Acad. Sci.* 121:321–349.
Raymond, S. 1964. "Acrylamide Gel Electrophoresis." *Ann. N.Y. Acad. Sci.* 121:350–365.
Righetti, P. G., and J. W. Drysdale. 1976. *Isoelectric Focusing.* North-Holland.
Sanger, F., and A. R. Coulson. 1975. "A Rapid Method for Determining Sequences in DNA by Primed Synthesis with DNA Polymerase." *J. Mol.Biol.* 94:441–448.
Schuster, L. 1971. "Preparative Gel-Density Gradient Electrophoresis," in *Methods in Enzymology*, vol. 22, edited by W. B. Jakoby, pp. 434–440. Academic Press.
Shapiro, A. L., E. Vinuela, and J. B. Maizel. 1967. "Molecular Weight Estimation of Polypeptide Chains by Electrophoresis in SDS-Polyacrylamide Gels." *Biochem. Biophys. Res. Commun.* 28:815–820. This is the first report of SDS-PAGE electrophoresis.
Smithies, O. 1953. "Zone Electrophoresis in Starch Gels: Group Variations in the Serum Proteins of Normal Human Adults." *Biochem. J.* 61:629–641. First use of starch gels.
Steinhardt, J., and J. A. Reynolds. 1969. *Multiple Equilibria in Proteins.* Academic Press. The binding of SDS by proteins is described in detail in this book.
Weber, K., and M. Osborn. 1969. "The Reliability of Molecular Weight Determined by Dodecyl Sulfate-Polyacrylamide Gel Electrophoresis." *J. Biol. Chem.* 244:4406–4412. The definitive paper on SDS-PAGE electrophoresis.
Wright, G. L., K. B. Farrell, and D. B. Roberts. 1973. "An Evaluation of Gradient Acrylamide Gel Electrophoresis and Acrylamide Gel Isoelectric Focusing for the Primary Separation of Complex Mixtures of Proteins." *Biochim. Biophys. Acta.* 295:396–411.
Wu, R. 1978. "DNA Sequence Analysis." *Annu. Rev. Biochem.* 47:607–634.
Zwaan, J. 1967. "Estimation of Molecular Weights by Polyacrylamide Gel Electrophoresis." *Anal. Biochem.* 21:155–168.
Zweig, G., and J. R. Whitaker. 1967. *Paper Chromatography and Electrophoresis*, vol. 1: *Electrophoresis in Stabilizing Media.* Academic Press.

PROBLEMS

9–1. Why is electrophoresis done in solutions having low salt concentration?

9–2. In two-dimensional chromatography and electrophoresis, does it matter which is done first?

9–3. In determining the conditions for maximum separation of two components, what parameters (e.g., pH, ionic strength, temperature, etc.) should be varied? What effect might the variation of each have on the resolution?

9–4. Will two molecules having the same molecular weight and charge have the same mobility? Explain.

9–5. A mixture of proteins is electrophoresed in a polyacrylamide gel at three different pH values. In each case five bands are seen. Can you reasonably conclude that there are only five proteins in the mixture? Explain.

9–6. An enzyme has been extensively purified. By a variety of criteria it is thought to be pure—that is, it shows a single peak when chromatographed, electrophoresed, or centrifuged in a variety of ways. When subjected to sodium dodecyl sulfate (SDS) gel electrophoresis, two bands result, one twice the area of the other. What information does this give about the protein? Because purity is always difficult to prove, how could you prove that your hypothesis is correct? (*Hint*: Use gel chromatography.)

9–7. Proteins A and B having a molecular weight of 16,500 and 35,400 move 1.3 cm and 4.6 cm, respectively, when electrophoresed through a gel. What is the molecular weight of protein C, which moves 2.8 cm in the same gel?

9–8. A virus contains 256 proteins, 64 having a molecular weight of 1800 and 192 with a molecular weight of 26,000. If the virus were disrupted and analyzed by SDS gel electrophoresis, what would be the relative distances migrated and the relative areas of the bands?

9–9. Suppose that you have isolated a protein that seems to have two enzymatic activities. This makes you suspect that you may have two proteins that copurify. To check this, you electrophorese the preparation in a polyacrylamide gel at a variety of pH values. In each case, a single band results, but the band is sufficiently broad that you suspect that it is really two bands which do not resolve. In a sodium dodecyl sulfate gel, a single, broad band is also found. Because a protein with two enzymatic activities is rare, it is necessary to try a little harder to see if the breadth is due to the presence of two bands. What parameters could you vary to improve resolution by electrophoresis? What other methods (nonelectrophoretic) might you try?

9–10. A DNA molecule has the restriction map shown below when digested by the *Eco*RI restriction endonuclease (numbers refer to relative distance of each cut from the left end of the molecule). When analyzed by gel electro-

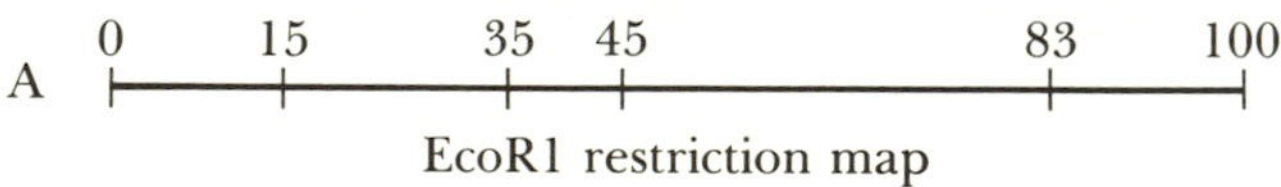

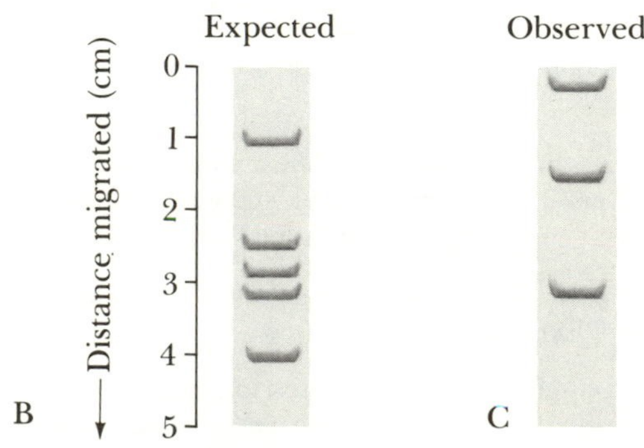

phoresis (after heating the digested sample to 65°C), instead of the expected five bands, the second pattern shown is observed. Explain.

9–11. A phage DNA molecule has short, complementary, single-stranded ends and circularizes when the phage infects a bacterium. The restriction map and the gel band pattern obtained after treatment of a free DNA molecule with the *Bam* restriction endonuclease are shown at A and B below. You are studying the properties of DNA isolated from an infected bacterium. Under certain conditions, the pattern of fragments obtained after enzymatic digestion and gel electrophoresis is that shown in panel C. What is the structure of the DNA under these conditions?

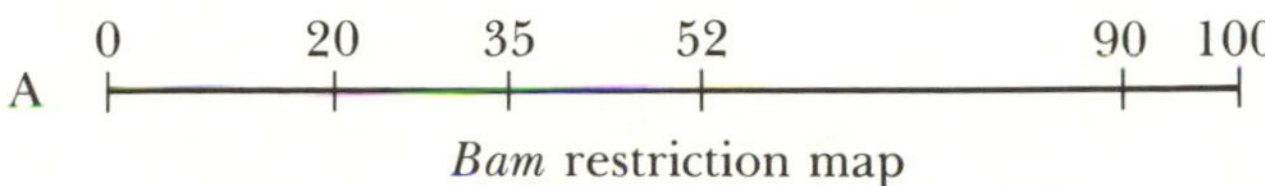

Pattern after gel electrophoresis

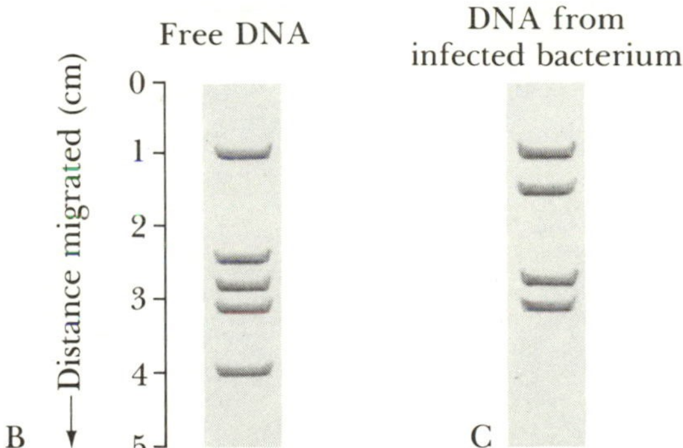

9–12. A linear DNA molecule is digested with a restriction enzyme and then electrophoresed in a gel. The sizes of the fragments are 0.1, 0.16, 0.18, 0.26, and 0.30. If the DNA was in its circular form prior to treatment with the enzyme, the fragments obtained are 0.18, 0.26, and 0.30. If the linear molecule is used but the enzyme is allowed to react for a fairly short time, the fragments obtained are 0.16, 0.18, 0.26, and 0.40. You also have in your refrigerator a related DNA molecule that differs only in that it is known to have a deletion that amounts to 10% of the size of the undeleted molecule and this deletion is very near the center of the molecule. If this is treated with the restriction enzyme, the fragments are 0.1, 0.16, 0.18 and 0.30. Draw the original molecule and show the sites at which the restriction enzyme cuts.

9–13. A DNA molecule is sequenced according to the Maxam-Gilbert procedure. The sample is divided into two parts and electrophoresed for a short time and a longer time. The gels are shown below. Columns 1, 2, 3, and 4 are the **G-only**, **A + G**, **T + C**, and **C-only** samples. What is the sequence of the molecule?

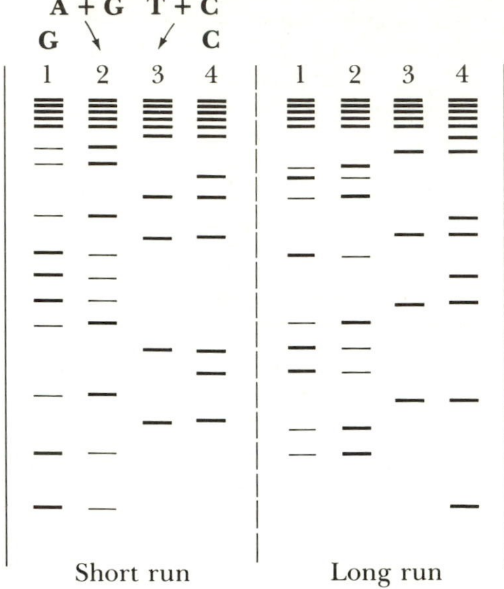

10

Immunological Methods

The identification and assay of biological material is most commonly done by chemical tests, spectroscopic methods, and the determination of various physical parameters such as sedimentation coefficients, electrophoretic mobilities, chromatographic constants, and so forth. If very small quantities of material must be assayed, traditional chemical methods fail, and an approach has been to use radiochemical tracers and the technology described in Chapter 5. Unfortunately, in the assay of complex systems such as cells in culture and certainly of whole animals or plants, it is not always possible to add enough radioactive material to label detectable quantities of a particular biochemical—either because of dilution of the label or because of danger to the organism.

Immunological procedures provide the solution to these difficulties because they make it is possible to assay minute amounts of nonradioactive material in complex mixtures.

The Immunological System

In response to the injection of a foreign substance into a higher animal, an *antibody* (Ab) is produced that can react with the substance. Antibodies are proteins found in the bloodstream and are part of a class of serum proteins called *immunoglobulins.* Any substance that can elicit antibody production is called an *antigen* (Ag). An antibody produced by exposure to an antigen has the important property of reacting specifically with the antigen that stimulated its production and not with most other antigens. Similarly, the antigen fails to react with any antibody other than that which it elicited.

A *hapten* is a substance that cannot by itself stimulate antibody synthesis but can react with a hapten-specific antibody. Most haptens are molecules of low molecular weight (<1000); an antibody to a hapten is usually prepared by chemically coupling the hapten (by covalent bonds) to an immunogenic substance such as the protein serum albumin and injecting this conjugated protein into an animal (e.g., a rabbit). An antibody prepared against a substance X (which may be either an antigen or a hapten) is commonly called *anti-X*. An antibody-antigen complex is often written Ab-Ag.

In addition to the original antigen, there are other substances that react with a specific antibody, though often with a somewhat lower efficiency. This weaker reaction is called a *cross-reaction*. One kind of cross-reaction is that which takes place when antigen A reacts partially with anti-B and antigen B partially with anti-A. Asymmetric cross-reactivity also occurs—that is, antigen A reacts with anti-B, but antigen B does not react with anti-A. Cross-reaction occurs when there is chemical similarity but not identity.

Antibodies, being proteins, are also antigenic in other animals. For example, an antibody obtained from rabbit serum can be injected into a goat and elicit an antibody. The goat antibody does not react with the antigen that stimulated the production of the rabbit antibody but to the rabbit antibody itself; it is usually called *goat anti-(rabbit antibody)*.

If the biological activity of a substance is destroyed by an antibody, the substance is said to be *neutralized*.

There are many types of antibody proteins found in serum. For the purpose of analytical immunological procedures, the most important family of antibodies is the immunoglobulin G or IgG class. (IgG is often called gamma globulin. However, the gamma globulin fraction of blood that is used in disease prevention usually contains other proteins also.) These proteins, whose basic structure is shown schematically in Figure 10-1 consists of three principal regions. Two of these regions are identical and are termed F_{ab} (F stands for fragment and ab for antigen-binding). The third section is called F_c. The central portion of an IgG molecule is a flexible region known as the *hinge*. Each F_{ab} branch contains a terminal antigen-binding site, as shown in the figure. *Thus, each IgG molecule can bind two antigen molecules.*

Preparation of Antibody

Because antibodies are produced in the bloodstream of an animal in response to the injection of a foreign substance, antibody can be obtained by bleeding an animal that has been repeatedly injected with the same antigen or conjugated hapten. Because of the specificity of the Ag-Ab

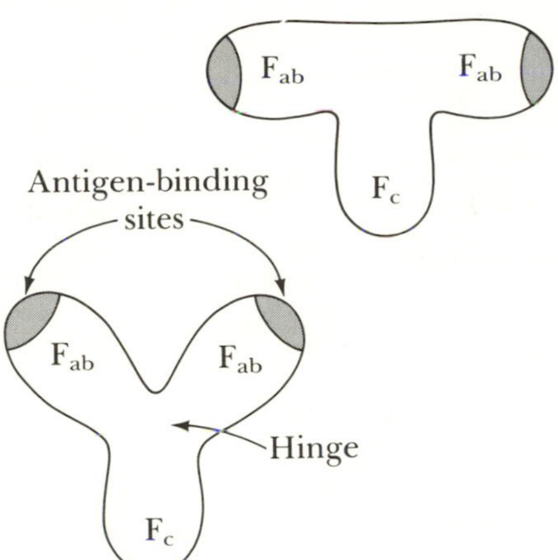

Figure 10-1
Immunoglobulin G has the shape of the letter Y. It contains a hinge for flexibility and two antigen-binding sites. [From *Biochemistry*, 2nd edition, by Lubert Stryer. W. H. Freeman and Company. Copyright © 1981.]

reaction, it is rarely necessary to isolate the specific antibody, or even the immunoglobulin fraction. Hence, in most immunological work, blood serum from which all cells have been removed by centrifugation is used. Serum known to contain a particular antibody is called *antiserum*. There are some occasions when an antiserum must be partially purified. For example, when a hapten conjugated to serum albumin is injected into an animal to prepare antihapten, the animal also makes anti-serum albumin. or if antibody to ultraviolet-irradiated DNA is made, an antibody will also be made that is active against unirradiated DNA. These secondary antibodies may be unacceptable in a particular experiment and must be removed. This is done by *adsorption*. That is, the antigen to the unwanted antibody is added to the serum, and the antigen-antibody complex is allowed to form. Under appropriate conditions, this complex can be removed by centrifugation. The repeated addition of antigen usually results in eliminating the unwanted antibody. On rare occasions, a pure antibody is needed; it can be prepared by means of affinity chromatography (see Chapter 8) using the antigen or hapten coupled to a matrix.

The preparation of an antihapten requires special procedures. Because the hapten itself is not antigenic, it is coupled to a strongly antigenic molecule such as bovine serum albumin or certain synthetic polypeptides and then injected into a rabbit. The coupling is sometimes a problem because the hapten must contain at least one functional group that can be coupled to the available functional groups (e.g., amino, carboxyl, phenolic, imidazolyl, sulfhydryl, indolyl, and guanidino) of the protein under chemical conditions that destroy neither the hapten nor the protein structure. Otherwise, a chemical derivative of the hapten must be synthesized to allow this coupling to be done. Table 10-1 lists haptens to which antibodies have been made. The ability to prepare antihaptens is essential to the radioimmunoassay that is described in a later section.

Table 10-1
A Variety of Haptens to Which Antibodies Have Been Made.

Peptide hormones	Insecticides
Nonpeptide hormones	Carbohydrates (e.g., glucose, maltose, lactose)
Coenzymes	Catecholamines
Vitamins	Lipids
Drugs	Steroids
Toxins	Nucleic acid constituents
Carcinogens	Plant hormones

NOTE: For a complete description, see V. P. Butler and S. M. Beiser, *Adv. Immunol.* 17(1973):255–310.

Reaction of Antibody and Antigen

After an antibody has been prepared, the antibody-antigen reaction is carried out merely by mixing the two substances in a buffer and incubating until the reaction is complete. The buffer is usually at neutral pH and is isotonic to prevent denaturation of the antibody protein, although a range of ± 2 pH units and a factor of 2 in ionic strength can be tolerated.

The reaction is remarkable in that it usually proceeds in complex mixtures containing other proteins, small molecules, polysaccharides, lipids, and so forth, with virtually no effect on the rate or extent of reaction. In quantitative studies, control experiments are always performed to insure that there are no interfering substances, because in isolated cases a reaction mixture may contain substances that degrade or metabolize either the antigen or the antibody.

Immune Reactions Useful in Bioassays

Precipitin Reaction

It is very common for an antigen to contain two or more sites for binding antibody. Such sites are called *antigenic determinants*. For example, a polynucleotide consists of repeating units, each of which contains an antigenic determinant, and a bacterial cell wall may contain hundreds of repeating regions. Individual protein molecules, even those without subunits, also seem to have more than one antigenic determinant; in this case, the antibody sample usually consists of a mixture of distinct IgG molecules, each of which reacts with a particular determinant. Let us consider the state of a solution containing a single type of IgG molecule and an antigen that has two determinants. If the molar ratio of IgG to antigen is one, all interacting molecules will be part of a gigantic

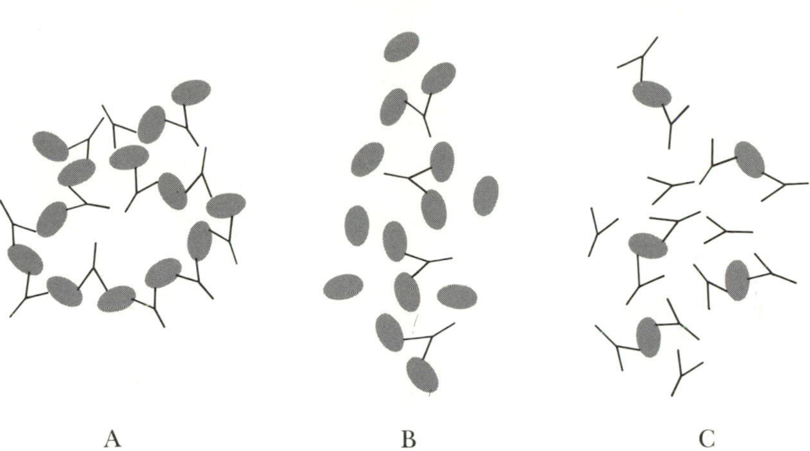

Figure 10-2
A. Antibody and antigen molecules are in roughly equal numbers and form a complex lattice called precipitin. B. Antigen is in excess. Each antibody binds two antigen molecules. Many antigens are free. C. Antibody is in excess. Each antigen binds two antibodies. Many antibody molecules are free.

lattice; each IgG molecule serves as a bridge between two antigen molecules and each antigen joins two IgG molecules. This is shown in Figure 10-2. If the antibody and antigen concentrations are high enough, the lattice can be so large that a visible precipitate forms. This precipitate is called *precipitin.* If there is a great excess of antigen, all of the antibodies will on the average have two bound antigen molecules (Figure 10-2B). In this case, there is very little opportunity to form a lattice and little precipitin is formed. If the antibody is in great excess instead, most of the antibody molecules will not carry any antigen so that a common configuration of those antibodies binding antigen is three antibodies bound to two antigens to form a short linear array, which is not precipitable (Figure 10-2C). Thus, with excess antibody there is also less precipitin formed than when the molar ratio is one. These facts are summarized by a graph of the amount of antibody precipitated by various amounts of antigen; this graph is called a *precipitin curve* (Figure 10-3). Several reactions that will be discussed are based on the fact that precipitin is a *reversibly dissociating* antigen-antibody complex. The reversibility is made evident by the fact that, even after precipitin has formed, the amount present can be varied by changing the ratio of antigen to antibody. This will be seen in the following section. In the regions marked *antibody excess* and *equivalence*, the antigen is totally precipitated; hence, precipitation with excess antibody can be used to remove particular substances from solution. For example, in a mixture of radioactive

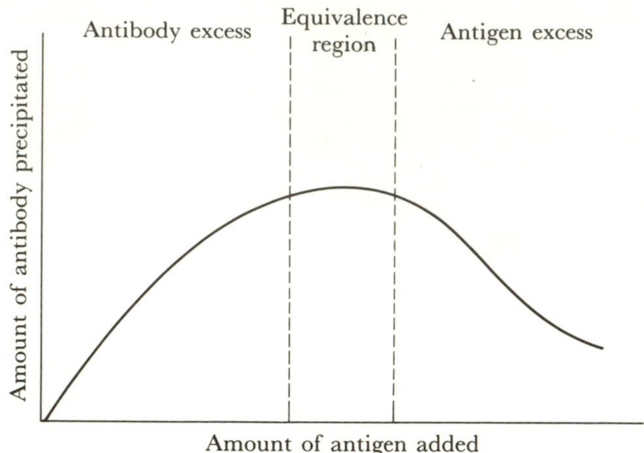

Figure 10-3
Precipitin curve. To a fixed amount of antibody various amounts of antigen are added. After a suitable period of incubation, the mixture is centrifuged and the amount of antibody precipitated is determined by measuring the amount of antibody remaining in the supernatant. In the antibody-excess region, all antigen is precipitated—that is, the supernatant contains the excess antibody. In the antigen-excess zone, much of the antibody does not form precipitin and the supernatant contains a mixture of antigen and antibody. At equivalence, all of the antigen and antibody are precipitated.

proteins, the amount of a particular protein (e.g., A) can be determined by adding anti-A and sedimenting the precipitate. The ratio of sedimentable-to-total radioactivity yields the weight fraction of the protein mixture that is A. That the precipitin reaction is an important part of the radioimmunoassay will be seen later.

Gel-Diffusion Precipitin Reactions and Immunodiffusion

The precipitin reaction can also occur in agar gels. Consider the arrangement (and method) shown in Figure 10-4A. A test tube is partly filled with agar containing antiserum, and this is overlaid with antigen. The antigen diffuses downward into the agar and a precipitate forms in the agar at the advancing, diffusing front. The precipitate is a band due to the reversibility of the precipitin reaction described in the preceding section. That is, in the upper part of the tube, there is little precipitin because of excess antigen (see Figure 10-3). As downward diffusion of the antigen continues, the precipitate appears to move downward. This is because at the leading side, more antibody is encountered, whereas, at the trailing side, more antigen is arriving; the excess antigen solubilizes the precipitate (Figure 10-3). If both antigen and antibody are mixtures of several reacting species, several bands will be present, presumably one for each antigen-antibody complex. The separation of these bands depends on the diffusion coefficients of the antigens and the relative concentrations of antigens and antibodies; hence, if two antigens have the same diffusion coefficients and the concentrations of antibodies and antigens are the same, only one band will appear. Thus, the number of observed bands is always a minimum value for the number of antigens present.

Additional information can be obtained by means of the *Ouchterlony double-diffusion method*, which is shown in Figure 10-4B–D. A Petri dish of agar contains three wells (holes in the agar), two of which are filled with antigen and the third with antiserum. Diffusion proceeds in all directions from all wells; when the equivalence point is reached in the space between an antibody and an antigen well, a precipitin line appears. If the two antigen wells contain the same antigen, a curved line such as that shown in Figure 10-4B is the result. If the antigens are different, the two lines are independent and will intersect as shown in Figure 10-4C. If there is cross-reaction between the two, the pattern of Figure 10-4D will result. The small region projecting above the curve in Figure 10-4D is called a *spur* and always points toward the less reactive antigen. Two different weakly reacting but cross-reacting antigens can appear to give intersecting bands as in Figure 10-4C and could be mistaken for independent substances. However, the crossovers are spurs only if they are less dense than the main line and show pronounced curvature. The Ouchterlony procedure can be used to determine whether an unknown material is the same as some known substance because, if it is, the pattern of Figure 10-4B results.

Complement-Fixation Assay

Blood sera contain a class of proteins that are not immunoglobulins and are collectively called *complement*, or C′. Complement reacts not with antigen or antibody alone but with a variety of antigen-antibody complexes. For reasons that are not clear, complement does not react with hapten-antibody complexes. If haptens are to be measured, inhibition assays (see next section) are usually used.

The complement-fixation assay for an antigen-antibody reaction makes use of the fact that complement is consumed ("fixed") by the antigen-antibody complex, making less complement available. A quantitative measure of the decrease in the amount of complement is made possible by the fact that the addition of complement to an antigen-antibody complex consisting of red blood cells and anti-(red cell surface polysaccharide) leads to cell *lysis* (bursting of the cells). This lysis is

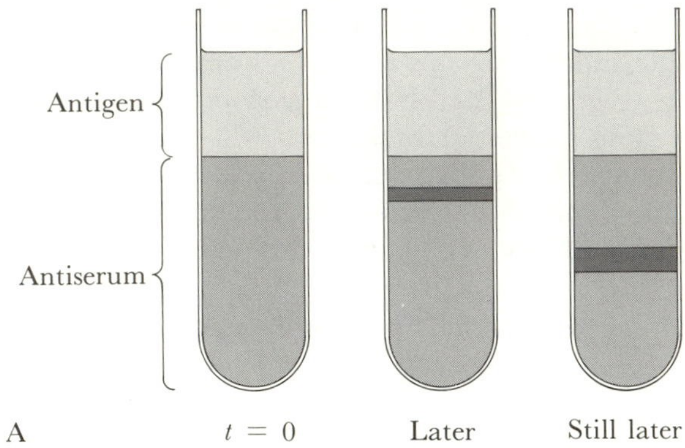

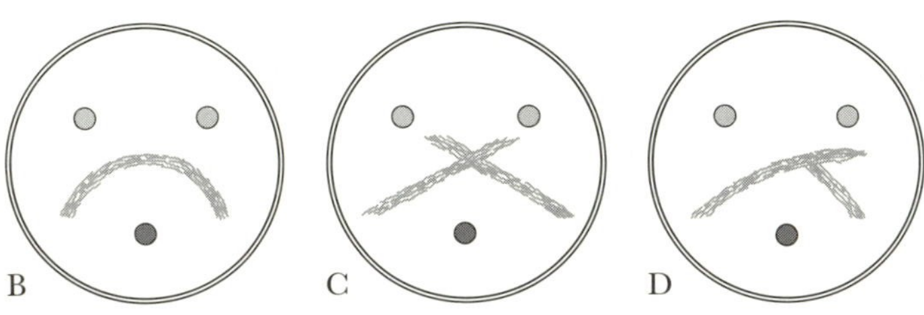

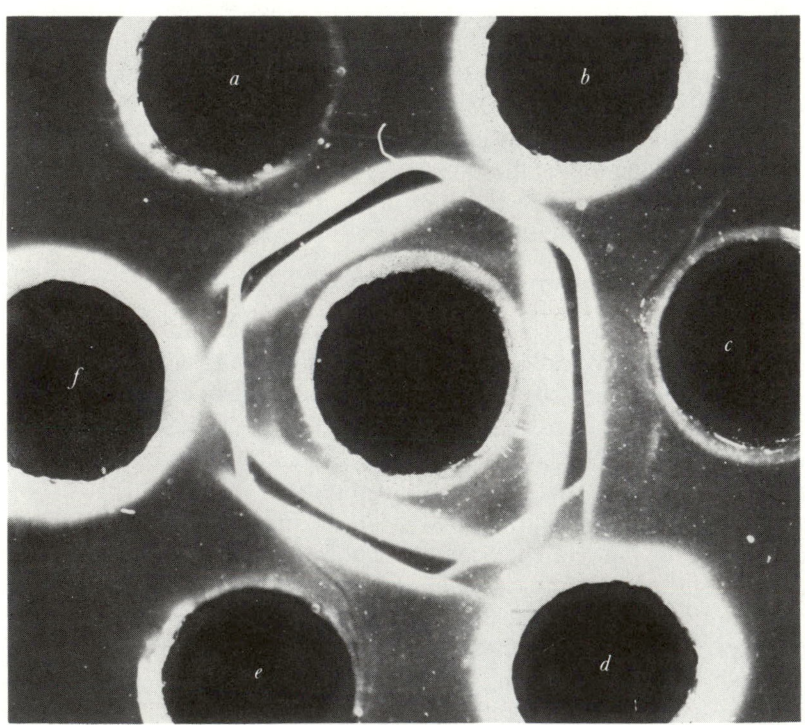

E

Figure 10-4 (*opposite*)
Gel diffusion precipitin reaction. A. Single diffusion in one dimension (Oudin method). The test tube is partly filled with agar containing antiserum. After the agar has hardened, it is overlaid with agar containing antigen; this agar also hardens. Antigen then diffuses into the antiserum. B–D. Double diffusion in two dimensions (Ouchterlony method). A Petri dish is filled with agar in which three cylindrical wells are cut. One is filled with antiserum (solid spot) and the others with antigen. In part B, both upper wells are filled with the same antigen. In part C, each upper well is filled with a different antigen; the lower well contains antibodies to both antigens. In part D, a single antibody is in the lower well; the upper left is the antigen and the upper right is a cross-reacting antigen. E. Photograph of an Ouchterlony plate. Consider the outer bands only because the inner band is a contaminating protein. Wells *a*, *c*, and *e* contain *E. coli* phage λ exonuclease. Well *b* contains a protein that is immunologically indistinguishable because the bands are continuous. Wells *d* and *f* contain altered exonucleases that are cross-reacting (note the spurs). [From C. M. Radding, *Methods Enzymol.* 21(1971):458–462.]

easily seen with red blood cells; if intact cells are centrifuged, the red hemoglobin pigment is carried to the bottom of the centrifuge tube; if lysis (hemolysis) has occurred, the hemoglobin remains in the supernatant and can be detected quantitatively by its absorbance at 541 nm (see Chapter 14). The standard test uses guinea-pig serum as a source of complement, sheep red cells, and rabbit antibodies to sheep red cells. The assay shown in Figure 10-5 is normally done in two steps.

1. Antibody, antigen, and a measured amount of complement are mixed and incubated until the binding of complement is complete. In this step, the complement is fixed by the antigen-antibody complex.

2. Sheep erythrocytes precoated with specific rabbit antibody are then added to the mixture. In this step, the erythrocyte-antibody complex binds that complement which was not consumed in step 1. After a period of incubation, during which lysis occurs, the mixture is centrifuged and the optical density of the supernatant is recorded.

Because hemolysis requires free complement (i.e., that not consumed in the first antibody-antigen reaction), the amount of hemoglobin in the supernatant measures the amount of complement remaining. Note that an *increased* antigen-antibody reaction means *decreased* lysis.

A variation of sheep cell hemolysis is sometimes used. Sheep cells are incubated in a buffer containing radioactive Cr^{3+} ions. These cells are then washed and used in the complement-fixation reaction. Lysis is detected by release of the radioactivity, which is precipitated as an insoluble chromium salt.

Complement fixation can be used to measure the amount of a particular antigen present in a complex mixture with great sensitivity.

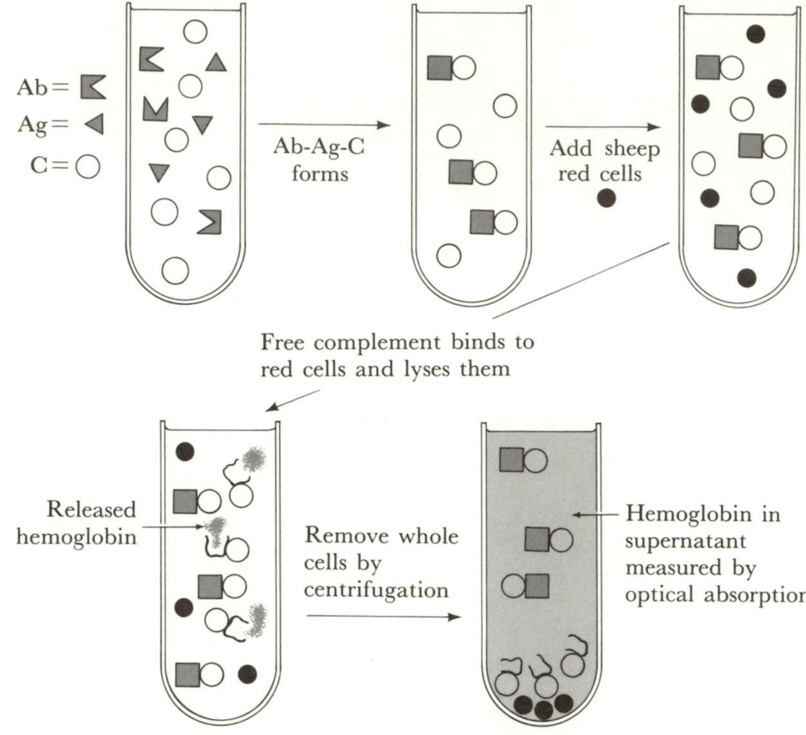

Figure 10-5
Complement fixation. Note that there are equal amounts of antibody and antigen and an excess amount of complement for the purpose of illustration. Normally, the amounts of antibody and complement are adjusted so that most of the complement is consumed at equivalence. The sheep red cells are precoated with rabbit antibodies to sheep red cells; this has been omitted from the illustration for clarity.

Figure 10-6 shows a curve relating the amount of complement fixed (i.e., decrease in lysis) as a function of added antigen in the presence of a fixed amount of antibody. Typically, the amount of complement fixed increases with the amount of antigen in the antibody-excess region and decreases in the antigen-excess region. Hence, to measure the amount of a particular antigen, the following protocol is used. First, a standard curve such as that shown in Figure 10-6 is obtained, using known amounts of the antigen. Then the reaction is carried out with several dilutions of the antigen sample. This is necessary because, if a single dilution were used, it would not be known whether the amount of complement fixed corresponded to the antibody- or antigen-excess side of the curve. After this has been done, the amount of antigen can be read from the curve.

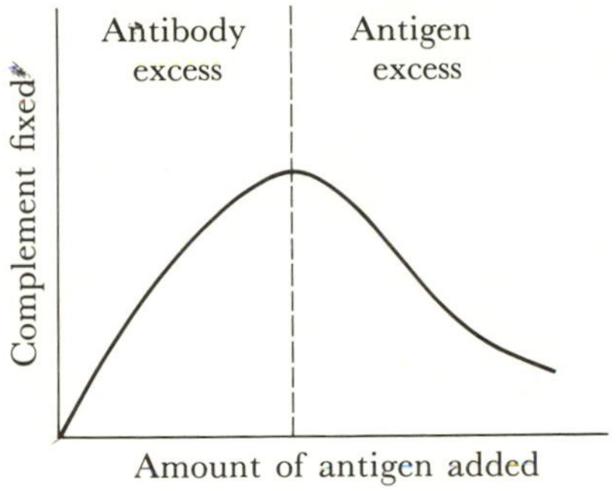

Antibody excess

Antigen excess

Complement fixed

Amount of antigen added

Figure 10-6
Standard complement-fixation curve. To a fixed amount of antibody and complement are added various amounts of antigen. After a period of incubation, sheep erythrocytes and anti-(sheep erythrocytes) are added. After a second period of incubation, the mixture is centrifuged to remove unbroken erythrocytes. The absorbance of the supernatant is measured to determine the amount of hemoglobin released. If no complement is fixed, the maximum amount of hemoglobin is found in the supernatant. Note that two different amounts of antigen can fix the same amount of complement. However, to measure the amount of an antigen whose concentration is unknown, it is necessary to make measurements with dilutions of the sample. If diluting the antigen decreases the amount of complement fixed, the amount fixed corresponds to the antibody-excess part of the curve.

The complement-fixation assay is roughly from fifty to one hundred times as sensitive as the precipitin reaction; it is possible to detect as little as 0.01 μg of certain antigens. Its main disadvantage is that, in some crude mixtures being assayed for the presence of a particular antigen, substances may be present that inactivate complement (anticomplementary substances) or stimulate hemolysis. Complement fixation as an assay is no longer in much use, having been replaced by the more sensitive and simpler radioimmunoassay. At present, it is used only on occasion for detecting substances that cannot be made radioactive.

Radioimmunoassay

A highly sophisticated immunoassay, the radioimmunoassay (RIA), is capable of detecting extraordinarily small amounts of nonradioactive material and can do so in mixtures of huge numbers and amounts of extraneous materials. Its sensitivity equals or surpasses all known chromatographic and spectrophotometric assays. Its main use is to

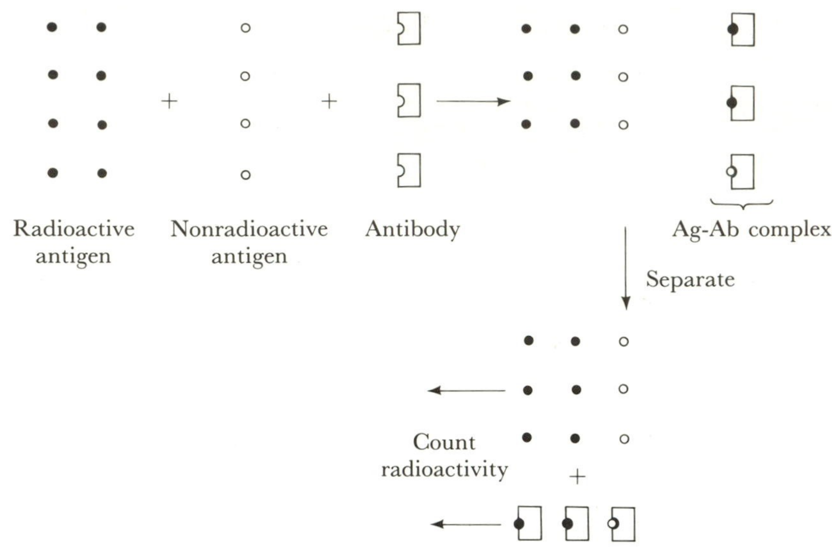

Figure 10-7
Principle of the radioimmunoassay. See text for details.

detect molecules (1) that cannot be radioactively labeled in vivo to a suitable specific activity or without also labeling other compounds; (2) that cannot fix complement when combining with specific antibody (e.g., haptens); and (3) whose identity is unknown but that can cross-react and thereby compete with known antigens.

Because the radioimmunoassay is such a powerful method, it is worthwhile examining more carefully how the inhibition procedure is used. The principle is shown schematically in Figure 10-7. Antibody (Ab) directed against the antigen (Ag) is saturated with radioactive Ag (i.e., Ag*), using excess Ag*. If nonradioactive Ag is added to Ab along with Ag*, less Ag* will be found in the antibody-antigen complex as the ratio Ag/Ag* increases (i.e., there will be an increasing amount of Ab-Ag and a decreasing amount of Ab-Ag*). For example, if Ag/Ag* = 1, only $\frac{1}{2}$ of Ag* will be bound; if Ag/Ag* = 2, only $\frac{1}{3}$ will be bound, and so forth. If the Ab-Ag* complex can be physically separated from Ag*, the amount of Ag can be determined.

To measure Ag, a standard curve must be constructed. This is done by mixing a fixed amount of Ab and Ag* and placing the mixture in a set of test tubes. Known amounts of Ag are added to each. When the reaction is complete, the Ab-Ag* is separated from Ag* (how this is done is described on page 335). A graph is then made that relates the radioactivity in the collected Ab-Ag* to the amount of added Ag (Figure 10-8). To determine the amount of Ag in an experimental sample, an aliquot of the sample is added to the same Ab-Ag* mixture used to obtain the standard

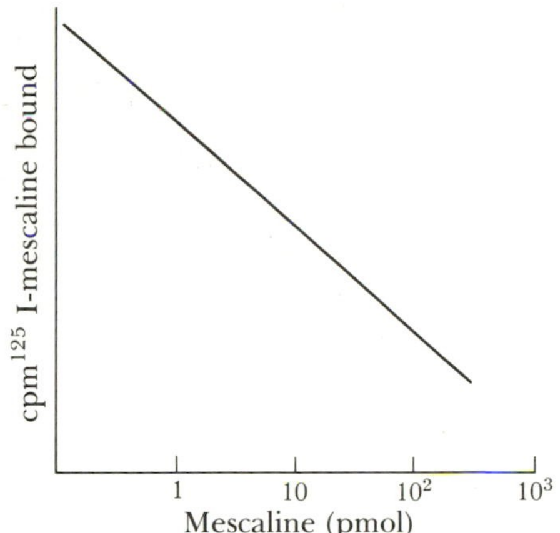

Figure 10-8
A typical standard curve for the radioimmunoassay. Antibody to mescaline and [^{125}I]-mescaline were mixed with various quantities of unlabeled mescaline. The inhibition of precipitation of radioactivity by added mescaline provides the standard curve. Note that the x-axis is a logarithmic scale. [Courtesy of Helen Van Vunakis.]

curve, Ab-Ag* is collected and the radioactivity measured, and the amount of Ag is read from the standard curve. This is possible with any sample (no matter how complex) as long as nothing in the mixture interferes with the Ab-Ag* reaction.

The separation of Ab-Ag* from Ag* is in practice performed in two ways (although other methods are also possible).

Dextran-coated activated charcoal method. Activated charcoal has the widely used property of adsorbing many small molecules virtually instantaneously. Proteins and the Ab-Ag* complex also adsorb, though more slowly and to a lesser degree. However, when charcoal is coated with the cross-linked polysaccharide dextran (a molecular sieve that cannot be penetrated by large molecules, see Chapter 8), both Ab and Ab-Ag* fail to adsorb. Hence, if dextran-coated charcoal is added to a mixture containing free Ag* and Ab-Ag*, immediate centrifugation results in pelleting radioactivity not bound to Ab (i.e., Ag*), and Ab-Ag* radioactivity remains in the supernatant. This is a convenient and rapid technique and is shown in Figure 10-9A.

Double-antibody method. Antibody is normally prepared in a rabbit. Rabbit IgG is antigenic in other animals and has been used to generate anti-(rabbit IgG) in the goat, sheep, horse, and cow. The addition of goat antiserum to rabbit IgG results in the formation of precipitin, which can be collected by centrifugation (Figure 10-9B). Hence, the following procedure is used. To each of the tubes used to generate the standard curve and to the sample tubes (all of which have been previously incubated sufficiently long to form Ab-Ag*) is added goat anti-(rabbit Ab) at the equivalence point of the precipitin reaction (see Figure 10-3). The mixture is then incubated (typically, for a period ranging from 16 to 48 hours) and, after incubation, the sample is centrifuged. Ab-Ag* is found in the pellet; free Ag* is in the supernatant. This general method is highly satisfactory but requires maintaining large animals as a source of antiserum.

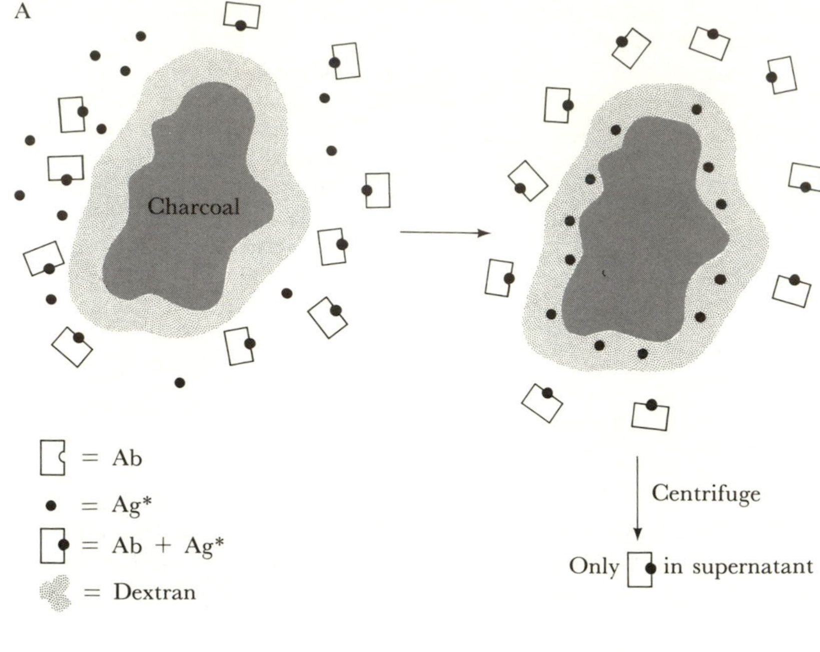

Ab(g) = goat anti-(rabbit IgG)

Ab(r) = rabbit anti-antigen

Ab(r) + Ag* ⟶ Ab(r)Ag* + Ag*

Excess ... Less

↓ Add Ab(g)

Ab(g)[Ab(r)Ag*] + Ag*

↓ Centrifuge

Ab(g)[Ab(r)Ag*] in pellet

Ag* in supernatant

B

Figure 10-9
Methods for separating Ab-Ag* from Ag*: (A) dextran-coated charcoal method (because Ag* is a small molecule, it can penetrate the dextran matrix); (B) double-antibody method (the Ab-Ag* complex forms precipitin in box when goat antibody is added).

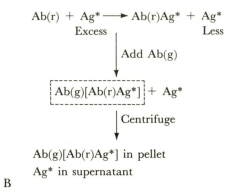

Preparation of Labeled Antigen

The radioimmunoassay differs from other immunoassays in that radioactive antigen is required. Occasionally, the appropriate ^{3}H- or ^{14}C-labeled compound can be purchased or synthesized from labeled precursors. More commonly, the antigen or hapten is iodinated with ^{125}I or ^{131}I; there is a special advantage in using ^{125}I because this isotope is a strong γ-ray emitter. As explained in Chapter 5, the β emitters have to be counted by liquid scintillation counting and because the sample in a radioimmunoassay frequently contains a substantial amount of protein, quenching in the scintillation system is a severe problem. With ^{3}H and ^{14}C, the quenching can be quite strong. With γ emitters, quenching rarely occurs because of the high energy of the radiation.

The preparation of iodinated antigens is not always simple, especially because it is necessary that the substance does not suffer any damage that might affect its antigenicity. Iodination of proteins or an amino group, or of haptens containing an amino group, can be accomplished by reaction with ^{125}I-3-(4-hydroxyphenyl)propionic acid *N*-hydroxysuccinimide ester. For proteins containing tyrosine or haptens with a phenolic group, the aromatic ring can be iodinated in a variety of ways. Nucleic acids can also be iodinated by a reaction with cytosine.

In some cases, there are no functional groups that can be directly iodinated; in these cases, the compound is first conjugated with tyrosine or histamine and the conjugate is iodinated. The alteration of a substance of interest so that it can be made radioactive is called *derivatizing*. It is, of course, always necessary to demonstrate that a derivative of a substance X remains an antigen for anti-X.

Radioiodination of protein is best done with a solid-phase enzymatic system. One commercially available system is called Enzymobeads (Bio-Rad Labs). These are very fine hydrophilic beads to which the enzymes glucose oxidase and lactoperoxidase are coupled. In a mixture containing Enzymobeads, proteins, $Na^{125}I$ and glucose, the glucose oxidase generates a tiny amount of H_2O_2 and this is used by the lactoperoxidase to convert $^{125}I^-$ to ^{125}I, which rapidly reacts with the protein (Figure 10-10) The main advantage of this procedure is that the concentration of all reactants is so low that chemical damage, which often occurs with other methods, is minimal, if it occurs at all. Furthermore the beads can be removed from the reaction mixture by centrifugation so that purification of the radioactive protein only requires removal of the $Na^{125}I$, which is easily accomplished by gel chromatography.

The Immunoradiometric Assay

In some cases, there is no satisfactory way to prepare a radioactive antigen—which precludes use of the radioimmunoassay. A variation of this assay, the immunoradiometric assay (IRA), allows the assay of such substances. It is best described by the diagram in Figure 10-11. Purified antigen is adsorbed or coupled to a stable substance such as cellulose (or Sepharose treated with cyanogen bromide, see Chapter 8). Purified immunoglobulin (IgG) obtained from sera containing an antigen-specific antibody is added. Unadsorbed antibodies are removed by washing and the adsorbed antigen-specific antibody is iodinated with radioactive iodine (note that here, unlike the radioimmunoassay, the antibody is radioactive instead of the antigen). The radioactive antibody is then removed from the adsorbant and used in the assay. For assay, Ab* is mixed with the antigen to be determined, using antibody excess (in the radioimmunoassay, antigen excess is used). The mixture is then added

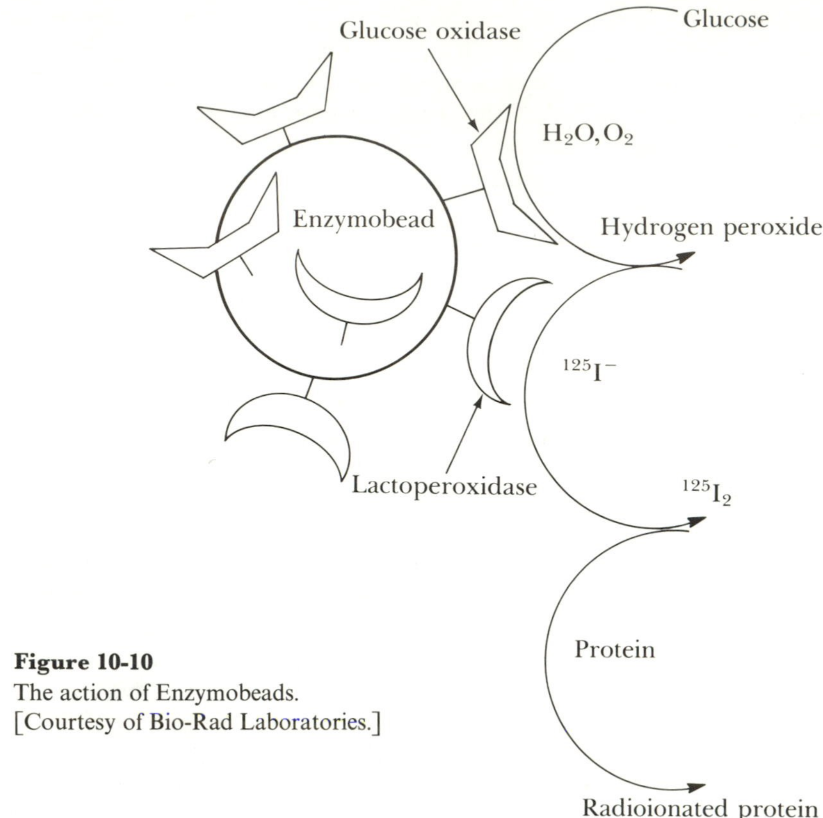

Figure 10-10
The action of Enzymobeads.
[Courtesy of Bio-Rad Laboratories.]

to the cellulose, which contains Ag; only unreacted Ab* can bind to the cellulose. Reacted Ab* is washed off and counted and is a measure of the amount of Ag.

Examples of Immunological Procedures Used in Bioassays

The following examples show the great variety of systems that can be examined by immunological procedures. Note that, in some cases, the immunoassay is used qualitatively and its value lies in its extraordinary sensitivity. In other cases, it is a quantitative and precise analytic tool.

Example 10-A □ Identification and classification of various bacteria by agglutination.

Many bacteria isolated in nature or obtained from patients can be identified by precipitation or agglutination with specific antibodies directed against known strains (Figure 10-12). The reaction can be detected either by observation of clumping with the light microscope or by changes in turbidity of a suspension measured spectrophotometrically.

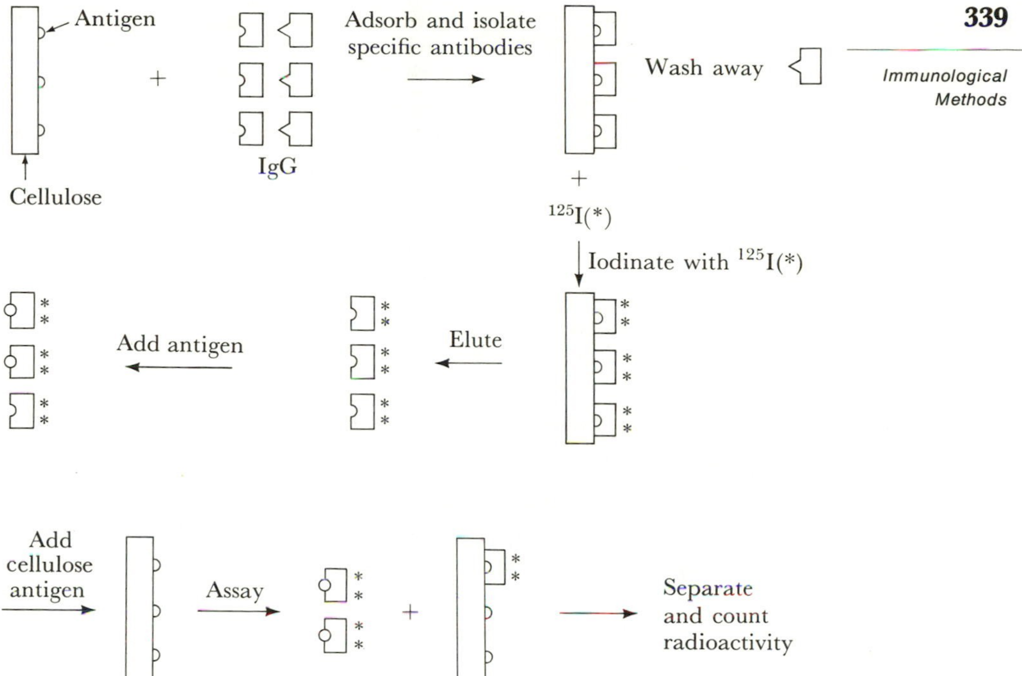

Figure 10-11
Principle of the immunoradiometric assay. See text for details.

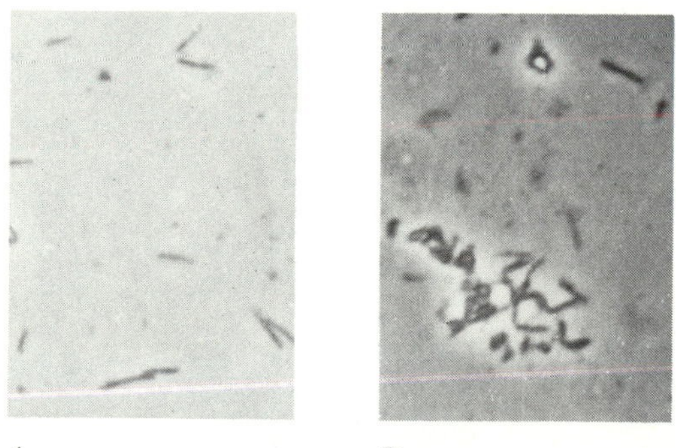

Figure 10-12
E. coli bacteria agglutinated by the addition of specific antiserum: (A) no antiserum; (B) with antiserum.

Example 10-B ☐ Identification of viruses by inhibition of virus-induced hemagglutination.

It is often necessary both in the laboratory and clinically to identify viruses. This can be easily done with a simple inhibition assay. Certain classes of viruses have the property of causing the clumping of red cells (*hemagglutination*). If a particular virus is causing agglutination, the addition of an antibody directed against that virus will bind to the virus and thereby prevent hemagglutination, but other antibodies will not. Therefore, if one is armed with a collection of antibodies directed against many viruses, the particular virus can be identified.

Example 10-C ☐ Detection of gonadotrophins in the urine of pregnant women by an inhibition assay.

Small polystyrene particles can be coated with human gonadotrophin, a hormone found in the urine of pregnant women. The addition of anti-gonadotrophin causes these coated particles to clump (Figure 10-13); this clumping is easily seen with a light microscope. If urine from a pregnant woman is added to the coated particles before adding antibody, urinary gonadotrophin competes with the coated particles for sites on the antibody so that the aggregates do not form. This is a useful pregnancy test.

Example 10-D ☐ Measurement of the synthesis of bacteriophage tail fibers in infected *E. coli* bacteria, using an inhibition assay.

Antibodies prepared against *E. coli* bacteriophage T2 are directed primarily against T2 tail fibers; hence, the addition of anti-T2 inac-

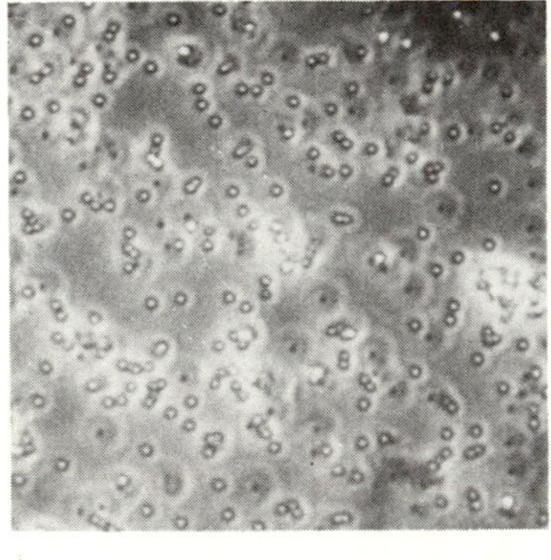

A

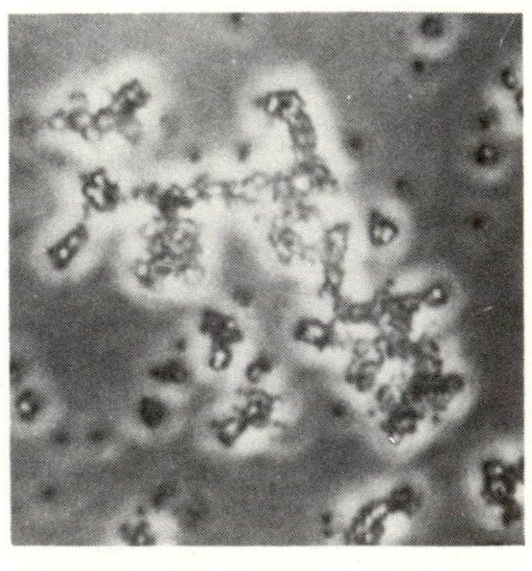

B

Figure 10-13

Clumping of gonadotrophin-coated polystyrene particles by anti-gonadotrophin: (A) no antiserum; (B) with antiserum.

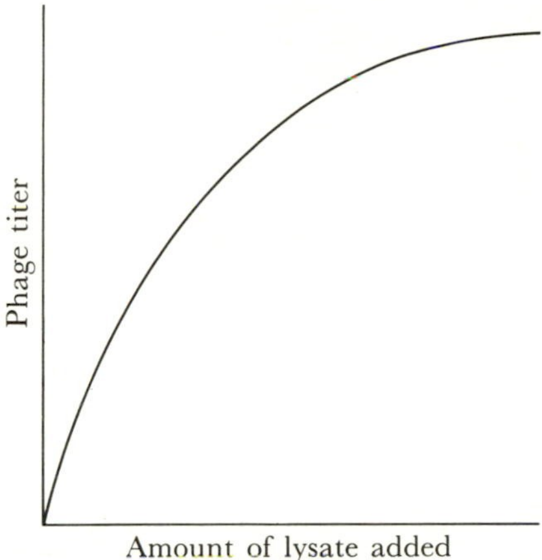

Figure 10-14
Serum-blocking-power assay. Anti-T2 is added to phage T2, along with various amounts of an extract of bacteria infected with a phage mutant that fails to make phage heads. In the absence of the extract, the antiserum inactivates the phage. If phage tails (to which the antiserum is directed) are present in the extract, they compete with the phage for the antibody so that phage survive. This can be used to determine the number of phage equivalents of tails present in the extract.

tivates the phage by preventing adsorption to the host. This is an example of the neutralizing power of antibodies. The intracellular synthesis of tail fibers can be followed by preparing cell extracts at various times after infection, mixing aliquots with anti-T2, and then adding phage. If tail fibers are present in the cell extract, they will be bound to the anti-T2, which will therefore be titrated (i.e., prevented from binding to the added phage) and the added phage will survive. Hence, the number of viable phage remaining after such treatment measures the amount of tail fibers in the extract (Figure 10-14). This inhibition assay is known in phage research as the *serum-blocking-power assay*.

Example 10-E

☐ Measurement of T4 DNA synthesis in infected *E. coli* by complement fixation.

Antibodies to denatured, glucosylated DNA can be prepared. *E. coli* bacteriophage T4 DNA contains glucosylated 5-hydroxymethylcytosine instead of cytosine and is therefore distinguishable from *E. coli* DNA in that T4 but not *E. coli* DNA will react with the antibodies. Using complement fixation, the course of synthesis of T4 DNA during infection has been followed by preparing cell extracts, denaturing the DNA, and assaying for glucosylated, denatured DNA (Figure 10-15).

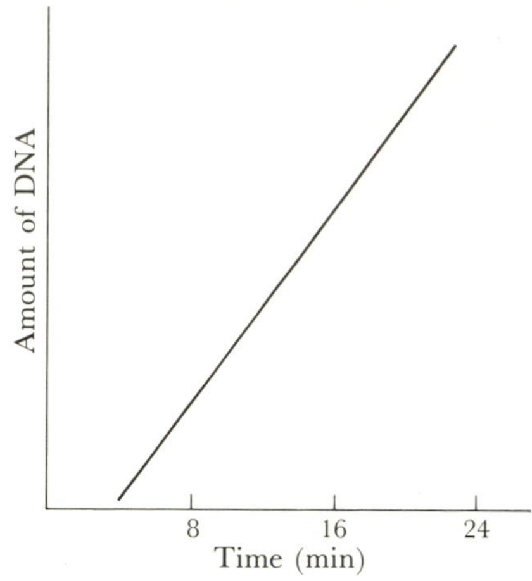

Figure 10-15
Measurement of *E. coli* phage T4 DNA synthesis by complement fixation. *E. coli* was infected with five phages per bacterium. Samples were taken at various times and assayed for the presence of glucosylated DNA, using complement fixation with antibody to glucosylated DNA. The virtue of this method is that *E. coli* DNA is nonglucosylated and does not react with the antibody, whereas T4 DNA contains glucosylated 5-hydroxymethylcytosine. Therefore, the antibody detects the T4 DNA and ignores the huge amount of bacterial DNA. [From L. Levine, W. T. Murakami, H. Van Vunakis, and L. Grossman, *Proc. Natl, Acad. Sci. U.S.A.* 46(1960): 1038–1043.]

Example 10-F □ Effect of denaturants on DNA.

The denaturation of DNA is usually studied by measuring the increase in optical density at 260 nm (see Chapter 14). If a potential denaturant also absorbs light of this wavelength, this simple assay cannot be carried out. However, denaturation can be detected immunologically, because several DNA antibodies react only with denatured DNA. In fact, the effects of a large number of destabilizing substances have been assayed by complement fixation, the extent of denaturation being determined as a function of denaturant concentration. An example of some of the data obtained is given in Table 10-2.

Example 10-G □ Identification of the *E. coli* phage λ *redβ* protein.

If *E. coli* is infected with phage λ, a large amount of an exonuclease is synthesized. This nuclease was purified and anti-(λ exonuclease) was prepared. When tested for homogeneity by Ouchterlony immu-

Table 10-2
Reagent Concentration Giving 50% Denaturation at 73°C and Ionic Strength of 0.043 M of *E. coli* Bacteriophage T4 DNA, Assayed by Complement Fixation.

Reagents	Concentration (M)
Methanol	3.50
Ethylene glycol	2.20
Glycerol	1.80
1,4-Dioxane	0.64
Phenol	0.08
Formamide	1.90
Urea	1.00
Thiourea	0.41

SOURCE: Data from L. Levine, J. A. Gordon, and W. P. Jencks, *Biochemistry* 2(1963): 168–175.
NOTE: DNA was placed in solutions containing various concentrations of the reagents and incubated at 73°C. The solutions were then diluted and assayed by complement fixation, using a DNA antibody that detects only single-stranded DNA. Fifty-four different reagents were tested.

nodiffusion (Figure 10-16), two bands were found, indicating that at least two proteins were present in the "pure" sample. One band resulted from the product of a recombination gene called *exo*. The other proved to be the product of another gene called *redβ*. The *redβ* protein complexes with the exonuclease but has no detectable enzymatic activity in vitro. Its synthesis and genetic control has been studied, using immunodiffusion as an assay because no other assay for the *redβ* protein exists.

Example 10-H

☐ Detecting *E. coli* DNA polymerase III in the presence of polymerase I by neutralization.

If two enzymes participate in similar reactions, it is often difficult to assay one in the presence of the other. However, in assaying DNA polymerases, polymerase I activity can be eliminated from a cell extract by the addition of anti-polymerase I before adding the extract to the reaction mixture. The neutralizing power of an antibody is employed as a biochemical tool, thus allowing polymerase III to be assayed unambiguously.

The following two examples make use of cross-reaction to detect substances that could not otherwise be detected.

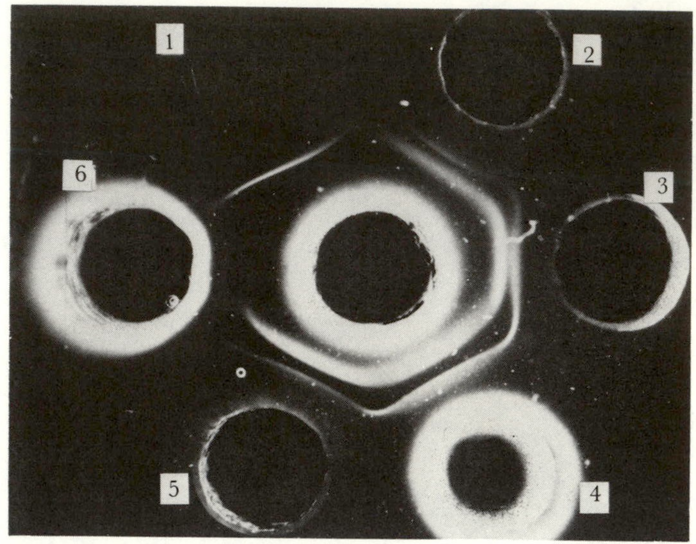

Figure 10-16
Ouchterlony analysis of the *E. coli* phage λ exonuclease and β protein. The center well contains antibody directed against presumably pure λ exonuclease. Well 1 contains highly purified λ exonuclease and a single precipitin line is seen. Well 3 is the same as well 1 but less pure. Well 2 contains a partially purified extract of a cell infected with wild-type λ. The exonuclease band (product of the *exo* gene) and the inner band (β protein) are present. Well 4 is the same as well 2 but less pure. Well 5 contains the result obtained with a λ mutant called *t11*, which overproduces exonuclease fails to make all replicative and structural phage proteins. Beta is present, indicating that it is made in the part of the genome containing the *exo* gene but that it is not in the replicative or structural region. Well 6 contains an extract from a λ N^- mutant. Both bands are absent indicating that the exonuclease and β are both under the control of gene N. [From C. M. Radding and D. C. Shreffler, *J. Mol. Biol.* 18(1966):251–261.]

Example 10-I ☐ Detection of mutant proteins or of protein fragments by cross-reaction.

Proteins made from mutant genes contain either a new amino acid replacing the one found in the wild-type protein or, with chain-termination mutations, fragments of the normal protein. In both cases, the activity of the protein (e.g., as an enzyme) is lost and there is no longer a bioassay. If the altered protein or the fragment still contains the same antigenic determinant (e.g., an amino acid sequence) as that of the wild-type protein, it can be detected immunologically. Complement fixation, immunodiffusion, or the radioimmunoassay would be the methods of choice. An example of this approach is shown in Table 10-3.

Table 10-3
Immunological Cross-Reaction Between Different Mutants of *E. coli* β-galactosidase.

Region of genetic map in which mutant occurs	Amount of protein (μg) for 50% complement fixation*
Wild-type	0.06
1	NF†
2	NF
3	NF
4	NF
5	4.0
6	1.0
7	2.0
8	NF
9	NF
10	1.4
11	0.5
12	1.4
13	0.6
14	1.3
15	0.5
16	0.5

Conclusions

1. The enzyme β-galactosidase can be detected immunologically even if no enzymatic activity is detectable.
2. Regions 1–4 and 8–9 either are the antigenic sites or more effectively alter the three-dimensional conformation of the antigenic site, because mutations in those regions eliminate all antigenic activity.

SOURCE: Data from A. V. Fowler and I. Zabin, *J. Mol. Biol.* 33(1968):35–47.
* The nearer the value is to the wild-type figure of 0.06, the more closely the antigenic determinant of the mutant protein resembles that of the wild type.
† No fixation detectable.

☐ Detecting relatedness between proteins by cross-reaction. **Example 10-J**

Certain proteins—for example, those found in the glial cells of the brain—are widespread in nature. Antibody to a particular rabbit glial protein has been found by means of complement fixation to react with proteins found in nerve tissues of species ranging through-

Table 10-4

Cross-Reaction Between the Brain 100,000 × *g* Supernatant (S-100) Proteins from Various Organisms and Antibody to Cattle S-100 Protein.

Animal	Relative antiserum concentration required for 50% fixation of complement*
Cow	1.0
Sheep	1.0
Rat	1.0
Guinea-pig	1.2
Mouse	1.3
Pig	1.8
Rabbit	1.9
Human	2.0
Chicken	2.2
Pigeon	2.3
Bullfrog	2.5

SOURCE: Data from D. Kessler, L. Levine, and G. Fasman, *Biochemistry* 7(1968): 758–764.
* The nearer the value is to 1, the more closely related is the protein.

out the evolutionary scale from frogs to humans. Representative data, which can be taken as evidence for related structure, are given in Table 10-4.

Examples of the Use of the Radioimmunoassay

Table 10-5 gives a variety of substances that have been assayed using the radioimmunoassay. At the present time, more than one thousand substances have been measured in this way and the catalog of such substances increases daily. In many cases, no other assay is possible at present.

One particularly important application of the radioimmunoassay is in the study of hormones. In this procedure, cross-reacting material is examined and this has led to the discovery of many unexpected hormone precursors. The reasoning used is the following.

Consider two cross-reacting substances A and B. If antibodies to A and radioactive A can be prepared, the radioimmunoassay can be used to measure the concentration of A in a complex mixture such as blood or urine. If a substance reacting with anti-A is found, there is no way, from the assay alone, to know whether it is A or a cross-reacting sub-

Table 10-5
Substances That Have Been Used in the Radioimmunoassay.

Hormones	
Gastrointestinal (glucagon, gastrin, enteroglucagon, secretin, pancreozymin, vasoactive intestinal peptide, gastric inhibitory peptide, motilin, insulin)	Prolactin Thyrocalcitonin Parathyroid hormone Human chorionic gonadotrophin
Corticotrophin Follicle-stimulating hormone	Posterior pituitary peptides (oxytocin, vasopressin, neurophysin)
Antidiuretic hormone Thyroid-stimulating hormone	Bradykinin Thyroid hormones
Pharmacologic agents	
Morphine and opiate alkaloids Cardiac glycosides Prostaglandins Lysergic acid and derivatives Amphetamines	Tetrahydrocannabinol Barbiturates Nicotine and metabolic products Phenothiazines
Vitamins and cofactors	
D, B12, folic acid, cyclic AMP*	
Hematological substances	
Fibrinogen, fibrin, and fibrinopeptides Plasminogen and plasmin Antihemophilic factor	Prothrombin Transferrin and ferritin Erythropoietin
Virus antigens	
Hepatitis antigen Herpes simplex Vaccinia Several Group A arboviruses	Polio Rabies Q fever Psittacosis group
Nucleic acids and nucleotides	
DNA, RNA, cytosine derivatives	

* Adenosine 3′,5′-monophosphate.

stance such as B.* However, this can be determined by fractionating the mixture by gel chromatography and examining each fraction for A using the radioimmunoassay. In general, A and B would separate in gel chromatography. Thus if two peaks of activity are found, there must be at least one cross-reacting substance (or possibly a dimer of A) in the mixture. If only one peak is found, it may contain either A or B. The peak

* Immunodiffusion or immunoelectrophoresis could in principle demonstrate whether A, another substance, or a mixture is present but, in general in biological fluids, there is not enough hormone to produce a visible precipitin band. A bioassay for A, if it were a hormone, could be carried out but the cross-reacting material might also have biological activity; also, the concentration is often too low for a bioassay.

may be identified by adding authentic A to another aliquot of the mixture, fractionating, and determining by the radioimmunoassay again whether there is one or two peaks of activity. If only one is found, the original fluid probably contained A only. If two peaks are found, only a cross-reacting substance is present. Note that even if the cross-reacting material has no biological activity (for instance, if it were a precursor to A), it would be detected in this way.

This technique has been used to demonstrate the existence of precursors (which we call pro-H) to many hormones. In most cases studied the pro-H elutes from a gel column before the hormone so that it is a larger molecule than the hormone. To determine whether pro-H is a precursor that is modified to yield a hormone, it is purified and treated with a protease to produce several fragments. The fragments are isolated and antibodies to each of the fragments are prepared. It is usually found that antibodies to some, but not all, of the fragments are active against the pure hormone H. Therefore, pro-H contains amino acid sequences that are not present in H so that pro-H cannot simply be a dimer of H. If pro-H is converted to H by proteolytic cleavage in body fluids, then these fluids should contain both H and fragments F_1, $F_2, \ldots, F_n$ that do not cross-react with anti-H. Antibodies to pro-H [anti-(pro-H)] are also prepared. If pro-H is a precursor to H, then the fragments $F_1, F_2, \ldots, F_n$ should react with anti-(pro-H) but not with anti-H.

Such fragments have been detected for many hormones and in many cases this provided the principal evidence for the existence of hormone precursors. Also, since the precursor can be assayed, even though it may have no biological activity before processing, a means has been provided for purifying the substances. Several dozen hormone precursors have been detected and purified in this way.

The two advantages of the radioimmunoassay are its ability to detect substances for which there is no other assay and its extraordinary sensitivity. Several examples of its use when great sensitivity is required follow.

Example 10-K □ Tumor diagnosis.

Some tumors produce tumor-associated antigens that can be found in blood or urine. In some cases—for example, human chorionic gonadotrophin, human placental lactogen, and placental alkaline phosphatase—these antigens are not present in normal blood but are found in serum from a certain percentage of patients with particular kinds of carcinoma.

If the tumor is in tissue that normally controls or makes a hormone, often associated with the tumor is an increased level of the particular hormone—for example, high insulin with pancreatic cancer. Each of these substances can be detected by the radioimmunoassay, with far greater sensitivity than with any known chemical or bioassay.

☐ Assay of clinically important substances in children or in other situations where only small amounts are available. **Example 10-L**

Many clinically important substances can be assayed in large blood samples by conventional chemical tests. However, for babies, small children, seriously ill patients, and small animals, the required sample size may be greater than that which can be safely withdrawn. The great sensitivity of the radioimmunoassay avoids this problem.

☐ Measurement of pyrimidine dimers in ultraviolet-irradiated DNA. **Example 10-M**

In the study of mechanisms of repair of irradiated DNA in vivo, it is almost always necessary to measure the number of pyrimidine dimers (the principal product of ultraviolet irradiation of DNA). This can be done by paper chromatography or electrophoresis if the intracellular DNA can be labeled to high specific activity with ^3H. However, this cannot be done with animal or plant cells. Furthermore, it is difficult to detect a very small number of dimers—for example, 1 per 10^8 base pairs—by conventional means, whereas it can be done by the radioimmunoassay (Figure 10-17).

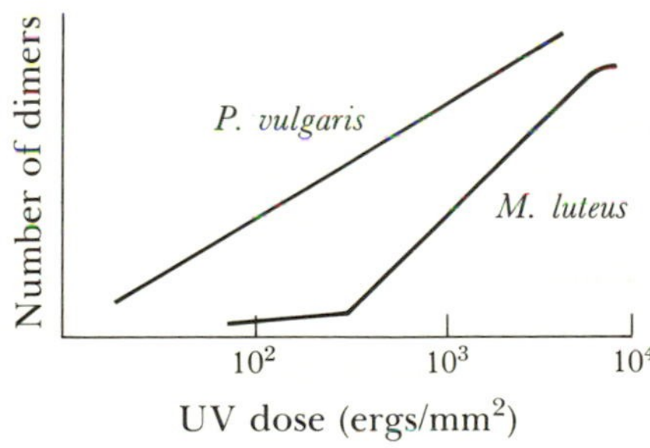

Figure 10-17
Detection of thymine dimers ($\widehat{TT}$) in DNA by the radioimmunoassay, using an antibody directed against ultraviolet-irradiated DNA, that is, anti-$\widehat{TT}$. A mixture is made of anti-$\widehat{TT}$ and iodinated UV-irradiated DNA and the amount of radioactive iodine precipitated by goat anti-(rabbit serum) is measured. The decrease in precipitated radioactivity as a function of added UV-irradiated DNA is measured. From a standard curve, the percentage inhibition can be related to the number of thymine dimers. Each curve represents DNA obtained from a different bacterium. *Micrococcus luteus* DNA has a lower thymine content than *Proteus vulgaris* DNA and therefore requires a higher dose to yield the same number of dimers. [From E. Seaman, H. Van Vunakis, and L. Levine, *J. Biol. Chem.* 247 (1972):5709–5715.]

Example 10-N ☐ The role of prostaglandins in humans.

Human venous blood contains from 10 to 100 picograms per milliliter of various polyunsaturated fatty acids called prostaglandins. Understanding their significance has been hampered by the inability to assay these substances quantitatively and with high sensitivity; the radioimmunoassay provides this opportunity as well as the ability to analyze a large number of samples in a short time. This becomes especially important (i.e., to monitor blood levels) because many prostaglandins have recently been shown to have pharmacological activity—but with a narrow therapeutic margin of safety.

The Fluorescent Immunoassay

The fluorescent immunoassay has about one-tenth the sensitivity of the radioimmunoassay but does not need radioactive material. However, it does require that purified specific antibody be available. Since purification of an antibody directed against a particular antigen may be tedious, this procedure is most useful when the same substance must be assayed repeatedly.

A fluorescent immunoassay of a substance X is done as follows (Figure 10-18). Purified anti-X is isolated from rabbit serum and covalently coupled to small beads to form an immunoadsorbent. The sample to be assayed for the concentration of X is mixed with an excess amount of the immunoadsorbent. All X binds to the beads. Then more anti-X, which has been chemically coupled to a fluorescent compound, is added; this combines with the X that is bound to the beads (as long as X has two antigenic determinants and hence can bind to two antibody molecules). The amount of fluorescent antibody bound is directly proportional to the amount of bound X. The beads are then removed from the mixture by centrifugation, resuspended in a fixed volume of buffer, and the fluorescence is measured. The amount of fluorescence is compared to a curve obtained by adding known amounts of X to a suspension of beads.

Beads capable of binding antibody are commercially available (e.g., Immunobeads, Bio-Rad Laboratories) as are beads to which purified antibodies to specific substances are already bound.

The Protein A Immunoassay

The cell wall of almost all species of the bacterium *Staphylococcus* contains a substance called protein A. This protein binds tightly to the F_c segment of an IgG molecule but does not interfere with the antigen-antibody reaction. Furthermore, protein A can be labeled with ^{125}I without affecting its interaction with IgG. ^{125}I-labeled protein A can be

Step I: Couple anti-X to beads.

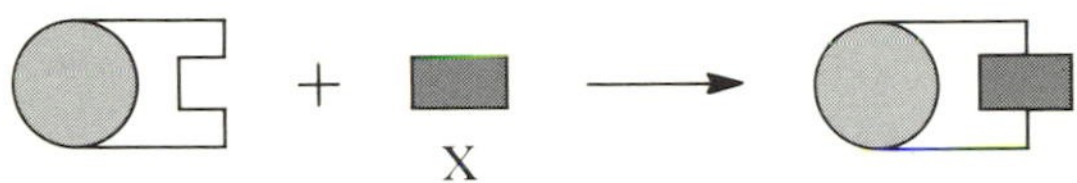

Step II: Add X.

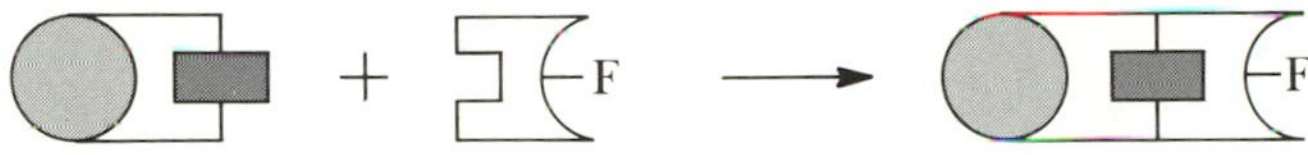

Step III: Add fluorescent (F) anti-X.

Step IV: Wash beads, resuspend, and measure fluorescence.

Figure 10-18
The fluorescent immunoassay using Immunobeads. The beads are washed free of unbound material between each step.

used as an assay for IgG. The following shows how it can be used in an immunoassay for a particular substance X.

Beads containing bound X (X_b) are prepared. If anti-X is added to the beads, it will react with X_b; if free X (X_f) is also present, X_b and X_f compete for sites on anti-X. Thus, the amount of X_f (which is the substance to be assayed) that is present determines the amount of anti-X that binds to the beads. This could be measured with ^{125}I-anti-X, which could be prepared by iodination of the antibody. However, the use of ^{125}I-protein A avoids the necessity for preparing iodinated antibody because if ^{125}I-protein A is added to the reaction mixture containing beads, X_f, and antiserum, the amount of ^{125}I bound to the beads is a measure of the amount of bound anti-X. Thus, X_f is measured by a standard inhibition curve—that is, as X_f increases, bound ^{125}I decreases.

It is not necessary to have the antigen coupled to beads; it is sufficient that the antigen is immobilized onto some object that is easily separated from the reaction mixture. For example, cell-surface antigens (S) have been assayed using whole cells as the source of immobilized antigen. If a reaction mixture is prepared containing cells, anti-S, and ^{125}I-protein A, ^{125}I-labeled cells can be centrifuged from the mixture. If free S is also present, less anti-S and thus less ^{125}I-protein A is bound to the cells so that the amount of cell-associated ^{125}I is a measure of the concentration of free S.

The advantage of the protein A assay is that neither radioactive antigen nor radioactive antibody has to be prepared to assay for each substance. The ^{125}I-protein A suffices for the assay of all antigens. Many examples of its use can be found in the reference by Van Vunakis and Langone given at the end of the chapter.

Immunological Techniques for Localizing Substances in Cells, Tissues, and Molecules: The Fluorescent Antibody and the Ferritin-Conjugated Antibody

The fluorescent dye fluorescein can be chemically coupled to most antibody molecules without loss of antibody activity. This material can be used to localize structures in cells and tissues by fluorescence microscopy (see Chapter 2). An example is the identification of cells to which a particular virus is bound at the surface, using fluorescein-conjugated antivirus. The presence of the virus is indicated by the fluorescence even though the size of the virus is beyond the limit of resolution of the microscope (see Chapter 2, page 43).

The iron-containing protein ferritin can also be coupled to antibody. If specific ferritin-conjugated antibody is reacted with cells and the cells are examined by electron microscopy, the antibody molecules can be localized by the position of the dark dots corresponding to the ferritin. This allows localization of particular molecules at high resolution. An example is shown in Figure 10-19.

Another very useful method, which is applicable to both light and electron microscopy, is the peroxidase–anti-peroxidase (PAP) procedure. This method uses antibodies to link the visualizable enzyme, horseradish peroxidase, to a tissue antigen X (Figure 10-20). In the first step, rabbit anti-X is allowed to bind to X. Then bivalent goat anti-(rabbit immunoglobin) is added in excess so that one of the combining sites on the goat antibody is bound to rabbit immunoglobulin and the other site remains free. Then a complex consisting of peroxidase and rabbit anti-peroxidase is added and this complex combines with the remaining goat anti-(rabbit immunoglobulin) site. In this way peroxidase, which remains enzymatically active, is bound to the tissue antigen. If hydrogen peroxide and diaminobenzidine are then added, a deep brown material is deposited at the position of the tissue antigen against which the rabbit antibody was directed. This brown substance is easily seen by ordinary light micro-

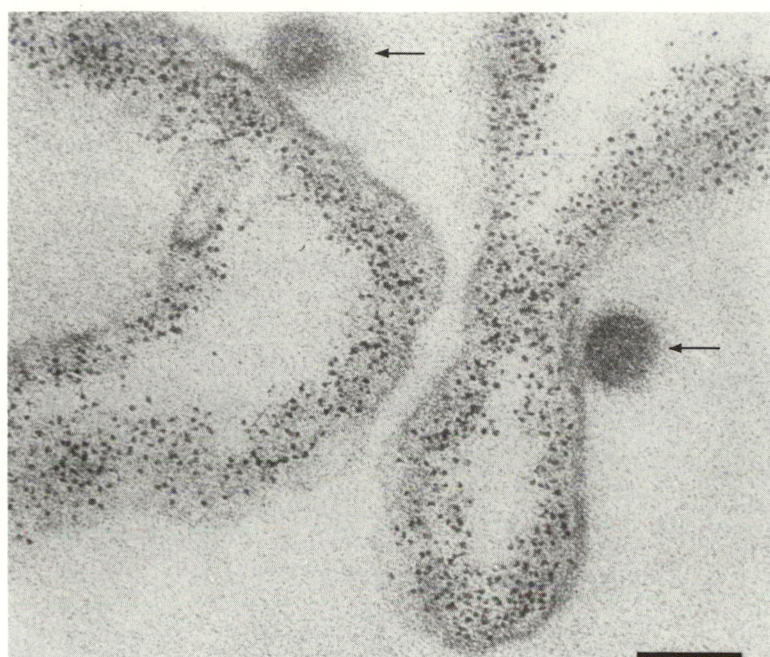

Figure 10-19
An example of the use of ferritin labeling. A human red cell has been broken open and incubated with a mixture of influenza virus (arrow) and ferritin-conjugated antibody directed against the protein spectrin. The influenza virus adsorbs to the outer membrane of the red cell. The ferritin is seen as dark dots and is observed only on the opposite side of the membrane from the virus. This shows that spectrin is primarily, if not solely, on the inner surface of the red cell membrane. Bar = 0.01 μm [From G. L. Nicolson, V. T. Marchesi, and S. J. Singer, *J. Cell Biol.* 51(1971):365–272.]

scopy. For electron microscopy OsO_4 is then added and the brown product reduces the OsO_4 to metallic osmium, which is readily visible by electron microscopy.

Immunoelectrophoresis

Immunoelectrophoresis includes a set of techniques for both identifying and quantitating various substances. Each method is performed in an agarose gel. They differ in how the sample is applied to the gel, whether the antibody is mobile, and whether the technique is one- or two-dimensional.

Step I: Add rabbit anti-X.

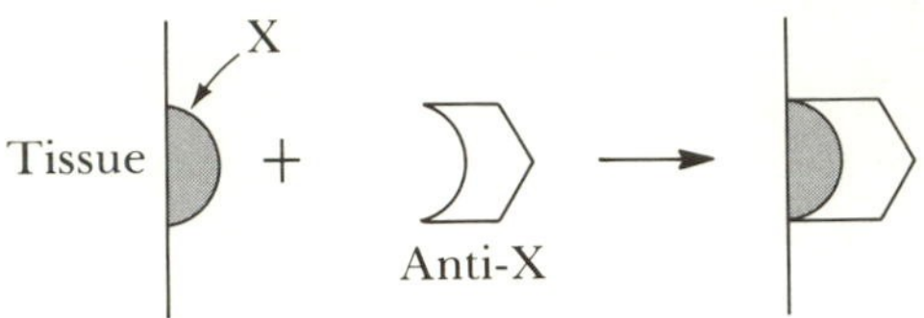

Step II: Add bivalent goat anti-(rabbit immunoglobulin).

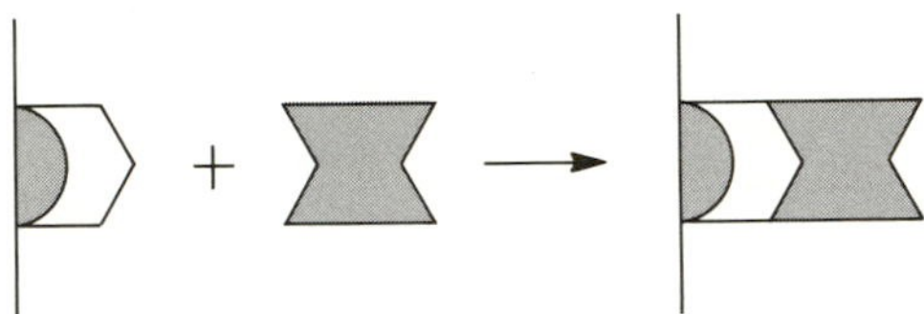

Step III: Add peroxidase–anti-peroxidase complex

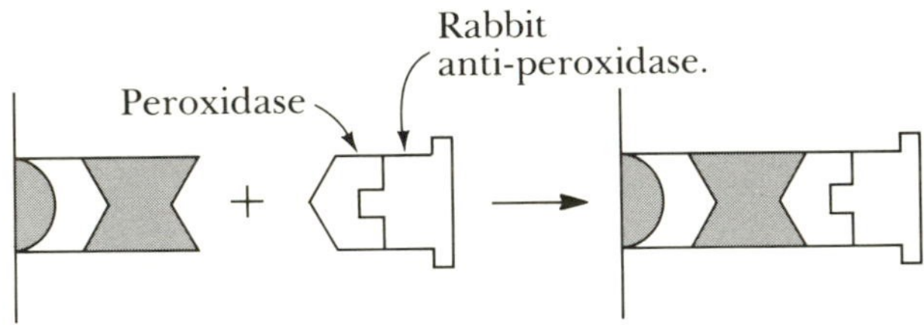

Step IV: add H_2O_2 + diaminobenzidine to generate brown material.

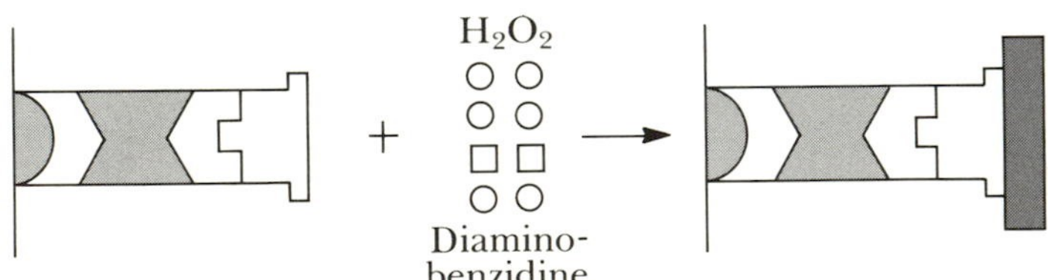

Figure 10-20
The steps in the peroxidase–anti-peroxidase procedure for locating a cellular or tissue antigen X. See text for details.

The simplest and oldest procedure is that of P. Grabar and C. A. Williams and is shown in Figure 10-21. A uniform layer is prepared and a well and trough are cut for antigen and antibody respectively. The antigen mixture is placed in the well and the slab is submitted to electrophoresis. During this period, molecules migrate in both directions, the

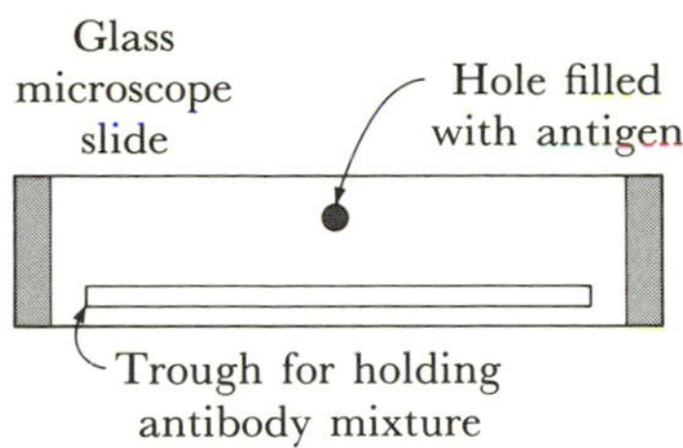

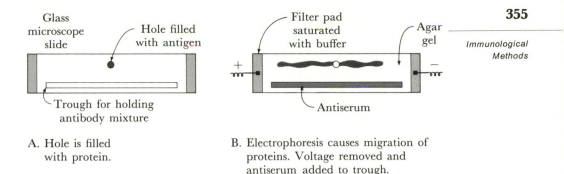

A. Hole is filled with protein.

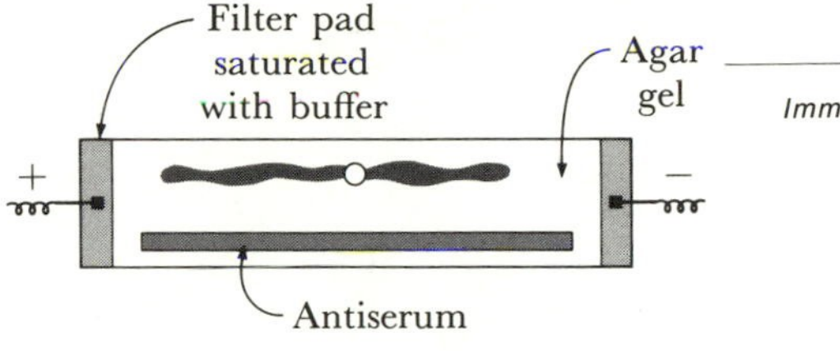

B. Electrophoresis causes migration of proteins. Voltage removed and antiserum added to trough.

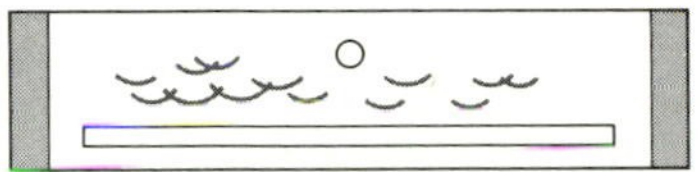

C. Immunodiffusion forms precipitin lines.

Figure 10-21
Arrangement for immunoelectrophoresis using the technique of Grabar and Williams.

distances depending on their mobilities and the sign of the charge. When electrophoresis is complete, antiserum is added to the trough. Both antiserum and antigens diffuse through the gel. If an antigen-antibody reaction occurs where they meet, a visible, curved precipitin line forms. This technique allows the identification of the number of antigens producing antibody and the relative amounts of each, and it allows the assay of materials for which only immunoassay is possible (e.g., cross-reacting, inactive fragments of enzymes).

A variant of this procedure, *crossed immunoelectrophoresis*, is more sensitive than the technique just described; it allows precise measurement of the amount of each antigen present in a mixture. The procedure is shown in Figure 10-22. In the first step of the procedure, an antigen mixture is placed in a slot in an agarose gel and electrophoresed, as shown in the figure. Bands of each antigen are present, although at this time they are not visible. Next a narrow longitudinal strip (panel A) is cut out along the middle of the gel. This strip contains bands of each antigen. This strip is then placed on a second gel (panel B), which contains a low concentration of antiserum that contains antibodies to the antigens present in the mixture applied to the first gel. This second gel is now electrophoresed, but at a 90° angle with respect to the movement of the antigens in panel A. During this period of electrophoresis both the antigens in the strip and the antibodies in the lower plate migrate in the electric field. Precipitin zones form for each antigen; these zones move as

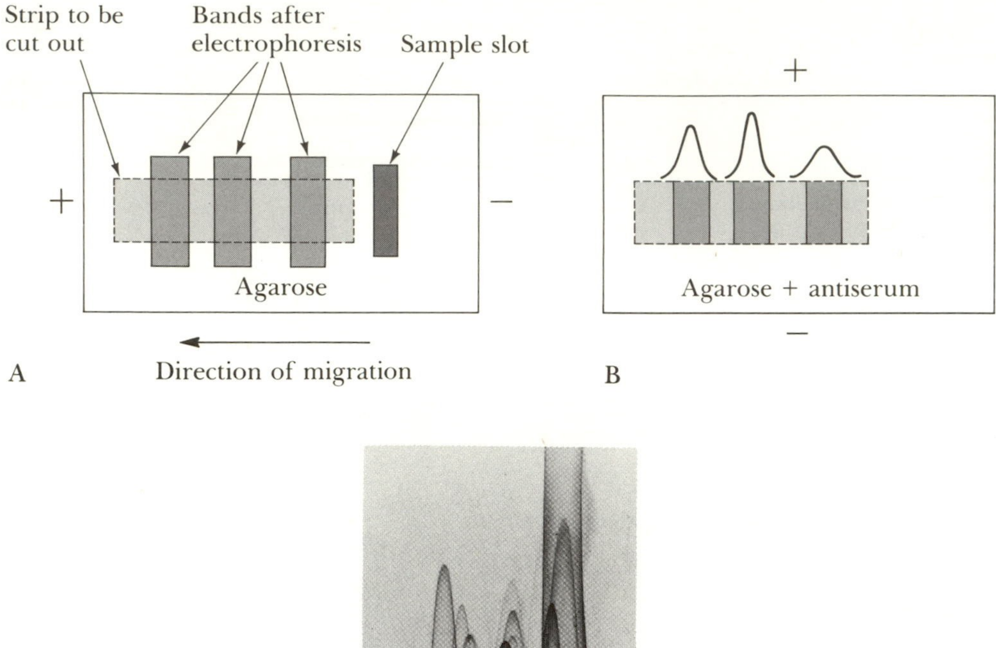

Figure 10-22
Crossed electrophoresis. Panel A shows the first stages. Antigens have been placed in the sample slot and electrophoresed in agarose. Three bands have formed. A strip of agar (outlined by dashed lines) is then cut out of the agarose. This strip is placed on a second slab of agarose that contains antiserum, as shown in panel B. Electrophoresis is carried out at a 90° angle with respect to that in panel A. Precipitin curves form. Panel C shows a photograph of the result of crossed electrophoresis. [Courtesy of Bio-Rad Laboratories.]

the antigen is forced by the electric field to migrate. Panel C shows that the resulting pattern of precipitin lines differs significantly from that obtained by the Grabar-Williams technique. Each antigen produces a long narrow peak rather than a small curve. This results in less overlap of the components of the mixture and allows a much less ambiguous count of the number of components.

Another important method is called *rocket electrophoresis* (Figure 10-23). This procedure allows the determination of the concentration of a particular antigen in a mixture. It is necessary, though, that the antiserum contains antibodies for the particular antigen and not for any other antigen in the mixture. A slab of agar containing a low concentration of antiserum is prepared. A hole is punched in the agar at one end of the slab and it is filled with the antigen. During electrophoresis, precipitin

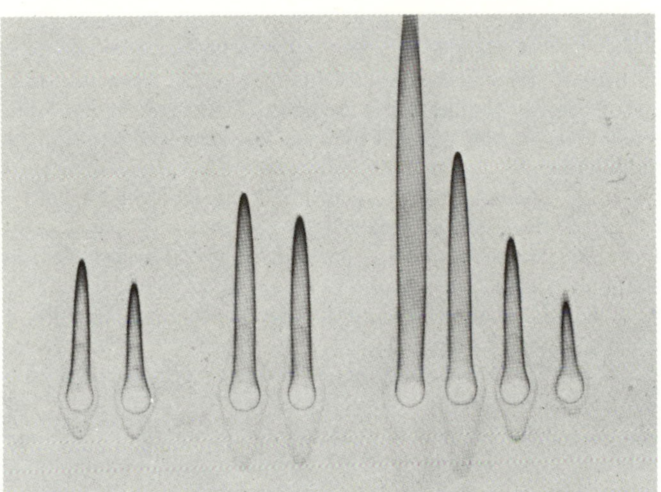

Figure 10-23
Rocket electrophoresis. [Courtesy of Bio-Rad Laboratories.]

lines form and then lines move along the direction of migration of the antigen. Owing to diffusion, a point is reached at which antigen excess cannot be induced ahead of the moving zone; when this occurs, the leading edge of the precipitin band stops moving. For a particular antibody concentration, the distance that the leading edge moves is proportional to the initial concentration of the antigen. Thus, the amount of a particular substance in a mixture can be measured in the following way. A set of holes is made in the gel. Each hole except one is filled with a different, but known, concentration of the antigen. The remaining well is filled with the mixture. After electrophoresis the distance moved can be plotted against the initial concentration of the standard wells. The concentration of the antigen in the mixture can then be determined from this graph.

Many other variations of the basic immunoelectrophoretic procedure have been developed. These are described in some of the references which follow.

SELECTED REFERENCES

Axelsen, N. H., J. Krøll, and B. Weeke. 1973. "A Manual for Quantitative Immunoelectrophoresis. Methods and Applications." *Scand. J. Immunol.* 2, suppl. 1.

Berson, S. A., and R. S. Yalow. 1971. "Radioimmunoassay: A Status Report," in *Immunobiology*, edited by R. A. Good and O. W. Fisher, pp. 287–293. Sinauer Associates. An excellent description of the radioimmunoassay.

Brewer, J. M., A. J. Pesce, and R. B. Ashworth. 1974. *Experimental Techniques in Biochemistry*, ch. 4. Prentice-Hall.

Clausena, J. 1969. "Immunochemical Techniques for the Identification and Estimation of Macromolecules," in *Laboratory Techniques in Biochemistry and Molecular Biology*, vol. 1, edited by T. S. Work and E. Work, pp. 405–556. North-Holland.

Crowle, A. J. 1973. *Immunodiffusion*. Academic Press. Hudson, L., and F. C. Hay. 1976. *Practical Immunology*. Blackwell.

Kabat, E. A. 1968. *Structural Concepts in Immunology and Immunochemistry*. Holt, Rinehart and Winston.

Kirkham, K. E., and W. M. Hunter. 1971. *Radioimmunoassay Methods. European Workshop*. Churchill Livingstone.

Kwapinski, J. B. 1972. *Methodology of Immunochemistry and Immunological Research*. Wiley-Interscience.

Sternberger, L. A. 1974. *Immunocytochemistry*. Prentice-Hall.

Van Vunakis, H., and J. J. Langone. 1980. "Immunological Techniques." *Methods in Enzymology*, vol. 70. Academic Press. This is the most up-to-date methods book available at present and is the best source for the use of the protein A immunoassay.

Wasserman, E., and L. Levine. 1961. "Quantitative Micro-complement Fixation and Its Use in the Study of Antigenic Structure by Specific Antigen-Antibody Inhibition." *J. Immunol.* 87:290–295. The definitive paper on micro-complement fixation.

Weir, D. M. 1978. *Handbook of Experimental Immunology*. Blackwell.

Williams, C. A. 1967. *Methods in Immunology and Immunochemistry*. Academic Press.

PROBLEMS

10–1. A complement fixation assay is performed, using an antigen X and anti-X. A standard curve is constructed by adding various amounts of X. The data are:

X added (μg)	OD	X added (μg)	OD
0.00	1.15	0.05	0.17
0.01	0.82	0.06	0.19
0.02	0.50	0.07	0.23
0.03	0.22	0.08	0.50
0.04	0.18	0.09	0.75

You wish to determine the amount of X in a cell extract. This is done by adding 1 μl of the extract and then 1:1 and 1:2 dilutions of the extract to the mixture (without adding any X); the OD readings are 0.20, 0.66, and 0.85. What is the concentration of X in the extract?

10–2. Suppose that you have a mixture of two proteins and antiserum prepared against the mixture. If the two proteins have the same molecular weight and shape, they will have nearly the same diffusion coefficient (see Chapter 12). If this mixture is subjected to the Ouchterlony diffusion test, the probability is very high that two bands will result. Why? Under what circumstances would only one band appear? Suppose that the two proteins had very different diffusion coefficients; under what circumstances might the bands fail to resolve?

10–3. Would you expect the reaction between specific antibody and a protein antigen (assayed in any way) to depend on whether the protein is native or denatured? Explain.

10–4. Why should single amino acid changes in a protein (a) eliminate all antigenicity, (b) have no effect, or (c) produce cross-reacting material?

10–5. Suppose that you wanted to measure the level of a particular drug in the bloodstream of an animal. The standard chemical assay for the drug cannot be carried out in the presence of the amount of protein in blood. How would you go about setting up an assay using immunological methods? Describe all necessary steps.

10–6. A radioimmunoassay is performed for substance X using anti-X and [^{3}H]X. A standard curve is constructed using 1250 cpm of [^{3}H]X in each sample. The data are:

Added X (μg)	^{3}H precipitated (cpm)
0	1250
0.001	940
0.002	748
0.004	535
0.008	346
0.010	287

A blood sample is then taken and the amount of X in 0.1 ml is measured.

a. If the precipitated radioactivity is 15 cpm, what can you say about the concentration of X in the blood?
b. How would you improve the data?
c. Addition of 0.1 ml of a fiftyfold dilution yields 625 cpm. What is the concentration of X in urine?

10–7. A sample of fluid F is extracted from a plant and injected into a rabbit in order to make antiserum to F. Cross-electrophoresis is then carried out with F and anti-F and seventeen bands are seen. What can be said about the number of substances present in F?

Four

HYDRODYNAMIC METHODS

11

Sedimentation

One of the more common techniques in use today for the characterization of macromolecules is sedimentation.* By using the appropriate variant of the technique, the molecular weight, density, and shape of a macromolecule can be obtained, changes in these parameters can be detected, and any of them can be used as the basis of the separation of the components of a mixture for preparative or analytic purposes. Furthermore, the ease with which measurements can be made with modern automatic instruments makes ultracentrifugation especially useful.

Basically, only one thing is done with an ultracentrifuge: particles are made to move by centrifugal force and the distribution in concentration of the particles along the length of the centrifuge tube is determined at one or more times. A measurement made while the molecules are moving along the centrifugal axis is called a *sedimentation velocity determination* and the result is a *sedimentation coefficient*, a number that gives information about the molecular weight and the shape of the particle. When the concentration distribution is measured under conditions such that the distribution no longer changes with time, the particles are said to have reached *sedimentation equilibrium*; this type of measurement yields data about molecular weights, density, and composition.

In principle, sedimentation is very simple. However, in practice, certain complications must be taken into account to obtain meaningful information.

* *Sedimentation* is a general term for motion in a centrifugal field. If sedimentation analysis is performed using a high-speed centrifuge called an ultracentrifuge, the word *ultracentrifugation* is commonly used.

In this chapter, both the principles and the complexities of many ultracentrifugal procedures are described, and the kinds of measurements that are made are shown by example.

Simple Theory of Velocity Sedimentation

If a particle is in a centrifugal field generated by a spinning rotor with angular velocity, ω, it will experience a centrifugal force, $F_c = m\omega^2 r$, in which m is the mass of the particle and r the distance from the center of rotation. If this particle is not in vacuum but in a solvent,* the solvent molecules will, of course, be displaced by the motion of the particle. Their resistance to this displacement constitutes a buoyant force, opposed to the centrifugal force. This *buoyancy* reduces the net force on the macromolecule by $\omega^2 r$ times the mass of the displaced solution; this mass is simply the volume of the particle multiplied by the density, ρ of the solvent. The particle volume is $m\bar{v}$, in which $\bar{v}$ is the partial specific volume† of the particle, so that the buoyant force is $\omega^2 rm\bar{v}\rho$.

Clearly, the particles and the solvent molecules cannot slip by one another without experiencing friction. This frictional force—which opposes the motion of both—is proportional to the difference between the velocities of the particle and of the solvent molecules and is expressed as fv, in which f is the frictional coefficient and v is the velocity relative to the centrifuge cell, which holds the solvent. Because the velocity is constant when the net force is zero, the velocity of the particle is

$$v = \frac{\omega^2 rm(1 - \bar{v}\rho)}{f} \tag{1}$$

Equation (1) says that (all other things being equal);

1. A more massive particle (or molecule) tends to move faster than a less massive one.
2. A denser particle (i.e., small partial specific volume) moves faster than a less dense one.
3. The denser the solution, the more slowly the particle will move.
4. The greater the frictional coefficient, the more slowly the particle will move.

These four statements constitue the basic rules of sedimentation and apply to all particles whether they are large structures, macromolecules, or small molecules.

---

* Solvent is used in the general sense of the suspending medium. That is, if the particle is in water, the solvent is water; if it is in 0.1 M NaCl, the NaCl solution is considered to be the solvent.

† A detailed discussion of partial specific volume is the subject of Chapter 12.

Because the velocity of a molecule is proportional to the magnitude of the centrifugal field (i.e., $\omega^2 r$), it is common to discuss sedimentation properties in terms of the velocity per unit field, or

$$s = \frac{v}{\omega^2 r} = \frac{m(1 - \bar{v}\rho)}{f} \qquad (2)$$

in which s is the *sedimentation coefficient*; the determination of s is the immediate goal of most sedimentation velocity experiments (see page 443 for method).

The units used in equations (1) and (2) often cause confusion. The following statements should eliminate the difficulty.

1. m is the mass and not the molecular weight; m is expressed in grams. When referring to the molecular weight, the symbol M will be used; $M = N_A m$, in which N_A is Avogadro's number (6.023×10^{23}).
2. ω is expressed in radians per second; one revolution equals 2π radians.
3. The unit of s is seconds (as can be seen by dimensional analysis of equation 2). Most values of s are between 10^{-13} and 10^{-11} second. To eliminate the exponents, the value of s is usually reported in svedbergs (S). 1 svedberg $= 10^{-13}$ second.

Thus, typical values of s range from 1 to 100 svedbergs. A particle whose s value is 60 svedbergs is usually called a 60-S particle. It must be remembered that in equations (1) and (2), the unit of s is not the svedberg but the second, as just stated.

Ideally, at a single temperature s would be a constant for a particle in a given solvent. However, this is not the case for macromolecules, owing to their great size, and the theory must be expanded. This is an appropriate place to do so but, because it will be more understandable through reference to real sedimentation data, the instrumentation of ultracentrifugation and the kinds of data obtained will be explained first.

Instruments for Ultracentrifugation

Centrifugation experiments require instruments that operate at accurately known speeds with small variation and without temperature fluctuations.* Modern ultracentrifuges operate at forces as great as $600{,}000 \pm 100g$ and with temperature control within approximately 0.1°C. Two types of instruments exist—*analytical* and *preparative*. Analytical centrifuges are equipped with *optical systems* designed to determine concentration distributions at any time during the measurement,

* The field of ultracentrifugation was born when the first ultracentrifuge was designed and built by The (pronounced "Tay") Svedberg in 1923 (see the Selected References near the end of the chapter).

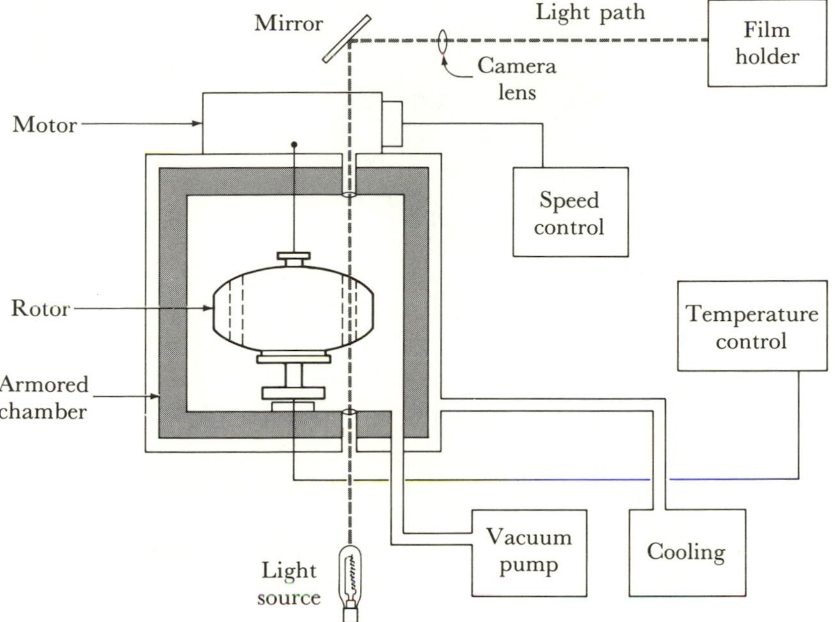

Figure 11-1
A Beckman analytic ultracentrifuge. The rotor is in an evacuated and cooled chamber and is suspended on a wire coming from the drive shaft of the motor. The tip of the rotor contains a thermistor for measuring temperature. Electrical contact of the thermistor to the control circuit is by means of a pool of mercury, which the rotor tip touches. The rotor chamber contains an upper and a lower lens. The lower lens collimates the light so that the sample cell is illuminated by parallel light. The upper lens and the camera lens focus the light on the film. In more advanced instruments, the film is replaced by an electronic scanning system that at any time can provide a plot of the concentration of the molecules at all points in the sample cell. In addition, a monochromator is placed between the light source and the cell so that any wavelength may be selected for determining the concentration of the sedimenting molecules.

whereas preparative centrifuges require fractionation of the contents of the centrifuge cell and measurement of the concentration in each fraction to obtain a concentration distribution. Preparative centrifuge is certainly a misnomer because, in addition to being used in the preparation and purification of various macromolecules and cell organelles, it is used at least as frequently to analyze mixtures in a quantitative way (i.e., as an analytical instrument).

Figure 11-1 is a schematic diagram of a Beckman analytical ultracentrifuge. This instrument consists basically of a motor; a centrifuge rotor, which is contained in a protective armored chamber; and a photographic system for recording the distribution in concentration of the sample in the centrifuge cell (see figure legend for details).

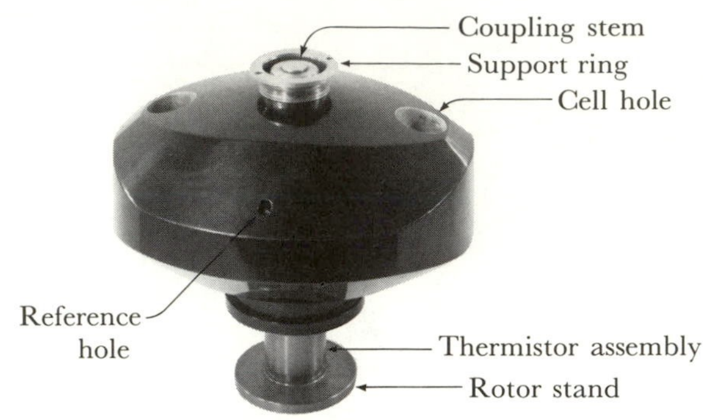

Figure 11-2
An analytical rotor. [Courtesy of Beckman Instruments.]

Figure 11-2 shows a standard rotor for the Beckman analytical centrifuge. The rotor is suspended in the center from the drive motor by a wire. One of the holes in the rotor contains the centrifuge cell; the other contains a counterbalance and reference hole for determining distances from the center of rotation. Figure 11-3 shows a diagram of one of the simplest sample holders—a single-sector cell. The walls of the cell are designed so that, if the cell is carefully oriented in the rotor, the walls will be parallel to the lines of centrifugal force. This insures that there is no pile-up of material against the walls. The cell consists of a centerpiece, which contains the liquid sample; two windows; a housing to hold the centerpiece and the windows; and a filling port. The windows are usually made of quartz, although sapphire is sometimes used for very high speed operation because it deforms less at high forces. The centerpiece may be made of metal (usually aluminum) or plastic (either Kel-F or an epoxy resin). Metal centerpieces have the advantage that thermal equilibrium is rapidly achieved, thus minimizing convection produced by thermal gradients. On the other hand, the metal frequently interacts with the solvent or solute, in which case the inert plastic centerpieces must be used. Centerpieces are also made of titanium and have the advantages of both, but they are limited to lower speeds because the high density of the metal increases the pressure and hence the stress on the rotor.

For some purposes, a double-sector cell is used; one of the sectors contains the solution and the other contains the solvent, which provides a baseline for the optical system.

Concentration distributions within the cell are determined by passing light through the moving cell and recording the intensity of the transmitted light either on photographic plates or by means of an electronic scanner. The electronic scanner, though exceedingly expensive, provides accuracy not achievable with the photographic system. The complete scanning system includes a monochromator between the light source

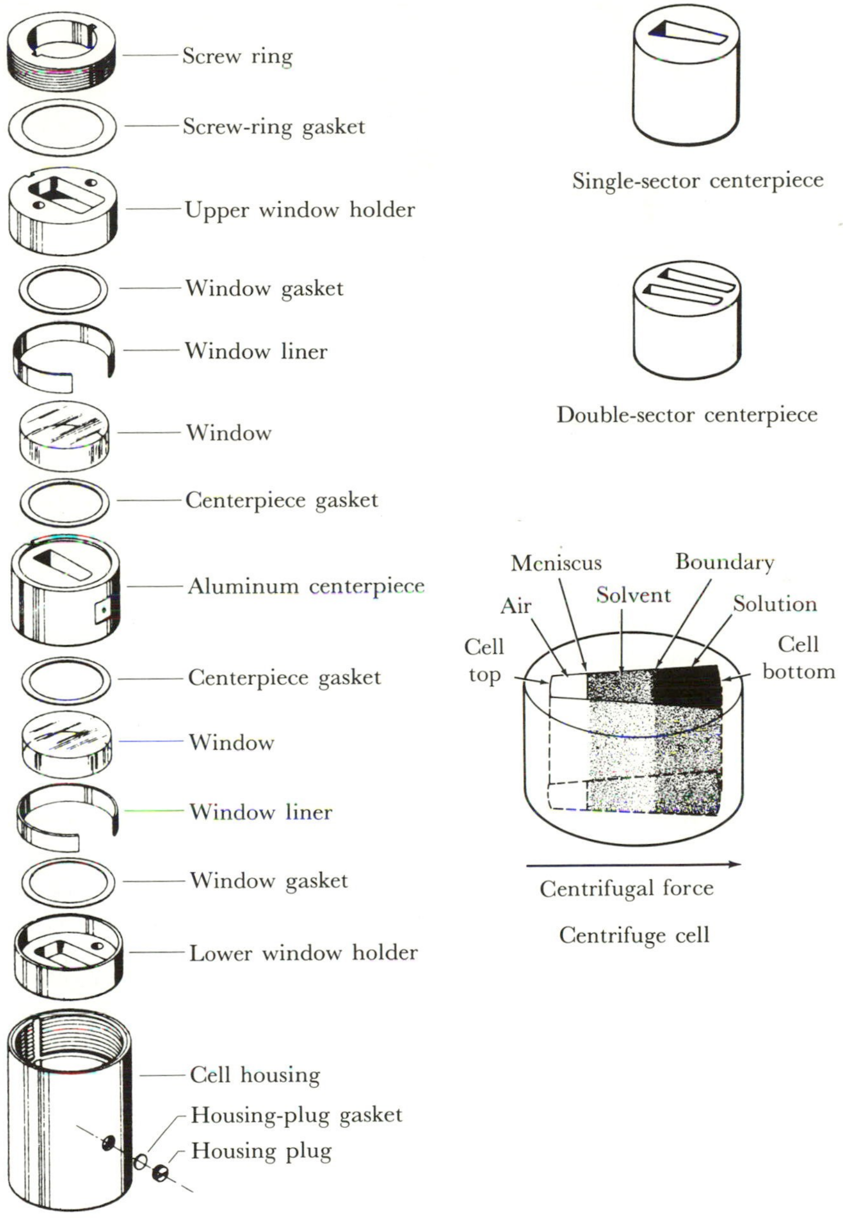

Figure 11-3
A centrifuge cell and the single- and double-sector centerpieces. [Courtesy of Beckman Instruments.]

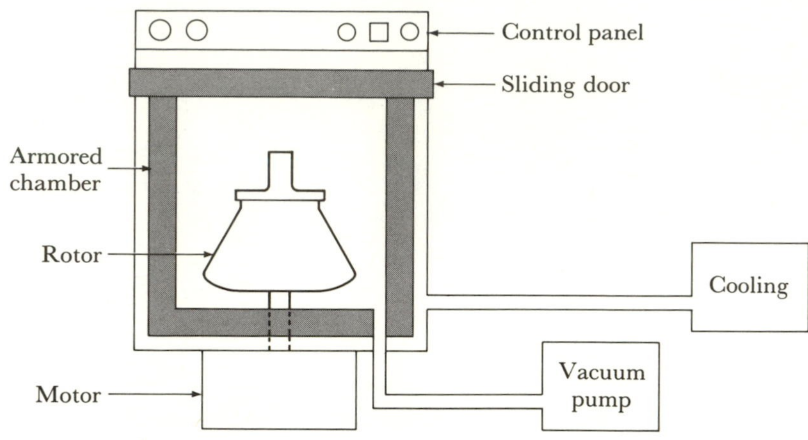

Figure 11-4
A Beckman preparative ultracentrifuge.

and the centrifuge cell; this allows the selection of particular wavelengths of light for viewing the sample. This is of especial value when studying substances that absorb light. Furthermore, when sedimenting a mixture of substances, the movement of different components can be examined separately if each one has a unique wavelength that is absorbed.

Figure 11-4 is a schematic representation of a Beckman preparative ultracentrifuge. This instrument is much simpler than the analytical instrument.

Figure 11-5 shows two types of rotors for the preparative instrument. The angle rotors are primarily used for pelleting materials.* They are very efficient for this purpose because the material moves only a short distance to the wall of the centrifuge tube; the resulting accumulated material then rapidly slides down the wall to the tube bottom. The swinging bucket rotors are used mostly for analytical equilibrium and zonal centrifugation (to be described shortly).

If the purpose of centrifugation is to pellet material on the tube bottom, the procedure is merely to pour off the supernatant liquid after the rotor has stopped spinning and the tube has been removed. If the centrifuge operation is for analytical purposes, the contents of the tube can be fractionated by puncturing a hole in the tube bottom and collecting the material by drops. If the rate of dripping is sufficiently slow that turbulence is not produced, each drop represents a single lamella from the tube as shown in Figure 11-6.

* Pelleting means sedimenting particles until they are tightly packed at the bottom of a centrifuge tube.

Figure 11-5
Beckman swinging-bucket rotor (left) and angle-head rotor (right). These rotors are designed to generate forces of 420,000 and 368,000 g respectively, at their maximum speed of 65,000 rpm. [Courtesy of Beckman Instruments.]

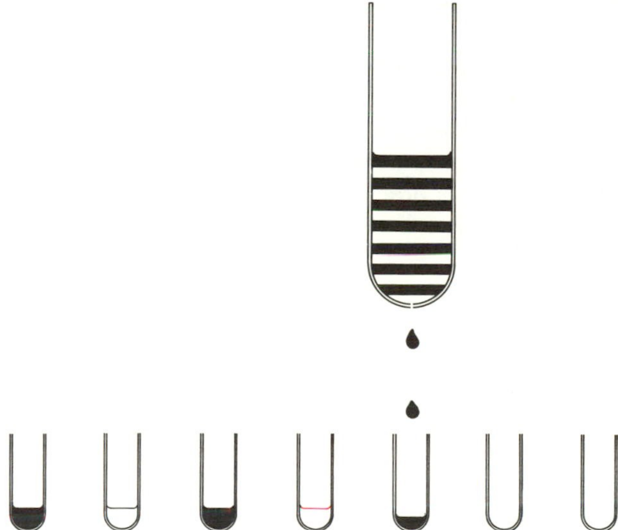

Figure 11-6
Fractionating the contents of a centrifuge tube by drop collection. The bottom of the tube is pierced with a needle. As long as the system is stabilized against convection (e.g., by a concentration gradient), the drops represent successive layers of liquid. These layers are shown schematically as alternating black and white.

It should be noticed that the walls of the tubes used with the preparative rotors are parallel, in contrast with the sector-shaped cell of Figure 11-3. This produces convection and will be discussed later in the section on zonal centrifugation. To date, no one has designed a satisfactory preparative rotor tube that is sector-shaped.

Meaning of Solute Concentration in Sedimentation Analysis

Usually when studying chemical phenomena the word concentration refers to molarity or molality. However, all of the methods used to detect macromolecules in centrifuge cells measure the mass per unit volume (for example, mg/ml) and not molarity because the molecular weight is often unknown. Thus, *in this chapter concentrations of macromolecules will always be expressed in mass per milliliter*. In contrast, the concentration of the ions and solutes that are small molecules are expressed in molarity, as is always the case in chemistry.

Distribution of Concentration in a Boundary-Sedimentation-Velocity Experiment Performed in a Sector-Shaped Cell

Two types of initial conditions are used in sedimentation experiments. In one, the particles are distributed uniformly throughout the solution. As sedimentation proceeds, the particles move through the solution, leaving behind a region of pure solvent. This is called *boundary sedimentation* because all information is obtained from the parameters of the boundary between solvent and solution. The other type is *zonal sedimentation*, in which the sample is layered on top of a denser solvent. This section will confine itself to a discussion of boundary sedimentation.

Figure 11-7 shows the concentration distribution in sector-shaped centrifuge cells filled with different solutions of macromolecules after a period of centrifugation sufficient to move the material roughly halfway down the cell. The cell in series A contains material with a single sedimentation coefficient (it is common to use the phrase "homogeneous with respect to sedimentation coefficient" or the term "monodisperse"). It also has a low diffusion coefficient so that it gives a sharp boundary. The cell in series B contains a mixture of equal amounts of two components, each having a unique sedimentation coefficient and sharp boundaries, as in series A. The cell in series C contains a mixture of components whose sedimentation coefficients are slightly different from one another. The second row consists of plots of the concentration of the material that sediments in each cell. The region in which concentration changes is called the *boundary*; the upper flat portion is the *plateau*.

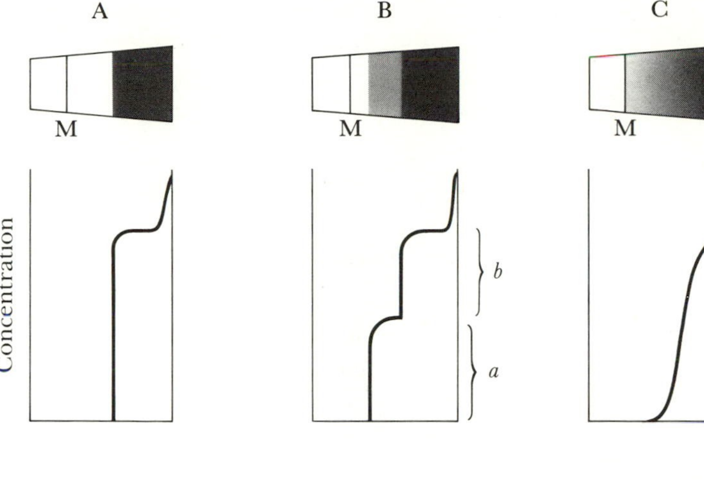
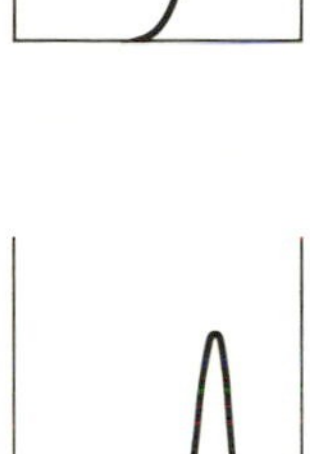

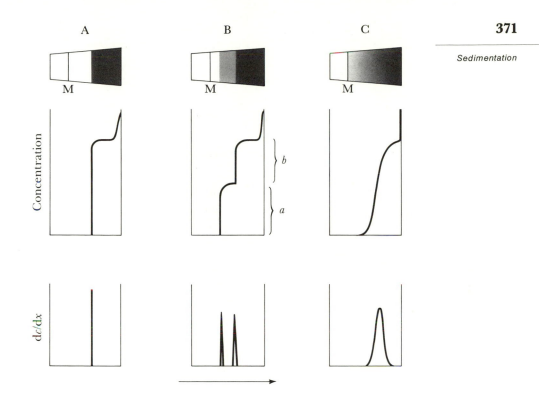

Figure 11-7
Concentration distribution after a given period for cells filled with three different solutions: (A) a single type of solute molecule having only one sedimentation coefficient; (B) two components, each having a distinct sedimentation coefficient; and (C) a heterogeneous mixture having a range of *s* values. Note the material accumulating on the cell bottom. The bottom row shows the concentration gradient; this is equivalent to a schlieren photograph. The letter M indicates the meniscus.

The third row consists of plots of the derivative of the concentration curve, that is, the *concentration gradient.*

The particular points to be noticed are that (1) a homogeneous material gives a very sharp boundary, (2) a mixture of two components gives two sharp boundaries that look like steps, (3) a mixture of many components gives a boundary consisting of many steps that produces a broad, smooth curve, and (4) in each case, particles are pelleting on the tube bottom.

Let us consider the plot of concentration versus distance for series B in a little more detail. The curve has two measurable features—the distance sedimented (from which *s* is determined) and the relative heights of the two boundaries. Because concentrations are additive, the heights

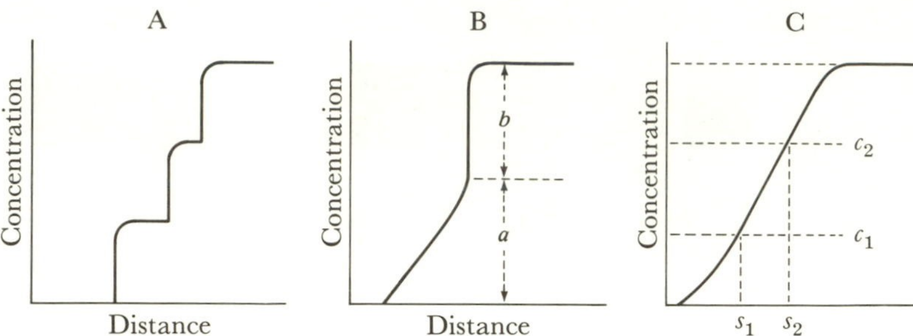

Figure 11-8
Determination of the composition of a solution by sedimentation velocity: (A) the solution contains three components, each at the same concentration; (B) the solution contains a major fraction having a single s value, the remainder being a slower heterogeneous mixture, and $a/(a + b)$ is the fraction of the material that is heterogeneous, whereas $b/(a + b)$ is the homogeneous fraction; (C) the solution is a heterogeneous mixture. c_1 and c_2 are the concentrations that have s values less than or equal to s_1 and s_2, respectively, and $c_2 - c_1$ is the concentration having $s_1 < s < s_2$.

a and b tell the ratio of the two components. Clearly, if the two components are present in equal concentrations and if they sediment independently, $a = b$. In general, if the slower material a is a fraction, f, of the mass of the total material, then $f = a/(a + b)$. Let us now consider a system with many components. If there were three components, the boundary shown in Figure 11-8A would result. In general, if a mixture consists of a weight fraction f of a heterogeneous mixture and $1 - f$ of a fast component, the pattern of Figure 11-8B will result with $a/(a + b) = f$. Hence, because a plot of concentration versus distance can easily be converted into a plot of concentration versus sedimentation coefficient, the weight fraction of total material with s between s_1 and s_2 is $(c_2 - c_1)/c$, as shown in Figure 11-8C.

Examination of the middle panel of Figure 11-7 and of Figure 11-8A shows an anomaly of sedimentation of homogeneous material; this anomaly has been called the *curvature at the top*. Ideally a pure substance having a very low diffusion coefficient should have a boundary that forms a right angle both with the baseline and with the plateau: This is because the sample should have only a single s value. The figure shows that there is a baseline right angle but that the upper part of the boundary is curved. The curvature is much too great to be a result of diffusion and in fact no satisfactory explanation has ever been given for its existence. Usually it is of no concern, although with mixtures it limits the ability to detect a small amount of material whose s value is slightly larger than that of the major fraction of material. This is discussed further in a later section, in which boundary sedimentation is compared with band sedimentation.

The analysis just given shows how the amount of material is measured from the heights of the components of the boundaries.

This same line of reasoning applies to plotting the concentration gradients, as shown in the lowest panel of Figure 11-7. However, in this case, the measurement is of areas rather than distances.

How these simple observations can yield significant information about a real molecule can be seen by the following example.

Example 11-A

☐ Structure of *E. coli* phage T7 DNA.

E. coli bacteriophage T7 DNA gives the boundary shown in Figure 11-9A. This DNA has a single, sharp boundary at all concentrations so that the DNA can be considered to be homogeneous with respect to molecular weight. If the DNA is sedimented instead at pH 12.5, at which pH the two strands of the DNA separate, the resulting boundary is that shown in Figure 11-9B. This boundary consists of two parts—one sharp (homogeneous) and the other broad (heterogeneous). The distance $a/(a + b)$ is 0.5, which means that 50% of the single strands sediment more slowly than the homogeneous component. It can be concluded, then (because in this case it is known

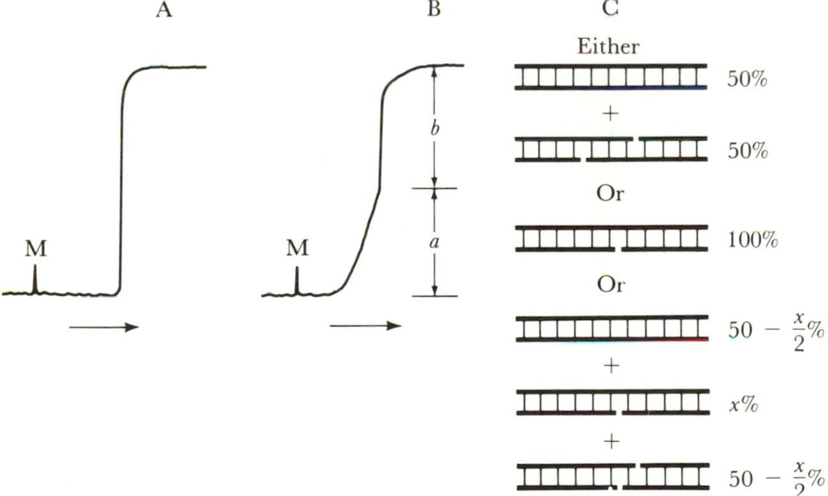

Figure 11-9
Sedimentation analysis of *E. coli* phage T7 DNA; M is the meniscus and the arrow indicates the direction of sedimentation. A. Native DNA has a single, sharp boundary, indicating that no molecules are broken. B. Sedimentation in alkali to separate the strands, which shows that some single strands are intact (b) and some are broken (a). Because $a = b$, the concentration of intact molecules equals the concentration of broken molecules. The number of broken molecules is, of course, greater than the number intact. C. Three possible interpretations of the data.

that a low sedimentation coefficient means a lower molecular weight), that 50% of the DNA single strands must be broken. Because the slowly moving material sediments quite heterogeneously, the broken strands are of a large range of sizes and therefore the breaks must be at a large number of places in the molecule—perhaps randomly distributed. Because half of the single strands are broken, either all the molecules in the population of double-stranded DNA molecules contain one single-strand break, or half the molecules contain no breaks and half have breaks in both strands, or there is a mixture of those having no broken strands, one broken strand, or two (see Figure 11-9C). These models are not distinguishable from these data alone.

Measurement of the Concentration Distribution in an Analytical Cell

In all the methods of analyzing distributions of molecules in the ultracentrifuge, it is necessary to measure the concentration of the components. Existing analytical ultracentrifuges utilize three different optical methods: *schlieren*, *interference*, and *absorption* (usually ultraviolet absorption). It is not of great importance to understand in detail how these optical systems form the image from which the concentration is obtained (this can be found in the manufacturer's manual), but it is worthwhile understanding how the image is related to the concentration distribution in the centrifuge cell.

The schlieren system depends on the fact that light passing through the regions of uniform concentration is undeviated but that passing through the regions of varying concentration is deviated because of the change in index of refraction (which varies with concentration). The optical system of modern centrifuges converts this deviation into a curve that shows the concentration gradient, as in Figure 11-10.* The optical system can be adjusted so that the curve is more or less sharp; however, in making this adjustment the area bounded by the schlieren curve and the baseline must be constant and proportional to concentration. By measuring this area and by making use of certain optical constants of the instrument, the concentration can be determined. As the concentration of material in the cell decreases, the area decreases until the peak is barely resolved above the baseline; in practice, the lower limit for the schlieren system is a concentration of a few milligrams per milliliter. The value of the schlieren system is its great ability to examine boundary

* Devising this ingenious optical system, which actually plots a graph of concentration gradient versus distance was a major accomplishment in the development of ultracentrifugal technique, without which the method might never have become useful. The system was designed by J. St. L. Philpot and modified by H. Svenson. Details about the optical systems can be found in the Selected References near the end of this chapter.

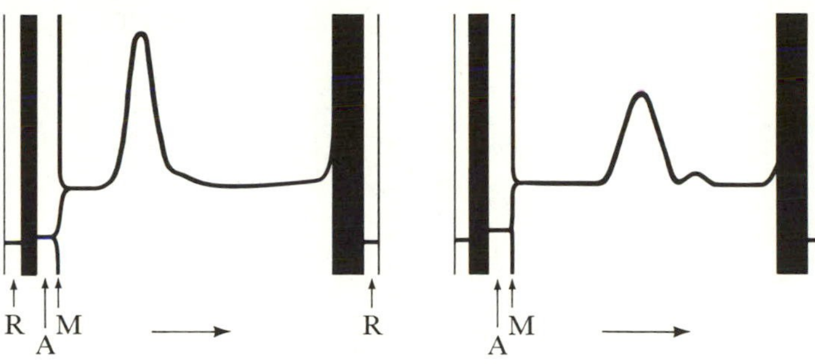

Figure 11-10
Schlieren patterns of a mixture of two proteins. The centrifugation time is different for each illustration, that on the right being the longer. Note that, with the longer time, a small amount of a fast-moving material is resolved. The blackened areas are the top and the bottom of the cell; A is the air space above the solution but in the cell; R stands for reference holes in the rotor that are at fixed distances from the axis of rotation; and M is the meniscus. After a longer time of sedimentation, the main peak is broader owing to diffusion of the protein molecules; the area is unchanged, however.

shape and thereby to detect the presence of inhomogeneity in s. It has been widely used especially for the determination of the sedimentation coefficients of proteins.*

For a variety of techniques (especially equilibrium and approach-to-equilibrium methods) the schlieren system is not sufficiently sensitive to detect small changes in concentration. For this reason, the *interference system* was developed, in which a double-sector cell (Figure 11-3) is used, one sector containing solvent only and the other the solution. The optical interference pattern is produced by light passing through both of the sectors. (The mechanism by which the interference pattern is produced is described in the texts listed in the Selected References near the end of this chapter. For the present purpose, suffice it to say that the optical system is based on the Rayleigh interferometer and that the fringe pattern is a function of the index of refraction at each point in the cell. Whereas the schlieren system plots refractive index *gradient* versus distance along the cell, with interference optics, each fringe traces a curve of the index of refraction versus distance.)

Figure 11-11 shows the interference patterns obtained for various situations. In part A, both sectors are filled with solvent only and the

* As will be seen later, the single problem with the schlieren system is that the required concentrations are so high that there is often strong concentration dependence. This does not affect work on proteins, but it severely affects that on nucleic acids (Figure 11-13). Almost no information could be gained about the sedimentation properties of DNA until the more sensitive, UV-absorption optical system was developed.

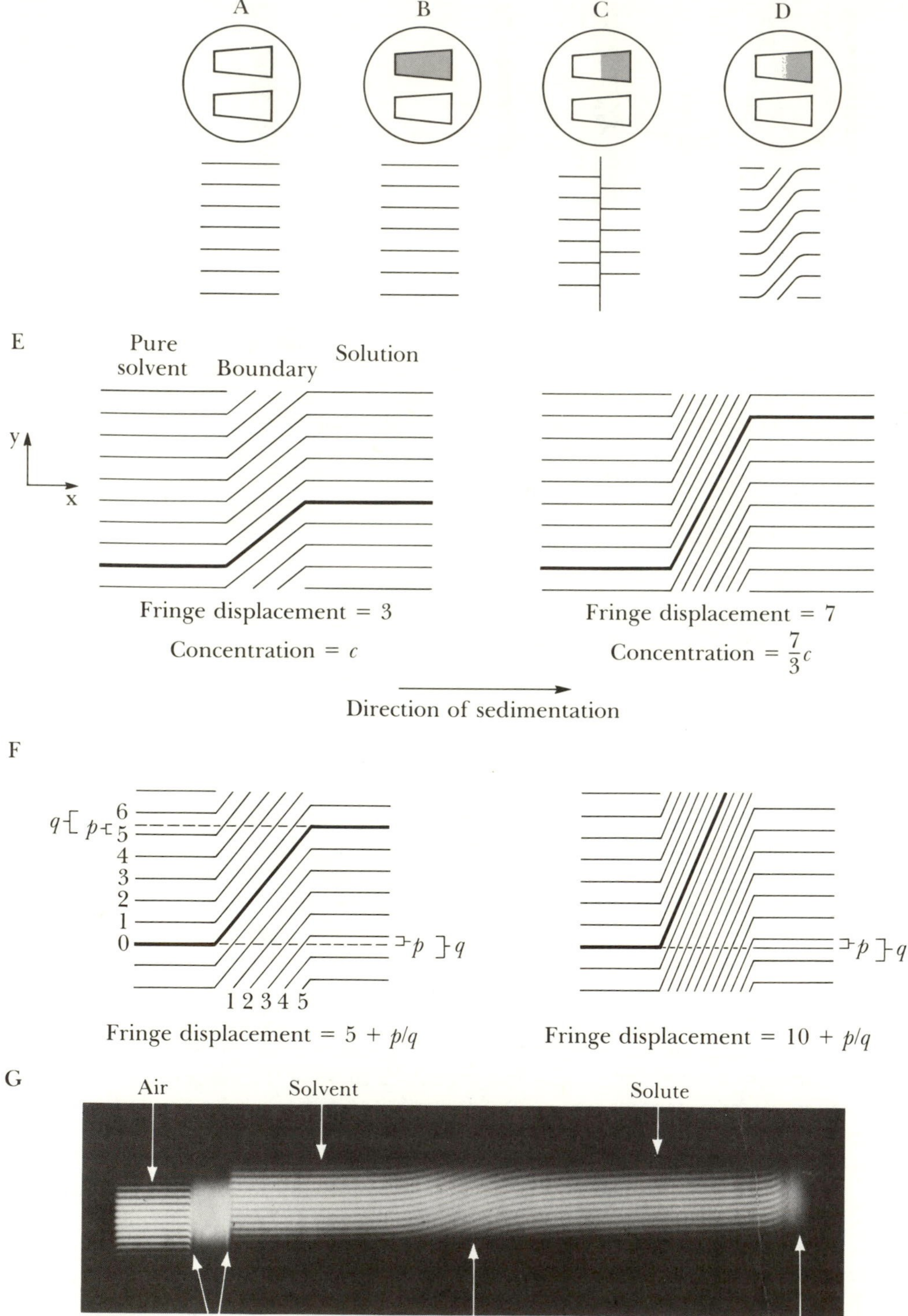

A
B
C
D
E
Pure solvent
Boundary
Solution
y
x
Fringe displacement = 3
Concentration = c
Fringe displacement = 7
Concentration = $\frac{7}{3}c$
Direction of sedimentation
F
q p
6
5
4
3
2
1
0
1 2 3 4 5
p q
Fringe displacement = 5 + p/q
Fringe displacement = 10 + p/q
G
Air
Solvent
Solute
M
B
Tube bottom

Figure 11-11 (*opposite*)
Generation and analysis of interference patterns from double-sector cells: (A) interference pattern resulting if both sectors contain the same solution; (B) one sector contains solvent, the other solution (but before sedimentation has begun); (C) same as part B, but after a period of centrifugation (the sample is homogeneous and has a very low diffusion coefficient); (D) same as part B, but the solute has a very high diffusion coefficient; (E) method for counting an integral number of fringes (see text); (F) method for counting a non-integral number of fringes (see text); (G) interference optical photograph of bovine serum albumin in H_2O sedimenting at 60,000 rpm for one hour. In part F, sedimentation is from left to right. With interference optics, a double-sector cell is used, one sector filled with solvent, the other with solution. In this photograph, the two sectors were not filled with the same volume; hence, two menisci (M) can be seen. The boundary (B) is indicated by the curvature of the fringes. Note that the fringes curve upward at the bottom of the cell because material is pelleting. Because the position of the fringes is determined by solute concentration, the concentration of solute along the cell can be plotted.

fringes are undeflected across the cell. In part B, one sector contains solute and the fringes are actually deflected a distance proportional to concentration, although this is not easily detectable because all the individual fringes look the same. If the material under study is homogeneous and has a small diffusion coefficient so that the boundary is narrow (as in Figure 11-7), the pattern shown in part C results, in which the fringes have shifted upward. In a real situation in which the boundary has finite width, the pattern shown in part D is obtained. Part E shows how the fringe pattern depends on concentration. Several points should be noted.

1. Fringes are equally spaced. The spacing is determined by the instrument optics, not by the contents of the cell.

2. Concentration is constant in any region of the cell in which the fringes are horizontal (parallel to the centrifugal direction or the x-direction).

3. Concentration changes where the fringes are at an angle. The vertical displacement of the fringes (perpendicular to the centrifugal direction or in the y-direction) is proportional to the solute concentration.

In order to measure the concentration at any point in the cell, one fringe is arbitrarily selected (the blackened ones in the left and right panels of part E) and traced in the x-direction. The y-displacement is proportional to concentration so that from the y-displacement for any value of x, the concentration at any point x along the cell can be determined, if the proportionality constant is known. It is common to measure the y-displacement using fringe spacing as a unit. Thus in the left and right panels of part E, the displacements are three and seven fringe spaces respectively. Assuming that centrifugation has proceeded so that no solute is present to the left of the boundaries in both panels, the concentrations in the left and right panels are three and seven units respectively. The constant of proportionality relating fringe number and concentration is determined simply by examining the fringe pattern early in the centrifugation run as soon as the boundary is free of the meniscus. Since the initial solute concentration is known (and determined by the experiments), this concentration divided by the fringe displacement equals the concentration per fringe. As part E has been drawn, the displacements in both panels are whole numbers of fringes. This is not generally true.

How to handle nonintegral numbers of fringes is shown in the left panel of part F of Figure 11-11. The two dashed lines are extensions of the horizontal portions of the heavy fringe. Note that the vertical fringe displacement is 5 + some fraction. If the parallel fringe displacement is q and our heavy line is displaced by p above fringe 5, then $5 + p/q$ is equivalent to the number of fringes *crossed* by the lower horizontal line. Referring again to the left pattern of part F, consider the lower horizontal dashed line. Note that the number of fringes *crossed* by this line is also 5 and that the fringe space on the right side of the boundary is also divided into p and $q - p$. Thus, *fringe displacement can be measured both horizontally and vertically.*

It is also possible that the concentration is so high that no fringe can be followed completely across the boundary. However, this situation can be analyzed using the technique described above for fractional fringes, where it is clear from reading horizontally that the fringe displacement is $10 + p/q$.

Part G shows an actual photograph of an interference pattern of a protein sample sedimenting from left to right. Note that the fringes in the boundary region (B) are curved rather than sharply angled as in the idealized drawings in the previous parts of the figure.

The methods just described allow the conversion of any fringe pattern to a plot of concentration versus distance along the cell merely by multiplying the y-displacement by the proportionality factor. Thus, a fringe line is actually itself a graph of concentration versus distance. This process of measuring concentration will be examined again in a later section when sedimentation equilibrium is discussed.

The interference system is usable to a concentration of a few hundred micrograms per milliliter, approximately one order of magnitude better than the schlieren system.

For nucleic acids, the dependence of s on concentration is so great that a method was sought so that concentrations of $<20\ \mu$g/ml could be used. Fortunately, because the molar absorption coefficient (see Chapter 14) of nucleic acid is very high at 253.7 nm (a strong emission line from a Hg arc light source), an absorption optical system was easily developed. To understand, refer to Figure 11-7, which shows a plot of concentration along the cell. If the cell is photographed with ultraviolet light, the blackening of the film will decrease as nucleic acid concentration increases. Figure 11-12A shows a typical photograph of sedimenting

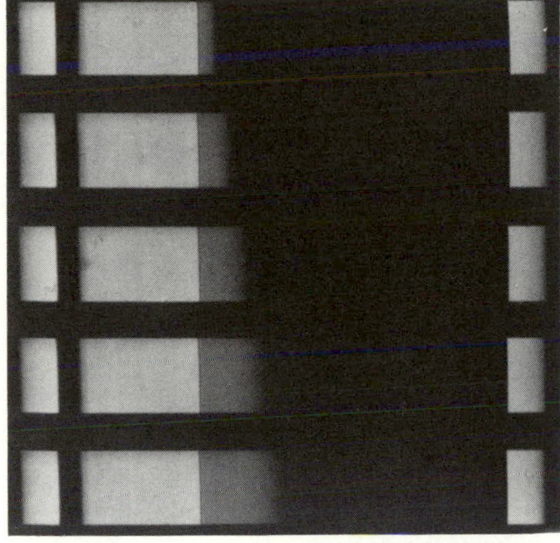

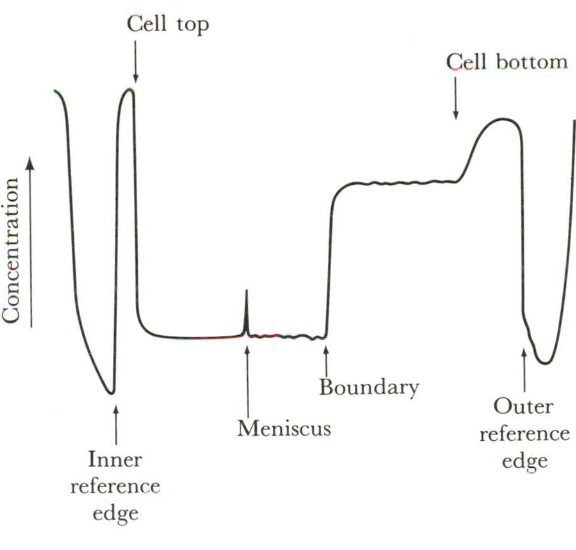

Figure 11-12
A. An ultraviolet-absorption pattern of sedimenting *E. coli* phage DNA. Sedimentation from left to right. Note that the picture shown is a positive print rather then the negative referred to in the text. B. A photometric trace of the blackening of the second frame of part A.

DNA. If the film is traced with a photometer,* a plot of concentration versus distance is easily obtainable, as shown in Figure 11-12B. Many instruments in current use eliminate the photographic intermediate by measuring the transmitted light directly and converting this by photocells

* A photometer measures light intensity. Thus if light passes through the film, the blackening of the film decreases the amount of transmitted light, which is measured by a photometer. If the photometer is coupled to a chart recorder, a plot is drawn of the intensity of the light transmitted by the film.

and appropriate circuitry into a plot of concentration versus distance, which appears on a chart recorder. Such instruments are called scanners.

Some analytical ultracentrifuges are now equipped with a monochromator for wavelength selection, which permits the detection of molecules by absorption optics, using a wavelength corresponding to their absorption maximum. This allows much lower concentrations of proteins to be used if the protein has strong absorption (e.g., cytochrome *c* or hemoglobin).

Factors Affecting Sedimentation Velocity

Before discussing the complexities of sedimentation velocity, it is worthwhile to have an intuitive understanding of the problems. Imagine yourself walking across a room. Then imagine walking through a swimming pool with water up to your neck; then through a pool of molasses. The effect of solvent viscosity on your speed should be clear. Now consider walking through the pool with your arms extended. Your frictional coefficient has increased and it will take longer to walk through the pool as equation (1) states. If there are one hundred people in the pool, it will be harder to get across and even worse if your arms are extended; you should now understand why s is lower at higher concentrations and that the concentration dependence is greater with very asymmetric molecules. Suppose that you tried to run though the water with your arms partly extended. With twice the energy expended, would you get across in half the time? You should now appreciate speed dependence. Later you will see that although intuition tells the direction of the effects, the causes at the molecular level may differ from those in the pool.

The sedimentation coefficient of macromolecules depends on a variety of factors because (1) the shape of the molecules depends on solvent composition, (2) many of the molecules may be charged, (3) they are very large compared with solvent molecules, and (4) they become deformed as they move. The manifestation of these effects is a dependence of s on the concentration, the centrifugal speed, and the ionic strength of the solvent.

Concentration Dependence of s

Consider a solution of macromolecules of such a size that they frequently collide with one another. This is characteristic of very large or extended molecules (such as proteins and nucleic acids) in solution because, as they rotate, they effectively occupy a relatively large volume of the solution. When they approach one another, it is more difficult for solvent molecules to move in the opposite direction—that is, the solvent viscosity is effectively increased in the vicinity of the macromolecule. This reduces the forward velocity of the particle in a given centrifugal field and therefore reduces s. Because the probability of collision (or close approach) increases both with molecular volume and with the degree of extension the molecule, the magnitude of the concentration dependence also

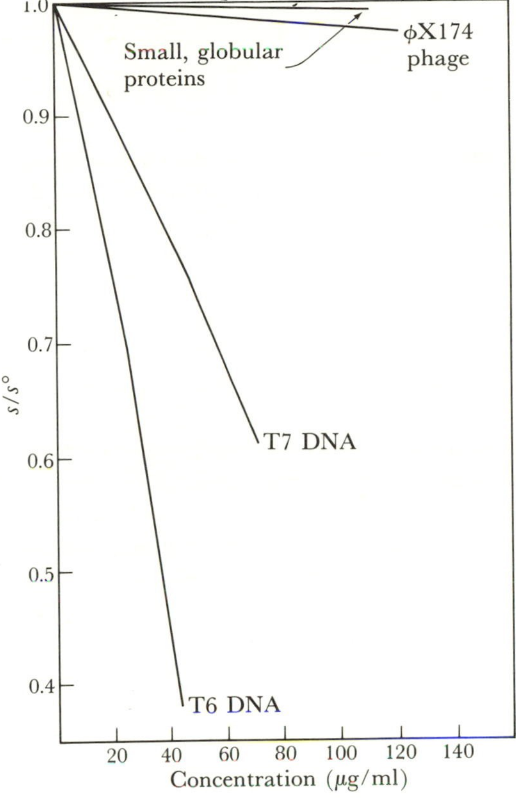

Figure 11-13

Dependence of s on concentration plotted as s/s° where s° is the value of s extrapolated to zero concentration. Note that the concentration dependence becomes more severe as the molecular weight increases and as the molecule is more extended.

increases with these parameters. This is exemplified in Figure 11-13, which shows several representative curves for s as a function of the concentration of DNA or protein. It can be seen that the greatest concentration dependence is with the large, extended molecules (i.e., the DNAs) and that it is very slight with small spheres (phage ϕX174) and globular proteins.*

There is no detailed theory to describe concentration dependence.† However, it has been found empirically that, for most systems, the concentration dependence is best described by the equation

$$s_c = s^\circ/(1 + kc) \tag{3}$$

* Very slight does not mean inconsequential. For example, proteins are usually studied at such high concentrations (such as 0.5–5.0 mg/ml) that the *total* decrease in s may be substantial.

† An attempt has been made to improve on equation (4). For example, it has been calculated that $s_c = s^\circ/(1 + k[\eta]c)$, in which $[\eta]$ is the intrinsic viscosity described in Chapter 13 and k is a constant having a value of 1.2 for rodlike molecules and 1.6 for spheres. However, this theory must still be considered semiempirical because $[\eta]$ cannot be readily calculated for most proteins and polynucleotides.

or

$$\frac{1}{s_c} = \frac{1}{s^\circ} + \frac{kc}{s^\circ} \tag{4}$$

in which s° is the s value at zero concentration, s_c is the s value at concentration c, and k is a constant for the particular molecule. The value of s° is obtained experimentally by measuring s at various concentrations, making a plot of $1/s$ versus c and determining $1/s^\circ$ by extrapolation to $c = 0$. Clearly, s° is a more useful parameter to describe a molecule and should always be determined. This has not always been done in published work and it is important to realize that there may be substantial differences between s° and s measured at standard concentrations, as is clear from Figure 11-13.

A particular problem arises if the macromolecules in a mixture have different s values and if the concentration is relatively high. In this case, the molecules interfere not only with the sedimentation of their own species, but also with that of the other species. Furthermore, the molecule with the higher s and greater concentration dependence must sediment through a solution of the slower-moving molecule. At high concentration, the faster molecules are so impeded that they sediment near or at the same rate as the slower molecules, which can disguise the fact that there is really a mixture and gives the misinformation that the material is homogeneous. This is known as the Johnston-Ogston effect. In fact, with large nucleic acids it is often necessary to use very low concentrations before the two species can be resolved; even at these concentrations, the data frequently indicate that there is less of the faster component than is really present. An example of this is given in Figure 11-14. As will be seen in a later section (page 396), this problem is minimized in the technique of zonal centrifugation.

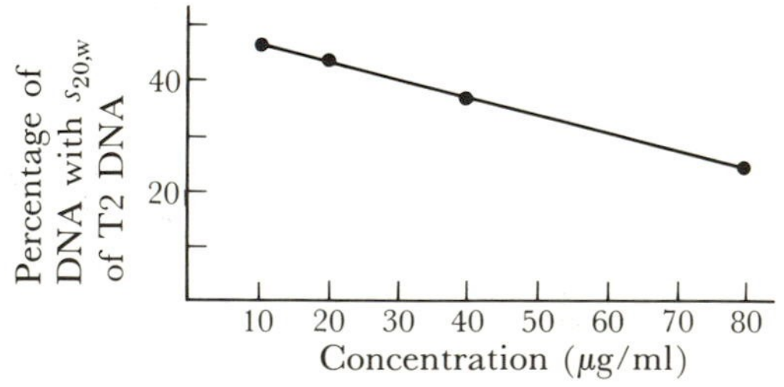

Figure 11-14
Amount of T2 DNA seen in a 50:50 (by weight) mixture of T2 and T7 DNA at various total DNA concentrations—the Johnston-Ogston effect. The molecular weights of T2 and T7 DNA are 106 million and 25 million, respectively.

Speed-dependent sedimentation refers to two independent phenomena—a speed-dependent aggregation of molecules that occurs at high concentration and an actual reduction of s° at high speed. In neither case is the phenomenon completely understood, but, owing to the work of Bruno Zimm and his co-workers, we have some ideas about the causes.

Speed-dependent aggregation refers to an apparent loss of material from the bulk of the solution, which decreases the effective concentration of the macromolecule (and therefore increases the apparent s). The explanation currently given for this phenomenon is that at high velocity a molecule leaves behind it a wake, which enhances the speed of the molecule just behind it. This results in the formation of molecular clusters that have very high s values and rapidly form a pellet at the bottom of the centrifuge tube. This process continues until the concentration is too low for clusters to occur. Figure 11-15 shows data in which this speed-dependent aggregation is occurring. The concentration of DNA in a centrifuge cell is plotted for two different times and three different speeds of centrifugation. At the higher speed, the DNA concentration decreases with time because of the pelleting. At the lower speed, it remains fairly constant. Note that the measured s values are 50 S and 37 S at the higher and lower speeds, respectively. Such an apparent loss of material is a strong indication of speed dependence and demands that measurements be done at lower speeds.

The second type of speed dependence occurs with very large molecules at low concentrations. With T4 DNA (molecular weight $= 106 \times 10^6$),

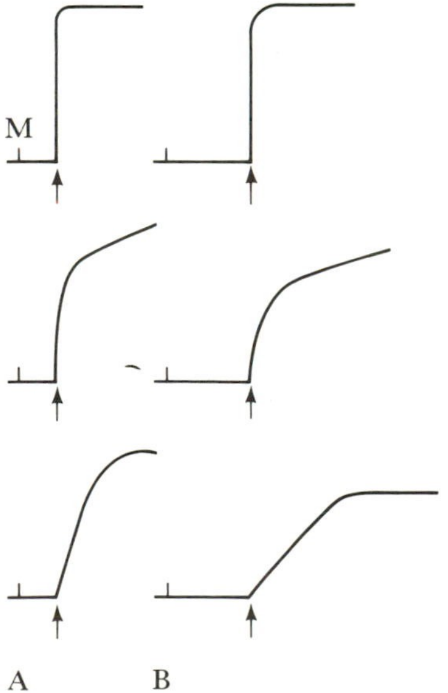

Figure 11-15
Photometric traces of ultraviolet-absorption patterns of *E. coli* phage T4 DNA sedimented at three speeds. In column A, all the arrows at the base of the boundaries are roughly equidistant from the meniscus, M. This is true of column B also, but the distance is greater. Note that as the speed increases, a greater fraction of the DNA seems to sediment ahead of the s value calculated from the position of the arrows. This s value is that expected for the molecular weight of the DNA. The rapid sedimentation is due to the speed-dependent aggregation described in the text. Sedimentation is from left to right.

the increase in s is 9% in going from 65,000 to 10,000 rpm. With larger DNA molecules (e.g., molecular weight greater than 500×10^6), the speed dependence becomes significant and can lead to three- to eightfold errors in estimates of molecular weight. This is an important phenomenon because the chromosomes of bacteria and eukaryotic cells possess DNA molecules having molecular weights ranging from 2×10^9 to 10×10^9. This phenomenon is indicated in Figure 11-16, which shows that *above a certain speed the dependence of s on molecular weight is lost*. Zimm has explained this surprising result as being due to a change in shape (an extension) of the DNA molecule at high speed because of enhanced frictional drag at the ends of the molecule.* Clearly, to distinguish molecular-weight differences for very large DNAs, it is necessary to use very low centrifuge speeds (or to make measurements of their visco-elastic properties—see Chapter 13).

Effect of Charge

A strong solvent effect that does not involve a change in the shape of a macromolecule arises if the molecule is charged (which is usually the case with biological macromolecules) and sedimentation is carried out in solutions of low (<0.01 M) ionic strength. Because of the large s value of the macromolecule compared with that of the neutralizing ions (the "counterions" such as Na^+ or Mg^{2+}), the ionized macromolecule sediments more rapidly than the counterions. Thus, in the case of a negatively charged macromolecule, negative charges move down the centrifuge cell leaving positive charges behind. This separation of charge results in an electrical potential gradient against the direction of sedimentation, which tends to reduce the s value of the macromolecules. This complication is simply avoided by the use of excess counterions—for example, in practice, by always using ionic strength in excess of 0.05 M.

Standard Sedimentation Coefficients

If sedimentation velocity studies are carried out in different solvent–solute systems, the measured s value is affected by the solution density (i.e., by the $1 - \bar{v}\rho$ term), by the solution viscosity (i.e., a molecule will move more slowly in a viscous medium), and by the temperature (primarily, by its effect on density and viscosity). To be able to compare experimental values obtained in different solvent systems and at different temperatures, observed values are converted into a standard solvent and temperature. In so doing, it is assumed that the partial specific volume

* One would expect that circular molecules would show no speed dependence because there are no free ends. A more sophisticated theory indicates that some speed dependence would be expected. In a recent experiment using a circular DNA molecule having a molecular weight of 220×10^6, a small speed dependence was observed. Presumably this results from the fact that at very high speeds a circle can be deformed so that it has ends of the sort possessed by a rubber band.

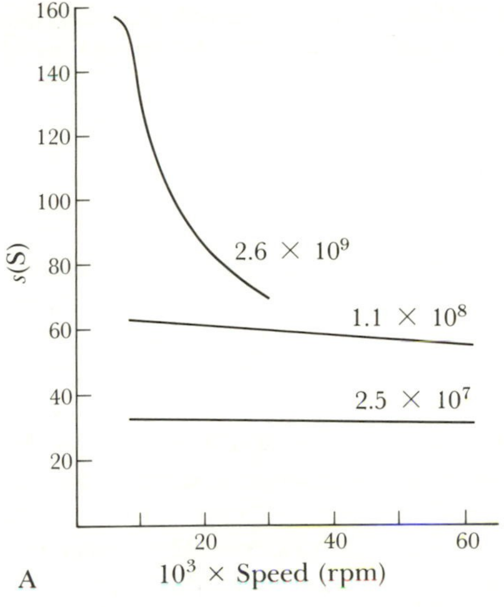

Figure 11-16
Speed dependence of *s* seen in two ways: A. Decrease in *s* as a function of speed showing that the effect is greater with higher molecular weight. The DNAs are intact chromosomes of *E. coli* and are from the phages T2 and T7. The molecular weights are as indicated. B. The *s* versus *M* curve as a function of speed (rpm) as indicated.

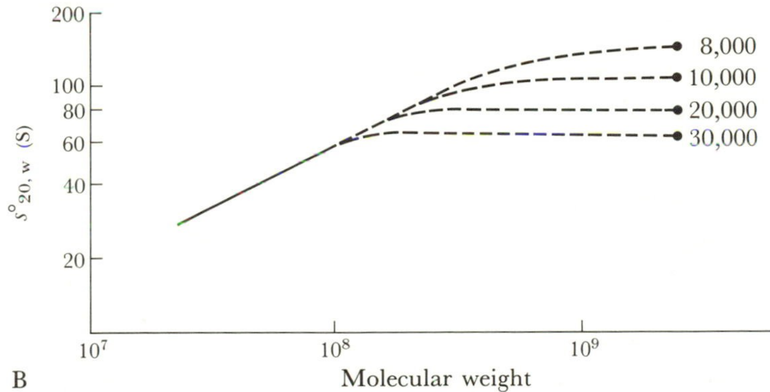

and the friction factor are unaffected by the change in the system. This is probably a reasonable assumption (although not always justified) as long as the counterion concentration is not very high and solute molecules that bind to the macromolecule are not present.* (However, see *Solute Effects* in the next section.)

* The problem is that almost all equations used in centrifugation have been derived by making the assumption that the system has only two components, the solvent molecules and the molecules whose sedimentation properties are being studied. However, in practice this is rarely the case, because of the addition of counterions. In fact, the concentration of the counterions usually ranges from 0.01 M to 1.0 M, whereas the molecule of interest is between 1.0×10^{-5} M and 1.0×10^{-3} M. In practice, this complexity is usually ignored because, if the salt concentration is <0.05 M, nothing anomalous is detectable. However, in 1 M NaCl, a common solvent for DNA, some corrections have to be made. This is rarely done because in fact the hydrodynamic theory of DNA sedimentation has not yet achieved the degree of sophistication such that these corrections would make much of a difference. Subtle problems do arise, though, if proteins are sedimented in 5 M guanidinium chloride or 7 M urea.

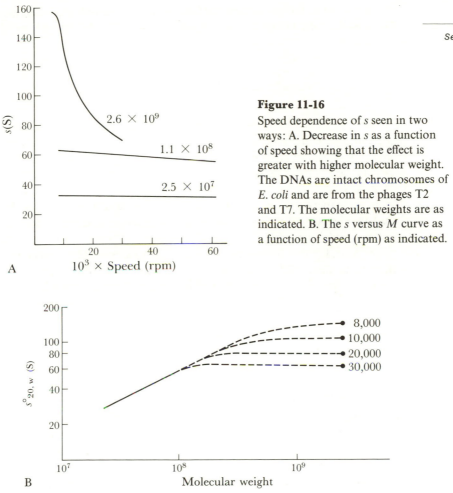

By agreement, sedimentation coefficients are corrected to a reference solvent having the viscosity and density of water at 20°C by means of the equation

$$s_{20,\mathrm{w}} = s_{\mathrm{observed}} \cdot \frac{1 - \bar{v}\rho_{20,\mathrm{w}}}{1 - \bar{v}\rho_T} \cdot \frac{\eta_T}{\eta_{20}} \cdot \frac{\eta}{\eta_0} \tag{5}$$

in which $s_{20,\mathrm{w}}$ is the standard s value, $\bar{v}$ is the partial specific volume, $\rho_{20,\mathrm{w}}$ and ρ_T are the densities of water at 20°C and of the solvent, respectively, η/η_0 is the relative viscosity of the solvent to that of water, and η_T/η_{20} is the relative viscosity of water at temperature T compared with that at 20°C.

The appropriate values of η for many solvent–solute systems can be found in various references; others can be measured directly by viscometry (see Chapter 13).

Note that, for this correction as well as for calculations of molecular weight to be described in later sections of this chapter, the value of $\bar{v}$ is necessary. How this value is obtained is described in Chapter 12.

Factors Affecting Standard Sedimentation Coefficients

Friction: The Shape Factor

Let us consider by a simple example the factors affecting the frictional coefficient. If a ball and a stick of the same mass and density are centrifuged through a viscous medium, it is intuitively obvious that the ball, which is more compact, will move faster. (If the rod could be oriented so that it moved only along its axis, this need not be true. However, Brownian motion keeps the rod—and a particle—rotating so that the statement is correct.) The general rule is that the more extended an object is, the greater will be its resistance to motion and therefore the greater its frictional coefficient. This simple consideration enables one to understand why the sedimentation coefficients of extended semirigid molecules increase markedly if the intramolecular structure giving rise to the rigidity is destroyed. A striking example of this can be seen in studies with DNA.

Example 11-B ☐ Detection of DNA denaturation by ultracentrifugation.

The rigidity of DNA is produced primarily by its double-stranded helical structure in which the nucleotide bases tend to stack one on top of the next (see Chapter 16) and in which bases on opposite strands are held in contact by hydrogen bonds. As the temperature of a DNA solution is increased, hydrogen bonds break and the stacking tendency of the bases decreases. When a region of bases loses its rigidity because

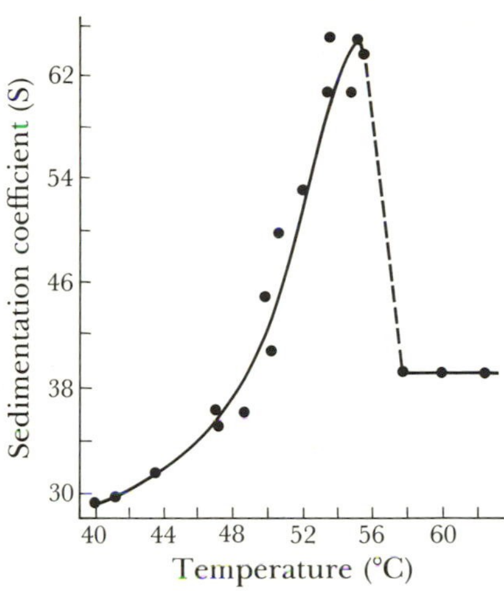

Figure 11-17
Sedimentation coefficient of the sharp boundary of T7 DNA fully or partly denatured by heating at the indicated temperature in 0.1 M PO_4, pH 7.8, containing formaldehyde. The formaldehyde prevents the reformation of hydrogen bonds when the solution is cooled to 20°C for centrifugation. The dashed line represents separation of the strands. [From D. Freifelder and P. F. Davison, *Biophys. J.* 3(1963): 49–63.]

of a breakdown of hydrogen bonds and of stacking, the DNA molecule will then have points at which bending can occur. This decreases the extendedness of the molecule and therefore increases *s*. As more and more intramolecular disruption occurs, the molecule becomes more and more flexible, less rigid, and less extended, and it has a higher *s* value. This continues until the last hydrogen bond is broken. At that point, the two DNA strands separate, thus producing two units having half the molecular weight of the original structure. This phenomenon can be clearly seen in Figure 11-17, which plots the *s* value of a DNA having a molecular weight of 25 million as a function of the temperature to which the DNA has been heated. The value of *s* increases by 220% during the period of disruption and then drops precipitously at the point of strand separation (because the molecular weight is halved). It is interesting that this simple measurement was one of the earliest indications that the strands of DNA could be separated.

Molecular Weight

Equation (2) shows that *s* increases with *M* if the partial specific volume ($\bar{v}$) and the frictional coefficient (f) are constant. Values of $\bar{v}$ do not usually change significantly with *M*, although they vary with the amino acid composition of proteins. The frictional coefficient, on the other hand, is a strong function of *M*. Many attempts have been made in the past twenty years to calculate the relation between *s* and *M* (which in fact means calculating that f-M relation). For solid spheres and rigid rods, this is a relatively simple task and, insofar as proteins can be treated as globular structures, the theory is also adequate. For flexible rods such as DNA, these attempts have not been very successful, although recently some headway has been made.

For the more common purpose of using s as an assay of M or of changes in M, the approach to this problem is to obtain empirical relations between s and M. This approach has been successful for DNA, although the equations in common use have been continually modified as the measured values of M for various DNAs have become more precise.* The best equation to date for double-stranded NaDNA at neutral pH in 1 M NaCl sedimented in a sector-shaped cell (Figure 11-3) is

$$s^{\circ}_{20,\mathrm{w}} = 2.8 + 0.00834\, M^{0.497} \tag{6}$$

This equation is, unfortunately, not based on a very large number of values of M and may be changed slightly in the future. An important point to be noticed is that s varies more slowly than M so that s alone is not a good measure of M unless a significant error can be tolerated. For example, a 5% error in s yields a 10% error in M. The value of s for DNA can be measured to 2% accuracy, but this is rarely done. Because the s-M relation is strictly empirical, it is important that M values are not evaluated from these equations if the s lies outside of the s values (10–60 S) used to obtain the equation.

For globular proteins, s varies roughly as $M^{\frac{2}{3}}$ so that a 5% error in s yields a 7% error in M. This is fairly reliable because s for proteins can be measured to 1% accuracy. However, the empirical equation for proteins suffers from the fact that $\bar{v}$ varies with amino acid composition.

As will be seen in a later section, there are better ways to determine precise values of M and s is usually measured to assay *changes* in M or in shape.

Solute Effects

Bound solute molecules can affect s because they can affect the molecular weight and sometimes also the shape of the molecule. For example, DNA in NaCl solution is NaDNA whereas in CsCl it is CsDNA, which has a higher molecular weight and hence a higher s value. The shape effect is very severe for charged flexible molecules. A highly charged flexible molecule, such as RNA or single-stranded DNA, which carries a negative charge (the phosphate) on each nucleotide, is very extended in solutions of very low ionic strength because the charges repel one another. In a higher ionic strength, the charges may be neutralized by counterions and, owing to its flexibility, the molecule may assume a compact configuration such as a random coil. Thus, there may be a huge difference in s value in 0.001 M NaCl and 0.05 M NaCl, for which the buoyancy and viscosity corrections may be very small. This solute

* Many other equations can be found in the literature. Unfortunately, they have been based on values of M that are considerably in error. The source of these errors is discussed in several of the references listed near the end of the chapter.

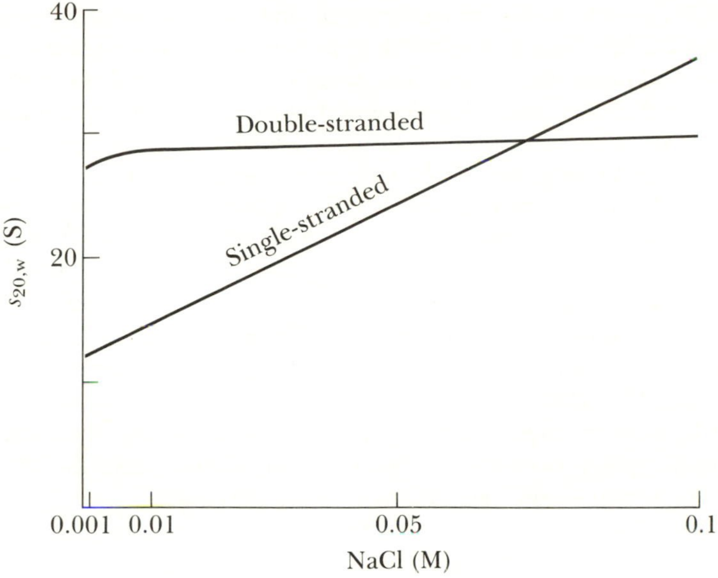

Figure 11-18
The dependence of $s_{20,w}$ of a single- and double-stranded DNA molecule, each having a molecular weight of 12×10^6, on the concentration of NaCl.

effect is shown in Figure 11-18. Note that s value of the single-stranded DNA varies with the NaCl concentration but the s value of double-stranded DNA, which is much less flexible than single-stranded DNA, shows virtually no dependence.

Procedure for the Accurate Measurement of Sedimentation Coefficient

Now that some constraints have been placed on the means of measuring s, it is worthwhile defining a procedure that will result in meaningful s values. First, to avoid the effects of charge, ionic strength should be maintained in the range of 0.05–1.0 M and pH should be controlled by a suitable buffer. Measurements should be performed at several speeds, each differing by about 50%, to ascertain that the effects of speed are absent. At a given speed, at least four concentrations should be used (including the lowest possible concentration) to ascertain that either a single species is present or that the true ratio of several components is being observed and to obtain s°. Finally, $s^\circ_{20,w}$ should be calculated. A sample calculation of $s^\circ_{20,w}$ is given in the appendix to this chapter.

At this point, it must be remembered that $s^{\circ}_{20,w}$ has been determined not for the bare macromolecule but for the molecule *with bound counterions and other ligands.* This fact cannot be emphasized too strongly. Unless there is a strong effect of counterion concentration on shape or on $\bar{v}$ (as described in an earlier section), it is usually the case that $s^{\circ}_{20,w}$ is the same in, for example, 0.1 M and 1.0 M NaCl and the same in 0.1 M and 1.0 M CsCl. However, for DNA, for example, it is certainly not the same for identical concentrations of NaCl and CsCl because NaDNA has an average nucleotide molecular weight of $336 + 23 = 359$ whereas CsDNA has an average nucleotide molecular weight of $336 + 137 = 473$. (The average molecular weight for a nucleotide is 336; Na and Cs have atomic weights of 23 and 137, respectively.) In both solute systems, s will probably reflect the actual molecular weight, but this conclusion must be made cautiously because it must be realized that the empirical equation (6) was obtained for NaDNA only. This is also the case for proteins, which bind either cations or anions or both.

Examples of the Use of Boundary Sedimentation

In the following examples, boundary sedimentation is used to determine the relative amounts of material having different values of s. It is interesting to note that strong conclusions are made without the necessity of knowing the actual s values precisely.

Example 11-C ☐ Mechanism of inactivation of bacteriophages by X irradiation.

If bacteriophages are irradiated with X rays, they lose viability. A possible explanation is that the DNA within the phages is broken by the radiation. To investigate this, phages were irradiated and samples were taken after various doses for determination of viability and for analysis of the DNA by ultracentrifugation. Figure 11-19A shows sedimentation diagrams for the DNA of phages given two X-ray doses. As the dose increased, more of the DNA sedimented more slowly; this represents broken molecules. The fraction of unbroken molecules was determined by the procedure shown in Figure 11-8. When this fraction and the fraction of surviving phages is plotted against dose (Figure 11-19B), it can be seen that phage killing is more rapid than DNA breakage and that DNA breakage occurs at about half the rate of killing. If the irradiation is carried out in the absence of oxygen (i.e., in a nitrogen atmosphere), the phages become more resistant to radiation, although the rate of DNA breakage is unchanged. If a plot such as that shown in Figure 11-19B is made, it is found that the rate of killing is the same as the rate of breakage. Hence, from this simple analysis, the conclusion can be drawn that

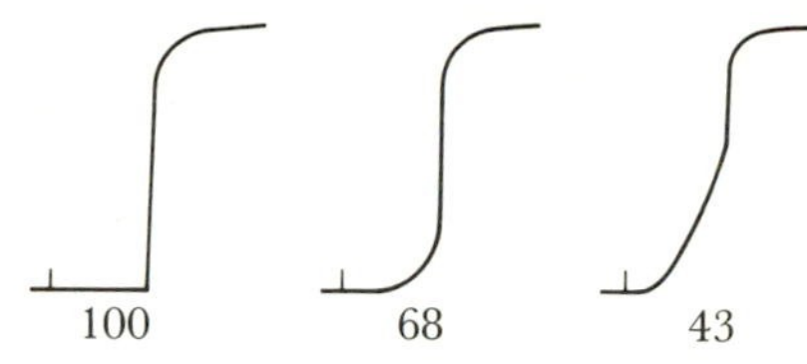

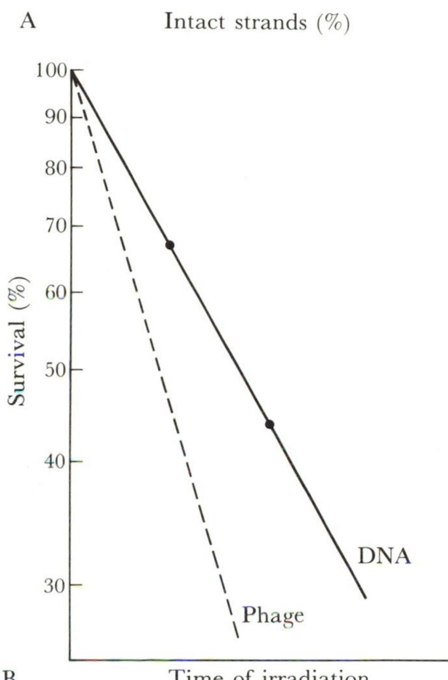

Figure 11-19
A. Sedimentation diagram of *Pseudomonas aeruginosa* phage B3 DNA after two doses of X rays. B. Plot of the percentage of intact double strands as a function of time of X irradiation compared with the phage survival. A comparison of the inactivation and strand-breakage rates requires measuring the relative doses that give the same survival level.

there are two modes of killing: an O_2-independent process, which is DNA breakage, and an O_2-dependent process, which does not include DNA breakage.

Example 11-D

☐ Binding of a cofactor to an enzyme.

The cofactor, reduced nicotinamide-adenine dinucleotide (NADH), binds to the enzyme chicken heart lactic dehydrogenase. To understand the mechanism of action of this enzyme, the number of binding sites and the dissociation constants characterizing the various equilibria must be known. Hence, an assay for binding must be done. This can be done with the ultracentrifuge. Using absorption optics and light at 340 nm (which is absorbed by NADH but not lactate dehydrogenase), the sedimentation of NADH can be detected (Figure 11-20). When only NADH is present, a boundary having $s = 0.2$ S is seen. When lactate dehydrogenase is added, a boundary appears

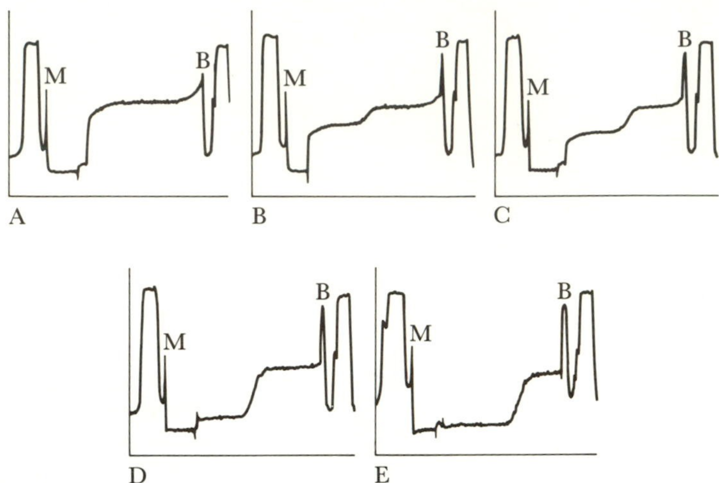

Figure 11-20
Sedimentation velocity patterns of various mixtures of lactic dehydrogenase and the coenzyme reduced nicotinamide-adenine dinucleotide (NAD). The cell was photographed with light of wavelength 314 nm, which is absorbed by NADH but not by the dehydrogenase, and the absorbance was plotted by a scanner. Thus the sedimentation of NADH is observed. (A) NADH with no enzyme; (B–E) NADH/lactate dehydrogenase = 12, 8, 4 and 2, respectively. Sedimentation is from left to right; M and B are the meniscus and bottom of the cell, respectively. [From H. K. Schachman and S. I. Edelstein, *Methods Enzymol.* 27(1973):3–58.]

at s = 7 S, the value for pure enzyme. Hence, NADH is sedimenting with the enzyme and must be bound to it. As the amount of enzyme increases, the relative amount of NADH at s = 0.2 S and s = 7 S can be measured, and this yields the fraction bound. Using this simple assay, the ratio of bound to unbound NADH can be determined as a function of NADH and lactate dehydrogenase concentration, and of pH, ionic strength, and so forth.

Example 11-E ☐ An indication of the strength of base pairs of DNA.

When DNA is treated with an endonuclease or with X rays, single-strand breaks are produced. As these breaks accumulate, ultimately two single-strand breaks in separate strands become close enough to constitute a double-strand break, and the double-stranded DNA is fragmented. Because the strength of hydrogen bonds depends on temperature and ionic strength, it might be expected that the minimum separation of two single-strand breaks that would not cause double-strand breakage would also depend on these agents. To measure single-strand breaks requires only the sedimentation of DNA in alkali because, at high pH, the double helix unwinds to form two single

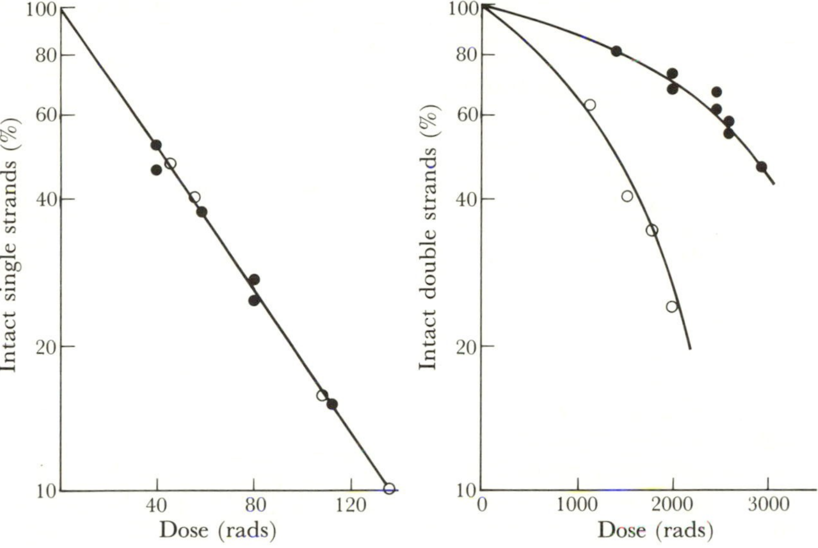

Figure 11-21
The percentage of intact single and double strands of *Pseudomonas aeruginosa* phage B3 DNA following introduction of single-strand breaks by X irradiation. Irradiated DNA was sedimented at alkaline pH (left) and at neutral pH (right) to measure single- and double-strand breakage. Solid circles, high ionic strength; open circles, low ionic strength. [From D. Freifelder and B. Trumbo, *Biopolymers* 7(1969):681–693.]

polynucleotide strands. To investigate this, DNA was X-irradiated and single- and double-strand breaks were determined by sedimentation in alkali and at neutral pH (Figure 11-21). The fraction of unbroken single- and double-stranded DNA molecules was measured for each dose, using the procedure shown in Figure 11-8. From a relatively simple equation,

$$h = [L(1 - F^{1/b})/2b] - \tfrac{1}{2}$$

in which h is the minimum number of base pairs required to maintain the structure, L is the number of nucleotides per single strand, F is the fraction of double-stranded molecules that are unbroken, and b is half the average number of breaks per single strand, the minimum number of base pairs required to keep the double-stranded molecule intact was calculated. It was found to be 15.8 and 2.6 in 0.01 M and 1.0 M NaCl at 25°C, respectively.

☐ Structure of ribosomes. **Example 11-F**

In isolating ribosomes from the bacterium *E. coli*, it is found that there are at least four species having s = 30, 50, 70, and 100 S (Fig.

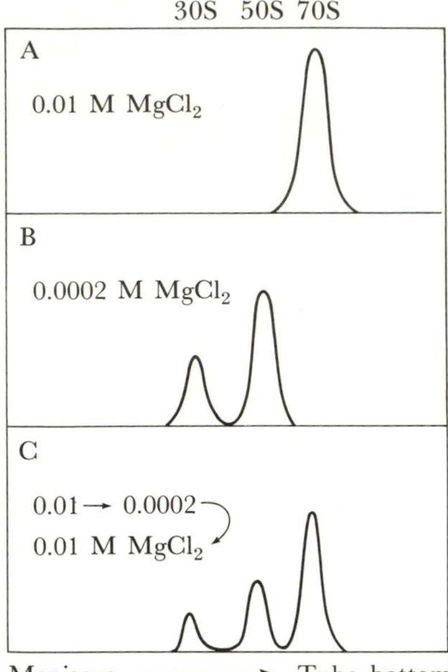

Figure 11-22
A sedimentation velocity study of dissociation and association of *E. coli* ribosomes. Drawings of schlieren patterns of ribosomes and subunits sedimented in different Mg^{2+} concentrations: (A) in 0.01 M $MgCl_2$, the 70S ribosome is stable; (B) in 0.0002 M $MgCl_2$, it dissociates to form one 50S and one 30S subunit (the areas are in a 2:1 ratio because the 50S subunit has about twice the molecular weight of the 30S); (C) partial reassociation by restoring the sample in part B to 0.01 M $MgCl_2$. At higher Mg^{2+} concentration, a 100S peak appears.

11-22). It can be shown by sedimentation analysis that these species constitute an associating system dependent on Mg^{2+} concentration in which a 30S and a 50S particle combine to form a 70S particle and two 70S particles combine to form a 100S species. This can be seen in Figure 11-22, in which peaks disappear and reappear at new positions and in which the relative areas reflect the concentration of material having each *s* value.

Centrifugation of Associating Systems

If all molecules in a mixture are independent of one another the number of *s* values observed equals the number of sedimenting species. This is not always the case for an associating system.

A mixture containing monomers and dimers that are in equilibrium may give a single peak if the following conditions are met: (1) the interaction is freely reversible in the sense that at equilibrium monomers are forming dimers and dimers are dissociating to form monomers; (2) the time required for association and dissociation is so short that equilibrium exists at all points in a centrifuge cell. Consider such a mixture at the instant sedimentation begins. The dimers will move more rapidly than the monomers but will dissociate to form monomers. Thus, a molecule that is a component of a dimer will first move fast and then move more slowly. A monomer initially present will encounter another monomer and then associate; the dimer will move more rapidly until it dissociates. Thus, each molecule alternates between having two different s values. If this change in state occurs a very large number of times, the molecules will have an average s value that is greater than that of a monomer and less than that of a dimer. The sedimentation boundary will be asymmetric though with an increasing fraction of the molecules trailing behind the main boundary. This is because a monomer left behind is at a lower concentration than those ahead of the boundary and therefore has a lesser tendency to dimerize as sedimentation proceeds. Figure 11-23 shows the kind of boundary that is observed for the system just described.

If the association-dissociation reaction is very slow, two asymmetric boundaries result, one having the s value of the monomer and the other having the s value of the dimer. With an intermediate rate of association and dissociation, a three-component boundary is sometimes observed. These components have the monomer and dimer s values and the intermediate values.

Many types of interacting systems have been studied and analyzed mathematically. The message derived from each of these studies is the following. *The number of components in a sedimentation boundary should not be taken as the number of species present, unless additional experiments have proved that the system is not an association-dissociation system.*

The existence of association-dissociation reactions can be demonstrated by studying the effect of total concentration in the observed s values. With a single substance s decreases as the concentration increases (equation 3). However, in the mixture, higher concentration favors

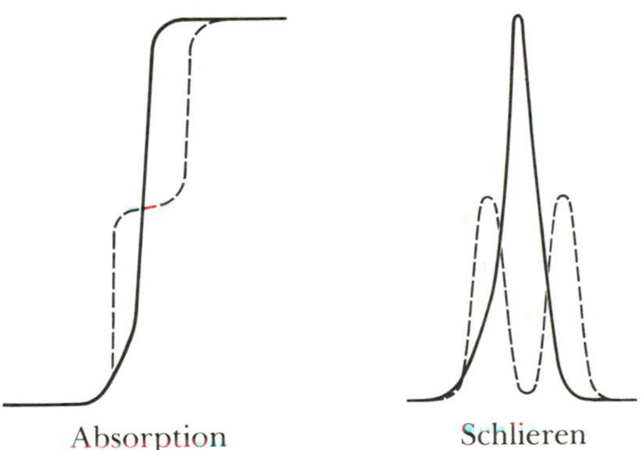

Figure 11-23

The effect of association and dissociation on a monomer-dimer equilibrium system. The dashed lines show the curves for a 50:50 mixture of monomers and dimers in which the monomers either are unable to associate or do so very slowly. The solid lines are the curves for an association-dissociation system.

association so that the apparent s value increases with increasing concentration; such an observation can always be taken as an indication that an association-dissociation system is present.

The interactions just described also have effects on sedimentation equilibrium. Because of the various causes of non-ideality in solutions of macromolecules, the molecular weight determined by any of the sedimentation equilibrium procedures decreases with increasing concentration. However, in an equilibrating mixture, higher concentration favors dimerization so that the molecular weight appears to rise with increasing concentration. Straightforward but tedious methods (which will not be described here) allow the degree of association and the dissociation constants to be calculated from studies of sedimentation equilibrium at several concentrations.

Zonal Centrifugation Through Preformed Density Gradients

In the centrifugation procedures described in the preceding sections, the starting solution consists of macromolecules uniformly distributed throughout the cell and all information is obtained from observations of the trailing boundaries of the molecules as they sediment down the centrifuge tube. Whereas this method has many advantages and is ideally suited to the analytical ultracentrifuge, it has the disadvantage that molecules having different s values never separate from one another so that faster molecules are always sedimenting through a solution of slower molecules. This has two effects: (1) concentration dependence is always at work to decrease resolution and (2) if a fast component is to be physically separated from a slow one, the fast one is always contaminated by the slower; the level of contamination can only be reduced by repeated sedimentation. An ideal technique would allow each molecular species to sediment only through the solvent. This situation can be approximated by *zonal*, or *band*, centrifugation. In the description given here, it is carried out in the preparative centrifuge, using swinging-bucket rotors (Figure 11-5). A later section will show how it can be performed in special cells in an analytical centrifuge.

The methods used in zonal centrifugation are shown in Figure 11-24 A small volume of a solution containing the molecules to be characterized is layered on top of a preformed concentration gradient contained in a centrifuge tube. The solution being layered always has a density less than that at the top of the gradient (otherwise it would sink into the gradient). The tube is then centrifuged and the molecules in the starting layer sediment through the gradient. If the molecules have the same sedimentation coefficient, they will sediment within a narrow zone as shown in Figure 11-25. If there are molecules having several sedimentation coefficients, they will separate from one another as sedimentation proceeds. The

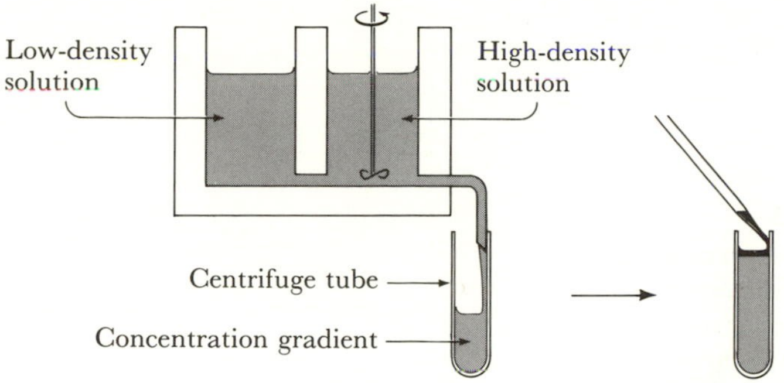

A. Formation of gradient

B. The sample is layered on top of the gradient.

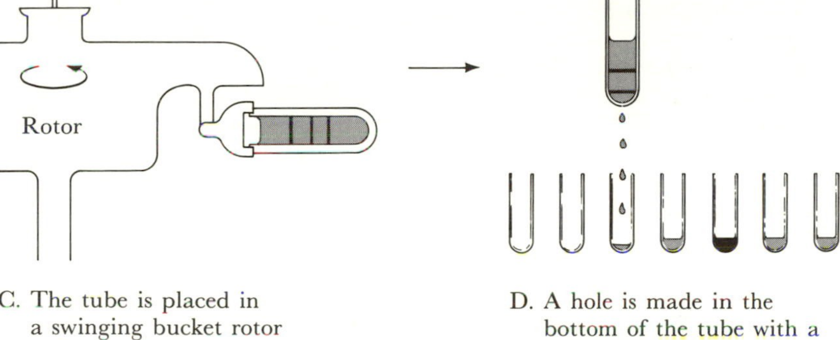

C. The tube is placed in a swinging bucket rotor and centrifuged. The components of the sample separate according to their *s* values.

D. A hole is made in the bottom of the tube with a needle, and the drops are collected in a series of tubes.

Figure 11-24
Operations in zonal centrifugation.

different components will then be resolved into a series of zones or bands, which then sediment only through solvent and independently of one another. After centrifugation is complete, the contents of the tube are fractionated, most commonly by drop collection from the tube bottom (Figure 11-6). If the dripping rate is sufficiently slow so that there is no turbulence, each drop represents a single lamella in the tube. The fractions then consist of one or more drops each. A great variety of techniques can be used to assay the materials and thereby determine the concentration distribution. In analytical centrifugation, optical techniques (ultraviolet absorption, schlieren, or Rayleigh interference) are relied on entirely to detect the molecules. However, the fractions obtained from

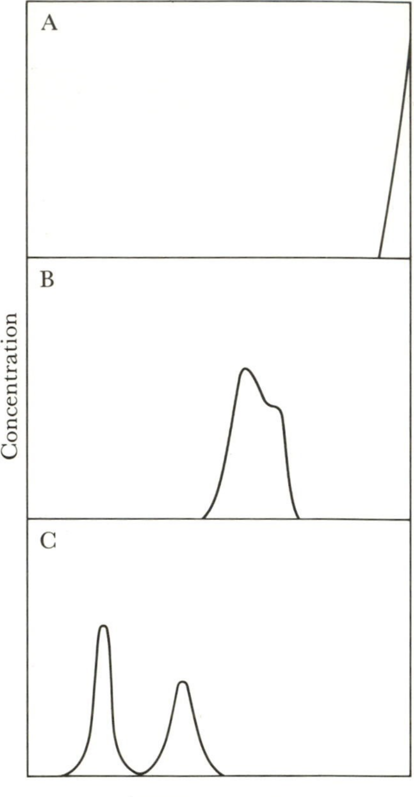

Figure 11-25
Separation of two components by zonal centrifugation in a sucrose density gradient—sedimentation is from right to left: (A) before centrifugation; (B and C) at successively longer periods of centrifugation. Note that it is not possible to determine the relative amounts of the two components from the curve in panel B; once they are separated, as in panel C, the relative mass is easily measured as the relative areas of the two zones.

a drop-fractionated tube can be assayed by radioactivity, chemical tests, enzymatic activity, absorption, fluorescence, or combinations of these. Several examples of these methods are shown in Figure 11-26.

Why is a concentration gradient needed for zonal centrifugation? Let us consider the consequence of layering a low-density solution on top of a higher-density solution in which there is no concentration gradient. Initially, the system is stable because of the density difference at the boundary—that is, the upper solution is simply floating on the lower. Suppose now that there are small fluctuations in temperature in the solution (as there certainly will be). Because the density of most solutions decreases with increasing temperature, there fluctuations will create local density inversions (i.e., higher-density regions above lower), which will result in a local flow of the liquid called *convection.* This will have little effect on the initial boundary because the density difference at the boundary is usually great enough that mixing across the boundary does not occur (unless the temperature difference is huge—10–20°C). However, after sedimentation, the molecules to be studied will have moved into the denser lower layer, and such convection can destroy any band

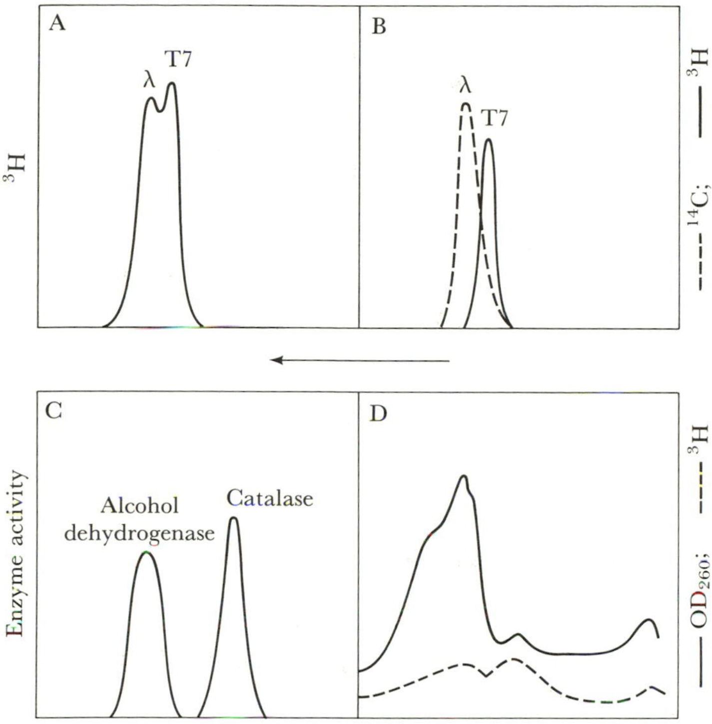

Figure 11-26
Several ways to assay a zonal gradient to get the concentration distribution: (A) T7 [^{3}H]DNA and λ [^{3}H]DNA; (B) T7 [^{3}H]DNA and λ [^{14}C]DNA; (C) a mixture of two enzymes; and (D) the incorporation of [^{3}H]uracil into RNA (OD_{260} indicates the amount of total RNA).

that might exist. The introduction of a steep density gradient ensures that temperature changes have to be very great to create a density difference sufficient to cause flow within the gradient. A second important function of the density gradient is to prevent mixing due to mechanical disturbances; any perturbation would be counteracted by the attempt to return to the situation in which low density is above high density. The gradient also serves a third purpose. Consider a system without a gradient and free of all temperature fluctuations and mechanical disturbances in which the sedimenting molecules have entered the lower layer and formed a zone. In this zone, the presence of the molecules increases the density of the solution by their own contribution to the density (normally a very small effect but, when high concentrations are used, it is significant). Hence, the density of the zone is also greater than the solution just below it, and this results in the convective flow of the zone toward the bottom of the cell. If instead sedimentation were through a preformed concentration gradient, the sedimenting molecules would continually be entering a region of higher density. This would still increase the density of that region but, if the gradient were sufficiently steep, the contribution of the molecules to the density would be insufficient to cause a density inversion and the system would remain stable. The material most commonly used to form density gradients is sucrose because of its purity, low cost, and lack of interference with most chemical, enzymatic, and optical assays. If the macromolecule to be studied is an enzyme or an unstable protein, glycerol is frequently used because many proteins are more stable and less easily denatured in the presence of glycerol. H_2O-D_2O gradients have also been used successfully; this is the method of choice if the particle under study requires constant osmotic pressure. Examples of the use of these systems will be described in a later section.

Determination of the Sedimentation Coefficient by Zonal Centrifugation in Preformed Density Gradients

Zonal centrifugation is often used to determine *s* by calibration with markers for which *s* values have been measured in an analytical ultracentrifuge. However, even if the concentration is low enough to eliminate concentration-dependent sedimentation (which is of great value if the material is not pure), the determination of *s* by means of zonal centrifugation has pitfalls not encountered with the analytical ultracentrifuge. There are three reasons for this: First, the density and viscosity of the solution increase with distance along the tube so that the net force on the molecules is not a simple function of distance. Second, the walls of the preparative tubes are parallel and therefore not aligned with the centrifugal direction, and this results in sedimentation against the walls and accumulation of material. This accumulation causes a local density inversion and some convection. Third, *s* is calculated from only two points, the position at zero time and at one later time (when the centrifuge has stopped) so that the rate of movement is not really known. The first reason can be dealt with theoretically in a complicated way; the second has been considered theoretically but the effect on *s* is not clearly understood. To avoid the problems of reason 1, the common practice is to use isokinetic gradients—that is, the concentration and viscosity gradient is selected so that the molecules move at constant velocity at all distances from the center of rotation. The currently standard 5% to 20% sucrose gradient was empirically selected to have this property. It is important to realize that the isokinetic property of a concentration gradient depends on the change in concentration as a function of *distance* along the tube; therefore, a 5% to 20% sucrose gradient is isokinetic in a particular centrifuge tube (e.g., the tubes for the Beckman SW39, SW50, and SW65 rotors) but not for longer tubes (e.g., SW25, SW41). Thus, gradients of different composition must be selected for tubes of different length if isokinetic conditions are needed. The student is cautioned to keep this in mind when reading the literature because this has rarely been appreciated and is one of the reasons that different values of s are reported from different laboratories.

Because of the second and third reasons, absolute determination of s is difficult, and the usual practice is to determine relative s by cosedimentation with molecules of known s, for which s has been determined in the *analytical* ultracentrifuge. However, the following considerations show that this practice is not valid. $s^{\circ}_{20,w}$ for double-stranded DNA varies with $M^{0.497}$ (equation 6) if measured in a sector-shaped cell. This equation would also be valid if the cell contained an isokinetic gradient, in which case the distance traveled (d) per unit time would also vary as $M^{0.497}$. However, if in a 5% to 20% sucrose isokinetic gradient, the relation between d and M is measured in a preparative centrifuge, it is found to be

$$\frac{d_1}{d_2} = \left(\frac{M_1}{M_2}\right)^{0.38} \tag{7}$$

in which the subscripts 1 and 2 refer to two independently sedimenting components. Because s is proportional to distance in an isokinetic gradient, the s value of DNA in a 5% to 20% sucrose gradient in a preparative cell (with parallel walls) will not be the same as that obtained in a sector-shaped cell because the exponents of M are not the same. (It is thought that this is due to the convection described in reason 2.) Therefore, s cannot be determined from the relative distance traveled. There is no cause to believe that this situation is any better for proteins. However, the problem is not really serious because normally s is only wanted for an estimation of M. Because the s-M relation is already an empirical one for most substances, an empirical d-M relation can be obtained just as well, using relative distance traveled at a given time. Equation (7) describes such a relation for DNA. However, equation (7) is valid *only* for an isokinetic gradient and, because of reason 3, it can never be known whether it is isokinetic.* This problem can be avoided by the procedure of cosedimenting two or three substances of known molecular weights that bracket that of the unknown and measuring relative d for two different periods of centrifugation. A simple plot of relative d versus log M will give an accurate M value. This cosedimentation procedure is the only one that gives an unambiguous result yet it is rarely used.

* I do not mean to imply that an isokinetic gradient cannot be prepared but rather that, since the position is determined at only one time, one does not know whether a supposedly isokinetic gradient is in fact isokinetic.

Examples of the Use of Zonal Centrifugation in the Preparative Ultracentrifuge

In the following examples of zonal centrifugation, very small amounts of material that cannot be detected by optical methods are studied. Note the variety of the assays and the strength of the conclusions.

Example 11-G ☐ Demonstration that the *E. coli* Rho factor terminates RNA synthesis.

When RNA polymerase synthesizes RNA from a DNA template, synthesis begins at base sequences known as promoters and ends at terminator sequences. However, at one time there was no evidence that termination sequences are recognized in vitro. Later Jeffrey Roberts isolated a protein called Rho, which decreased the amount of RNA synthesized in vitro. Evidence that Rho supplies the information for the recognition of termination sites came from analyzing the synthesized RNA by zonal centrifugation in sucrose gradients. One of the relevant experiments is shown in Figure 11-27. If Rho

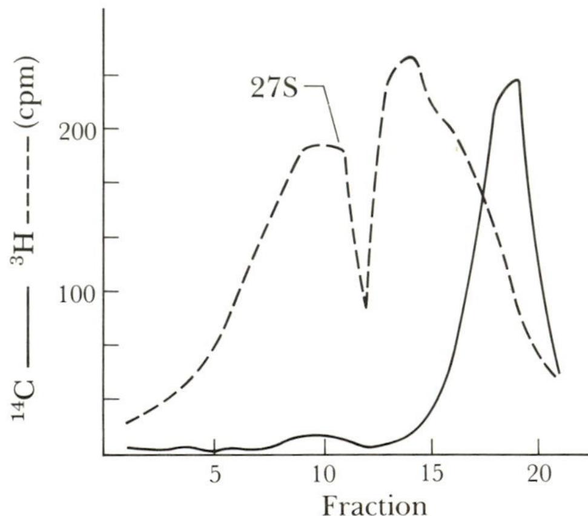

Figure 11-27
Zonal sedimentation (5%–20% sucrose gradient) of RNA synthesized from an *E. coli* phage λ DNA template. Two reaction mixtures were prepared containing λ DNA, *E. coli* RNA polymerase, and appropriate buffers. The dashed line represents RNA synthesized when [^{14}C]uridine triphosphate (UTP) is incorporated. The solid line represents RNA made with [^{3}H]UTP in the presence of the factor. Rho decreases the size of the RNA. Sedimentation is from right to left. [From J. W. Roberts, *Nature* (*Lond.*) 224(1969):1168–1174.]

decreased either the rate of initiation of synthesis or of polymerization, after long periods of synthesis (when many molecules had been synthesized), the size distribution of the RNA molecules would be the same whether Rho were present or not. If, however, it induced termination, the molecules would be smaller. Because the overall shape of all RNA molecules is the same, this would mean that, when Rho is present, the synthesized RNA would sediment more slowly as shown in Figure 11-27.

☐ Measurement of the change in size of a mutant enzyme. **Example 11-H**

Many of the methods used to determine molecular weights of enzymes either require highly purified samples (e.g., sedimentation equilibrium, as will be described) or, if impure samples can be used, yield subunit weights (e.g., sodium dodecyl sulfate gel electrophoresis, Chapter 9). The size of an active enzyme can be determined from a crude lysate, using zonal sedimentation and enzyme activity as an assay. For example, *E. coli* DNA polymerase I sediments at $s = 5.4$ S. The polymerizing and $5' \rightarrow 3'$ exonuclease activities associated with this protein cosediment so that either activity can be used to localize the enzyme in a sucrose gradient. A mutant form of this enzyme lacks polymerase activity but retains $5' \rightarrow 3'$ exonuclease activity. As shown in Figure 11-28, the s value of the mutant enzyme is 2.8 S. Because the particular mutation is a chain-termination mutation, the reduced s value indicates that the enzyme is a small fragment of the wild-type protein, which has 0.4% activity.

☐ Determination of the molecular weight of T7 DNA by end-group labeling. **Example 11-I**

The enzyme T4 polynucleotide kinase can be used to label the 5′ ends of a double-stranded DNA molecule (by transferring ^{32}P from [γ-^{32}P]ATP to 5′-OH groups). This can be used to determine molecular weights if the amount of DNA and the number of ^{32}P atoms coupled are known. However, if the DNA sample contained broken DNA molecules, a low value of M would result because too much ^{32}P would be bound—for example, if 1% of the molecules were broken into one hundred fragments, the observed M would be half the true M. To determine the number of ^{32}P atoms per intact DNA molecule, the reaction mixture can be sedimented through a sucrose gradient (Figure 11-29). Figure 11-29 shows that, as is often the case, a significant fraction of the radioactivity barely sediments and remains at the top of the gradient. However, the main peak represents intact molecules so that, from the ratio of ^{32}P to DNA concentration in the peak, M can be calculated.

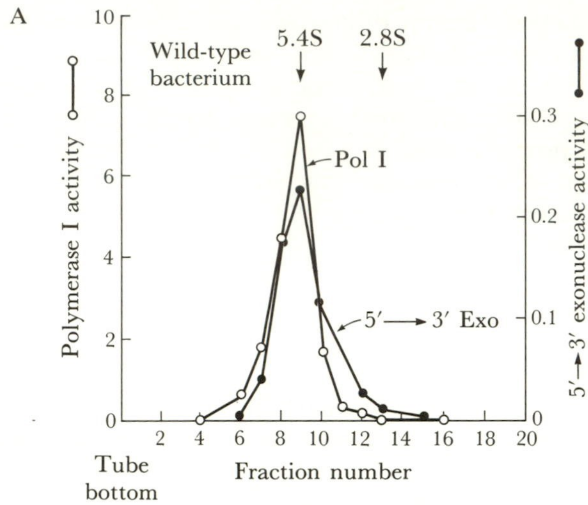

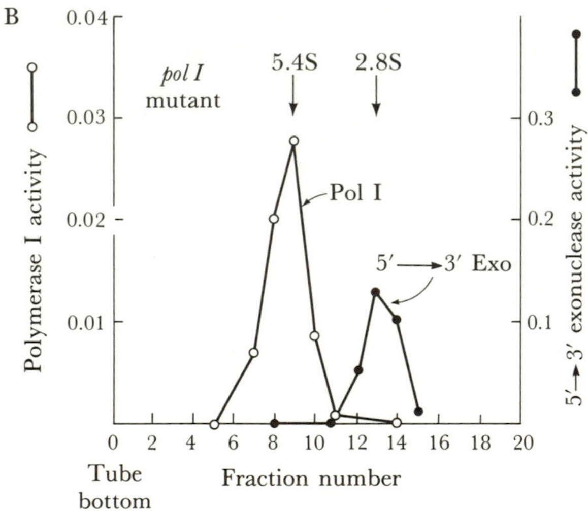

Figure 11-28

Zonal sedimentation through a 5%–20% sucrose gradient of *E. coli* DNA polymerase I isolated from (A) a wild-type strain and (B) a strain making only a fragment of the enzyme. The wild-type enzyme possesses two activities: the polymerase activity is measured by the incorporation of a ^{32}P-containing nucleotide into acid-insoluble material (left axis); the 5′ → 3′ exonuclease is assayed as the removal of ^{32}P-labeled polynucleotide (right axis). In part A, the two activities cosediment because they are possessed by a single protein. In part B, the mutant enzyme has 0.4% of the polymerising activity, but the more slowly moving fragment retains the 5′ → 3′ exonuclease activity. [From I. R. Lehman and J. Chien, *J. Biol. Chem.* 248(1973): 7717–7723.]

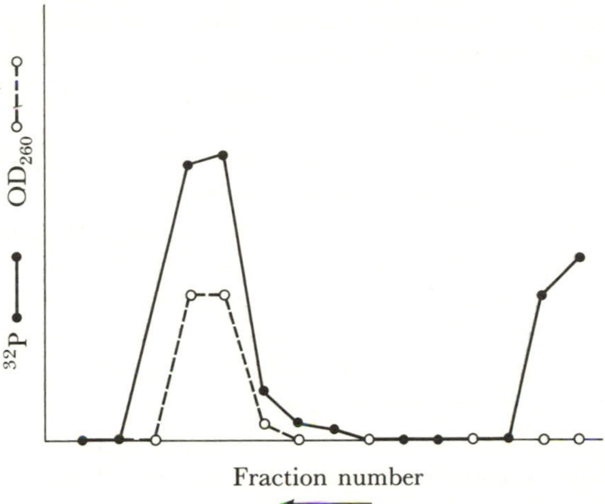

Figure 11-29
Sedimentation pattern of T7 DNA labeled at its termini with ^{32}P, using T4 polynucleotide kinase: OD_{260} is a measure of the total DNA concentration; ^{32}P measures number of termini. Note that, at the top of the gradient, no OD_{260} is detectable. Hence, the DNA concentration is very low and the ^{32}P measured represents the labeling of tiny fragments.

Example 11-J

☐ Partial purification of an osmotically fragile structure containing a hormone.

A hormone called substance P can be isolated from the hypothalamus of a rat. Dorothy Freifelder and Susan Leeman found that the hormone can be pelleted from macerated tissue at very low centrifugal force if the tissue were suspended in 20% sucrose before disrupting the cells; in the absence of sucrose, it was not sedimentable even at very high forces. This suggests that the hormone is contained in osmotically fragile particles. It is not possible to sediment the particles through a 5% to 20% sucrose gradient because the sample layer would immediately sink to the bottom. A 25% to 70% gradient is not useful because the particles are apparently destroyed at very high osmotic strength. To avoid these problems a gradient was prepared from 20% sucrose in H_2O and 20% sucrose in D_2O. This provides a stable gradient because D_2O is 11% denser than H_2O, and the constant sucrose concentration maintains the necessary osmotic strength. Sedimentation through this gradient yielded partly purified particles.

Band Sedimentation in Self-Generated Density Gradients

It is often desirable to carry out zonal sedimentation with the analytical ultracentrifuge and ultraviolet-absorption optics. This method is preferable if only very small samples are available that cannot be labeled with radioisotopes or assayed by chemical or enzymatic means. Because of difficulties in preparing density gradients in cells of the analytical ultracentrifuge, Jerome Vinograd developed a procedure for zonal centrifugation without using a preformed gradient.

As indicated in an earlier section, if a sample is layered onto a denser solution without a preformed density gradient, convection will ultimately disrupt or destroy the boundary. Vinograd realized that, if the sample is in a solution of very low density, because of the diffusion of the solute used to increase the density of the denser solution, in time the density discontinuity generated at the interface will slowly be propagated down the centrifuge tube, forming a shallow gradient. Below the boundary generated by the propagating gradient, there is no gradient and convection can occur. However, at the propagating gradient and above it, there is stabilization against convection. Clearly, as long as the material being sedimented does not move faster than the gradient being generated, the material will be in a density gradient and sedimentation will be normal. Furthermore, if the solute of the denser solution has a fairly high density (which is the case for CsCl, the material most frequently used), the salt will also sediment slightly and form an additional stabilizing gradient. This technique has been termed *band sedimentation* in self-generated density gradients.

Figure 11-30 shows the type of centrifuge cell that is used with this method. The cell is prefilled with a solution of high density (usually 2 M CsCl) and the sample hole is filled with the solution of the material to be studied. Shortly after acceleration, material leaks across the scratch on the face of the cell and is layered on top of the denser solution. As the speed of rotation increases, sedimentation occurs. As long as the speed is not too high, the zone is undisturbed. Figure 11-31 shows representative concentration distributions (i.e., ultraviolet-absorption patterns) of DNA sedimenting through concentrated CsCl solutions. Note that the boundary is asymmetric with material moving faster in the leading direction. This is clearly understood from concentration dependence; material on the leading side of the boundary is at the lowest concentration and therefore moves fastest. The trailing side of the boundary is sharp because any molecule that diffuses in the centripetal direction finds itself in a region of zero concentration and therefore sediments at maximum rate. As a matter of fact, the complete relation between s and concentration for a given molecule can be derived from the shape of the boundary.

The area of a band is proportional to the amount of material in the band (The height at any point is proportional to the concentration at that point.) This is of course always true of zonal centrifugation. Thus in a mixture of two components that completely resolve from one another the relative amounts of the components is the ratio of the areas of the bands.

Band centrifugation is the procedure to use for determining whether a sample contains a very small amount of *rapidly* sedimenting material.

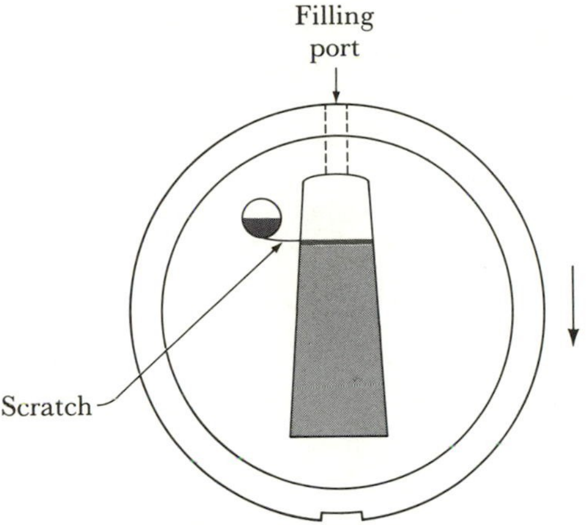

Figure 11-30
Band-forming centerpiece viewed from the top. The solvent (shaded area) is placed in the sector through the filling port. The sample is placed in the circular well. During acceleration, the sample flows along the scratch and forms a layer on the solvent. The sample in this figure is in the process of layering. Note that the sample flows only through the scratch because the entire surface of the centerpiece is in close contact with an end window. [Centerpiece designed by Jerome Vinograd, California Institute of Technology.]

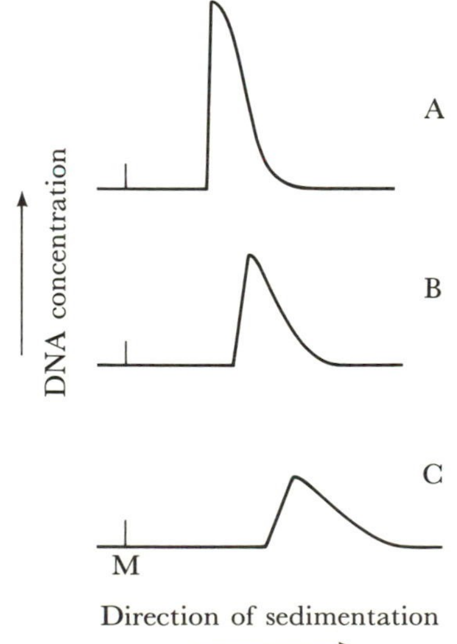

Figure 11-31
Sedimentation of double-stranded T7 DNA in a band-forming cell like that shown in Figure 11-30. The time of centrifugation increases in going from part A to part C; M is the meniscus. The peak becomes lower and broader with time because of concentration dependence, as described in the text. The areas remain constant though.

In standard boundary sedimentation, the rapidly sedimenting material moves through the slower material and is slowed down by the Johnston-Ogston effect; therefore, it is frequently not observable. In band centrifugation, this small fraction is sedimenting in a region of zero concentration and is easily observable.

Band sedimentation is rapidly becoming the method of choice for analytical centrifugation of DNA but it has not yet had widespread use in the study of proteins. However, one important application in the study of enzymes is the following.

An ingenious new technique combining band sedimentation and absorption optics (using the Beckman photoelectric scanner described on page 380) has been developed for detecting the active form of an enzyme [B. L. Taylor, et al. *J. Biol. Chem* 253(1978):3062–3069]. Many enzymes have several oligomeric forms of which only one has enzymatic activity; the structure of the active form can be determined in the following way. A mixture of the forms is allow to form a thin layer on a buoyant solution that contains the assay mixture for the enzyme. As the enzyme sediments through the assay mixture, the enzyme reaction will occur but only in the region of the cell occupied by the enzyme. If the reaction product absorbs light of a wavelength that is not absorbed by the assay mixture, a band of absorption will appear to sediment through the mixture and the *s* value of this band can be measured. Alternatively, if the substrate has a characteristic absorbance and the product does not, a band of reduced absorbance will sediment through the cell. The *s* value of oligomeric forms can be measured independently and matched up to the observed *s* values; in this way the active oligomeric form can be identified. The kind of data that may be observed is shown in Figure 11-32.

A Comparison of Boundary and Band Sedimentation

Operationally the main difference between boundary and band sedimentation is the amount of material that is used. Whereas the concentrations required are the same for both techniques, boundary sedimentation requires that the cell is *filled* with the solution; thus, 1.0–1.7 milliliters are needed, the amount depending on the cell that is used. However, in band sedimentation only a *layer* of 0.010–0.025 milliliter is required.

Another important difference is that the solvent used in band sedimentation is more limited because the solvent density must be sufficiently high that the sample will form a floating layer. The most usual solvents are 2 M CsCl and 2 M NaCl, which have a very high ionic strength. Lower ionic strength is obtainable if the density is increased by sucrose or glycerol. However, band sedimentation in a dilute solvent is not possible.

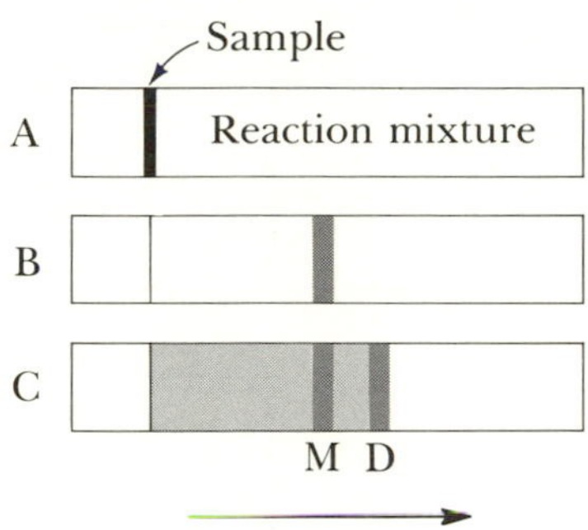

Figure 11-32
Determination of the active form of an enzyme by analytical band sedimentation. Panel A shows the cell before sedimentation begins; the sample is layered on a buoyant reaction mixture that does not absorb light. Panels B and C show the result after centrifugation. M and D refer to the bands of sedimenting monomers and dimers, respectively. The shaded area represents a region of the reaction mixture in which a light-absorbing product has been produced. In B, conditions are adjusted so that only monomer is present; there is no absorbing material. In C, monomers and dimers are both present and the boundary of the absorbing material sediments with the dimer. Thus, the dimer but not the monomer is the active form of the enzyme.

A final difference, which is extremely important, is in the ability to detect minor components and the manner of calculating the relative amounts of each component. Figure 11-33 shows boundary and band patterns for a sample containing a single component and for mixtures of a major (95%) and a minor (5%) component. Note that in panel B, in which the minor component has a lower *s* value, both procedures are able to resolve the two components. Furthermore, there is no difficulty determining the relative amounts of each—in boundary sedimentation the value is obtained from the relative heights and in band sedimentation it is from the areas. However, in panel C, in which the minor component has the higher *s* value, this component is more easily seen by band sedimentation because in boundary sedimentation, the minor component is buried in the anomalous curvature at the top (see page 372). Of course, if the difference in *s* values had been much greater both methods would have provided good resolution. Great care must be taken when the bands do not resolve, as serious quantitative errors have been made in the analysis of overlapping bands. Figure 11-34A,B shows the patterns observed for the two modes of centrifugation for a sample containing a 30:70 mixture of components whose *s* values are 23 S and 27 S, respectively. In panel A, the ratio of the two components is easily measured from the relative heights. However in panel B, there is no

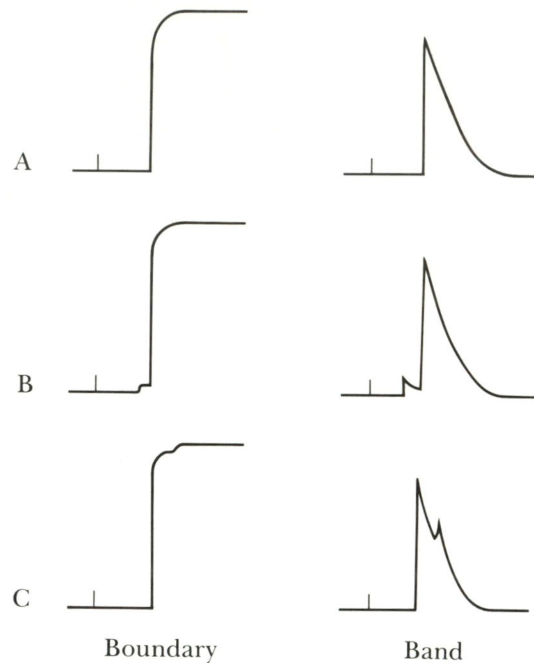

Figure 11-33
Comparison of boundary and band sedimentation. Panel A shows the result when a single component is present. In B, 5% of the material sediments more slowly than the bulk of the molecules. In C, 5% sediments more rapidly.

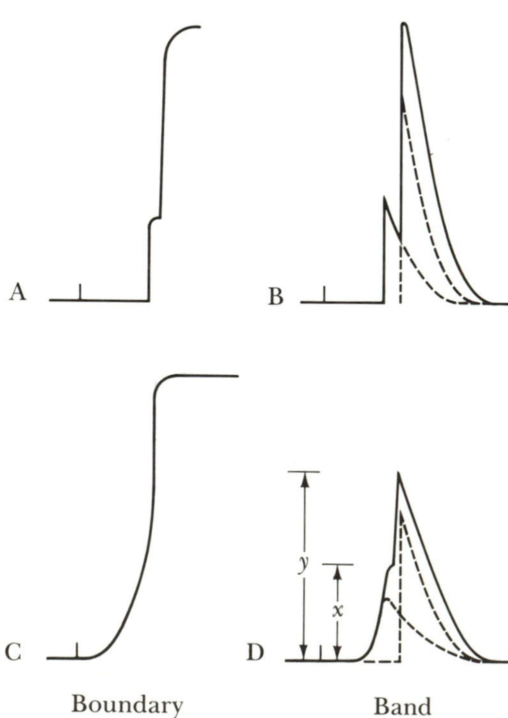

Figure 11-34
Comparison of boundary and band sedimentation of mixtures. In panels A and B, 30% of the mixture contains two types of molecule; the slower molecules comprise 30% of the mixture by weight. In panels C and D, 50% of the molecules are the same size and the remaining 50% is a heterogeneous mixture moving more slowly. In panels B and D, the dotted lines represent the sedimentation profile for the individual components: the solid line is the observed sum of all of the components.

simple way of determining the areas of the two component bands from the sum of the two that is observed. An estimate could be made by assuming that both bands have the same shape. Nonetheless, measurement of the boundary in panel A is much more precise. If the sample consists of a single homogeneous fraction and a more slowly moving heterogeneous fraction (a very common situation with DNA samples, in which a fraction of the molecules are broken to fragments of many sizes), the band gives virtually no quantitative information. Panel C of Figure 11-34 shows the boundary pattern for a sample in which 50% of the molecules are fragmented. The fraction that is fragmented is again determined from the heights of the heterogeneous and homogeneous components. Panel D shows the band pattern for the same sample. Note that there is material moving behind the main band. With boundaries such as these, an error has been made repeatedly in the scientific literature, namely, to calculate the fraction of the molecules moving slowly as the ratio of the heights x/y. This is certainly incorrect, since in band sedimentation the amount of material is proportional to the area of the band. In such a band there is no way to calculate the fraction of each component because one has no idea of the shape of the band for the heterogeneous component. The only information obtainable from panel D is that there is a heterogeneous fraction.

In summary, band sedimentation is the method of choice for analyzing mixtures of discrete components whose bands completely separate from one another. However, boundary sedimentation is preferable for samples such as those in Figure 11-34C and D.

Sedimentation of DNA at Alkaline pH

DNA is denatured at high pH. Hence, by preparing alkaline sucrose gradients (i.e., 5%–20% sucrose in 0.1–0.3 M NaOH), denatured DNA can be studied.*

Sedimentation through alkali has two important uses: to identify single-strand breaks in double-stranded DNA and to distinguish various forms of DNA. The identification of single-strand breaks in linear molecules is based on the following considerations. If the strands of a double-stranded molecule that contains no interruptions are separated in alkali, the two strands that result have the same size and therefore the same s

* The only complication with this technique is that sucrose at high concentration is a reasonably good buffer in the pH range of 10–11. Furthermore, its buffering capacity increases substantially as the temperature is lowered. Hence, alkaline sucrose gradients cannot be prepared by making a dilute pH 12.5 buffer and then adding sucrose. To insure denaturation, the pH must be adjusted after sucrose is added, and the value must be checked at the temperature at which it will be used. The use of 0.3 M NaOH eliminates this problem. All alkaline sucrose solutions should contain a chelating agent such as EDTA (ethylenediaminetetraacetate) to avoid a metal-ion–catalyzed alkaline hydrolysis of phosphodiester bonds.

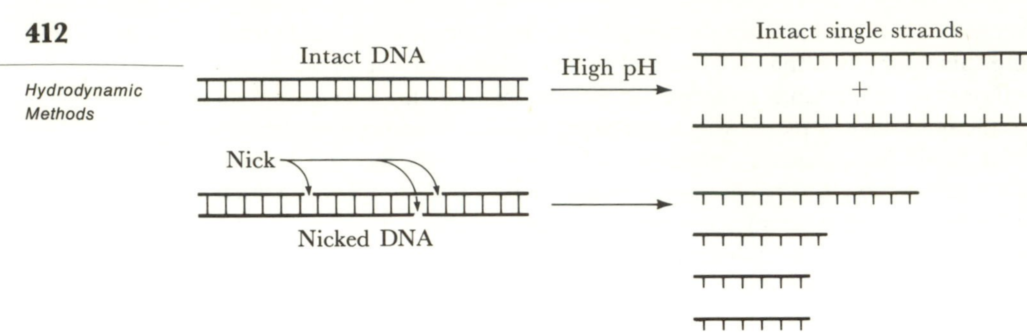

Figure 11-35
Result of the denaturation of intact and nicked double-stranded DNA.

value (Figure 11-35). However, if there is a break—let us assume it is in a unique position in one strand—the result is one intact single strand and two fragments of unique size, that is, three *s* values. If the break is randomly situated, the fragments can be of all sizes. This reasoning can of course be extended to one or more breaks in one or both strands. The main point is that a single-strand break is made evident in alkali by the appearance of DNA that sediments more slowly than intact strands. The examples that follow demonstrate the use of this principle.

Example 11-K ☐ Demonstration of naturally occurring single-strand breaks, in phage DNA.

The DNA of *E. coli* phage T7 DNA shows a single sharp boundary when sedimented at neutral pH. However in alkali the boundary shown in Figure 11-9B (page 373) is observed. Half of the DNA sediments move more slowly than the main boundary, whose *s* value is that expected of intact single-stranded T7. This indicates that half of the single strands are broken. Possible interpretations of the data are given in Figure 11-9C.

Example 11-L ☐ Measurement of the rate of depurination of DNA.

When DNA is incubated at pH 4 for a period of several days, purine bases are slowly removed from the DNA. If the number of bases removed is large, they can be separated by chromatography or electrophoresis and measured by chemical or spectral analysis. However, by centrifugation it is possible to detect the removal of one base in a DNA molecule, although whether the base is adenine or guanine cannot be determined. DNA can be denatured and maintained in single-stranded form either by adding alkali or by heating in the presence of formaldehyde. The two procedures are equally effective as long as the DNA is not heated for too long and at too high a temperature (in which case thermal hydrolysis of phosphodiester bonds occurs). Alkaline hydrolysis occurs at a very low rate and is usually no problem. If either a purine or a pyrimidine base is missing, the two procedures give very different results. This is because one of the phosphodiester bonds adjacent to the base-free deoxyribose is very rapidly (in seconds) hydrolysed above pH 12 whereas the bonds are stable in the heat–formaldehyde procedure. Thus if a base has been removed, it will appear as a single-strand break in alkali but will have no effect on the sedimentation diagram following heat–formaldehyde denaturation.

The presence of fragments when DNA is denatured by both procedures is evidence that native DNA molecules possess single-strand breaks and is a condition that must be met before that conclusion can be drawn. This essential test was used in the experiment that was described in Example 11-K.

Example 11-M

☐ Identification of small fragments of newly synthesized DNA: the Okazaki fragments.

If the bacterium *E. coli* is grown for many generations in a growth medium containing [^{3}H]thymidine, the synthesized DNA is labeled with ^{3}H. If this DNA is isolated and sedimented at neutral pH, its *s* value is characteristic of DNA whose molecular weight is very high. However, if sedimented in an alkaline sucrose gradient, a small fraction of the DNA sediments very slowly (Figure 11-36A). This implies that somewhere in the double-stranded DNA there are closely spaced single-strand breaks. If nonradioactive cells are grown for only 5 seconds (0.2% of a generation) in radioactive medium and then collected and analyzed in an alkaline sucrose gradient, most of the radioactivity (i.e., all DNA made during the 5-second period) has a very low *s* value (Figure 11-36B). Hence, recently-made DNA must contain many closely spaced single-strand breaks. If the cells labeled for 5 seconds are transferred to nonradioactive medium and allowed to grow for 10 minutes ($\frac{1}{4}$ generation), sedimentation analysis in alkaline sucrose shows that all radioactivity has a high *s* value (Figure 11-36C); that is, "old" DNA does not contain single-strand breaks. These simple results lead to the conclusion that DNA is synthesized in small fragments, which are then joined to form a larger unit.

Example 11-N

☐ Detection of radiation damage in bacterial DNA.

If bacteria labeled with [^{3}H]thymidine are treated with the enzyme lysozyme, which removes part of the bacterial cell wall, and then layered directly on an alkaline sucrose gradient, the alkali will complete the lysis of the bacteria and denature the DNA. If this is then

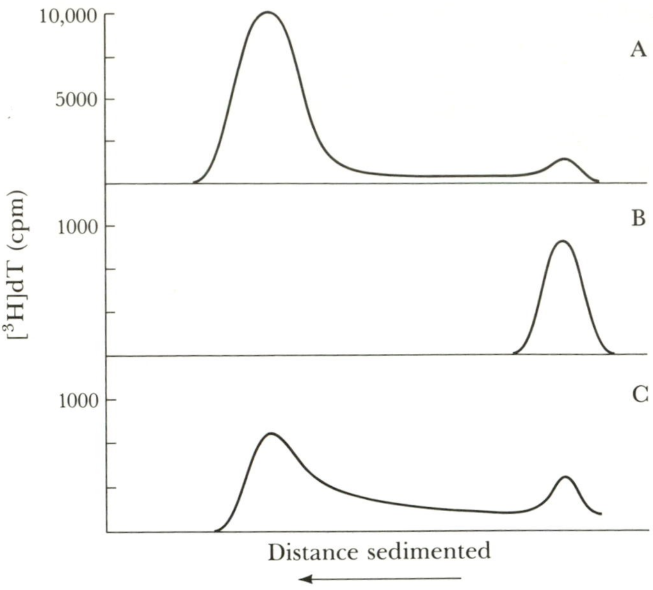

Figure 11-36
Detection of short single-strand DNA fragments in *E. coli* by sedimentation in alkaline sucrose: (A) the bacteria were grown in [^{3}H]thymidine for many generations; (B) the bacteria were pulse-labeled; (C) pulse-labeling followed by growth in the absence of label (see text for details). Note the change in scale (cpm) between part A and parts B and C. [Data from the laboratory at Brandeis University.]

centrifuged, the sedimentation pattern indicates that most of the DNA has a very high s value (Figure 11-37). If the cells are X-irradiated before lysis, the s value is much lower, indicating that the X rays cause single-strand breaks. From an empirical relation between s and M, the number of single-strand breaks can be calculated and compared with the X-ray dose. In that way, the relation between strand breakage and X-ray-induced killing can be investigated.

The distinction between various conformations of DNA is based on the following considerations. As mentioned in Chapter 1, some types of DNA may be either linear, an open circle (with at least one single-strand break), or a covalent circle (containing no interruptions). Combinations of these are also possible—for example, two circles may be linked as in a chain; such linked circles are called *catenanes*. Each of these structures has a distinct s value as shown in Figure 11-38 although the differences are not always large. For instance, the ratio of the s values for a linear molecule, a nicked circle, and a naturally occurring supercoil is 1.0:1.14: 1.4 respectively; furthermore a nicked circle and a non-supercoiled

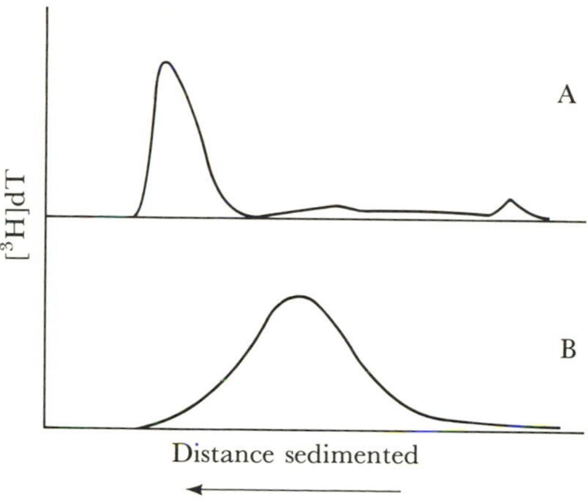

Figure 11-37
Detection of single-strand breaks in *E. coli* DNA by the technique of Richard McGrath and R. W. Williams [*Nature* 212(1966): 534–535]. (A) Unirradiated bacteria; (B) X-irradiated bacteria. The cells were treated with lysozyme and layered on an alkaline sucrose gradient. Lysis and the release of DNA occurred in the layer. [Data from the laboratory at Brandeis University.]

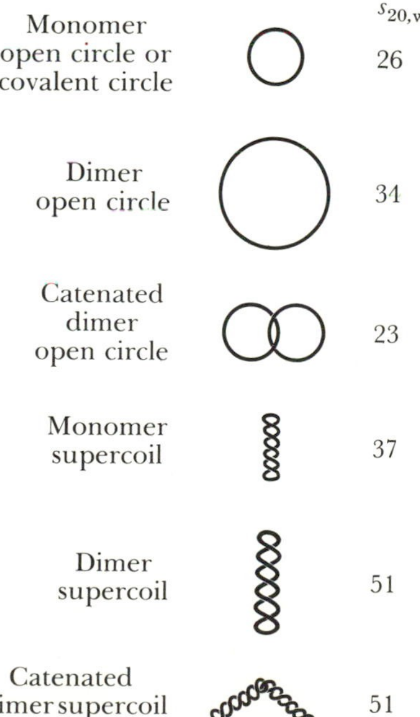

Figure 11-38
Sedimentation coefficients of different circular structures of the DNA isolated from the mitochondria of human leucocytes. [From B. Hudson, D. A. Clayton, and J. I. Vinograd, *Cold Spring Harbor Symp. Quant. Biol.* 33(1968): 435–442.]

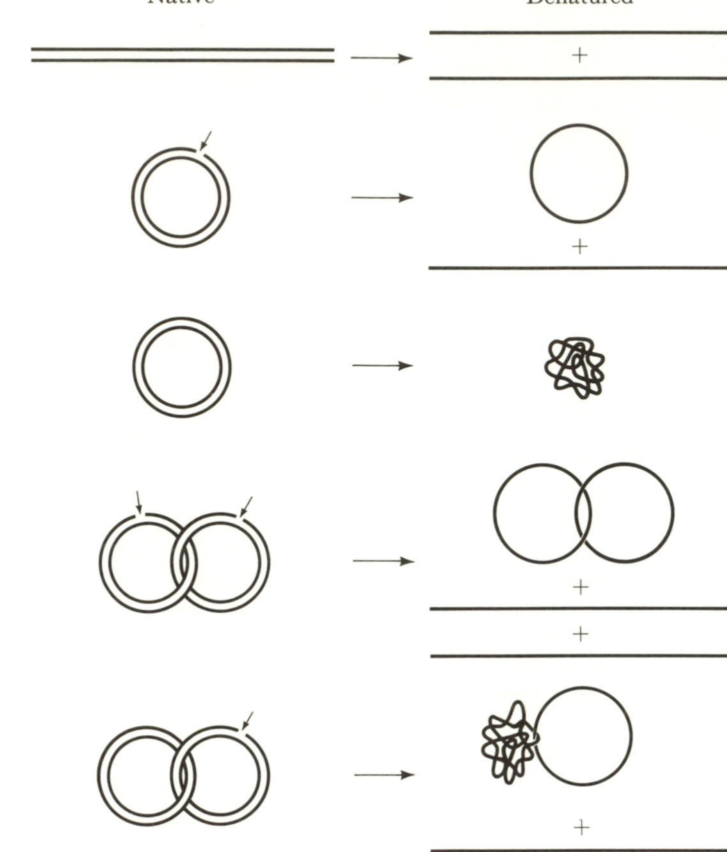

Figure 11-39
Products of the denaturation of different forms of DNA.

covalent circle have the same *s* values. Consider, however, the effect of placing each of these structures in alkali (Fig. 11-39). A linear molecule is dissociated into two single strands; a circle with one single-strand break separates into one linear molecule and one single-strand circle. However, the two single-strand circles of a covalent circle cannot separate because they are intertwined. The result is that, in alkali, a denatured covalent circle collapses to a tangled, compact mass. Because the molecular weight is unchanged and the molecule becomes compact, its *s* value becomes very large. Supercoiling does not contribute significantly to the *s* value of a covalent circle in alkali. That is, the *s* values in alkali of a nonsupercoiled and a supercoiled covalent circle are not detectably different. This is because once the molecule has collapsed, the small additional twisting of the helix introduced by supercoiling (about 1 twist per 25 turns of the helix) does not affect the shape of the tangled mass.

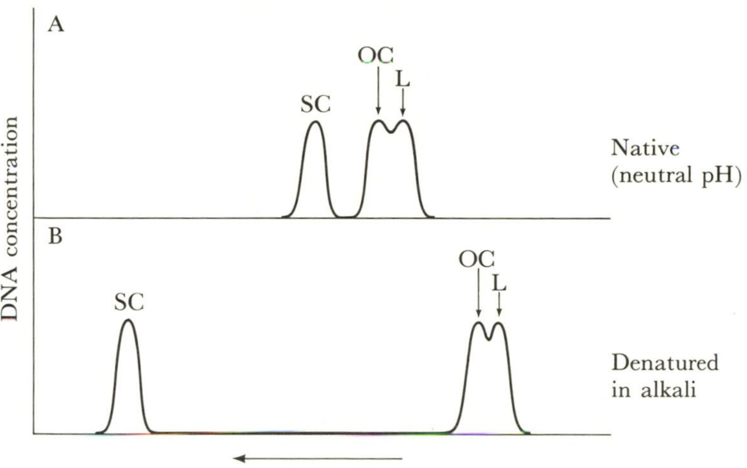

Figure 11-40
Separation of supercoils (SC), open circles (OC), and linear (L) molecules by sedimentation at neutral pH (native DNA) or in alkali (denatured DNA). The time of centrifugation of part B is half that of part A. Note the improved separation of supercoils in alkali.

The relative *s* values of the various forms usually enable one to determine the form of a particular molecule or the composition of a mixture. Figure 11-40 shows two hypothetical sedimentation patterns for a mixture of linear molecules, open circles, and supercoils. The very rapid sedimentation of the supercoils is evident. If panel A had contained non-supercoiled covalent circles, they would not have been seen because they cosediment with open circles. However, the result in alkali would still be that shown in panel B. This provides the criterion for a non-supercoiled covalent circle, namely cosedimentation with a nicked circle at neutral pH and very rapid sedimentation in alkali.

The extraordinary separation of covalent circles in alkali is a useful property that is exploited in some of the examples that follow.

☐ Determination of the structure of the DNA of the *E. coli* F plasmid. **Example 11-O**

The *E. coli* F plasmid is responsible for the ability of a male strain to transfer its DNA to a female strain. Physical analysis has shown that it is a piece of DNA distinct from that of the chromosome and only about 1% as large. If a male strain of *E. coli* is grown for many generations in a medium containing [^{3}H]thymidine, all DNA is labeled. If these cells are lysed and the DNA is sedimented through alkaline sucrose, a rapidly sedimenting fraction (approximately 1% of the total DNA) is observed (Figure 11-41) whose *s* value is greater than that of the intact *E. coli* chromosomal DNA. Such rapid sedimentation is usually indicative of a covalent circle. A covalent circle has the

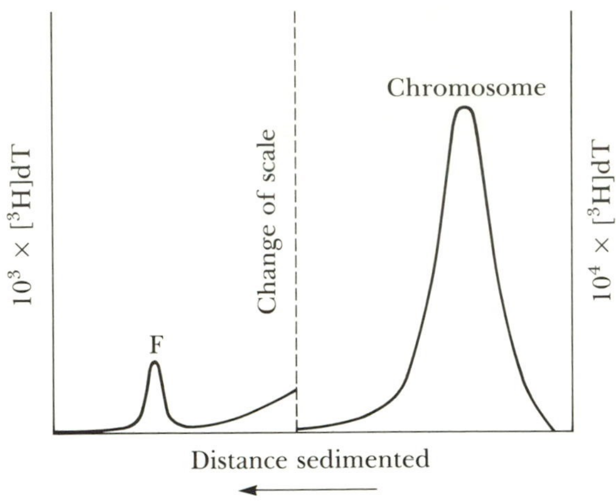

Figure 11-41
Detection of covalent circles of *E. coli* F plasmid DNA. Bacteria were grown in [^{3}H]thymidine, lysed, and sedimented through an alkaline sucrose gradient. The F covalent circles account for approximately 1% of the total DNA but are easily resolved because of their high *s* values in alkali.

property, described earlier, that one single-strand break—converting it into an open circle—allows the strands to separate in alkali and thereby sediment much more slowly. Indeed, several agents that are known to produce single-strand breaks result in the disappearance of this rapid peak. The kinetics of disappearance are first order, indicating that loss of the peak is a result of one single-strand break. Hence, it can be concluded that F is a covalent circle. From an empirical *s* versus *M* relation (for covalent circles in alkali), *M* can also be calculated.

Example 11-P ☐ Measurement of the rate of production of single-strand breaks by the measurement of the survival of covalent circles.*

In Example 11-N, single-strand breaks are detected by breakage of linear DNA molecules. This can be done more accurately by analyzing covalent circles. Bacteria containing the *E. coli* sex plasmid F′*lac* were labeled, X-irradiated, and sedimented, as in Example 11-0. The survival of F′*lac* covalent circles as a function of X-ray dose was measured by comparing the amount of radioactivity in the rapidly sedimenting fraction. The curve of log(survival) versus dose is shown in Figure 11-42. Because the curve is linear, the dose yielding a survival of 1/e represents that producing an average of one single-

* See Appendix 2 at the end of this chapter for a discussion of the analysis of survival curves.

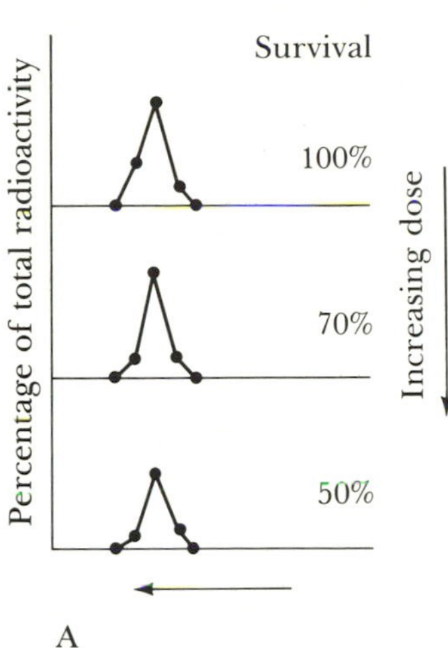

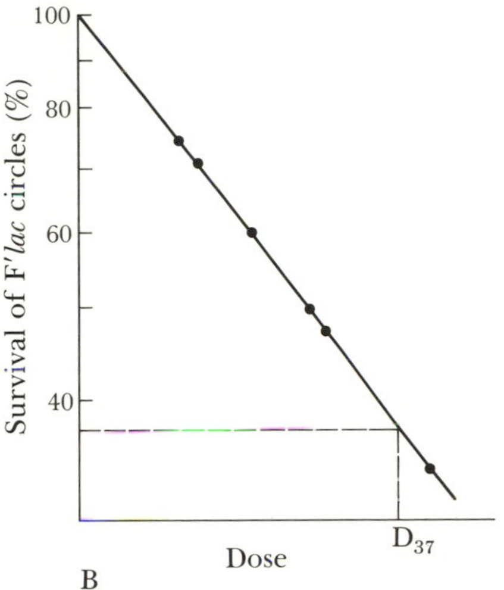

Figure 11-42
Effect of X-irradiating *E. coli* that contain *F′lac*. Part A shows the portion of the sucrose gradient containing the F′*lac* covalent circles, for three different X-ray doses. Part B shows a semilogarithmic plot of the survival of F′*lac* covalent circles; survival was determined from gradient analysis, as in part A. D_{37} is the dose giving a survival of 1/e, or 37%. At this dose, there is an average of one single-strand break per F′*lac* molecule.

strand break per covalent circle. Because *M* for F′*lac* is known, a value of the number of breaks per dose per unit of molecular weight can be calculated.

Determination of Molecular Weight *M* by the Sedimentation-Diffusion Method

Earlier in this chapter, it was shown that

$$s = m(1 - \bar{v}\rho)/f$$

in which f is the frictional coefficient. The frictional coefficient is very difficult to calculate theoretically but can be shown to be related to the diffusion coefficient, D, by the equation

$$D = kT/f \qquad (8)$$

in which k is the Boltzmann constant and T is the absolute temperature. These equations can be rearranged to yield

$$m = skT/D(1 - \bar{v}\rho) \qquad (9)$$

Hence, by measuring s and D (both extrapolated to zero concentration), m can be calculated.

To evaluate M, one need only substitute $M = N_A m$, in which N_A is Avogadro's number. Thus

$$\begin{aligned} M &= sN_A kT/D(1 - \bar{v}\rho) \\ &= sRT/D(1 - \bar{v}\rho) \end{aligned} \tag{10}$$

in which $R = N_A k$ is the gas constant. The coefficients s and D are either measured in the same solvent and at the same temperature (the better procedure) or corrected to water at 20°C. This is a reliable method.

The diffusion coefficient is not difficult to measure with modern instruments and is discussed in Chapter 12. However, for very large and extended molecules such as DNA, D is too small to be measured accurately.

In a procedure developed by Robert Baldwin and Kensall Van Holde (see Selected References and Chapter 12), D is measured directly from the spreading of the sedimentation boundary with time. In this way, M can be measured from a single sedimentation experiment. The procedure is only effective for compact molecules such as globular proteins and viruses, which have a high diffusion coefficient.

Measurement of Molecular Weight by Sedimentation Equilibrium

The sedimentation equilibrium method allows a direct determination of m because it eliminates the need to determine D. With this method, centrifugation is carried out at relatively low speeds so that sedimentation of the molecule is slow enough to be counterbalanced by diffusion—that is, the tendency of the centrifugal force to cause a decrease in concentration at the meniscus and an increase at the bottom of the cell is antagonized by diffusion, which tends to maintain the same concentration everywhere in the cell. At equilibrium the concentration is indeed lower at the meniscus and greater at the bottom but is unchanging. A description of the state of affairs at equilibrium is provided by the well-known Boltzmann distribution, which in terms of a system in a centrifugal field is

$$\frac{c_1}{c_2} = e^{-(E_1 - E_2)/kT} \tag{11}$$

in which c_1 and c_2 are the concentrations of solute molecules at distances r_1 and r_2 from the axis of rotation at which points the molecules have potential energies E_1 and E_2 and k is the Boltzmann constant. The potential energy difference, $E_1 - E_2$, is the work necessary to move a molecule of mass m from r_2 to r_1, that is, *against* the force field, or

$$E_1 - E_2 = -\int_{r_2}^{r_1} m(1 - \bar{v}\rho)\omega^2 r dr$$
$$= \tfrac{1}{2}m(1 - \bar{v}\rho)\omega^2(r_2^2 - r_1^2) \quad (12)$$

Combining equations (11) and (12) and replacing k by the gas constant R so that the units of m (which are now given as M) will be molecular weight units (daltons) instead of grams produces the following equation, which can be used to calculate M directly from centrifugation data:

$$M = \frac{2RT}{(1 - \bar{v}\rho)\omega^2} \ln \frac{c_r}{c_a} \cdot \frac{1}{r^2 - a^2} \quad (13)$$

in which c_r is the concentration of the solute at a distance r from the axis of rotation and a is the distance of the meniscus from the axis of rotation. Hence, a plot of ln c_r versus r^2 is a straight line from whose slope M can be calculated (Figure 11-43).

In the derivation, the system was assumed to consist of only two components and the variation in density due to changes in c and pres-

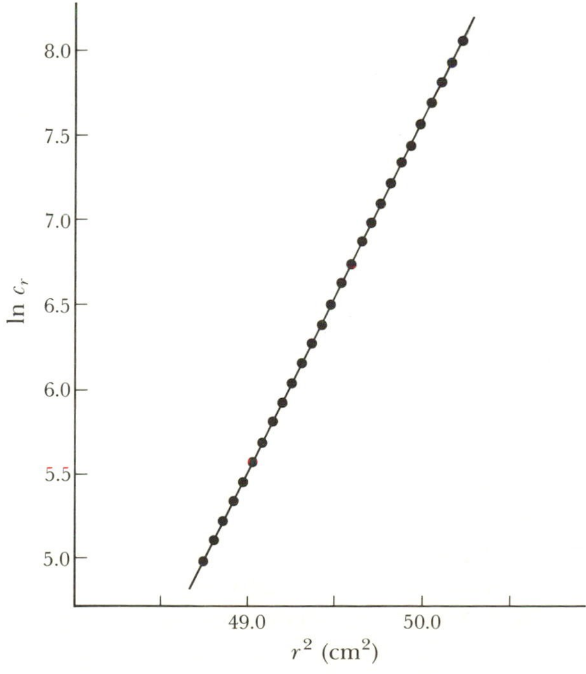

Figure 11-43
Sedimentation equilibrium data for bovine serum albumin in 0.1 M NaCl at 27,690 rpm and 20°C. Initial protein concentration was 650 μg/ml.

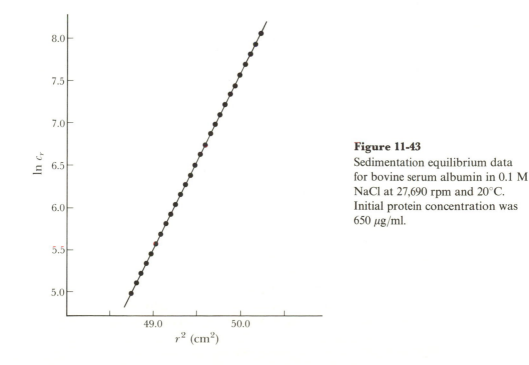

sure* was assumed to be negligible. Pressure effects can in fact be ignored as long as the speed of centrifugation is low; the variation in ρ will be small as long as the concentrations used are low. However, the counterion concentration must be kept low enough to satisfy the two-component assumption. One might ask why counterions should even be included. They are necessary because usually the solute molecule is charged so that an electrical potential gradient would be produced by the distribution of charged molecules. The counterion effectively neutralizes this gradient and eliminates the enormous complexities that would result if this were not possible.

The requirement for a two-component system also introduces a complexity when proteins are studied in denaturing conditions because often the denatured state is achieved by the addition of a very high concentration of a denaturant—for example, 6 M guanidinium chloride. Such a study might be carried out if the molecular weight of the monomer of a self-associating system (for example, a protein that dimerizes) is needed because association does not occur in denaturing conditions. Detailed theoretical corrections of equation (13) have been carried out. However, in the case of denaturants that do not bind to proteins, of which guanidinium chloride is one example, these corrected equations are not needed. Equation (13) remains applicable as long as the value of $\bar{v}$ (Chapter 13) is determined in 6 M guanidinium chloride. Even simpler, it has been shown [J. Lee and S. N. Timasheff, *Methods Enzymol.* 61 (1979): 49–57] that it is possible to calculate $\bar{v}$ in guanidinium chloride from the value of $\bar{v}$ in dilute salt solution and the amino acid composition of the protein.

The presence of associating monomers would be indicated by an upward curvature of a plot of the type shown in Figure 11-43. This curvature is a result of greater accumulation of dimers, trimers, and so forth, at the bottom of the cell. In some cases it is possible to determine the monomer: dimer ratio from analysis of this curvature.

The determination of c_r in sedimentation equilibrium is done by an analysis of an interference diagram. The procedure is related to that shown in Figure 11-11, parts E and F, but requires one modification. Figure 11-44A shows a schematic interference diagram for a protein at equilibrium. As in Figure 11-11, a single fringe is arbitrarily selected (the heavy fringe) and the y-displacement as a function of x is plotted. In order to calculate c_r, it is necessary to determine the constant of proportionality between concentration and fringe width. This cannot usually be done by examining the boundary shortly after centrifugation begins because the concentration is too low. This low concentration is needed in order to avoid having excessive concentration at the bottom of the cell at equilibrium. The following technique, shown in Figure 11-44B and C, is used. The centrifuge cell is filled with a high concentration of protein in a buffer. A small volume of the same buffer (prepared

* Pressure is a factor caused by the action of the centrifugal field.

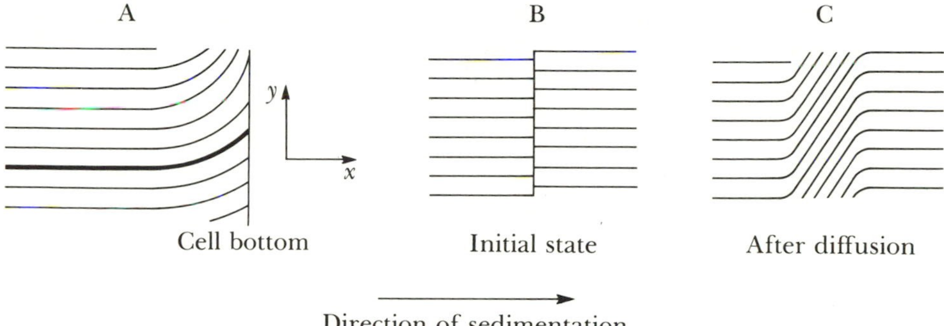

Figure 11-44
Calculation of concentrations in the sedimentation of equilibrium technique, using interference fringes. A. A single fringe (heavy line) is selected. B. The fringes in a synthetic boundary cell at the start of centrifugation. C. The same cell as in panel B, after diffusion has spread the boundary.

by exhaustive dialysis of the solution against the buffer) is layered on the protein solution. The layering is accomplished in a special centrifuge cell called a synthetic boundary cell (see Figure 12-7). The cell is centrifuged at a speed so slow that the molecules barely sediment. Diffusion occurs, however, so that the initially sharp boundary shown in panel B spreads to yield the pattern in panel C. Since the initial concentration is known, the relation between vertical fringe displacement and concentration can be measured as has been done in Figure 11-11E.

The advantages of the equilibrium method are that (1) it has a firm theoretical basis; (2) it requires only a small amount of material; and (3) the precision of measurement is very high (error $<1\%$). Its only minor disadvantage is the length of time required to reach equilibrium (16–24 hours).

Molecular Weights by Approach to Equilibrium (Archibald Method)

In the approach-to-equilibrium method (often called the "Archibald" method, after its inventor), as in the equilibrium method, molecular weight is obtained directly from the concentration distribution but in considerably less time. This is often an advantage if unstable molecules are being studied.

The equilibrium method was originally developed because the mathematical equations describing the transport of molecules in a centrifugal field become solvable for the condition of equilibrium in which there is no net flow at any point in the centrifuge cells. The Archibald method instead makes use of the fact that at all times (not only at equilibrium) there is no net flow of material at either the meniscus or the cell bottom. Although it is clear that, before equilibrium is reached, the macromolecule being studied is certainly moving away from the meniscus and accumulating at the cell bottom, this statement simply means that material neither leaves nor enters the cell. The mathematical consequences of this are that M can be determined by measuring the depletion of the macromolecules from the meniscus and the accumulation at the cell bottom *at any time*. This is accomplished by use of the equations

$$\frac{(\partial c/\partial r)_{\mathrm{a}}}{\omega^2 r_{\mathrm{a}}} = -\frac{M_{\mathrm{a}}(1 - \bar{v}\rho)(c_0 - c_{\mathrm{a}})}{RT} + \frac{M_{\mathrm{a}}(1 - \bar{v}\rho)c_0}{RT} \tag{14}$$

and

$$\frac{(\partial c/\partial r)_{\mathrm{b}}}{\omega^2 r_{\mathrm{b}}} = -\frac{M_{\mathrm{b}}(1 - \bar{v}\rho)(c_0 - c_{\mathrm{b}})}{RT} + \frac{M_{\mathrm{b}}(1 - \bar{v}\rho)c_0}{RT} \tag{15}$$

in which c is the concentration of the macromolecule at a distance r, c_0 is the initial concentration, r is the distance from the axis of rotation at which c is measured, and the subscripts a and b refer to the meniscus and bottom, respectively.

The advantage of the Archibald method is that only very short periods of sedimentation are required. A major drawback of the method is that measurements are made very near the meniscus and cell bottom where the precision of the data is poorest.

Sedimentation Equilibrium in a Density Gradient

So far, the molecules under consideration have had a density that is much more than that of the solvent. In this section, the important technique of sedimentation equilibrium in a density gradient (developed by Matthew Meselson, Franklin Stahl, and Jerome Vinograd) in which the density of the solvent is almost precisely that of the molecule under study is examined. This technique requires a third component (typically a cesium salt) of low molecular weight and high density. When the solution is centrifuged, the salt redistributes (according to equation 13) and reaches equilibrium, thus forming a concentration gradient that is less dense at the top and more dense at the tube bottom—that is, a density gradient. The macromolecules also sediment, but the material at the top of the cell moves centrifugally through the gradient and that at the bottom moves centripetally (see Figure 11-45). This continues until the macromolecules form a band at a position in the concentration gradient at which the density of the macromolecule equals the density of the solution (Figure

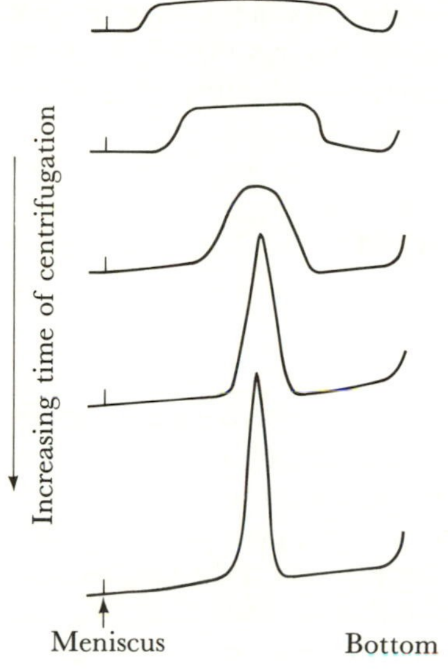

Figure 11-45
Movement of DNA molecules to the equilibrium position during centrifugation in a CsCl density gradient. The height of the curve is the concentration of DNA at each point in the cell. The area of the curve remains constant.

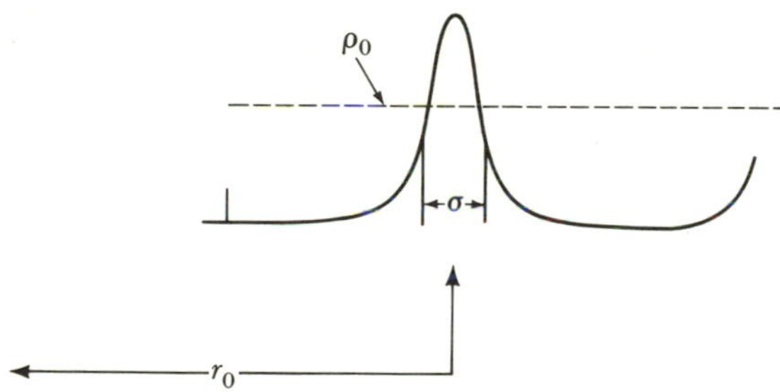

Figure 11-46
Parameters of a band in a CsCl gradient: the solid and dashed lines show the DNA concentration and the solution density, respectively; r_0 is the distance from the center of the band to the axis of rotation; ρ_0 is the density at the center of the band; σ is the band width at 1/e the height.

11-46). (In this case, the density of the macromolecule refers to the density of the molecular unit including water and the cesium salt.) The band width is determined by a balance between the centrifugal force (which tends to make the band narrow), diffusion (which tends to broaden the band), and the density gradient (a steeper gradient will give a narrower band).

To date, this technique has been used almost exclusively for DNA and bacteriophages; in a few cases, it has been used for RNA and ribosomes but for proteins its use is rare. The discussion herein will pertain primarily to DNA and phages. In a later section, a brief description of the use of density gradient centrifugation in the separation of lipoproteins will be given.

Measurement of Molecular Weight by Sedimentation Equilibrium in a Density Gradient

The theory of sedimentation equilibrium in a density gradient is complicated because (1) the salt used to generate the gradient is usually at a very high concentration (5–7 M), so that the mathematical simplicity accompanying the assumption that the system consists of only two components cannot be applied; (2) the positive and negative ions of the salt, because of differences in molecular weight and diffusion coefficient, do not distribute in the same way so that an electrical potential gradient is set up, and (3) the centrifugal speeds are so high that there are pressure effects. For these reasons, it has not been common to use this method to determine molecular weights. However, because it is possible to do so if the data are handled appropriately, the method will be described briefly.

If the fact that the system has three components is ignored, the following equation can be derived:

$$M = \frac{8RT\rho_0}{\sigma^2(\mathrm{d}\rho/\mathrm{d}r)\omega^2 r_0} \tag{16}$$

in which ρ_0 is the density at the center of the band, σ is the width of the band at 1/e the height, $\mathrm{d}\rho/\mathrm{d}r$ is the density gradient, ω is angular velocity in radians per second, and r_0 is the distance from the center of the band to the center of rotation. It has been repeatedly observed that this equation gives values of M that are too low (usually about twofold), the error increasing with increasing M. However, Carl Schmid and John Hearst showed that this is primarily a result of inaccuracies in measuring $\mathrm{d}\rho/\mathrm{d}r$ and of the effects of the concentrations of the macromolecules in the band. The problem of measuring $\mathrm{d}\rho/\mathrm{d}r$ is solved by determining, for a given macromolecule whose density is known, the separation of two bands, which differ in density by virtue of isotopic substitution (e.g., ^{14}N for ^{15}N or ^{13}C for ^{12}C). Then equation (16) can be used to obtain accurate values of M by measuring M at several concentrations c_i and then extrapolating to zero concentration, using the empirical equation

$$\ln M = \ln M(c_0) + kc_0 \tag{17}$$

in which $M(c_0)$ is the value of M measured at c_0 (the concentration at the center of the band), k is a constant, and c_0 is calculated from

$$c_0 = \frac{0.20 c_i L}{\sigma} \tag{18}$$

in which c_i is the initial concentration (before centrifugation) and L is the length of the liquid column, and σ is defined in equation (16).

The reader is cautioned to discount all measurements of the molecular weight of DNA made by density gradient equilibrium centrifugation before 1969 when Schmid and Hearst presented their work. At the present time, however, this method* and the direct measurements of length by electron microscopy are probably the most reliable means of determining M for DNA.

All of these considerations are valid only for materials that are homogenous with respect to density. Density heterogeneity will clearly result in a broader band (owing to a superposition of many nearby bands). This will give an artifically large value of σ, which will produce lower values of M. This heterogeneity can be eliminated if Cs_2SO_4 is used (see page 435); however, to date Cs_2SO_4 has not been used in molecular weight measurements.

When determining the molecular weight of a macromolecule by any procedure, one must know whether the value obtained includes the mass of the counterions and of the bound water molecules. In the case of equilibrium centrifugation, it is that of the unhydrated form but including the counterions—in this case, the Cs^+ ion. Thus, the molecular weight is that of CsDNA. As we have discussed elsewhere, for DNA, values of the molecular weight are by agreement given for NaDNA so that for consistency the measured value should be corrected to NaDNA.

Sedimentation in Density Gradients to Determine Density

The greatest value of equilibrium centrifugation in density gradients is to resolve material according to density. For example, DNA containing the isotope ^{15}N can be separated from [^{14}N]DNA. If sedimented to equilibrium in a buoyant CsCl solution, the density difference of 0.014 g/cm^3 results in a separation of [^{14}N]- from [^{15}N]DNA of about 0.5 mm in a standard centrifuge cell at 40,000 rpm. Other isotopes produce even greater separation, as indicated in Table 11-1.

The value of this ability to separate DNA molecules by density differences can best be seen in the first example of its use by Matthew Meselson and Franklin Stahl (the method was actually devised to do this experiment).

* However, this method is not commonly used because of the memory of the twofold error just discussed.

Table 11-1
Density Increments in CsCl for DNA Containing Various Isotopes or a Substitute Base.

Isotope*		Density (g/cm^3)
^{15}N	(100%)	0.014
^{13}C	(100%)	0.049
^{13}C	(60%)	0.030
^{2}H	(95%)	0.056
^{15}N^{2}H	(95%, 95%)	0.072
^{15}N^{13}C	(95%, 60%)	0.046
Substitute base: 5-bromouracil	(100%)	0.119

* Number in parentheses refers to percentage substitution for normal isotope.

Example 11-Q □ The Meselson-Stahl experiment demonstrating semiconservative replication of DNA.

Bacteria grown in a medium containing ^{15}N for many generations were shifted to one containing ^{14}N for two generations. DNA samples were isolated at various times and analyzed in CsCl density gradients. As the bacteria grew, the density of the isolated DNA shifted first to that of the average of [^{14}N]- and [^{15}N]DNA (this is called the hybrid density), and then a band at the density of [^{14}N]DNA appeared (Figure 11-47). The shift in density as a function of the number of generations of growth gave clear indication that DNA had an even number of subunits and was replicated semiconservatively (i.e., the single strands but not the double strands remain intact).

Another use of isotopic labeling to study replication follows.

Example 11-R □ Isolation of replicating DNA molecules.

An elegant procedure was used by Jun-ichi Tomizawa and Hideyuki Ogawa to isolate intact, replicating DNA molecules. *E. coli* phage λ, containing the naturally occurring isotopes ^{1}H and ^{14}N, were used to infect bacteria in a growth medium containing ^{2}H and ^{15}N. The infected cells remained in the ^{2}H, ^{15}N medium for a short time. All bacterial DNA had the ^{2}H,^{15}N density; however, if the λ DNA had replicated once, it would have had a density corresponding to one strand of [^{2}H,^{15}N]- and one strand of [^{1}H,^{14}N]DNA, as in Example 11-Q. If, however, the λ had not replicated once but had partially replicated, its density would have shifted to a slightly higher value. By centrifugation of the DNA isolated from the infected cells and selection of fractions containing DNA with a small density shift from

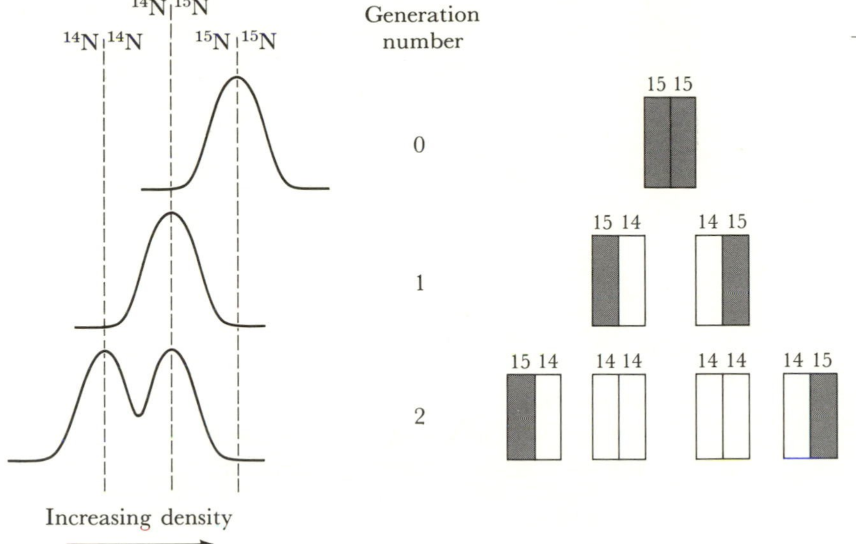

Figure 11-47
The type of data obtained in the Meselson-Stahl experiment [*Proc. Natl. Acad. Sci. U.S.A.* 44(1958):671–682]. *E. coli* were grown for many generations in ^{15}N medium and then at zero time transferred to ^{14}N medium. After one generation, all DNA had hybrid density. DNA of less than hybrid density is not seen before the first generation because the chromosome is fragmented into about 250 pieces during isolation. At the right, the state of the isolated DNA at various times is shown.

the position of unreplicated λ DNA, partially replicated molecules were isolated. These were then examined by electron microscopy and the first photographs of replicating DNA were obtained.

Density differences in CsCl solutions also arise from differences in the percentage of guanine·cytosine (G·C) pairs in the DNA (Figure 11-48). This has been used to identify minor species of DNA. For example, the mean G + C content of the DNA of the crab *Cancer borealis* is 42%. However, if the isolated DNA is centrifuged in CsCl, it forms two bands (Figure 11-49), one corresponding to the major fraction of chromosomal DNA (58% G + C) and the other to the minor species (3% G + C). Such minor bands are called *satellite bands* and are widespread in nature. Satellite bands represent several different things. Frequently they consist of the DNA of mitochondria and chloroplasts whose G + C content is different from that of nuclear DNA. In one bizarre case a satellite band was found in the DNA of a particular mouse, which was found to have a systemic infection with a DNA-containing virus. Another major class of satellite DNA consists of iterated sequences—that is, a particular

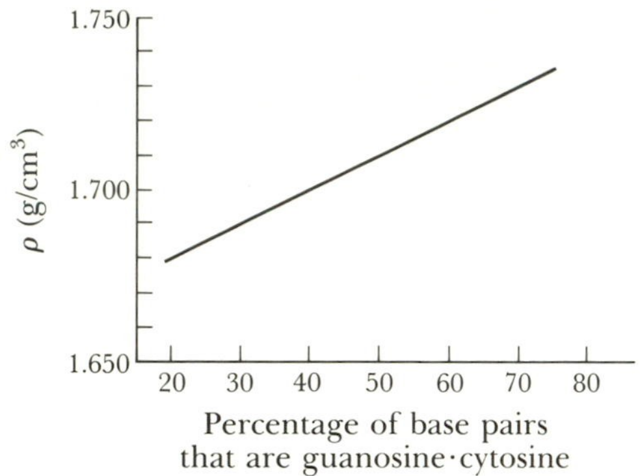

Figure 11-48
Density (ρ) of DNA in CsCl as a function of guanine + cytosine content. The equation that describes the curve is $\rho = (0.098)[G + C] + 1.660\ g/cm^3$, in which [G + C] is the mole fraction of G + C in native DNA. [From C. L. Schildkraut, J. Marmur, and P. Doty, *J. Mol. Biol.* 4(1962):430–443.]

Figure 11-49
Photometric trace of the result of equilibrium centrifugation in CsCl of the DNA of the bacterium *E. coli* and the crab *Cancer borealis*. As is common with bacteria, the DNA has a narrow range of densities. The crab DNA consists of two discrete fractions, one of very low density. This minor band is called *satellite DNA*. [From N. Sueoka, *J. Mol. Biol.* 3(1961):31–40.]

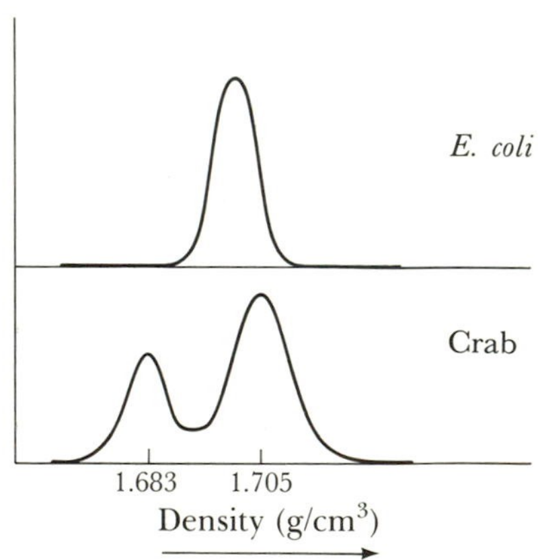

sequence of bases, corresponding to one or a small number of genes, that is tandemly repeated hundreds or thousands (and occasionally millions) of times.

The first successful isolation of plasmid DNA made use of the dependence of density on G + C content.

The densities of the DNA of the bacteria *E. coli* and *Serratia marcescens* are 1.708 g/cm³ and 1.718 g/cm³ respectively. Julius Marmus isolated the F′*lac* plasmid (whose density is also 1.708 g/cm³) of *E. coli*

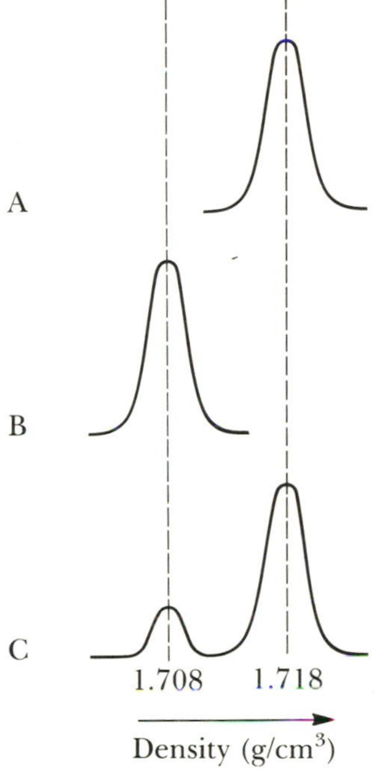

Figure 11-50
Banding patterns in CsCl of DNA isolated from (A) *Serratia marcescens*, (B) *E. coli*, and (C) *S. marcescens* containing the *E. coli* F′*lac* plasmid.

in the following way. Using genetic techniques the F′*lac* plasmid was transferred to *Serratia*; a large batch of these bacteria were grown and the DNA was prepared and centrifuged in CsCl. The pattern shown in Figure 11-50 was observed. Two bands were seen—the minor band had the density of *E. coli* DNA. The centrifuge tube was fractionated by drop collection (Figures 11-6 and 11-24) and the F′*lac* DNA was thereby isolated.

In CsCl solutions, single-stranded DNA has a density that is from approximately 0.015 g/cm^3 to 0.020 g/cm^3 greater than that of double-stranded DNA. This fact can be used either as an analytical tool or to purify either single- or double-stranded DNA. This is shown in the following example.

Example 11-S

☐ Demonstration of an early step in the infection of *E. coli* by phage ϕX174.

The phage ϕX174 contains single-stranded DNA. Shortly after infecting the bacterium *E. coli* with ϕX174, whose DNA is radioactive, analysis of radioactive intracellular DNA by centrifugation in CsCl indicated that the radioactivity at the position of single-stranded DNA decreases in density to the position of double-stranded DNA. This provided evidence that ϕX174 DNA serves as a template for the synthesis of a double-stranded replicating form, which was later isolated and studied further in the laboratory of Robert Sinsheimer.

Note that radioactive ϕX174 DNA was used in order to distinguish this DNA from the great excess of *E. coli* DNA, whose density is close to that of ϕX174 DNA.

The densities of DNA and protein are quite different in CsCl solutions—roughly 1.7 g/cm^3 and 1.3 g/cm^3. Hence, banding in CsCl can be used to separate these molecules. This has been a useful procedure for preparing protein-free DNA. Another important consequence of this density difference is that nucleoproteins such as phages and viruses have a density that reflects their ratio of nucleic acids to protein. For example, a phage that is half protein and half DNA has a density of $\frac{1}{2}(1.3) + \frac{1}{2}(1.7) = 1.5$ g/cm^3, and one that is 60% protein and 40% DNA has a density of $0.6(1.3) + 0.4(1.7) = 1.46$ g/cm^3. The density of viruses is sufficiently different from all macromolecular components of cells that banding in CsCl whose density is 1.45–1.55 g/cm^3 has become the standard procedure for purifying phages and viruses. With this procedure, free DNA and RNA molecules are found on the tube bottom (usually as a gelatinous pellet) and proteins float on the top surface of the liquid forming a scum that is easily removed with forceps.

Using this principle, bacteriophage λ mutants that contain less DNA by virtue of a genetic deletion have been isolated by CsCl banding by Jean Weigle and Grete Kellenberger. Because the protein content of a λ phage is unchanged by a reduction in nucleic acid content, a deletion phage has a lower density than the original phage. Phage particles containing excess DNA have also been isolated because they have a higher density. This not only provides a tool for isolating interesting mutants, but allows different phages to be distinguished in physical experiments—that is, relative DNA content becomes a density label. This can be seen in the following example.

Example 11-T ☐ Recombination studies with *E. coli* phage λ.

Phage λ contains a site in its DNA called *att* at which a recombinational exchange occurs, catalyzed by an enzyme called integrase. It was believed that this exchange involves the breakage of two DNA molecules and the joining of the fragments. This hypothesis was tested by performing a cross between a wild-type phage and a mutant that contains two deletions (b2 and b5) as shown in Figure 11-51. These deletions are large enough to have a significant effect on the phage density in CsCl. In the experiment to be described, all other recombination systems were eliminated by mutation so that the integrase system alone was active. As shown in the figure, λb2b5 labeled with [^{3}H]thymidine was crossed with wild-type nonradioactive phage in *E. coli.* After the infected cells lysed, the released phages were analyzed in CsCl. Four phage types can result from such an infection: b2b5 (parent), b2$^+$b5$^+$ (parent), and the recombinants, b2b5$^+$ and b2$^+$b5. Furthermore, they either will contain all

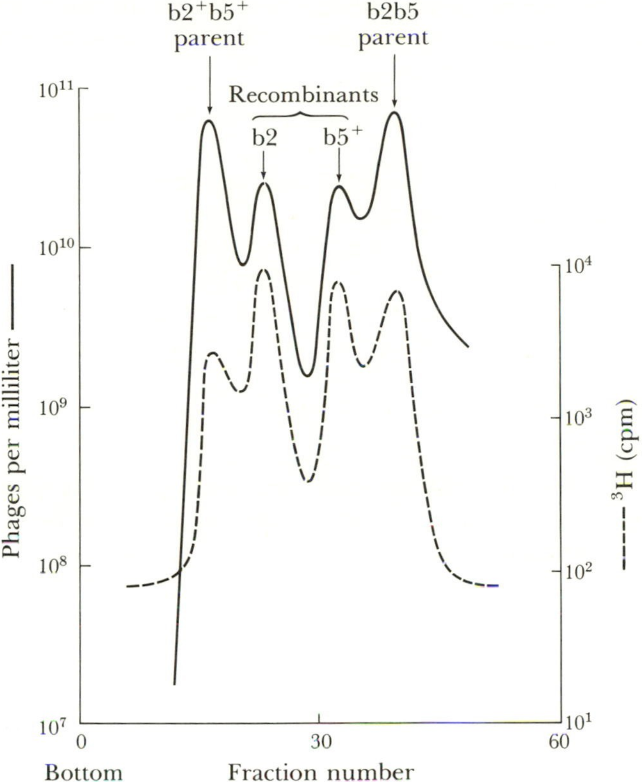

Figure 11-51
Resolution of parental and recombined λ phage by centrifugation in CsCl. Phage $\lambda b2b5$ labeled with ^{3}H was crossed with unlabeled λb^+. The results show that there is ^{3}H associated with the recombinant DNA, indicating that parental DNA can get into the recombinants. [From F. Kellenberger-Gujer and R. A. Weisberg, in *The Bacteriophage Lambda*, edited by A.D. Hershey, pp. 407–415, Cold Spring Harbor Laboratory, 1971.]

parental DNA (i.e., unreplicated) or will be replicas. Although a great deal of information can be obtained from this experiment, the question here is simply whether parental DNA is ever found in the recombinants. Because all four phage types have different densities, they are readily distinguished and it is necessary to note only whether ^{3}H is present in the bands corresponding to the recombinants. Figure 11-51 verifies that it is by showing that the DNA from the b2b5 parent appears in the recombinants. If the experiment is repeated with the ^{3}H in the $b2^+b5^+$ parent, ^{3}H is again found in the recombinant. This shows that both parents contribute DNA to the recombinants.

In many experiments, the natural density difference between two DNA molecules is inadequate and the density difference must be enhanced. This can be done in several ways.

Separation of DNA molecules in solutions of Cs_2SO_4 *containing* Hg^{2+} or Ag^+ *salts.* The Hg^{2+} ion binds strongly to the adenine·thymine (A·T) base pair. Because this ion is very dense, the density of the Hg^{2+}-A·T complex is even greater than that of a G·C pair. With the proper concentration of Hg^{2+}, the density difference can be substantially increased. The Ag^+ ion binds to G·C pairs and can be used in a similar way. Cs_2SO_4 is used rather than CsCl to avoid precipitation of $HgCl_2$ or AgCl. This technique is used in the following example.

Example 11-U □ Isolation of selected regions of phage λ DNA by centrifugation in Cs_2SO_4 containing Hg^{2+}.

If *E. coli* phage λ DNA is broken into small pieces by shearing or sonication (see Chapter 19), the fragments can be separated into different density classes by centrifugation in Cs_2SO_4 containing Hg^{2+}. This fractionation occurs because different regions of λ DNA have different ratios of A·T to G·C pairs. The extraordinary separation produced by this method is shown in Figure 11-52.

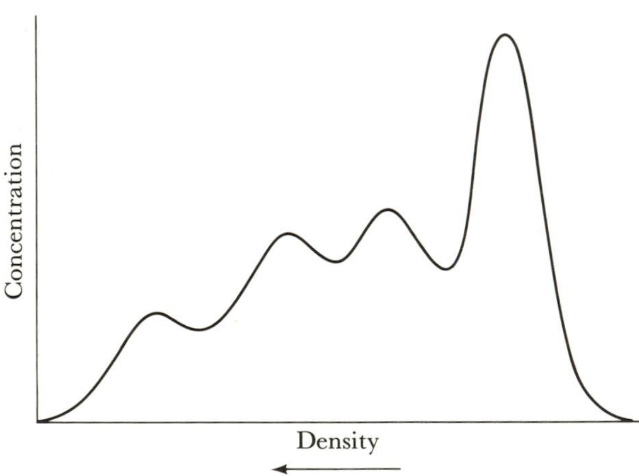

Figure 11-52
Equilibrium sedimentation of λ DNA fragments in Cs_2SO_4 containing Hg^{2+}: (A) various bands are seen that correspond to different regions of the DNA of *E. coli* phage λ. Note that high-(G + C) DNA has a lower density than low-(G + C) DNA in Cs_2SO_4–Hg^{2+} in contrast with CsCl. By similar studies of fragments derived from purified left and right halves and of quarters of λ DNA, the various density classes can be assigned to specific regions of the λ DNA. [From A. M. Skalka, A. D. Hershey, and E. A. Burgi, *J. Mol. Biol.* 34(1968): 1–16.

Separation of the individual polynucleotide strands of DNA by complexing with polyribonucleotides and centrifuging in Cs_2SO_4. The density difference between individual polynucleotide strands is rarely great enough to allow separation in density gradients. However, these strands frequently contain purine or pyrimidine tracts (i.e., base sequences consisting of only purines or only pyrimidines in a single strand). If a DNA molecule contains predominantly purine tracts on one strand (and therefore pyrimidine tracts on the other), the addition of a poly(ribopyrimidine nucleotide) will result in binding to one of the strands and not to the other. Polyribonucleotides have a very high density in Cs_2SO_4 so that the density of one of the DNA strands is markedly increased compared with the other. After separation in Cs_2SO_4, the polyribonucleotide can be easily removed by an enzyme such as RNase or by digestion of the polyribonucleotide with alkali. In this way, pure, separated strands of DNA can be obtained.

Example 11-V

□ Fractionation of the two complementary strands of phage λ DNA.

If λ DNA is heated and denatured to separate the strands, mixed with poly(I,G) (a mixed polymer of inosinic and guanylic acids), and then sedimented to equilibrium in Cs_2SO_4, the two complementary strands separate, as shown in Figure 11-53. In this case, both single strands of λ DNA bind poly(I,G). This is evident because with the bands in the figure have a higher density than the denatured DNA. Separation occurs because the two strands have different numbers of pyrimidine tracts so that the density shift of one strand is greater than that of the other strand.

Separation of density-labeled DNAs having different A+T *content with* Cs_2SO_4. Occasionally, it is necessary to distinguish DNA molecules that not only have a density label but also differ in base composition. However, if a DNA whose A+T content produces a density in CsCl of 1.700 g/cm^3 and another whose density is 1.715 g/cm^3 were labeled with ^{15}N and ^{14}N, respectively, their densities would be 1.714 g/cm^3 and 1.715 g/cm^3, which would be inseparable. However, in Cs_2SO_4 solution, the density is (for unknown reasons) insensitive to base composition. Hence, in Cs_2SO_4, only the isotopic differences will be detected.

Separation of glucosylated from nonglucosylated DNA in Cs_2SO_4 *gradients.* Some phage DNA molecules contain glucosylated bases—for example, the *E. coli* phages T2, T4, and T6 contain 5-hydroxymethylcytosine instead of cytosine, and this base is glucosylated. The glucose decreases the density in CsCl slightly (0.002 g/cm^3), but in Cs_2SO_4 gradients (for reasons that are unclear) the difference is considerably greater (0.020 g/cm^3). This difference has been used to follow the synthesis of glucosylated phage DNA in hosts whose DNA is nonglucosylated.

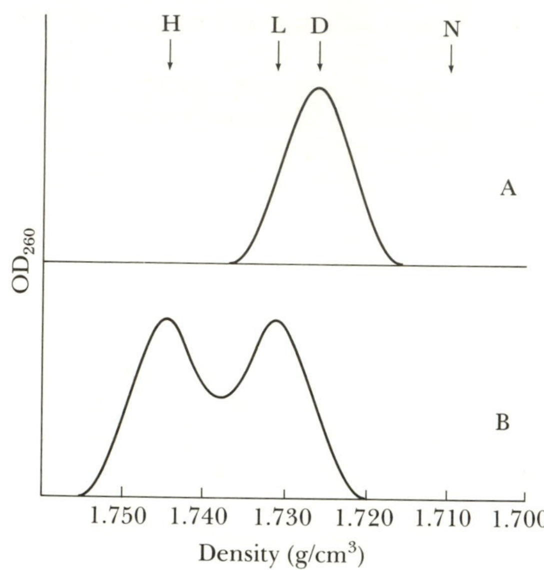

Figure 11-53
Separation of the complementary strands of phage λ DNA by complexing with poly(I,G) (mixed copolymer of inosinic acid and guanylic acid). Phage λ DNA was denatured by heating to 95°C. In part A, this DNA is sedimented to equilibrium in CsCl and forms a single band at position D (denatured); N is the position for native DNA. In part B, the denatured DNA is mixed with equimolar amounts of poly(I,G) before centrifugation in CsCl. The density increased because poly(I,G) has a higher density than DNA. The two strands, H (heavy) and L (light), bind different amounts of poly(I,G) and therefore band at different densities. The two bands can be collected by drop collection and freed from poly(I,G) by enzymatic digestion with pancreatic RNase to yield pure H and L strands. In the early literature, the H and L strands are labeled C (Crick) and W (Watson). This widely used technique was developed by Waclaw Szybalski.

Separation of covalently closed, circular DNA molecules from linear DNA or DNA containing single-strand breaks by centrifugation in CsCl *containing ethidium bromide or propidium iodide.* Ethidium bromide (Figure 11-54) binds very tightly to DNA in concentrated salt solutions and, in so doing, decreases the density of the DNA by approximately 0.15 g/cm^3. It binds by intercalation between the DNA base pairs and thereby causes the DNA molecule to unwind as more of the ethidium bromide is bound. A covalently closed, circular DNA molecule has no free ends so that, as it unwinds, the entire molecule twists in the opposite direction to compensate. For example, an O-shaped molecule that has bound enough ethidium bromide to produce one clockwise turn will twist in the counterclockwise direction to produce a molecule shaped like the figure eight. As more and more of the molecules intercalate, the

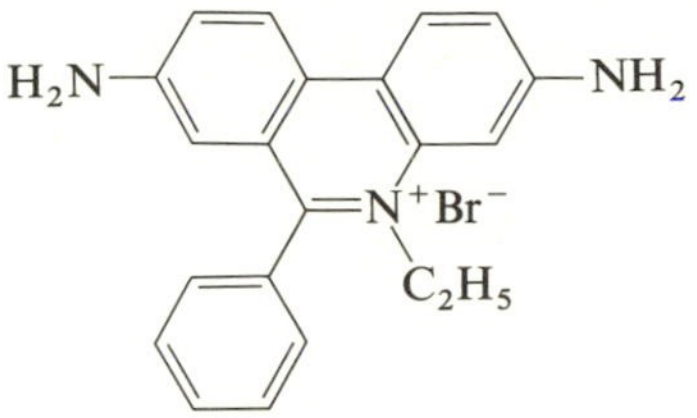

H_2N — NH_2 — N^+Br^- — C_2H_5

Ethidium bromide

Figure 11-54
Chemical formula for ethidium bromide.

8-shaped molecule will become more twisted. Ultimately, the DNA molecule is unable to twist any more so that no more unwinding is possible. Therefore, no more ethidium bromide molecules can be bound. On the other hand, a linear DNA molecule or a circular DNA with one or more single-strand breaks does not have the topological constraint of reverse twisting and can therefore bind more of the ethidium bromide molecules. Because the density of the DNA and ethidium bromide complex decreases as more is bound and because more ethidium bromide can be bound to a linear molecule or an open circle than to a covalent circle, the covalent circle has a higher density at saturating concentrations of ethidium bromide. Therefore, covalent circles can be separated from the other forms in a density gradient as shown in Figure 11-55. Propidium iodide introduces a greater density difference.

This method, developed in the laboratory of Jerome Vinograd, has become one of the most widely used tools in biochemistry and molecular biology both as a means of purifying covalent circles and as an analytic tool.

Example 11-W

☐ Demonstration that resistance transfer factors are plasmids.

Many strains of *E. coli* are simultaneously resistant to several antibiotics. This capability is transmitted from a resistant cell to a sensitive one, which suggests that the resistant cells contain a transferable DNA plasmid. If DNA is isolated from both a resistant and a sensitive cell and analyzed in CsCl containing ethidium bromide, it is found that the resistant cell contains a small amount of denser material not present in the sensitive cell. If a sensitive strain becomes resistant, it acquires this extra band. Therefore, the information for drug resistance is carried on a covalent circle of DNA. From the relative amount of material in the denser band, a minimum value can be calculated for the fraction of total DNA that is plasmid DNA. It is a minimum value because, in the course of DNA isolation, single-strand breaks might have been introduced in some covalent circles, which would result in the movement of the plasmid DNA to the main band. If

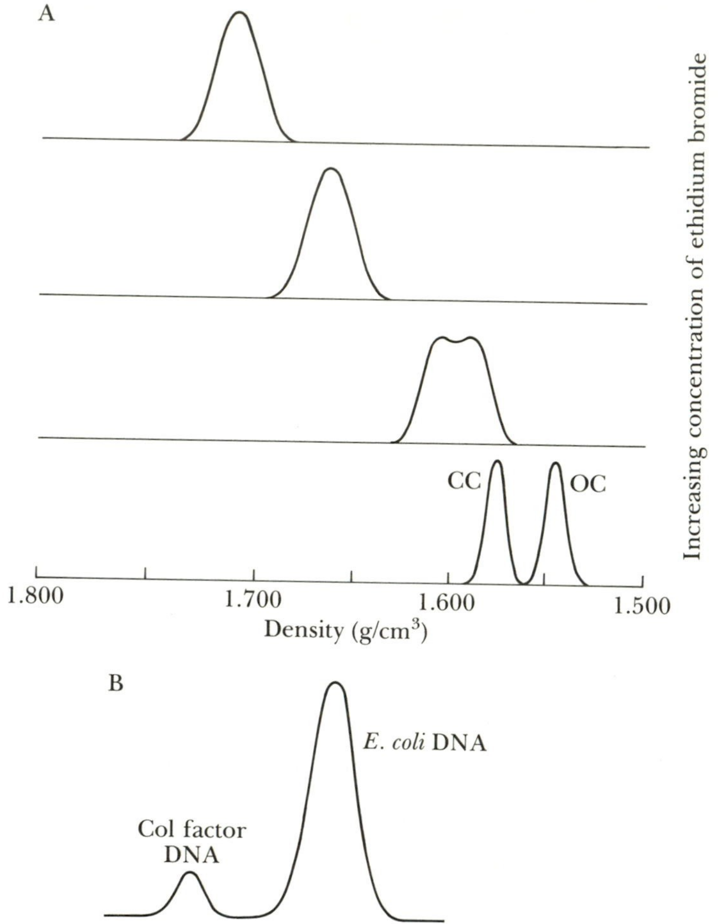

Figure 11-55
Effect of ethidium bromide on the density of DNA in CsCl. In part A, a mixture of equal amounts of open circles (OC) and covalent circles (CC) is centrifuged in CsCl containing various amounts of ethidium bromide. The density decreases until, at saturation, the two components separate. The covalent circles bind less ethidium bromide and are therefore at a higher density. Part B shows a photometric tracing of the DNA extracted from an *E. coli* strain containing ethidium bromide. Note that this differs from Figure 11-49, in which satellite DNA is separated from crab DNA, in that the Col factor DNA and the *E. coli* have the same density and separate here only because the Col factor DNA is a covalent circle.

the plasmid band is removed from the gradient, the DNA molecules can be visualized by electron microscopy and the molecular weight determined. Because the molecular weight of the *E. coli* chromosome is known, it is then possible to calculate the minimum number of plasmid DNA molecules per cell. This procedure is an alternative to the method described in Example 11-O.

☐ Distinction between DNA dimers and catenanes. **Example 11-X**

DNA molecules sometimes interact to form circular structures having twice the molecular weight of the monomer. Such a dimeric structure might be a true circular dimer or it might be two linked monomers (a catenane). Both will band at the same position in CsCl containing ethidium bromide. However, consider the effect of introducing one single-strand break into either structure. If the molecule is a dimeric covalent circle, it will shift in the gradient to the position of a nicked circle (Figure 11-56). However, if it is a catenane, one single-strand break will convert one of the units into an open circle; hence, the density will be an average of the densities of a covalent and a nicked circle—that is, half-way between the two positions. If the catenanes

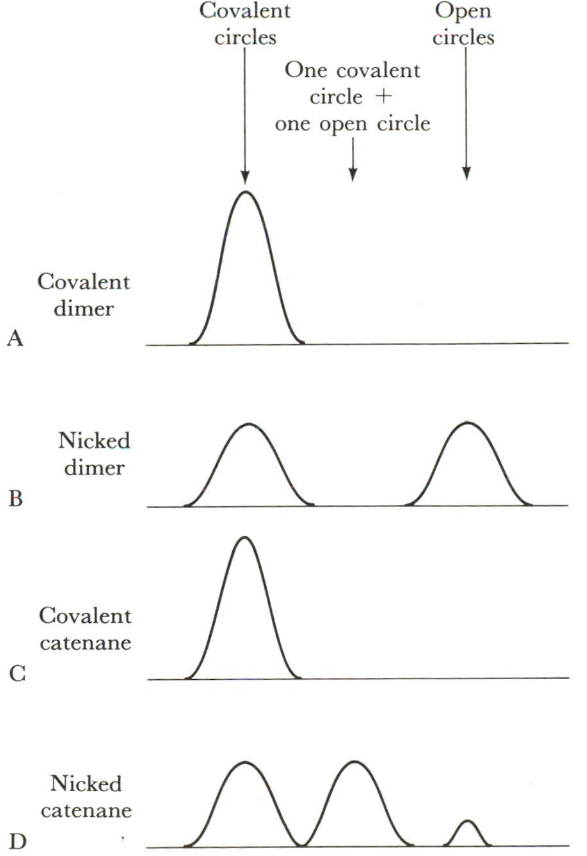

Figure 11-56
Method for distinguishing a covalent dimer from two linked covalent monomers (a covalent catenane) by centrifugation in CsCl containing ethidium bromide. The covalent dimer and the catenane band at the same position (A, C). However, if a single-strand break is introduced into half of the molecules (by X rays or, enzymatically, with DNase), the covalent dimer becomes an open-circular dimer (B), but the covalent catenane will be a mixture of, mainly, one covalent circle linked to one open circle and a few linked open circles (D).

consist of two linked units of unequal molecular weight, the introduction of single-strand breaks will produce two bands of intermediate density, corresponding to the two combinations of covalent and open circles.

Matthew Meselson and I used this technique to show that prophage DNA is inserted into bacterial DNA. A strain of *E. coli* that contains a circular sex factor with the attachment site for phage λ was lysogenized. The sex factor with the attached prophage was isolated and by a combination of centrifugation in alkaline sucrose gradients and in CsCl containing ethidium bromide was shown to be a larger circular structure. However, these tests did not reveal whether the lysogenic form was a larger circle or a catenane consisting of a linked sex factor and a λ circle. When single-strand breaks were introduced by X rays, the circles shifted in density to that of linear DNA—no intermediate density bands were found. Hence, it was concluded that the lysogenized form is a continuous circle.

Use of Density Gradients in the Study of Lipoproteins

The use of equilibrium centrifugation in a density gradient is rather uncommon for proteins for several reasons: (1) the density of most proteins is very nearly the same; (2) it is difficult to prepare a buoyant solution having the appropriate density without having an ionic strength that exceeds a value that can be tolerated by most proteins. This is not the case for lipoproteins, whose densities are usually quite low (< 1.1 g/cm^3) and for which the density depends upon the amount of liquid. It is possible to prepare buoyant solutions using solutions containing various amounts of NaCl and sucrose. Solutions of these molecules do not easily form density gradients as do CsCl solutions so that it is necessary to prepare concentration gradients, as in zonal centrifugation using the apparatus shown in Figure 11-24. A solution of lipoproteins is layered on the concentration gradient. After about 48 hours at 30,000–40,000 rpm, the lipoproteins have usually come to rest at their equilibrium positions.

Purification of Specific Types of Animal Cells by Zonal and Equilibrium Centrifugation

Many cells, such as those in blood, have different sedimentation coefficients and are separable by zonal centrifugation. However, when centrifuging intact cells, it is necessary that the suspending fluid have the same osmotic pressure as the cells because shrinkage and swelling (even to the point of cell rupture) can occur in hypertonic and hypotonic solutions respectively. Furthermore, the suspending medium should not contain molecules that can penetrate and thereby alter the structure of the cells. Except for rare cases, these requirements eliminate the use of sucrose and glycerol to form a density gradient.

Early attempts to separate cells by zonal centrifugation used concentration gradients of bovine serum albumin (BSA). This was reasonably successful in separating the cells of peripheral blood. The principal disadvantage in the use of bovine serum albumin was the difficulty of maintaining sterility of the solutions. The material of choice at present is Ficoll (Pharmacia), a polysaccharide that cannot penetrate biological membranes. Solutions of the appropriate concentrations have a very low osmotic pressure and thereby permit the construction of density gradients whose osmotic pressure is the same as that of the cells.

Buoyant density gradients have also been prepared with another polysaccharide, Percoll (also made by Pharmacia). Solutions of Percoll are harmless to cells, have a very low viscosity, and can be prepared at physiological pH and osmotic pressure. Equilibrium sedimentation has been done at unit gravity (That is, without a centrifuge) or, rapidly, at 10–100 *g*.

Both Ficoll and Percoll have been widely used in the separation of the cells of peripheral blood and in the isolation of particular cell types from complex tissues. They have been particularly useful in the purification of osmotically sensitive intracellular organelles such as chloroplasts, mitochondia, nuclei, and secretory vesicles.

Centrifugation of Small Volumes—The Airfuge

Most centrifuge rotors are not capable of handling volumes less than four milliliters although with special adaptors one-milliliter samples can be centrifuged in swinging-bucket rotors. For rapidity of sedimentation, angle rotors are more effective because of the shorter distance of sedimentation. To accomodate small volumes in angle rotors, Beckman Instruments has made commercially available the Airfuge, a table-top instrument utilizing an air-driven, magnetically suspended rotor. This instrument employs a rotor that is 3.7 cm in diameter and holds six samples whose volume is at most 0.24 ml. The rotor reaches 95,000 rpm in thirty seconds.

The main use of the Airfuge has been to pellet material rapidly either to collect the sediment or to produce a clarified supernatant. The instrument is very effective because of the high speed and short sedimentation distance. It has also been adapted for clinical assays in which only small samples of body fluids or tissues are available. In addition, two very important new techniques have been developed—one for particle counting and the other for measuring the molecular weights of proteins. A description of these procedures follows.

The counting of particles such as viruses has always been difficult owing to a variety of technical problems. With the development of a specially adapted Airfuge rotor, particle counting has become straightforward. This rotor contains six horizontal 0.1-ml cavities, at the end of which can be placed a support for microscopy, namely, a small piece of a membrane filter, a glass cover slip, or an electron-microscope grid. The cavity is sector-shaped so that when sedimentation is complete, all of the particles are spread uniformly on the bottom of the cavity. Since the area of the bottom is accurately known, a microscopic count of the number of particles per unit area provides the value of the total number of particles sedimented. The volume of added solution is measured exactly so that the initial particle concentration is also known.

The determination of molecular weights of proteins has been discussed in the section concerned with equilibrium centrifugation. The technique is accurate and straightforward and has three disadvantages—it requires a somewhat high concentration of protein, the sample must be pure, and a difficult measurement of the partial specific volume (See Chapter 12) must be made. A new technique developed in the laboratory of Howard Schachman eliminates these disadvantages although at a small reduction in precision [M. A. Bothwell, G. J. Howlett, and H. K. Schachman, "A sedimentation equilibrium method for determining molecular weights of proteins with a tabletop high speed air turbine centrifuge," *J. Biol. Chem.*

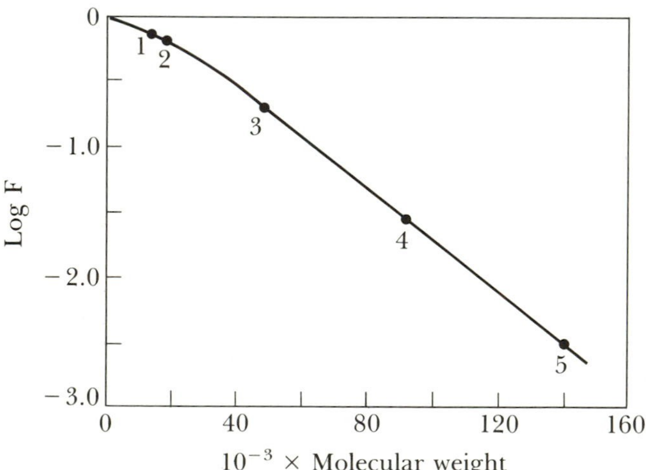

Figure 11-57
The relation between molecular weight of proteins and the fractional depletion of material in the upper portion of an Airfuge tube after centrifugation to equilibrium. The proteins are: (1) cytochrome *c*; (2) lysozyme; (3) peroxidase; (4) alkaline phosphatase; (5) glyceraldehyde-3-phosphate dehydrogenase. [Data taken from the paper by Bothwell et al., referenced in the text, by permission of the author and the publisher.]

253 (1978): 2073–2077]. A sample of a particular protein is prepared and its concentration determined. It is not necessary to know the concentration in grams per milliliter—any assay such as radioactivity, enzyme activity or mass per milliliter is acceptable. A 100-microliter sample is centrifuged in the Airfuge until equilibrium is achieved (usually 8–18 hours). At the end of the period of centrifugation, the uppermost 40 microliters of the solution are removed and assayed for protein concentration by the same technique used to measure the initial concentration. The result is recorded as the logarithm of the fraction F of the initial concentration that is present in these 40 microliters. Several of the tubes contain proteins who molecular weight is known. A graph of log F versus molecular weight is prepared for these known proteins. This empirical plot is used to determine the molecular weight of an unknown protein. The kind of data obtained is shown in Figure 11-57. For enzymes for which there is a sensitive assay it is possible to determine the molecular weight using as little as 10^{-10} g (or possibly less) of protein whereas in traditional equilibrium centrifugation, 10^{-4}–10^{-3} g is needed.

APPENDICES

1. Determination of s

The definition of s (from equation 2) is

$$s = \frac{v}{\omega^2 r} = \frac{1}{\omega^2 r} \cdot \frac{\mathrm{d}r}{\mathrm{d}t}$$

in which $\mathrm{d}r/\mathrm{d}t$ is the rate of movement of the particle. If the boundary is at r_0 at $t = t_0$ and r_1 at some time t_1, then

$$\int_{t_0}^{t_1} s\mathrm{d}t = \frac{1}{\omega^2} \int_{r_0}^{r_1} \frac{\mathrm{d}r}{r}$$

or

$$s(t_1 - t_0) = \frac{1}{\omega^2} [\ln(r_1) - \ln(r_0)]$$

or

$$s = \frac{1}{\omega^2} [\text{slope of a plot of } \ln(r) \text{ versus time}]$$

The following discussion shows how the slope is determined.

The photometric traces from a sedimentation velocity experiment are shown in Figure 11-58. Note that the diagram at the left shows the reference mark (in the centrifuge rotor), which is 5.7 cm from the axis of rotation, and the meniscus,

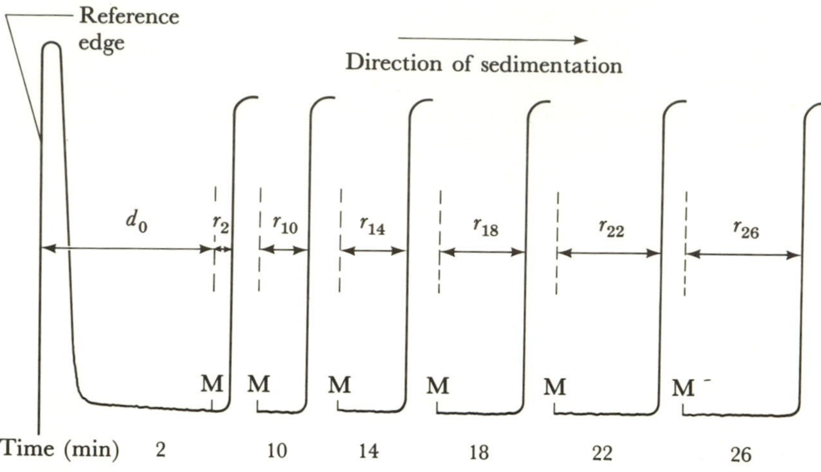

Figure 11-58
Photometric traces of sedimenting *E. coli* phage T7 DNA photographed by the ultraviolet-absorption system at six successive times of centrifugation (increasing from left to right). M is the meniscus and d_0 is the distance from the reference edge to the meniscus. Note that the distance between the meniscus and the boundary increases with time. See page 443 for calculation of r from d_0 and r_x.

and each subsequent diagram also shows the meniscus. The combined magnification of the photographic and photometric systems is known so that distance as a function of time can be measured. This is done by determining the distance r_x from the meniscus to the boundary as indicated and then adding the distance d_0 from the reference edge to the meniscus. Dividing this sum by the magnification factor and adding the result to 5.7 cm yields the following data:

Time after reaching 33,000 rpm (min)	Distance (r) from axis of rotation (cm)	ln r
2	6.225	1.8284
10	6.310	1.8421
14	6.358	1.8493
18	6.400	1.8561
22	6.445	1.8632
26	6.491	1.8705

These numbers are then plotted, as shown in Figure 11-59. The line is straight and the slope is $0.0521/(24 \times 60)$ sec^{-1} $= 3.62 \times 10^{-5}$ S^{-1}. The centrifuge speed was 33,000 rpm $= (33{,}000 \times 2\pi)/60 = 3.45 \times 10^3$ radians/sec. Therefore, ω^2 is 1.19×10^7. Thus,

$$s = 30.1 \times 10^{-13} \text{ sec}$$

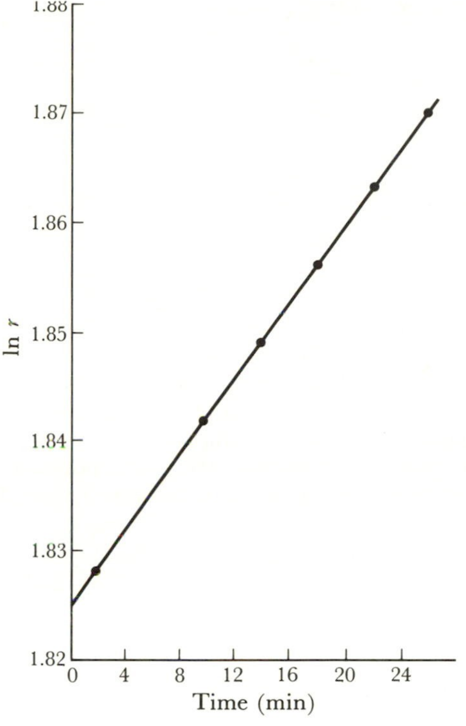

Figure 11-59
Plot of the data (given on page 444) obtained from the photometric traces of Figure 11-58.

Because one Svedberg unit (S) $= 10^{-13}$ sec,

$$s = 30.1 \text{ S}$$

This can be converted into $s_{20,\text{w}}$ by the use of equation (5).

2. *The analysis of radiation survival curves*

Consider a population of N identical organisms or molecules that is exposed to a dose D of some radiation that causes inactivation. It is further assumed that there are one or more radiosensitive sites in each organism or molecule and that inactivation requires that n sites be damaged. The effect on the population is expressed in terms of the fraction of the population that has not sustained some particular type of damage, as a function of the dose. As will be seen, the data are most conveniently plotted as $\log_e$(surviving fraction) or ln S versus the dose D. The resulting curve is called a *survival curve* or a *dose-response curve*.

First, we analyze a population of organisms (there is only a single site in each organism that, if damaged or "hit" by a radiation photon, will cause inactivation. (Note that simple absorption of the photon is not sufficient—the photon must cause damage.) The number dN damaged by a dose dD is proportional to the number N that existed prior to receiving that dose or

$$-\frac{\mathrm{d}N}{\mathrm{d}D} = kN \qquad (1)$$

which can be integrated from $N = N_0$, at $D = 0$ to yield

$$N = N_0 e^{-kD} \tag{2}$$

The surviving fraction $S = N/N_0$ is

$$S = \frac{N}{N_0} = e^{-kD} \tag{3}$$

Clearly a plot of $\ln S$ versus D gives a straight line having a slope of $-k$. Curves of this type are called *single-hit* curves and are commonly observed with irradiated biological systems. An example of such a curve was shown in Figure 11-42. The constant k is a measurement of the effectiveness of the dose and is proportional to the fraction of the incident photons that cause an inactivating hit—in other words, the probability that a single photon can cause such a hit.

Let us now consider a population of organisms in which each organism contains n units, *each* of which must be hit if the organism is to be inactivated. Thus, inactivation requires at least n hits—"at least", because it is assumed that two hits on a single unit are no more effective than one hit. The probability of one unit being hit by a dose D is $1 - e^{-kD}$ so that the probability P_n that all n units become inactivated is

$$P_n = (1 - e^{-kD})^n \tag{4}$$

Thus, the surviving fraction S of the population is $1 - P_n$ or

$$S = 1 - (1 - e^{-kD})^n \tag{5}$$

This equation can be expanded to yield

$$S = 1 - (1 - ne^{-kD} + \cdots + e^{-nkD}) \tag{6}$$

As D increases, the terms containing e^{-2kD}, e^{-3kD}, and so forth, become negligible compared to ne^{-kD} so that at high dose, $S = ne^{-kD}$ or

$$\log S = \log n - kD \tag{7}$$

A plot of equation (5) for $k = 1$ and various values of n shows that for small values of D, $\log S$ changes slowly. (Figure 11-60). At large D, equation (7) takes over and the curve becomes linear. Extrapolation of this linear part to $D = 0$ gives $S = n$ as the y-intercept. Thus, as long as experimental data are obtained for high enough values of D, n can be determined. Curves of this sort are called n-*hit curves* and a system showing such a curve is said to have n-hit inactivation kinetics. An example of a system that shows this effect of hit number is the inactivation of baker's yeast by X rays. This yeast exists in both a haploid (one copy of each chromosome) and a diploid (two copies of each) state. Damage to any chromosome is sufficient to kill the haploid variety. However, inactivation of a diploid requires that both copies of any chromosome be inactivated. Therefore, as long as damage to a single chromosome follows single-hit kinetics, then the haploid and diploid strains should have single- and double- hit inactivation kinetics; this is indeed the case.

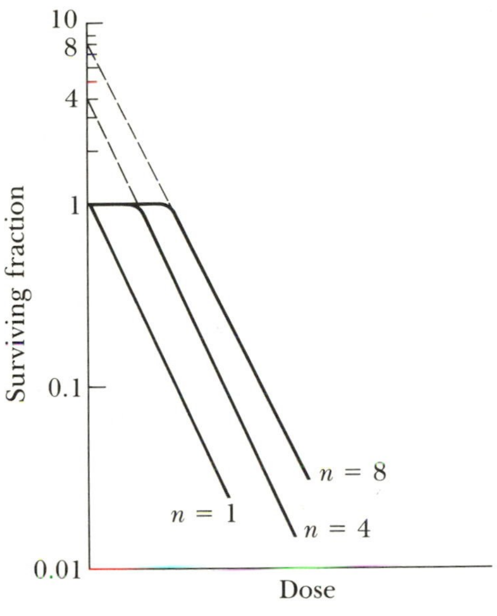

Figure 11-60
Survival curves for various values of n showing that at high dose each curve becomes linear and that extrapolation to the y-axis yields n as the intercept.

More complicated situations can also be imagined. For instance, an organism might contain n targets, each of which might have to be hit m times. In this case, the relevant equation is

$$S = [1 - (1 - e^{-kD})^n]^m \tag{8}$$

which at high dose is $S = n^m e^{-mkD}$ and at $D = 0$, the linear portion of the log S versus D curve has a slope $-mk$ and extrapolates to $S = n^m$.

SELECTED REFERENCES

Aune, K. C. 1978. "Molecular Weight Measurements by Sedimentation Equilibrium: Some Common Pitfalls and How to Avoid Them," in *Methods in Enzymology*, vol. 48, edited by C. H. W. Hirs and S. N. Timasheff, pp. 163–185. Academic Press.

Baldwin, R. L. 1957. "Boundary Spreading in Sedimentation-Velocity Experiments. 5. Measurement of the Diffusion Coefficient of Bovine Albumin by Fujita's Equation." *Biochem. J.* 65:503–512. Details of the measurement of *s*.

Brewer, J. M., A. J. Pesce, and R. B. Ashworth. 1974. *Experimental Techniques in Biochemistry*, ch. 6. Prentice-Hall.

Chervenka, C. H. 1980. "The Air-Driven Centrifuge: A Rediscovered Wonder," in *Fractions*, no. 1. Beckman Instruments Publication.

Freifelder, D. 1970. "Molecular Weights of Coliphages and Coliphage DNA. 4. Molecular Weights of DNA from Bacteriophages T4, T5, and T7 and the General Problem of Determination of M." *J. Mol. Biol.* 54:567–577.

Freifelder, D. 1973. "Zonal Centrifugation," in *Methods in Enzymology*, vol. 27, edited by C. H. W. Hirs and S. N. Timasheff, pp. 140–150. Academic Press.

Fujita, H. 1975. *Foundations of Ultracentrifugal Analysis.* Wiley.

Hearst, J. E., and C. W. Schmid. 1973. "Density Gradient Sedimentation Equilibrium," in *Methods in Enzymology*, vol. 27, edited by C. H. W. Hirs and S. N. Timasheff, pp. 111–127. Academic Press.

Hinton, R., and M. Dobrota. 1976. *Density Gradient Centrifugation*. North-Holland.

Howlett, G. J., E. Yeh, and H. K. Schachman. 1978. "Protein-Ligand Binding Studies with a Table-top, Air-driven, High-speed Centrifuge." *Arch. Biochem. Biophys.* 190:809–819.

Meselson, M., and F. W. Stahl. 1958. "The Replication of DNA in *Escherichia coli*." *Proc. Natl. Acad. Sci. U.S.A.* 44:671–682.

Meselson, M., F. W. Stahl, and J. Vinograd. 1957. "Equilibrium Sedimentation of Macromolecules in Density Gradients." *Proc. Natl. Acad. Sci. U.S.A.* 43:581–588. The development of equilibrium centrifugation in CsCl.

Radloff, R., W. Bauer, and J. Vinograd. 1967. "A Dye-Buoyant-Density Method for the Detection and Isolation of Closed Circular Duplex DNA: The Closed Circular DNA in HeLa Cells." *Proc. Natl. Acad. Sci. U.S.A.* 57:1514–1521. The ethidium bromide method.

Schachman, H. K. 1959. *Ultracentrifugation in Biochemistry*. Academic Press.

Schumaker, V., and B. H. Zimm. 1973. "Anomalies in Sedimentation. 3. A Model for the Inherent Instability of Solutions of Very Large Particles in High Centrifugal Fields." *Biopolymers* 12:877–894. Speed-dependent aggregation.

Svedberg, T., and K. O. Pedersen. 1940. *The Ultracentrifuge*. Oxford University Press.

Szybalski, W., H. Kubinski, Z. Hradeĉna, and W. C. Summers. 1971. "Analytical and Preparative Separation of Complementary DNA Strands," in *Methods in Enzymology*, vol. 21, edited by L. Grossman and K. Moldave, pp. 383–413. Academic Press.

Szybalski, W. 1967. "Use of Cs_2SO_4 for Equilibrium Density Gradient Centrifugation," in *Methods in Enzymology*, vol. 12B, edited by L. Grossman and K. Moldave, pp. 330–360. Academic Press.

Van Holde, K. E. 1975. "Sedimentation Analysis of Proteins," in *The Proteins*, 3rd edition, vol. 1, edited by H. Neurath and R. L. Hill, pp. 228–253. Academic Press.

Van Holde, K. E., and R. L. Baldwin. 1958. "Rapid Attainment of Sedimentation Equilibrium," *J. Phys. Chem.* 62:734–743. The low-speed equilibrium method.

Vinograd, J., R. Bruner, R. Kent, and J. Weigle. 1963. "Band-centrifugation of Macromolecules and Viruses in Self-generating Density Gradients." *Proc. Natl. Acad. Sci. U.S.A.* 49:902–910. Development of the band-centrifugation method.

Yphantis, D. 1964. "Equilibrium Ultracentrifugation of Dilute Solutions." *Biochemistry* 3:297–317. The high-speed equilibrium method.

Zimm, B. 1974. "Anomalies in Sedimentation. 4. Decrease in Sedimentation Coefficients of Chains at High Fields." *Biophys. Chem.* 1:279–291. Theory of speed-dependent sedimentation.

The manual for the Beckman Model E ultracentrifuge contains a wealth of information.

PROBLEMS

11–1. Which would have the higher *s* value, a rigid rod or a flexible rod, both having the same molecular weight, thickness, and length? A solid or a hollow sphere, both having the same radius and the same mass?

11–2. A small enzyme has a sedimentation coefficient of 3.4 S. When it binds to its substrate (a small organic molecule), its s value changes to 2.9 S. Explain this change.

11–3. A flexible molecule having many surface charges of the same sign decreases in s value as the salt concentration is reduced and the molecule becomes extended. Single-stranded DNA has this property. However, very rigid molecules carrying a surface charge of single sign also have lower s values in 0.0001 M NaCl than in 1 M NaCl—even when there is no change in shape. There are two reasons for this. Name them.

11–4. If a boundary moves halfway down a centrifuge cell in 20 minutes at 20,000 rpm, how long would it take to reach the same position if the speed is 40,000 rpm?

11–5. The sedimentation coefficient of a protein molecule is 5 S at pH 7.0 in 0.5 M NaCl. At pH 10.5, it is denatured and the sedimentation coefficient (still in 0.5 M NaCl) is 8 S. Is the native molecule more likely to be rodlike or a compact sphere?

11–6. A particular single-stranded DNA has an s value in solution A of 39 S and in solution B of 11 S. One of the solutions contains 0.01 M NaCl and the other is 1 M NaCl. Which is which?

11–7. The sedimentation coefficient of a particular DNA molecule is 22×10^{-13} sec. How far will a molecule move at 40,000 rpm in 20 minutes at a distance 6.0 cm from the axis of rotation? (Note: the s value is not in svedbergs.)

11–8. The following questions concern the relation between shape, molecular weight, and the sedimentation coefficient.

a. If a protein sample gives a very sharp sedimentation boundary that clearly indicates that only a material having a single sedimentation coefficient is present, can one conclude that all molecules have the same molecular weight?

b. If instead two components are observed, can one conclude that material having two different molecular weights is present?

11–9. Draw the concentration distribution in the centrifuge cell for a mixture of four components having s = 18, 26, 45, and 52 S and being 15%, 25%, 50%, and 10% of the total concentration, respectively.

11–10. A DNA sample is prepared and you suspect that you may have broken $\frac{1}{2}$ of the molecules. If $\frac{1}{2}$ had been broken, which of the following centrifuge diagrams, A, B, or C, would you get?

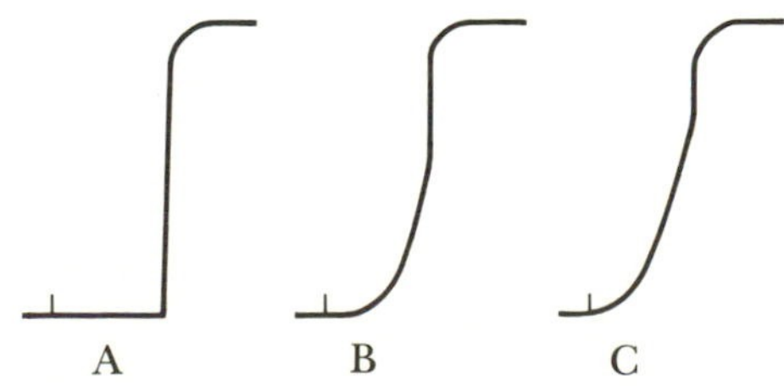

11–11. A DNA sample shows the centrifuge diagram A at neutral pH and B in alkali. What information does this give you about the DNA molecule? What would you conclude if instead diagram C had been obtained in alkali?

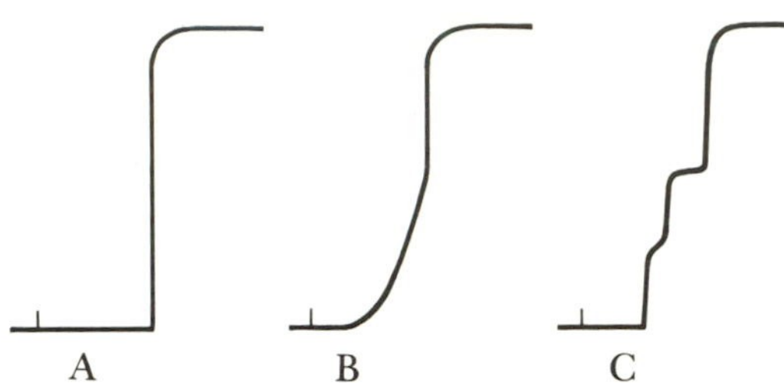

11–12. Two DNA samples are treated in two different ways but in each 50% of the molecules get broken. In one sample, the breaks are randomly arranged. In the other sample, the breaks tend to be near the center of the molecule but they can be at many positions; however, they are always in the central third of the molecule. Match up the diagrams A and B with the two samples.

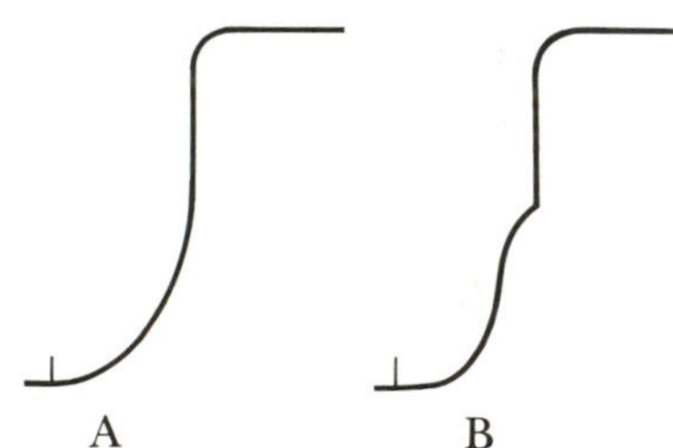

11–13. A protein having a molecular weight of 50,000 and a single, sharp sedimentary boundary is sedimented in 6 M guanidinium chloride. Two schlieren peaks are seen corresponding to molecular weights of approximately 5000 and 15,000. The area of the slower peak is two-thirds that of the faster. What is the subunit structure of the protein?

11–14. The sedimentation velocity properties of a supercoiled DNA are being studied as a function of the concentration of added ethidium bromide. It is found that s decreases, reaches a minimum, and then increases. Explain.

11–15. A lipoprotein is suspended in 1 M NaCl and centrifuged to determine its sedimentation coefficient. When the schlieren patterns are looked at, it is found that a peak appears at the tube bottom and moves upward. Explain what has happened.

11–16. The sedimentation of a DNA solution is being studied with ultraviolet optics. The DNA solution has been divided into three parts. To part A, nothing is added. To parts B and C, two different materials are added. Then each is centrifuged. In part A, typical boundaries are observed. In parts B and C, no boundaries are seen, neither at the time of reaching speed nor after several hours. In part B, the cell appears uniformly transparent to ultraviolet light, both above and below the meniscus. In part C, the cell remains nearly opaque. What have the reagents probably done to the DNA?

11–17. If DNA were labeled with [^{3}H]thymidine at a specific activity of 25 Ci/mmol, what shift in density in CsCl would result? Repeat for [^{14}C]thymidine at 1 mCi/mmol and for [^{32}P]phosphate at 1 mCi/μmol.

11–18. The $s_{20,w}$ of a pure protein differs by 10% when sedimented in 0.25 M NaCl versus NaI. What are possible explanations for this? How could the possibilities be distinguished by sedimentation equilibrium?

11–19. Why is more protein than DNA required to obtain a value of s? (Consider how the boundaries must be detected.) For what kind of protein might this not be true?

11–20. You are doing a sedimentation velocity experiment. When the schlieren peak has moved half way down the cell, you remember that you forgot to add something to the solution. You stop the machine, add the necessary material, shake up the contents, and recentrifuge. You notice that the area of the schlieren peak is smaller than in the first centrifugation. Can this be due to the material you added? Is the s value you obtain an accurate representation of what is in the cell?

11–21. Double-stranded DNA can be converted into single-stranded DNA either by heating in the presence of formaldehyde or by adjusting the pH to 12.5. The sedimentation patterns of DNA denatured by either procedure are normally identical. However, if DNA is mixed with acridine orange and then the mixture is irradiated with low doses of light absorbed by acridine orange, denaturation by the two procedures gives different patterns; that is, by the heat-formaldehyde treatment the sedimentation pattern of the DNA is that of unirradiated DNA—a single, sharp boundary with $s_{20,w}$ roughly 30% greater than that of native DNA—whereas by the alkaline treatment there is a small, sharp boundary with a great deal of material sedimenting more slowly. Explain this difference.

11–22. Describe how you would determine experimentally whether a particular density gradient would produce isokinetic sedimentation. Would a gradient that is isokinetic for DNA also be isokinetic for a protein?

11–23. What effect would ethidium bromide have on $s_{20,w}$ of *linear* DNA?

11–24. If proteins are suspended in buffers containing mercaptoethanol, disulfide bonds are broken. Would this be detectable by sedimentation? Consider native and denatured protein.

11–25. A ^{14}C-labeled particle is dissociated to release RNA molecules and then run in a sucrose gradient. Pattern A always appears. The area of the 12S peak is always twice that of the 8S peak. If sedimented in a gradient containing the denaturant dimethylsulfoxide and in which the NaCl concentration is very low, pattern B is observed. What can you say about the structure of the two RNA molecules and the relative numbers of each type in the particle?

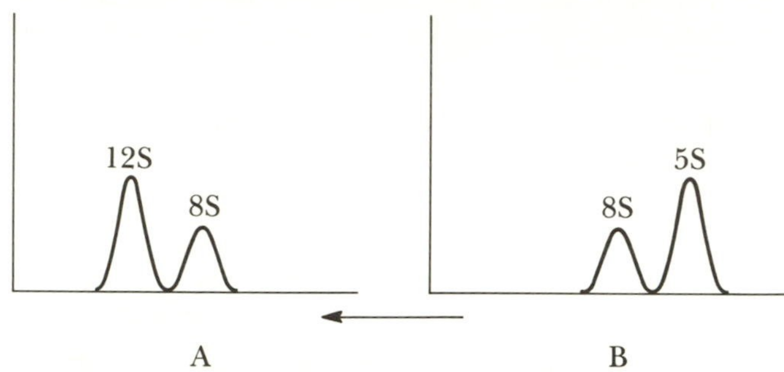

11–26. 10,000 cpm of a ^{14}C-labeled protein are layered on a sucrose concentration gradient, centrifuged, and later the tube is fractionated and the samples counted. Only 90% of the total radioactivity is recovered after a 1-hour centrifugation run. On another day you do a similar experiment but run the centrifuge for 2 hours. This time a smaller percentage is recovered. Why might this be?

11–27. A highly compact protein is treated with a reagent that causes the s value to decrease by approximately 40%. A single boundary results. What do you think the reagent has done if the boundary has become (a) narrower or (b) broader?

11–28. Most proteins have a density of 1.3 g/cm^3 in CsCl, although variations of up to 0.050 g/cm^3 exist. Would it be feasible to separate two proteins whose molecular weight is 10,000 having densities of 1.300 and 1.340 g/cm^3 by equilibrium centrifugation in buoyant CsCl? (This speed of the centrifuge usable for CsCl centrifugation ranges from 20,000 to 50,000 rpm. Although the density gradient is speed-dependent, for a rough calculation, 0.020 g cm^{-3} mm^{-1} can be used.)

11–29. *E. coli* [^{14}N]- and [^{15}N]DNA (50% G + C) is separated in a CsCl density gradient by 1.32 mm at a certain speed and temperature. What would be the distance from *E. coli* [^{14}N]DNA of two DNAs that are 30% and 70% G + C, centrifuged in the same cell?

11–30. The density of DNA in 7 M CsCl containing 0.1 M $MgCl_2$ is less than that in 7 M CsCl alone. Explain.

11–31. A phage DNA has a density of 1.700 g/cm^3 in CsCl. If labeled with both ^{13}C and ^{15}N, its density is 1.750 g/cm^3. What is the density of an unbroken molecule that is prelabeled with ^{13}C and ^{15}N and replicates for $\frac{1}{10}$ of a generation in a ^{12}C, ^{14}N medium?

11–32. A λ DNA molecule has a molecular weight of 30×10^6. A deletion mutant is smaller but the molecular weight is not known. You have some ^{15}N-labeled normal λ DNA and you link it via the single-stranded cohesive ends to a single deletion mutant DNA that has no density label.

This joint molecule is centrifuged to equilibrium in CsCl and the density is found to be 1.708 g/cm^3. The density of normal λ DNA is 1.700 g/cm^3. What is the molecular weight of the deletion DNA?

11–33. What would be the density of a bacteriophage that is $\frac{2}{3}$ by weight protein and $\frac{1}{3}$ by weight DNA? The densities of the protein and the DNA molecules are 1.3 g/cm^3 and 1.7 g/cm^3 respectively.

11–34. Under particular conditions the band width of a DNA molecule, whose molecular weight is 25×10^6, is 1.3 mm when centrifuged to equilibrium in CsCl. What would be the width of the band of a transfer RNA molecule, whose molecular weight is 25,000, under the same conditions?

11–35. The molecular weight of DNA is being determined by CsCl density gradient centrifugation. If bacterial DNA (whose molecular weight when intact is about 3×10^9) is broken down (during the isolation procedure) into pieces whose average molecular weight is about 5×10^6, the measured average value is usually about 30% too small. Explain why this is the case.

11–36. In CsCl equilibrium centrifugation

a. What would be the relation between the band widths of a sample of a nicked circle and that of a linear molecular having the same molecular weight?

b. What would be the relation between the densities of a nicked circle and a covalent (but not supercoiled) circle, if ethidium bromide were present?

11–37. One way to evaluate M from data obtained from centrifugation to equilibrium in CsCl is to plot $\ln(C_r/C_0)$ (where C_0 is the concentration at r_0) against $(r - r_0)^2$. This should give a straight line whose slope is $M\omega^2 r_0 \bar{v}(\Delta\rho/\Delta r)/2RT$. In such a plot, it is sometimes observed that the graph is curved concave upward, the departure from linearity being most noticeable as $(r - r_0)$ becomes large. Density heterogeneity is not the explanation when this is the case if the data are obtained from both sides of the band and if the band is symmetric. Propose an explanation.

11–38. From the s value of a particular DNA sedimented in 1 M NaCl, a molecular weight of 25×10^6 is calculated from an empirical relation between s and molecular weight. The molecular weight is also measured from the band width after centrifugation to equilibrium in CsCl but the value obtained is 33×10^6. Why are the values different?

11–39. A protein solution is sedimented to equilibrium at 20°C at a speed of 20,000 rpm. The protein concentrations at 6.425 cm and 6.703 cm from the axis of rotation are 1.012 mg/cm^3 and 6.905 mg/cm^3 respectively. The partial specific volume of the protein is 0.71 cm^3/g. The density of the solvent is 1.01 g/cm^3. What is the molecular weight of the protein?

11–40. A DNA molecule has a density of 1.700 g/cm^3 in CsCl and 1.60 g/cm^3 in CsCl containing ethidium bromide. For a typical circular DNA molecule the introduction of one single strand break by X rays shifts the density to 1.55 g/cm^3 (i.e., the density of linear DNA). However, this molecule shifts in density to 1.575 g/cm^3. (Some molecules are unnicked and remain at a density of 1.60 g/cm^3.) If the unnicked DNA is centrifuged in an alkaline sucrose gradient, a single peak is seen; after nicking, three peaks appear and the pattern shown below is seen.

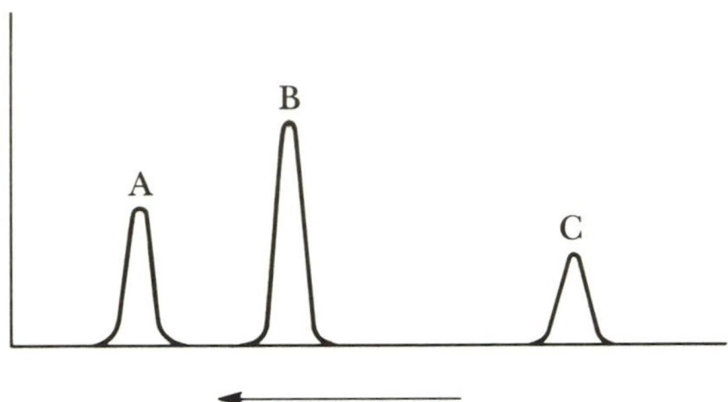

Peak A consists of unnicked molecules. The area of B is always 3 × that of C, no matter how large A is. What is the structure of the molecule?

11–41. A sample of circular DNA molecules is X-irradiated. This causes double-strand breaks; that is, the molecule becomes linear. Samples exposed to several doses are centrifuged and two species are observed. The ratio of the amount of slower moving material to the amount of the faster moving material is determined as a function of dose; the data are the following.

Dose (rads)	Fraction fast	Fraction slow
0	1.00	0
1000	0.61	0.39
2000	0.37	0.63
3000	0.23	0.77
4000	0.14	0.86

On the average, how many rads are required to convert a circular molecule to a linear molecule?

11–42. A sample of DNA gives one band in CsCl but two in CsCl containing ethidium bromide. The ratio of the areas of the denser to the lighter band is 2:1. The molecular weight of the DNA is 30×10^6. Suppose that the DNA were treated with an enzyme that produces on the average one single-strand break in each molecule. What would the ratio of the band areas be after such a treatment? (Remember to use the Poisson distribution to determine the fraction of molecules receiving no breaks.)

12

Partial Specific Volume and the Diffusion Coefficient

In the preceding chapter, it was mentioned that measurements of the partial specific volume and the diffusion coefficient are frequently necessary in using hydrodynamic methods to characterize macromolecules. How this is done is the subject of this chapter.

Measurement of Partial Specific Volume

In the theory of ultracentrifugation, a term that is often encountered is $1 - \bar{v}\rho$, in which ρ is the solution density and $\bar{v}$ is the partial specific volume. The volume increment produced in a solution when unit mass of solute is added is $\mathrm{d}v/\mathrm{d}m = \bar{v}$. (The partial specific volume is sometimes approximated as the reciprocal of the density of the solute, which is not exactly true but often a good approximation.) Common experience indicates that the volume increment in solution differs from the volume of the solid—for example, a cup of sugar dissolved in a cup of water makes much less than two cups of solution (nearer one cup). It is also important to know that $\bar{v}$ is not an invariant parameter of a particular macromolecule but varies with the solvent composition—that is, the salt concentration, pH, presence of other dissolved substances, and so forth.

An evaluation of $\bar{v}$ is essential in determining both molecular weight and sedimentation coefficients. Furthermore, $\bar{v}$ must be measured with great precision because, due to the range of values for biological macromolecules (0.6–0.75 cm^3/g), a 1% error in $\bar{v}$ gives about a 3% error in M or $s_{20,\mathrm{w}}$. In fact, the measurement of $\bar{v}$ is frequently the limiting factor in determining molecular weight. The three major methods are the summation of $\bar{v}$ values of the residues of a macromolecule, those that use density measurement, and parallel sedimentation equilibrium measurements in isotopically labeled solvents.

Summation of $\bar{v}$ of the Residues of a Macromolecule

If the amino acid composition of a protein is accurately known, $\bar{v}$ can be calculated from the $\bar{v}$ values of the individual amino acids as $n_i m_i \bar{v}_i / \sum n_i m_i$, in which n_i is the number of residues per mole of the ith amino acid in the protein, m_i is the residue molecular weight (the molecular weight of the amino acid minus the weight of one mole of water, because one mole of water is removed in the formation of a peptide bond), and $\bar{v}_i$ is the partial specific volume of the ith residue. Values of $\bar{v}$ for the residues can be found in various reference tables. If other groups such as lipids, carbohydrates, flavins, and so forth, are present, $\bar{v}$ for the group must be added in. This is an accurate method for proteins but has not been tested for nucleic acids.

Methods Using Density Measurement

Because $\bar{v} = \mathrm{d}v/\mathrm{d}m$, $\bar{v}$ can be determined from the variation of the density of a solution with solute concentration. Four methods for accurately measuring density are described first, and then the surprisingly formidable problem of precisely measuring concentration is discussed.

Pycnometry

A pycnometer is simply a container whose volume can be accurately measured and which can be filled with great precision (Figure 12-1). The

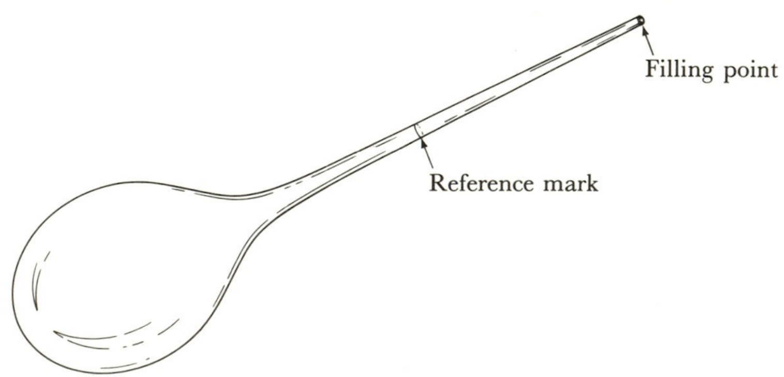

Figure 12-1
A pycnometer. The pycnometer is first weighed empty and then filled to the mark and reweighed. The volume up to the mark is accurately known.

volume is measured by filling with water and weighing, since the density of water is accurately known. It is then filled with the *solution* and reweighed. Because of the temperature dependence of volume, the temperature of both the water and the solution must be accurately controlled and known. Pycnometry is the most direct way to determine density. However, in order for the weight differences to be large enough to be measured with precision, large volumes of solution of high concentration (approximately 10 ml at 50 mg/ml) are needed; frequently it is very difficult to obtain so much material. Pycnometry usually fails with highly extended molecules such as high-molecular-weight DNA because at 10 mg/ml the solution is a semisolid gel and the pycnometer cannot be filled.

Linderstrøm-Lang density gradient column

A density gradient column of bromobenzene and kerosene, both of which are immiscible with water, is prepared (Figure 12-2). If a small droplet (usually 1 μl) of an aqueous solution is placed on the surface, it will fall through the column and come to rest at a point at which its density equals that of the column. If the column is calibrated with solutions of known concentration (usually of KC1), the density of the sample can be determined from its position relative to the standards. A large number of standards and sample drops are needed to define the density gradient and to determine the position of the sample with precision. To avoid thermal convection, which would result in the movement of the drops, the temperature of the gradient must be controlled to $\pm 0.01°$C. This method has the great advantage of using tiny amounts of material and is generally reliable although there have been a few instances of error caused by interaction with the solvents in the gradient.

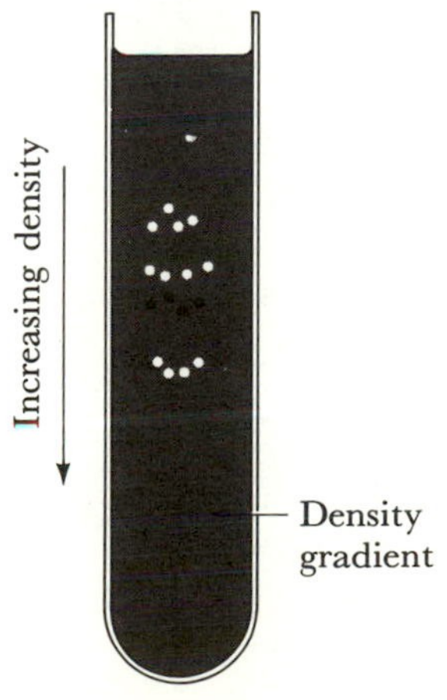

Figure 12-2
Linderstrøm-Lang density gradient column. A linear density gradient of two organic liquids (typically kerosene and bromobenzene, which are miscible with one another but immiscible with H_2O) is prepared using apparatus of the type shown in Figure 8-5. Droplets of solutions of known density (open circles) are introduced into the column for calibrating the density gradient. The position of the sample droplets (solid circles) is measured with respect to the reference drops.

Cahn electrobalance

Like a pycnometer, this instrument accurately weighs a solution of known volume but uses small (1 ml) volumes at relatively low ($\sim$10 mg/ml) concentration. This is possible because its accuracy is ± 0.1 μg—roughly, 1000 times as sensitive as standard laboratory balances. However, its cost ($7000) prohibits its widespread use.

Mechanical oscillator technique

This method uses a commercially available mechanical oscillator that can be filled with fluid (Figure 12-3); its resonance frequency is related to the density of the liquid. The advantage of the instrument is high precision with a sample volume of less than 1 ml. The instrument is gaining widespread use.

With each of the methods requiring density measurement, it is necessary to know concentration precisely. This seems trivial because to make a known volume of solution necessitates only weighing a sample and dissolving it. However, the weight must be the *anhydrous weight* and, unfortunately, proteins and nucleic acid invariably contain bound water. With inorganic materials, an anhydrous sample can be obtained by heating to a high temperature, but proteins and nucleic acids are degraded at temperatures higher than 100°C. Hence, a standard method is to dry the protein or nucleic acid sample at 60–80°C in a vacuum until the weight becomes constant and assume that constant weight indicates that all water has been removed. However, to remove all water from

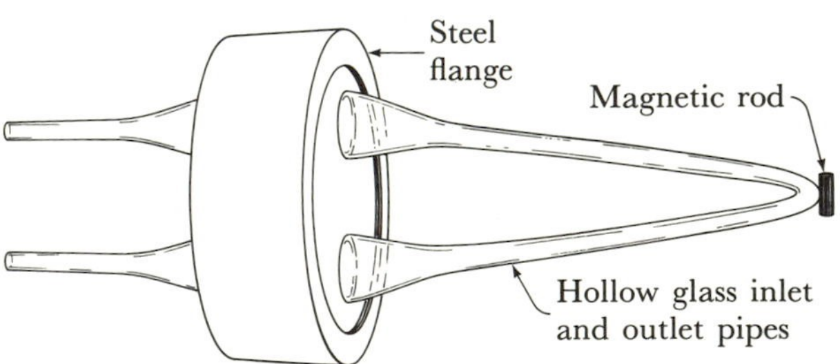

Figure 12-3
A mechanical oscillator for density measurement. An oscillating magnet elsewhere in the system causes the magnet rod and therefore the entire V-shaped tube to vibrate. The natural frequency of vibration is determined by the geometry and mass of the tube. The tube is filled with the sample whose density is to be measured, and the natural frequency changes. From the measured frequency, the weight of added liquid is calculated. Because the volume of the V-shaped tube is accurately known, the density can be determined.

proteins requires a temperature so high that degradation occurs. An alternative approach, which has been used only a few times, is to determine the dry weight from the elemental composition. For example, because the chemical formulas of all of the amino acids and nucleotides are known, the weight of amino acids or nucleotides can be calculated from the weight of nitrogen and phosphorus in a sample, both of which can be measured with an error of no more than 1%. Therefore, the solution used for density determination can be analyzed for nitrogen and phosphorus content and dry weight can be calculated from the amino acid or nucleotide composition of the protein or nucleic acid (which must, of course, be determined if not already known).

Example 12-A

☐ Calculation of $\bar{v}$ by pycnometry.

The first step in the use of a pycnometer is to determine its volume. This is done by measuring the weight of water that it can hold. A pycnometer weighs 14.3082 g when empty and 24.2651 g when it is filled with water at 20°C. The density of water at 20°C is 0.9982 g/cm so that the volume is (24.2651 − 14.3082)/0.9982 = 9.9749 cm^3. The next step is to determine the density of a solution containing a particular ratio of solute and solvent. Thus a solution is prepared from 3.7184 g of solute and 9.9582 g of water. The pycnometer is filled with this solution (that is, with 9.9749 cm^3 of solution, as calculated above) and it weighs 25.8237 g at 20°C. The weight of the solution in the pycnometer is 25.8237 − 14.3082 = 11.5155 g and the density is 11.5155/9.9749 = 1.1544 g/cm^3. Finally the change in volume per change in mass is calculated. The solution has a total weight of 3.7184 (solute) + 9.9582 (water) = 13.6766 g and a volume of 13.6766/1.1544 = 11.8474 cm^3. The volume of water used in preparation of the solution is 9.9582/0.9982 = 9.9762 cm^3. The increase in mass (by addition of the solute) is 3.7184 g and the increase in volume is 11.8474 − 9.9762 = 1.8712 cm^3. Thus $\bar{v}$ is 1.8712/3.7184 = 0.503 cm^3/g.

Note that for high precision the weights must be corrected for the buoyancy of air, when appropriate.

Parallel Sedimentation Equilibrium Measurement in H_2O and D_2O Solutions

At equilibrium, the distribution of materials in a centrifuge cell is described by

$$M_H(1 - \bar{v}\rho_H) = \frac{2RT}{\omega^2}\left(\frac{d\ln c}{dr^2}\right)_{H_2O} \tag{1}$$

in which M_H is the molecular weight in H_2O, ρ is the density of the aqueous solution, R is the gas constant, T is the absolute temperature, ω is the angular velocity in radians per second, c is the concentration, and r

is the distance from the axis of rotation in centimeters. A similar equation can be written for a D_2O solution. Dividing one equation by the other yields

$$\frac{M_H(1 - \bar{v}_H \rho_H)}{M_D(1 - \bar{v}_D \rho_D)} = \left(\frac{\mathrm{dln}c}{\mathrm{d}r^2}\right)_H \Bigg/ \left(\frac{\mathrm{dln}c}{\mathrm{d}r^2}\right)_D \tag{2}$$

in which the subscript D refers to D_2O. Howard Schachman and his colleagues have shown that these two equations can be solved to yield $\bar{v}$ because $\bar{v}$ is virtually the same in H_2O as in D_2O. The values of M_H and M_D are not the same because the amide hydrogens exchange with deuterium. However, the ratio M_H/M_D is easily calculated from the chemical formula. Thus, simultaneous equilibrium centrifugation analyses, one in H_2O and one in D_2O, which allow the cancellation of the $2RT/\omega^2$ terms in equation (1) to obtain equation (2), yield the value of $\bar{v}$, since the $\mathrm{dln}c/\mathrm{d}r^2$ terms are measurable quantities. This method has the great advantages that only a tiny amount of material is needed and $\bar{v}$ is measured at the same time that M is being measured. It is the only method available if material is limiting; yet, in principle, it is not as accurate as the methods using the measurement of density because the difference in the two differentials in equation (2) is small. However, because dry weight determination is generally inaccurate, this method probably gives better precision in practice than the others. A considerable increase in accuracy can be achieved by the use of $D_2^{18}O$, which has recently become available.

Measurement of the Diffusion Coefficient

As discussed in Chapter 11, molecular weight, M, can be determined quite accurately from a measurement of the sedimentation coefficient, s, and the frictional coefficient, f. The direct determination of f is very difficult; fortunately, this can be bypassed by measuring the diffusion coefficient, D.

Diffusion is the net flow of molecules from a region of high concentration to one of low concentration *if there is no driving force*—that is, the result of random movement (Figure 12-4). The diffusion coefficient, D, of a molecule can be simply defined by Fick's first law of diffusion, which states that the number of molecules, $\mathrm{d}n$, passing through an area, A, in time $\mathrm{d}t$ is related to the concentration gradient, $\mathrm{d}c/\mathrm{d}x$ by this equation:

$$\frac{\mathrm{d}n}{\mathrm{d}t} = -DA\left(\frac{\mathrm{d}c}{\mathrm{d}x}\right) \tag{3}$$

The minus sign is introduced so that D is positive. From the fact that molecules will move more slowly if the frictional coefficient, f, is large,

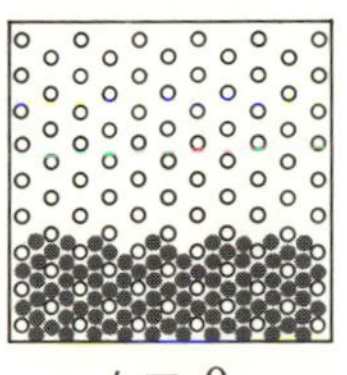

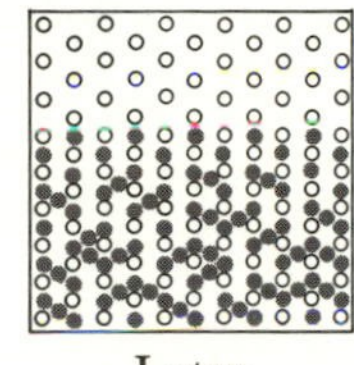

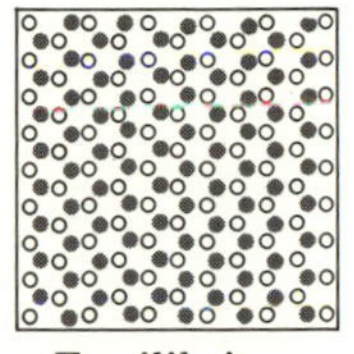

Figure 12-4
The mechanics of diffusion. The solid circles are originally located at the bottom of the box. They diffuse upward until they are distributed uniformly throughout the system.

it can be shown that

$$D = \frac{kT}{f} \tag{4}$$

in which k is the Boltzmann constant and T is the absolute temperature. Hence, the Svedberg equation (Chapter 11, equation 1) can be written with D instead of f as

$$M = \frac{RTs}{D(1 - \bar{v}\rho)} \tag{5}$$

in which R is the gas constant, T is the absolute temperature, $\bar{v}$ is the partial specific volume of the macromolecule, and ρ is the solution density, and M can be calculated if s and D are known.

The diffusion coefficient for macromolecules is usually measured by creating a boundary between a buffer and a solution of macromolecules of known concentration and observing the spreading of the boundary with time. The theory behind the measurement is simple, but in practice the measurement is filled with potential error. For example, D for a macromolecule is so small, from 10^{-8} to 10^{-5} cm^2/sec, that it normally takes a day for measurable spreading to occur. During this time, the system must be free of all mechanical disturbances, and thermal convection must be avoided by accurate temperature control. Furthermore, if a molecule is charged (e.g., a net positive charge) the extra OH^- ions in solution, which diffuse more rapidly than the macromolecule, create an electric potential gradient that drives the charged molecules to the region of low concentration. Hence, it is necessary to conduct diffusion experiments at or near the isoelectric point and in the presence of sufficiently high ionic strength to neutralize or eliminate the effect of the developed electric field. In addition, because in theory the molecules must move independently of one another (i.e., they must not collide), it is necessary to extrapolate to infinite dilution.

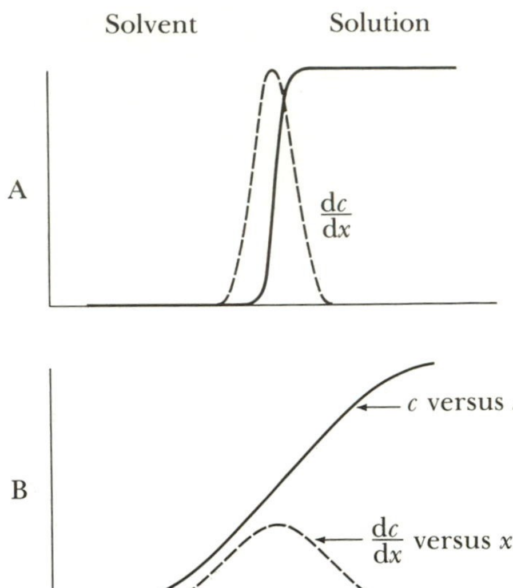

Figure 12-5
Measurement of diffusion: (A) initial concentration distribution in which the solvent and the solution are in contact; the curve c versus x shows the concentration of the solution across the cell and $\mathrm{d}c/\mathrm{d}x$ versus x is the concentration gradient or the schlieren pattern; (B) at a later time.

Each of the methods that make use of boundary spreading starts with a concentration distribution such as that shown in Figure 12-5A, and the change in time is measured as shown in Figure 12-5B. The measurement is facilitated if the concentration gradient is observed by schlieren optics (Chapter 11). If A and H are the area and height of the schlieren curve,

$$4\pi Dt = \left(\frac{A}{H}\right)^2 \tag{6}$$

and a plot of $(A/H)^2$ versus t gives a straight line of slope $4\pi D$ (Figure 12-6).

The initial boundary can be prepared in several ways. In a standard diffusion cell, the solvent is layered onto the solution of macromolecules or a very thin zone is formed between two solutions having two different densities.

These procedures are, however, frought with experimental difficulties. This is because diffusion occurs very slowly. (The time required for a molecule to move a particular distance d is proportional to d^2.) For example, it takes several hours for a typical protein to produce a boundary 1 mm wide and about a week to reach 1 cm. A typical diffusion experiment requires several days during which time many external agents may disturb the spreading boundary. Small temperature changes cause convective flow within the solution; to reduce this, the temperature variation must be kept to less than 0.01°C. Mechanical vibrations are a major problem although they may be minimized by using special "vibration-free" mounts for the apparatus. Both problems are reduced somewhat by using a stabilizing density gradient. The usual criteria for

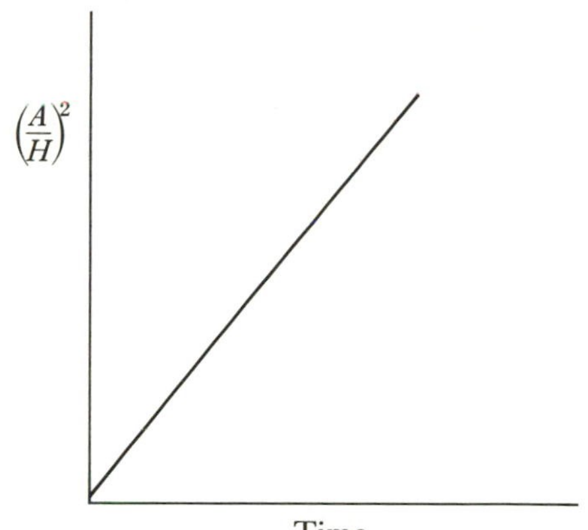

Figure 12-6
Typical curve for determining the diffusion coefficient; A and H are the area and height of the dc/dx curve of Figure 12-5. For small proteins, the time scale is of the order of several hours; for large viruses, it can be several days. Note that the curve does not pass through the origin. This is because the boundary is rarely perfect at zero time. The slope of the curve is $4\pi D$.

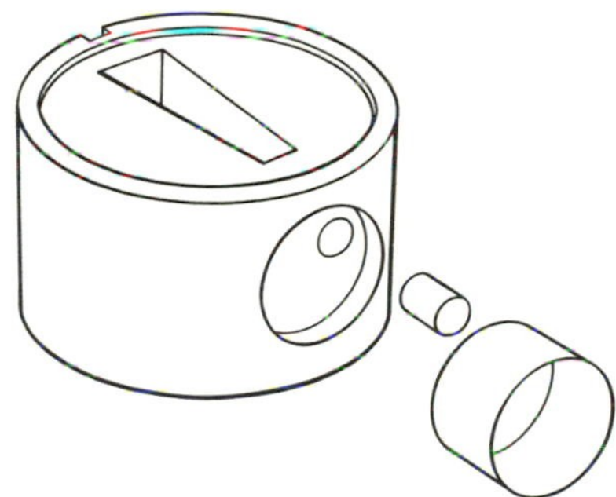

Figure 12-7
One type of synthetic boundary cell (valve type) for the Beckman ultracentrifuge: an aluminum centerpiece with a flat-bottomed cavity in place of the usual filling hole (see aluminum centerpiece in Figure 11-3). There is a small, round hole, which has a groove on its perimeter (not shown), on the bottom of the cavity in which a rubber plug is placed. A large cup containing a hole in which the plug fits snugly is placed in the cavity. This cup contains the solvent. The sector of the centerpiece contains the solution. At rest, the plug prevents entry of the solvent, but, during centrifugation, the plug is compressed and solvent leaks from the cup along the groove into the sector. [Courtesy of Beckman Instruments.]

a "disturbance-free" measurement is linearity of the curve shown in Figure 12-6.

The initial boundary can also be formed in a special analytical centrifuge cell—a synthetic boundary cell as shown in Figure 12-7, in which, at a critical speed, solvent passes through a fine capillary into the solution. The centrifuge is then operated at a speed sufficiently low that sedimentation does not occur appreciably. This method is useful if samples of only small size are available. In a variation of the centrifugal method, the cell is filled with the solution only and centrifuged: a boundary forms as a result of the sedimentation and the spreading of the boundary as sedimentation proceeds is measured, but a different equation is needed to analyze the data.

A remarkable method—optical mixing spectroscopy—involving the scattering of a laser light beam, has been developed. Because the molecules in solution undergo translational motion (i.e., diffusion), there is a Doppler shift in the scattered light so that the scattered light has a slightly different frequency. This frequency shift is measured by mixing the scattered and unscattered light (i.e., beating the scattered light against the incident light) and measuring the beat frequency (i.e., the frequency difference). This frequency can be simply related to D. This method gives a 1% error, uses small samples (a few hundred μg), is very rapid, and is not affected by convection or electrical effects. Its only disadvantage is that it often fails with very highly extended molecules. The technique has not yet become popular.

To date, there has been no satisfactory measurement of D for large nonspherical molecules such as DNA, because D is so small for such molecules.

In Chapter 11, the utility of the value of D in determining M was explained. Another important use is to give information about the shape of a macromolecule. How this is done is shown by the following argument. In any arrangement in which diffusion is occurring, molecules are moving from a region of high concentration to one of low concentration. The driving force is determined by the concentration difference; however, as soon as the movement begins, the driving force is counteracted by the frictional drag of the particles moving through the solvent. Stokes' Law expresses the frictional force on a sphere of radius a moving through a fluid of viscosity η at constant velocity, and a simple calculation yields the value of D for a sphere—namely,

$$D = kT/6\pi a\eta \tag{7}$$

The units of this equation are poise for η (=0.01 for water at 20°C), cm for a, and a value of 1.38×10^{-16} erg deg^{-1} for k; D, therefore, is in cm^2/sec. The mass m of a sphere is $\frac{4}{3}\pi a^3\rho$, in which ρ is the density. Thus,

$$D_{\text{sphere}} = \frac{kT}{6\pi\eta}\left(\frac{4\pi\rho}{3m}\right)^{\frac{1}{3}} \tag{8}$$

The density can be approximated as the reciprocal of the partial specific volume $\bar{v}$ so that

$$D_{\text{sphere}} = \frac{kT}{6\pi\eta}\left(\frac{4\pi}{3m\bar{v}}\right)^{\frac{1}{3}} \tag{9}$$

If the mass (in grams) of a molecule is known, from an independent measurement, it is possible to calculate the value of D that the molecule would have *if the molecule were spherical.* The difference between that value and the experimentally observed value is a measure of the departure of the shape of the molecule from that of a sphere.

It is common to rewrite these expressions in terms of the frictional coefficient f, using equation (4) and the notation f_0 for the frictional coefficient of a sphere. Thus,

$$f_0 = 6\pi\eta\left(\frac{3m\bar{v}}{4\pi}\right)^{\frac{1}{3}} \tag{10}$$

If f is a measured value, the frictional ratio f/f_0 (also called the Perrin factor) is a measure of the departure of the molecule from spherical shape.

Several quantitative theories have been developed for relating the value of f/f_0 to molecular dimensions. In each theory, the value of f is calculated for simple regular shapes, just as has been done for the sphere. Since molecules never have a really simple shape, the information provided by these theories is primarily suggestive. The two shapes examined in greatest detail are the oblate and prolate ellipsoid. An oblate ellipsoid is a disc-shaped ellipsoid generated by rotating an ellipse about its short axis. A prolate ellipsoid is rodlike (actually more like a cigar or an American football) and is generated by rotating an ellipse about its long axis. The ratio of the long axis of either ellipsoid to its short axis is called the *axial ratio.* A commonly used relation between the axial ratio and f/f_0 is shown in Table 12-1. Note that it is not really necessary to distinguish the oblate and prolate ellipsoids until the axial ratio is greater than ten.

The following examples show how to use equation (10) and Table 12-1.

☐ Estimation of the shapes of two proteins from a diffusion measurment. **Example 12-B**

Protein P has a molecular weight of 40,000, $D = 9.3 \times 10^{-7}$ cm^2/sec, and $\bar{v}$ is 0.75 g/cm^3 (a typical value for a protein). The value of m is $40{,}000/6 \times 10^{23} = 6.67 \times 10^{-20}$ g. Assuming that D is measured in water at 20°C so that $\eta = 0.01$ poise (usually η would be measured for the solvent used in the diffusion measurement), then $f_0 = 4.31 \times 10^{-8}$. From the value of D, $f = kT/D = (1.38 \times 10^{-16})(293)/(9.3 \times 10^{-7}) = 4.35 \times 10^{-8}$ g cm^{-1} sec^{-1}. Thus $f/f_0 = 1.01$ and the shape of the molecule is not very different from a sphere. Actually one means by this that the molecule could fit neatly in a sphere.

Myosin, a muscle protein, has a molecular weight of 493,000, $\bar{v} = 0.728$ cm^3/g, and $D = 1.1 \times 10^{-7}$ cm^2/sec. A calculation similar to that just performed yields $f/f_0 = 3.65$. Referring to Table 12-1 the axial ratio of myosin is very large. Other measurements indicate that it is more like a prolate ellipsoid than an oblate one so that its

Table 12-1
Values of f/f_0 for Prolate and Oblate Ellipsoids Having Various Axial Ratios.

	f/f_0	
Axial ratio	prolate	oblate
1	1.00	1.00
2	1.04	1.04
3	1.11	1.11
4	1.18	1.17
5	1.25	1.22
6	1.31	1.28
8	1.43	1.37
10	1.54	1.46
15	1.78	1.64
20	2.00	1.78
30	2.36	2.02
40	2.67	2.21
50	2.95	2.38
60	3.20	2.52
80	3.66	2.77
100	4.07	2.93

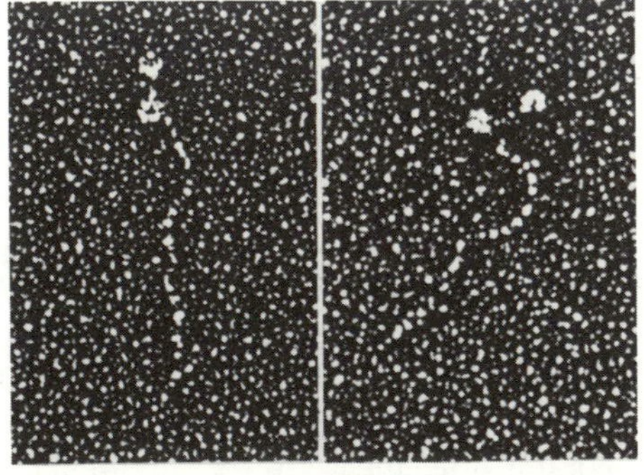

Figure 12-8
An electron micrograph of two myosin molecules prepared by the replica method. Note the two-headed region of the protein attached to the long tail. [From S. Lowey, H. S. Slayter, A. G. Weeds, and H. Baker, *J. Mol. Biol.* 42(1969):1.]

axial ratio is about 80. Figure 12-8 shows an electron micrograph of myosin taken many years after the diffusion measurement was done; the predicted high axial ratio is clearly correct although the diffusion measurement could not have indicated the dual beads at the end of the polypeptide strand.

Table 12-2
Several Results of Diffusion Measurements.

Molecule	M	D (cm^2/sec)	$\bar{v}$ (cm^3/g)	f/f_0	Maximum possible axial ratio
Pancreatic ribonuclease	13,683	1.19×10^{-6}	0.728	1.14	3.4
Egg-white lysozyme	14,100	1.04×10^{-6}	0.688	1.32	6.1
Bovine serum albumin	66,500	6.1×10^{-7}	0.734	1.31	6.0
Human fibrinogen	330,000	1.98×10^{-7}	0.706	2.35	31.0
DNA	2.5×10^7	$<10^{-8}$	0.55	$\gg 12$	known to be 6×10^4
Tobacco mosaic virus	4×10^7	4.4×10^{-8}	0.73	2.19	24.0

Values of D and the axial ratios for several molecules are listed in Table 12-2.

Diffusion measurements can also give information about the arrangement of subunits in a protein because different arrangements would have distinct shapes. This is shown in the following example.

☐ The arrangement of subunits in hemoglobin. **Example 12-C**

Hemoglobin consists of four subunits, each having a molecular weight of 16,125; the value of D for a subunit is 1.13×10^{-6} cm^2/sec. Repeating the calculation in Example 12-B yields a value of $f/f_0 = 1.1$ so that the axial ratio of a subunit is about 3 (Table 12-1). The value of D for the tetramer is 6.9×10^{-7} cm^2/sec so that $f/f_0 = 1.16$. Thus, the axial ratio of hemoglobin is about 3.5. This means that the subunits cannot be arranged linearly but must form a compact cluster; if there were a linear array, the axial ratio of hemoglobin would be about 12 and f/f_0 would have to be about 1.6.

SELECTED REFERENCES

Bancroft, F. C., and D. Freifelder. 1970. "Molecular Weights of Coliphages and Coliphage DNA. 1. Measurements of the Molecular Weights of Phage Particles by High Speed Equilibrium Centrifugation." *J. Mol. Biol.* 54:537–546. An example of $\bar{v}$ measurement by pycnometry and measurement of nitrogen and phosphorus.

Bauer, N. 1949. "Determination of Density," in *Physical Methods of Organic Chemistry*, vol. 1, part 1, edited by A. Weissberger, pp. 253–296. Wiley. A treatise on pycnometry.

Dubin, S. B., G. B. Benedeck, F. C. Bancroft, and D. Freifelder. 1970. "Molecular Weights of Coliphages and Coliphage DNA. 2. Determination of Diffusion Coefficients Using Optical Mixing Spectroscopy and Measurement of Sedimentation Coefficients." *J. Mol. Biol.* 54:547–566.

Dubin, S. B., J. H. Lunacek, and G. B. Benedeck. 1967. "Observation of the Spectrum of Light Scattered by Solutions of Biological Macromolecules." *Proc. Natl. Acad. Sci. U.S.A.* 57:1164–1171. The use of optical mixing spectroscopy to determine *D* is described in this paper and in the preceding one.

Edelstein, S. J., and H. K. Schachman. 1973. "Measurement of Partial Specific Volume by Sedimentation Equilibrium in H_2O–D_2O Solutions," in *Methods in Enzymology*, vol. 27, edited by L. Grossman and K. Moldave, pp. 83–98. Academic Press.

Einstein, A. 1956. *Investigations on the Theory of Brownian Movement*. Dover. The classic on diffusion.

Gosting, L. J. 1956. "Measurement and Interpretation of Diffusion Coefficients of Proteins." *Adv. Protein Chem.* 11:429–554. A good review of the technology of diffusion measurement.

Kratky, O., H. Leopold, and H. Stabinger. 1973. "The Determination of the Partial Specific Volume of Proteins by the Mechanical Oscillator Technique," in *Methods in Enzymology*, vol. 27, edited by L. Grossman and K. Moldave, pp. 98–110. Academic Press.

Kupke, D. W. 1973. "Density and Volume Change Measurements," in *Physical Principles and Techniques of Protein Chemistry*, part C, edited by S. J. Leach, pp. 1–75. Academic Press.

Kupke, D. W., and T. E. Torrier. 1978. "Protein Concentration Measurements: The Dry Weight," in *Methods in Enzymology*, vol. 48, edited by C. W. H. Hirs and S. N. Timasheff. pp. 155–162. Academic Press.

PROBLEMS

12–1. The partial specific volumes of amino acids have different values if the amino acids are in LiCl and KCl. Would you expect this to be true of proteins? Explain and estimate the magnitude of the effect.

12–2. Would $\bar{v}$ of RNA differ if the RNA were in NaCl rather than $MgCl_2$?

12–3. A pycnometer is being used to measure $\bar{v}$ of a solute. The pycnometer weighs 14.2056 g if empty and 24.1305 g if filled with H_2O at 20°C. A solution is prepared by dissolving 3.5921 g of the solute in 9.9413 g of H_2O. The pycnometer is filled with this solution at 20°C and weighs 25.5307 g. What is $\bar{v}$ for the solute? (The density of H_2O at 20°C is 0.9982 g/cm^3.)

12–4. A macromolecule is known to bind the Hg^{2+} ion. How is $\bar{v}$ for the molecule to which the Hg^{2+} ion is bound related to $\bar{v}$ of the molecule lacking the metal ion?

12–5. If in determining *D* from measurements of *A* and *H*, $(A/H)^2$ plotted against *t* produces a curve rather than a straight line, what conclusion might you draw? Suppose that the curve has two distinct components, each asymptotic to a straight line. What conclusion might you draw?

12–6. A macromolecule with $\bar{v} = 0.74$ cm^3/g is sedimented in H_2O at 20°C; $s^{\circ}_{20,w}$ is 14.2 S; $D^{\circ}_{20} = 5.82 \times 10^{-6}$ cm^2/sec. What is the molecular weight?

12–7. In the absence of convection, the width of a sedimentation boundary, observed in an analytical ultracentrifuge, is determined almost entirely by diffusion. Indeed, the rate of boundary spreading can be used to determine D. Would you expect the measurement of D to be more accurate at high or at low centrifugal speed? Explain. Would it ever be reasonable to perform such a measurement with a mixture of two components? When?

12–8. Will D increase or decrease as axial ratio increases? Which has the greater D, a rigid rod or a flexible rod, both having the same length and cross section and made of the same material?

12–9. Answer the following:

a. What is the diffusion coefficient in water at 20°C of a virus particle whose molecular weight is 5×10^7 and which is spherical? The particle is assumed to be 50% protein (density = 1.3 g/cm^3) and 50% DNA (density = 1.7 g/cm^3). The viscosity of water is 0.01 poise.

b. Many phages have long protein tails used for attachment to bacteria. If this phage had a tail, would D be greater or smaller than the value for a spherical, tailless phage?

12–10. A macromolecule whose molecular weight is 22,600,000 and density is 1.79 g/cm^3 has a value of D of 2.1×10^{-8} cm^2/sec in water at 20°C. What may you conclude about the shape of the molecule?

12–11. A protein has a molecular weight of 366,000 and a diffusion coefficient of 2.58×10^{-7} cm^2/sec in water at 20°C. After heating to 75°C and restoring the temperature to 20°C, the molecular weight is found to be 61,000 and the diffusion coefficient is 1.07×10^{-6} cm^2/sec. What can be said about the structure of the molecule?

12–12. A protein is sedimented at 25°C in pure water (density = 0.998 g/cm^3). Its sedimentation coefficient is 8.6×10^{-13} sec and the diffusion coefficient is 6×10^{-7} cm^2/sec. What is the molecular weight? $\bar{v} =$ 0.74 cm^3/g.

12–13. The sedimentation coefficient and the diffusion coefficient for a particular protein are found to be 18.3×10^{-13} sec^{-1} and 4.62×10^{-7} cm^2/sec at 20°C. The partial specific volume of the protein is 0.73 cm^3/g. What is the molecular weight of the protein? At 20°C the density of water is 0.998 g/cm^3.

13

Viscosity

Solutions containing macromolecules have greater viscosities than does the solvent alone. The viscosity increment over the solvent alone is a function of several parameters of the molecule, each of which increases this increment. These parameters are the volume of the solution that is occupied, the ratio of length to width of the molecule (the *axial ratio* or the ratio of the axes of the smallest ellipsoid of revolution in which the molecules could fit), and the rigidity of the molecule. For globular molecules such as many of the proteins, the principal effect is through molecular volume and this is simply related to molecular weight. For very rigid, thin molecules, such as DNA, the major effect is due to the axial ratio and this is also a function of molecular weight. Hence, viscometry can be used for the determination of M; on the other hand, if M is known even approximately, information about the overall shape of the molecule can be obtained. These are the two main uses of viscometry.

Simple Theory of Viscosity

When any substance moves across a surface, the motion is impeded by friction. If the substance is a liquid, this friction generates the effect called viscosity.

Consider a liquid between two large parallel plates (Figure 13-1A), one of which is stationary and the other moving in the x direction with velocity v. The infinitesimal layer of liquid next to each plate encounters frictional resistance to the motion. Hence, the moving plate carries liquid in the x direction at a velocity nearly equal to v and the layer next to the

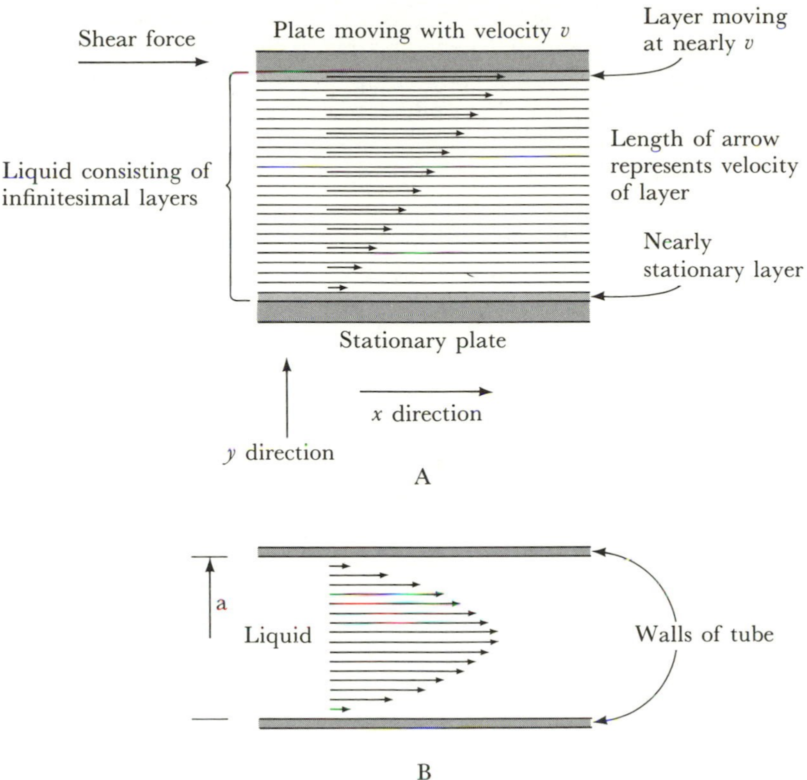

Figure 13-1
A. Shearing of a liquid between two parallel plates, one stationary and one moving at velocity v. The position of the tips of the arrows defines the velocity gradient. B. Velocity profile of a liquid flowing through a cylinder with radius "a".

stationary plate moves very slowly. If the liquid is thought of as consisting of a large number of layers, each layer will slide along the adjacent one, and the frictional resistance between the adjacent layers generates a velocity gradient (Figure 13-1A). The kind of deformation of a liquid produced by a velocity gradient is called *shear*. Newton showed that the frictional force f between the layers is proportional to the area A of the layers and to the velocity gradient dv/dy between them; that is,

$$f = \eta A\left(\frac{dv}{dy}\right). \tag{1}$$

One usually calls η the *coefficient of viscosity*, or simply the *viscosity*; $f/A = F$, the *shear stress*; and $dv/dy = G$, the *shear gradient*, or *shear rate*. If η is a constant, the fluid is called Newtonian; if η is a function of F or G, the solution is called non-Newtonian.

If, instead of being between plates, the liquid is flowing through a cylindrical tube, the friction is encountered at the walls of the tube; in this case, the velocity is maximum along the axis of the tube and minimum adjacent to the walls (Figure 13-1B) so that the velocity gradient is parabolic instead of linear.

All viscometers in use in physical biochemistry employ either the parallel plate or the tube configuration.

The conditions of flow described in Figure 13-1A and B are called *laminar* flow and persist as long as the shear gradient is not too great; at very high shear gradients, *turbulence* sets in and the situation becomes difficult to treat both theoretically and experimentally. Turbulence will not be discussed here.

Effect of Macromolecules on the Viscosity of a Solution

The addition of macromolecules to a solvent* with viscosity η_0, yields a solution of higher viscosity, η. This can be thought to result from increased friction between adjacent unimolecular liquid planes (see Figure 13-1) caused by the fact that the macromolecules are larger than the solvent molecules and hence extend through several of these hypothetical planes. The change in viscosity is usually expressed as a ratio, η/η_0, called the relative viscosity, η_r. Einstein showed that η_r is a function of both the size and the shape of the macromolecules and derived the equation

$$\eta_r = \eta/\eta_0 = 1 + a\phi + b\phi^2 + \cdots \qquad (2)$$

in which a is a shape-dependent constant ($a = \frac{5}{2}$ for spheres), ϕ is the fraction of the solution volume occupied by the molecules, and b is a second shape-dependent constant. This equation can be rewritten in terms of the concentration, c, of the macromolecules by defining V as the specific volume of one molecule, so that $\phi = Vc$, to give

$$\eta_r = \eta/\eta_0 = 1 + aVc + bV^2c^2 + \cdots \qquad (3)$$

Viscosity is frequently expressed as the *specific viscosity*, η_{sp}, which is the fractional change in viscosity produced by adding the solute, that is,

$$\eta_{sp} = \frac{\eta - \eta_0}{\eta_0} = \frac{\eta}{\eta_0} - 1 = \eta_r - 1 = aVc + bV^2c^2 + \cdots \qquad (4)$$

Neither η_r nor η_{sp} can be simply related to molecular parameters (i.e.,

---

* *Solvent* as used here refers to either a pure solvent or a dilute solution of small molecules such as salts.

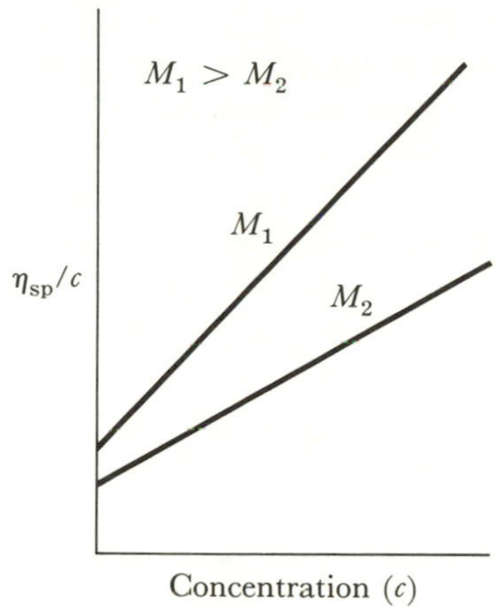

Figure 13-2
Plot of η_{sp}/c versus c for two different DNA molecules. Typically, the curves are linear and less steep as the molecular weight (M) decreases. Although not shown, the curve is much flatter for spherically symmetric molecules than for rods.

shape and volume) because of intermolecular interactions (e.g., collision, entanglement). To avoid this problem, one must consider the situation at very low (i.e., zero) concentration. To do this, the *intrinsic viscosity* $[\eta]$ is defined as

$$[\eta] = \lim_{c \to 0} \frac{\eta_{sp}}{c} = \lim_{c \to 0} aV + bV^2c + \cdots \simeq aV \tag{5}$$

which depends only on the shape-dependent constant, a, and the specific volume, V. Operationally, this means that $[\eta]$ is determined by measuring η_{sp} at several concentrations, plotting η_{sp}/c versus c and extrapolating to $c = 0$. An example of the kind of data usually obtained is shown in Figure 13-2.

As defined above, $[\eta]$ is still not always a useful parameter because of the possible non-Newtonian dependence of $[\eta]$ on G. When a solution of macromolecules having a large axial ratio (i.e., long and thin, such as DNA) is sheared, the molecules tend to become oriented so that their long axes are parallel to the direction of flow of the solution. This orientation decreases η_r because the resistance contributed by the macromolecules to the sliding of the layers past one another decreases as the molecules are confined to fewer layers. The strength of this dependence of η_r on the shear gradient increases with axial ratio because a long, thin molecule is more easily oriented than a short, thick one. This phenomenon can be seen in Figure 13-3B, which shows $[\eta]$ versus G for DNA molecules of different molecular weight—clearly, the greater M is (which for DNA means a greater axial ratio), the lower $[\eta]$ is for a given value of G.

If shape is dependent on the shear gradient (i.e., if the molecule is deformed by the shear), $[\eta]$, which depends on shape (see equation 5), will also decrease with increasing G. Therefore, to determine molecular parameters from $[\eta]$, it is necessary to measure $[\eta]$ as a function of G

and extrapolate to $G = 0$. Unfortunately, there is not a special notation for $[\eta]$ at $G = 0$ (although $[\eta]_0$ seems reasonable); however, in the scientific literature, $[\eta]$ invariably means the value extrapolated to $G = 0$ if there is dependence on G. Figure 13-3A shows why $[\eta]$ must be measured at low shear rather than simply extrapolated from high values.

These considerations lead to two qualitative rules-of-thumb, which can be used to estimate whether a given macromolecule has a large or a small axial ratio: (1) if η_r is large at low concentration, the macromolecule must have a high axial ratio and, similarly, if η_r is small at high concentrations, the molecule must be somewhat compact; (2) if η_r decreases significantly with increasing shear gradient, the axial ratio must be high.

Two other effects of shear on η_r are worth mentioning : *degradation* and *rheopexy*.

Degradation refers to the fact that at high shear stress, long, thin molecules are broken. Figure 13-4 shows the state of such a molecule in a velocity gradient (illustrated by Figure 13-1B). Note that, because the velocity of flow is not constant across the tube, the ends of any molecule at an angle to the streamlines will move at two different velocities. This

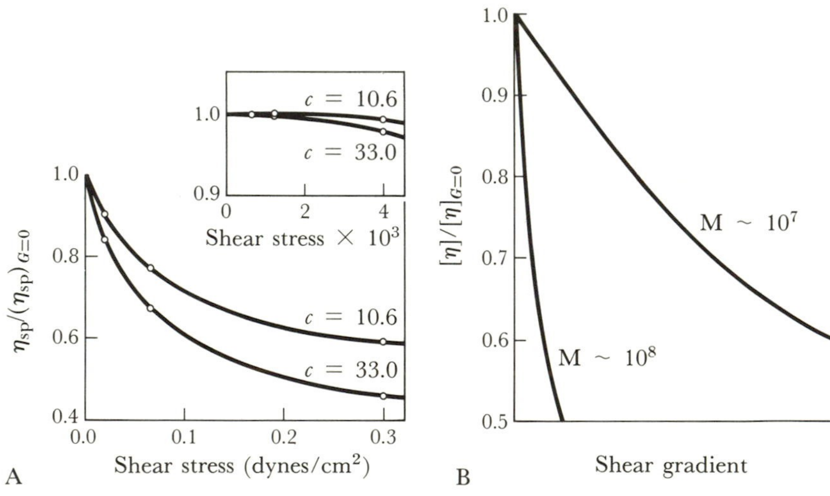

Figure 13-3
A. Dependence of η_{sp} of T2 DNA ($M = 106 \times 10^6$) on shear stress, expressed as a percentage of the value at zero shear. Curves for two different concentrations are shown in μg/ml). The inset shows the change in shape at very low shear. The zero slope shows why low shear viscometers are important in the study of native DNA because the extrapolation to zero shear from large values can cause great errors. [From D. M. Crothers and B. H. Zimm, *J. Mol. Biol.* 12(1965):625–536.] B. Curves showing the effect of molecular weight on the dependence of $[\eta]$ on G. (Data only approximate; collected from a variety of unrelated experiments.)

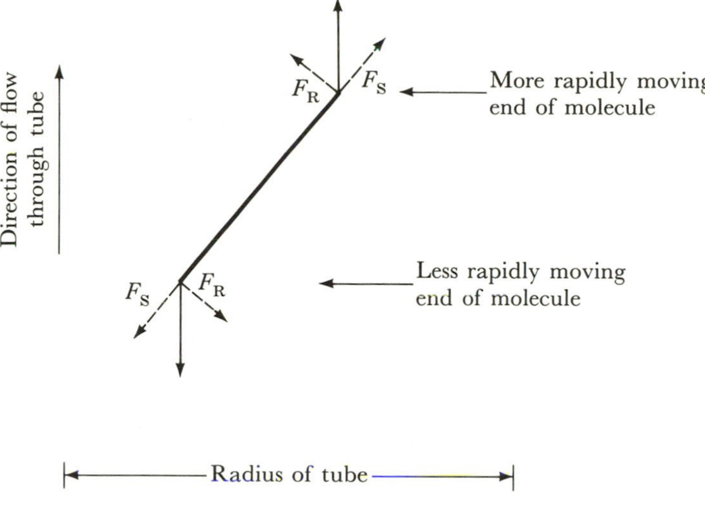

Figure 13-4
A rodlike molecule in a velocity gradient generated by flow through a tube. The streamlines move more rapidly near the center of the tube than near the walls. Hence, the molecule is subjected to a driving force at one of its ends and a retarding force at the other. These can be resolved into the forces F_R, which tends to rotate the molecule, and the stretching force, F_S, which causes the molecule to break near the center.

will tend to rotate the molecule until it is aligned with a streamline, in which case it experiences no force. Hence, when a molecule is at an angle to the streamlines, the forces at either ends of the molecule are not the same. These forces can be resolved into a perpendicular force (which produces rotation) and a parallel force, which produces stretching (Figure 13-4). It can be shown that the stretching force is maximal (1) in the center of the molecule and (2) when the molecule is at an angle of 45° to the streamlines. Because the force is greatest at the midpoint, the molecule will have the greatest probability of breaking in half. Several interesting studies of this halving phenomenon for DNA have been made; some of the results are shown in Figure 13-5.

Experimentally, degradation is recognized by either a sharp drop in η_r at some value of G in an η_r versus G curve such as that shown in Figure 13-3B or a decrease in η_r in repeated measurements of η_r at a given value of G. Shear degradation is of great concern with high-molecular-weight DNA (see Chapter 19).

Rheopexy refers to the fact that at relatively high shear stress and concentration, η_r increases with time. This is observed only with material of very high molecular weight, such as the DNA of eukaryotic chromosomes, and the phenomenon is not clearly understood. Because η_r

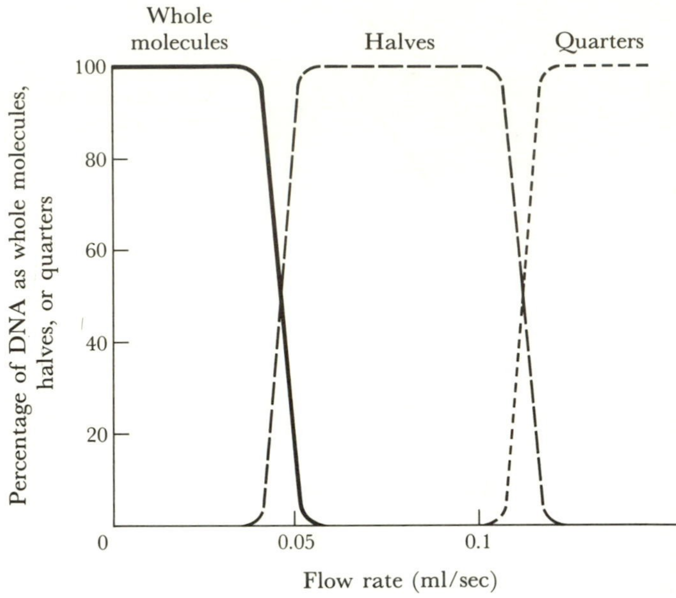

Figure 13-5
Breakage of T2 DNA by passage through a fine tube of 0.25 mm diameter at 0.1 μg/ml and various flow rates. The size of the DNA was determined by zonal centrifugation in an H_2O–D_2O concentration gradient. At a critical rate, the halves are further broken into quarters. [Somewhat idealized curves are extrapolated from the data of C. Levinthal and P. F. Davison, *J. Mol. Biol.* 3(1961):674–683.]

returns to the normal value after shearing is stopped, rheopexy is normally seen only if η_r is measured by repeated determinations under conditions of continuous shear, as with the Zimm-Crothers viscometer (see next section).

Measurement of Viscosity

Many instruments have been developed for measuring viscosity. A description of those used with biological polymers follows. Viscosity varies by about 2% per °C at 20°C. Thus, precise temperature control (±0.01°C) is needed and is done with a thermostat water bath.

Ostwald Capillary Viscometer

This viscometer consists of a capillary tube of radius r and length L through which a volume V passes (Figure 13-6). The instrument is used in the following way. The solution is added at opening 1 until the liquid level at rest is at scratch C. Suction is then applied at opening 2 until

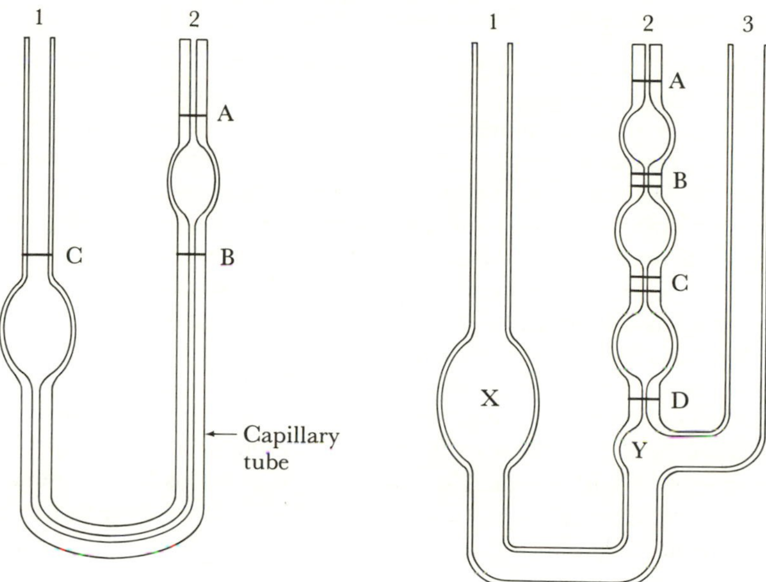

Figure 13-6
Capillary visometers: (left) Ostwald type; (right) Ubbelohde type. See text for details of operation. In the Ubbelohde type, B and C are pairs of scratches. The upper scratch is for timing the movement of the meniscus from A to B and the lower from B to C. The double scratch is to allow time to restart a stopwatch. The relative shear gradients are usually calculated from constants provided by the manufacturer.

the liquid level is above scratch A. The suction is removed and the liquid falls owing to the difference in height between the two arms (i.e., the hydrostatic head). The time t required for the meniscus to move between scratches A and B is measured. Because of the change in the relative liquid heights, the flow rate is not constant. The viscosity η and *average* G are

$$\eta = \frac{\pi hg\rho r^4 t}{8LV} \text{ and } G = \frac{8V}{3\pi r^2 t} \tag{6}$$

in which h is the average liquid height, g the gravitational constant, and ρ the density. Precise evaluation of h, r, and L is avoided by measuring

$$\eta_r = \frac{\eta}{\eta_0} = \frac{t}{t_0} \cdot \frac{\rho}{\rho_0} \tag{7}$$

This instrument has the advantage of low cost but the disadvantage that the shear gradient cannot be varied, and solutions must be relatively dust-free to avoid clogging the fine capillary.

Ubbelohde Capillary Viscometer

As indicated earlier, to measure $[\eta]$ at $G = 0$, one must be able to vary both concentration and shear. The Ubbelohde viscometer is designed with this in mind. As shown in Figure 13-6, this viscometer has several bulbs so that, with reduced pressure head, the average shear gradient decreases. The instrument is used in the following way. Liquid is added through opening 1 to fill bulb X. Opening 3 is closed and suction is applied at opening 2, until the liquid has been drawn above A. Opening 3 is then opened with opening 2 closed, which allows bulb Y to drain. Opening 2 is then opened and the times for the meniscus to pass scratches A, B, C, and D are determined. This gives the viscosity at three different values of G. For varying concentration, the liquid in bulb X can be diluted because the amount of liquid in X does not determine the volume of liquid contained between A and the bottom of the capillary. This instrument is fairly breakable and easily clogs but is highly useful as long as very low values of G are not required.* The Ubbelohde viscometer is no longer in common use, having been replaced by the Zimm-Crothers viscometer.

Couette Viscometer

This instrument, designed for use at relatively low shear gradients, consists of two concentric cylinders separated by a narrow annulus, which is filled with the sample (Figure 13-7). One cylinder is fixed and the other rotates. The situation is equivalent to the two parallel plates shown in Figure 13-1A. From the angular velocity and the dimensions of the cylinder, the terms in equation (1) can be calculated as follows. The velocity of the liquid layer next to the stationary cylinder is nearly zero; for the rotating cylinder, it is the speed of rotation. At intermediate layers, the velocities are proportional to the radial distance from the stationary cylinder. Therefore, the shear gradient is constant and is

$$G = \frac{\pi RS}{30d} \tag{8}$$

in which R is the average radius of the cylinders, S is the rotor speed in rpm, and d is the annular distance. The shear stress F is $T/2\pi R^2 h$, in which h is the height of the cylinder and T is the torque necessary to maintain the speed S. Hence, this instrument operates at a low, controllable, and easily measured value of G, which is of great value with

* The limitation on G is a result of the fact that G is primarily determined by the length and diameter of the capillary. As the diameter is increased, the flow rate would increase to the point that it would be too great to measure—that is, the meniscus would move too rapidly. To compensate for this, the length of the capillary would have to be increased. For very low shear gradients, the length would be unmanageably great.

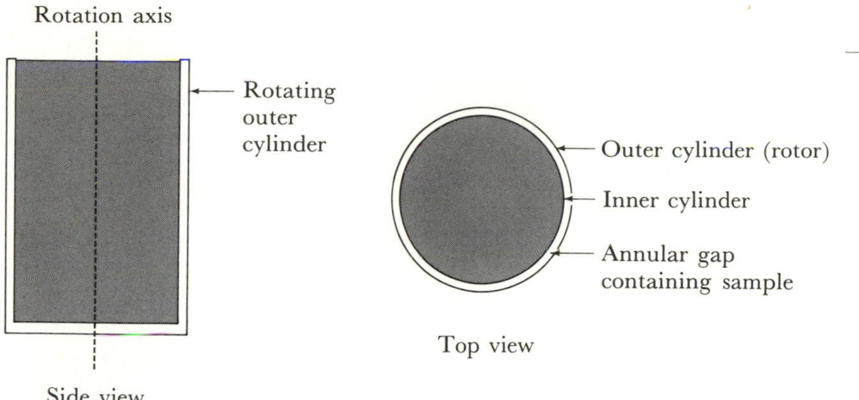

Figure 13-7
A Couette viscometer: the outer cylinder usually rotates; the inner cylinder is stationary.

non-Newtonian liquids for which an extrapolation of η to $G = 0$ is necessary. A complication of the Couette viscometer arises from the forces at the ends of the viscometer (the "end effects"). Note that the top surface of the liquid experiences the force of capillarity or surface tension; at the bottom, the liquid is between a stationary and rotating surface. In practice, a Couette viscometer is made sufficiently long that end effects become proportionally small.

Zimm-Crothers Floating-Rotor Viscometer and the Cartesian-Diver Modification

This modification of the Couette viscometer allows high-precision work at very low shear gradients (Figure 13-8). It differs in principle from the Couette type in that, with a Couette, the shear gradient is fixed and the shear stress is measured, whereas, with this viscometer, the shear stress is fixed and the shear gradient is measured. The solution is placed between a stationary outer cylinder and a potentially rotating inner cylinder, which floats in the liquid. The inner cylinder contains a steel pellet. External to the outer cylinder are rotating magnets, which cause the steel pellet, and therefore the inner cylinder, to rotate. The speed of this rotation depends on the amount of steel in the inner cylinder and the viscosity of the fluid and is independent of the speed of the rotating magnet and the density of the sample. The shear rate is varied by changing the amount of steel in the inner cylinder (by adding more pellets). Because the torque is proportional to the amount of metal in the rotating magnetic field, the rotor will turn more rapidly and the shear gradient will increase. The shear gradient is equal to the tangential velocity divided by

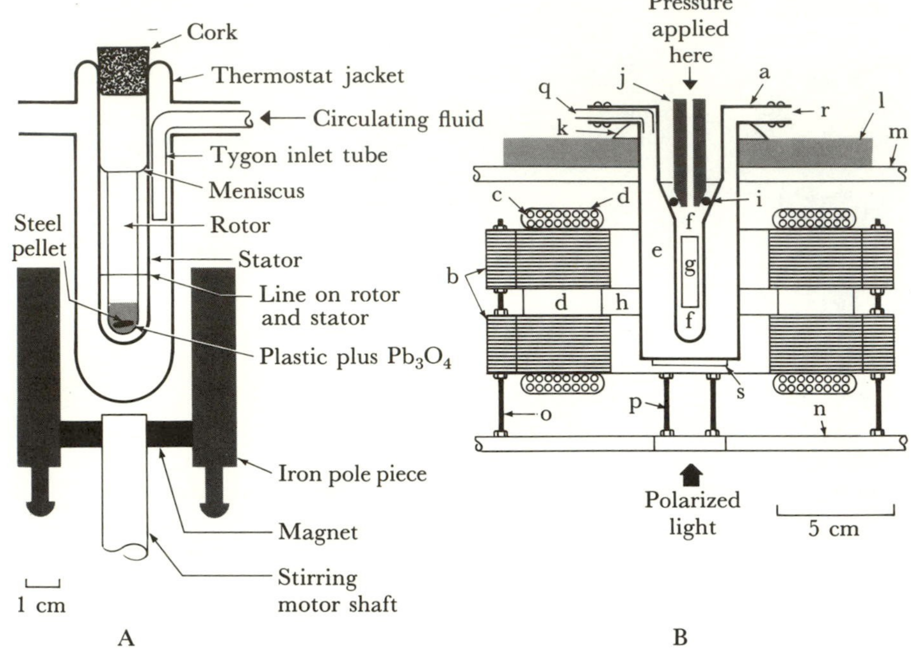

Figure 13-8

A. The Zimm-Crothers viscometer. [From B. H. Zimm and D. M. Crothers, *Proc. Natl Acad. Sci. U.S.A.* 48(1962):905–911.] B. The Cartesian-diver modification: (a) the glass, thermostated sample chamber (black section lining), (b) electromagnet laminations (horizontal section lining), (c) magnet wire, (d) insulating tape around outside of magnet wire, (3) circulating water, (f) DNA solution, (g) Cartesian-diver rotor, (i) rubber O-ring used to form pressure seal, (j) "pressure bar," (k) plaster, (l) Bakelite block, (m) upper aluminum support plate, (n) middle aluminum support plate, (o, p) nuts and bolts for attaching magnet poles to frame, (q, r) circulating water inlet and outlet, respectively, (s) frosted glass plate. All parts are drawn to the scale at the bottom of the figure except the magnet laminations (b) and the magnet wire (c). [Reprinted with permission from L. C. Klotz and B. H. Zimm, *Macromolecules* 5(1972):471–481. Copyright by The American Chemical Society.]

the radial distance between the cylinders. This instrument simply measures the viscosity relative to a standard, such as water, by determining the relative amounts of time required for the inner cylinder to make a given number of rotations.

A problem with all rotating-cylinder viscometers is the effect of surface films at the meniscus (the end effects referred to in the preceding section). The Cartesian-diver rotating-cylinder viscometer avoids this problem. In this modification of the Zimm-Crothers instrument, the inner rotor contains a trapped air bubble, which is compressed by exerting pressure on the solution from above. This converts the inner cylinder into a Cartesian diver, which can be adjusted to be totally immersed in the liquid. This instrument is by far the best available for precise measurement at low shear rate. A modification of the Cartesian-diver viscometer, called a viscoelastometer (used for another purpose), is described in a later section.

Relation Between Intrinsic Viscosity and Molecular Weight

For polymers whose configuration is near that of a random coil, the relation between $[\eta]$ and M is of the form

$$[\eta] = KM^a \tag{9}$$

in which K and a are constants depending on the solvent. These constants are usually determined empirically for each solvent-solute system, using molecules of known M, because the theory is not yet adequate for calculating them. Equations of this type have not been derived for very compact molecules such as viruses because $[\eta]$ depends in a complicated way on the ratio of molecular volume to molecular weight. In general, for a given value of M, $[\eta]$ is smaller for the compact form than for a random coil of the same M. Actually, viscometry is rarely used to characterize highly compact molecules because other methods such as centrifugation are of greater value.

For proteins denatured in solutions of 6 M guanidinium chloride (a substance that breaks all hydrogen bonds so that a protein is a random coil if there are no intrastrand disulfide bridges), a and K are well known from extensive data collection. The relation is usually written as

$$[\eta] = 0.716n^{0.66} \tag{10}$$

in which n is the number of amino acid residues in the protein. The average molecular weight per residue can be determined from the amino acid composition so that M can be calculated from η. Because multi-subunit proteins dissociate in 6 M guanidinium chloride, M is the molecular weight of a subunit and not that of the intact protein. This relation can be used to determine other properties of proteins, as will be seen in Examples 13-A and 13-B.

For double-stranded linear DNA molecules, the relation between $[\eta]$ and M has been found to be

$$0.665 \log M = 2.863 + \log([\eta] + 5) \tag{11}$$

This strictly empirical equation can be used to calculate M if the DNA sample is homogeneous with respect to molecular weight. This precaution must be borne in mind because of the great sensitivity of DNA to degradation by shearing induced by handling and isolation procedures (see Chapter 19).

Examples of the Use of Viscometry

Viscometry can be used in a quantitative way to determine molecular weight or semiquantitatively to estimate shape or to detect changes in molecular weight or in shape. Molecular weight is calculated using data of the type shown in Figures 13-2 and 13-3; that is, η_r is measured at several values of c and G, and $[\eta]$ is determined for each value of G by extrapolating η_{sp}/c to $c = 0$ and then again extrapolating these values to $G = 0$. This is straightforward and will not be explained further. The examples that follow show the semiquantitative uses of viscometry. The basic rule used is that stated on page 474—an increase or a decrease in viscosity indicates an increase or a decrease in axial ratio, respectively. It should be noted in the examples that this type of information can often be obtained by measuring η_r only.

Example 13-A ☐ Estimation of the overall shape of proteins.

As discussed in Chapter 1, a protein can be roughly categorized as being highly compact, a random coil, helical, or semirigid (or a combination of the last three). A statement about the overall shape can be made by comparing the viscosity of native and denatured proteins, if the protein consists of a single subunit. (Denaturation can be accomplished by acid, high temperature, or the addition of denaturants such as guanidinium chloride.) Because the denatured form is usually a random or near-random coil (see Example 13-B for this distinction), one can determine whether the native form is more or less compact than a random coil by noting whether the viscosity increases or decreases on denaturation. For example, η_r of ribonuclease increases markedly on thermal denaturation in acid, indicating that it has a compact native structure (which is the case for most globular proteins). On the other hand, the viscosity of poly(γ-benzyl-L-glutamate) in the rigid rod form decreases fourfold if placed in conditions in which it is random coil. Similarly, the viscosity of the protein collagen and of DNA, highly rigid triple- and double-stranded helices respectively, decreases markedly on denaturation. (Note, however, the implicit assumption that the conditions of denaturation do not introduce any degree of depolymerization and that the decrease in viscosity is due solely to a change in configuration. This is not always the case—for example, low pH, which denatures DNA, also produces single-strand breaks and therefore is to be avoided if simple denaturation is desired.)

Example 13-B ☐ Detection of intrastrand disulfide bonds in proteins.

The cysteine moieties of proteins are frequently coupled by means of disulfide bonds. These disulfide bonds prevent a protein from assuming a completely random coil configuration when denatured, especially if the cysteines are separated by a large number of amino acids.

Such a molecule might be described as a nearly random coil because it is more compact than a true random coil. Disulfide bonds are broken by reduction with 2-mercaptoethanol and reformation of these bonds from the resulting sulfhydryl (SH) groups can be prevented by *S*-carboxymethylation with iodoacetamide. Hence, if disulfide bonds are present in the native structure, the viscosity in 6 M guanidinium chloride will be *greater* after treatment with mercaptoethanol, because the molecule will be less compact. For example, calf brain tubulin has $[\eta] = 36.0$ ml/g before reduction and 44.0 ml/g after reduction, indicating that disulfide bridges are present.

From the values of $[\eta]$ for the reduced and nonreduced form, information can be obtained about the distance between the cysteines participating in the S—S bond. It has been calculated that the ratio of $[\eta]$ for a single polymer in a straight-chain random-coil configuration to that of a ring structure is 1.6. Hence, in the tubulin example, because the value of 44.0 ml/g represents the straight chain (i.e., no S—S bonds), a circle (i.e., when the S—S bonds are between the two terminal amino acids) would have $[\eta] = 44.0/1.6 = 27.5$ ml/g. Because the observed value is 36.0 ml/g, the S—S bonds are clearly not between terminal amino acids. If the bonds were between adjacent amino acids, they would have little effect and $[\eta]$ should be very near (and probably indistinguishable from) 44.0 ml/g. Hence, the S—S bonds are not between amino acids that are located close to one another. Because $[\eta]$ is about halfway between 27.5 and 44.0 ml/g, it may be concluded that, if there is one or a small number of S—S bonds, the cysteines must be reasonably far apart along the polypeptide chain but not too near the termini.

Example 13-C

☐ Circularity versus linearity in DNA molecules.

In 1961, it was shown that the genetic map of *E. coli* phage T2 is circular. Because it was not possible to observe DNA with the electron microscope at that time, to test for circularity, the change in η_r of a single sample of T2 DNA, which was continuously being digested by pancreatic DNase, was studied. This enzyme produces single-strand phosphoester breaks that accumulate and ultimately match to form double-strand breaks (Figure 13-9A). The single-strand breaks alone have no effect on DNA viscosity (Figure 13-9B) because of the rigidity of the molecule conferred by base stacking (see Chapter 16). Hence, for a linear molecule, η_r is constant during the period in which there are only single-strand breaks and decreases when matching occurs, owing to decreasing molecular weight. However, when the first double-strand break occurs in a circular DNA molecule, the molecule becomes linear and η (Figure 13-9C) increases; because the axial ratio is increased, η_r does not begin to decrease until two double-strand breaks occur. For T2 DNA, η_r is constant for several hours of DNase treatment and then decreases, indicating that it is a linear molecule.

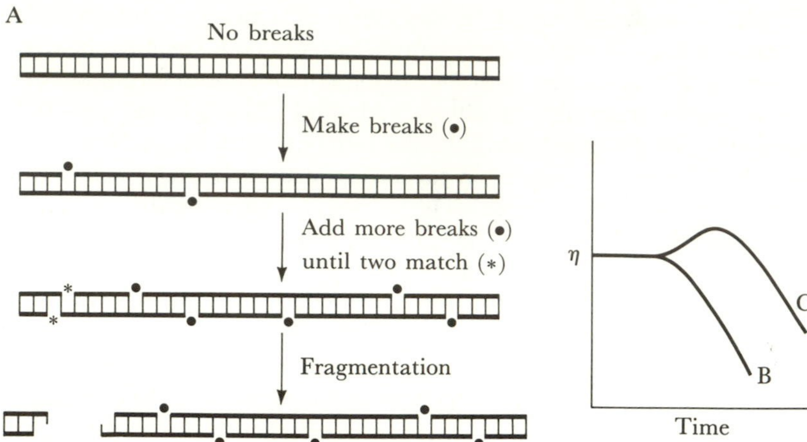

Figure 13-9
A. Process of the matching of single-strand breaks in DNA to make double-strand breaks. The dots indicate the positions of the breaks. The asterisks indicate a pair of breaks that result in a double-strand break: they are separated by only one base pair, which is insufficient to maintain the integrity of the molecule. B. Viscosity of T2 DNA as a function of time of digestion with pancreatic DNase. The DNase produces single-strand breaks, which ultimately match to produce double-strand breaks, resulting in a decrease in the viscosity. The curve is flat at first because single-strand breaks do not affect the viscosity. C. Expected curve (as in curve B) if the DNA were circular. At the time that double-strand breaks occurred, the circle would become linear and the viscosity would increase. Viscosity would not decrease until a second double-strand break had formed.

Example 13-D ☐ Viscometric evidence that certain substances can intercalate between nucleotide bases of DNA.

The dye acridine orange binds tightly to double-stranded DNA. The sedimentation coefficient (Chapter 11) of the complex *decreases* compared with that of the free DNA; this decrease could result either from depolymerization or from an increase in the frictional coefficient due to an increase in the axial ratio. If the decrease is due to depolymerization, $[\eta]$ should decrease; if due to an increase in the frictional coefficient, it will increase. Experimentally, $[\eta]$ of the complex is greater than that of free DNA, so that the axial ratio of the complex must be greater than that of the free DNA. Hence, the DNA must increase in length when the dye is bound. Fluorescence polarization studies (Chapter 15) have shown that the dye is immobilized and is in the same plane as the base pairs. Hence, it has been inferred that the dye intercalates between the DNA bases, thus lengthening the molecule (Figure 13-10). This inference has been confirmed by electron microscopy of the complex; DNA becomes longer and less flexible when the dye is bound.

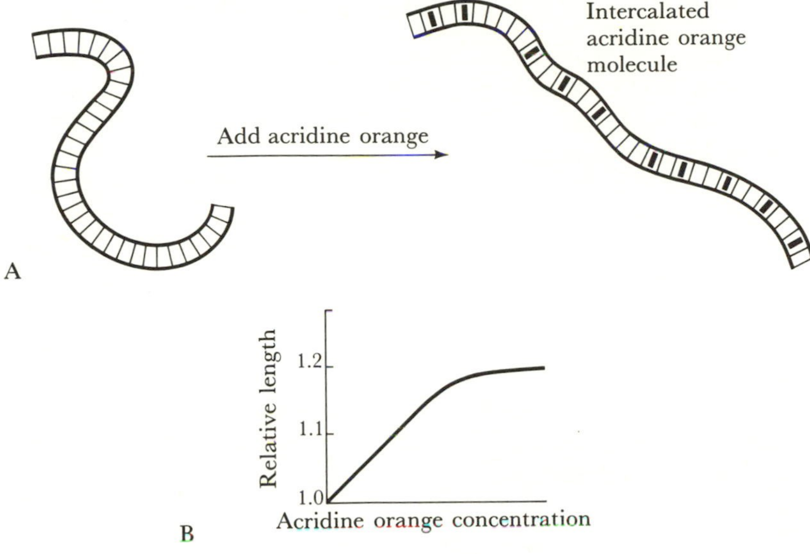

Figure 13-10
A. Acridine orange molecules intercalating between the base pairs of DNA. The molecule is extended and becomes more rigid. B. The length of *E. coli* phage λ DNA as a function of acridine orange concentration, as determined by electron microscopy. [From the laboratory at Brandeis University.]

Example 13-E

☐ Detection of the enzymatic polymerization of DNA from mononucleotides.

If a mixture of DNA and radioactive nucleoside triphosphates is incubated with the enzyme *E. coli* DNA polymerase I, some radioactive material becomes acid precipitable (see Chapter 5, Example 5-A). This could result from net synthesis leading to *an increase* in the amount of high-molecular-weight DNA or to nucleotide *exchange* into the original molecules. If there is net synthesis, η_r should increase because of an increase in the concentration of the DNA; if there is only exchange, the amount of DNA will be unchanged and η_r will be constant. At the time the work with polymerase I was beginning, DNA could be distinguished from nucleotides by light scattering (using purified protein-free samples with DNA at high concentrations), by ultracentrifugation or ultraviolet absorbance (at low concentration and in the absence of excessive UV-absorbing material), or by viscometry. The enzyme reaction mixture contains DNA at low concentration and both protein and UV-absorbing nucleotides at high concentrations, thus eliminating the first three possibilities. By viscometry, it was seen that η increased with time of incubation, indicating either that more DNA was present or that molecular weight was increased; either case implies DNA synthesis.

Example 13-F ☐ Conformation of DNA-histone (chromatin) complexes.

Histones are proteins contained in chromosomes and bound to DNA. To understand the role played by the histones in chromosomal structure and in the regulation of DNA expression, the structure of the DNA-histone complex has been studied.

As histone is added to DNA, the sedimentation coefficient of the DNA increases. An increase in s means either an increase in M or a decrease in the frictional coefficient (or axial ratio). The intrinsic viscosity also increases, indicating either an increase in molecular weight or an increase in axial ratio. Qualitatively, these two results would suggest that M is increasing. However, as more histone is added, the ratio $s/[\eta]$ increases markedly, which indicates that the axial ratio also decreases with increasing bound histone. Hence, the increase in M by histone binding is probably accompanied by some folding of the DNA.

Example 13-G ☐ Detection of the injection of DNA by phages.

Phages are highly compact and therefore have a very low $[\eta]$ value for their molecular weight. Treatment with certain reagents (e.g., alkaline buffers, $NaClO_4$, and guanidinium chloride) lead to extraordinary increases in $[\eta]$ due to the release of DNA. This can be used to measure the extent of the injection of DNA from the particles. It should be noticed though that, because viscometry measures the average property of a solution, the following cases cannot be readily distinguished: 100% of the phages injecting half of their DNA and half of the phages injecting all of their DNA. Hence, if a set of conditions results in a value of η less than that obtained when the phages are totally disrupted, the observed η cannot be readily interpreted. These cases could, of course, be distinguished by sedimentation experiments because the sedimentation coefficient would depend on the amount injected. Hence, in the first case, a single sedimentation coefficient whose value would be less than that of the intact phage and greater than that of free DNA would be observed; in the second case, two sedimenting species would be seen—the phage and the free DNA.

Measurement of the Viscoelasticity of DNA Solutions

The DNA molecules of bacteria have a molecular weight greater than 2×10^9; that of animal cells approaches 10^{11}. In this range of M, hydrodynamic methods suffer from the complex phenomena indicated in Table 13-1. Note that the viscoelasticity method described herein is insensitive to these artifacts.

Table 13-1
Problematic Phenomena Occurring with Various Hydrodynamic Techniques.

Phenomenon	Sedimentation	Viscometry	Visco-elasticity
Speed-dependent sedimentation	+	−	−
Stress-dependent aggregation	+	+	−
Shear degradation	+	+	−
Variation with shear stress	−	+	−

NOTE: A plus indicates that a problem exists; a minus, that it does not.

If solutions of DNA are subjected to a shear stress, the DNA molecules are extended as shown in Figure 13-11. After the stress has been removed, the molecule returns to a relaxed configuration in which the molecule, if it is very long, approximates a random coil. A solution of an extendable molecule shows viscoelasticity when sheared in a Couette-type viscometer. With the Zimm-Crothers Cartesian diver viscometer (Figure 13-8), the application of a shear stress causes the inner cylinder to rotate. When the shear stress is removed, if the viscometer contains a pure solvent, the angular velocity gradually approaches zero. However, if the rotor is suspended in a solution of high-molecular-weight DNA, the rotor stops and then the direction of rotation reverses owing to the relaxation of the previously stretched DNA molecules; ultimately, of course, the rotor

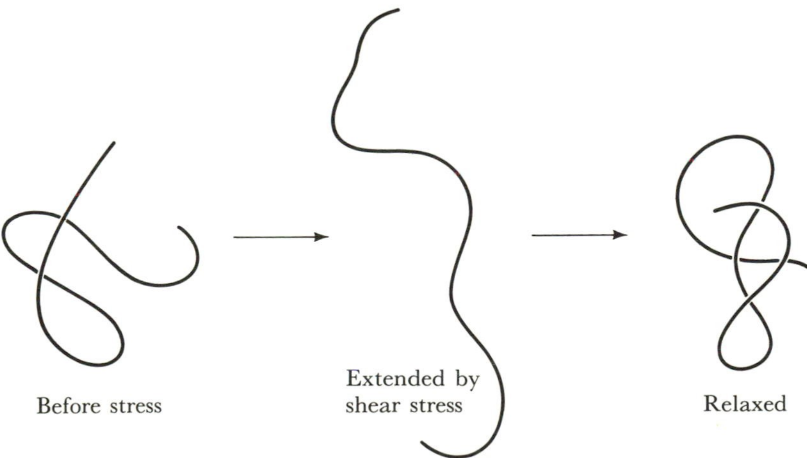

Figure 13-11
Extension of a DNA molecule followed by visoelastic relaxation after the stress has been removed.

stops owing to friction. The exponential decay rate at which the rotor comes to rest can be characterized by a time constant τ_R called a *retardation time* (i.e., the time required to reach 1/e the total angular displacement occurring during reverse motion). When τ_R is extrapolated to zero DNA concentration, it is called the *relaxation time*, τ_R°.

The molecular weight of a linear DNA molecule is determined in a somewhat complex way from the specific viscosity and the value of the retardation time extrapolated to zero shear rate and zero DNA concentration. Details of the measurement are beyond the scope of this book but can be found in some of the Selected References. A theory applicable to circular DNA molecules has been developed also but has not yet been tested experimentally.

The great usefulness of this method is that, with solutions containing molecules of different sizes, the observed value of M corresponds to that of the *largest* molecules present (unlike $[\eta]$ measurements, which yield an average value). Hence, with very fragile molecules, the system is unaffected by the degradation of a fraction of the molecules.

The instrumentation for viscoelasticity measurements is not yet commercially available so that work has been confined to laboratories of Bruno Zimm and his colleagues. To date, the method has been used successfully to evaluate M for the DNAs of large bacteriophages, of several bacteria, algae, and yeasts, and of the chromosomes of several species of the fruit fly *Drosophila* (Table 13-2). This technique should find even greater applications in the future.

Study of the viscoelasticity of DNA molecules has led to the development by Ken Dill and Bruno Zimm of a novel method for separating

Table 13-2
Size of Various DNA Molecules Determined by Viscoelastic Measurements.

DNA	Molecular-weight viscoelasticity	Other methods*
E. coli phage T7	$25 \pm 2 \times 10^6$	25×10^6
E. coli phage T2†	$126 \pm 5 \times 10^6$	111×10^6
E. coli	2.7×10^9	2.6×10^9
Bacillus subtilis	2.0×10^9	$1\text{–}4 \times 10^9$
Drosophila melanogaster	$41 \pm 3 \times 10^9$	–
Drosophila americana	$79 \pm 10 \times 10^9$	–

SOURCE: Data from L. C. Klotz and B. H. Zimm, *J. Mol. Biol.* 72(1972):779–800; R. Kavenoff and B. H. Zimm, *Chromosoma* 41(1973):1–27; R. Kavenoff, L. C. Klotz, and B. H. Zimm, *Cold Spring Harbor Symp. Quant. Biol.* 38(1973):1–8; B.C. Bowen and B. H. Zimm, *Biophys. Chem.* 7(1978):235–252.
* A dash indicates that the measurement has not been made by any other method.
† The discrepancy between the two values has not been explained.

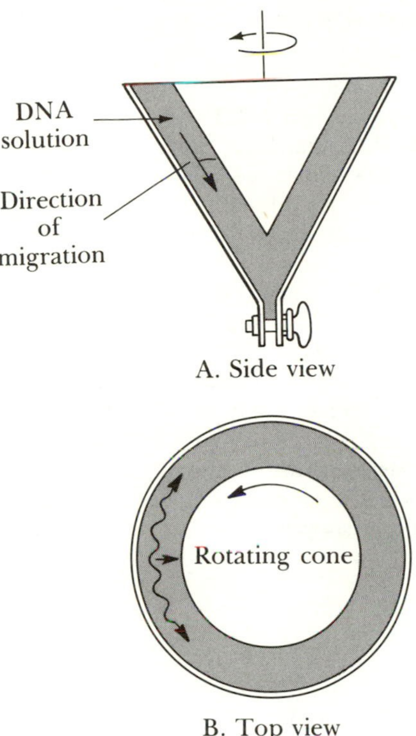

Figure 13-12
Two views of the radial migration separator developed by Ken Dill and Bruno Zimm. [*Nucleic Acids Res.* 7(1979):735–749.]

very large DNA molecules according to molecular weight. This method is called *radial migration separation*. The apparatus and the principle of separation are shown in Figure 13-12. Two concentric cones are arranged as shown in panel A. The space between the cones is filled with a DNA solution and the inner cone is rotated slowly (approximately 1 rotation per 5 seconds). The elastic molecules are aligned and stretched as shown in panel B. Because the molecules are stretched in a curve, there is a small force (central arrow) directed toward the axis of rotation. This makes the molecules migrate toward the center of the system. The rate of migration is proportional to the $\frac{5}{2}$ power of the molecular weight so that separation occurs during the migration. (The separation would presumably be improved if the DNA solution were layered, as in zonal centrifugation, but this has not been tested.) After several hours of migration, the rotation is stopped and the stopcock at the base of the outer cone is opened; the solution is then fractionated by drop collection. The largest molecules emerge first. This method is being used to concentrate and separate intact DNA molecules of chromosomes from the fragments that are produced during isolation of the DNA from tissue.

Bowen, B. C., and B. H. Zimm. 1978. "Molecular Weight of T2 NaDNA from Viscoelasticity." *Biophys. Chem.* 7:235–252.

Bowen, B. C., and B. H. Zimm. 1979. "Improvements in Instrumentation for Viscoelastometry of DNA Solutions." *Biophys. Chem.* 9:133–136.

Eigner, J. 1968. "Molecular Weight and Conformation of DNA," in *Methods in Enzymology*, vol. 12B, edited by L. Grossman and K. Moldave, pp. 386–429. Academic Press.

Uhlenhopp, E. L., and B. H. Zimm. 1973. "Rotating Cylinder Viscometers," in *Methods in Enzymology*, vol. 21, edited by C. H. W. Hirs and S. N. Timasheff, pp. 483–491. Academic Press.

Yang, J. T. 1961. "The Viscosity of Macromolecules in Relation to Molecular Conformation." *Adv. Protein Chem.* 16:323–400. Viscometry as applied to proteins.

Zimm, B. H. 1971. "Measurement of Viscosity of Nucleic Acid Solutions," in *Procedures in Nucleic Acid Research*, vol. 2, edited by G. C. Cantoni and D. R. Davies. pp. 245–261. Harper and Row.

PROBLEMS

13–1. A DNA has an intrinsic viscosity $[\eta]$ of 500 ml/g. When heated to 95°C in 0.01 M NaCl, $[\eta]$ drops to 200 ml/g. When heated to 90°C in 0.5 M NaCl, $[\eta]$ drops to 30 ml/g. Explain the effect of NaCl concentration. What differences might be expected if the heating was performed with 1 M formaldehyde in the NaCl solutions?

13–2. If bacteria are treated with low concentrations of some detergents, the suspension becomes visibly viscous. If the suspension is briefly sedimented, a pellet results that contains all of the DNA and RNA. The supernatant fluid is still viscous. What is the probable explanation for this viscosity?

13–3. Which in the following pairs will produce the greater viscosity when suspended in H_2O: (a) two spheres of identical mass but different radii; (b) a solid sphere and a porous sphere through which solvent can flow; (c) a rigid rod and a flexible rod; (d) a sphere and a sphere with a linear branch (e.g., a lollipop)?

13–4. Which will produce the greater relative viscosity, a molecule in H_2O or in 20% glycerol?

13–5. Which of the following pairs will be more resistant to degradation by hydrodynamic shear: (a) a linear DNA molecule and a circular molecule, both having the same total length; (b) a native linear DNA molecule and a denatured molecule in 1 M NaCl, both having the same total length: (c) denatured linear DNA in 1 M NaCl and 0.01 M NaCl, both having the same molecular weight; (d) a covalent circle and a twisted circle of DNA, both having the same molecular weight?

13–6. What might you conclude about the structure of a protein molecule if $[\eta]$ increases when placed in 6 M guanidinium chloride? If it decreases? What is the principal source of uncertainty in drawing these conclusions?

13–7. In general, the addition of various salts to H_2O increases viscosity, that is, $\eta_r > 1$. However, in a few cases, in dilute solution $\eta_r \leq 1$. Propose a possible explanation.

13–8. Under certain conditions of ionic strength and pH, a particular polynucleotide has a fairly small variation of η_{sp}/c with c, and $[\eta]$ is relatively independent of shear. On changing pH, the concentration and shear dependence, as well as the actual value of $[\eta]$, increase markedly. What is a possible effect of the pH change?

13–9. Normally, native DNA solutions show a relatively small dependence of $[\eta]$ on ionic strength, owing to the rigidity of the molecule. However, a particular DNA sample isolated from bacteria has the property that, at low ionic strength, $[\eta]$ becomes substantially smaller. Furthermore, if the solution is diluted before lowering the ionic strength, η_r is higher than that observed if the ionic strength is decreased before dilution. This is not true of all DNA preparations from these bacteria. What is a possible cause of this effect with the particular sample?

13–10. A sample of supercoiled DNA is subjected to a treatment that introduces single-strand breaks at a constant rate of one every 30 minutes. On the average, it takes ten single-strand breaks before a double-strand break forms. Assume that the measurement can be made sufficiently rapidly that the viscosity does not change significantly during the measurement. Draw a curve showing η_r versus time.

13–11. Some linear DNA molecules possess single-stranded termini that can join together. Would η_r of a solution of these molecules be higher or lower after joining than the value before joining? Would $[\eta]$ be different?

13–12. A DNA sample has the melting curve shown below. On the graph draw the curve for viscosity as a function of temperature.

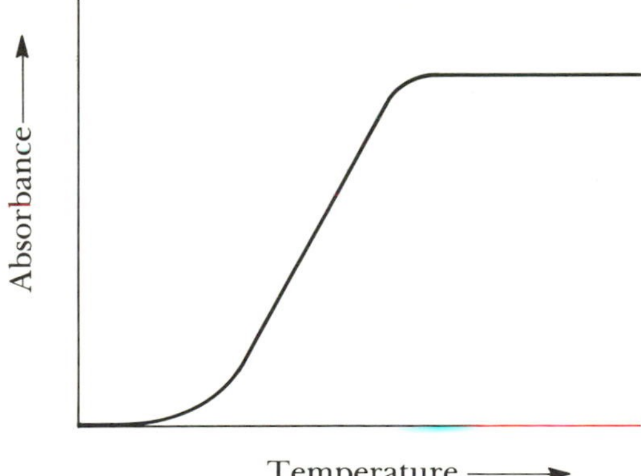

Five

SPECTROSCOPIC METHODS

14

Absorption Spectroscopy

Molecules absorb light. The wavelengths that are absorbed and the efficiency of absorption depend on both the structure and the environment of the molecule, making absorption spectroscopy a useful tool for characterizing both small and large macromolecules.

Simple Theory of the Absorption of Light by Molecules

Light, in its wave aspect, consists of mutually perpendicular electric and magnetic fields, which oscillate sinusoidally as they are propagated through space (Figure 14-1).

The energy E of the wave is

$$E = \frac{hc}{\lambda} = h\nu \tag{1}$$

in which h is Planck's constant, c is the velocity of light, λ is the wavelength, and ν is the frequency. When such a wave encounters a molecule, it can be either *scattered* (i.e., its direction of propagation changes) or absorbed (i.e., its energy is transferred to the molecule). The relative probability of the occurrence of each process is a property of the particular molecule encountered. If the electromagnetic energy of the light is absorbed, the molecule is said to be *excited* or in an *excited state*. A molecule or part of a molecule that can be excited by absorption is called a *chromophore*. This excitation energy is usually converted into

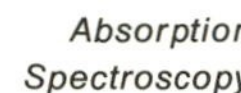

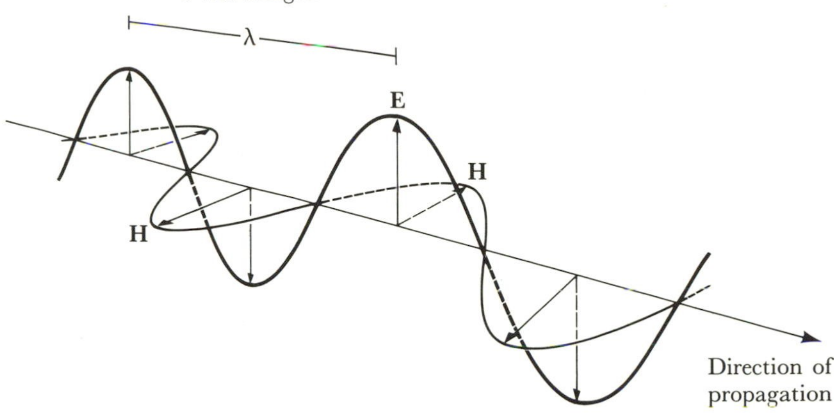

Figure 14-1
Propagation of an electromagnetic wave through space. The **E** and **H** vectors are mutually perpendicular at all times.

heat (kinetic energy) by the collision of the excited molecule with another molecule (e.g., a solvent molecule). With some molecules it is reemitted as *fluorescence*, which is discussed in more detail in Chapter 15. In both cases, the intensity of the light transmitted by a collection of chromophores is less than the intensity of the incident light.

An excited molecule can possess any one of a set of discrete amounts (quanta) of energy described by the laws of quantum mechanics. These amounts are called the *energy levels* of the molecule. The major energy levels are determined by the possible spatial distributions of the electrons and are called *electronic energy levels*; on these are superimposed *vibrational levels*, which indicate the various modes of vibration of the molecule (e.g., the stretching and bending of various covalent bonds). (There are even smaller subdivisions called *rotational levels*, but they are of little importance in absorption spectroscopy and will not be discussed.) All these energy levels are usually described by an *energy-level diagram* (Figure 14-2). The lowest electronic level is called the *ground state* and all others are excited states.

The absorption of energy is most probable only if the amount absorbed corresponds to the difference between energy levels. This can be expressed by stating that light of wavelength λ can be absorbed only if

$$\lambda = \frac{hc}{E_2 - E_1} \tag{2}$$

in which E_1 is the energy level of the molecule before absorption and E_2 is an energy level reached by absorption.

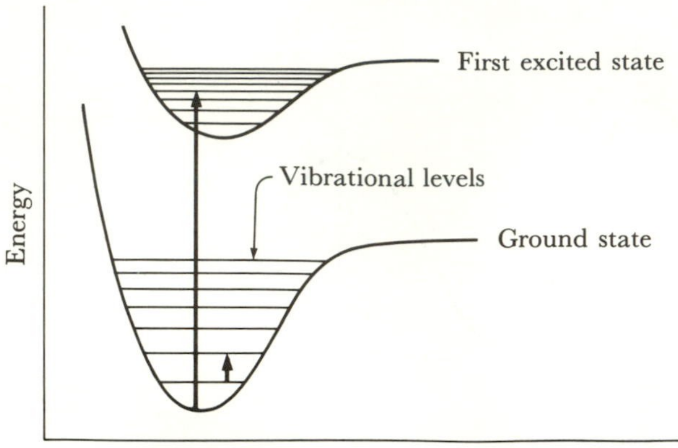

Figure 14-2
Typical energy-level diagram showing the ground state and the first excited state. Vibrational levels are shown as thin horizontal lines. A possible electronic transition between the ground state and the fourth vibrational level of the first excited state is indicated by the long arrow. A vibrational transition within the ground state is indicated by the short arrow.

A change between energy levels is called a *transition.* Mechanically, a transition between electronic energy levels represents the energy required to move an electron from one orbit to another. Transitions are represented by vertical arrows in the energy-level diagram.* A plot of the probability of absorption versus wavelength is called an *absorption spectrum* and absorption spectroscopy refers to the gathering and analysis of absorption data. If all transitions were between only the lowest vibrational levels of the ground state and the first excited state, then an absorption spectrum would consist of narrow, discrete lines. However, this is not the case for the following reason. The transition from one electronic level to the next level can occur in many ways. Because energy differences between vibrational levels are small compared to the minimum energy difference between electronic levels (namely, between the lowest vibrational levels of each), the electronic transition consists of a cluster of very closely spaced spectral lines. For complex reasons, each line has significant width and the width is comparable to the spacing between the lines. This has the effect that the lines overlap so much that a broad peak, called an *electronic absorption band,* results

* All transitions do not occur with high probability; those that do are determined by the so-called *selection rules* of quantum mechanics, which will not be discussed here.

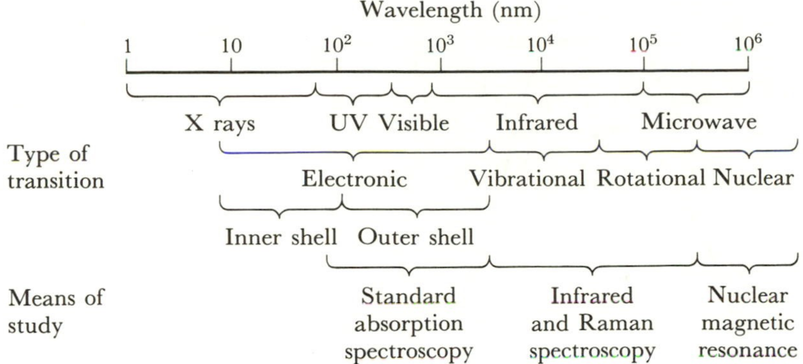

Figure 14-3
The part of the electromagnetic spectrum that is relevant to physical biochemistry.

(see Figure 14-6). For most molecules, the wavelengths corresponding to transitions between the ground state and any vibrational level of the first excited state fall in the range of ultraviolet and visible light. Low-energy transitions are also possible between vibrational levels within a single electronic level. These transitions produce radiation in the *infrared* range. Figure 14-3 shows the part of the electromagnetic spectrum relevant to the work of biochemists and the transitions producing radiation in different frequency ranges.

The probability of absorption at a single wavelength is characterized by the *molar absorption coefficient* at that wavelength. This is most easily defined in terms of how it is measured. If light of intensity I_0 passes through a substance (which may be in solution) of thickness d (in cm) and molar concentration c, the intensity I of the transmitted light obeys the Beer-Lambert law:

$$I = I_0 10^{-\varepsilon dc}, \quad \text{or} \quad \log_{10}\left(\frac{I}{I_0}\right) = -\varepsilon dc \qquad (3)$$

in which ε is the molar absorption coefficient*. Absorption data are reported either as % *transmission* ($100 \times I/I_0$) or, more commonly, as the *absorbance*, A, ($\log I/I_0$). When $d = 1$ cm, A is commonly called OD_λ or *optical density*, in which the subscript λ tells the wavelength at which the measurement was made. Optical density is convenient because it equals $\varepsilon \times c$. In some cases, if c is high, ε appears to be a function of c and it can be said that Beer's law† is violated. The two kinds of violations

* ε is also frequently referred to as the molar extinction coefficient.
† The Beer-Lambert law is almost universally called Beer's law. Although this is not strictly correct (there being another law of Beer), this convention will be followed herein.

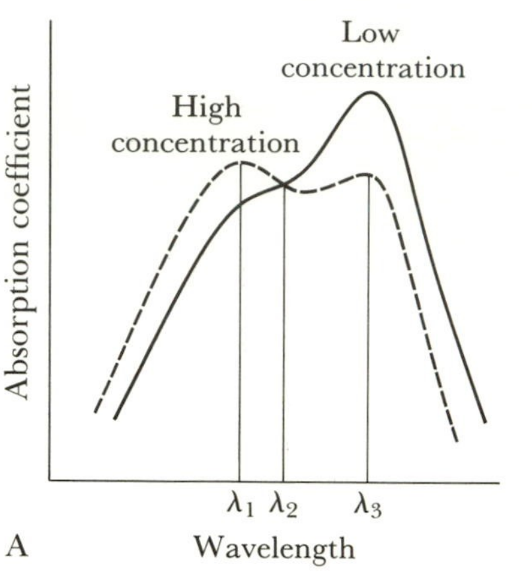

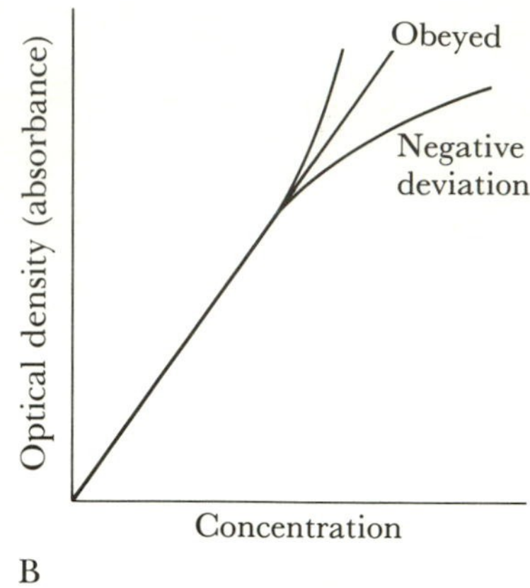

Figure 14-4

Positive and negative deviation from Beer's law and the causes. At the left is a spectral shift associated with increasing concentration—often a result of polymerization. Note that at one wavelength, λ_2, there is no change in molar absorption coefficient with change in concentration. This wavelength is called the isosbestic point. At the right is a curve showing deviation from Beer's law. At λ_1, the deviation is positive and, at λ_3, it is negative. At the isosbestic point, λ_2 it is always obeyed.

are shown in Figure 14-4. There are several causes of deviation from Beer's Law. For example, in solution many molecules form dimers or higher polymers as the concentration increases. The spectrum of the dimers may differ from that of the monomer. This could lead to either positive or negative deviation as indicated in Figure 14-4A. Another effect at high concentrations is aggregation; this frequently leads to scattering of light, thus decreasing the amount of light transmitted and hence causing positive deviations. Aggregation might also produce electronic interactions that could either increase or decrease ε. Other causes of departure from Beer's law are denaturation of proteins at low concentrations and chemical reactions at high concentrations.

Instrumentation for Measuring the Absorbance of Visible and Ultraviolet Light

Absorbance measurements are made by a *spectrophotometer*. (Because in biochemistry almost all studies of molecules are done with molecules in solution, the discussion that follows refers to samples of that type.) Although they vary in design, all spectrophotometers consist of a light source, a monochromator (for wavelength selection), a transparent sample holder called a cuvette, a light detector, and a meter or recorder for measuring the output of the detector (Figure 14-5). In a typical opera-

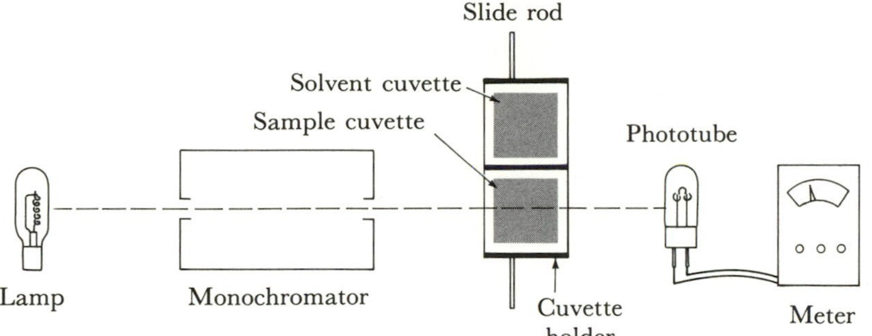

Figure 14-5
A spectrophotometer. Light from a lamp passes through a monochromator for wavelength selection. Sample and solvent are contained in two cuvettes in a cuvette holder. Light passes through a cuvette and falls on a phototube whose output is recorded on a meter. The cuvette holder is on a slide so that each cuvette can be separately placed in the beam.

tion at a single wavelength, a measurement is made of the light transmitted by the solvent alone (which may be a buffer or a solution of small molecules), followed by a measurement of that transmitted by the sample when dissolved in the same solvent; the first value is then subtracted from the second to give the absorbance of the solute. In practice, this subtraction is not done arithmetically; rather, the instrument is adjusted to read zero absorbance when the solvent alone is measured (this is called zeroing the instrument). Then, with the instrument so adjusted, the absorbance of the sample is read directly. To obtain a spectrum this operation is repeated at many wavelengths. Certain instruments, called automatic double-beam recording spectrophotometers, scan a range of wavelengths and simultaneously measure the absorbance of the sample and solvent (contained in separate cuvettes) and electronically subtract the two values at each wavelength. The spectrum is then plotted on a chart recorder. These instruments are very expensive but essential if a great deal of spectral analysis is to be done.

Parameters Measured in Absorption Spectroscopy

Figure 14-6 shows ultraviolet–visible spectra for two biological molecules. The parameters usually measured are optical density or ε. The wavelength corresponding to a peak of maximum absorption is called λ_{max}, and it is at this wavelength that ε is usually measured. Some of the absorption bands consist of multiple peaks and the wavelengths corresponding to the peaks having smaller molar absorption coefficients are

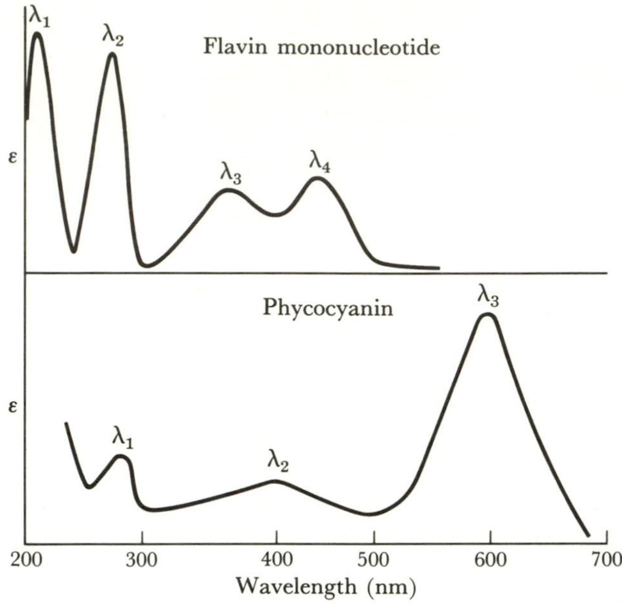

Figure 14-6
Spectra of two biological molecules, flavin mononucleotide and phycocyanin, indicating that spectra can be very different, which often allows the identification of compounds from their spectra.

frequently recorded. These wavelengths are sometimes also called λ_{max}, or it is stated that a substance has absorption maxima at $\lambda_1, \lambda_2, \ldots, \lambda_n$.

Sometimes the width of a band is measured, although this is not common.

A useful list of λ_{max} and ε for common biological chromophores is given in Table 14-1.*

Factors Affecting the Absorption Properties of a Chromophore

The absorption spectrum of a chromophore is primarily determined by the chemical structure of the molecule. However, a large number of environmental factors produce detectable changes in λ_{max} and ε. Environmental factors consist of pH, the polarity of the solvent or neighboring molecules, and the relative orientation of neighboring chromophores. It is precisely these environmental effects that provide the basis for the use of absorption spectroscopy in characterizing macromolecules.

* Note that wavelength is expressed in nanometers (nm). In the older literature and today in certain disciplines, the nanometer is called a millimicron (mμ). Some of the old instruments are labeled in angstrom units (Å). 1nm = 1 mμ = 10 Å = 10^{-9} meter.

Table 14-1

Absorption Maxima (λ_{max}) and Molar Absorption Coefficients (ε) for Various Substances at Neutral pH Encountered in Biological studies.

Molecule	λ_{max} (nm)	$10^{-3} \times \varepsilon$ at λ_{max} (M^{-1} cm^{-1})
Tryptophan*	280	5.6
	219	47.0
Tyrosine*	274	1.4
	222	8.0
	193	48.0
Phenylalanine*	257	0.2
	206	9.3
	188	60.0
Histidine*	211	5.9
Cysteine*	250	0.3
Adenine	260.5	13.4
Adenosine	259.5	14.9
Guanine	275	8.1
Guanosine	276	9.0
Cytosine	267	6.1
Cytidine	271	9.1
Uracil	259.5	8.2
Uridine	261.1	10.1
Thymine	264.5	7.9
Thymidine	267	9.7
DNA	258	6.6
RNA	258	7.4

* Other amino acids show insignificant absorption.

The general features of these environmental effects are the following.

pH effects. The pH of the solvent determines the ionization state of ionizable chromophores. An example is shown in Figure 14-7, which illustrates the effect of pH on the tyrosine spectrum.

Polarity effects. For polar chromophores, the value of λ_{max} for $n \rightarrow \pi^*$ transitions occurs at a shorter wavelength in polar hydroxylic solvents (H_2O, alcohols) than in nonpolar solvents. The shift is toward longer wavelengths for $\pi \rightarrow \pi^*$ transitions. The $\pi \rightarrow \pi^*$ transition is the more common transition for biological molecules. An exception are the amino acids that are usually studied in conformation analysis; these are $n \rightarrow \pi^*$ transitions so that the spectrum is shifted toward the blue as shown in Figure 14-8.

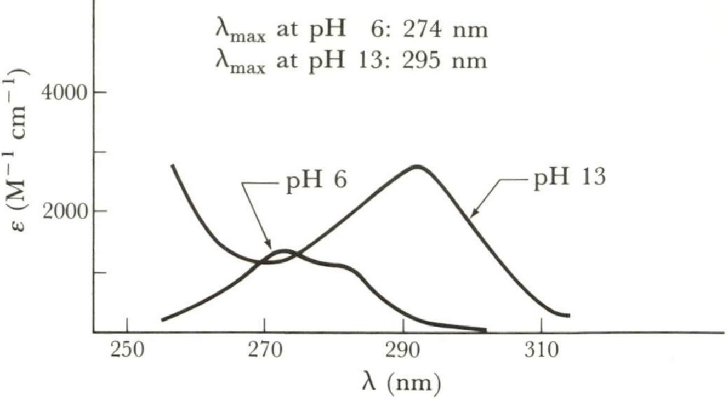

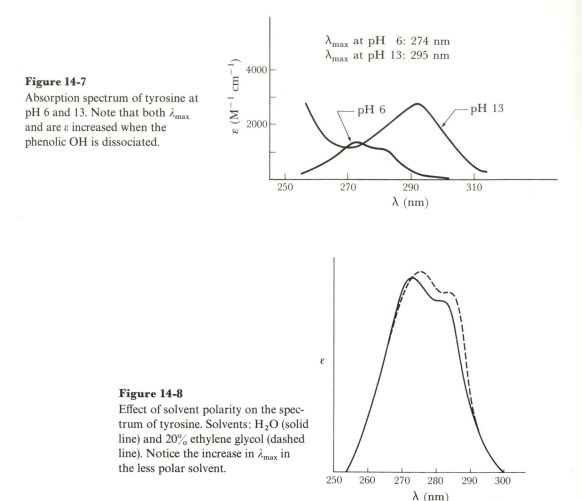

Figure 14-7
Absorption spectrum of tyrosine at pH 6 and 13. Note that both λ_{max} and are ε increased when the phenolic OH is dissociated.

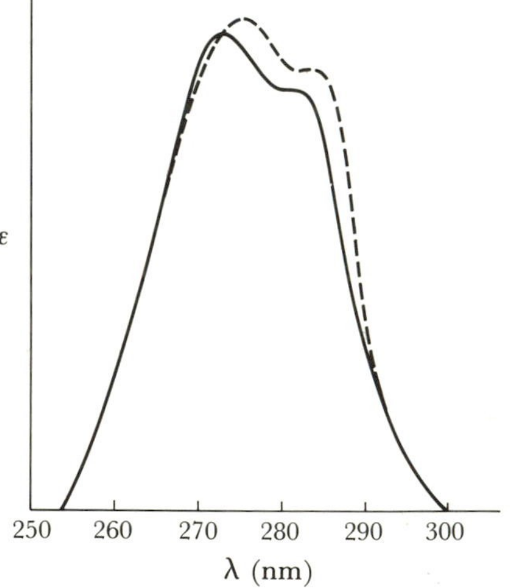

Figure 14-8
Effect of solvent polarity on the spectrum of tyrosine. Solvents: H_2O (solid line) and 20% ethylene glycol (dashed line). Notice the increase in λ_{max} in the less polar solvent.

Orientation effects. Geometric features frequently have strong effects on λ_{max} and ε. The best known is the *hypochromism* of nucleic acids. That is, the absorption coefficient of a nucleotide decreases when the nucleotide is contained in a single-stranded polynucleotide in which the nucleotide bases are in proximity. There is a further decrease with a double-stranded polynucleotide because the bases are arranged in an even more orderly array. This is shown in Figure 14-9.

As a result of a large number of studies with biological compounds and macromolecules whose structures are well known in a variety of conditions, a set of empirical facts has been assembled that may be called the working rules of biochemical absorption spectroscopy. They are given in Table 14-2. Examples of the application of these rules will be given in parts of the following section.

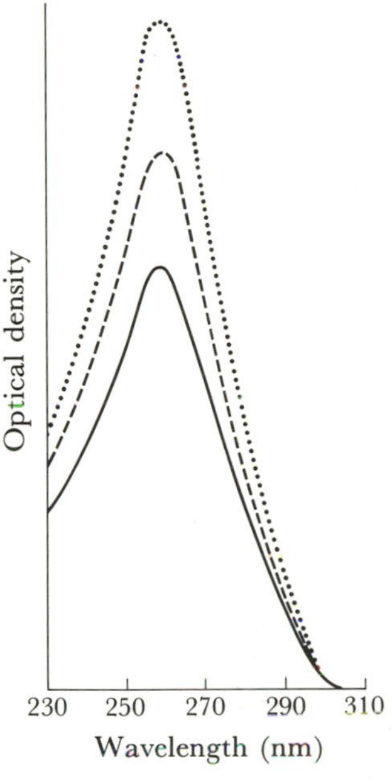

Figure 14-9
Spectra of T7 DNA as a double-stranded DNA (solid line), as a single-stranded DNA (dashed line), or after hydrolysis to free nucleotides (dotted line), showing the decrease in optical density (hypochromicity) that accompanies the formation of a more ordered structure. All spectra were obtained at the same concentration.

Table 14-2
Empirical rules for the interpretation of the absorption spectra of biological macromolecules.

1. If the amino acids tryptophan, tyrosine, phenylalanine, and histidine are shifted to a less polar environment, λ_{max} and ε increase. Hence:
   a. If the spectrum of an amino acid in a protein in a polar solvent shows that λ_{max} and ε are higher than they are for the free amino acid in the same solvent, then that amino acid must be in an internal region of the protein ("buried") and surrounded by nonpolar amino acids.
   b. If the spectrum of a protein is sensitive to changes in the polarity of the solvent, the amino acid showing the change in λ_{max} and ε must be on the surface of the protein.

   To make use of this rule, it must be ascertained that the change in polarity does not cause a conformational change that could bring an internal amino acid to the surface. Such changes do not usually occur with the solvents used.

2. For amino acids, λ_{max} and ε always increase if a titratable group (e.g., the OH of tyrosine, imidazole of histidine, and SH of cysteine) is charged. Hence, when the pH is changed:
   a. If no spectral change is observed for one of these chromophores and if the pH change is such that titration of a free amino acid would have occurred, the chromophore must be buried in a nonpolar region of the protein.
   b. If the spectral change as a function of pH indicates that the ionizable group has the same pK as it would if free in solution, then the amino acid is on the surface of the protein.
   c. If the spectral change as a function of pH indicates a very different pK, then the amino acid is likely to be in a strongly polar environment (e.g., a tyrosine surrounded by carboxyl groups).

3. For purines and pyrimidines, ε decreases as their ring systems become parallel and nearer to one another (more stacked). The value of ε decreases in the following series: free base > base in an unstacked single-stranded polynucleotide > base in a stacked single-stranded polynucleotide > base in a double-stranded polynucleotide.

Chemical Analysis by Absorption Spectroscopy Using Visible and Ultraviolet Light

Absorbance measurements allow the determination of the concentration of a substance, assay of certain chemical reactions, and the identification of materials. Examples of each follow.

Example 14-A ☐ Concentration measurement.

The most common use of absorbance measurements is to determine concentration. This can be done if the absorption coefficient is known and Beer's law is obeyed. For example, for double-stranded DNA, an $OD_{260\ nm} = 1$ corresponds to 50 μg/ml. Because Beer's law is obeyed to at least OD = 2 (which approaches the limit for most spectrophotometers), concentration is easily calculated—that is, OD = 0.5 corresponds to 25 μg/ml, OD = 0.1 to 5 μg/ml, and so forth. Note that the optical method for determining concentration is nondestructive. Similarly if an amino acid mixture is fractionated by ion-exchange chromatography, the concentration of tryptophan in a particular tube can be determined from the absorbance at 280 nm using the value $\varepsilon = 5600\ M^{-1}/cm^{-1}$ given in Table 14-1. Thus if the absorbance in a cuvette having a 4-cm path length is 0.26, the concentration is $0.26/4(5600) = 1.16 \times 10^{-5}$ M.

Sometimes the material being measured consists of light-absorbing particles in suspension rather than in solution—for example, bacteriophages whose absorbance is determined almost entirely by their DNA content. Bacteriophages not only absorb but also scatter light (by Rayleigh scattering) and therefore appear to have an artificially high absorbance. However, a scattering correction can be made by measuring the absorbance at a series of wavelengths far from the λ_{max}. Because scattering varies as λ^{-4}, a plot of measured absorbance versus λ^{-4} can be made and the linear part of this curve (where all observed absorption is due to scattering) can be extrapolated to λ_{max} to correct the measured absorbance for the amount due to scattering. An example of such a correction is shown in Figure 14-10. This is a successful method for determining the nucleic acid content of bacteriophages and viruses, using the relation that OD_{260} (corrected for scattering) = 1 means 50 μg DNA/ml.

The concentration of bacteria is also often determined by spectrophotometry, although at the wavelengths used, scattering accounts for all of the apparent absorbance. In this case, instead of correcting for scattering, a calibration curve is constructed by comparing observed optical density with viable cell count. This is a precise way to measure cell concentration or dry mass per unit volume, as shown in Figure 14-11. It is important to appreciate that the calibration curve is an empirical one and applies only to a particular suspending

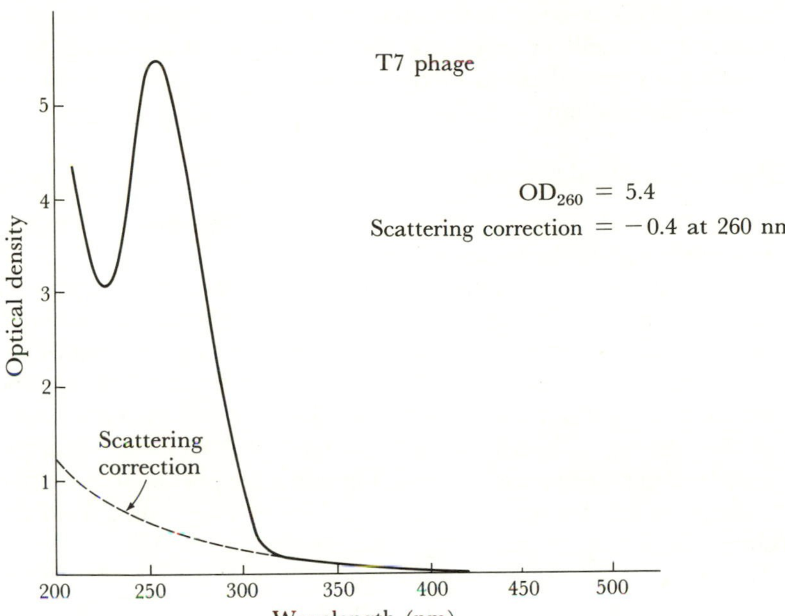

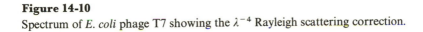

Figure 14-10
Spectrum of *E. coli* phage T7 showing the λ^{-4} Rayleigh scattering correction.

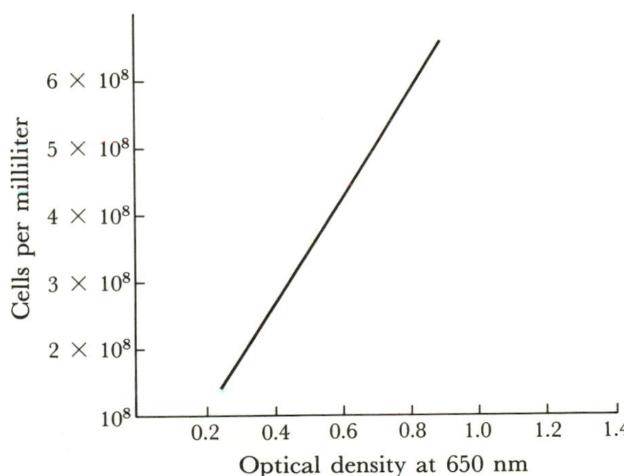

Figure 14-11
Concentration of bacteria determined from measurement of optical density.

medium. This is because for small particles the optical density is primarily a result of scattering and the degree of scattering by a suspended object depends in part on the index of refraction of the suspending medium.

Example 14-B □ Assay of chemical reactions.

Many chemical reactions can be assayed if one of the reactants changes in absorbance during the course of the reaction.

An example from enzymology that had a tremendous impact on molecular biology in studies of the lactose operon is the measurement of the activity of the enzyme β-galactosidase. This enzyme can cleave *o*-nitrophenyl galactoside to form *o*-nitrobenzene, which can be detected by its absorbance at 420 nm. The OD_{420} is then a measure of the amount of hydrolysis of *o*-nitrophenyl galactoside. Because, for a certain range at least, the reaction rate is proportional to enzyme concentration, the amount of enzyme can be determined from the slope of a plot of optical density versus hydrolysis time (Figure 14-12). This assay is an example of an especially useful trick. A colored substance, which loses its color when coupled to other compounds, is chemically linked to the substrate of of an enzyme so that production

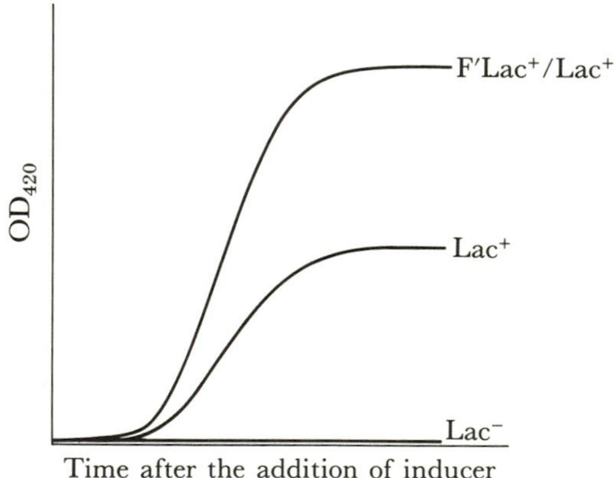

Figure 14-12
Synthesis of β-galactosidase assayed by OD_{420} as a measure of hydrolysis of *o*-nitrophenyl galactoside. Three cultures, one Lac^-, one Lac^+, and a diploid containing two copies of the *lac* genes were growing in a glycerol–salts medium. At $t = 0$, an inducer of β-galactosidase synthesis was added. At various times, samples were removed and treated with toluene to release the enzyme and kill the cells; then *o*-nitrophenyl galactoside was added. Thirty minutes later, the OD_{420} was measured. The experiment shows that the Lac^- cell makes no enzyme and that the diploid makes roughly twice as much as the haploid.

of the colored compound accompanies enzymatic activity. For example, *o*-nitrobenzene can be coupled to proteins to yield a colored complex useful in the assay of proteases.

A different type of example is the enzymatic hydrolysis of polynucleotides. Rule 3 in Table 14-2 states that ε(free base) > ε(base in a polynucleotide). Hence, if a polynucleotide is hydrolyzed to mononucleotides, ε_{260} will increase. Therefore, if a nuclease is added to a polynucleotide sample and the OD_{260} is measured as a function of time, the OD_{260} will increase and this increase is an indicator of hydrolysis. Like that of β-galactosidase, the amount of nuclease can be determined from the slope of the curve of OD_{260} versus time.

A third example, not involving enzymes, is the means of determining the dose rate of an X-ray machine. If a solution of $FeSO_4$ in H_2SO_4 is X-irradiated, the Fe^{2+} ion is converted into Fe^{3+}, which can be detected by its absorbance at 305 nm. From OD_{305}, the total dose received can be calculated. This is the method of choice for precise quantitation of all ionizing radiation; it is called *ferrous sulfate dosimetry*.

Example 14-C

□ Identification of substances by spectral measurement.

Most substances have characteristic spectra and can be identified thereby. This can be done either by measuring a complete spectrum or by measuring the ratio of absorbance at different wavelengths. For example, in the early work in determining the base composition of DNA, the DNA was hydrolyzed and the bases were separated by chromatography. The individual bases could be identified by obtaining their spectra and noting their λ_{max}: adenine, 260.5 nm; thymine, 264.5 nm, guanine, 275.0 nm, and cytosine, 267.0 nm. To avoid determining the complete spectrum (which would normally be done with a recording spectrophotometer), the bases could be distinguished by measuring the ratio OD_{250}/OD_{280}. These values are: adenine, 2.00; thymine, 1.26; guanine, 1.63; and cytosine, 0.31. After the bases had been identified, the amount of each was determined as in Example 14-A by measuring optical density because the values of ε at λ_{max} (13.4×10^3, 7.9×10^3, 8.1×10^3, and $6.1 \times 10^3\ M^{-1}\ cm^{-1}$, respectively) are known. (For example, if the OD_{260} of adenine were 0.65, the molar concentration would be $0.65/(13.4 \times 10^3) = 4.9 \times 10^{-5}$ M.)

Structural Studies of DNA Using Absorption of Ultraviolet Light

The absorbance of various forms of DNA differs as described in Table 14-2, item 3. This allows absorbance measurements to be used to study DNA structure, as shown in the following examples.

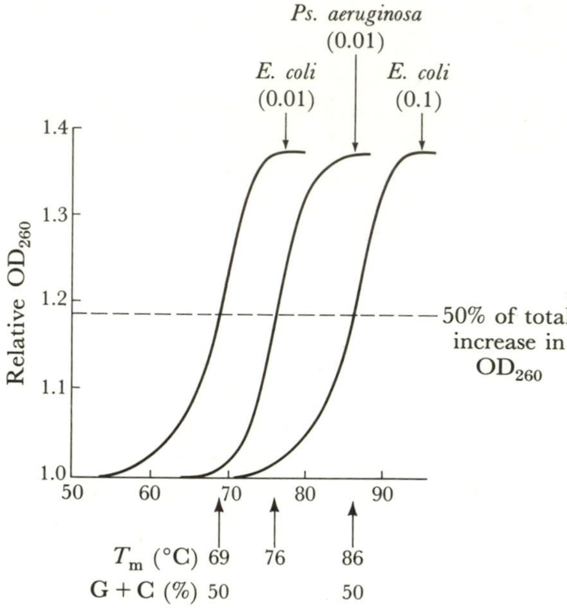

Figure 14-13
Optical density of three DNA solutions as a function of temperature: *E. coli* DNA (40% G + C) in 0.01 M PO_4, pH 7.8; *Pseudomonas aeruginosa* DNA (68% G + C) in 0.01 M PO_4, pH 7.8. The temperature at which the absorbance change is 50% complete is T_m, the melting temperature. Note that T_m increases with ionic strength and with G + C content.

Example 14-D ☐ The helix-coil transition of double-stranded DNA: denaturation and renaturation.

The OD_{260} of DNA increases if the DNA is heated through a particular temperature range (Figure 14-13). This so-called *hyperchromicity* is a measure of denaturation or a helix-coil transition and results from the unstacking of the DNA bases associated with the continuous separation of the two polynucleotide strands (see rule 3 in Table 14-2). This simple optical assay allows the determination of the stability of DNA in relation to temperature, pH, ionic strength, and added small molecules, and in a variety of polar and nonpolar solvents. Some of the most important properties of DNA were elucidated from this powerful assay in the following ways. First, the thermal stability of DNA increases with guanine + cytosine content (Figure 14-14). Hence, a guanine·cytosine base pair is probably hydrogen-bonded more strongly than an adenine·thymine pair. Second, if the temperature is raised (but not to the point of maximum

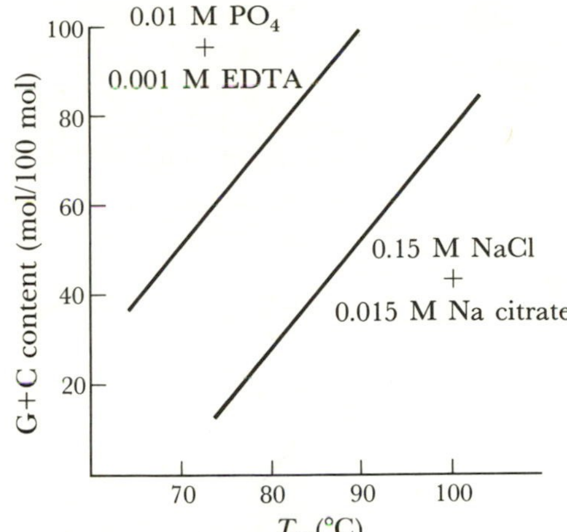

Figure 14-14
Plot of T_m versus G + C content of various DNA molecules. The values of T_m for two different ionic strengths are shown. Note that the curves have the same slope. [From J. Marmur and P. Doty, *J. Mol. Biol.* 5(1962): 109–118.]

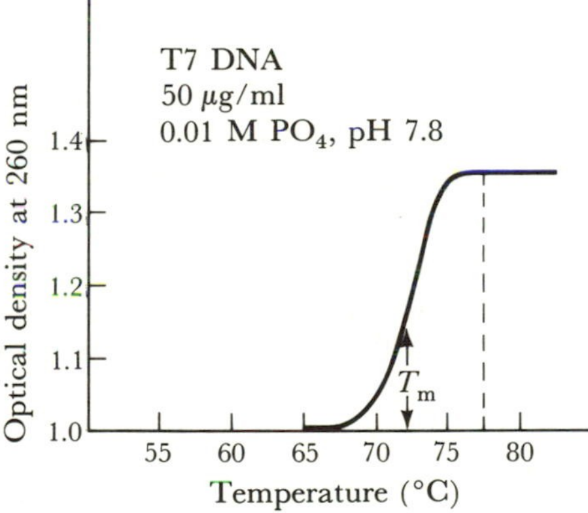

Figure 14-15
Difference between melting curves of a DNA solution obtained by measuring OD_{260} at the indicated temperature (solid line) or by heating the DNA to the indicated temperature, cooling to 25°C, and then measuring the OD (dashed line). The temperature at which strand separation occurs is at the intersection of the dashed line and the *x*-axis.

absorbance) and then lowered to a temperature below which no increase in optical density is observed, the optical density immediately drops to its original value, indicating that, if strand separation is not complete, the native structure is restored (Figure 14-15). A special case is that of an interstrand cross-link in which the absorbance always drops to the normal value. Third, strand separation does not occur until well past the optical-transition region (e.g., 77.5°C, Figure 14-15) because separation of the last few base pairs has only a tiny effect on the total absorbance change. Fourth, if DNA is heated past the point of strand separation and then cooled, the increase in absorbance drops from a value of 37% (i.e., the maximum value) to 12% in high ionic strength (0.1 M), because hydrogen bonds (intra- and inter-

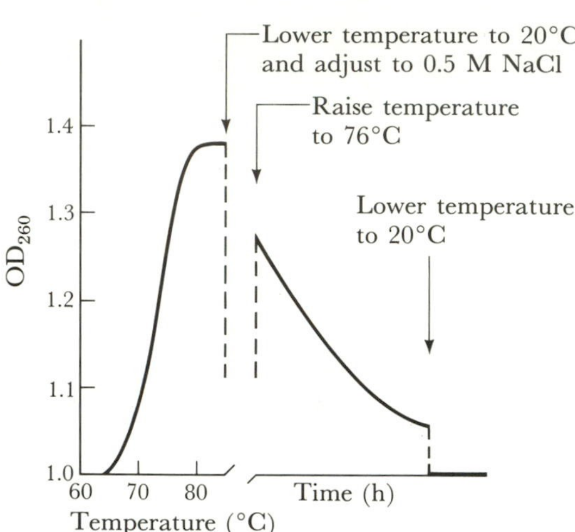

Figure 14-16
Detection of the renaturation of T7 DNA by measuring OD_{260}. DNA is first denatured in 0.01 M NaCl. When adjusted to 0.5 M NaCl, random hydrogen bonds form and the OD increases. Correct base pairing begins. Because the temperature is below that for strand separation, the OD decreases. However, DNA is partly denatured at that temperature. When the temperature is lowered to 20°C, all hydrogen bonds reform and renaturation is complete.

strand) reform at random. Hence, in high ionic strength, denatured DNA is aggregated. If the ionic strength is low (0.01 M), the optical density remains at the maximum value because charge repulsion of the phosphate groups keeps the strands separated. Fifth, if the DNA with a 12% increase in optical density is put at an ionic strength ranging from 0.5 to 1.0 M and at a temperature above the midpoint (T_m) of the transition (Figure 14-16) and the DNA concentration is high, in several hours the optical density returns to the original value. This is called *renaturation* and represents reformation of the native double-stranded structure (see Chapters 1 and 19).

Example 14-E ☐ Solvent perturbation of nucleic acids.

A change in solvent from H_2O to 50% D_2O causes characteristic spectral changes in mononucleotides but not in base pairs. Thus, spectral changes of a DNA sample in 50% D_2O can be used to determine the fraction of bases that are not base-paired. This is of great value in the study of such substances as transfer RNA (tRNA). In tRNA the sequence does not allow all bases to be engaged in hydrogen bonds; three-dimensional models in which there is partial hydrogen bonding can be constructed (Figure 14-17). Different models require different fractions to be hydrogen bonded so that some of the possible models can be ruled out by spectral data. (See Chapter 17 for a discussion of the role of nuclear magnetic resonance in distinguishing possible tRNA models by measuring the number of hydrogen-bonded base pairs.)

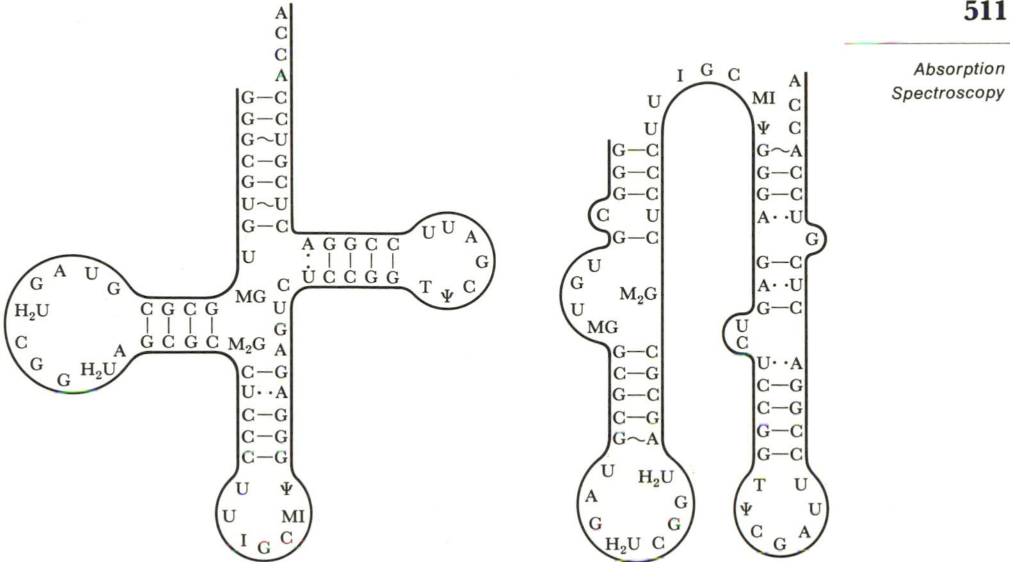

Figure 14-17
Two of several possible structures of yeast alanine tRNA arranged so that a large fraction of the bases participate in hydrogen bonding. The hydrogen bond of a G·C pair is indicated by —; that of an A·U pair is indicated by ··; the weaker hydrogen bonds found in other base pairs are indicated by ~. An NMR analysis yielding the number of base pairs of various types indicates that the cloverleaf structure (left) is the correct one. The symbols are: G, guanosine; C, cytidine; A, adenosine; U, uridine; I, inosine; H_2U, dihydrouridine; T, ribosylthymine; MG methylguanosine; M_2G, dimethylguanosine; Ψ, pseudouridine; MI, methylinosine. [The base sequence is that reported by R. W. Holley, J. Apgar, G. A. Everett, J. T. Madison, M. Marquisee, S. H. Merrill, J. R. Fenswick, and A. Zamer, *Science* (*Wash., DC*) 147(1965):1462–1465.]

Structural Studies of Proteins Using Absorption of Ultraviolet Light

In Table 14-2, items 1 and 2, various spectral changes accompany alteration of the charge or the environment of certain amino acids. Thus, absorbance measurements can give an idea of the location of particular amino acids in a protein. Detailed information about protein structure is not obtained; nonetheless a strong statement that an amino acid is either internal or external can simplify the job of the X-ray diffractionist, who is attempting to determine the complete structure of the molecule. Furthermore, information about the amino acids that are in a binding site is invariably valuable to the enzymologist attempting to determine the reaction mechanism of an enzyme. Several elegant techniques are used, which are presented in the examples that follow.

Example 14-F ☐ Spectrophotometric pH titration of proteins.

Many studies of protein structure require the determination of pK values for proton dissociation from ionizable amino acid side chains, because these values give an indication of the location of the amino acid in the protein (rule 2 in Table 14-2). This can often be done spectrophotometrically because dissociation often changes the spectrum of one of the chromophores (e.g., tyrosine); see Figure 14-7. Let us consider a hypothetical tyrosine-containing protein, and use rule 2 to determine the number of external tyrosines. Suppose that this protein has five tyrosines. If all are on the surface and they are ionized by increasing the pH, the entire tyrosine spectrum will shift to that seen in Figure 14-7 for free tyrosine at high pH (rule 2a). In other words, a plot of OD_{295} (λ_{max} for the ionized form) versus pH would look like curve A in Figure 14-18. If, instead, three tyrosines were internal, and in a nonpolar environment, the curve would be like curve B; the ratio of the first plateau value to the final value would be $\frac{2}{5}$ (rule 2b). Note

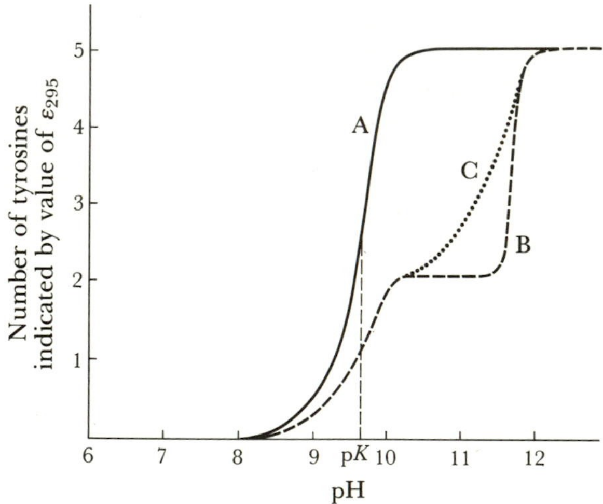

Figure 14-18
pH titration curves for tyrosine, using ε_{295} as an assay. The hypothetical protein contains five tyrosines. In curve A, all five are on the surface. In curve B, two are on the surface and the remaining three are internal—in a nonpolar environment and therefore not titratable. In curve C, the three internal ones are in a polar environment and accessible to the solvent.

that the curve shows a large rise in OD_{295} at very high pH. This indicates that the internal tyrosines have become exposed to the solvent—that is the protein has unfolded (become denatured).

If the three internal tyrosines were in a polar environment, a curve similar to curve C might be obtained, which indicates that these three have a pK value different from that of the exposed groups (rule 2c).

Example 14-G

☐ Determination of some aspects of the conformation of proteins by the solvent-perturbation method and by difference spectroscopy.

Rule 1 in Table 14-2 indicates that the spectra of chromophores depend on the polarity of their environment. This fact can be used in two ways: (1) to determine whether certain chromophores in a protein are internal or external and (2) to determine the polarity of the environment of an internal amino acid.

Point 2 is described first because it is easier to follow. A protein is placed in a polar solvent and the spectrum of an internal chromophore is obtained. This is compared with known spectra of the same chromophore in polar and nonpolar solvents. If the spectrum resembles that of the chromophore in a nonpolar solvent, then the chromophore must be in a nonpolar region in the protein, and vice versa; the extent of the shift in λ_{max} allows an estimate of the degree of polarity.

The determination of whether an amino acid is internal or external by measuring the spectra of a protein in a polar and nonpolar solvent is called the *solvent-perturbation method.* Because one is generally interested in knowing the structure of a protein in aqueous salt solutions (as the protein would be if in a living cell), it is necessary that the nonpolar solvent itself does not introduce conformational changes and this must always be checked by other methods. In fact, proteins are rarely studied in completely nonpolar solvents because most proteins are either insoluble or denatured in these solvents. The usual practice is to use a solvent that is 80% water and 20% a substance of reduced polarity. Some standard mixtures are given in Table 14-3. These solvents of reduced polarity are called *perturbing solvents.*

The most convenient procedure for applying the perturbation method is *difference spectroscopy* because it eliminates the necessity

Table 14-3
Solvents Commonly Used in the Solvent-Perturbation Method.

Liquid additive (20 volumes/100 volumes final solution in H_2O*)*	
Dimethylsulfoxide	Ethylene glycol
Dioxane	Glycerol
Solid additive (20 g/100 g final solution in H_2O*)*	
Arabitol	Polyethylene glycol
Erythritol	Sucrose
Mannitol	Urea

of determining the spectrum twice—that is, in the presence and absence of the perturbant. In standard absorption spectroscopy, a spectrum is determined by measuring the absorbance of a solution and subtracting the absorbance of the solvent (or, as described earlier, by adjusting the spectrophotometer to read zero for the solvent). In difference spectroscopy, the two sample holders contain solutions that are identical except that one contains a perturbant. Instead of measuring the spectra of each, the absorbances at each wavelength are subtracted from one another (or course, it is necessary that both solvents have identical absorbances at all wavelengths used—this is the case for the solvents in Table 14-3).

A nonzero *difference spectrum* will appear only if the spectrum of the sample is affected by the perturbant. The parameters obtained from such spectra are λ_{max} and $\Delta\varepsilon$ at λ_{max}. Figure 14-19 shows the difference spectra for tryptophan and tyrosine (the amino acids most commonly examined in the solvent-perturbation method) in 20% ethylene glycol.

An example of perturbation analysis, using tryptophan as a chromophore, and difference spectroscopy follows. Consider a hypothetical protein with five tryptophans. The difference spectrum between H_2O and 20% ethylene glycol shows a peak at 292 nm, as shown in Figure 14-19. From the OD_{280} and ε_{280} obtained in H_2O, the amount of tryptophan in the sample can be determined; and from $\Delta\varepsilon_{292}$ for free tryptophan in H_2O and 20% ethylene glycol, the expected ΔOD_{292} for the sample—if all tryptophans were perturbed—is known. Let us assume that this value is 1.00. If ΔOD_{292} is measured for this protein, it is found to be 0.60. Because only those on the surface are perturbed, $0.60/1.00 \times 5 = 3$ are on the surface. If the amino acid sequence of the protein is known, the particular tryptophans that are on the surface can be identified by a simple trick. If a protein is treated by various oxidizing agents, the indole group of tryptophan is oxidized and becomes nonabsorbing. Presumably, only those tryptophans on

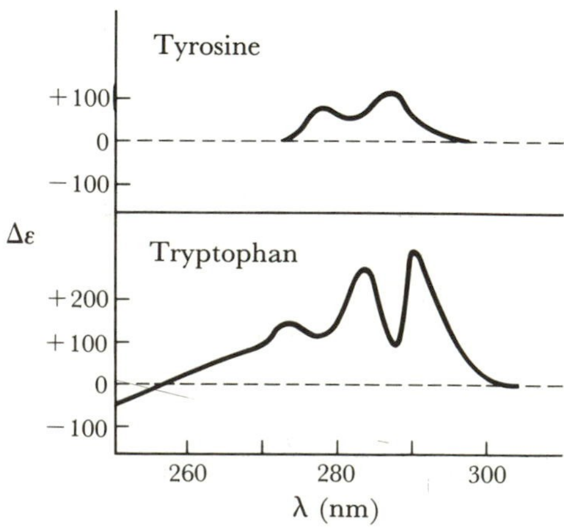

Figure 14-19
Difference spectra for tyrosine and tryptophan in 80% H_2O, 20% ethylene glycol. Notice that difference spectra can have negative values of $\Delta\varepsilon$.

the surface will be oxidizable, but this must be checked. This also can be done by the solvent-perturbation method and difference spectroscopy. If $\Delta OD_{292} = 0$ after oxidation, the oxidized tryptophans must be on the surface. After this has been ascertained, the sequence of amino acids in the oxidized protein can be redetermined and the positions of the oxidized tryptophans noted. Hence, the particular tryptophans that are on the surface can be identified. Such information may be useful in the elucidation of the three-dimensional structure of the molecule and in locating regions of the polypeptide chain.

In the preceding example, the ΔOD_{292} expected if all tryptophans are on the surface was calculated from the absorbance and absorption coefficient of the pure protein in the absence of perturbant. This may not be possible if the protein sample contains a contaminating absorbing material. The measurement can, however, be made in another way because if a protein is denatured, all the amino acids are in contact with the solvent. Hence, the maximum ΔOD_{292} can be determined from a difference spectrum of the denatured protein in a perturbing solvent versus the native protein in a polar solvent.

We have so far attempted to distinguish external from internal amino acids. However, this distinction is not always clear because, in some cases, a chromophore is not totally buried but is in a deep crevice so that its spectrum is affected only if the perturbing molecules are below a critical size necessary to enter the crevice. This also provides information about molecular conformation. The substances most commonly used, together with their mean diameters, follow: D_2O (2.2 Å), dimethylsulfoxide (4.0 Å), ethylene glycol (4.3 Å), glycerol (5.2 Å), arabitol (6.4 Å), glucose (7.2 Å), and sucrose (9.4 Å). For example, let us reconsider the protein that contains five tryptophans that was just analyzed. The value of ΔOD_{292} of 0.60 when the solvent is in 20% ethylene glycol allowed us to conclude that three tryptophans are on the surface. However, if in 20% glycerol the value of ΔOD_{292} is 0.40, then only $(0.40/1.00) \times 5 = 2$ tryptophans are available to the glycerol. Thus, by this measurement one gains the added information that one of the tryptophans is in a crevice whose diameter is between 4.3 Å (ethylene glycol) and 5.2 Å (glycerol).

☐ Observation of the helix-coil transition in proteins: denaturation. **Example 14-H**

Because a buried chromophore becomes exposed to the solvent during denaturation, by monitoring the absorbance of these chromophores, one can observe the helix-coil transition for proteins in the same way that denaturation is studied by examining the hyperchromicity of DNA. For example, if a protein contains tryptophans, some of which are internal, the unfolding as a function of temperature could be detected by measuring $\Delta\varepsilon_{292}$. This could then be used to examine the effects of other agents such as NaCl concentration on the thermal stability. The kind of data obtained for a hypothetical protein is shown in Figure 14-20.

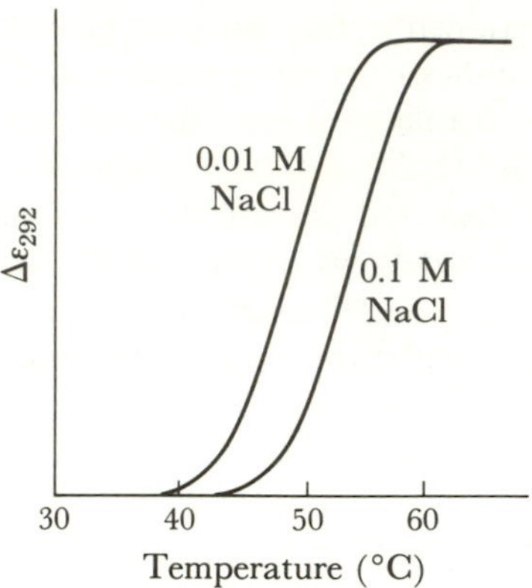

Figure 14-20
Helix-coil transition of a hypothetical protein assayed by perturbation difference spectroscopy at a single wavelength using 80% H_2O, 20% ethylene glycol containing two different NaCl concentrations. The reference solution for the difference spectrum is the protein in the ethylene glycol–NaCl solution at 20°C. Note that this protein is more stable at higher NaCl concentration since a higher temperature is needed for unfolding in 0.1 M NaCl.

Example 14-I ☐ Detecting the binding of small molecules to proteins.

The binding of an enzyme substrate to the active site of an enzyme frequently produces spectral changes in chromophores in or near the active site by affecting the polarity of the region or the accessibility to solvent. By comparing the observed changes with those obtained by solvent perturbation, information about the structure of the active site can be obtained. For example, the addition of various substrates to the enzyme lysozyme produces a shift in λ_{max} for tryptophan to longer wavelength. The magnitude of the change is that expected from the transfer of one tryptophan from a polar to a nonpolar environment. This suggests that a tryptophan is in the binding site. Furthermore, solvent-perturbation studies of lysozyme, such as that in Example 14-F, show that there are four tryptophans on the protein surface (i.e., a certain fraction of the known number of tryptophans are perturbable); if the enzyme-substrate complex is studied, the analysis indicates that only three are on the surface. Hence, one tryptophan is no longer in contact with the solvent when the substrate is added. Again the simple interpretation is that the active site contains tryptophan. This simple optical analysis is important because it is performed in solution and therefore confirms that the structure determined by X-ray diffraction analysis (which is done with dry samples), that is, there is a crevice in the molecule containing tryptophan in which substrate is bound, is probably valid for lysozyme in solution.

In many cases, chromophores in the enzyme, the substrate, or both, change their absorption during complex formation. Hence, a spectral assay (i.e., $\Delta\varepsilon$ as a function of substrate or enzyme concentration) can be used to determine dissociation constants. This is in fact a very useful method applicable to the binding of any material (e.g., small molecules or metal ions) to a protein whenever spectral changes are observed.

Example 14-J

☐ Protein-protein association.

Spectral changes can accompany protein-protein association either because chromophores on the surface become inaccessible to the solvent by being buried in the region in which binding takes place or because a conformational change that buries or exposes a chromophore in another part of the molecule can accompany binding. Hence, as in Example 14-I, the spectral changes can be used to monitor interaction and thereby determine conditions, kinetics, and so forth (Figure 14-21).

This can be studied especially well using difference spectroscopy, in which case the difference spectrum is produced between two solutions that are identical except for concentration.

Reporter Groups

A molecule of interest may often have many chromophores but none will be in or near a region that participates in the biological function of the molecule. This situation can sometimes be corrected by the addition of an artificial chromophore in the relevant region. Such a chromophore is called a *reporter group*. For successful use, the reporter group must have

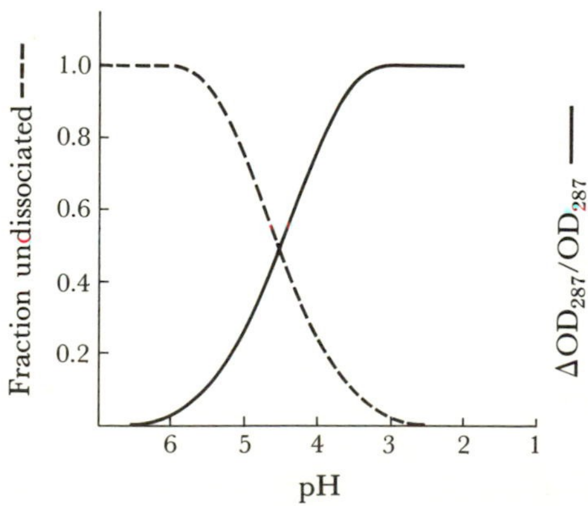

Figure 14-21
Detection of the pH-induced dissociation of ovomacroglobulin into its subunits by difference spectroscopy. The OD_{287} at various pH values was measured against OD_{287} at pH 7.

a spectrum distinguishable from the remainder of the macromolecule, it must react at a single position, and its insertion must not affect the interaction. Useful reporters are dimethylaminoazobenzene and arsanilic acid. An example of the use of reporter groups follows.

Example 14-K □ Location of a zinc atom in an enzyme.

An enzyme contains a tyrosine in the active site and a zinc atom whose location is unknown. Arsanilic acid can be coupled to the tyrosine in the active site by diazotization. The absorption spectrum for free arsanilazotyrosine is significantly altered by binding zinc. In the protein in solution, the spectrum is that of the zinc complex. Hence, the protein must be folded so that the zinc-binding site and the active site are close together. An economical explanation is that the Zn atom is in the active site.

The similarity of this procedure to the method of extrinsic fluorescence (see Chapter 15) should be noted.

Absorption of Polarized Light

The plane of polarization of an electromagnetic wave is defined as the plane of the **E** vector (Figure 14-1; see Chapter 16 for a more complete description of polarization and the production of plane-polarized light). A typical beam of light is nonpolarized because it can be thought of as a collection of waves whose planes of polarization are randomly oriented. However, plane-polarized light can be obtained either by passing light through various materials or by the reflection from a surface at a critical angle.

All chromophores have at least one preferred axis of absorption of plane-polarized light. For example, the probability of absorption by the symmetric planar ring of hexamethylbenzene

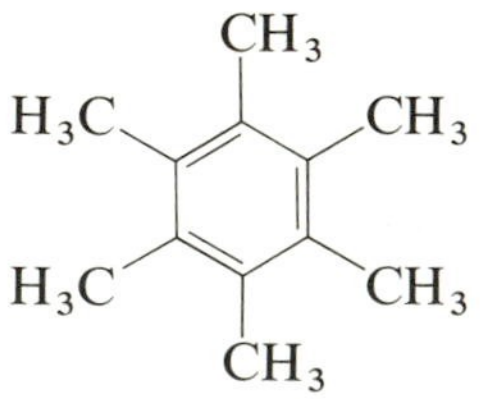

CH_3

H_3C CH_3

H_3C CH_3

CH_3

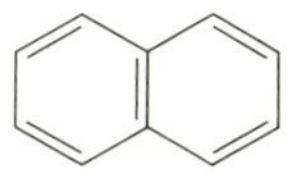

is maximum if the **E** vector is in the plane of the ring and zero if perpendicular to this plane. If the molecule has axes of symmetry of different lengths, as in the planar molecule naphthalene, absorption occurs only if the plane of the **E** vector is in the plane of the ring, but the probability of absorption depends on whether the **E** vector is parallel to the long or the short axis of the molecule. This effect is called *dichroism* and requires that ε be defined for particular directions. It is useful in determining the orientation of molecules in biological systems and is described in greater detail in the section concerning polarization microscopy in Chapter 2. In studies of macromolecules, it has been used most frequently in infrared spectroscopy to determine the orientation of particular bonds with respect to molecular axes. This will be described in the next section.

Infrared Spectroscopy

Transitions between vibrational levels of the ground state of a molecule (Figure 14-2) result from the absorption of light in the infrared (IR) region: from 10^3 nm to 10^5 nm (Figure 14-3). These vibrational levels and, hence, infrared spectra are generated by the characteristic motions (bond stretching, bond bending, and more complex motions) of various functional groups (e.g., methyl, carbonyl, amide, etc.). The value of infrared spectral analysis comes from the fact that the modes of vibration of each group are very sensitive to changes in chemical structure, conformation, and environment and, from this point of view, infrared is not different from visible and ultraviolet spectroscopy. It is thought of as being different principally because it has a somewhat different technology and because it is used to examine chemical groups not accessible to ultraviolet and visible-light absorption spectroscopy.

Technology of Infrared Spectroscopy

Infrared spectrophotometers are in principle no different from ultraviolet and visible-light spectrophotometers. The source of radiation is an object heated to 1500–1800 K; a monochromator is used for wavelength selection, but a thermocouple (instead of a photocell) is the detector.

The principal complication of infrared spectroscopy is that it is usually not possible to use aqueous solutions because of the powerful absorption of infrared by water (because it has a high absorption coefficient and is at a concentration of 55 M). This problem can be circumvented in part by the use of D_2O or of H_2O–D_2O mixtures. Chloroform is also sometimes used because it dissolves many polar molecules, but chloroform-induced conformational changes present many difficulties. The most common method with macromolecules is to use thin, fairly dry films. They are prepared by dissolving the macromolecule in a volatile solvent, placing the solution on a flat plate, allowing the solvent to evaporate, and lifting the film from the plate. These films can, if desired, be stretched to prepare films in which all molecules are oriented in the same direction. They are useful in studies with polarized light in which the orientation of particular groups with respect to the molecular axis can be determined. Of course, the possibility that the structure of a macromolecule in a dry film is not the same as that in solution must always be considered.

Information in Infrared Spectra

First, it should be noted that infrared spectra are conventionally plotted in terms of wave numbers ($1/\lambda$) or frequency (ν) rather than wavelength. In frequency terms, each band in an infrared spectrum can be characterized by the frequency of an absorption maximum (ν_{max}), its band width at half maximum height ($\Delta\nu_{\frac{1}{2}}$), the absorbance at ν_{max} (A_{max}), and the band shape.* In oriented films, the dichroic ratio (R), that is, the ratio of the band area if the electric vector is parallel to and perpendicular to the axis, is measured. By extensive studies on many monomers with known structure, the identity of the groups and the type of vibration corresponding to each band in a spectrum are well known. For simple compounds, infrared spectra differ from spectra for visible and ultraviolet light in that they generally consist of fairly narrow lines (Figure 14-22). However, for macromolecules, each bond type exists in such great numbers and in so many different configurations that each band is shifted—to an extent that depends on where it is in the molecule—so that all bands overlap and therefore the spectrum appears to contain a few relatively broad bands (Figure 14-22).

However, certain bonds that are prevalent in biological macromolecules (for example, hydrogen bonds) cause characteristic shifts in the absorption bands of the groups involved in the bonding. These shifts may be used in the study of macromolecules. An example of the study of hydrogen bonds is shown in Figure 14-23. These spectra demonstrate an interaction between the C=O group of uracil (U) and the NH_2 group of adenine (A) in the double-stranded polymer poly(A)·poly(U) consisting of one strand each of poly(A) and poly(U). In panel A, a small part of the spectrum of poly(U) is shown. The principal band is due to the C=C group. Panel B shows a portion of the spectrum of poly(A) in which the band is that of the C—N bond. The monomers uridylic acid (UMP) and adenylic acid (AMP) presumably cannot maintain hydrogen bonds at room temperature so that the spectrum of a mixture of equimolar quantities of UMP and AMP should be the sum of the spectra of each, as shown in panel C. Poly(A)·poly(U) is presumably held together by hydrogen bonds. Panel D compares the measured spectrum of poly(A)·poly(U) with a spectrum that is just the sum of the spectra of poly(A) and poly(U). The C=O band has moved to a higher frequency, as might be expected by the greater inertia of the oxygen atom resulting

* The wave number is $1/\lambda = \nu/c$, in which c is the velocity of light in vacuum. The units of ν and $1/\lambda$ are s^{-1} and cm^{-1}, respectively. The wave number is frequently abbreviated as $\overline{\nu}$ or k.

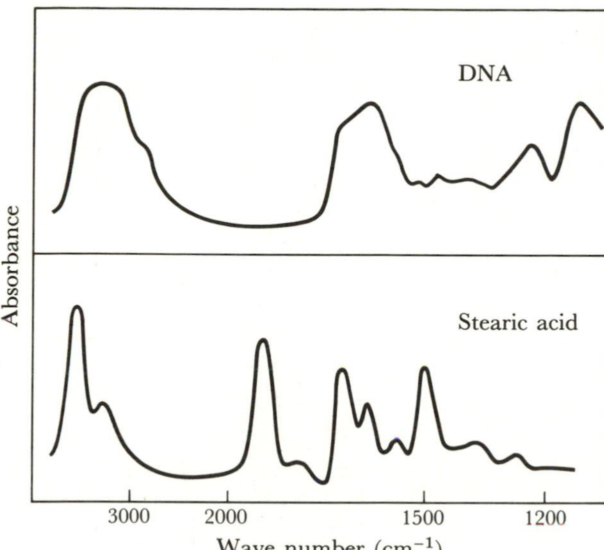

Figure 14-22
Infrared spectra for DNA and for stearic acid. Note the broad bands in the DNA spectrum.

from being constrained to the hydrogen bond. The C—N vibration is also shifted to a higher frequency but to a lesser extent than the C=O shift. Thus, the spectrum gives supporting evidence for hydrogen bonding. Although not shown in the figure, the two C=O bonds in uracil, that from C_2 and that from C_4 have different absorption frequencies. The band shown is for the C_4=O bond; only this band shifts so the spectrum indicates that the C_4=O bond is the one that is hydrogen bonded. Hydrogen bonding could also have been studied in other regions of the spectrum, for instance, the band corresponding to the N—H stretch in the adenine amino group. A shift is also seen here but to lower frequencies indicating that the covalent N—H bond is weakened when the H is in a hydrogen bond. This fact has also been confirmed by other measurements.

There are many modes of vibration (bending, stretching, and so forth) of the amide group of the peptide bond and at least ten different bands result, which are designated amide I, amide II, and so forth. The amide I band has been of great value because its frequency depends upon the environment of the bond and thus can be used to learn something about protein structure. In particular, its frequency depends on whether the particular peptide bond is in the α helix, β sheet, or random coil configuration. The wave numbers corresponding to each of these structures are the following:

α helix	1650 cm^{-1}
β sheet	1632, 1685 cm^{-1}
Random coil	1658 cm^{-1}

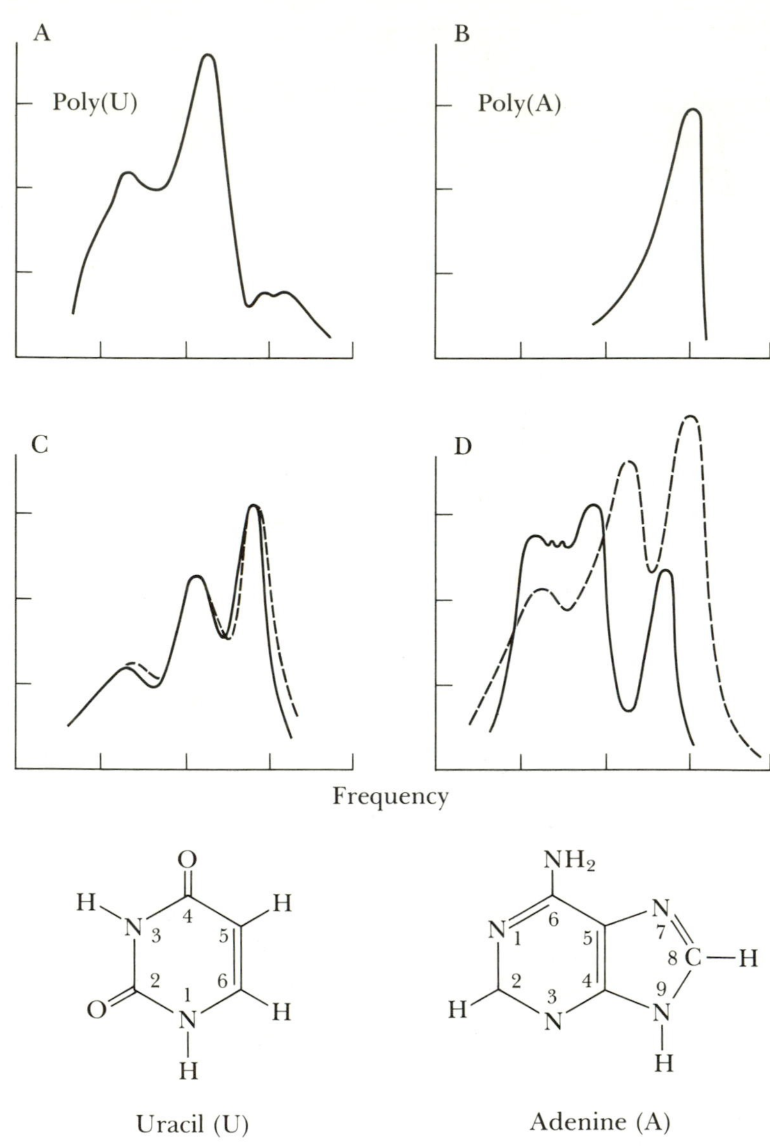

Figure 14-23
Infrared spectra indicating hydrogen bonding. Panels A and B show the spectra of poly(U) and poly(A), respectively. Panel C shows the spectrum for an equimolar mixture of poly(U) and poly(A) (solid line) and the expected curve obtained by summing the curves in panels A and B (dashed line). Panel D shows the spectrum of poly(A)·poly(U) (solid line); the dashed line is the curve of panel C. [From H. T. Miles, *Biochim. Biophys. Acta* 30(1958):324–328. With permission of the publisher.]

These bands are all well resolved. If a protein contains peptide bonds in all three configurations, all four bands will appear. Thus, simply by measuring the intensity of absorption in these bands, the relative proportion of the amino acids that are in the two configurations can be determined. Of course, since the measurement is done in D_2O and not H_2O, it is not always clear that the values apply also to aqueous solution. This problem is avoided with Raman spectroscopy, as will be seen in a later section.

A few other applications of infrared spectroscopy follow.

Identification of exchangeable hydrogen. Many bands change frequencies when deuterium is substituted for hydrogen. In general, the functional group responsible for a given band (i.e., carbonyl, hydroxyl, amino) is known; thus, by observing the bands that have shifted, the groups in which exchange is possible can be identified. Because some functional groups (e.g., hydroxyl) normally exchange rapidly, a delayed shift in ν_{max} indicates a slow exchange; this generally means that the group is buried.

Identification of the number of hydrogen bonds and the functional groups engaged in hydrogen bonding and measurement of their breakage during denaturation. This can be done, for example, by dissolving the macromolecule in D_2O, denaturing the sample, and observing which bands corresponding to deuterated groups appear during denaturation.

Identification of tautomeric forms by the appearance of unexpected bands. Suppose that a molecule contains a hydroxyl group, but its infrared spectrum indicates that a carbonyl is present. This can be taken as strong evidence for tautomerization. If the molecule is part of a macromolecule, the relative intensities of the hydroxyl and carbonyl bands may change compared with the free molecule—one band might even disappear. This would indicate the chemical structure of the molecule in the polymer. This has been of special importance in understanding nucleotide structure and the effect of various nucleotides on mutation frequency.

Interaction between small molecules, such as riboflavin and adenine, and protein–ligand binding. This produces characteristic shifts in ν and intensity, as in ultraviolet-absorption spectroscopy.

Determination of the ratio of $A \cdot U$ to $G \cdot C$ pairs in transfer RNA. The two types of base pairs give bands at different ν. This has been used to distinguish the structures shown in Figure 14-17.

Titration of protein carboxyls. Some "buried" carboxyls titrate in a pH region that is far from normal and in a range of titration of other ionizable groups (see Example 14-D for the analogous experiment with ultraviolet light). Such titrations can be followed by spectral changes in D_2O.

Determination of the orientation of hydrogen bonds in stretched films of proteins and polypeptides by measuring the orientation of the constituent C=O *and* NH— *groups, using polarized infrared radiation.* This is possible because the maximum absorption by these groups occurs when the **E** vector is parallel to the group. This can be used to identify an α-helix because the theoretical ratio of the absorbance when the **E** vector is parallel to and perpendicular to the axis of a protein (the dichroic ratio) is 44 at 3300 cm^{-1} (the stretching vibration for NH—). If the measured ratio is near this value, the structure has a high probability of being α-helical. The theoretical value itself will rarely be achieved because the polypeptide chains cannot be perfectly oriented in a film. Also, from the measured dichroic ratios at 3300 cm^{-1} and 1660 cm^{-1} (the stretching vibration for C=O), the maximum angle that these groups could make with the axis can be calculated. This kind of information is frequently of great value in the interpretation of X-ray diffraction patterns.

Raman Spectroscopy

Infrared spectroscopy has the disadvantage that the powerful absorption of infrared radiation by 55 M H_2O makes the analysis of substances in aqueous solution nearly impossible. Raman spectroscopy, which also probes the vibrational energy levels of a molecule, eliminates this problem; this is because incident light of almost any wavelength can be used even though the same transitions used in the infrared are examined. How this can be done is explained in the following.

A fundamental rule of the quantum theory is that absorption and emission of radiation cannot occur unless the energy of the radiation precisely equals the difference between two energy levels. However, "cannot occur" is a term that is inconsistent with quantum mechanical thinking and it must be replaced with the phrase "occurs with low probability". Raman scattering is an example of such a low-probability event.

Consider a population of molecules illuminated with monochromatic light, whose frequency is such that it is not absorbed—that is, $h\nu$ is not the difference between any pair of energy levels. If no light is absorbed, most of the light passes right through the sample. However, some of the incident light is scattered and this light can be observed by viewing at 90° away from the direction of the incident beam. The frequency of this scattered light will predominantly be the same as that of the incident light—that, is, no energy is lost and the scattering is elastic. However, with very low probability, a small amount of energy of an incident photon (that is, a part of a quantum!) can be transferred to one of the molecules so that the molecules will go to a higher vibrational state. The amount of energy lost by the incident photon is exactly equal to the energy of a transition from one vibrational state to another;

thus a small fraction of the scattered light has a frequency that is decreased by the frequency associated with the vibrational state. Similarly, with equally low probability, if the molecule is not in the ground state, vibrational energy can be transferred *from* the molecule to *raise* the frequency of the scattered light. Of course, there are many more molecules in the ground state than in an excited vibrational state so that a frequency loss, albeit rare, is much more common than a frequency gain. The result of these frequency shifts is that at right angles with respect to the incident beam a collection of spectral lines, whose frequencies are determined by vibrational transitions, can be observed. These lines constitute the Raman spectrum. Several important points should be noted about this spectrum.

1. Raman scattering occurs with very low probability so that the intensity of the lines is very low. In fact it is very difficult to detect the lines unless a high-intensity light source such as a laser is used. A typical experimental arrangement is shown in Figure 14-24.

2. The intensity of the Raman lines is so low that a highly concentrated sample (usually about 2% w/v) is needed. This can create problems in the interpretation of the spectra because at such high concentrations many molecules aggregate, often in specific arrays such as dimers, stacks, and so forth, whose spectra differ from that of the

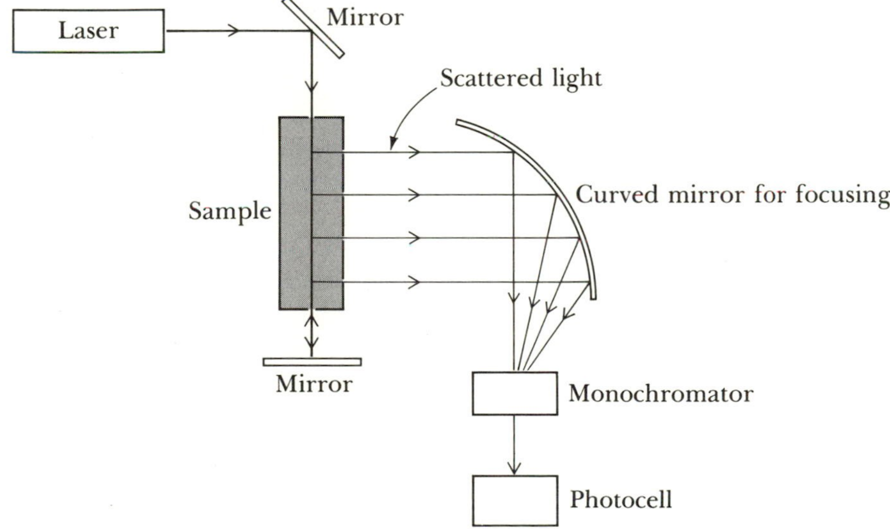

Figure 14-24
Raman spectrometer. The lower mirror reflects the light back through the sample in order to double the intensity of the scattered light. The curved mirror gathers as much scattered light as possible.

individual molecules. This problem is avoided in resonance Raman spectroscopy, which is described in a later section.

3. The line resulting from a gain of energy is so weak that it is usually not seen by ordinary spectrometers.

4. Any wavelength can be used to produce the Raman spectrum; furthermore, the spectrum is independent of the wavelength used. If the frequency of a vibrational transition normally observed in the infrared is ν, then the observed frequency, using incident light of frequency ν_0, is $\nu_0 - \nu$. Thus *the observed line is not in the infrared but in the same wavelength range as the incident light* because in general $\nu \ll \nu_0$. How Raman and infrared spectra are related is shown in Figure 14-25. Note that just as a C=O or a C—H group has a characteristic absorption frequency in their infrared spectra,

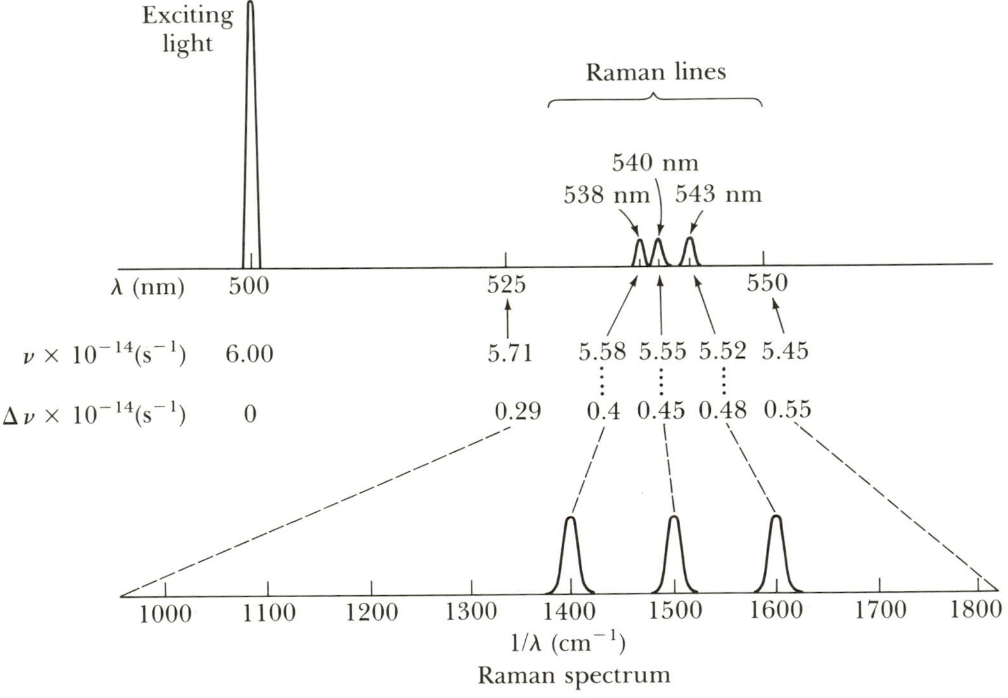

Figure 14-25
The relation between the wavelengths of the actual Raman lines and the spectrum expressed in terms of wave numbers in the infra-red range. The wavelengths of the exciting light and the Raman lines are converted to frequencies. The difference $\Delta\nu$ between the exciting light and a Raman line is calculated and then converted to a value expressed in wave numbers (lowest set of numbers). This value, which is actually the difference in wave numbers between the Raman line and the exciting light, is the value plotted.

they have a characteristic frequency *shift* in the Raman spectrum. It is important to appreciate that whereas the Raman spectrum is a result of transitions that occur normally in the infrared, there is no infrared radiation detected. If visible light is used for excitation, it is visible light, whose frequency is only slightly shifted, and therefore is still visible light, that is detected. However, the spectrum is normally plotted with a frequency or wave number along the x-axis whose values are those of the responsible infrared transition. This conversion from the observed spectrum to the infrared spectrum is accomplished electronically and mechanically by the spectrometer and is not an intrinsic property of the Raman phenomenon.

5. There are vibrational transitions that do not produce emission of radiation because radiation can never be absorbed or emitted unless the change in vibrational energy level is accompanied by a change in dipole moment. Thus diatomic, homonuclear molecules such as O_2 and N_2, which neither have a dipole moment nor gain one by vibration, have no infrared spectra. However, Raman scattering does not have this requirement so that these molecules do have Raman spectra.*

6. Molecules having certain types of symmetry do not produce Raman spectra. This is not a common occurrence for biochemicals, however.

It is important to understand that the information content of a Raman spectrum is no greater than that of an infrared spectrum; however, it is obtained without the need for dry films or D_2O solutions. In actual use, one does not even need to understand any of the points that have just been made. This is because, as in all spectroscopic techniques, a set of empirical rules provides the basic information.

The interpretation of Raman spectra is no different from that of infrared spectra and each of the studies described in the preceding section could have been carried out by Raman spectroscopy. However, Raman spectroscopy is far superior because numerous spectral lines, which are inaccessible in infrared spectroscopy owing to the strong water absorption, are easily detected. Furthermore, since scattering rather then absorption is measured, it is possible to perform Raman spectroscopy with dry fibers or crystals, which allows a ready comparison with the results of X-ray diffraction analysis. In fact, many conclusions derived from X-ray studies have been confirmed and, in some cases, altered by Raman analysis. The application of this procedure to the study of protein structure is described in the following section.

---

* The requirement for production of transitions in the infrared is that the induced vibrations cause a change in the permanent dipole moment of the absorber. The requirement for the Raman effect is that the polarizability of the absorber changes with molecular motion. Thus, a transition may have low intensity with one technique and high intensity with the other.

Use of Raman Spectroscopy in the Elucidation of the Structure of Proteins

Raman spectra provide valuable information about protein structure. Studies with model compounds, whose structures were already known from X-ray diffraction analysis, have demonstrated that many features of protein structure produce characteristic spectral lines. Raman spectra provide information about the peptide bond, the geometry of disulfide bonds, and the environment of the side chains of tyrosine, tryptophan, and methionine. Furthermore, since Raman spectra can be obtained from crystals, powders, gels, or aqueous solution, it is possible to compare the structure determined by X-ray diffraction, which uses either solid material or a gel, with the structure of the molecule in the biologically relevant state, namely, in aqueous solution.

The observations most commonly made and the structural features to what they correspond are described in Table 14-4.

It was pointed out earlier that with Raman spectroscopy it is possible to observe vibrational transitions that do not give infrared spectra—for instance, transitions of the homonuclear diatomic molecules, such as O_2. Since O_2 participates in many biological processes, Raman spectroscopy has been of great value. An example of this follows.

Example 14-L ☐ The state of oxygen in oxygenated hemoglobin and hemerythrin.

The protein hemoglobin, found in blood, carries oxygen. If the Raman spectra of pure hemoglobin and of hemoglobin carrying oxygen

Table 14-4
The rules of Raman Spectroscopy of proteins.

1. A strong peak at $1650 \pm 5\ \text{cm}^{-1}$ is generated by the α helix. It is a result of a stretching vibration of the peptide bond and is known as the amide I band.
2. A very sharp amide I band at $1665 \pm 5\ \text{cm}^{-1}$ indicates a β-sheet structure. If the band is very broad, there are peptide bonds in a random-coil configuration.
3. The amide III band, which is also a result of a vibration of the peptide bond, is broad and ranges from $1265\ \text{cm}^{-1}$ to $1300\ \text{cm}^{-1}$ for the α helix, sharp and strong at $1235 \pm 10\ \text{cm}^{-1}$ for β-sheet structures, and sharp at $1248 \pm 10\ \text{cm}^{-1}$ for a random coil.
4. The stretching vibration for the disulfide bond is affected by the rotation about the C—C and C—S bonds. The three different conformations of the disulfide bonds, namely, *gauche-gauche-gauche*, *trans-gauche-gauche*, and *trans-gauche-trans*, have sharp bands at 510, 525, and $540\ \text{cm}^{-1}$, respectively.
5. The relative intensities of the lines at $828\ \text{cm}^{-1}$ and $853\ \text{cm}^{-1}$ are determined by the environment of tyrosine. If the ratio is 0.7, the tyrosine is totally exposed to water; if it is 0.8, the tyrosine is buried. This ratio is informative only if the protein contains only one tyrosine.
6. A tryptophan totally removed from water yields a sharp peak at $1361\ \text{cm}^{-1}$. As the tryptophan becomes more and more accessible to water, the band becomes weaker and broader. The size and shape of this peak is informative only when the protein contains one tryptophan.

are compared, a strong absorption band at a frequency of 3.321×10^{13} sec^{-1} is seen when oxygen is present. The frequencies corresponding to molecular O_2 and to other oxygen-containing groups such as superoxide, O_2^-, and peroxide, O_2^{2-} have also been determined by Raman spectroscopy of compounds containing these groups. The absorption frequency of the superoxide group, O_2^-, is also 3.321×10^{13} sec^{-1} suggesting that this is the form of oxygen in oxygenated hemoglobin. The protein hemerythrin is the oxygen carrier in many invertebrates. The spectrum of the oxygenated molecule shows absorption at 1.532×10^{13} sec^{-1}, corresponding to peroxide, so that peroxide is the form of oxygen in this molecule.

Other Uses of Raman Spectroscopy

Raman spectroscopy is rapidly becoming an important and versatile technique. Following is a list of several of its applications.

Distinguishing adenosine monophosphate, diphosphate, and triphosphate and their ionized forms in solution. This has been useful in studying certain enzyme reactions because the Raman spectrum of a reaction mixture indicates the relative proportions of each molecular species during the reaction or at equilibrium.

Determination of the number of S—S *bonds in proteins.* The S—S and —SH bonds each give characteristic ν_{max}.

Determination of the number of paired and unpaired bases in RNA. Each base is distinguishable by its Raman spectrum so that base composition can be determined by the relative intensity of peaks. There is also a spectral change on deuteration so that sites of slow and rapid hydrogen-deuterium exchange are identifiable. A base showing slow exchange is though to be in a base pair because hydrogen bonding delays exchange (see Example 14-G).

Identification of intermediates in enzymatic reactions. Most simple organic compounds and functional groups have characteristic Raman spectra in aqueous solution. Thus, if their concentration is sufficiently high, reaction intermediates can be detected in reaction mixtures.

Resonance Raman Spectroscopy

In the kind of Raman spectroscopy that has been discussed, the exciting light is at a wavelength that is far from an absorption band. If the wavelength of the exciting light lies in an absorption band of an electronic transition for a particular chromophore, there is a large increase in the intensity of certain lines in the Raman spectrum. These lines result from vibrational transitions involving motion of those atoms that participate in the electronic transition. The meaning of this statement can be made clearer by looking at Figure 14-26, which shows the visible-

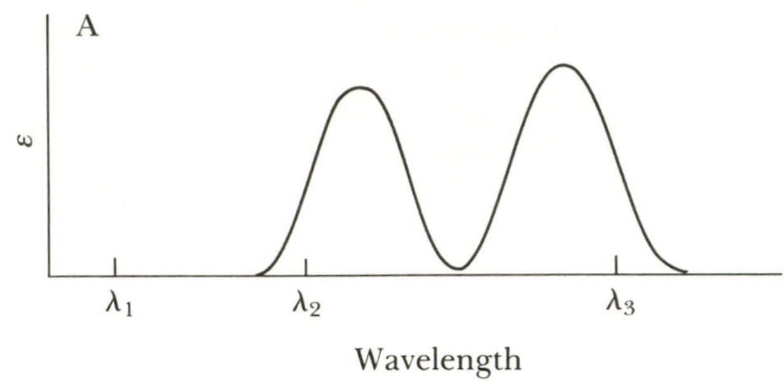

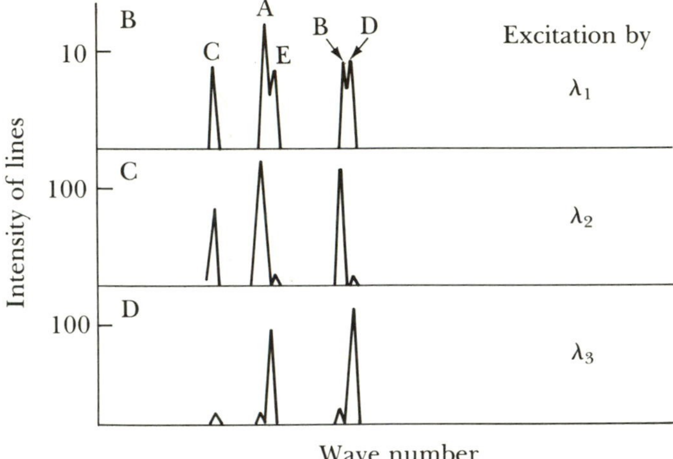

Figure 14-26
(A) Visible light absorption spectrum of a hypothetical molecule. (B) Raman spectrum resulting from excitation with λ_1, which is not absorbed. (C, D) Resonance Raman spectra using exciting wavelengths λ_2 and λ_3, which are absorbed. Note that the vertical scale is tenfold greater than in B.

light absorption spectrum for a hypothetical molecule. The spectrum has two bands, each of which corresponds to a particular electronic transition. Part of the breadth of each electronic absorption band is a result of the superposition of many vibrational modes. Let us assume for simplicity that the left band includes three vibrational transitions, whose wave numbers in the infrared are $\bar{\nu}_A$, $\bar{\nu}_B$, and $\bar{\nu}_C$, and the right band includes two vibrational transitions, whose wave numbers are $\bar{\nu}_D$ and $\bar{\nu}_E$. If the molecule is illuminated with light that cannot be absorbed, for instance λ_1, the Raman spectrum will consist of the five weak lines having wave numbers $\bar{\nu}_A$ through $\bar{\nu}_D$. However, if the wavelength of the incident light is λ_2, which can be absorbed in the left band, the vibrational modes corresponding to the transition whose electronic band contains

λ_2 will produce Raman lines that are much more intense than those resulting from illumination with λ_1. In particular, the Raman spectrum will consist of two very weak lines, namely, $\bar{\nu}_D$ and $\bar{\nu}_E$, and three very intense ("enhanced") lines, $\bar{\nu}_A$, $\bar{\nu}_B$, and $\bar{\nu}_C$. Similarly, if the incident light light has the wavelength λ_3, the Raman spectrum will consist of three very weak lines, $\bar{\nu}_A$, $\bar{\nu}_B$, and $\bar{\nu}_C$ and two very strong lines, $\bar{\nu}_D$ and $\bar{\nu}_E$. The technique of obtaining Raman spectra using incident light that can be absorbed is called *resonance Raman spectroscopy*. It has two principal advantages over ordinary Raman spectroscopy. First, the lines are sufficiently intense that much lower concentrations of the molecule of interest can be used. Second, when the concentration is reduced, only the enhanced lines are seen that the spectrum is considerably simplified. There are also four disadvantages. (1) The Raman scattered light is in the range of the wavelengths that are absorbed. This requires the use of samples of very small volume in order to minimize loss of this light. Usually the sample is contained in a fine capillary. (2) The absorption of the incident light sometimes leads to fluorescence, whose intensity may be as great or greater than the Raman lines. Addition of fluorescence quenchers (see Chapter 15) sometimes eliminates this problem; when it does not, several specialized techniques still in the developmental stages (such as the use of pulsed lasers and coherent anti-Stokes Raman spectroscopy) are used. (3) The absorption of the high-intensity laser light causes heating of the sample, and sometimes photodegradation or photoisomerization occurs. Thus, after the spectrum is obtained, it is necessary to demonstrate by some means that the sample is unchanged. (4) The instrumentation required to perform resonance Raman spectroscopy is not generally available.

In order to perform resonance Raman spectroscopy, a chromophore that absorbs in a spectral region that can be studied with available laser frequencies is needed. At the present time this means a chromophore that absorbs visible light. The use of ultraviolet is possible but presents problems that have not yet been fully worked out. The chromophore may be part of the molecule of interest, that is, an inherent chromophore, or can be a resonance Raman label (similar to a report group). Examples of molecules of biological interest that have inherent chromophores are heme proteins (hemoglobin, cytochromes), copper and non-heme-iron metalloproteins, carotenoids and visual pigments, chlorophyll, and vitamin B_{12} derivatives. The resonance Raman labels used to date are colored enzyme substrates and inhibitors (used to probe the structure of active sites) and azo dyes that bind to particular sites on proteins.

In infrared spectroscopy there are characteristic bands, such as amide I vibrations, which are present in all proteins. However, in resonance Raman spectroscopy, the types of vibrations depend only on the chromophore so that there is not a collection of commonly occurring lines in proteins. Thus, for each chromophore it is necessary to identify the particular atoms responsible for each observed Raman line. However, the utility of the procedure is the same as that in infrared spectroscopy.

That is, the frequency of these lines is affected by molecular interactions so that it is possible to study interactions involving particular atoms, namely, those contained in the chromophore.

One way to identify the atoms responsible for particular lines is by isotopic substitution. For instance, the copper-containing protein hemocyanin (the oxygen carrier of some molluscs and crustaceans) has a resonance Raman spectrum consisting of several narrow lines. The frequencies of two of these lines are reduced when $^{18}O_2$ is substituted for $^{16}O_2$ in the oxy form of the protein. On this basis these lines have been assigned to vibrations of the oxygen. There are many other methods for identifying lines; these are described in the references given at the end of the chapter.

The identification of a line is often sufficient to give structural information about a protein. For example, the groups to which the metal atom in several copper proteins is bound have been identified in this way. Resonance Raman spectra have been obtained for simple model copper-peptide complexes and the few lines observed have been identified. The frequencies of certain lines were independent of the nature of the side chains, indicating that they represent vibration of the copper and the peptide bond. Other data showed that the binding is probably to the N atom. These lines are found in resonance Raman spectra of the copper proteins suggesting that in these proteins the copper is bound to the peptide N also. Studies with other model compounds showed that another line is due to a Cu—S stretching vibration. Thus, it was concluded that in the copper proteins the copper is bound both to a peptide N and a cysteines. This information plus other data have led to detailed models of the copper binding site.

Published experiments using resonance Raman spectra are somewhat limited at present because of the small number of instruments capable of producing spectra. However, the great potential of the system, especially when ultraviolet excitation becomes possible, and the construction of more instruments should increase its use markedly in the next few years.

SELECTED REFERENCES

Beaven, G. H., E. A. Johnson, H. A. Willis, and R. G. J. Miller. 1961. *Molecular Spectroscopy*. Heywood.

Brewer, J. M., A. J. Pesce, and R. B. Ashworth. 1974. *Experimental Techniques in Biochemistry*, ch. 7. Prentice-Hall.

Bush, C. A. 1974. "Ultraviolet Spectroscopy, Circular Dichroism, and Optical Rotatory Dispersion," in *Basic Principles in Nucleic Acid Chemistry*, edited by P. O. P. Ts'o, pp. 92–169. Academic Press.

Donovan, J. W. 1973. "Ultraviolet Difference Spectroscopy: New Techniques and Applications" and "Spectrophotometric Titration of the Functional Groups of Proteins," in *Methods in Enzymology*, vol. 27, edited by C. H. W. Hirs and S. N. Timasheff, pp. 497–525; 525–548. Academic Press.

Edisbury, J. R. 1965. *Practical Hints of Absorption Spectrometry*. Hilger-Watts.
Fraser, R. D. B., and E. Suzuki. 1970. "Infrared Methods," in *Physical Principles of Protein Chemistry*, vol. B, edited by S. J. Leach, pp. 213–273. Academic Press.
Gratzer, W. B. 1967. "Ultraviolet Absorption Spectroscopy of Polypeptides," in *Poly-α-Amino Acids*, edited by G. D. Fasman, pp. 177–238. Dekker.
Herskovitz, J. T. 1967. "Difference Spectroscopy," in *Methods in Enzymology*, vol. 11, edited by C. H. W. Hirs, pp. 748–775. Academic Press.
Horton, H. R., and D. E. Koshland. 1967. "Environmentally Sensitive Groups Attached to Proteins," in *Methods in Enzymology*, vol. 11, edited by C. H. W. Hirs, pp. 856–870. Academic Press. A description of the use of reporter groups.
Kronman, M. J., and Robbins, F. M. 1970. "Buried and Exposed Groups in Proteins," in *Fine Structure of Proteins and Nucleic Acids.*, edited by G. D. Fasman and S. N. Timasheff, pp. 271–416. Dekker.
Mathieu, J. P. 1973. *Advances in Raman Spectroscopy*. Heydon.
Spiro, T. G., and B. P. Gaber. 1977. "Laser Raman Scattering as a Probe of Protein Structure." *Annu. Rev. Biochem.*, 46:553–572.
Timasheff, S. N. 1970. "Some Physical Probes of Enzyme Structure in Solution," in *The Enzymes*, vol. 2, edited by P. D. Boyer, pp. 371–443. Academic Press.
Van Wart. H. E., and Scheraga, H. A. 1978. "Raman and Resonance Raman Spectroscopy," in *Methods in Enzymology*, vol. 49, edited by C. H. W. Hirs and S. N. Timasheff, pp. 67–148. Academic Press.

PROBLEMS

14–1. A molecule has a spectral line at 2989 cm^{-1}.
  a. What are its frequency and its wavelength?
  b. In what part of the spectrum is it located?

14–2. You have just prepared two concentrated solutions of the amino acids tyrosine and isoleucine. You neglect to label the bottles immediately and become confused. You have no more so you cannot make up fresh solutions. What spectral measurement might you make to distinguish the solutions?

14–3. The molar absorption coefficient of benzene is 100 $M^{-1}cm^{-1}$ at 260 nm.
  a. What concentration would give an absorbance of 1.0 in a 1-cm cell at 260 nm?
  b. What concentration would allow 1% of 260 nm light to be transmitted through a 1-cm cell?
  c. If the density of liquid benzene is 0.8 gm/cm^3, what thickness of benzene would give an absorbance of 1.0 at 260 nm?

Assume that the absorption coefficient is independent of solvent and that Beer's law is obeyed over the entire range of concentration.

14–4. A solution at a concentration of 32 μg/ml of a substance having a molecular weight of 423 has an absorbance of 0.27 at 540 nm measured in a cuvette with a 1-cm light path. What is the molar absorption coefficient at 540 nm? Assume that Beer's law is obeyed.

14–5. A solution of a molecule (A) has an $OD_{260} = 0.45$ and $OD_{450} = 0.03$. A solution of a second molecule (B) has $OD_{260} = 0.004$ and $OD_{450} =$

0.81. Two milliliters of A are mixed with 1 milliliter of B. The resulting "optical densities" of the mixture are $OD_{260} = 0.30$ and $OD_{450} = 0.46$. Is there an interaction between A and B? Explain. What assumption is made to justify this conclusion?

14–6. A particular molecule has a molar absorption coefficient of 348 at 482 nm. A solution of this molecule has an $OD_{482} = 1.6$. When diluted 1:1, 1:2, 1:3, 1:4, 1:5, and 1:6, the values of OD_{482} are 1.52, 1.42, 1.05, 0.84, 0.70, and 0.61, respectively. What is the molarity of the original solution?

14–7. The molar absorption coefficients of substance A at 260 nm and 280 nm are 5248 and 3150 M^{-1} cm^{-1}, respectively. In isolating A, a reagent B is used whose molar absorption coefficients at 260 and 280 nm are 311 and 350 M^{-1} cm^{-1}. After isolating A, $OD_{260} = 2.50$ and $OD_{280} = 2.00$. What is the concentration of A?

14–8. Solutions containing tryptophan and tyrosine have characteristic absorption spectra. The molar absorption coefficients at 240 nm and 280 nm are the following:

Wavelength (nm)	ε_{Tyr} (M^{-1} cm^{-1})	ε_{Trp} (M^{-1} cm^{-1})
240	11,300	1960
280	1,500	5380

A 10-mg sample of a protein is hydrolyzed to its constituent amino acids and diluted to 100 ml. The absorbance of this solution, using a 1-cm light path, is 0.717 at 240 nm and 0.239 at 280 nm. Estimate the content of tryptophan and tyrosine using units of μmol/g protein. (Hint: set up two simultaneous equations.)

14–9. Suppose that you have just prepared two DNA samples, one native and one denatured, and have dialyzed each against 0.01 M NaCl. You then add a very small amount of an enzyme to each and in so doing mix up the samples. How can you determine the identity of each sample by an absorbance measurement? You may assume that a small part of each sample is consumed in the testing and that the enzyme will not interfere with the test. Design a second test that does not require the introduction of an agent or condition that causes denaturation.

14–10. Suppose that you would like to know much cytochrome *c* is contained in one *E. coli* cell. The molar absorption coefficient at its absorption maximum is known. How would you make this measurement?

14–11. The OD_{260} of denatured DNA (measured at a temperature at which the maximum OD_{260} is reached—see Figure 14-13) is invariably 37% higher than that of native DNA (e.g., at 20°C). However, in a solution of 6 M sodium trifluoroacetate, the OD_{260} of a DNA sample is only 16% higher than at 20°C. Propose an explanation for this.

14–12. Two proteins A and B are isolated from a bacterium. You are interested in proteins that act on DNA so you mix an aliquot of each protein with a DNA solution whose concentration is 10 μg/ml ($OD_{260} = 0.2$). After a one-hour incubation time, the OD_{260} of the solution containing protein A is 0.24 and that containing B has an OD_{260} of 0.32. Both solutions are then boiled. The OD_{260} of the solution containing A rises to 0.26 and that containing B remains at 0.32. Suggest possible activities of proteins A and B.

14–13. The growth of a bacterial culture is monitored by reading the OD_{450}. Another person measures bacterial number by counting the cells with a microscope. Explain why the curves shown below have different shapes.

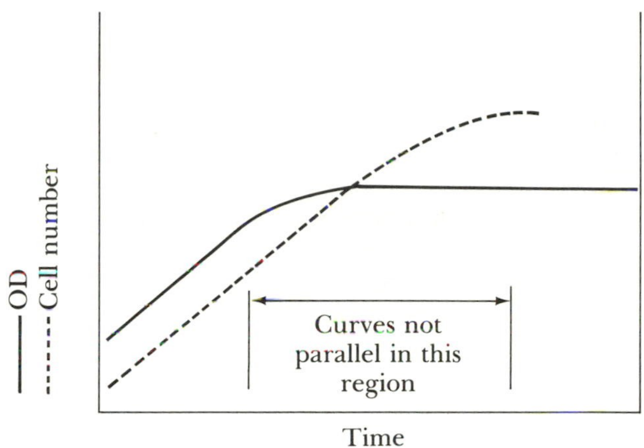

14–14. What spectral changes would probably accompany complete enzymatic hydrolysis of a compact protein in 0.1 M NaCl?

14–15. What structural features of a protein containing tryptophan as the only ultraviolet-absorbing amino acid might lead to no increase in absorbance when the protein is heated above the denaturation temperature in a solution of 0.5 M NaCl, pH 7?

14–16. A protein is known to have a single tryptophan and a single tyrosine. At a protein concentration of 100 μg/ml, if put in 20% ethylene glycol, both λ_{max} and ε increase compared to the value in 0.1 M NaCl. If the concentration is raised to 10 mg/ml and the protein is still in 0.1 M NaCl, both λ_{max} and ε increase. At this high concentration of protein, there is no further change in λ_{max} and ε, if ethylene glycol is present. What two things have you learned about this protein?

14–17. A protein is placed in a 20% ethylene glycol solution and its difference spectrum determined against the solution without ethylene glycol. At all wavelengths $\Delta\varepsilon = 0$. What can you conclude about the structure of the protein?

A second protein is also studied in both 20% dimethylsulfoxide and 20% sucrose. In sucrose, $\Delta\varepsilon = 0$. In dimethylsulfoxide, there is a characteristic difference spectrum of tryptophan. The observed value of $\Delta\varepsilon$ is doubled if the protein is heated to 60°C in dimethylsulfoxide. In 20% sucrose, after heating to 60°C, the value of $\Delta\varepsilon$ is equivalent to eight tryptophans. What can you say about the protein?

14–18. In performing a pH titration of a protein at 295 nm, it is found that ε_{295} increases sharply at pH 9.6. At pH 11.7, ε_{295} increases again, the latter increase being one-third that of the first rise. If the pH is then reduced to 6 and then gradually increased, the curve for ε_{295} versus pH is nearly identical with that of the protein before titration except that the increase in ε_{295} is only two-thirds that originally observed. The protein is known to have eight tyrosines. What can you say about the protein structure?

14–19. DNA is isolated from an *E. coli* bacterial culture. The mean G + C content of *E. coli* is 50%. The DNA is dissolved in a mixture of 0.15 M NaCl and 0.015 M Na citrate. A melting curve is obtained. Instead of a smooth curve, there are two steps having T_m of 80°C and 88°C. The first step accounts for 20% of the total increase. What possible explanation can be given for this?

DNA from another bacterium is isolated in the same way but by accident is heated to 75°C for 1 minute. It is later used in a melting experiment. The melting curve is surprising in that, although the major transition is between 82°C and 88°C, there is a gradual increase in OD_{260} of 10% in the temperature range of 32°C to 55°C. Explain this finding.

15

Fluorescence Spectroscopy

In the preceding chapter, we saw that certain types of information about both the properties of macromolecules and their interactions with other molecules could be obtained from studies of absorption spectra. The underlying principle is that the absorption spectrum of a chromophore is significantly affected by the physical and chemical environment. In this chapter, we consider a more sensitive spectroscopic probe, fluorescence.

With some molecules, the absorption of a photon is followed by the emission of light of a longer wavelength (i.e., lower energy). This emission is called fluorescence (or phosphorescence, if the emission is long-lived). As is true for absorption spectroscopy, there are many environmental factors that affect the fluorescence spectrum; furthermore, fluorescence efficiency is also environmentally dependent. Because these parameters of fluorescence are more sensitive to the environment than are those of absorbance and because smaller amounts of material are required, fluorescence spectroscopy is frequently of greater value than absorbance measurements (although absorption spectroscopy is simpler to perform). With macromolecules, fluorescence measurements can give information about conformation, binding sites, solvent interactions, degree of flexibility, intermolecular distances, and the rotational diffusion coefficient of macromolecules. Furthermore, with living cells, fluorescence can be used to localize otherwise undetectable substances.

As with other physical methods, the theory of fluorescence is not yet adequate to permit a positive correlation between a fluorescent spectrum and the properties of the immediate environment of the emitter; hence, once again the utility of the procedure is based on establishing empirical principles from studies with model compounds.

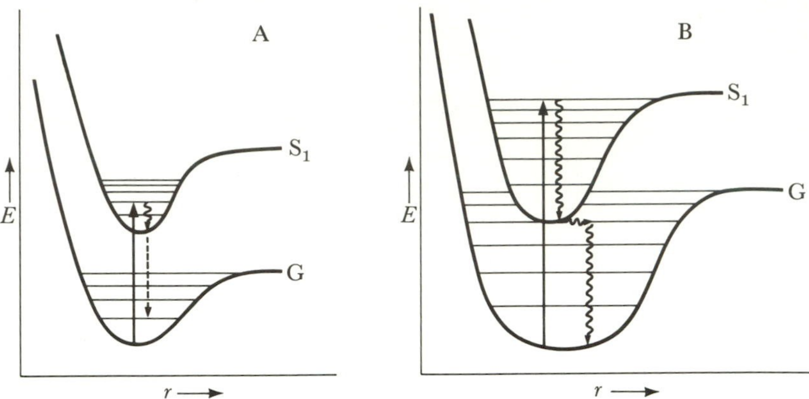

Figure 15-1
Energy-level diagram of two chromophores; G and S_1 indicate the ground and first excited states, respectively (heavy lines). The vibrational levels are the thin lines. A. This molecule is capable of fluorescing by the transition (solid arrow) indicated in the diagram. After excitation, there are vibrational losses (wavy arrow) to the lowest level of the excited state and then emission from this state (dashed arrow). B. This molecule fails to fluoresce because the vibrational levels of G are higher than the lowest level of S; hence, there can be a nonradiative transition (horizontal wavy arrows) from S_1 to a vibrational level of G followed by nonradiative losses to the bottom of G (vertical wavy arrow).

Simple Theory of Fluorescence

In Chapter 14, it was seen that a molecule can possess only discrete amounts of energy. The potential energy levels of the molecule are described by an energy-level diagram (Figure 15-1). This figure shows two electronic levels, the lower or ground state (G) and one upper or first excited state (S_1) and some of the vibrational levels of each (see Chapter 14 for a discussion of vibrational levels). As explained in Chapter 14, light energy can be absorbed only when the molecule moves from a lower to a higher energy level.* Such transitions are indicated on an energy-level diagram by vertical lines. If the molecule is initially unexcited (i.e., at its lowest energy level or ground state, G) and the absorbed energy is greater than that required to reach the first electronic excited state, S_1, the excess energy can be absorbed as vibrational energy and the molecule will be at one of the vibrational levels shown in the figure. This vibrational energy is rapidly dissipated as heat by collision with solvent molecules (if the excited molecule is in solution), and the molecule drops to the lowest vibrational level of S_1. The excited molecule

* Not all transitions are possible. Allowable transitions are defined by the selection rules of quantum mechanics.

then returns to G either by emitting light (fluorescence) or by a nonradiative transition (described in the discussion of quantum yield). Because energy is lost in dropping to the lowest level of S_1, the emitted light will have less energy (i.e., longer wavelength) than the absorbed light. Therefore, fluorescent light always has a longer wavelength than the exciting light. However, in returning to G, the molecule may arrive at one of the vibrational levels of G instead of the absolute ground state; this vibrational energy will also be dissipated as heat. Hence, if there are many absorbers, the light emitted will have many wavelengths (all of course greater than that of the absorbed light); the probability of dropping from the first excited state to each vibrational level of the ground state determines the shape of the fluorescence spectrum. A typical absorption and fluorescence spectrum is shown in Figure 15-2.

As stated earlier, the excited molecule does not always fluoresce. The probability of fluorescence is described by the *quantum yield*, Q; that is, the ratio of the number of emitted to absorbed photons. (A photon is a unit of light having energy $E = h\nu$, in which h is Planck's constant and ν is the frequency of the light wave.) Several factors determine Q; some of these are properties of the molecule itself (internal factors) and some are environmental. The internal factors derive mostly from the distribution of vibrational levels between G and S. For example, if a vibrational level (V_G) of the ground state has the same energy as a vibrational level of a lower order (V_S) of the first excited state, there can be a nonradiative transition (which will not be explained here) from V_S to V_G (Figure

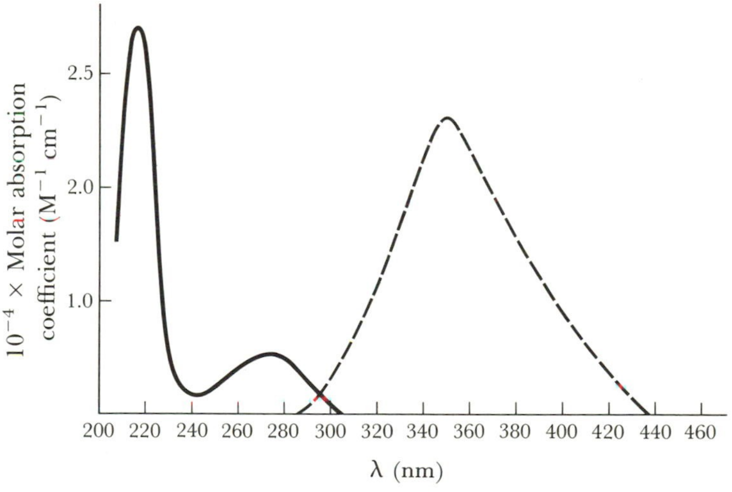

Figure 15-2
Absorption and fluorescence spectrum of tryptophan. Solid line: molar absorption coefficient as a function of wavelength. Dashed line: emission spectrum in arbitrary units.

15-1B), followed by the conversion of the energy of V_G into heat. This is what usually happens with flexible molecules because they have very high vibrational levels of G (see Figure 16-1B). In fact, this is the most common way for excitation energy to be dissipated and accounts for the fact that fluorescent molecules (fluors) are rare and that those which do fluoresce are almost invariably fairly rigid aromatic rings or ring systems.

The internal factors are not generally of interest to biochemists concerned with the properties of macromolecules; environmental factors are more important. The effect of the environment is primarily to provide radiationless processes that compete with fluorescence and thereby reduce Q; this reduction in Q is called *quenching*. In biological systems, quenching is usually a result of either collisional processes (either a chemical reaction or simply *collision* with exchange of energy) or a long-range, radiative process called *resonance energy transfer* (which will be discussed in a later section). These three factors are usually expressed in an experimental situation involving solutions as an effect of the solvent or dissolved compounds (called *quenchers*), temperature, pH, neighboring chemical groups, or the concentration of the fluor. How to make use of these environmental effects in studying macromolecules will be discussed in a later section.

Instrumentation for Measuring Fluorescence

Figure 15-3 shows a standard arrangement for measuring fluorescence. A high-intensity light beam passes through a monochromator for the selection of an excitation wavelength (i.e., a wavelength efficiently absorbed by the fluor). The exciting light beam then passes through a cell containing the sample. To avoid detecting the incident beam, use is made of the fact that fluorescence is emitted in all directions so that observation of the fluorescence can be made at right angles to the incident beam. The fluorescence then passes through a monochromator for wavelength analysis and finally falls on a photosensitive detector (usually a photomultiplier tube). Many modern instruments have scanning systems and chart recorders that automatically vary the wavelength detected and plot the emitted intensity as a function of the wavelength of the emitted light.

The intensity of light is a measure of the energy E per unit area per unit time. Because the response of the usual light detectors, (i.e., photomultiplier tubes), is wavelength dependent and because $E = hc/\lambda$, the ratio of the outputs (electrical currents) produced by a photomultiplier when two different wavelengths fall on it is not the same as the ratio of the intensities. However, in most experiments, a simple measurement of the photomultiplier output is sufficient because usually only *relative* intensities at each wavelength are being measured—for example, the measurement of fluorescence in the presence and absence of an agent if

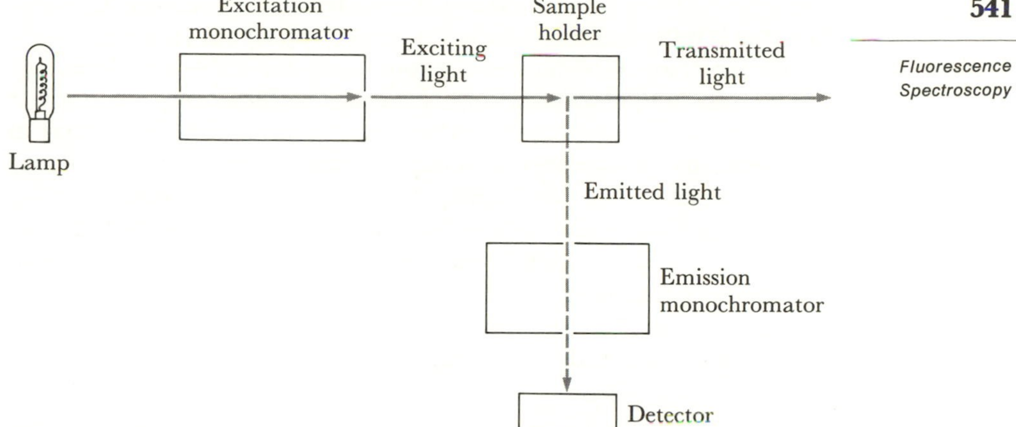

Figure 15-3

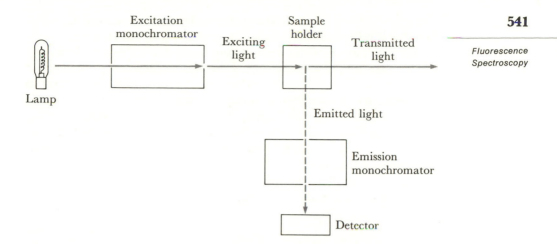

A spectrofluorimeter. Fluorescence is emitted in all directions by the sample, but most systems look at only that emitted at 90° with respect to the exciting light. Note that this differs from an absorption spectrophotometer only in that, with that instrument, the transmitted light would be detected. Modern automatic spectrofluorimeters have a chart recorder and plot either detector output versus emitted wavelength or detector output at a single wavelength or a collection of wavelengths versus wavelength of the exciting light.

the agent does not affect the efficiency of absorption of the exciting light. However, in some experiments, it is necessary to measure Q or to determine the absolute energy distribution of a fluorescence spectrum; both of these require measuring absolute intensity. The usual methods for measuring fluorescence intensity require calibrating the photomultiplier system with a *thermopile*, an instrument whose ability to measure the energy of incident light is wavelength-independent. This calibration may be supplied by the manufacturer of the spectrofluorimeter; if not, another instrument, called a quantum counter can be used.

It is important to know that the distinction between a corrected spectrum and an uncorrected one is not often made in the presentation of fluorescence spectra in journal articles. It is common to plot a spectrum as the photomultiplier output versus wavelength. This is an *uncorrected* spectrum. Plotting fluorescence intensity or quantum yield produces a *corrected* spectrum. Invariably, when photomultiplier output is plotted, it is incorrectly called fluorescence or fluorescence intensity. It is probably safe to assume that a given spectrum is uncorrected unless stated otherwise.

To measure Q requires the counting of photons because

$$Q = \frac{\text{photons emitted}}{\text{photons absorbed}} \tag{1}$$

Note that Q is a dimensionless quantity. Because the energy, E, of one photon is related to the frequency, ν, of the light by the relation $E = h\nu$, a measurement of the number of photons requires measuring the energy of the radiation and correcting for frequency. Measuring the absolute quantum yield is a difficult and tedious process and is rarely done in biochemistry. (Procedures for doing so are given in the Selected References near the end of this chapter.) This usual method for determining Q requires a comparison with a fluor of known Q; two solutions are prepared—one of the sample and one of the standard fluor—and, with the same exciting source, the integrated fluorescence (i.e., the area of the spectrum) of each is measured.

The quantum yield, Q_x, of a sample X is

$$Q_x = \frac{I_x Q_s A_s}{I_s A_x} \tag{2}$$

in which Q_s is the quantum yield of the standard, I_x and I_s are the integrated fluorescence intensities of the sample and the standard, respectively, and A_x and A_s are the percentage of absorption of each solution at the exciting wavelength. Usually, the solutions are adjusted so that $A_x = A_s$.

Once again, it is necessary to understand a convention used in presenting data. Very often published spectra are plotted as Q versus λ, or fluorescence intensities at a particular wavelength are stated as a value of Q. In fact, it is rare that Q has been measured. This inconsistency derives from the fact that, in biochemistry, one is usually measuring the relative intensities in two different situations and, of two spectra (e.g., a quenched and unquenched), the relative values of Q have the same ratio as the fluorescence intensities.

Intrinsic Fluorescence Measurements for Studying Proteins

Two types of fluors are used in fluorescence analysis of macromolecules—intrinsic fluors (contained in the macromolecules themselves) and extrinsic fluors (added to the system, usually binding to one of the components). Intrinsic fluorescence will be discussed first.

For proteins, there are only three intrinsic fluors—tryptophan, tyrosine, and phenylalanine (listed in order of Q, from larger to smaller). The fluorescence of each of them can be distinguished by exciting with and observing at the appropriate wavelength. In practice, tryptophan fluorescence is most commonly studied, because phenylalanine has a very low Q and tyrosine fluorescence is frequently very weak due to quenching. The fluorescence of tyrosine is almost totally quenched if it is ionized, or near an amino group, a carboxyl group, or a tryptophan. In special situations, however, it can be detected by excitation at 280 nm.

The principal reason for studying the intrinsic fluorescence of proteins is to obtain information about conformation. This is possible because the fluorescence of both tryptophan and tyrosine depends significantly on their environment (i.e., solvent, pH, and presence of a quencher, a small molecule, or a neighboring group in the protein).

The use of measurements of intrinsic fluorescence in proteins is based on empirical "rules" obtained from studies of model compounds whose structure and conformation are well known. The rules in common use are presented in Table 15-1. How some of these rules are used in practice is described in the following examples.

Information About the Conformation and Binding Sites of Proteins Obtained from Studies of Intrinsic Fluorescence

In the following examples, relatively strong conclusions can be made about structural features of proteins, using simple measurements of fluorescence. It should be noted that in many cases measurement of relative fluorescence is sufficient. ("Free amino acid" always refers to an amino acid dissolved in water.)

Example 15-A

☐ Conformational change in a hypothetical enzyme induced by the binding of a cofactor.

The three tryptophans of a hypothetical enzyme have a much greater fluorescence intensity and a shorter λ_{max} than does free tryptophan (Figure 15-4). Hence, they must be in a very nonpolar environment (rule 2 in Table 15-1). When the cofactor of the enzyme is added, the fluorescence intensity decreases and λ_{max} becomes longer. Hence, the conformation of the enzyme must have changed (rule 4) either to make one or more of the tryptophans available to the polar solvent, water, or to bring them near charged groups. The addition of either the iodide or cesium ion has no effect on the fluorescence spectrum of the enzyme either before or after the cofactor is bound. Hence, no tryptophans were on the surface before additon of the cofactor (and hence are not in the binding site) nor have they been brought to the surface (rule 3); they have probably been moved to a more polar, internal region of the protein.

Note that, because of these changes in fluorescence, the value of one of the parameters (e.g., intensity at a particular wavelength, the value of λ_{max}, or Q) could be used as an assay for binding. Hence, a plot of fluorescence versus cofactor concentration can be used to determine both the stoichiometry of binding and the dissociation constants by the standard methods of ligand-binding analysis (Chapter 18).

Table 15-1

Empirical Rules for Interpreting Fluorescent Spectra of Proteins.

1. All fluorescence of a protein is due to tryptophan, tyrosine, and phenylalanine unless the protein is known to contain another fluorescent component.
2. The λ_{max} of the tryptophan fluorescence spectrum shifts to shorter wavelengths and the intensity of λ_{max} increases as the polarity of the solvent decreases.
   a. If λ_{max} is shifted to shorter wavelengths when the protein is in a polar solvent, the tryptophan must be internal and in a nonpolar environment.*
   b. If λ_{max} is shifted to shorter wavelengths when the protein is in a nonpolar medium, either the tryptophan is on the surface of the protein or the solvent induces a conformational change that brings it to the surface.*
3. If a substance known to be a quencher (i.e., it quenches the fluorescence of the *free* amino acid), such as the iodide, nitrate, or cesium ions, quenches tryptophan or tyrosine fluorescence, the amino acids must be on the surface of the protein. If it fails to do so, there are several reasons:
   a. The amino acid may be internal.
   b. The amino acid may be in a crevice whose dimensions are too small for the quencher to enter it.
   c. The amino acid may be in a highly charged region and the charge might repel the quencher. For example, the iodide ion (a negative quencher) fails to quench tryptophan fluorescence if the tryptophan is in a negative region; the Cs^+ ion is ineffective if the fluor is in a positive region. The neutral quencher, acrylamide, disregards the charge.
4. If a substance that does not affect the quantum yield of the free amino acid affects the fluorescence of a protein, it must do so by producing a conformational change in the protein.
5. If tryptophan or tyrosine are in a polar environment, their Q decreases with increasing temperature, T, whereas, in a nonpolar environment, there is little change. Hence, deviation from a monotonic decrease in Q with increasing T indicates that heating is inducing a conformational change because the polarity of the regions to which the tryptophans are being exposed must be changing. An increase with T of the temperature dependence of Q when the protein is in a polar solvent such as water indicates that more tryptophan molecules are being exposed to the solvent—that is, the protein is unfolding.
6. The Q for both tryptophan and tyrosine are decreased if the α-carboxyl group of these amino acids is protonated. In order to be protonated during a titration, the amino acid must be accessible to the solvent, that is, it must be on the surface or in a crevice.
7. Tryptophan fluorescence is quenched by neighboring protonated acidic groups. Hence, if the pK measured by monitoring tryptophan fluorescence is the same as the pK for a known ionizable group (e.g., the imidazole of histidine or the SH bond of cysteine), then that group must be very near a tryptophan. This rule applies only if it can be shown independently that the pH change does not introduce a conformational change.
8. If a substance binds to a protein and tryptophan fluorescence is quenched, either there is a gross conformational change as a result of binding or some tryptophan is in or very near the binding site. Furthermore, because a decrease in the polarity of the solvent causes a shift in λ_{max} to shorter wavelengths, such a shift associated with binding indicates that water is excluded in the complex.
9. If the absorption spectrum of a small molecule overlaps the emission spectrum of tryptophan and the distance is small, there is quenching. Hence, if the binding of such a molecule to a protein quenches tryptophan fluorescence, tryptophan must be in or near the binding site.

NOTE: Two important points must be made about these rules: (1) fluorescence is so sensitive to environmental factors that other interpretations must always be sought for the results described in these rules; (2) if a protein contains several tryptophans, as is usually the case, each may have a different quantum yield. Therefore, the absolute magnitude of changes cannot be used to determine the fraction in a given environment—for example, internal versus external.

* A shift of λ_{max} in a macromolecule is, in standard terminology, in relation to λ_{max} of the free amino acid in water.

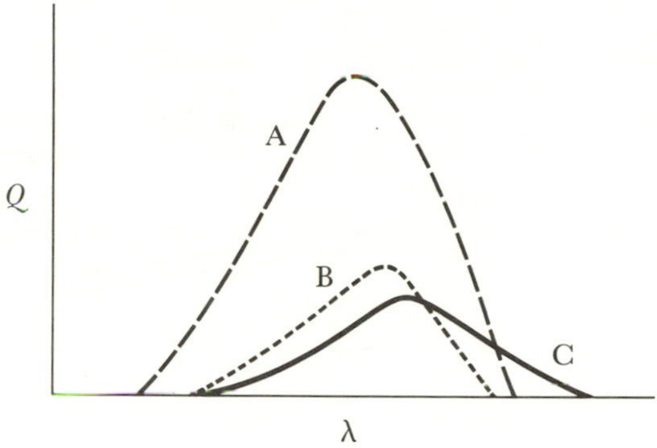

Figure 15-4
Fluorescence spectrum of a hypothetical protein in solution: (A) without added cofactor; (B) with added cofactor. Curve C is the spectrum of free tryptophan in water.

☐ Properties of the active site of an enzyme. **Example 15-B**

A hypothetical enzyme is known to contain a single tryptophan and λ_{max} is nearly the same as that for the tryptophan, suggesting that the tryptophan is in a polar environment (rule 2). The Q is low, suggesting that a quencher might be nearby. Acid titration of the enzyme decreases Q and the pK of the transition is that of a carboxyl. Hence, the low Q is caused by proximity to a carboxyl (rule 7). The addition of the substrate shifts λ_{max} to shorter wavelengths and increases Q. The shift in λ_{max} indicates that the tryptophan is in a less polar environment (rules 2 and 4). This could mean either that there has been a conformational change moving the tryptophan to a nonpolar region or that the tryptophan is in or near the binding site and that the substrate binding excludes water. The increase in Q suggests that the quenching effect of the carboxyl has been eliminated. A likely but not exclusive interpretation is that tryptophan is in the active site and the substrate binds either to the tryptophan or to the carboxyl.

☐ Studies on the denaturation of a protein. **Example 15-C**

The Q of tryptophan of a hypothetical protein drops slowly and continually with temperature from 20°C to 63°C. Hence, the tryptophan molecules are in a polar environment (rule 6). Between 63°C and 65°C, Q increases markedly (Figure 15-5) so that, in the range from 63°C to 65°C, there must be a conformational change resulting in a change in the exposure of many of the tryptophan molecules to the solvent (rule 6a). This change in Q can be used as an assay for the helix-coil transition so that various parameters (e.g., salt concentration, pH, etc.) affecting the process can be studied.

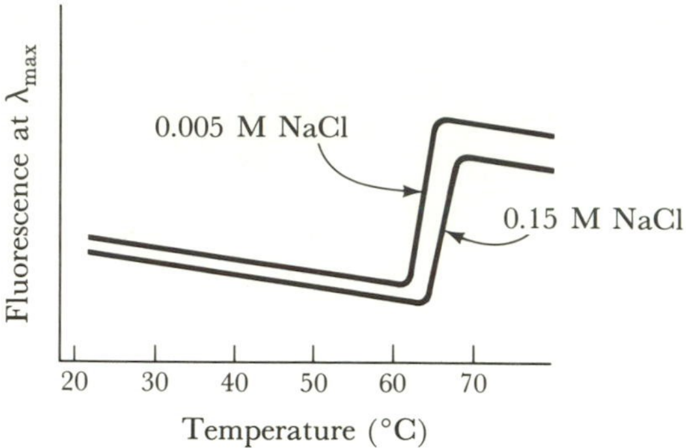

Figure 15-5
A helix-coil transition of a protein measured by changes in fluorescence. Note that, in 0.15 M NaCl, the protein is more stable in that a higher temperature is required for the transition.

Example 15-D □ Location of tryptophans in a hypothetical enzyme.

An enzyme is known to have five tryptophans. The Q is higher than that for free tryptophan, indicating that some are in nonpolar regions (rule 2). If the protein is heated from 20°C to 55°C, the decrease in Q is only 35% that found for free tryptophan, suggesting that 0.35×5 or ~2 tryptophans are on the surface (rule 5). If a high concentration of iodide is added, 30% of the fluorescence is quenched, again consistent with two tryptophans (rule 3). In neither case is 40% quenched, because the three internal tryptophans are in a slightly nonpolar environment and therefore contribute *more* than three-fifths of the value of Q. If a substrate of the enzyme is added, there is no change in Q, suggesting either that there is no tryptophan in the binding site or that, if tryptophan is in the binding site, the bound molecules must create a polar environment. Iodide quenches only 15% of the fluorescence when the substrate is bound; hence, $0.15/0.30 = \frac{1}{2}$ of the tryptophans on the surface are not exposed to the solvent when the substrate is bound. Hence, $\frac{1}{2} \times 2 = 1$ tryptophan is in the active site.

If the enzyme is titrated, 18% of the fluorescence is quenched. The pK associated with this quenching is that of histidine. Hence, there must be a histidine near one of the tryptophans (rule 7). If the substrate is bound and then the enzyme is titrated, there is no change in Q, indicating that in the presence of substrate, the histidine cannot be titrated. Hence, the binding site contains both a histidine and a tryptophan and the substrate may bind directly to the histidine.

In Example 15-D, the assumption was made that the Q for all tryptophans is nearly the same. This certainly need not be the case (e.g., one or more could be near carboxyls). If this assumption were incorrect, the calculations of the number of tryptophans in each location would be invalid. However, the value of this naive analysis is to provide a working hypothesis for the structure of the protein; this hypothesis must then be tested in other ways.

☐ Assay by iodide quenching of the dependence of conformation of a protein on NaCl concentration. **Example 15-E**

The conformation of many proteins in solution is affected by the NaCl concentration of the solvent. This can be studied by examining the degree of quenching produced by the iodide ion. For example, consider a protein for which, as the NaCl concentration is decreased to zero, the concentration of iodide required to produce a given degree of quenching decreases (Figure 15-6). Hence, at low salt concentration, the structure of the protein becomes more open so that more of the tryptophans become available for collision with iodide. Furthermore, if in the absence of NaCl the maximum quenching is only 50% of the quenching of free tryptophan produced by iodide, then it might be concluded that, if no NaCl is present, the structure cannot be a random coil because 50% of the tryptophans are still not exposed to the solvent. Note that iodide quenching is a general method for studying the helix-coil transition induced by any agent that does not affect the ability of iodide to quench.

Extrinsic Fluorescence

Nature does not always supply the investigator with a fluorescent group in the appropriate place in a macromolecule. However, in many cases, a fluor can be introduced into the molecule to be studied either by chemical coupling or by simple binding (as in the use of reporter groups in absorption spectroscopy). The use of such added molecules in fluorescence analysis is called the method of extrinsic fluorescence. Several requirements of the fluor must be met for using extrinsic fluorescence: (1) the fluor must be tightly bound at a unique location; (2) its fluorescence should be sensitive to environmental conditions; and (3) it should not itself affect the features of the macromolecule being investigated. These three criteria must always be verified. For proteins, the most common extrinsic fluors are 1-anilino-8-naphthalene sulfonate (ANS); 1-dimethylaminonaphtha-

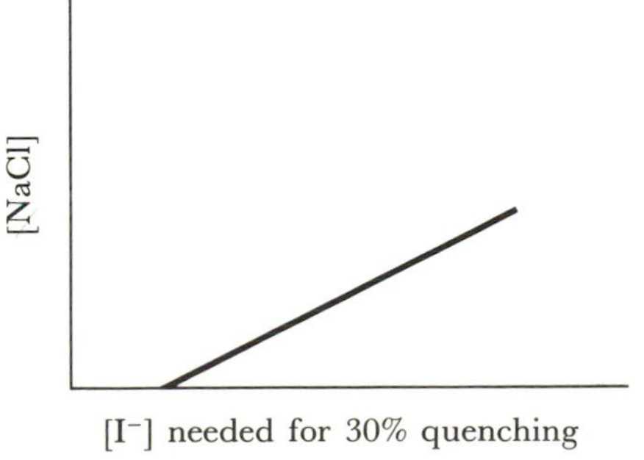

Figure 15-6
Plot of the concentration of iodide needed to reduce tryptophan fluorescence of a protein by 30% as a function of NaCl concentration.

lene-5-sulfonate (DNS) and its chlorinated derivative, dansyl chloride; 2-*p*-toluidylnaphthalene-6-sulfonate (TNS); rhodamine; fluorescein; and the isothiocyanates of rhodamine and fluorescein. For nucleic acids, various acridines (acridine orange, proflavin, acriflavin) and ethidium bromide are used (Figure 15-7).

ANS, DNS, and TNS have the valuable property that in aqueous solution they fluoresce very weakly. However, in a nonpolar environment, *Q*

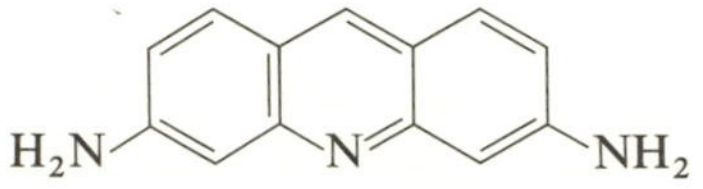

Proflavin

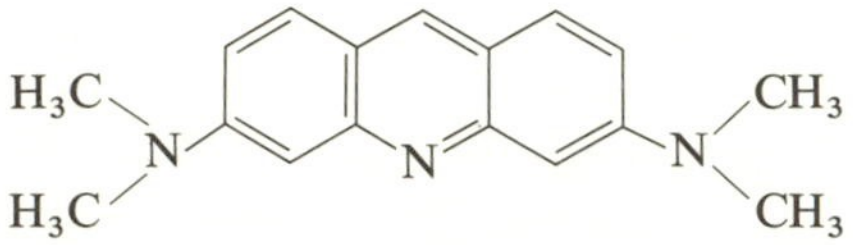

Acridine orange

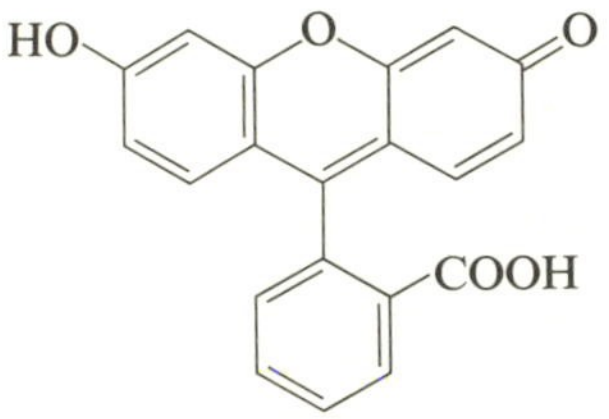

Fluorescein

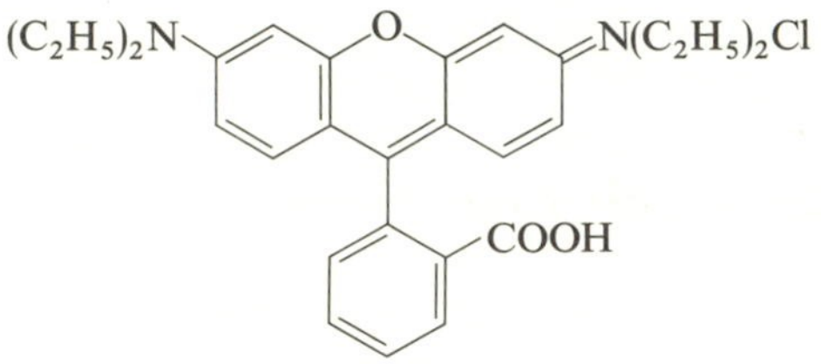

Rhodamine B

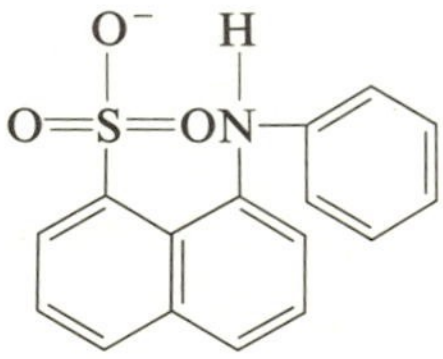

1-Anilino-8-naphthalene sulfonate (ANS)

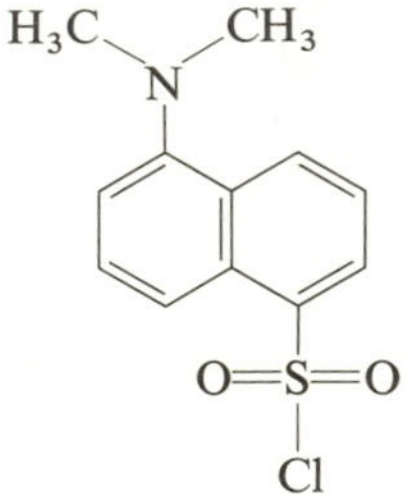

Dansyl chloride

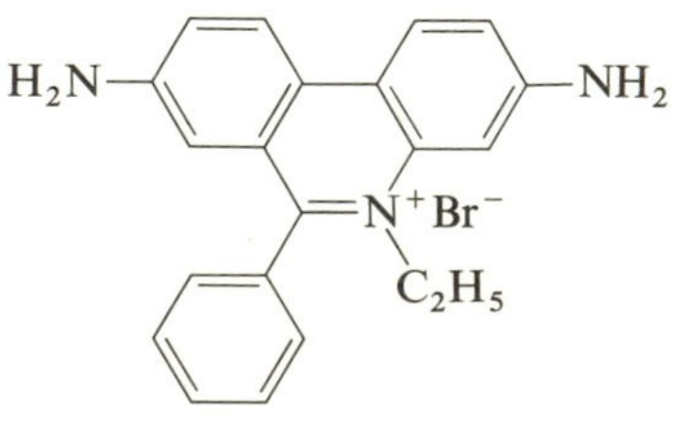

Ethidium bromide

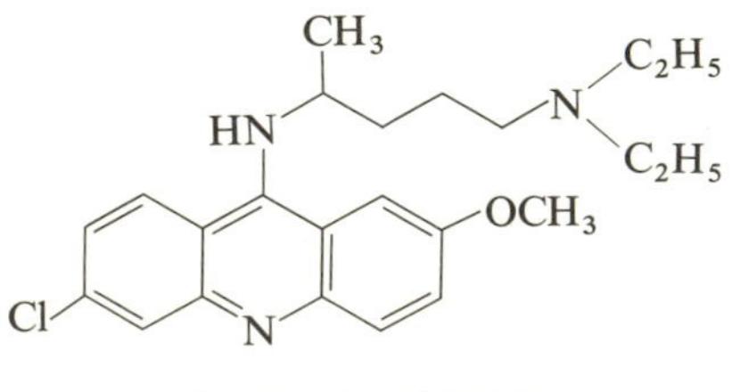

Quinacrine chloride

Figure 15-7
Structures of common extrinsic fluors.

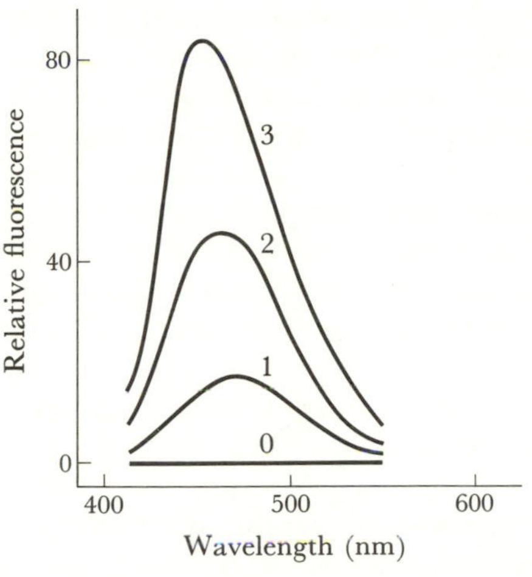

Figure 15-8
Effect of the addition of bovine serum albumin to ANS: 0 means no bovine serum albumin; 1, 2, and 3 indicate increasing amounts of bovine serum albumin.

increases markedly and the spectrum shifts toward shorter wavelength, both effects increasing as polarity decreases (Figure 15-8). Studies with model compounds have provided a rough relation between Q and the degree of nonpolarity. Dansyl chloride and the isothiocyanates have the useful property of reacting with specific amino acids in proteins. Hence, these substances are widely used to detect nonpolar regions in proteins because, on binding to such a region, their fluorescence increases.

How these extrinsic fluors can be used is shown in the following examples.

Example 15-F

☐ Determination of the properties of the heme-binding site in hemoglobin.

Hemoglobin is a complex of a small prosthetic group with the protein apohemoglobin. The extrinsic fluor ANS fluoresces when added to solutions of apohemoglobin but not with hemoglobin. The addition of heme to the apohemoglobin-ANS complex eliminates the fluorescence by displacing the ANS, so that ANS and heme must bind at the same site, which must be nonpolar. From the exact value of Q and the spectral shift, the degree of nonpolarity can be estimated.

A study of this sort can give valuable clues in the interpretation of X-ray diffraction patterns. For example, one would know that certain configurations of amino acids (e.g., a highly polar cluster) could not be the point of interaction with a prosthetic group.

Example 15-G

☐ Detection of a conformational change in an enzyme when the substrate is bound.

TNS fluoresces when added to α-chymotrypsin and therefore must bind to it (TNS fluoresces only if bound). TNS does not competitively inhibit the enzymatic hydrolysis of various substrates so that it cannot bind to the substrate-binding site. However, the addition of a substrate decreases the fluorescence. This decrease could be a result of decreased binding or of increased polarity of the TNS-binding site. By using fluorescence as an assay of binding and studying the fluorescence as a function of TNS concentration (in the presence and absence of substrate), it was shown that the binding constants are the same in both conditions. Therefore, there is not less TNS bound and the decrease in fluorescence indicates an increase in the polarity of the binding site. The change in polarity could result from a polar group moving near the binding site or from the uncovering of a polar group—in either case, a conformational change is associated with substrate binding. Note also that, with this fluorescence assay, the binding parameters of the substrate and the effects of various agents both on binding and the conformational change can be determined.

Example 15-H ☐ Presence of bound fatty acids in bovine serum albumin.

If different samples of bovine serum albumin are prepared and ANS is added, ANS fluorescence is observed, but the values of Q and λ_{max} differ from one sample to the other. The variation in Q is too great to be due to variations in the polarity of the binding site and hence is due to the number of binding sites—that is, less ANS is bound in the samples whose fluorescence is weak. If the bovine serum albumin samples are treated with a lipid solvent, the quantum yields and spectral shifts become the same for all samples. The extraction with the lipid solvent apparently makes some nonpolar sites available to the ANS. However, analysis of the solvent demonstrates that fatty acids are extracted from the samples whose fluorescence is weak. Hence, so-called pure samples of crystalline bovine serium albumin may contain bound fatty acids. Clearly, ANS fluorescence can be used both to assay the purity of these samples and to study the binding of fatty acids to bovine serum albumin—for example, to determine the binding constants.

Example 15-I ☐ Determining the strandedness of polynucleotides.

The fluor acridine orange shows an increase in Q and a shift in λ_{max} when bound to polynucleotides (Figure 15-9). When saturating amounts of acridine orange are added, the values of λ_{max} are significantly different for double- (green fluorescence) and single- (red fluorescence) stranded polynucleotides. Hence, the structures can be distinguished in this simple way. If a sample contains both double- and single-stranded polynucleotides, the fluorescence spectrum will have two peaks, one for each value of λ_{max}.

More examples of the use of extrinsic fluoresence will be given in the sections on energy transfer and polarization.

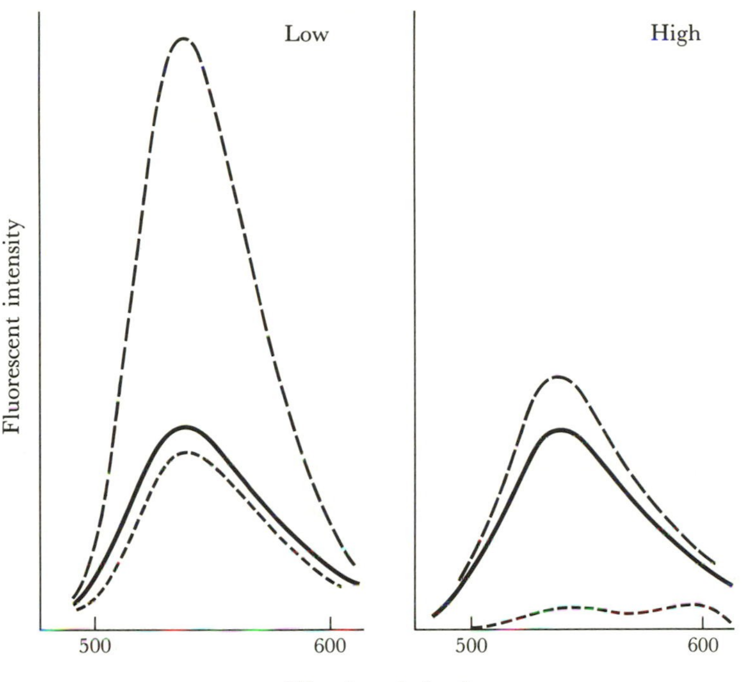

Figure 15-9
Effect of a double-stranded (long dashes) polynucleotide (e.g., DNA) and a single-stranded (short dashes) polynucleotide (e.g., denatured DNA or RNA) on the fluorescence of acridine orange. The solid line is the fluorescence spectrum for free acridine orange. Low and high refer to ratio of dye molecules to nucleotides. Note that DNA enhances fluorescence without affecting λ_{max} and that the enhancement decreases as more dye is added. Single-stranded material quenches fluorescence and shifts the spectrum toward the red. This accounts for the fact that, if a high concentration of acridine orange is added to a cell, the nuclear material fluoresces green whereas the RNA-containing cytoplasm appears orange.

Extrinsic Fluorescence and Energy Transfer

Consider a system containing two fluors (numbered 1 and 2) whose absorption and emission spectra are as shown in Figure 15-10. A convenient wavelength (e.g., λ_3) is selected for detecting the fluorescence of fluor 2 to determine the *excitation* (or *action*) spectrum for producing fluorescence. By excitation spectrum is meant the set of wavelengths that, when absorbed, will produce the fluorescence. It is clear that, in general, the excitation spectrum will be the same as the absorption spectrum

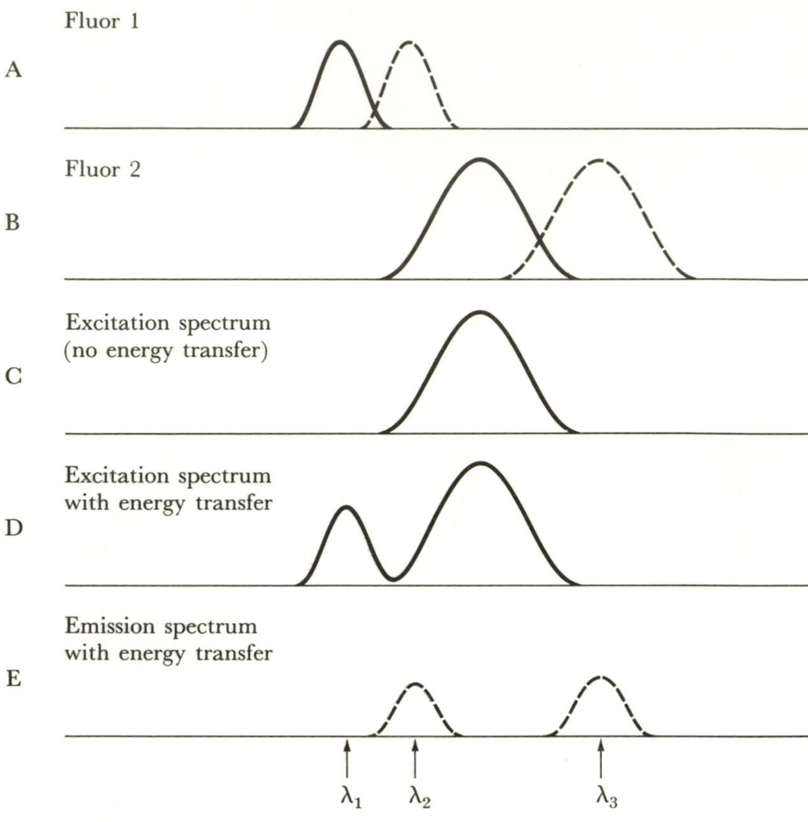

Figure 15-10
Spectra for two fluors (the solid curves represent absorption and the dashed lines emission): (A and B) absorption and emission spectra of fluors 1 and 2, respectively; (C) excitation spectrum for emission at λ_3, in which there is no energy transfer between fluors 1 and 2; (D) same as part C, but with energy transfer; (E) emission spectrum with excitation at λ_1 when there is energy transfer between fluors 1 and 2.

(Figure 15-10C).* However, on occasion the excitation spectrum will be very different from the absorption spectrum (Figure 15-10D) in that fluorescence is produced by absorption of much shorter wavelengths. This is almost invariably due to the fact that under certain circumstances energy absorbed by one molecule (a donor) can be transferred to another fluor (an acceptor) at some distance away. This is called *resonance energy transfer*, a necessary but not sufficient condition of which is that the

* This statement assumes that the absorption spectrum is due to a single electronic transition. This is not always the case, but the assumption is acceptable for our purpose.

emission spectrum of the donor overlaps the absorption spectrum of the acceptor (Figure 15-10).*

Note that, if energy transfer occurs, there is a decrease in the amount of the fluorescence of fluor 1 when excitation is by a wavelength in the absorption spectrum of fluor 1. This is clearly a kind of quenching, especially if the acceptor molecule is not a fluor. However, energy transfer can be easily detected if the entire fluorescence spectrum is determined, because it will be found (Figure 15-10E) that the fluorescence spectrum resulting from excitation at λ_1 contains a new band at long wavelengths. In other words, energy transfer is to be suspected whenever the observed emission spectrum cannot be accounted for by the wavelength used to excite a particular fluor.

So far, energy transfer between different fluors has been considered, but energy transfer can also occur between identical fluors because the absorption and emission spectra of a fluor usually overlap. Such transfer is indicated when it is observed that, as increasing amounts of a fluor are added to a sample, the fluorescence intensity decreases without a decrease in the absorbance at the exciting wavelength—that is, Q decreases. (For this to occur not only must the usual conditions for energy transfer be satisfied, but also Q must be less than 1.) This kind of energy transfer is clearly a kind of quenching.

The importance of the energy-transfer phenomenon to biochemistry is that the efficiency of transfer is a function of the separation of the fluors and can therefore be used to measure molecular distances. The efficiency of transfer, $\mathscr{E}$, is described by the following equation:

$$\mathscr{E} = \frac{R_0^6}{(R_0^6 + R^6)} \tag{3}$$

in which R is the distance between donor and acceptor, and R_0 is a constant related to each donor-acceptor pair that can be calculated from certain parameters of the absorption and emission spectra of each. The utility of energy transfer becomes apparent if equation (3) is rewritten as

$$R = R_0\left(\frac{1 - \mathscr{E}}{\mathscr{E}}\right)^{\frac{1}{6}} \tag{4}$$

(See Chapter 17 for a discussion of spin labeling, a method of measuring molecular distance by nuclear magnetic resonance.)

* A simplistic explanation is that the donor emits a photon, which is then absorbed by the acceptor, and the probability of energy transfer is merely the product of the probabilities of emission and of absorption. However, although emission and reabsorption is certainly a possibility, this is not what occurs in resonance energy transfer and the probability of transfer is described by an expression called the *overlap integral.* [This difference is described in some detail in a paper by T. Förster, *Disc. Faraday Soc.* 27(1959): 7–17.]

To understand how $\mathscr{E}$ is measured, let us refer to Figure 15-10, which shows the absorption and emission spectra of a donor and an acceptor. We will consider two methods, both of which utilize three wavelengths, λ_1, λ_2, and λ_3, selected as follows: λ_1 is chosen so that absorption by the donor is efficient but inefficient by the acceptor; λ_2 is a wavelength in the emission spectrum of the donor but not in that of the acceptor; and λ_3 is a wavelength emitted only by the acceptor.

In the first method of measuring $\mathscr{E}$, the quenching effect on the donor fluorescence is determined—that is, the quenching of the fluorescence at λ_2 if the donor is excited by λ_1. Let $f_{1,2}$ be the fluorescent intensity at λ_2 if excited at λ_1 and f^{D} and $f^{\mathrm{D,A}}$ represent the fluorescence if only the donor (D) is present or if both donor and acceptor (A) are present, respectively. The fraction of the donors that remains excited is $1 - \mathscr{E}$, or

$$\mathscr{E} = 1 - \frac{f_{1,2}^{\mathrm{D,A}}}{f_{1,2}^{\mathrm{D}}} \tag{5}$$

In the second method, the intensity of emission at λ_3 is measured. In this case, the relevant equation is

$$\mathscr{E} = \left(\frac{\varepsilon_1^{\mathrm{A}} C^{\mathrm{A}}}{\varepsilon_1^{\mathrm{D}} C^{\mathrm{D}}}\right)\left(\frac{f_{1,3}^{\mathrm{D,A}}}{f_{1,3}^{\mathrm{A}}}\right) - 1 \tag{6}$$

in which $\varepsilon_1^{\mathrm{A}}$ and $\varepsilon_1^{\mathrm{D}}$ are the molar absorption coefficients of acceptor and donor, respectively, at λ_1; C^{A} and C^{D} are the concentrations of acceptor and donor, respectively; and $f_{1,3}^{\mathrm{A}}$ and $f_{1,3}^{\mathrm{D,A}}$ are the fluorescence intensities at λ_3 when excited by λ_1, if either the acceptor or the donor-acceptor pair, respectively, are present.

After $\mathscr{E}$ has been measured, R can be evaluated if R_0 is known. The determination of R_0 requires a straightforward measurement of three things: the absorption coefficient of the donor, the fluorescence intensity, in arbitrary units, of the acceptor as a function of λ, and the value of Q of the donor in the absence of the acceptor. However, there is also an important multiplicative geometric factor resulting from the fact that the fluorescence of the donor is polarized and the plane of polarization is in general at some angle with respect to the axis of maximal absorption by the acceptor. This geometric factor is known as the orientation factor. Usually it cannot be evaluated precisely because the relative orientation of the donor and the acceptor is not known. A good discussion of the orientation factor is presented in the 1978 article by Lubert Stryer listed in the Selected References. The conclusion of Stryer's analysis is that if it can be confirmed by studies of fluorescence polarization (as described later in this chapter) that the donor and the acceptor are in rapid motion with respect to one another, the orientation factor will not be more than 1.20 and will in fact usually be about 1.07. Thus, the value of this factor is often estimated, with little resulting error.

To use energy-transfer measurements for distance determination, the following conditions must be met: (1) there must be a single donor and acceptor and each must be at a known chemical site, (2) the value of R_0 for the donor-acceptor pair must be known and it should be near that of the distance being measured, and (3) the addition of the fluors should not alter the structure of the macromolecule being studied.

The validity of the method for determining distance has been demonstrated by studies of a linear polymer in which oligomers of known length of poly-L-proline serve as spacers between a dansyl and an α-naphthyl group (Figure 15-11). These molecules were excited by a wavelength absorbed by the α-naphthyl group and the fluorescence of the dansyl group was measured as a function of the number of proline residues. Efficiency of transfer was calculated as previously described. When $\mathscr{E}$ was plotted as a function of the length of the polyproline chain, it was found to obey the $1/R^6$ law (Figure 15-12).

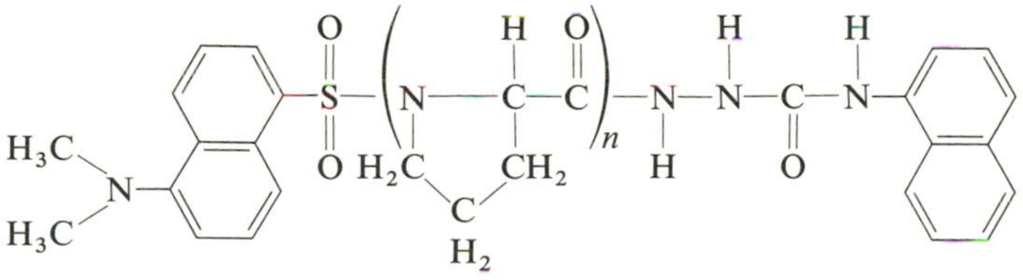

Dansyl L-Prolyl α-Naphthyl

Figure 15-11
Formula for poly-L-proline separating a dansyl (acceptor) and an α-naphthyl (donor) group (n varied from 1 to 12).

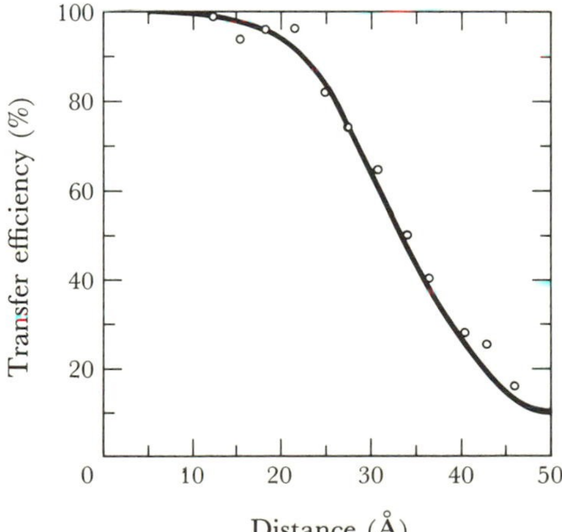

Figure 15-12
Efficiency of energy transfer as a function of a distance in dansyl-(L-prolyl)$_n$-α-naphthylene, in which n varies from 1 to 12. The distances (in Å) for each value of n were independently determined from other types of measurements. The solid line corresponds to a $1/R^6$ dependence. [From L. Stryer and R. P. Haugland, *Proc. Natl. Acad. Sci. U.S.A.* 58(1967): 719–726.]

There is a variety of strategies for carrying out energy-transfer measurements. One method is to use an intrinsic fluor as either donor or acceptor and couple another fluor at a defined site in the molecules of interest. For the study of proteins, tryptophan is very useful. For example, if a protein contains a single tryptophan, fluors can be attached to various amino acids and one by one the distance of each amino acid from tryptophan can be determined. Fluorescent analogues of molecules that bind to proteins are also very useful. At present there are available fluorescent analogues of various adenine nucleotides, vitamins, fatty acids, heme, phospholipids, the Ca^{2+} ion, and a variety of substrates and inhibitors of enzymes and binding proteins. Many of these are acceptors when tryptophan is excited. Another approach is the covalent attachment of one or more fluors at specific sites in macromolecules. For example, several fluors are available that react with sulfhydryl and amino groups of proteins, and terminal sugars of polysaccharides, glycoproteins, and nucleic acids.

The energy-transfer procedure is extremely valuable in determining the proximity of two regions of the same molecule, whether two different molecules occupy or are part of the same spatial unit such as a vesicle, membrane or organelle, and whether distant parts of a molecule move with respect to one another. The technique can easily indicate whether two sites are separated by 40 Å or 70 Å although it cannot readily distinguish 40 Å from 45 Å. The various kinds of information obtained from energy-transfer measurement can best be seen in the following examples.

Example 15-J ☐ Proximity of tryptophan to the active site of carbonic anhydrase.

The enzyme carbonic anhydrase contains a zinc atom in the active site. If a donor, *m*-acetylbenzenesulfonamide, is bound to this zinc atom, excitation of this fluor causes tryptophan fluorescence. Hence, a tryptophan must be near the active site.

Example 15-K ☐ Spacing between hydrophobic groups of bovine serum albumin.

As increasing amounts of ANS are added to bovine serum albumin, the induced fluorescence of ANS first increases and then decreases. The decrease is caused by energy transfer between nearby ANS molecules. The magnitude of the effect gives a rough idea of the spacing between nonpolar regions.

Example 15-L ☐ Determination of the shape of the visual pigment rhodopsin.

Rhodopsin is the primary light absorber in the discs of rod cells in the eye. It is a complex between a protein, opsin and a fluor, *cis*-retinal, which absorbs given light. The molecular weight of rhodopsin

(40,000) implies that if the molecule were spherical, its diameter would be 45 Å. A fluor, which served as an acceptor, was bound to a particular sulfhydryl group in opsin. Energy-transfer experiments showed that this sulfhydryl group is 75 Å away from the *cis*-retinal. This measurement not only suggested that rhodopsin is an elongated protein but also indicated that it is long enough to traverse the membrane of the retinal disc. It was then proposed that rhodopsin might serve as a light-sensitive gate that controls the ion flow across the membrane. Subsequent X-ray-scattering experiments confirm the length of the protein and several lines of evidence suggested that the idea of rhodopsin as a gate is essentially correct.

□ The localization of metal ions in metalloproteins. **Example 15-M**

Two useful substituents for metal ions are Tb^{3+} (terbium), which can serve on as acceptor of tryptophan and tyrosine fluorescence, and Co^{2+}, which is an acceptor from many organic donors.

In one experiment, the protein transferrin was studied. Transferrin contains two binding sites for the Fe^{3+} ion and both of these can be filled with Tb^{3+}. Transferrin was prepared containing one Tb^{3+} and one Fe^{3+} ion. Tyrosine was excited and Fe^{3+} fluorescence was observed when Tb^{3+} was both present and when it was absent. Tb^{3+} fluorescence was also measured when no Fe^{3+} was present; the high efficiency of transfer of energy from tyrosine to Tb^{3+} showed that the tyrosine and the Tb^{3+} ion are very close. When both Tb^{3+} and Fe^{3+} were present, Fe^{3+} fluoresced; from the efficiency of transfer it could be calculated that the Tb^{3+} and Fe^{3+} ions are 25 Å apart.

Carboxypeptidase A is a Zn^{2+}-containing enzyme. The Zn^{2+} ion, which is in the active site, was replaced by the Co^{2+} ion and the dansyl group was coupled to the amino terminus of a series of peptide substrates. Carboxypeptidase A binds the carboxyl end of the peptides so that energy-transfer measurements, in which dansyl was excited and the Co^{2+} ion fluoresced, indicated the shape of the substrate when bound. If the substrate folds so that the amino and carboxyl groups are near one another, the dansyl group will remain near the Co^{2+} ion. However, the distance ranged from 7 Å with dansylglycine-tryptophan to 12 Å for dansyl(glycine)$_3$-tryptophan. Since the value increases with chain length, the substrate apparently maintains an extended configuration.

□ Conformational changes of enzymes during substrate binding. **Example 15-N**

In studies of this sort, a donor and an acceptor are bound to an enzyme and the distance between them is measured by the energy-transfer method. It is frequently found that this distance changes when the substrate of the enzyme is bound, indicating that the enzyme changes shape on binding the substrate.

Example 15-O ☐ The structure of *E. coli* ribosomes.

In the bacterium *E. coli*, a ribosome, the nucleoprotein particle on which protein synthesis occurs, contains two subparticles known as the 30S and 50S subunits. The 30S particle consists of 21 protein molecules and one 16S RNA molecule. *E. coli* ribosomes are easily dissociated and then reassembled. The ribosomes were dissociated and each of the 21 proteins was purified. One was labeled with a donor and each of the other twenty was labeled with an acceptor. Twenty different kinds of ribosomes were then separately reassembled; in each case there was one donor and a particular receptor-labeled protein. Energy-transfer experiments gave information about the separation of each protein pair.

Example 15-P ☐ Three-dimensional structure of yeast phenylalanine tRNA.

The three-dimensional structure of yeast phenylalanine tRNA has been determined by X-ray diffraction analysis of crystalline tRNA; the molecule has a clover-leaf configuration such as that in Figure 14-17. In the anticodon loop of phenylalanine tRNA there is a naturally occurring fluorescent base that can serve as a donor. An acceptor fluor was attached at the C-C-A terminus of the tRNA and the energy-transfer technique was used to measure the distance between these two regions of the molecule in solution. The data showed that the distances were comparable, suggesting that the structure of the molecule is the same in solution as that determined by X-ray diffraction.

In another series of experiments, other tRNA molecules were studied. In these experiments, a donor and an acceptor were placed in the tRNA molecule at five different pairs of sites. This allowed the distance between the various loops and stems to be measured and again confirmed the data from the X-ray analysis.

Example 15-Q ☐ The distribution of particular peptides in membranes.

There are numerous proteins and peptides that can enter channels in cell membranes. In several cases, the number of peptide molecules per channel can be measured by the energy-transfer technique. The peptide of interest is labeled with either a donor or an acceptor. Both donor(D)-labeled or acceptor(A)-labeled peptides are allowed to enter the membrane channels. If only a single molecule can enter a channel, energy transfer is not possible. If two molecules can enter a channel, a collection of saturated channels will contain $\frac{1}{4}$ two As, $\frac{1}{4}$ two Ds and $\frac{1}{2}$(D + A) and energy transfer can occur. Furthermore, the measured separation of D and A will give information about the lateral dimension of the channel.

Excimers

On occasion, an excited molecule can form a complex with an identical unexcited molecule and the complex will exist until the excited molecule emits. This can be described by the following scheme, in which A is an absorber, A* is an excited absorber.

$$A + \text{photon} \rightarrow A^* \quad \text{(excitation)}$$

$$A^* + A \rightarrow A^*A \quad \text{(formation of excited dimer)}$$

$$A^*A \rightarrow A + A + \text{photon} \quad \text{(fluorescence of dimer)}$$

Such a complex (A*A) is called an *excimer* and is recognized by the production of a new fluorescent band at a longer wavelength than the usual emission spectrum. It is distinguishable from resonance energy transfer in that the excitation spectrum is identical with that of the monomer; with energy transfer, the excitation spectrum is composed of the absorption spectra of the donor and the acceptor (Figure 15-10).

Excimer formation is a concentration-dependent phenomenon and can be used to indicate high local concentrations of a fluor. It has had very little use in biochemistry to date but will certainly find applications in the future. For example, the presence of a vesicle or some unit that can concentrate certain substances could be studied by studying the excimers of an extrinsic fluor.

Polarization of Fluorescence

The intensity of polarized light transmitted by a polarizer depends on the orientation of the polarizer—the transmission is maximum when the plane of polarization is parallel to the axis of the polarizer and zero when it is perpendicular (see Chapter 16, Figure 16-1B). For a light beam that is only partially polarized, a polarization, P, can be defined as

$$P = \frac{I_{\parallel} - I_{\perp}}{I_{\parallel} + I_{\perp}} \qquad (7)$$

in which $I_{\parallel}$ and $I_{\perp}$ are the intensities observed parallel and perpendicular to an arbitrary axis. Polarization can vary between -1 and $+1$. It is zero when the light is unpolarized; otherwise it is called *partially polarized.* The polarization of fluorescence can be determined using the standard experimental arrangement for measuring fluorescence (Figure 15-3) with a simple modification (Figure 15-13): a polarizer is placed in the path of the exciting light in order to excite the sample with polarized

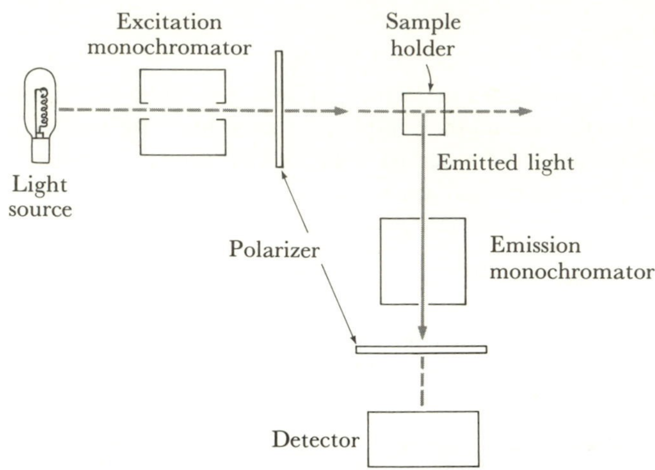

Figure 15-13
Experimental arrangement for measuring the polarization of fluorescence.

light and a second polarizer is placed between the sample and the detector with its axis either parallel or perpendicular to the axis of the excitation polarizer. If the exciting light is polarized, it is usually observed that the fluorescence is either only *partially* polarized or completely unpolarized. The cause of this loss of polarization is explained next, and it will be seen that the magnitude of the change in polarization gives information about the physical state of the emitter.

To begin with, it must be remembered that the absorption of polarized light by a chromophore is maximal when the plane of polarization is parallel to a particular axis of the chromophore, called the electric dipole moment (see Chapter 16). In general, the chromophores will be randomly oriented (this is certainly true in solution), and the probability of absorption of the exciting polarized light is proportional to $\cos^2 \theta$, in which θ is the angle between the plane of polarization and the electric dipole moment. Furthermore, the plane of polarization of the emitted light is not determined by the absorption dipole moment but by the transition dipole moment (which is generally not parallel to the absorption dipole moment), and the probability of emission of fluorescence with the plane of polarization at an angle ϕ with respect to the transition dipole moment is proportional to $\sin^2 \phi$. The result of these probabilities is that, if the absorbers are randomly arranged (*but stationary*) and the two dipoles are not parallel, the polarization, P, of the fluorescence is $<\frac{1}{2}$. (In fact, even if the absorbers were perfectly aligned with the plane of polarization of the exciting light, P would be <1 because of $\sin^2 \phi$.) The fact that P is always <1 is called *fluorescence depolarization.*

Many other factors can increase the extent of depolarization; the two most important from the point of view of biochemistry are (1) the motion of the absorber and (2) energy transfer between like chromophores. If the emitter is rotating very rapidly (i.e., if there is Brownian motion) so that there is a substantial change in orientation during the lifetime of the excited state, the polarization will be further reduced. This is an important phenomenon because the extent of this type of depolarization is affected by temperature, solvent viscosity, and the size and shape of the molecule containing the emitter. The second factor, energy transfer between identical chromophores, results from the fact that, although resonance energy transfer occurs with highest probability between molecules having parallel dipoles, it also occurs when they are nonparallel, with resulting depolarization. Because the efficiency of energy transfer falls off with the sixth power of the distance between the donor and acceptor, this type of depolarization is highly concentration dependent.

The *intrinsic polarization*, P_0, is that which would be observed if the absorber were immobilized and far from all other molecules (i.e., if there were no effect of either molecular motion or energy transfer). In practice, this is the polarization observed if the fluor is in a solvent of very high viscosity and at very low concentration. Note that low concentration can mean either a low concentration of the macromolecule-fluor complex or a large separation between the fluors bound to a single macromolecule. At very high viscosity and high concentration, the effect of energy transfer is primarily detected and, at low viscosity and low concentration, the effect of molecular motion. These effects can be used to obtain information about macromolecules not readily obtained by other techniques.

Applications of Fluorescence Polarization to Proteins and Nucleic Acids

The principal applications in the study of proteins make use of the changes in polarization due to the changes in mobility of either the entire molecule or a part of the molecule.*

To observe depolarization due to motion, it is necessary that the lifetime of the excited state be sufficiently long that reorientation can occur before emission. For both proteins and nucleic acids, which on a molecular scale move quite slowly, the lifetimes of intrinsic fluors are usually too short, although, with small proteins ($M < 20{,}000$), some information can be obtained from studies of the fluorescence depolarization of tryptophan. In general, polarization analysis utilizes extrinsic fluors.

---

* Note that the kind of *mobility* under discussion is *rotation* because translational motion will not cause depolarization. However, if a molecule is free to move and moves by collision with solvent molecules (i.e., by Brownian motion), it is very unlikely that it could move without rotating also. For this reason, mobility can also be thought of in the general sense as simply the ability to move.

Like other procedures, the method of fluorescence polarization uses a set of rules that govern the interpretation of data. The basic rule and some of its general consequences are given in Table 15-2.

The utility of the fluorescence polarization method is described in the following examples.

Example 15-R ☐ Measurements of binding using fluorescence polarization as an assay.

If an antigen labeled with a fluor reacts with an antibody, the polarization of the fluorescence increases owing to the decreased mobility of the antigen in the antigen-antibody complex (consequence 3 in Table 15-2). This has been used as an assay for the binding of a variety of fluorescein-labeled antigens—for example, ribonuclease, bovine serum albumin, and the insulin β chain—to their respective antibodies. From such simple titrations, equilibrium constants and stoichiometry have been obtained that have enhanced our understanding of the mechanism of the antigen-antibody reaction.

The binding of a fluorescent substrate by an enzyme enhances polarization by reducing the mobility of the fluor (consequence 2 in Table 15-2). This is a sensitive assay for enzyme-substrate binding and can be used to determine those parameters whose measurement depends on knowing precisely the amount of bound substrate (e.g., binding constants, stoichiometry).

Table 15-2

Basic Rule Used in Interpreting Fluorescence Polarization Data and Some of the Consequences of This Rule.

Rule

With increasing mobility, polarization decreases.

Consequences

1. If free in solution, extrinsic fluors show virtually no polarization because they are rapidly and freely rotating. However, if coupled to a low-molecular-weight molecule, their polarization becomes greater owing to somewhat reduced mobility.
2. If an extrinsic fluor is bound to a marcromolecule, the polarization can become substantial owing to greatly decreased mobility. The degree of polarization depends not only on the mobility of the entire macromolecule but also on the mobility of the binding site. For example, if a fluor intercalates between the bases of DNA, it is held rigidly in place so that its mobility is that of the DNA molecule; if it is bound to a long amino acid side chain, its polarization will reflect both the mobility of the entire protein and the freedom of movement of the side chain with respect to the polypeptide chain.
3. If a fluor is bound to a macromolecule and then the macromolecule either aggregates (e.g., dimerizes) or binds a large molecule, the polarization increases owing to the reduced mobility of the structure.
4. If a fluor is bound to a macromolecule and the macromolecule undergoes a conformational change (e.g., a helix-coil transition), the polarization decreases as the molecular structure becomes more disordered and increases as it becomes more ordered.

☐ Detection of the association and dissociation of proteins. **Example 15-S**

The polarization of dansyl derivatives of α-chymotrypsin, chymotrypsinogen, and lactic dehydrogenase increases when the proteins self-associate because the increased size of the macromolecule reduces the mobility (consequence 3 in Table 15-2). Hence, self-association and dissociation can be assayed by measuring changes in polarization. The effects of pH, ionic strength, and composition of the solvent can be examined with this assay because they have little or no effect on the polarization itself (i.e., if the enzymes are so dilute that no association is possible, no changes in polarization are observed).

The hybridization of beef muscle (M) and beef heart (H) lactate dehydrogenase, known tetramers, has been studied elegantly using fluorescence polarization of dansyl derivatives of each. Certain conditions (e.g., low pH) lead to a *decrease* in both polarization and enzyme activity. Because decreased polarization indicates increased mobility or decreased molecular size, these conditions presumably cause dissociation of the tetramer to monomers. Having reached their presumably monomeric state, both the H and M enzymes can be treated in such a way that polarization and enzyme activity increase, indicating reassociation. This will also happen if the dissociated H and M lactate dehydrogenases are mixed before reassociation, indicating that in the mixture one of them does not hinder the other from reassociating. If the reassociated material is then subjected to low pH again and the polarization is measured as a function of time to determine the kinetics of dissociation, it is found that the kinetics of dissociation of the reassociated mixture indicate that more than two components are present. Fractionation of the reassociated mixtures has indicated that hybrid tetramers are present (e.g., a tetramer having 3 H subunits and 1 M subunit) and that each of these has different dissociation kinetics. Note that the value of P is merely used as an assay of the dissociation-reassociation reaction and that this is a general way to detect hybridization.

☐ Studies of the helix-coil transition in proteins. **Example 15-T**

For some proteins, there is no simple optical method for observing the helix-coil transition. If an extrinsic fluor (e.g., ANS) can be coupled to a protein without affecting the conformation, the helix-coil transition can be observed by measuring the polarization of the fluorescence because, as the protein unfolds, the polarization will decrease owing to the increasing flexibility of the protein (consequence 2 in Table 15-2).

The polarization method can also be used to detect S—S bonds in a denatured protein. The principle is that *the limiting state of a structureless random coil should have maximum flexibility* and therefore minimum polarization. Hence, a fully denatured molecule containing S—S bonds will not have minimum polarization until the S—S bonds are broken. This means that, if the addition of an agent known to break S—S bonds (e.g., mercaptoethanol) decreases the polarization, S—S bonds must have been present. Polarization is probably the most sensitive way to detect these subtle changes.

Example 15-U □ Structure of a fluor-DNA complex: an example of the determination of orientation from fluorescence polarization.

The fluor acridine orange binds tightly to DNA. Because it is an effective mutagen, the structure of the complex is of some interest. If bound to DNA, acridine orange becomes inaccessible to certain chemical treatments (e.g., diazotization), which suggests that it is somehow within the DNA double helix. If bound to DNA, acridine orange fluoresces when it is excited not only by wavelengths in its own absorption band but also by light absorbed only by the DNA bases, thus indicating energy transfer from the bases. The quantum yield of this energy transfer is very high, so that the acridine orange (which is a planar structure) must be very near the bases. The plane of polarization of the fluorescence of acridine orange is known with respect to both the plane of its rings and its long axis so that, if the system were immobilized, the orientation of the acridine orange molecules with respect to the DNA could be determined by measuring the plane of polarization of the fluorescence with respect to the DNA molecules (using light absorbed by acridine orange but not by the bases). The fluorescence of acridine orange is highly polarized, indicating greatly reduced mobility. If it were bound in such a way that it is not free to move with respect to the DNA helix, this low mobility would be expected because DNA is so highly extended that it has a very low diffusion coefficient. The inaccessibility to diazotization, the proximity to the bases, and the apparent rigidity of the binding suggests that a weak binding to an external group such as the phosphates of the polynucleotide chain is not likely. When the complex of acridine orange and DNA is oriented by flow through a capillary tube (see Chapter 13) and the plane of polarization of acridine orange fluorescence is determined, it is found that the plane of the bound acridine orange ring is perpendicular to the helix axis. Because the bases are also perpendicular to the helix axis and the resonance energy transfer observed previously indicates that the acridine orange molecules and the bases are very close together, a reasonable interpretation of the data is that acridine orange is intercalated between the base pairs. This is consistent with the weak binding of acridine orange to single-stranded DNA and with numerous other physical measurements.

Other Applications of Fluorescence Polarization

Polarization of fluorescence can be used to determine the orientation of molecules in *rigid* systems and the hydrodynamic properties of certain solutions of molecules. Some examples of these uses follow.

☐ Orientation of chlorophyll in chloroplasts by polarization of intrinsic fluorescence. **Example 15-V**

The polarization of the red fluorescence of chlorophyll in a single chloroplast can be measured with a fluorescence polarization microscope (see Chapter 2). From measurements of artificially immobilized chlorophyll, the angles between the plane of polarization of the fluorescence to the axes of the chlorophyll is known. Hence, from the observed plane of polarization of the fluorescence of the chloroplasts, the orientation of the chlorophyll molecules in the chloroplasts can be determined.

☐ Orientation of DNA in chromosomes by polarization of extrinsic fluorescence. **Example 15-W**

In Example 15-U, it was shown that the fluor acridine orange intercalates between the base pairs of DNA and has its plane perpendicular to the helix axis. This dye also binds intracellularly to eukaryotic chromosomes. The polarization of acridine orange fluorescence with respect to the axis of stretched chromosomes has been measured with a fluorescence polarization microscope and used to determine the orientation of DNA in the chromosomes.*

☐ Determination of viscosity within living cells. **Example 15-X**

If a fluorescent molecule is reorienting very rapidly compared with its lifetime, its fluorescence will show little or no polarization (consequence 1 in Table 15-2). However, if it is in a very viscous medium so that its movement is greatly impaired, polarization will be observed. If the lifetime of the fluorescence is roughly known (and indeed lifetimes are measurable), the viscosity can be determined. Note that here we are not concerned with the properties of a macromolecule but merely with determining the viscosity of a liquid that might be inaccessible to direct viscometry. Intracellular viscosity has been measured in both mouse ascites cells and in the bacterium *E. coli*, by allowing them to take up a fluorescent amino acid, aminonaphthylalanine. A small polarization appears; because it is known that this amino acid does not bind to any large molecules in the cells, the polarization can be used to calculate intracellular viscosity. This is of interest because it allows estimation of the rates of diffusion within cells and determination of whether certain diffusion-limited reactions may participate in various regulatory pathways.

Polarization Analysis with Pulsed Excitation

Fluorescence polarization analysis can be used in a quantitative way to determine rotational diffusion coefficients. For this purpose, the standard

---

* J. W. MacInnes and R. B. Uretz, *Science* (*Wash., DC*) 151(1966):689–691.

method of measuring polarization of fluorescence has two disadvantages: (1) it measures an average value, which is undesirable if a molecule has two rotational coefficients (e.g., a rodlike molecule can rotate about its long axis or tumble end-over-end); and (2) it requires measurement of the intrinsic polarization, which is accomplished by studying the polarization in solutions of increasing viscosity and extrapolating to infinite viscosity. Such extrapolation often fails because the substances added to increase viscosity sometimes affect conformation. These problems are avoided by the *nanosecond-pulse technique*, in which excitation is accomplished with a nanosecond (10^{-9}) pulse of polarized light, and the intensities of the emitted light in directions parallel ($I_{\parallel}$) and perpendicular ($I_{\perp}$) to the direction of excitation are recorded. With a rotating molecule, there is an initial difference in these intensities, which decays in a few nanoseconds as the molecule tumbles as a result of Brownian motion. The decay rate is related to the rotational diffusion coefficient.

The data are analyzed in the following way. The emission anisotropy, A, which is defined below, is measured as a function of time;

$$A(t) = \frac{I_{\perp}(t) - I_{\parallel}(t)}{I_{\perp}(t) + 2I_{\parallel}(t)} = A_0 e^{-3t/\rho} \tag{8}$$

in which A_0 is the anisotropy at the instant of excitation and ρ is the rotational relaxation time. A plot of log A versus t is usually linear and has a slope of $-3/\rho$. If the substance has two rotational relaxation times, this will show up as a log A versus t curve consisting of two connected lines (Figure 15-14). An example of the utility of this method follows.

Example 15-Y ☐ Determinants of ρ for chymotrypsin.

An anthraniloyl group can be coupled to the active site of α-chymotrypsin. Using the nanosecond-pulse technique, a single value of ρ of

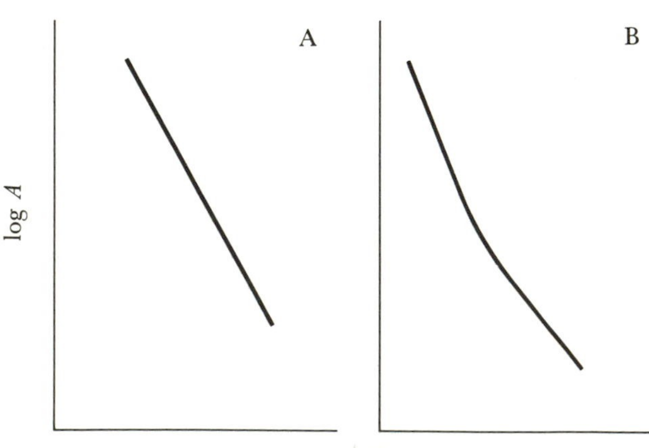

Figure 15-14
Plots of logarithm of the emission anisotropy versus time for a molecule having (A) one and (B) two rotational relaxation times.

52 ns is found. This value of ρ could reflect either flexibility of the site of binding of the anthraniloyl group or rotation of the entire protein (consequence 2 in Table 15-2). If it is assumed that the anthraniloyl group is moving virtually independently of the protein, ρ can be calculated (from the rotational relaxation time of anthranilic acid) to be approximately 1 ns, which is incompatible with the measured value. If it is assumed that the protein is a rotating unhydrated sphere, then $\rho = 3\eta V/kT$, in which η is the solvent viscosity, V is the volume of the sphere, k is the Boltzmann constant, and T is the absolute temperature. Using this equation and determining V from the molecular weight, a value of $\rho = 22$ ns, which is also too small, is calculated. Hence, chymotrypsin must have a larger V and must be either hydrated or nonspherical. This high value of ρ also shows that the active site must be rigid.

Special Uses of Fluorescence in Biology and Biochemistry

Fluorescence is of great use in obtaining information other than the conformation of macromolecules or the arrangement of molecules within cells (e.g., the localization within cells and tissue or the quantitative assay of minute amount of material). Examples of some of these applications follow.

☐ Fluorescence microscopy. **Example 15-Z**

Many fluors are localized intracellularly and can be detected by fluorescence microscopy. A specimen is illuminated with an exciting wavelength and observed through a filter that excludes the exciting light and transmits the fluorescence. Acridine orange binds to nucleic acids and fluoresces green and orange if bound to DNA and RNA, respectively. This has been used with eukaryotes to observe nucleic acids and chromosomes and to detect RNA in the nucleus; it has been used with prokaryotes to localize DNA. See Chapter 2 and Figure 2-19 and 2-21 for details.

An interesting procedure, using a fluor known as SITS, is used for distinguishing living from dead cells. SITS is taken up by living cells but is restricted to small vesicles, which thereby fluoresce. If the cell is not living, the vesicles are usually disrupted so that the SITS diffuses freely throughout the cell. It binds tightly to nuclear membranes and fluoresces brightly. Thus a fluorescent nuclear membrane is a criterion for cell death.

☐ Fluorescent antibody method. **Example 15-AA**

An antibody to a particular substance (e.g., a viral antigen or a cell-wall component) is conjugated to a fluorescent dye. If the antibody is incubated with cells containing the antigen and then washed away, examination of the cells by fluorescence microscopy will show fluorescence only where the antigen is present. This method is widely used to detect tumor antigens and to identify intracellular viruses. It is described more fully in Chapter 2 and an example is given in Figure 2-20.

Example 15-BB ☐ Assay of S—S bonds and SH groups.

In 1 M NaOH, the fluorescence of fluorescein mercuric acetate is quenched by disulfide bonds in protein. This is a relatively accurate means to assay these bonds. A protein with a known number of S—S bonds and SH groups (e.g., ribonuclease) must be used for calibration (Figure 15-15). At neutral pH, the fluorescence is quenched instead by SH groups but not S—S bonds.

Example 15-CC ☐ Enzyme assays.

Fluorescence assays, in which nonfluorescent substances are converted into fluorescent ones, or vice versa, exist for numerous hydrolytic enzymes (e.g., cholinesterase, lipase, hyaluronidase, and β-galactosidase), oxidative enzymes (e.g., aryl hydroxylases, peroxidases, and

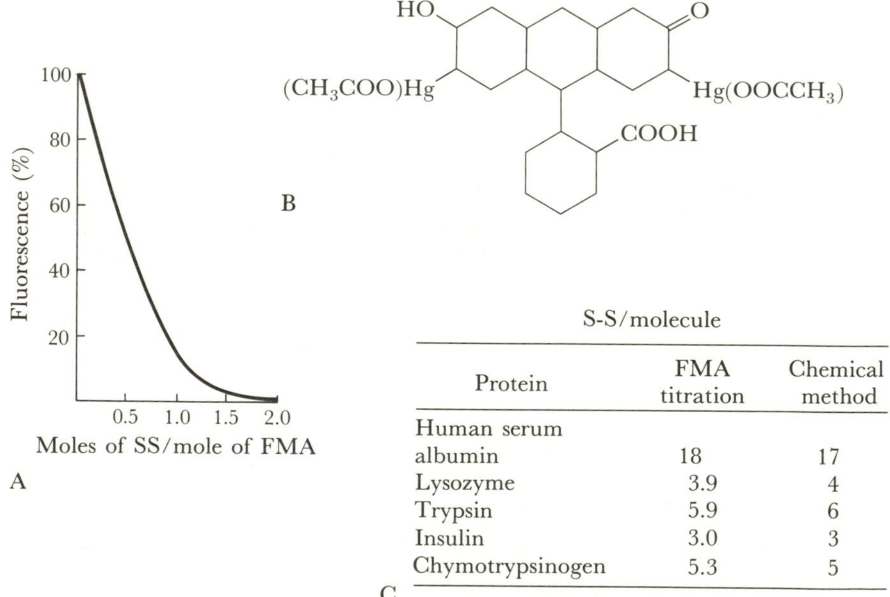

S-S/molecule

Protein	FMA titration	Chemical method
Human serum albumin	18	17
Lysozyme	3.9	4
Trypsin	5.9	6
Insulin	3.0	3
Chymotrypsinogen	5.3	5

Figure 15-15
A. Measurement of S—S bonds by the quenching of fluorescence of fluorescein mercuric acetate (FMA). B. The structure of the fluor. C. A comparison of the results with other methods. [From F. Karush, N. R. Klinman, and R. Marks, *Anal. Biochem.* 9(1964): 100–114.]

oxidases), transaminases, dehydrogenases, isomerases, kinases, and decarboxylases. An example of the extraordinary sensitivity of such assays can be seen in the work of Boris Rotman [*Proc. Natl. Acad. Sci. U.S.A.* 47(1961):1981–1991], who measured the activity of a *single* molecule of β-galactosidase. Purified β-galactosidase was diluted and dispersed into droplets of roughly 10^{-9} ml—each containing a known concentration of the nonfluorescent substance fluorescein di-(β-D-galactopyranoside). Hydrolysis of this substance yields fluorescein. The fluorescence of each droplet was measured with a microscope equipped with a photomultiplier and suitable apertures so that a single drop could be observed. In these droplets it was possible to detect fluorescein at 2×10^{-6} M, or 1.7×10^{6} molecules per droplet. Because one enzyme molecule hydrolyzes roughly 2×10^{5} molecules of substrate per hour, the fluorescein could be detected after approximately 10 hours. This method was then used to determine the number of enzyme molecules in a single *E. coli* bacterium.

☐ Quantitative measurement of DNA. **Example 15-DD**

The fluor ethidium bromide binds tightly to DNA; in so doing, the quantum yield increases substantially. This increase is linear throughout a wide range and the measurement of fluorescence intensity of an ethidium bromide solution containing small amounts of DNA can be used in a quantitative way to measure DNA concentration.

This enhancement of Q is also being used to detect DNA that has been electrophoresed in polyacrylamide and agarose gels (see Chapter 9, Figure 9-13). If a gel containing DNA is soaked in an ethidium bromide solution, it will take up the fluor. If the gel is then exposed to exciting light, the DNA bands become visible as regions of intense fluorescence.

☐ Detection of primary amines and peptides. **Example 15-EE**

In chromatography and electrophoresis, it is frequently necessary to detect primary amines such as amino acids and peptides. Several chemical techniques are available for this purpose but sometimes, when only a very small sample is available, these procedures are not adequate. When *o*-phthalaldehyde, a nonfluorescent substance, is mixed with primary amines, an intense blue fluorescence is produced. This is so sensitive that it has been possible to detect all of the spots of a protein fingerprint using only 10^{-5} g of protein.

SELECTED REFERENCES

Cantoni, G. C., and D. R. Davies, eds. 1971. *Procedures in Nucleic Acid Research*, vol. 2. Harper and Row. This contains several articles on nucleic acid fluorescence.

Chen, R. F., and H. Edelhoch. 1975. *Biochemical Fluorescence*, vol. 1 and 2. Dekker.

Chen, R. F., H. Edelhoch, and R. F. Steiner. 1973. "Fluorescence of Proteins," in *Physical Principles and Techniques of Protein Chemistry*, part A, edited by S. J. Leach, pp. 171–244. Academic Press.

Kronman, M. J., and F. M. Robbins. 1970. "Buried and Exposed Groups in Proteins," in *Fine Structure of Proteins and Nucleic Acids*, vol. 4, edited by G. D. Fasman and S. N. Timasheff, pp. 271–416. Dekker.

Lehrer, S. S., and P. C. Learis. 1978. "Solute Quenching of Protein Fluorescence," in *Methods in Enzymology*, vol. 49, edited by C. H. W. Hirs and S. N. Timasheff, pp. 222–236, Academic Press.

Pesce, A. J., C. G. Rosen, and T. L. Pasby. 1971. *Fluorescence Spectroscopy*. Dekker.

Stryer, L. 1968. "Fluorescence Spectroscopy of Proteins." *Science (Wash., DC)* 162:526–540. A good review of energy transfer.

Stryer, L. 1978. "Fluorescence Energy Tranfer as a Spectroscopic Ruler." *Annu. Rev. Biochem.* 47:819–846. A superb and very complete article, which is quite easy to read.

Timasheff, S. N. 1970. "Some Physical Probes of Enzyme Structure in Solution," in *The Enzymes*, vol. 2, edited by P. D. Boyer, pp. 371–443. Academic Press.

Udenfriend, S. 1962, 1969. *Fluorescence Assay in Biology and Medicine*, vol. 1 and 2. Academic Press.

Weber, G. 1952. "Polarization of the Fluorescence of Macromolecules: 1. Theory and Experimental Methods; 2. Fluorescent Conjugates of Ovalbumin and Bovine Serum Albumin." *Biochem J.* 51:145–155; 155–167. The classic papers on fluorescence depolarization.

Weber, G. 1972. "Uses of Fluorescence in Biophysics: Some Recent Developments." *Annu. Rev. Biophys. Bioeng* 1:553–569.

Weber, G., and F. W. J. Teale. 1966. "The Interaction Proteins with Radiation," in *The Proteins*, vol. 3, edited by H. Neurath, pp. 445–452. Academic Press.

PROBLEMS

15–1. Consider the following pair of spectra. One is an absorption spectrum and the other is a fluorescence spectrum. Which is which?

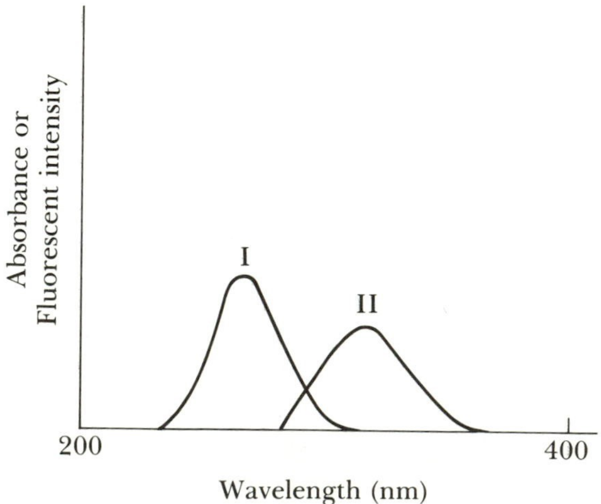

15–2. A protein causes ANS to fluoresce. If the protein concentration is increased before adding ANS, the fluorescence decreases. Give two possible explanations for this decrease.

15–3. Iodide quenching or tryptophan fluorescence can be used to determine whether tryptophans are exposed to the solvent. If a protein is known to contain only one tryptophan and iodide fails to quench, what possible explanations might be given to account for the lack of quenching? If the protein contains eight tryptophans and iodide quenches 25% of the fluorescence, it is tempting to assume that two tryptophans are accessible to the solvent. State several factors that could make this conclusion invalid.

15–4. When excimers are formed, the excitation spectrum is that of the monomer because the monomer is excited before dimerization. If many dimers are present and the dimer is excited, do you expect the excitation spectrum to match the absorption spectrum of the monomer? Explain.

15–5. Iodide quenching decreases fluorescence intensity. Would you expect there to be a change also in the shape of either the excitation or emission spectrum?

15–6. Give several possible mechanisms for an increase in the quantum yield of a fluor when bound to another molecule. Why might there sometimes be a shift in the excitation and/or emission spectra? Could the shift be to either longer or shorter wavelengths?

15–7. Is the excitation spectrum of a fluor always the same as the absorption spectrum? Explain.

15–8. A protein containing ten tryptophans shows fairly strong tryptophan fluorescence. A small molecule known to bind tightly to the protein produces virtually no change in the fluorescence, even though it is known that there are two tryptophans in the binding site. Give several possible explanations for this.

15–9. When the fluor acridine orange is bound to DNA and the mixture is irradiated with light absorbed by the acridine orange, irreversible chemical changes (e.g., broken purine rings, single-strand breaks) occur in the DNA. For these changes to occur, molecular O_2 must be present; hence, the process is called photosensitized oxidation, or photo-oxidation. These reactions occur with many acridine derivatives; the efficiency of the reaction is roughly proportional to the quantum yield of fluorescence. Propose a mechanism for photo-oxidation.

15–10. A fluor is covalently coupled to a protein. The polarization of the fluorescence is measured as a function of the ionic strength of the suspending buffer; it is found to decrease markedly as the ionic strength increases. What effect does increasing ionic strength have on the protein?

15–11. A protein has a fluor covalently attached at a unique site. When tryptophan is excited, tryptophan fluorescence is very weak but the fluor emits strongly. Explain each of the following possible observations. (*Note*: These are *alternative* and mutually exclusive observations.)

Iodide quenches the fluorescence of F but not of tryptophan.

Iodide quenches the fluorescence of F *and* of tryptophan.

Adjustment to pH 9 eliminates the fluorescence of F and enhances that of tryptophan (assume that there is no pH effect on free tryptophan or unbound F).

Adjustment to pH 9 eliminates all tryptophan fluorescence and markedly enhances the fluorescence of F.

15–12. A protein has two tryptophans. Three X-ray diffractionists are studying the protein. The first is sure that there is an internal tryptophan surrounded by a lysine, an arginine, and a tyrosine. The second person thinks that it is surrounded by glycine, two isoleucines, and a valine. The third person believes that both tryptophans are on the surface. You perform a fluorescence measurement and discover that (a) addition of I^- or Cs^+ decreases the tryptophan fluorescence to half of the initial value, and (b) addition of I^- causes a shift of λ_{max} to shorter wavelengths. Which of the three diffractionists is correct?

15–13. Two fluorescent compounds A and B are chemically attached to the end of the poly(amino acid) poly-L-proline. A is at one end and B is at the other end of the molecule. This particular poly(amino acid) differs from other poly(amino acid)s in that the bond joining the prolines is not a peptide bond but a more rigid bond. The result is that poly-L-proline is a much more extended molecule than is polyalanine. Let us use the notation A-(proline)$_n$-B to designate the former polymer, in which n denotes the number of prolines in the chain. A and B have the property that if A absorbs exciting light, B frequently fluoresces.

a. If solutions of A-(proline)$_8$-B and of A-(proline)$_{11}$-B are separately illuminated with light that excites A, for which solution will the fluorescence of B be brighter?

b. If A-(proline)$_7$-B and A-(alanine)$_7$-B are separately illuminated as in part(a), for which compound will B fluoresce more intensely?

16

Optical Rotatory Dispersion and Circular Dichroism

Another set of techniques for elucidating both the conformation of a macromolecule or macromolecular complex in solution and the interactions between macromolecules is described in this chapter. Although absorption spectroscopy is capable of supplying a great deal of useful information of this sort, by studying the absorption of *polarized* light—that is, the technique of optical rotatory dispersion (ORD) and circular dichroism (CD) spectroscopy, both of which satisfy the criteria of speed and applicability to solutions—even more information can be obtained (although at the cost of somewhat greater instrumental and theoretical complexity than in absorption spectroscopy). These methods measure the wavelength dependence of the ability of an optically active chromophore to rotate plane-polarized light (ORD) and the differential absorption of right and left circularly polarized light (CD). The physical basis of ORD and CD is the same and, in fact, they are merely two different ways of looking at the same interaction of polarized light with *optically active* molecules. Because a very large fraction of biological molecules contain optically active centers, ORD and CD have great applicability to their study.

In this chapter, it will be seen that, because the ORD and CD spectra of proteins and nucleic acids result primarily from the spatial asymmetry of the constituent amino acids and nucleotides, respectively, in the backbones of the macromolecules, ORD and CD are useful in structural studies with proteins, nucleic acids, and nucleoproteins. However, once again the theory connecting the spectra with molecular structure is not yet fully developed; thus, the working rules used to interpret spectra (like those of absorption spectroscopy) are for the most part empirical.

Simple Theory of ORD and CD

In the preceding chapter, it was explained that light is an electromagnetic wave consisting of an oscillating electric (E) field and a magnetic (H) field, both of which can be represented by mutually perpendicular vectors (Figure 16-1A). The *plane of polarization* is defined as the plane of the **E** vector. Because a light source usually consists of a collection of randomly

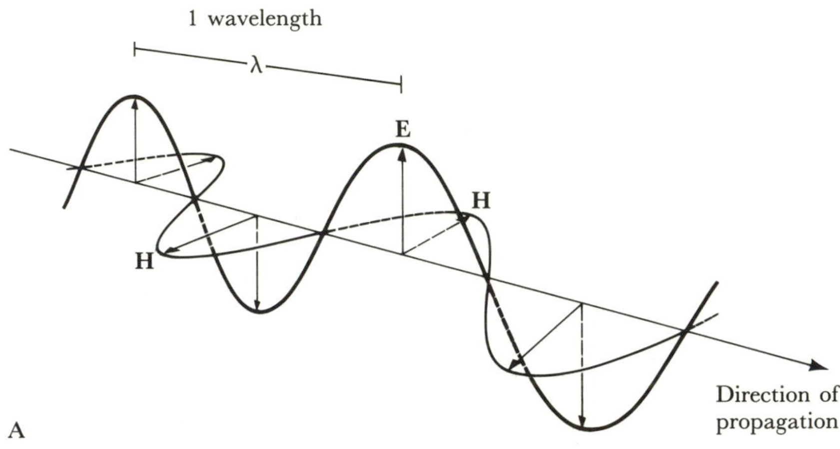

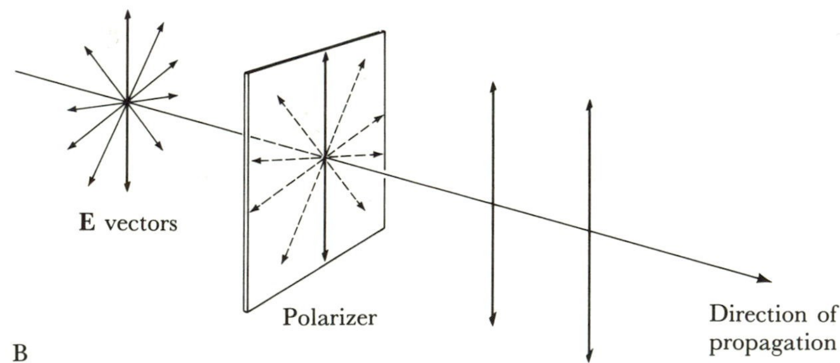

Figure 16-1
A. Propagation of an electromagnetic wave. The path of the tip of the **E** vector is indicated by the heavy line; that of the **H** vector by the light line. The **E** and **H** vectors and the direction of propagation are mutually perpendicular. The plane of the **E** vector is called the plane of polarization. B. Production of plane-polarized light. A collection of waves falls on the polarizer, which passes only those components of the **E** vector that are parallel to the axis of the polarizer.

oriented emitters, the emitted light is a collection of waves with all possible orientations of the **E** vectors. Plane-polarized light is usually obtained by passing light through an object (e.g., a Polaroid screen or a Nicoll prism) that transmits light with only a single plane of polarization (Figure 16-1B).

Suppose that two plane-polarized waves differing in phase by one-quarter wavelength (i.e., when one sine curve crosses the axis of propagation, the other is at a maximum or minimum), whose **E** vectors are perpendicular to one another, are superimposed. As the waves propagate forward, the result **E** vector rotates so that its tip follows a helical path (Figure 16-2). This is, of course, also true of the magnetic-field vector. Such light is called *circularly polarized* and is defined as *right* circularly polarized if the **E** vector rotates clockwise to an observer looking at the source.

If a right (R) and a left (L) circularly polarized wave, both of equal amplitude, are superimposed, the result is plane-polarized light, because at any point in space the **E** vector of each will sum as shown in Figure 16-3A. Similarly, plane-polarized light can be decomposed into R and L components. If the amplitudes of the two circularly polarized waves are not the same, the tip of the resultant **E** vector will follow an elliptical path and such light is said to be *elliptically polarized* (Figure 16-3B). A parameter called the ellipticity, θ, is often used to describe the elliptical polarization. This is the angle whose tangent is the ratio of the minor and major axes of the ellipse shown in Figure 16-3B—that is, $\theta = \tan^{-1}(b/a)$ (see equation 6).

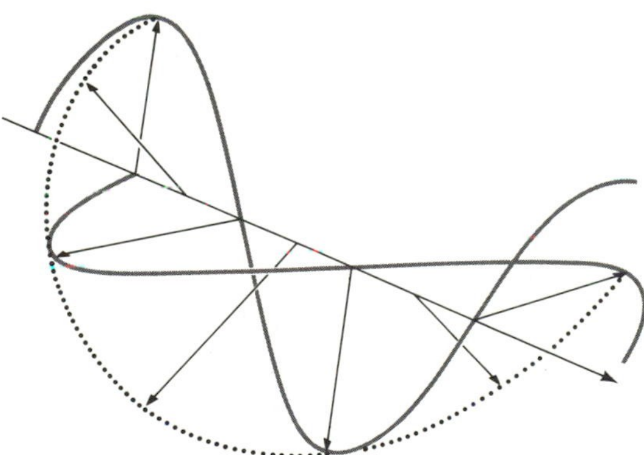

Figure 16-2

Generation of circularly polarized light. The **E** vectors of two electromagnetic waves are one-quarter wavelength out of phase and are perpendicular. The vector that is the sum of the **E** vectors of the two components rotates so that its tip follows a helical path (dotted line).

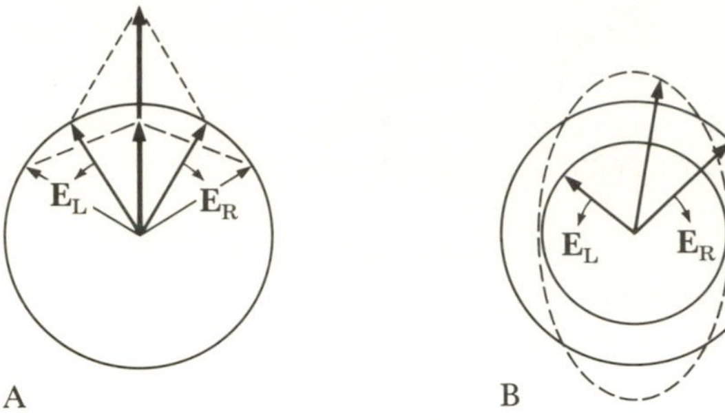

Figure 16-3
Diagrams showing how right and left circularly polarized light combine: (A) if the two waves have the same amplitude, the result is plane-polarized light; and (B) if their amplitudes differ, the result is elliptically polarized light—that is, the head of the resultant vector will trace the ellipse shown as a dashed line. The lengths of the major and minor axes of the ellipse are a and b.

When a beam of light (i.e., a propagating electromagnetic field) passes through matter, the electric (**E**) vector of the propagating wave interacts with the electrons of the component atoms. This interaction has the effect of reducing the velocity of propagation (also called *retarding* the light) and in decreasing the amplitude of the **E** vector. Reducing the velocity of propagation is called *refraction* and is described by the *index of refraction*, n, and decreasing the amplitude of the **E** vector is *absorption* and is described by the *molar absorption coefficient*, ε. Both n and ε depend on wavelength in a way that reflects the electronic structure and geometry of the molecules.

For most substances, simple refraction and absorption are the only detectable result of such an interaction even if the light is polarized. However, the behavior of some molecules is sensitive to the plane of polarization of the incident light. Such a molecule or chromophore is called *optically active* and is characterized by having distinct indices of refraction, n_L and n_R, and molar absorption coefficients, ε_L and ε_R, for left and right circularly polarized light, respectively. Optical activity is a characteristic of many organic and almost all biological molecules. The property that determines whether a chromophore is optically active is its *asymmetry*. If a molecule is asymmetric *in the sense that it cannot be superimposed on its mirror image*, it is optically active. Examples of optically active structures are given in Figure 16-4. The physical basis of optical activity is difficult to explain without the formalism of electrodynamics and quantum mechanics and will not be explained here. It can be found in the Selected References near the end of this chapter.

```
     H   Cl
     |   |
  H—C—C*—OH
     |   |
     H   H
```

A 1-Chloro-1-hydroxyethane

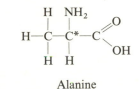

```
     H  NH2
     |   |      O
  H—C—C*—C⟨
     |   |      OH
     H   H
```

B Alanine

C

Figure 16-4
Examples of several optically active structures. In parts A and B, the asymmetric carbon is identified with an asterisk.

Let us consider the interaction of an optically active substance with polarized light. As indicated earlier, a plane polarized wave can be thought of as a mixture of L and R circularly polarized light. The entire interaction can be thought of in terms of these circularly polarized components and, in fact, it is easier to do so. If a substance retards both L and R equally (i.e., if the indices of refraction for L and R circularly polarized light, n_L and n_R, are the same), the L and R waves will recombine on leaving the substance to form plane-polarized light, with the plane of the transmitted beam being the same as that of the incident beam. However, if n_L and n_R are unequal, the transmitted L and R components are each retarded to a different extent so that on leaving the material the phases of the two sine waves differ (Figure 16-5). Henceforth, at any point in space, the **E** vectors of the L and R waves combine to form a beam of plane-polarized light whose angle differs from that of the plane of polarization of the incident wave; hence, the plane of polarization of the resulting wave will be rotated. For any substance that interacts with light in this asymmetric way, the extent of the rotation produced by a sample of a given volume depends on the number of chromophores with which the wave interacts—that is, on the concentration of the molecules multiplied by the path length, d, and on the wavelength, λ, of the light—because n is always a function of λ.

Quantitatively, the observed angle of rotation, α_λ, expressed in degrees, can be described by

$$\alpha_\lambda = \frac{180d}{\lambda}(n_L - n_R) \tag{1}$$

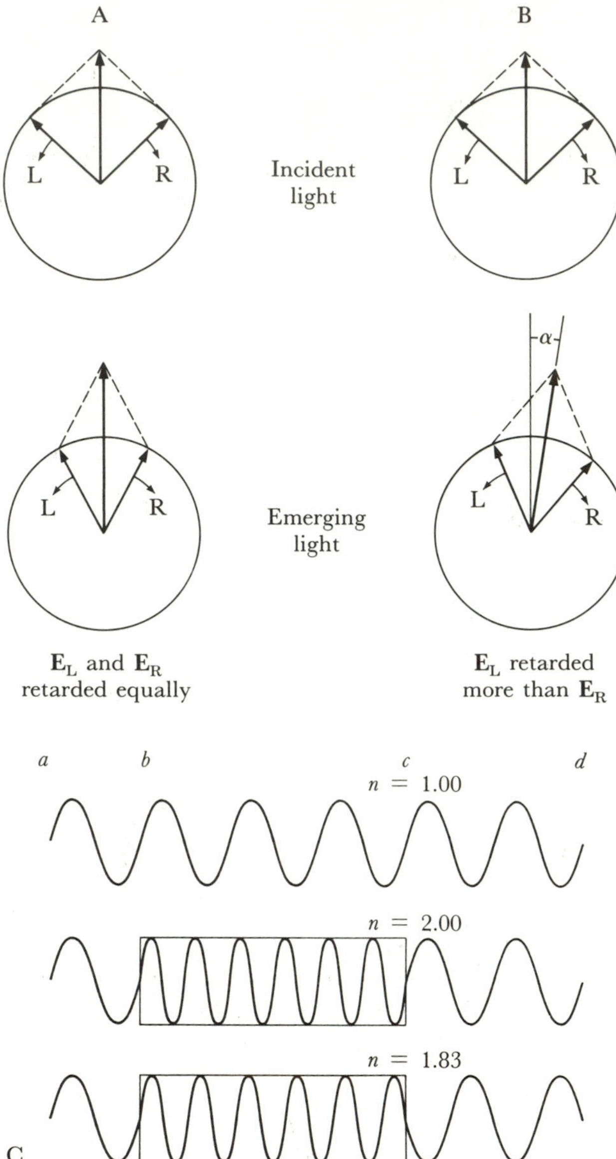

Figure 16-5
Rotation of the plane of polarization: (A) both right and left circularly polarized light are retarded equally so that the resultant **E** vector remains in the same plane; (B) the left circularly polarized light is retarded more than the right so that the resultant **E** vector changes orientation (because the amount of retardation is proportional to the distance the light travels, the **E** vector will rotate clockwise with increasing optical path length); (C) physical significance of the retardation of a light wave by passing through media of different indices of refraction and traveling equal distances *ad*. As a consequence of the lower velocity of light in refractive media and the invariant frequency of light, more cycles of vibration are squeezed in a given path length in the refractive media than in air ($n = 1.00$). Note that all three beams are in phase up to position *b*. In this example, if $n = 2.00$, the beam emerges at *c* in phase with the unretarded beam. If $n = 1.83$, it is exactly out of phase. [From E. Slayter, *Optical Methods in Biology*, Wiley-Interscience, 1970.]

For most work, the terms *specific rotation*, $[\alpha]_\lambda$, and *molar rotation*, $[M]_\lambda$, are used. These are defined as

$$[\alpha]_\lambda = \frac{\alpha_\lambda}{dc} \tag{2}$$

in which α_λ is the observed rotation in degrees, d is the light path in centimeters, and c is the concentration in grams per cubic centimeter; and

$$[M]_\lambda = \frac{\alpha_\lambda M}{10dc} \tag{3}$$

in which M is the molecular weight in grams per mole. The units of $[M]_\lambda$ are deg M^{-1} cm^{-1}.*If the substance under study is a polymer, as is often the case in biochemistry, it is more common to use the *mean residual rotation*, $[m]_\lambda$, *in which*

$$[m]_\lambda = \frac{\alpha_\lambda M_0}{10dc} \tag{4}$$

and M_0 is the mean residue molecular weight (i.e., the molecular weight of the polymer divided by the number of monomers).

A curve that shows the wavelength dependence of optical rotation, expressed either as α, $[\alpha]$, $[M]$, or $[m]$, is called an *optical rotatory dispersion spectrum.*

So far, only the *retardation* of R and L waves has been considered, but it is also of interest to know what happens to the *intensity* of each as these waves pass through matter. If the substance is optically *inactive*, the absorption of each is equal. If, on the other hand, the material is optically active, then in the range of wavelengths in which absorption occurs (the absorption band) there will be, for each wavelength, differential absorption of the L and R circularly polarized light. This difference is usually expressed in terms of the absorption coefficients for L and R light, ε_L and ε_R; that is,

$$\varepsilon_L - \varepsilon_R = \Delta\varepsilon \tag{5}$$

in which $\Delta\varepsilon$ is called the *circular dichroism*, or CD. It is positive if $\varepsilon_L - \varepsilon_R > 0$ and negative if $\varepsilon_L - \varepsilon_R < 0$. An important point is that, *if a given optically active molecule has positive CD, then its mirror image will have a negative CD of precisely the same magnitude.* Because differential absorption of the L and R waves means that the amplitudes

* In older texts d is measured in decimeters. The reader must be careful to note whether $[M]_\lambda$ is expressed in deg M^{-1} dm^{-1} or deg M^{-1} cm^{-1}.

of the transmitted waves will differ, the result, as shown in Figure 16-3B, is elliptically polarized light. Experimentally, it is usual to measure $\Delta\varepsilon$, but for historical reasons the ellipticity, θ, is plotted; θ is defined as

$$\theta = 2.303(A_L - A_R)180/4\pi = 33\,\Delta A \textit{ degrees} \tag{6}$$

in which A is the absorbance defined by equation (3) in Chapter 14. A curve showing the dependence of θ on wavelength is called a *CD curve* or *CD spectrum.*

For most work, the terms molar ellipticity and mean residue ellipticity are used. Unfortunately, in contrast with molar rotation and mean residual rotation in optical rotatory dispersion (equations 3 and 4), the same notation is used for both of these terms:

$$[\theta]_\lambda = \frac{M\theta_\lambda}{10dc} \tag{7}$$

in which θ_λ is the observed ellipticity in degrees, M is the molecular weight *or* mean residue molecular weight, d is the path length in centimeters, and c is the concentration in grams per milliliter. Combining equations (6) and (7) and using the relation between ΔA and $\Delta\varepsilon$, obtained from the Beer-Lambert law (equation 3 in Chapter 14), yields

$$[\theta]_\lambda = 3300\,\Delta\varepsilon \tag{8}$$

which is the most common form of equation (7).

It is useful to see the relation between an ordinary dispersion curve (i.e., the index of refraction, n, versus wavelength, λ), the ORD curve (which is, in fact, $\Delta n = n_L - n_R$ versus λ), a standard absorption curve (ε versus λ), and a CD curve ($\Delta\varepsilon = \varepsilon_L - \varepsilon_R$ versus λ). The types of curves obtained by examining a single absorption band of an optically active chromophore are shown in Figure 16-6. The solid curve of Figure 16-6A shows the spectrum for the absorption of nonpolarized light; for example, for a single electronic transition from the ground state to the first excited state. The wavelength corresponding to maximum absorption, λ_0, is often called the absorption peak or λ_{max}. This absorption band can also be characterized by a value, Δ, or the half-width at 1/e times the maximum height. The solid line of Figure 16-6B shows the index of refraction, n, for nonpolarized light as a function of wavelength. The principal features of this curve are that (1) $n = 1$ at $\lambda = \lambda_0$; (2) the curve to the left of λ_0 can be superimposed on that to the right by means of a 180° rotation; (3) the wavelengths corresponding to maximum and minimum n are approximately $\lambda_0 + 0.9\Delta$ and $\lambda_0 - 0.9\Delta$, respectively; and (4) at wavelengths outside of the absorption band, n approaches 1 asymptotically. If the chromophore is optically active, the curves for CD and ORD, respectively, are essentially the same except that the y-axes are $\Delta\varepsilon = \varepsilon_L - \varepsilon_R$

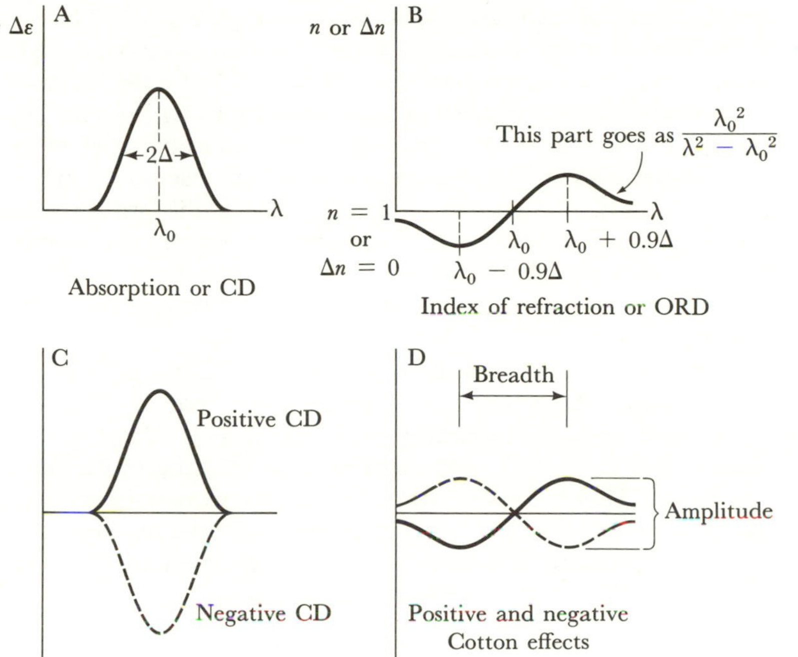

Figure 16-6

Relation between (A) an absorption spectrum (ε versus λ) and circular dichroism ($\Delta\varepsilon$ versus λ) and between (B) a dispersion curve (n versus λ) and optical rotatory dispersion (Δn versus λ) for an optically active substance (note that the symbol Δ is used to denote both the y-axis and the band width—a common notation, unfortunately; (C) the CD curve drawn as a solid line is called positive CD, and the dotted line is negative CD; (D) the ORD curve drawn as a solid line is called a positive Cotton effect, and the dotted line is a negative Cotton effect. If the substance is not optically active, the n-versus-λ and ε-versus-λ curves are of the same type, but $\Delta\varepsilon$ and n equal zero at all values of λ.

and $\Delta n = n_L - n_R$. The ORD curves are usually presented with the y-axis being some measure of rotation (i.e., $[\alpha]$ or $[M]$), but it must be remembered that the rotation is a result of the different indices of refraction of L and R circularly polarized light.

An ORD curve of the type shown in Figure 16-6B is called a *Cotton effect*. Such curves can be negative or positive, as shown in Figure 16-6D. (If an optically active molecule has a positive Cotton effect, its mirror image will have a negative Cotton effect of precisely the same shape.) Each Cotton-effect curve consists of two extremes called a *peak* and a *trough*; the magnitude of Δn at the wavelength corresponding to the maximum of a peak is always the same as that for a trough, except with the opposite sign. Two other parameters used to describe a Cotton effect are the amplitude and breadth, as shown in Figure 16-6D. The parts of the Cotton effect that tail outward from the absorption band are called *plain curves*. These curves are only rarely studied now but have the occasionally useful property that they extend very far from the absorption band and can therefore be used to determine approximately some of the parameters of a molecule in a solvent that absorbs strongly in the region near λ_0. For proteins and polypeptides, the principal Cotton effect occurs in the wavelength range near 200 nm and is caused by the absorption of the peptide bond. For nucleic acids, it occurs in the 250–275-nm range and is caused by electronic transitions of the nucleotide bases.

CD bands can also be positive or negative, as shown in Figure 16-6C, and a positive CD band always corresponds to a positive Cotton effect. A CD band can also be characterized by the height—that is, the magnitude of θ at λ_0—which is to some extent a measure of the degree of asymmetry. However, with CD it is more common to refer to the *rotational strength*, R, of a band—the area under the $\Delta\varepsilon$-versus-λ curve. Rotational strength is difficult to evaluate precisely and is normally calculated approximately for a Gaussian band as

$$R \sim 1.23 \times 10^{-42}[\theta_0]\frac{\Delta}{\lambda_0} \tag{9}$$

in which λ_0 is the wavelength of the peak of the CD curve, $[\theta_0]$ is $[\theta]$ at λ_0, and Δ is the half-width of the band at $1/e$ times the height. It can also be measured with a curve analyzer.*

The rotational strength describes the *intensity* of a CD band and physically tells something about the motion of the electrons when the absorbing center is raised from the ground state to an excited state by the absorption of light; it is not necessary to understand in detail the factors that determine the magnitude of R in order to interpret CD data. The main rule is the R is not zero in an optically active substance and that it generally increases with increasing asymmetry.†

Relative Values of ORD and CD Measurements

It should be apparent by now that ORD and CD are both manifestations of the same underlying phenomenon. In fact, as might be expected, ORD and CD spectra can be generated from one another. This is done by means of a mathematical conversion called the general Kronig-Kramers

* A curve analyzer is an electronic instrument that can (1) measure the area of a curve and (2) decompose a curve into a collection of Gaussian curves whose sum is the original curve.

† A corollary derived from this statement is that sometimes a substance either without optical activity or with a low R can achieve very high R by interacting with a polar asymmetric molecule.

transformation, which can be found in the specialized texts referred to near the end of the chapter.

The degree to which ORD and CD have been used to obtain information about macromolecules has been determined primarily by the availability of the necessary instrumentation. At first, only ORD instruments were available. Furthermore, before about 1960, these instruments were operable only in a wavelength region well above the absorption bands for proteins and nucleic acids. The result was that only the plain curve was available for study. These ORD curves were usually analyzed by a relation called the Drude equation:

$$[\alpha]_\lambda = \frac{A}{\lambda^2 - \lambda_0^2} \tag{10}$$

in which $[\alpha]_\lambda = \alpha_\lambda/dc$ (see equation 1), and λ is the wavelength at which $[\alpha]_\lambda$ is measured; λ_0 was called the *dispersion constant* (often written λ_c) and later proved to be the wavelength of the center of the absorption band; A was called the *rotatory constant* and was later shown to equal $2\ R\lambda_0^2/(0.696 \times 10^{-42}\pi)$, in which R is the rotational strength. It was clear that this equation could not be correct because it predicted infinite rotation at $\lambda = \lambda_0$; nonetheless, it was all that was available at the time and some information was actually obtainable from the determination of λ_0 and A. Various empirical studies showed that λ_0 and A could be related in certain ways to conformation of macromolecules. During this period of rather inadequate instrumentation, other equations were used to analyze the data. The most notable of these was the Moffitt equation, a multiparameter equation that attempted to relate the shape and magnitude of the plain curve to the fraction of amino acid residues in a protein that were in the α-helical and random-coil configurations. In later years, the instrumentation of ORD was improved so that the range of measurements extended to the far ultraviolet and allowed studies in regions of absorption—that is, observation of the Cotton effect. As will be seen in later parts of this chapter, analyses of Cotton effects have provided a large amount of information.

However, in the late 1960s, instruments for measuring CD became available. The simplicity of the CD curve compared with the Cotton-effect curve (see Figure 16-6) makes CD analysis vastly superior to ORD analysis so that at the present time, ORD is rarely used.

The principal advantage of CD analysis is its greater ability to resolve bands due to different optically active transitions. This is best seen by examining curves consisting of more than one band. First, referring to Figure 16-6, which shows the ORD and CD for a particular optically active substance, it should be noted that at no point is a flat baseline reached in an ORD curve; the curves are asymptotic. The CD curve, however, has a defined zero baseline outside of the absorption band. Furthermore, CD bands are narrow, allowing fairly good resolution of nearby bands.

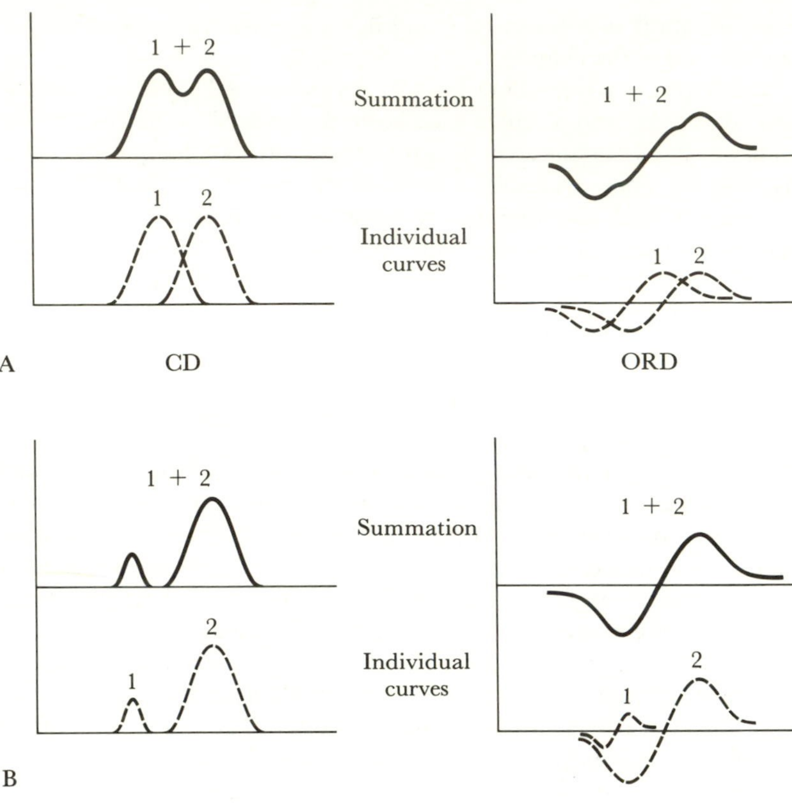

Figure 16-7
Comparison of the ability of ORD and CD to indicate that two optically active bands exist. In each case, the solid upper curve is the sum of the two dashed lower curves. In part A, the two curves have the same sign and amplitude but different λ_{max}. In part B, the amplitude differ. In part C, the signs differ but the amplitudes are nearly the same. Part D is a comparison of absorbance (upper solid line), CD (dashed line), and ORD (lower solid line) of a system containing four bands. Note that the CD curve tells which bands are optically active and distinguishes the sign.

Figure 16-7A shows the ORD and CD spectra that would result if the molecule contained two optically active centers giving rise to two Cotton effects very near one another. Note that the CD spectrum clearly shows two elements, whereas the ORD curve is somewhat complex. Even more striking is Figure 16-7B, which shows the results for a molecule with a strong Cotton effect at one wavelength and a very weak one at a more distant wavelength. Because ORD spectra totally overlap, the weak band is almost undetectable, whereas the CD spectrum with its narrow

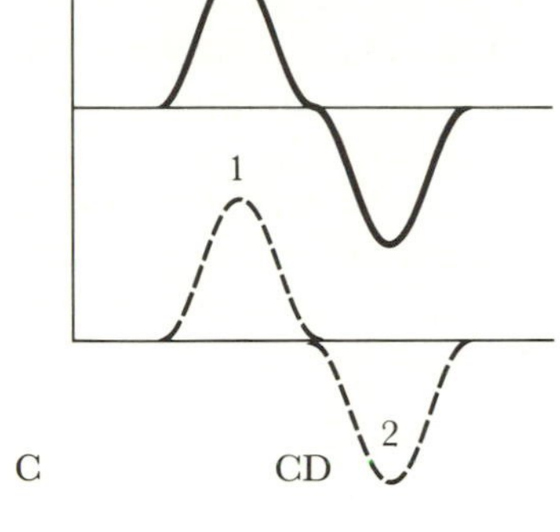

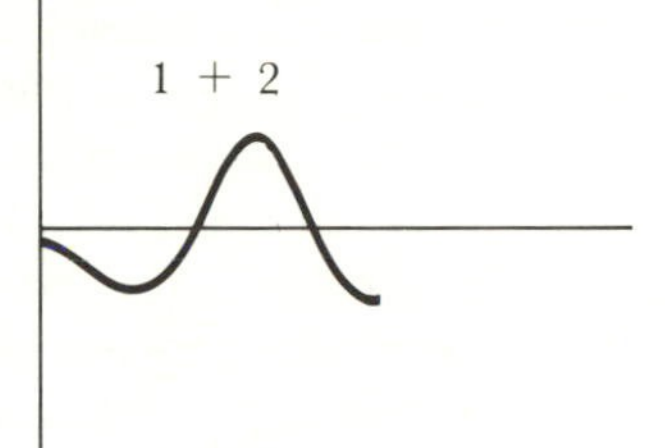

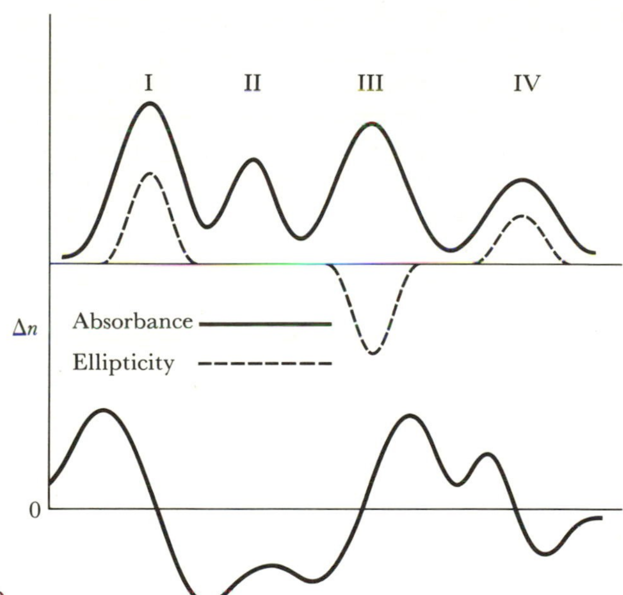

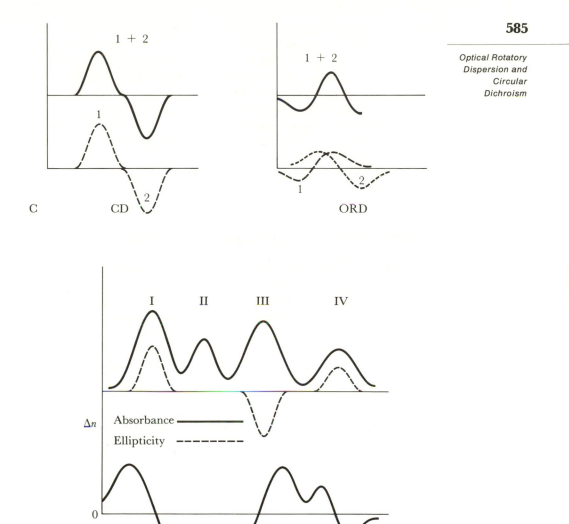

bands shows both bands quite clearly. In Figure 16-7C, the effect of the juxtaposition of positive and negative curves can be seen. These examples should convince the reader that the resolution of CD spectra is superior to ORD analysis.

A second diagram (Figure 16-7D) serves to reinforce this conclusion in addition to indicating that CD analysis also provides greater information than simple absorption spectroscopy with unpolarized light. Note that the absorption spectrum merely indicates that the system consists of four chromophores (i.e., there are four bands), whereas the CD spectrum shows that three of the four (I, III, and IV are due to optically active centers. The sign of the CD also describes the handedness of the chromophore. It can also be seen that the ORD fails to give clear evidence for whether band II is optically active because of the overlapping of the plain curves of the Cotton effects for bands I and III.

ORD measurements are presented in some of the examples of this chapter because many important results have been obtained in this way, despite the complexity of the spectra. However, in reading the recent scientific literature, the student will find that CD is now the method more frequently used.

Techniques for Measuring ORD and CD

Although films and solids are occasionally used, solutions are generally used for ORD and CD measurements. The solution is contained in a vessel called a cell. The basic instrumentation requires a light source whose wavelength can be varied, a system for polarizing the light, a system for measuring the polarization after the light has passed through the cell, and a detector by which the amount of light can be measured.

Figure 16-8 shows a simple model for an ORD instrument. A light source plus a monochromator is used to select the illuminating wavelength. This light passes through a polarizer, which produces plane-polarized light. An analyzer (which is, in fact, just another polarizer) is used to determine the angle of polarization. When the plane of the

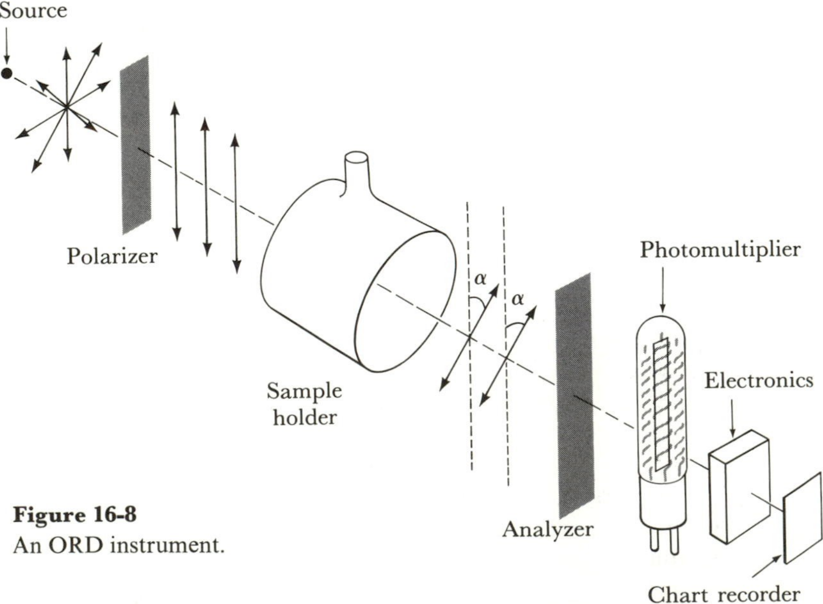

Figure 16-8
An ORD instrument.

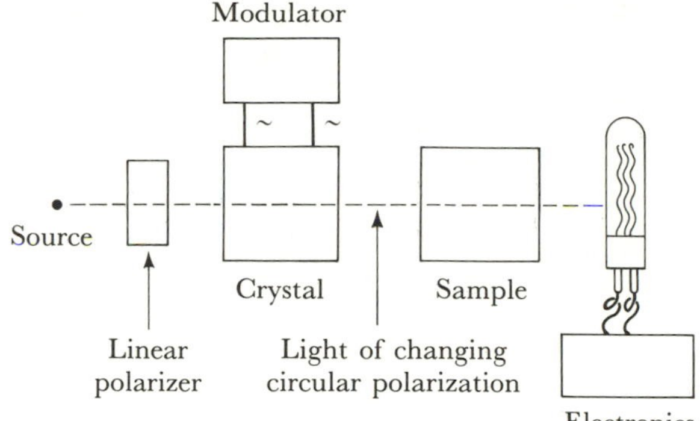

Figure 16-9
A CD instrument.

analyzer is parallel to that of the polarizer, the maximum intensity of light passes through. The transmitted light then falls on a photomultiplier tube, which converts the intensity into an electric current. If the cell is filled with a sample that rotates light, the analyzer must be rotated to allow the maximum passage of light. The angle at which the transmitted intensity is maximal defines the observed rotation in degrees, α_λ (see equation 1). In practice, modern automatic instruments simultaneously vary the wavelength, determine the rotation, and make a plot of α_λ versus λ. Different commercial instruments vary slightly in the way this is done, but the principle is as described.

For CD measurement (Figure 16-9), in principle two light sources are needed, one for L and the other for R circularly polarized light, each provided with a monochromator for wavelength selection. However, commercial instruments utilize a simple trick for generating L and R light from a single source. Plane-polarized light passes through a crystal that is subjected to an alternating electric field. This crystal (called an electrooptic modulator) has the remarkable property that the polarity of the field determines whether the L or R component of the light is transmitted. Because the field is alternating current, the beam continually modulates from the production of L to the production of R light. This beam then passes through the sample cell and falls on a photomultiplier. The output of the photomultiplier is then processed electronically in a fairly complex way to provide a voltage that is proportional to the ellipticity. This is automatically plotted as a function of wavelength to give the CD spectrum.

Interpretation of ORD and CD Curves

The theory of optical activity is not yet capable of yielding the precise structure of a protein from its CD spectrum, although, as will be seen later, it is somewhat better for nucleic acids. The complications are that very frequently it is not that the chromophore is asymmetric but that the chromophore is asymmetrically perturbed by neighboring groups.

Furthermore, with proteins, there is the additional complication that, because the peptide bond (which is the principal element whose spectrum is detected by CD) exists in many conformations depending on its precise location in the protein, the spectrum is a result of an average of the various conformation parameters. Hence, in practice, an empirical approach of obtaining an ORD or CD spectrum for molecules whose structure is accurately known from X-ray diffraction is used, and the spectrum is related to the structural features of the molecule. This spectrum is then compared with the spectrum of a protein of unknown structure. The principal problem of this approach is that the (rarely proven) assumption must be made that the structure of a macromolecule in solution (remember that ORD and CD are determined in solution) is nearly the same as that of a fiber, crystal, or dry powder (as is used in X-ray analysis) prepared from the same solvent. This equivalence can be tested by Raman spectroscopy as described in Chapter 14.

As a result of these problems, a set of working "rules" that are used to analyze spectra has been developed. Some of these rules are presented in Table 16-1. Rule 1 is primarily used to make an estimation of the basic conformation of a macromolecule (e.g., the fraction of a protein that is helical) or it may be used to confirm that a structure determined by X-ray analysis is valid in solution. Rules 2 and 3 are primarily used to assay interactions between macromolecules and small molecules. These applications will be seen in many of the examples later in this chapter.

Application of ORD and CD Analysis to Protein and Polypeptide Structure

Because of the lack of adequate theory discussed earlier, the approach to the elucidation of the secondary structure of a protein has been to determine empirically ORD or CD curves for model polypeptides. (A

Table 16-1

Empirical Rules for Interpreting ORD and CD Spectra.

1. An ORD or a CD spectrum is additive—that is, it is the simple sum of the spectra of its components (as in Figure 16-7). This is not always strictly true but is a good approximation.
2. The amplitude of an ORD curve or the rotational strength of a CD curve is a measure of the degree of asymmetry. An agent that increases or decreases these parameters usually does so by increasing or decreasing asymmetry (although other spectral features usually accompany the change in asymmetry).
3. A chromophore that is symmetric can become optically active when it is in an asymmetric environment (e.g., a helix). This may or may not be accompanied by a change in λ_0.
4. The value of λ_0 and the magnitude and sign of $\Delta\varepsilon$ at λ_0 allows the chromophore to be identified because it is always very near the value of λ_0 obtained from simple absorption spectroscopy.

model polypeptide has only a single conformation and its structure is known from X-ray scattering.) Then an attempt is made to construct from these "standard" curves a weighted sum that is the same as the observed curve of the sample.

For proteins, the principal standards are three forms of poly-L-lysine: α helix, β form, and the random coil, whose ORD and CD spectra are shown in Figure 16-10A. If it is assumed that no other conformations

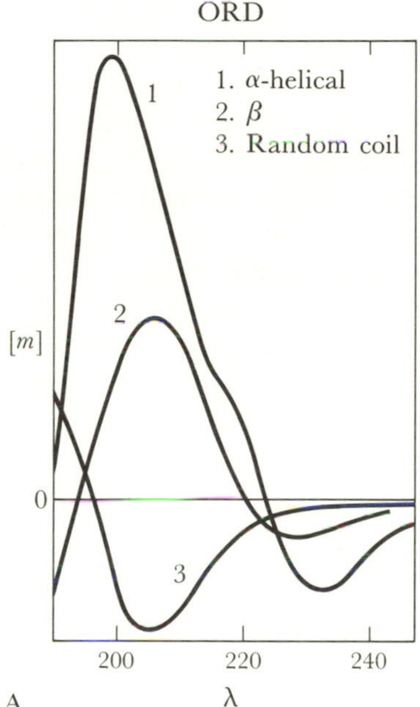

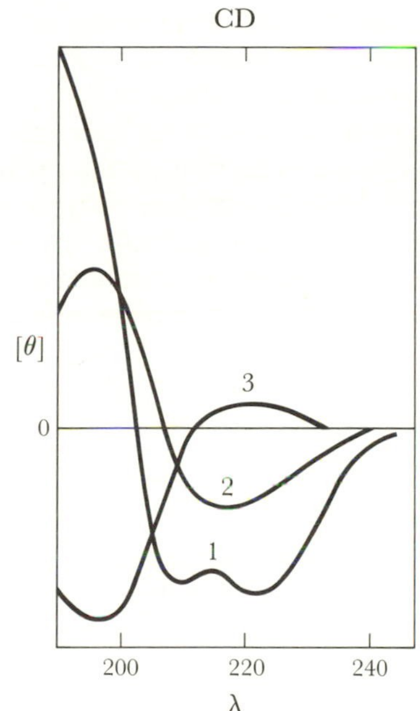

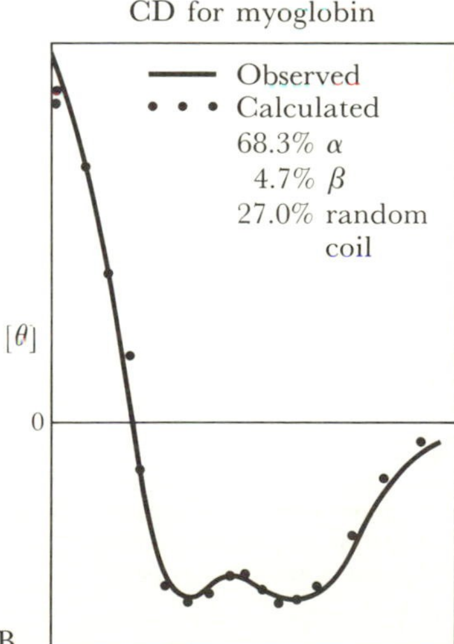

Figure 16-10
A. ORD and CD spectra for poly-L-lysine in the α-helical, β, and random-coil conformations. B. Observed CD curve for myoglobin and a curve calculated from the poly-L-lysine data of part A. [From N. Greenfield and G. Fasman, *Biochemistry* 8(1969):4108–4115.]

exist and that the amino acid side chains have no effect on the spectra, then it is possible to calculate the expected curve for a protein containing a mixture of the three conformations by simple graphic addition. An example is shown in Figure 16-10B, which shows the observed and calculated CD curve for myoglobin (the structure of which is accurately known from X-ray studies). As can be seen, the analysis is fairly good, although the fit is not exact. Such graphic addition for calculating a curve for a protein *with known structure* is simple. However, the more interesting problem of determining the fractional contribution of the three forms in a protein of unknown structure is a substantial task that must be done by curve fitting, using electronic computers. This is done by continually adjusting the fraction of each in the summation until the observed curve is obtained.

The problem with such an analysis is that, as noted for myoglobin, the calculated curve is rarely a perfect fit. As might be expected, this is due to violations of the assumptions stated in the preceding paragraph. The fact is that the spectra are somewhat affected by the presence of aromatic side chains, disulfide bridges, and prosthetic groups, and by chain length, and the peptide bond may exist in conformations other than α, β, or random coil* (e.g., the various kinds of β turns described in Chapter 1). The observations leading to this realization are the following.

1. The spectra for proteins claimed to be "true" α helices are not all the same; this is caused by a small effect of nonaromatic side chains on the rotatory strength of the peptide bond and by an occasional distortion of the α helix due to hydrogen bonding between two amino acids separated by *one* amino acid rather than three (the α-helical situation).

2. Long-chain homopolypeptides do not have the same rotational strength per residue as short ones. This is a problem because α-helical regions of proteins are usually from three to twenty amino acids in length; yet all model compounds that have α helicity are very long.

3. The random-coil "standards" are probably not totally in the random-coil configuration. This is also the case for random-coil segments of proteins, which probably contain β turns.

4. By X-ray diffraction, structures have been found in proteins that are unknown in the synthetic polypeptides used as model compounds.

5. Phenylalanine, tyrosine, histidine, and tryptophan side chains can contribute to CD spectra if in certain configurations. Polymers of

* It is also probably the case that no part of a native protein is ever in a true random-coil configuration because of a variety of subtle interactions between the side chains and because short regions of a protein chain bounded by structured regions can never have complete flexibility because the ends of the short regions are not completely free to move.

these amino acids give spectra that are different from the spectra of polypeptides of nonaromatic amino acids.

6. Proline and glycine can form left-handed helical structures, as in collagen.
7. Cysteine disulfide bridges give CD bands, like those in insulin and ribonuclease, which cannot be due to peptide absorption.
8. Nonprotein prosthetic groups (e.g., heme in hemoglobin) affect spectra.

By this time, the reader will have certainly decided that ORD and CD analyses are best dispensed with insofar as determining protein structure is concerned. However, as Figure 16-10 illustrates for myoglobin, these analyses are useful in some cases. In fact, for a general description of the helical content of proteins the ORD-CD analysis is adequate, as shown in Table 16-2. Furthermore, recent studies using CD analysis have begun to clarify some of the problems and to yield information that will make spectral features more easily interpreted. The important point to realize is that an experimenter cannot always defer experimentation until precise X-ray analysis is available, and the information gained from CD studies, if coupled with other methods such as fluorescence, nuclear magnetic resonance, and sedimentation, can often lead to a reasonable description of the conformation of a macromolecule.

The next important use of ORD and CD in the study of proteins is to assay conformational *changes* for which it is exquisitely sensitive, as will be seen in the following section.

Table 16-2
Helical Content of Proteins Determined by CD Compared with X-ray Analysis.

Protein	α helix (%) by CD	α helix (%) by X-ray analysis
Myoglobin	77	77
Lysozyme	29	29
Ribonuclease	18	19
Papain	21	21
Lactic dehydrogenase	31	29
α-Chymotrypsin	8	9
Chymotrypsinogen	9	6

SOURCE: Data from Y. H. Chen, J. T. Yang, and H. M. Martinez, *Biochemistry* 11(1972): 4120–4131.
NOTE: Similar data for β structure show poorer agreement because the CD analysis does not always adequately detect the various kinds of β turns.

Assay of Changes in Conformation by CD Analysis

Although CD analysis rarely gives absolute information about structure and it is always necessary to make comparisons with standards whose structures are known from X-ray diffraction, CD is extraordinarily sensitive to *changes* in conformation—if a CD spectrum changes in any way, there must be a conformational change. Hence, even if the structure is not known at all and the CD curve is virtually uninterpretable, CD analysis can still be a sensitive assay for any interaction or agent that causes a conformational change. Furthermore, this can be made semiquantitative by looking at changes of particular features of a spectrum. Hence, experiments can be designed with a CD change as the variable. How this is done is best seen by example.

Example 16-A □ Changes in enzyme structure caused by substances that bind to the enzyme.

The CD spectra of many enzymes change when the enzyme interacts with its substrate, a coenzyme, or an inhibitor. This can be used as an assay of binding. For example, binding constants can be determined by measuring rotational strength as a function of concentration of the bound molecules. Because each bound molecule will change R by a definite amount, the number of bound molecules can be determined from the magnitude of the total change. In cases in which the changes are small, they can be amplified by introducing at or near the active site an optically active chromophore, or one that becomes optically active when coupled to the protein, and studying the changes in CD of that chromophore. Active sites can also be identified in this way by introducing a chromophore at various sites and determining at which site the CD is affected by substrate binding. This is much like the reporter-group method of absorption spectroscopy (Chapter 14). Suppose that it is known that there is a histidine in the active site but that there are four histidines in the protein. A chromophore can be introduced (often with difficulty) next to one of the histidines and the CD of the chromophore can be examined before and after substrate binding. If there is no effect on the CD of the chromophore when adjacent to histidines 1, 2, and 4, but there is for 3, this can be taken as evidence that 3 is in the active site.

Example 16-B □ Protein denaturation.

The denaturation of proteins is always accompanied by CD changes, indicating the loss of α and β structure and the enhancement of the random-coil spectral components. Hence, denaturation can be followed by plotting ellipticity at a particular wavelength as a function of the denaturing conditions. Figure 16-11 shows an example for the thermal denaturation of myoglobin.

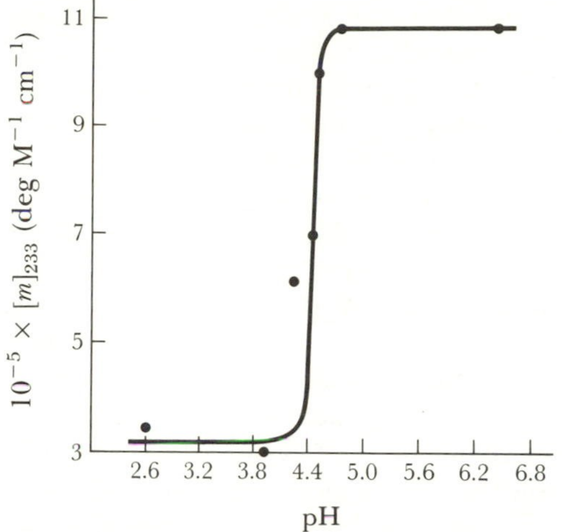

Figure 16-11
Detection of the denaturation of myoglobin by acid titration, using specific rotation at 233 nm as an assay. [From P. Appel and W. D. Brown, *Biopolymers* 10(1971): 2309–2322.]

☐ Studies of the folding of proteins. **Example 16-C**

The forces responsible for determining the three-dimensional structure of a protein are of great interest. This has frequently been studied by examining helix-coil transitions. CD adds another dimension to this analysis if the amino acid sequence is known, because it is possible to look directly at the environment of aromatic amino acids and of disulfide bridges because they have CD bands at characteristic wavelengths. Hence, it is possible to detect changes in the neighborhood of particular amino acids caused by the destruction of hydrophobic forces, or hydrogen bonds, the titration of particular amino acids, and so forth. An interesting example is that of a tyrosine in pancreatic ribonuclease far removed from the active site of the enzyme—its CD is changed when the inhibitor 3′-cytidylic acid, which binds to the active site, is added. This shows that binding to one part of a protein can cause a conformational change in a distant region.

ORD and CD Measurements for Polynucleotides and Nucleic Acids

For all polymers containing nucleotides, the optically active groups that are observable are the purines and the pyrimidines because the bonds in the sugars and the phosphoester linkages do not absorb in the wavelength range that is usually studied. Purines and pyrimidines themselves are examples of symmetric chromophores that become somewhat optically active when attached to a sugar by means of an *N*-glycosidic bond; furthermore, the optical activity increases substantially when they become part of a helical structure.

ORD and CD are extraordinarily sensitive probes of conformational changes in polynucleotides. For example, for many years, the most commonly used assay of conformational change in polynucleotides has been what is known as hypochromicity—the decrease in optical density that accompanies the formation of an ordered structure (see Chapter 14). The optical density at 258 nm of polyadenylic acid [poly(A)] is 26% smaller than that of the monomer adenylic acid. However, the CD for the same substances changes, as shown in Figure 16-12. Note the reversal of the sign and the increase in the value of θ at 260 nm. Clearly, these are huge changes compared with the change in absorbance. Changes such as these can be used to study (1) the loss of helicity of single-stranded polymers by various agents such as high temperature or extremes of pH [note in Figure 16-12 the large change in the CD spectrum of poly(A) when denatured]; (2) the transition from single- to double-stranded polynucleotides and vice versa; (3) structural changes introduced by the binding of cations, peptides, proteins, and so forth; and (4) the effect of the charging of tRNA with an amino acid.

For single-stranded polynucleotides, a reasonable theory (that of Ignacio Tinoco) exists relating ORD and CD spectra to structure. This theory has been nearly confirmed in all of its details and actually allows the calculation of a CD spectrum from molecular structure with some accuracy. The most important element of the theory is that the optical

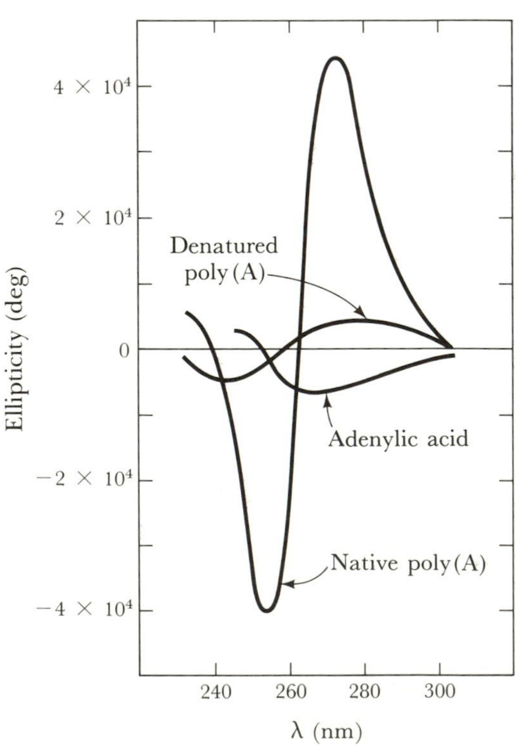

Figure 16-12
Circular dichroism of adenylic acid, denatured polyadenylic acid [poly(A)], and native poly(A). Note the change in sign when adenylic acid is polymerized and the great increase in ellipticity when in the native structure.

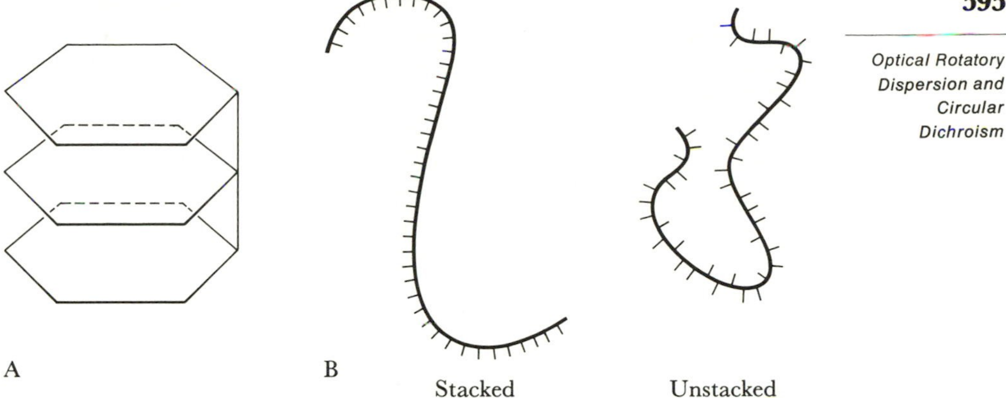

Figure 16-13
A. Three stacked bases. B. Structures of stacked and unstacked polynucleotides. The stacked polynucleotide is more extended because the stacking tends to decrease the flexibility of the molecule.

activity of oligonucleotides is a result of an interaction between two *adjacent bases—the nearest neighbors*—that are *stacked* one above the other (Figure 16-13). Second neighbors (i.e., two bases separated by the nearest neighbor of each) do not make a significant contribution. This has the result that the CD spectra of all nearest-neighbor pairs are different so that it is possible to calculate by direct summation the CD spectrum of a single-stranded polynucleotide if the nearest-neighbor frequencies are known.

The theory for double-stranded polynucleotides is not yet complete, although progress is rapidly being made. For double-stranded DNA, the CD spectra seems to be independent of base composition in the wavelength range usually studied, although there is such an effect in the far (vacuum) ultraviolet.

Examples of ORD and CD Analysis in Studies of Polynucleotide and Nucleoprotein Structure

Several examples of the use of ORD and CD in determining the structural features of polynucleotides follow.

Example 16-D

☐ Structure of polycytidylic acid, poly(C).

The ORD spectrum of poly(C) shows $[m]_{292} = 35{,}160$ deg M^{-1} cm^{-1}, which by rules 2 and 3 (Table 16-1) means that poly(C) is very asymmetric—for polynucleotides, this implies helicity. The forces responsible for the helicity are easily identified by spectral analysis. For example, the addition of formaldehyde–which titrates amino groups and thereby eliminates hydrogen bonding–leaves $[m]_{292}$ virtually unchanged. Hence, the helicity is not stabilized by hydrogen bonds. However, in 90% ethylene glycol, a reagent known to interfere with interactions between nonpolar (hydrophobic) groups, there is $[m]_{292} = 7223$ deg M^{-1} cm^{-1}, a fivefold decrease; this means that the asymmetry (i.e., the helicity) is sharply reduced. The $[m]_{292}$ value of the monomer, cytidylic acid, is approximately 8000 deg M^{-1} cm^{-1} in both H_2O and 90% ethylene glycol. Hence, the structure is stabilized by the so-called hydrophobic forces. Stabilization by a hydrophobic force means that the interacting molecules tend to become sufficiently near that water is excluded; hence, this result suggests that the bases are arranged one above the other and that they are relatively near—that is, the stacked-base conformation (Figure 16-13B).

Suppose that the effect of these reagents was examined instead by simply determining absorbance changes. The OD_{260} increases by about 10% when formaldehyde is added, but this could be due to the loss of helicity or to a change in the absorbance of the cytosine itself (because formaldehyde titration of free bases causes an increase in absorbance). Hence, no statement about the role of hydrogen bonding could be made. The OD_{260} increase when ethylene glycol (a weakly polar solvent) is added is 30%. This would suggest hydrophobic effects but, without the ORD results, that this is an effect on helicity could not be proved.

Example 16-E ☐ Strong evidence for base stacking.

Di- and trinucleoside phosphates show *strong* ORD and CD bands compared with the monomers, indicating that even short oligomers have a very asymmetric and probably a helical conformation. The rotational strength is vastly reduced by organic solvents so that in aqueous solution there must be helical arrangement stabilized by hydrophobic forces, as in Example 16-A.

Tinoco's theory provides a basis for understanding these data because CD spectra of dinucleotides agree almost precisely with those calculated on the basis of the stacked arrangement. From the magnitude of $[M]$ and $[\theta]$, the fractions of the bases that are stacked under various conditions can also be determined.

Uracil oligonucleotides show little or no optical activity; hence they are rarely in an asymmetric configuration so that uracil does not stack.

The studies on stacking confirmed and actually proved an early hypothesis that hydrophobic interactions are essential in stabilizing the double-stranded DNA structure. Many studies involving hyperchromicity and viscosity measurements in solvents of different polarity had

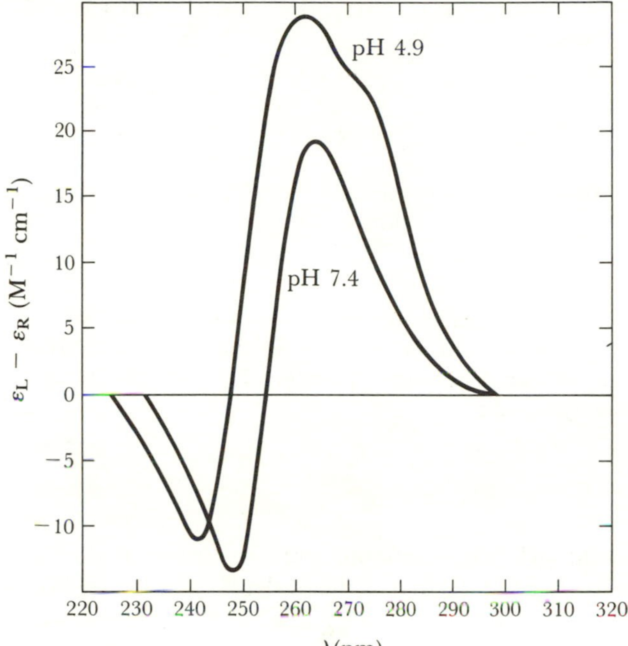

Figure 16-14
CD spectra of poly(A) at acid and neutral pH. [From J. Brahms, *Nature* (*Lond.*) 202(1964):797–789.]

shown that the rigid character of double-stranded DNA was lost as polarity decreased. The ORD and CD studies proved that the absorbance changes were due to changes in helicity.

Example 16-F

☐ Structure of various forms of polyadenylic acid, poly(A).

Single-stranded poly(A) at neutral pH has very strong CD bands and therefore possesses a highly ordered structure (Figure 16-14). A possible structure is a double-stranded helix held together by hydrogen bonds between the 6-amino group and the OH group of the ribose on the opposite strand. However, poly-(N^6-hydroxyethyladeninylic acid), which cannot form hydrogen bonds (Figure 16-15), has a similar CD spectrum; hence the hydrogen bond is not likely to be a part of the structure and the structure is probably not a double helix. Because the CD of the dinucleotide is that predicted for stacking and because, with increasing numbers (as many as 7 or 8) of A in oligo(A), the rotational strength of the CD band increases, poly(A) at neutral pH is probably an extended semirigid helical molecule whose structure is stabilized by base stacking.

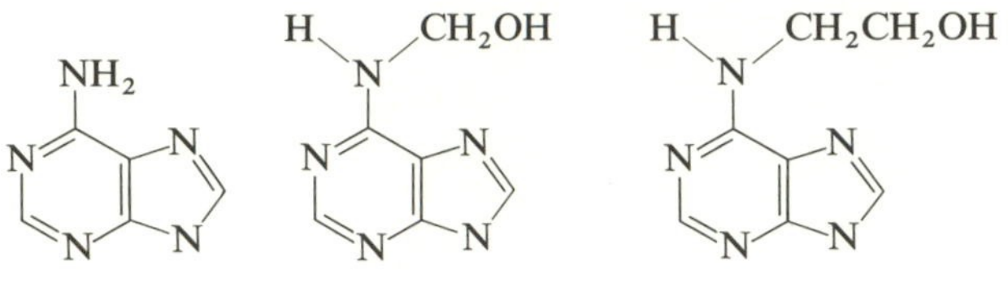

Adenine

Formylated adenine

N^6-Hydroxyethyladenine

Figure 16-15
Chemical structures of adenine, adenine reacted with formaldehyde (formylated adenine), and N^6-hydroxyethyladenine. Formylation blocks the NH group of adenine that participates in hydrogen bonding. This group is already blocked in N^6-hydroxyethyladenine.

At acid pH, the rotational strength (area) of the CD band is 1.5 times as great as that at neutral pH (Figure 16-14), indicating greater asymmetry at acid pH. Furthermore, λ_0 for the CD band shifts toward shorter wavelength compared with that at neutral pH and approaches λ_{max} of the simple absorption spectrum of a double-stranded helix. By rule 4 (Table 16-1), this suggests that poly(A) is also a double-stranded helix at acidic pH. Poly-(N^6-hydroxyethyl-adenylic acid) does not show either of these CD changes when shifted to acid pH. Because this polynucleotide cannot form hydrogen bonds, it is likely that the poly(A) structure at acid pH requires hydrogen bonding. This evidence suggests that acidic poly(A) is a hydrogen-bonded, double-stranded helix. This agrees with the results of X-ray analysis of poly(A) fibers prepared from acidic solution (i.e., that it is a double-stranded helix) and confirms that the structure determined by X-ray analysis also exists in solution.

Example 16-G ☐ Structure of the poly(A)·poly(U) complex.

If poly(A) is mixed with equimolar amounts of poly(U), a polymer is formed whose rotational strength is less than that obtained by summation of the CD spectra of poly(A) and poly(U). This lack of simple additivity shows that the spectrum of at least one of the components has changed (rule 1, Table 16-1). Furthermore, λ_0 is shifted to shorter wavelengths, as it is for acidic poly(A). Hence, again by rule 4, a reasonable conclusion is that poly(A)·poly(U) is a double-stranded helix. Note the simplicity of this observation (Figure 16-16).

Example 16-H ☐ Structure of single-stranded polydeoxynucleotides.

The polydeoxynucleotides generally show the same spectral features as the ribose forms except that the rotational strength is much lower.

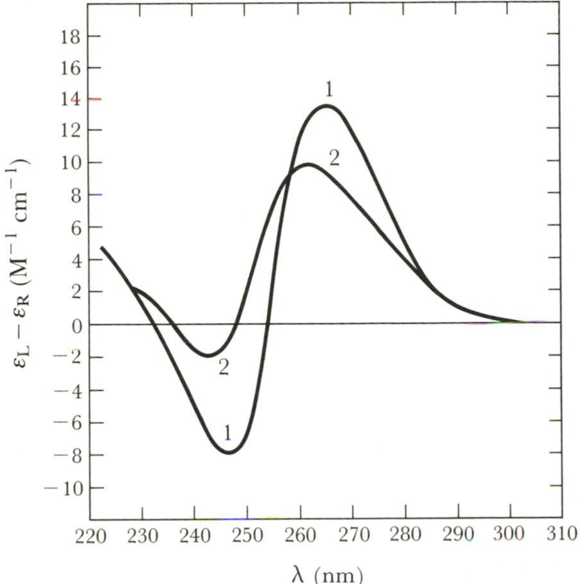

Figure 16-16
Circular dichroism of the poly(A)· poly(U) complex. Curve 1 is the sum of the spectra of poly(A) and poly(U) measured separately. Curve 2 is for the complex. Note that curve 2 is shifted to shorter wavelengths and that there is less negative CD. This is a characteristic of double-stranded polymers. [From J. Brahms, *J. Mol. Biol.* 11(1965): 785–801.]

Hence, these structures are less helical and probably less stacked. Furthermore, the OH group present in ribose but not in deoxyribose must somehow be a part of the helical structure.

CD spectra are of great value in studying nucleoproteins. The bands of protein and nucleic acids overlap but in the longer-wavelength region of DNA, there is no contribution. Hence, conformational differences between free nucleic acid and the nucleic acid in the complex can be monitored. This procedure has been used to study the properties of ribosomes and of chromatin, a complex of DNA and various histone proteins that is contained in chromosomes. Some of these experiments are described in the following examples.

☐ The structure of RNA in ribosomes. **Example 16-I**

The helix-coil transitions (see Example 14-D and Figure 14-15) of pure ribosomal RNA and *E. coli* ribosomes are indistinguishable. This suggests that there is little or no difference in the intrastrand hydrogen bonding of the RNA in the two forms. However, because of the insensitivity of absorbance measurements to small changes in structure, both ORD and CD spectra were also obtained. Again little or no difference was observed between RNA and ribosomes, which suggests that if there are conformational differences, they must be very small. Changes in the degree of helicity of the RNA would have a substantial effect on the spectra so that structural models, in which an RNA molecule is wrapped around protein molecules, can be eliminated. These observations have led to the view that ribosomes are assembled by proteins building onto the RNA.

Example 16-J ☐ Conformational changes associated with DNA-histone interactions in chromatin.

A very active current field of research is the study of the structure of chromatin. Several important conclusions came from a comparison of the CD spectra of free DNA and of chromatin. X-ray diffraction analysis suggested that there may be a small difference between the conformation of DNA in chromatin and of free DNA. The CD spectra showed that the CD band for chromatin has about half the maximum ellipticity and half the area of the spectrum of free DNA. Since the CD spectrum of DNA is primarily a result of base stacking, it was concluded that the DNA is less stacked in chromatin. This simple analysis does not give any detailed information about chromatin structure; however, CD analysis does provide a convenient tool for rapidly studying the conditions for interaction between histones and DNA. Indeed X-ray diffraction could do the same but the great length of time required to analyze each interacting mixture by this method clearly precludes the study of the effects of various parameters on the interaction—for example, pH, ionic strength, DNA and histone concentrations, and the ratio of DNA to histone. This was easily done with CD spectra. Note that the CD spectrum is not used to determine chromatin structure but only as an assay of binding, indicated by a conformational change of the DNA.

There are five major histones in animal cells. Each of these was allowed to interact with DNA singly, in pairs, in threes, and so forth. The CD spectra not only showed which histone is bound and in what combination but also showed (from a measure of the maximum magnitude of the ellipticity and the value of λ_{max}) that binding of different histones produces several conformational changes.

SELECTED REFERENCES

Adler, A. J., and G. D. Fasman. 1967. "Optical Rotatory Dispersion as a Means of Determining Nucleic Acid Conformation," in *Methods in Enzymology*, vol. 12B, edited by L. Grossman and K. Moldave, pp. 268–302. Academic Press. Dated but excellent.

Adler, A. J., N. J. Greenfield, and G. D. Fasman. 1973. "Circular Dichroism and Optical Rotatory Dispersion of Proteins and Polypeptides," in *Methods in Enzymology*, vol. 27D, edited by C. H. W. Hirs and S. N. Timasheff, pp. 675–735. Academic Press. An excellent, up-to-date review.

Bloomfield, V. A., D. M. Crothers, and I. Tinoco. 1974. *Physical Chemistry of Nucleic Acids*. Harper and Row.

Brahms, J., and S. Brahms. 1970. "Circular Dichroism of Nucleic Acids," in *Fine Structure of Proteins and Nucleic Acids*, vol. 4, edited by G. D. Fasman and S. N. Timasheff, pp. 191–270. Dekker. An excellent review.

Bush, C. A. 1974. "Ultraviolet Spectroscopy, Circular Dichroism, and Optical Rotatory Dispersion," in *Basic Principles in Nucleic Acid Chemistry*, vol. 2, edited by P. P. O. Ts'o, pp. 92–169. Academic Press.

Charney, E. 1979. *The Molecular Basis of Optical Activity*. Wiley.

Gratzer, W. B., and D. A. Cowburn. 1969. "Optical Activity of Biopolymers." *Nature* (*Lond.*) 222:426–431.

Jirgensons, B. 1969. *Optical Rotatory Dispersion of Proteins and Other Macromolecules*. Springer. Dated but good.

Kauzman, W. 1957. *Quantum Chemistry*. Academic Press. This contains both the classical and the quantum mechanical theory of optical activity.

Mommaerts, W. F. H. M. 1967. "Ultraviolet Circular Dichroism in Nucleic Acid Structural Analysis," in *Methods in Enzymology*, vol. 12B, edited by L. Grossman and K. Moldave, pp. 302–329. Academic Press. A good description of both experiment and theory.

Schellman, J. A. 1975. "Circular Dichroism and Optical Rotation." *Chem. Rev.* 75:323–331.

Sears, D., and S. Beychok. 1973. "Circular Dichroism," in *Physical Principles and Techniques of Protein Chemistry*, vol. C, edited by S. J. Leach, pp. 446–593. Academic Press. Detailed discussion of the theory of CD.

Tinoco, I., and C. R. Cantor. 1970. "Applications of Optical Rotatory Dispersion and Circular Dichroism to the Study of Biopolymers," in *Methods of Biochemical Analysis*, vol. 18, edited by D. Glick, pp. 81–203. Interscience.

PROBLEMS

16–1. Figure 16-2 shows how right circularly polarized light is generated. How should the figure be altered to show the production of left circularly polarized light? Of elliptically polarized light?

16–2. If two substances are found whose absorption spectra are identical and whose CD curves are identical except that one curve is positive and the other negative, what can probably be said about the structural relation between the two substances?

16–3. The structure of a protein has been determined by the X-ray diffraction of a protein crystal. It is found to contain 31% α helix, 58% β sheet, and 11% random coil. From CD analysis, the values are 60% α helix, 35% β sheet, and 5% random coil. What can you conclude about the structure?

16–4. Draw the expected and observed CD curves for the protein of Problem 16-3.

16–5. Why are amino acid side-chain effects easier to detect by CD than by ORD?

16–6. If a rotation of $\pi/12$ radians is observed for a certain substance, how can you be sure that it is not really $4\frac{1}{12}\pi$ radians?

16–7. In what wavelength range would you expect to find the CD spectrum for tyrosine in a protein?

16–8. Ethidium bromide is known to intercalate between the base pairs of DNA. This binding reduces the rotational strength of the CD curve for the DNA and substantially increases it for the ethidium bromide. Why?

16–9. The CD spectra of native DNA in 0.01 M NaCl and in 0.5 M NaCl are only slightly different. After boiling the DNA (to denature it), the rotational strength decreases substantially. In which solvent would the decrease be greater? Explain.

16–10. The CD spectra of a variety of phages differ markedly from that of free DNA. Furthermore, the spectra differ from one phage to the next. What can you conclude about the structure of DNA in phages?

16–11. The rotational strength of oligonucleotides increases with the number of nucleotides up to about 10; after that the changes are insignificant, if present at all. Explain.

16–12. The CD spectra of polyamino acids are not all the same—both rotational strength and λ_{max} differ. What are some of the factors that account for this?

16–13. A small molecule is added to a protein solution and the CD spectrum of the protein changes. What conclusions can you draw?

16–14. A DNA sample and a protein sample are mixed. In mixtures A and B, the NaCl concentrations are 0.01 M and 1 M, respectively. In sample A, the CD spectrum of the DNA is changed but that of the protein is not. In sample B, the CD spectrum of neither is changed. The CD spectra of pure DNA in 0.01 M NaCl and 1 M NaCl are the same. What can you conclude?

16–15. DNA molecules and histones are known to interact to form chromatin. The CD spectrum of DNA changes in a characteristic way when chromatin forms. Someone observes that in 0.01 M NaCl + 0.01 M $CaCl_2$, the CD spectrum of chromatin is different from that in 0.01 M NaCl. The DNA portion of the CD spectrum resembles, but is not identical to, that of free DNA. Someone hypothesizes that the Ca^{2+} ion has a specific effect on chromatin structure. To test this, CD spectra are obtained for chromatin in various solvents. The ellipticity at 260 nm is plotted as a function of molarity, as shown below. Is there a specific effect of the Ca^{2+} ion?

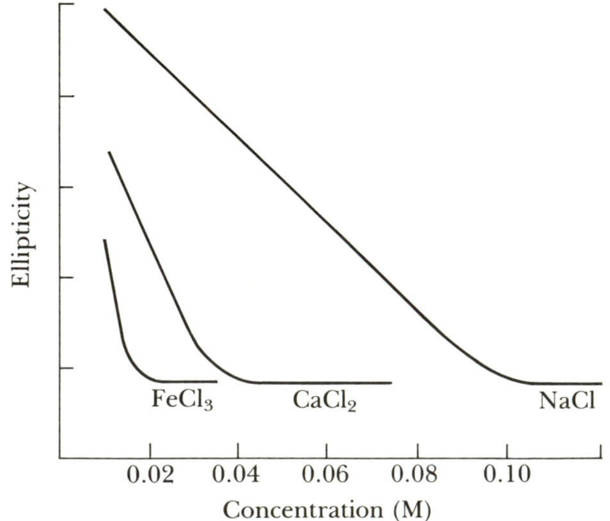

17

Nuclear Magnetic Resonance and Electron Spin Resonance

Nuclear magnetic resonance (NMR) is another spectroscopic method capable of yielding information about the structure of biological polymers, about interactions between molecules, and about molecular motion. Its special advantages are that (1) the theory is sufficiently good that, in principle, the detailed arrangement of individual atoms could be calculated from NMR spectra; (2) hydrogen atoms (which are beyond the resolution of X-ray diffraction analysis in large molecules) can be located; and (3) different atoms (e.g., H, N, C, and P) can be examined separately. For small molecules (e.g., molecular weight <500), NMR has been highly successful in determining structure. However, for macromolecules, the potential has not yet been realized because of the huge number of spectral lines, which are often poorly resolved and for which identification of the atom producing the line is frequently very difficult, and because of the large number of possible interactions for each atom.

How to approach the problem of structure determination is an important part of this chapter. Unfortunately, the discussion will be limited because of the great complexity of the subject. The intent is that the reader will be able to understand some of the biochemical literature employing NMR and will in the future know when to turn to NMR for aid in the solution of biochemical problems.

It will also become apparent that another difference between NMR and other methods for studying macromolecules is that the limitation of NMR is not primarily the complexity of the theory but often in instrumentation. This point is made because it seems that the great advances in the utility of the method have come principally from improvements in instrumentation.

For the present purpose, a nucleus consisting of a single proton (i.e., hydrogen will be considered although most of the concepts apply to more complex nuclei. In addition to charge and mass, a proton possesses angular momentum, or *spin*. A spinning charge generates a magnetic field and can be thought of as a tiny bar magnet oriented along the spin axis. The strength of the magnetic field is expressed as a magnetic moment, μ. Like a bar magnet, which has a north and south pole, μ has a direction. In an external magnetic field, a macroscopic bar magnet will become oriented with its magnetic moment along the lines of magnetic force of this field. However, in dealing with atomic and nuclear particles, the rules of quantum mechanics come into play, and it is found that, if a proton is in a magnetic field of strength H_0, the proton magnetic moment assumes one of two angles with respect to the direction of H_0—that is, the angles θ and 180—θ, in which θ is in degrees (Figure 17-1). These two orientations of the magnetic moment are referred to as being *aligned with and against* H_0, respectively. The potential energy of each of the orientations is $-\mu H_0 \sin \theta$ and $+\mu H_0 \sin \theta$, respectively, so that the difference in the energy levels is $2\mu H_0 \sin \theta$. As discussed in Chapter 14, transitions between electronic and vibrational levels can occur by the absorption of electromagnetic radiation (in the ultraviolet to infrared range) having energy equal to the difference (ΔE) between the two energy states. This is also true of nuclear energy levels except that the radiation

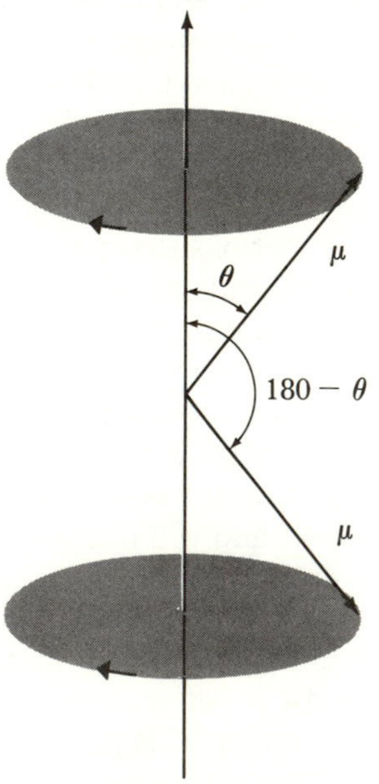

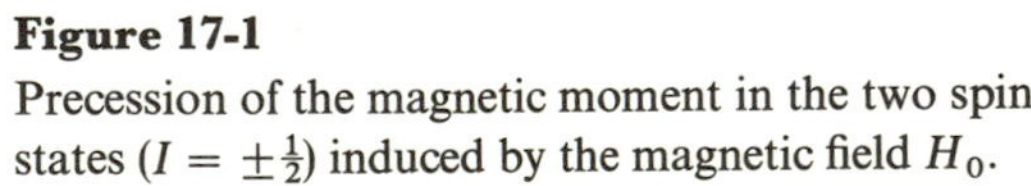

Figure 17-1
Precession of the magnetic moment in the two spin states ($I = \pm\frac{1}{2}$) induced by the magnetic field H_0.

is in the radiofrequency range. Because $\Delta E = h\nu$, in which ν is the frequency of the absorbed radiation and h is Planck's constant,

$$\nu = \frac{2\mu H_0 \sin\theta}{h} \tag{1}$$

In the jargon of NMR workers, absorption of radiation of this frequency causes the proton to *flip* from one orientation to the other.

To generalize to nuclei other than the single proton requires making use of the quantum mechanical spin number, I, which is a characteristic of each nucleus. (The factors determining the values of I can be found in the Selected References near the end of the chapter.) The number of different orientations and therefore energy levels of a given nucleus in a magnetic field is $2I + 1$, in which I is always an integral multiple of $\frac{1}{2}$. Hence, for protons, which have two levels, $I = \frac{1}{2}$. (In this chapter, only nuclei with $I = \frac{1}{2}$ will be considered.) The magnitude of μ also varies with the nucleus. It is difficult to predict the values of I and μ for a given nucleus so that in all cases they have been determined by actual measurement.

The parameters for several nuclei of interest to biologists are shown in Table 17-1. Nuclear energy levels differ from electronic and vibrational levels in that they are equally spaced and transitions can occur only between adjacent levels. This means that, *for a given nucleus, there is only a single transition frequency for each value of H.*

An understanding of the resonance phenomenon can also be gained by considering the theory from the viewpoint of classical electrodynamics.

Table 17-1
NMR data for Nuclei of Importance in Biological Systems.

Isotope	Natural abundance (%)	Spin, I	Sensitivity*	Relative sensitivity†	Approximate range of chemical shifts (ppm)
^{1}H	99.98	$\frac{1}{2}$	1.000	1.00	12
^{2}H	0.0016	1	0.0096	1.5×10^{-6}	12
^{13}C	1.1	$\frac{1}{2}$	0.016	0.0002	350
^{15}N	0.37	$\frac{1}{2}$	0.001	0.00037	1000
^{19}F	100	$\frac{1}{2}$	0.83	0.83	500
^{31}P	100	$\frac{1}{2}$	0.066	0.066	700

* Relative to ^{1}H, in terms of equal number of nuclei.
† Product of (natural abundance) × (sensitivity). For ^{2}H, ^{13}C, and ^{15}N, in which enrichment is possible (to >99% for ^{2}H and ^{15}N), the sensitivity is the more meaningful number.

If a spinning nucleus is placed in a magnetic field, H_0, the basic equations of electromagnetic theory show that the magnetic moment of the nucleus is subjected to a torque. Because the nucleus is spinning and therefore possesses angular momentum, the law of conservation of angular momentum requires that the magnetic nucleus precess about the field direction in the same way that a gyroscope responds to a gravitational field (Figure 17-1). Note that θ is not changed. The angular frequency of precession, ω_0 (called the Larmor frequency), is

$$\omega_0 = \frac{2\mu}{h} H_0 \sin\theta \qquad (2)$$

To flip the nucleus from one orientation to another, θ must be changed to $180 - \theta$. Increasing H_0 cannot do this because this increase affects only ω_0 and increases the rate of precession. However, if a small magnetic field, H_1, is added perpendicular to H_0 (Figure 17-2), the magnetic dipole will then be subjected to a second torque, tending to induce precession about this field direction. If $H_1 \ll H_0$, this torque is much smaller than that due to H_0 and, because the magnetic moment is precessing and therefore continually changing direction with respect to H_1, the torque due to H_1 will have little effect on the orientation of μ. To be effective H_1 would have to act continuously and therefore must also rotate in the plane indicated in Figure 17-2 and *at the same frequency as the rotating dipole*—that is, ω_0. If H_1 rotates at a different frequency, the force acting on the dipole will be continually changing in magnitude and direction and the precession due to H_0 will experience only a periodic wobbling (called nutation, with a gyroscope). Because there is only a finite number of orientations of μ with respect to H_0 (i.e., because the system is quantum mechanical), only if the frequency of oscillation of H_1 is ω_0 (the so-called resonance condition) can energy be transferred to reorient the magnetic

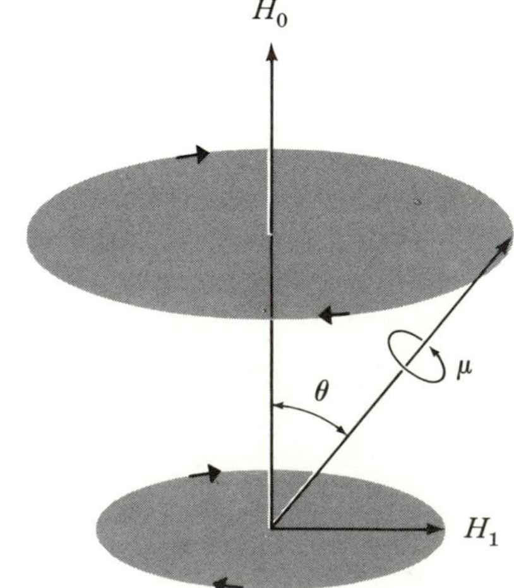

Figure 17-2
The magnetic moment μ rotates about H_0. The perpendicular field H_1 induces a torque on μ; if H_1 rotates at the Larmor frequency it can exert the torque continuously.

moment. Note that this is exactly what is expressed by equation (1) if the oscillating magnetic field, H_1, is simply the magnetic field of the electromagnetic radiation used to raise the nucleus from one energy level to the next.

The Chemical Shift

In the preceding discussion, it has been stated that only one transition can occur for a given nucleus in a particular magnetic field so that the spectrum should consist of only one resonance line. However, this is not the case because the magnetic field seen by the nucleus is not simply the applied field, H_0. For example, all nuclei are surrounded (shielded) by electrons and these electrons are also induced to circulate by H_0. Because moving electrons themselves generate a magnetic field, the nucleus sees an effective field, $H_{\text{eff}} = (1 - \alpha)H_0$, in which α is called the screening constant. Other factors (such as the magnetic field due to nearby nuclei) also affect the value of H_{eff}. What this means is that *the observed resonance frequency depends on the environment.* It is this environmental effect that makes NMR spectroscopy valuable (exactly as in other forms of spectroscopy) because, in a molecule, each nucleus of the same type will have a resonance frequency that depends on the chemical group in which the nucleus resides. For example, the hydrogen protons in a methyl group will have a different resonance frequency from that of an amino hydrogen proton, and, furthermore, the protons of toluene will differ from those of acetic acid. This shift in resonance frequency due to chemical environment is called the *chemical shift.*

In optical spectroscopy, every transition has a defined frequency. However, because nuclear resonance frequency is dependent on H_0, there is no natural basic scale unit. Furthermore, H_0 is difficult to measure accurately. Hence, the following approach has been adopted for reporting the resonance condition or the chemical shift. A reference material is added to the sample and its resonance line is assigned an arbitrary value of H or ν_0 (whichever is varied in the particular procedure used to obtain a spectrum), and chemical shifts are expressed as displacements from this value. An even better system is to use a dimensionless displacement from the reference standard—called parts per million (ppm). If H is varied and ν is constant,

$$\text{chemical shift (ppm)} = \frac{H_s - H_{\text{ref}}}{H_{\text{ref}}} \times 10^6 \qquad (3)$$

in which H_s and H_{ref} are the magnetic fields producing resonance for the sample and the reference material, respectively. If ν is varied,

$$\text{chemical shift (ppm)} = \frac{\nu_s - \nu_{\text{ref}}}{\nu_{\text{ref}}} \times 10^6$$

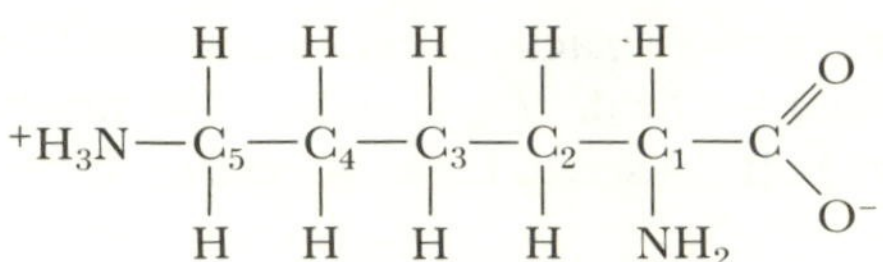

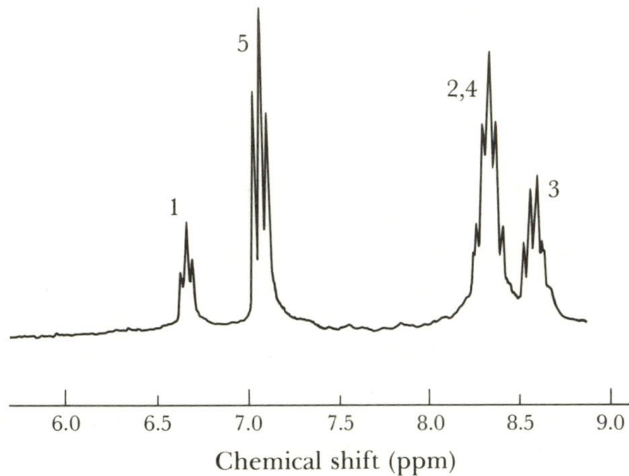

Figure 17-3
NMR spectrum of lysine in D_2O in zwitterion form. The structure of lysine is shown with its carbon atoms numbered. The numbers next to the peak indicate the carbon atom whose protons produce the cluster of lines. [From F. A. Bovey, *High Resolution NMR of Macromolecules*, Academic Press, 1972.]

This dimensionless scale has the advantage that chemical shifts are independent of the actual value of H_0 or the frequency of the radiofrequency signal, and spectra obtained with different NMR spectrometers are comparable. Nonetheless, it is still fairly common to see data reported as a frequency or field shift at fixed field or frequency, respectively.

Figure 17-3 shows an NMR spectrum for protons, clearly indicating the large number of resonances for a single nuclear type and showing the effect of the chemical group in which the proton resides.

Many factors affect the chemical shift, of which the following are important in biochemistry.

Intramolecular Shielding

This effect, which produces shifts on the order of 10–20 ppm, is the type briefly mentioned earlier—that is, nearby electrons are induced to move by the external magnetic field; this generates a field at the nucleus opposite in direction to that of the applied field so that $H_{eff} < H_0$. Somewhat larger effects are found if the nucleus is in an organic ring or a conjugated ring system. In rings, the π electrons are "delocalized" and free to move in a circular path in the plane of the ring. If a magnetic field is applied that is perpendicular to the plane of the ring, these electrons circulate in such a way that a magnetic field that is antiparallel to the applied field is induced in the center of the ring and a parallel field is induced external to the ring (Figure 17-4). This is analogous to what takes place when current flows through a wire. The direction of the resulting chemical shift therefore depends on where the nucleus is located with respect to the carbon atoms of the ring. The shift is in the opposite direction if the applied field is in the plane of the ring so that the chemical shift depends also on orientation. This type of chemical shift is called the *ring-current shift* and occurs in aromatic amino acids, purines, pyrimidines, porphyrins, and flavins.

Paramagnetic Effects

An unpaired electron (e.g., in free radicals or in certain metal ions) produces a large chemical shift because the electron itself has a significant magnetic moment. The shift is from about 20 to 30 ppm and can be detected when the unpaired electron is as much as 10 Å from the nucleus producing the resonance; furthermore, because the magnitude is inversely related to the cube of the distance, these effects can be used as a molecular yardstick.

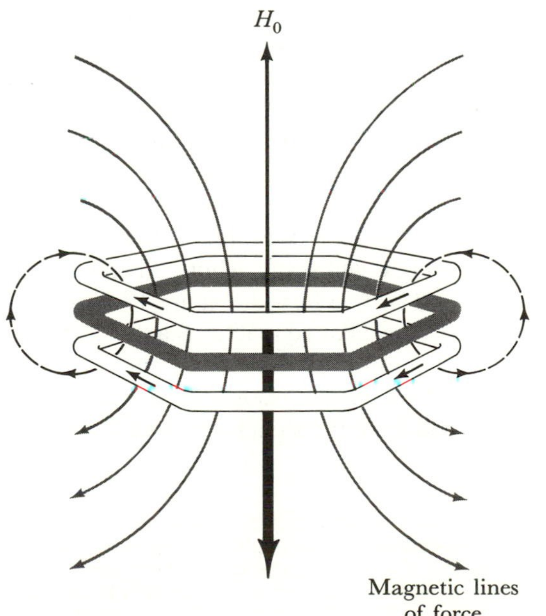

Figure 17-4
Electron density, ring currents, and magnetic lines of force about a benzene ring. The solid thin arrow indicates the applied field H_0. The solid thick arrow is the magnetic field induced by the circulation of electrons. [From R. A. Dwek, *Nuclear Magnetic Resonance in Biochemistry*, Clarendon Press, 1973.]

Intermolecular Effects

When two molecules interact, the electron distribution of one molecule can affect the chemical shifts of nuclei in the other. If the nucleus is in a polymer, such effects can be between different regions or residues that are distant within the linear sequence of molecules but nearby in three-dimensional space. That is, the folding and tertiary structure of a polymer can bring together distant chemical groups, thereby creating new environments so that new chemical shifts occur; these are generally smaller than the intramolecular effects described in the preceding paragraph. These effects are of great importance in biochemistry because they can be used to monitor the binding of a ligand to a macromolecule (e.g., interactions between an enzyme, a cofactor, and a substrate) and to describe the three-dimensional structure of a region of a polymer.

Chemical Exchange

If a nucleus is in a molecule that is rapidly undergoing reversible chemical or physical changes, the environment is continually changing. Hence, the effect on chemical shift depends on the time scale of the exchange. Let us consider a ligand that can be either bound to another molecule or free in solution—clearly, two distinct environments. If the ligand exchanges very slowly between the bound and unbound states, the greater fraction of a population of ligands will be either bound or unbound rather than changing from one state to the other. Therefore, there will be two chemical shifts for a nucleus in the ligand—that of the bound and that of the free environment (Figure 17-5A). If exchange is more rapid, there will be more nuclei entering and leaving each environment and the jumping between the two states causes the energy of the resonance to have an intermediate value.* This will have the effect of broadening the resonance lines (Figure 17-5B). As exchange becomes more rapid, there will be fewer bound or free and more that are jumping between the two states. Hence, the lines corresponding to the two states will not be present, and a single intermediate and broad band will appear (Figure 17-5C). If the exchange is very rapid, all nuclei will be jumping, and a narrow resonance line will result whose chemical shift is the average of that of the bound or free ligand (Figure 17-5D). This is a useful phenomenon for studying exchange reactions and the mechanism of enzyme reactions. For example, the number of nuclei in each state can be determined because the area of a resonance line is proportional to the number of nuclei producing the signal (see page 617).

* A classical but useful way to envision what is occurring is to think about a spin precessing first at one speed when it is in one state and then at another speed when in the other state. Thus, it alternates between moving slowly and rapidly and therefore, over a period of time, long compared to one revolution, it moves at an average and hence intermediate speed.

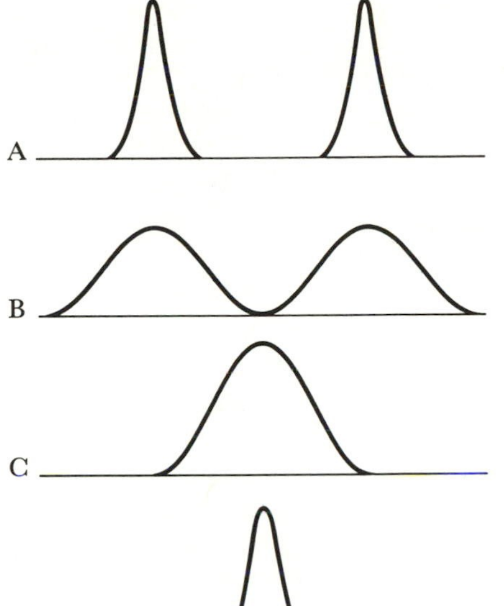

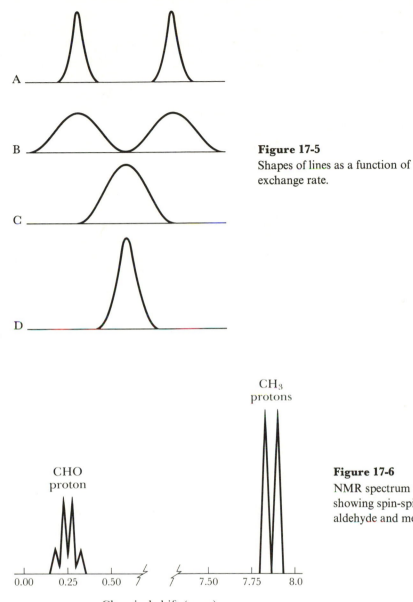

Figure 17-5
Shapes of lines as a function of exchange rate.

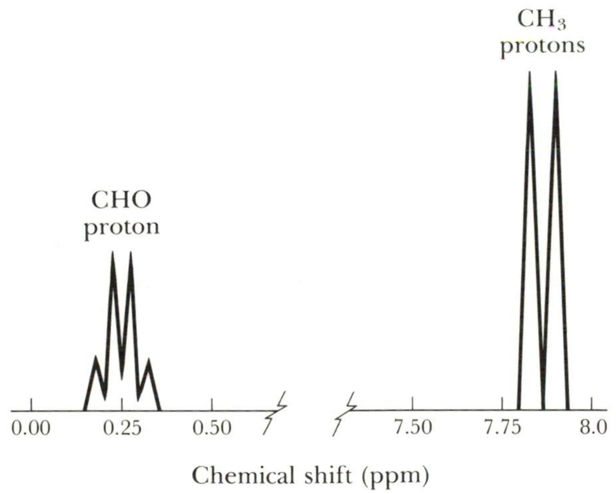

Figure 17-6
NMR spectrum for acetaldehyde, showing spin-spin splitting of the aldehyde and methyl protons.

Spin-Spin Interactions

In all of the preceding discussion, a single proton in a particular environment was assumed to produce a single resonance line. However, a peak will usually appear to consist of a cluster of lines (see Figure 17-6). This is called *line splitting* and results from *spin-spin coupling*. A simplified explanation of this follows. In any atom, the magnetic field of its nucleus will tend to orient its valence electron (which itself has a magnetic moment) so that the electron spin is antiparallel to that of the nucleus. If the atom is in a diamagnetic compound so that its valence electron is paired with an electron of a second nucleus, the valence electron of the second nucleus will be antiparallel to that of the first and therefore parallel to the spin of the first nucleus. The orientation of the second electron affects the orientation of the second nucleus, so that the first nucleus creates a magnetic field at the second. Similarly, the second nucleus affects the first. This mutual interaction is called *spin-spin coupling* and, according to the laws of quantum mechanics, increases the number of resonances. In general, if a nucleus is coupled to n identical nuclei of spin I, its absorption line will be split into $2nI + 1$ components. It is important to realize that spin-spin coupling comes about only if nuclei are coupled by covalent bonds (i.e., by means of paired electrons); it is not a result of simple proximity as in binding, adsorption, or protein folding.

Spin-spin coupling can best be understood by a simple example—that of the spectrum of the protons of acetaldehyde:

$$CH_3\text{—}C\overset{\displaystyle O}{\underset{\displaystyle H}{\lessgtr}}$$

This molecule has two classes of protons—the three methyl protons and one CHO proton. The molecule is small enough for the two sets of protons to be close enough together that they can mutually affect one another and cause splitting. Because each proton has roughly equal probability of having spin $+\frac{1}{2}$ and $-\frac{1}{2}$ [i.e., of tending to be oriented parallel (↑) or antiparallel (↓) to the applied field], the spin arrangements of the protons are like those shown in Figure 17-7. Note that, in the methyl group in which there are three identical protons, there are four distinct situations—all parallel; two parallel and one antiparallel; one parallel and two antiparallel; and all antiparallel. Note that each of the second and third configurations can be produced by three different arrangements of the three spins. Because there are two possible spin

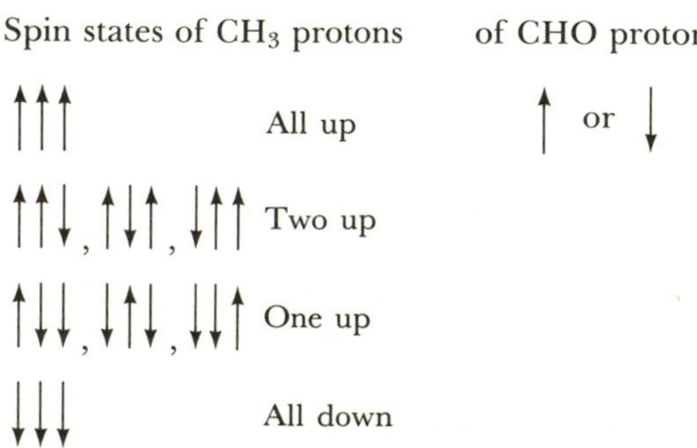

Figure 17-7
Possible spin arrangements in acetaldehyde for the CH_3 protons and the CHO proton.

arrangements of the CHO proton, the CHO proton can alter the field in the region of the methyl protons in two ways. Hence, the CHO proton results in the methyl proton resonance being split into two peaks. Because the parallel and antiparallel arrangements are equally probable, and because the area of a peak is proportional to the number of nuclei producing the signal (see page 617), the areas of the two peaks of the methyl doublet are the same. Conversely, because there are four possible arrangements of the methyl protons, the CHO proton line is split into four components (called a quartet). However, two of the methyl configurations are three times as probable as the others (because they are a result of three possible arrangements); hence, the ratio of the areas of the four components is 1:3:3:1. The spectrum of acetaldehyde is shown in Figure 17-6; the multiplets are easily seen.*

Note that whereas the effect of the CHO proton on the CH_3 protons and the effect of the CH_3 protons on the CHO proton have been considered, we have ignored the effect of the CH_3 protons on one another. Splitting might be expected because each proton is subject to a magnetic field that depends upon the spin of the other protons. However, this splitting does not occur although in fact the energy levels are split. The reason for the lack of splitting is derived from a quantum-mechanical selection rule that forbids transitions between certain energy levels of magnetically equivalent protons. In terms of NMR spectra the rule may be stated as follows. *There is no spin-spin splitting due to the interaction of magnetically equivalent nuclei, that is, nuclei whose chemical shifts are equal.* Thus the same considerations would apply to amino protons.

The presence of multiplets is sometimes a help in identifying the nuclei producing particular lines because the multiplicity defines the properties of adjacent nuclei. For instance, in propionaldehyde the resonances of both the CHO proton and the CH_3 protons would be triplets because of coupling to the two CH_2 protons. However, since there are three CH_3 protons and one CHO proton, the total area of the CH_3 triplet would be three times that of the CHO triplet. The remaining lines would be those corresponding to the CH_2 protons.

Splitting can be eliminated by a method known as *spin-spin decoupling*. With this method highly intense radiation is applied to a molecule at a frequency that can be absorbed by a particular proton. The effect of this treatment is (for reasons that are beyond the scope of this chapter) that splitting of the spin-coupled proton vanishes. That is, if proton A in a particular molecule is responsible for a particular peak, whose frequency is ν_A, and another proton B is responsible for a peak, whose frequency is ν_B, then irradiation of the sample with radiowaves having a frequency ν_A will eliminate splitting of the B resonances *only if protons*

* The statement that a cluster of lines is a methyl quartet often appears in the literature about NMR. It is important to realize that the four lines are not methyl resonances but the splitting of a resonance of another proton because of its being coupled with a methyl group.

A and B are coupled. Therefore, in the example shown in Figure 17-6, the four peaks of the CHO cluster will revert to a single peak if radiation absorbed by the CH_3 protons is applied and the pair of peaks in the CH_3 cluster will revert to one peak if radiation absorbed by the CHO proton is applied.

Spin-spin decoupling is primarily used as a means of identifying the nucleus responsible for a particular resonance. For proteins, it is of greatest value when the amino acid sequence of a protein is known.

Multiplets may also hinder analysis by decreasing resolution. This could be a consequence either of unresolved components having the appearance of a broad line or of components of one cluster of lines overlapping a second cluster, thus making it difficult to distinguish peaks. This is especially true for macromolecules.

Spin-spin coupling can produce spectra having a great deal of information (although the information is sometimes difficult to extract). This can be seen in a simple way by considering the spectrum of propionaldehyde $C_\alpha H_3C_\beta H_2C_\gamma HO$. In this case, the resonance of the proton on the γ-carbon is split into three peaks by the CH_2 group as are the protons on the α-carbon. However, the protons on the β-carbon are coupled to both the CH_3 protons and the CHO protons. The CH_3 protons produce a splitting to four lines and the CHO protons produce a splitting to two lines, as was seen in Figure 17-6. The protons on the β-carbon experience both splittings and the result is 4×2 or 8 lines—that is, each component produced by one mode of splitting is then split in the other mode. Such complex patterns are often a great help in identifying lines. However in a very large molecular having hundreds of lines, multiplets may be such that there is a great deal of overlap of lines, making it difficult to identify lines and to determine their width and area.

The spacing between the components of a multiplet (that is, the *coupling constant J*) depends on the degree of coupling; the stronger the coupling (that is, the closer the nuclei), the greater is the spacing. In favorable circumstances a great deal of information can be obtained from measuring J. The best known use of the value of J is to measure the torsional angle ϕ

H   H
 \C—C/
    ϕ

between protons on adjacent C, N, or O atoms. The relation between ϕ and J is expressed by the following set of empirical equations called the Karplus equations:

$$\text{H—C—C—H}', \quad J = 17\cos^2\phi + 1.1$$

$$\text{H—C—N—H}', \quad J = 12\cos^2\phi + 0.2$$

$$\text{H—C—O—H}', \quad J = 10\cos^2\phi - 1.0$$

in which J is expressed in Hertz. A valuable equation for the peptide protons

$$\begin{array}{c} \mathrm{H}\quad\mathrm{O} \\ -\mathrm{C}\!-\!\mathrm{N}\!-\!\mathrm{C}- \\ \mathrm{H}\quad\phi \end{array}$$

allows one to determine which of the protons are *cis* ($\phi < 90°$) or *trans* ($90° < \phi < 180°$) with respect to the C—N bond. The equations are

$$J = 8.5 \cos^2 \phi, \quad 0 < \phi < 90°$$

$$J = 9.5 \cos^2 \phi, \quad 90 < \phi < 180°$$

There are many complications and assumptions that must be satisfied if the Karplus equations are to be used. More detailed references on NMR should be consulted before they are applied.

Relaxation Processes

At the resonance frequency, nuclear magnetic moments are reoriented. Because the transition from the upper to the lower energy level occurs with the same probability as that from the lower to the upper, equal populations of the nuclei in the two energy levels would result in no net absorption of energy. However, because the difference in energy levels is very small (approximately 10^{-2} calorie or 0.04 joule), the distribution of nuclei in lower and upper levels is strongly dependent on temperature. At absolute zero, all nuclei are in the lower state. In the range usually used for measurement (0–25°C), thermal excitation has raised many to the upper state; however, there is still a small excess of nuclei in the lower state (1 in 10^5). When radiofrequency radiation of the resonance frequency is applied, energy is absorbed and the populations of nuclei in the upper and lower states are equalized. Once equal, no further absorption can be detected. For absorption to be continuous (as indeed it is), there must be a mechanism to restore the initially unequal distribution. The general term for any process that restores a system to its initial state is *relaxation*. In optical spectroscopy (involving electronic and vibrational levels), the status quo is restored either by degradation of the absorbed energy to heat (by collisional processes) or by fluorescence. There are basically two kinds of relaxation processes in NMR: *spin-lattice*, or *longitudinal*, relaxation and *spin-spin*, or *transverse*, relaxation. These complex processes will not be discussed in detail; in rather oversimplified terms, they result from the fact that the *motion* of molecules or the relative motion of parts of molecules produces fluctuating local magnetic fields, causing changes in resonance conditions and the dissipation of energy as heat owing to the induced motion of electrons.

These processes are characterized by relaxation times T_1 and T_2. For the depth of analysis in this chapter, there are only two important results concerning relaxation: (1) $\Delta\nu_{\frac{1}{2}} = 1/\pi T_2$, in which $\Delta\nu_{\frac{1}{2}}$ is the resonance line width at half-height, and (2) T_2 depends on molecular motion. The second point is that rapid motion decreases the ability to transfer energy by spin-spin relaxation (because the nuclei are rarely at the correct relative orientation); hence, with rapid motion, T_2 is long and line width is narrow.

Instrumentation of NMR

An NMR spectrometer needs both a fixed and a rotating radiofrequency magnetic field, a sample holder, and a detector of some sort. The basic instrument is shown in Figure 17-8. The sample is placed in a tube (A) that is situated between the poles (B) of a powerful electromagnet (usually 10^4–10^5 gauss). A coil of wire (C) in a plane perpendicular to the field H_0 of the electromagnet surrounds the sample. A radiofrequency transmitter generates a fixed, high frequency (approximately 10^8 Hertz or cycles per second) oscillating magnetic field in the coil of wire. The frequency in the coil is selected to be near the Larmor frequency corresponding to H_0. To achieve resonance, the magnetic field of a small

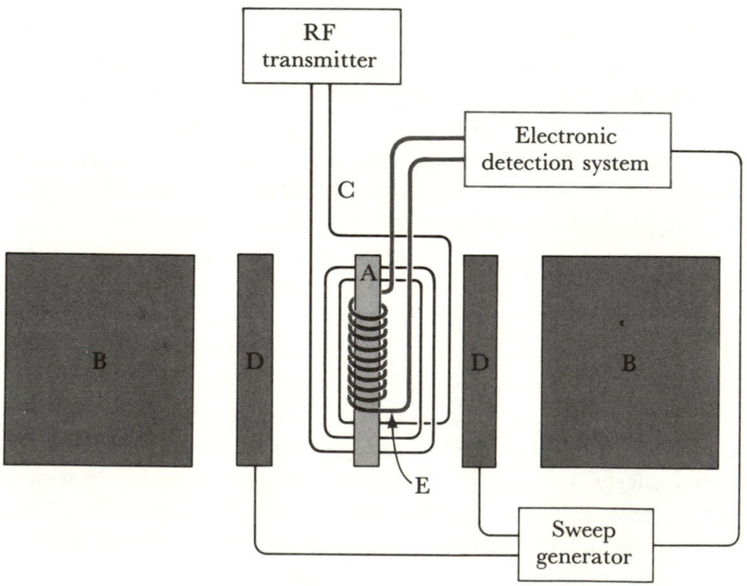

Figure 17-8
A sweep-type NMR spectrometer. See text for details.

accessory magnet (D) is increased slowly. (In some instruments, there is no accessory magnet and H_1 is varied by changing the frequency.) At resonance, the nuclear magnetic moments flip, and this sudden change in the magnetic field induces a current in a small coil (E) that is at right angles to both H_0 and H_1. Variations of this design exist, the most important being the Fourier transform NMR spectrometer, which will be discussed later.

Data Obtained from an NMR Spectrum

Four parameters of NMR spectra are used to obtain information about the molecules under study: the line position (i.e., its chemical shift), the area of the line, the band width, and the splitting. It is worth reviewing the information supplied by each parameter.

1. The *position* of the line or its chemical shift is determined by local magnetic fields (from other nuclei or unpaired electrons) and induced magnetic fields (produced by the surrounding electrons), resulting in the magnetic field in the region of a nucleus being different from the applied field. In other words, in a particular molecule, all nuclei of the same type (i.e., chemically identical) need not be in the same environment and hence can have different chemical shifts.

2. The area or intensity of a line is proportional to the *number* of nuclei in a given chemical environment. This means that a solution of molecules having a concentration that is twice that of another solution will have a peak with twice the area and that, if a molecule contains nuclei in two functional groups (e.g., methyl) in identical environments, or magnetically equivalent, the resonance will also have double area. This property differs from all other types of spectroscopy—that is, the area of an NMR line is *independent* of the electronic environment of the nucleus; there is no such thing as absorbance probability or quenching.

3. The *band width* (i.e., width at half-height) in spectra of macromolecules is primarily determined by molecular motion.

4. The *splitting* or clustering of lines is caused by the interactions of one nucleus that is covalently coupled with another. It is, however, often confused with a collection of nearby lines corresponding to different nuclei. The criterion for a multiplet is the relative areas of the components—for example, 1:1 for a doublet, 1:3:1 for a triplet, 1:3:3:1 for a quartet, 1:4:6:4:1, for a quintet, and so forth.*

* The relative intensities are proportional to the coefficients of the binomial expansion.

Rules for Interpreting NMR Spectra

Like those of other spectroscopic methods, the theory of NMR is not yet adequate for calculating the molecular parameters of complex molecules. However, if peaks are identified (how this is done is discussed in a later section), certain properties of the molecule can be deduced from measurements of the four parameters given in the preceding section. In practice, a collection of "rules" is used, some empirical and some theoretical. Those that are applicable to biological molecules are given in Table 17-2.

Technical Problems of NMR and Their Solutions

The first problem of NMR spectroscopy is obviously to obtain a spectrum. For proton resonance spectroscopy (still the most frequently done) of macromolecules in which aqueous solvents are required, it is necessary to prepare the sample in D_2O rather than in H_2O to eliminate a powerful H_2O-proton resonance that obscures most of the resonances of the protons of interest (because H_2O is 55 M). The necessity of using D_2O unfortunately eliminates the possibility of observing protons bonded to nitrogen or oxygen because these exchange rapidly with the solvent deuterons and are usually not observable (because of both the low sensitivity of the deuteron resonance and the fact that the lines are in distant parts of the spectrum). Hence, carbon-bound hydrogens are normally observed. To express all data as parts per million, a reference standard is frequently added. A common one is sodium 2,2-dimethyl-2-sila-pentane-5-sulfonate (DSS). Often, small amounts of H_2O are added as a standard, or the DHO, which contaminates all D_2O samples, can be used.

With special techniques using Fourier transform NMR, it is possible to suppress the H_2O resonance and perform the measurement in H_2O solution. In this case, D_2O is used as a standard. This method now has widespread use and is gradually supplanting work using D_2O. The main limitation is the availability of Fourier transform NMR spectrometers, which are exceedingly expensive.

For other nuclei (e.g., ^{13}C and ^{31}P discussed later), H_2O can be used as a solvent because the frequency range used is far from the H_2O resonance.

A second problem in NMR spectroscopy is the low sensitivity of detection. This difficulty results from the weakness of an NMR signal compared with the background noise of the instruments used. The method to reduce this problem is to sweep through the spectrum many times in succession, sum the results, and average the spectra (each sweep is usually called a *scan*). Because the noise is random, it will tend to cancel out, whereas the signal from the resonance will be enhanced. In

Table 17-2
Rules Used in the Interpretation of NMR Spectra.

1. The magnitude of the chemical shift for protons in a particular chemical group (e.g., methyl, ethyl, hydroxyl) varies with the molecule of which it is a part (e.g., a particular amino acid or nucleotide).
2. If a compound that contains a proton having a particular chemical shift is in a polymer, the chemical shift generally changes owing to the proximity of other molecules or groups.
   a. The largest shifts (20–30 ppm) are caused by the presence of unpaired electrons (i.e., paramagnetic centers).
   b. Shifts of approximately 2 ppm are usually the result of ring-current fields. The magnitude of the ring-current shift is roughly in the following order: flavins and porphyrins > tryptophan and nucleotides > histidine > tyrosine and phenylalanine.
   c. Electric fields from many charged groups (such as in aspartate, glutamate, lysine, arginine, and histidine) cause shifts. The shift is small if the group is on the exterior of a folded macromolecule and hence interacting primarily with the solvent but is larger if the group is internal.
3. A change in the chemical shift of a particular proton, following treatment of the macromolecule by a physical or a chemical agent, means a change in the structure of the macromolecule in the region surrounding the particular proton. If the treatment includes a change in pH, the shift may result from a change in the ionization state rather than a three-dimensional change.
4. Splitting implies that two sets of nuclei are *covalently* coupled to one another. The number of lines of the multiplet gives a rough indication of what group it might be. For example, a quartet often means that a methyl group is coupled. The separation of the components of a multiplet is related to the angle between the groups containing the nuclei. From a comparison with compounds whose structure is known from X-ray analysis, the bond angles for simple, small molecules can sometimes be determined. In a macromolecule, any agent that alters the separation of the components of a multiplet must alter the bond angle; this indicates what may be a subtle conformational change.
5. In a multicomponent, rapidly self-associating (e.g., dimerizing) system, the magnitude of the chemical shift, the width of the line, and the number of lines (Figure 17-5) is a result of the relative amount of time spent in each chemical or physical state. In such a system, a change in the equilibrium or the rate can change the width of a particular line or produce a splitting like that shown in Figure 17-5. Such changes can be used to study equilibria.
6. The width of a line is a measure of the relative mobility of a nucleus. If the nucleus is moving very rapidly, as it might in a small molecule, the peaks are very narrow. If the nucleus moves more slowly, as in a macromolecule diffusing or in a rigid part of a macromolecule, the lines are broader. Often band width can be used to estimate the mobility of a functional group or residue in a polymer, if the change in band width is not due to the cause given in rule 5 or to a change in position of a very close, unresolved and unrelated line.
7. The binding of a ligand to a site in a macromolecule usually affects the spectra of *both* the ligand and the macromolecule. Although such factors as the redistribution of a local charge, changes in orientation at a distance from the binding site, and so forth, may cause spectral changes, in many cases, the changes are effected by nuclei in the binding site. The usual effect is a shift in line position accompanied by a broadening of the line because there is likely to be decreased freedom of motion in the binding region.

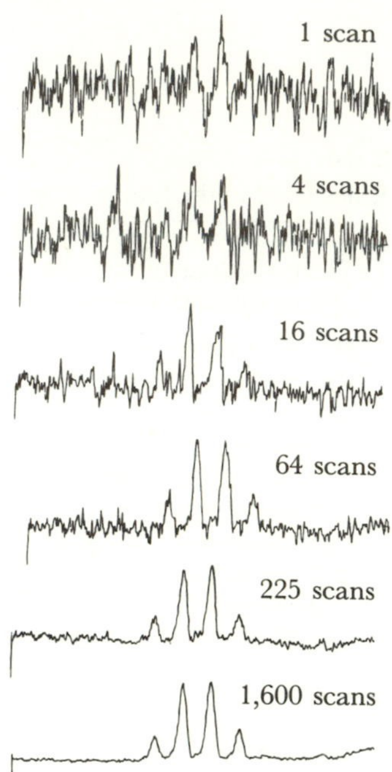

Figure 17-9
The methylene quartet of ethylbenzene as a function of the number of scans. [Courtesy of Varian Associates.]

practice, this is done with a small computer called a multichannel pulse-height analyzer. The success of this procedure is shown in Figure 17-9. After one scan, the signal is not discernible but, after several hundred, the resonances become clear. By this procedure, a particular proton can be detected at a concentration of 10^{-3} M. Other nuclei have lower sensitivity (see Table 17-1) and require higher concentration. Fourier transform NMR, which increases the signal-to-noise ratio tenfold, allowing detection of protons at 10^{-4} M, is described in a later section.

Resolution is the third major problem because of the huge number of peaks (100–1000) in the spectra of many macromolecules. However, because the chemical shift is a linear function of the size of H_0, the resolution problem has been attacked by constructing stronger magnets. The largest fields ($\sim$50 kG) are currently generated by superconducting magnets. At the same time, of course, it is necessary to increase the frequency of the electromagnetic radiation; radiofrequency generators of 3×10^8 Hertz are now available. With nuclei other than protons (e.g., ^{13}C and ^{31}P), the resolution problem is less severe because chemical shifts are much larger.

After a high-resolution spectrum has been obtained, the final problem is to identify the peaks—that is, to identify the nucleus producing a particular peak. For small molecules, this is not difficult; for large molecules, it is rarely simple, although the methods described in the following section are easy to understand.

Assignment of Lines for Proteins and Polynucleotides

The identification of the peaks in polymers is usually done by the following methods.

1. The spectra of amino acids, nucleotides, and certain functional groups have been tabulated. How these spectra vary with pH and certain solvents is also fairly well known. Many of these substances have sufficiently characteristic sets of lines at particular positions that they can be recognized in complex spectra by simple inspection. Examples of three such spectra are shown in Figure 17-10. This is probably one of the most common methods for peak identification.
2. If a protein has a known sequence, it can be enzymatically fragmented and the spectra of some of the purified fragments can be determined separately. In this way, peaks can sometimes be shown to come from particular regions of the molecule. The limitation of this method is that the chemical shifts are not always the same in the fragments as in the protein because the interaction with other parts of the molecule may not be the same. This method is usually not of great value for very large structures but is applicable to oligopeptides.
3. Chemical reactions with a particular residue or the binding of a ligand to the residue may result in the loss or displacement of particular lines. This is usually, accompanied by significant conformational changes so that the effect on the resonance may be indirect. Occasionally, however, the situation is more favorable and peaks can be identified in this way.
4. Some polymers may be selectively deuterated by the growth of an organism in a medium containing a particular deuterated residue. Proton resonances from the deuterated residue will vanish because of the weakness of the deuteron signal. Inversely, an organism can

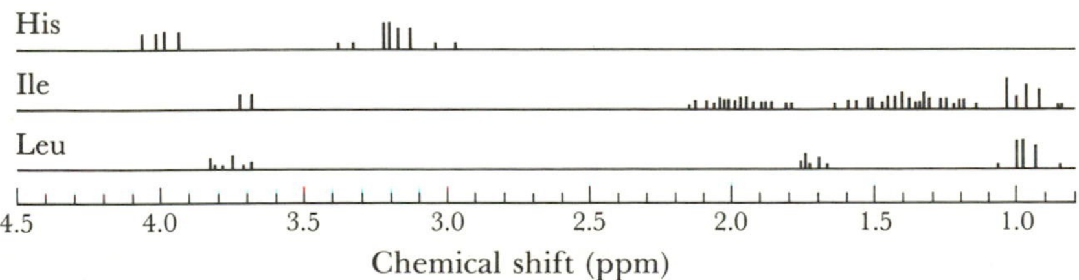

Figure 17-10
Spectra of histidine, isoleucine, and leucine in the zwitterion form. The relative heights of lines correspond to areas in true spectra.

be grown in a completely deuterated medium but with a single protonated residue; then, only peaks corresponding to the protonated species will be seen. This method is reliable but is very expensive owing to the high cost of deuterated compounds.

5. Spectra have been obtained for proteins and polynucleotides whose three-dimensional structure is accurately known from X-ray diffraction analysis. The spectra of samples can be compared with them. This is a useful method.

6. Comparison of the spectrum of a protein with that of a mutant in which there is a single amino acid substitution is a valuable technique. The lines corresponding to the missing amino acid will not be present in the mutant spectrum and thus are identified. Significant spectral changes also appear for those amino acids that interact with the original amino acid.

7. The spin-spin decoupling method described in page 614 is also used for line identification.

In summary, no single method is ever used to identify a peak; usually several must be used—along with a fair amount of ingenuity and reasoning.

Use of NMR to Study Protein Structure

NMR has the potential for supplying a great deal of information about protein structure because the parameters of the spectra are sensitive to changes in both sequence and conformation. For example, in the absence of any interaction, the spectrum of a polypeptide or protein would be the sum of the spectra of the constituent amino acids, all of which are well known. However, this is not the case, as seen in Figure 17-11, which shows the spectrum of the enzyme lysozyme in its native and denatured (i.e., random coil) forms and that calculated from the amino acid composition. It is clear that there are many differences between the observed and calculated spectra, which is the usual situation. Although each protein has its own differences, there are a few general features of the spectra that seem to be characteristic of most proteins. For example, it is generally observed that large chemical shifts occur if a proton is in an amino acid that is either amino-terminal, carboxyl-terminal, or next to the nitrogen atom of a peptide bond, and small shifts are produced if the proton is next to either the carbon atom of a peptide bond or a carbon or nitrogen atom of the nearest-neighbor amino acid; small shifts are also produced by titration of a carboxyl group or a nitrogen atom in the nearest-neighbor amino acid. Furthermore, if a protein is in a native configuration, there are characteristic chemical shifts of certain amino acid protons produced by being in an α helix. Unfortunately, the magnitude of the changes in chemical shift, band width, and intensity due to

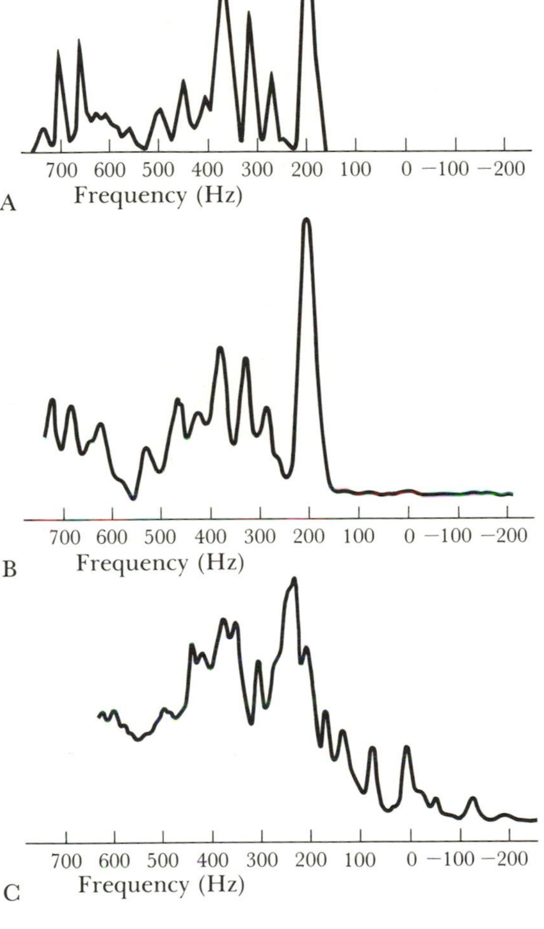

Figure 17-11
Part of the proton spectra of lysozyme: (A) calculated; (B) measured spectrum of denatured lysozyme; and (C) measured spectrum of native lysozyme. [Redrawn with permission from C. C. McDonald and W. C. Phillips, *J. Am. Chem. Soc.* 91(1969): 1513–1521. Copyright by the American Chemical Society.]

tertiary structure and the proximity to distant amino acids (e.g., because of ring-current shifts or hydrogen bonds) varies for each protein, and there are no truly general principles that might be stated as empirical rules.

Since the NMR spectra of amino acids are affected by the secondary and tertiary structures of proteins this theoretically allows the determination of three-dimensional structure; such a determination has not yet been done because of the great complexity of the analysis as shown by the following problems.

First, a typical protein has about 600 to 1000 lines, many of which overlap. Line width is affected by the mobility of the molecule (Table 17-2, rule 6), roughly increasing with decreasing diffusion coefficient, that is, with increasing molecular weight. Useful spectra (in which lines can be resolved) are therefore obtainable at present only for molecules whose molecular weight is less than 25,000. Second, the *unambiguous* identification of lines requires that the complete amino acid sequence be known. This is not always available and is not even sufficient owing to the interactions resulting from the tertiary structure. Observing changes in line position during the titration of particular amino acids or the effects of selective deuteration is an aid in identification but does not always work. The net result is that in a complex protein a significant number of lines remain unidentified. Nonetheless, a great deal of information can be obtained.

Currently, NMR is used successfully in five ways to study proteins: (1) to determine the fraction of amino acids in the α-helical configuration; (2) to monitor helix-coil transitions and especially to observe local changes during the transition; (3) to determine the conformation of selected regions of the protein (e.g., around a particular amino acid); (4) to observe the binding of small molecules and metal ions to selected regions of the protein (by observing either the ligand or the protein spectra); and (5) to study paramagnetic active sites in electron-transfer proteins.

The types of observations made and the method of analysis of the data can best be seen in a few examples.

Example 17-A ☐ Fraction of residues of poly-(γ-benzyl-L-glutamate) in the α-helical configuration.

The proton of the α-carbon of some amino acids has two different chemical shifts, depending on whether the carbon atom is in an α helix or a random coil (Figure 17-12). Because the area of a line is proportional to the number of protons giving that resonance and because there is one α-carbon proton per unit of the α helix, the relative areas of the two lines can be used to calculate the fraction of amino acids in a α-helical configuration. The fraction is simply

$$\frac{A_{\mathrm{H}}}{A_{\mathrm{H}} + A_{\mathrm{R}}}$$

in which A_{H} and A_{R} are the areas of the helix and the random-coil lines, respectively.

With real proteins rather than synthetic poly(amino acids), the error in measuring the amount of α helix may be substantial because not all amino acids show the clear effect depicted in the figure. Usually NMR provides a minimum value.

Example 17-B ☐ Unfolding of proteins.

The denaturation of proteins produces substantial changes in NMR spectra. Lines become narrower because freedom of motion is increased and the chemical shifts of identical amino acids approach

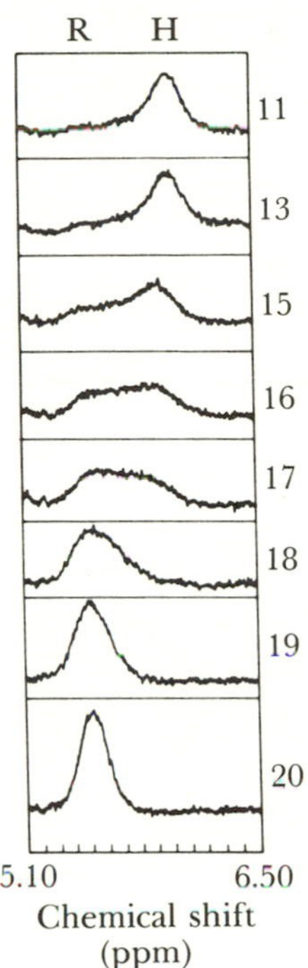

Figure 17-12
Selected region of the spectrum of poly-γ-benzyl-L-glutamate) in $CDCl_3$ with added trifluoroacetic acid: the peak represents the α-carbon proton only; H and R refer to the line positions when in the α-helical and random-coil configurations, respectively; the numbers indicate the percentage of CF_3COOH added. [Redrawn with permission from J. A. Ferretti, B. W. Ninham, and V. A. Parsegian, *Macromolecules* 3(1970):20–42. Copyright by the American Chemical Society.]

one another. By following line width and position as a function of pH, temperature, the concentration of various denaturing agents, and so forth, it is possible to follow the unfolding of various parts of the protein independently.

For example, let us consider the thermal denaturation of a hypothetical protein containing several histidine (His) lines and one tryptophan (Trp) line (Figure 17-13). The subscripts N and R denote line positions for native and random-coil configuration, respectively. At 25°C, the protein is in the native configuration, and various histidines have different chemical shifts owing to their various environments. The tryptophan line must represent many residues in the same environment because its area is greater than any of the histidine lines (remember that area is proportional to the number of protons). At 37°C, the histidine lines have shifted to the position they have in a random coil. Hence, the region of the protein containing the histidines must have relatively low thermal stability compared with that containing the tryptophan. At 45°C, a line appears at Trp_R. Because the Trp_N line remains, the tryptophans must be in at least two different parts of the protein; that is, one that is disrupted at 45°C and one that is not. From the relative areas of Trp_R and Trp_N (1:5), one-sixth of the tryptophans are in a region that is a random coil at 45°C. Note that the Trp_R line is as wide as the Trp_N line, suggesting that the tryptophan must have been in a highly flexible region of the molecule. By 65°C, the Trp_N line is gone so that all regions containing tryptophan are denatured.

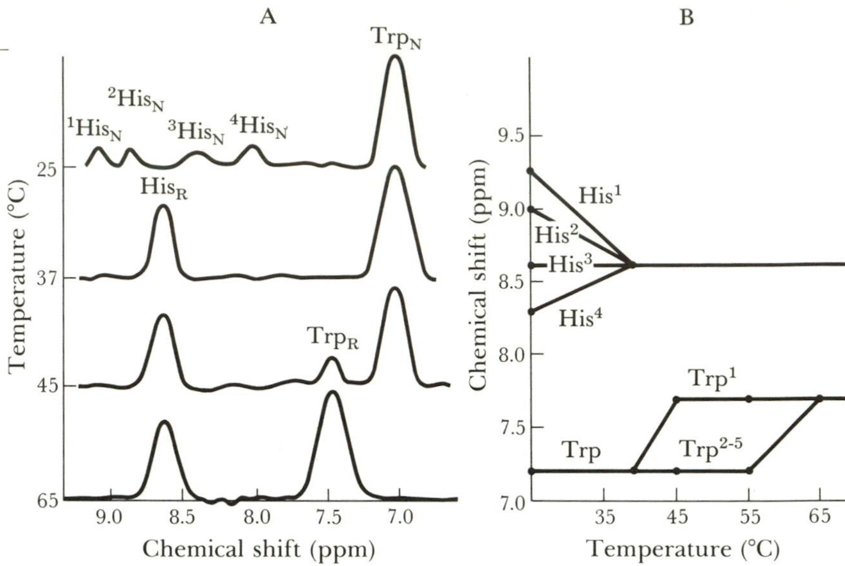

Figure 17-13

A. Spectra for a hypothetical protein as a function of temperature. Only the relevant lines have been drawn. In a real protein, as many as 50 additional lines might be in this region. B. Plot of line position as a function of temperature for each line of part A.

Note that this NMR study provides more information than would a hydrodynamic analysis of denaturation. For example, if this protein had been studied by viscometry (Chapter 13), one would have simply observed a continuous change in viscosity over the same temperature range and this would be a reflection of the *average* state of the molecule. However, the NMR analysis gives information about the state of *particular* regions of the molecule.

Example 17-C ☐ Properties of the active site of ribonuclease A.

The structure of the active site of pancreatic ribonuclease A has been elucidated by examining the spectrum of the C-2 histidine protons when the inhibitor 3′-cytidine monophosphate (3′-CMP), which binds to the active site, is added at various concentrations and over a range of pH values. This enzyme has several histidine residues but only three (12, 48, and 119)* show significant changes in chemical shift when 3′-CMP is added (Figure 17-14A and B), and histidine-119

* These numbers and their alternate designations, His-12, His-48, and so forth, refer to the position of the amino acid in the complete amino acid sequence.

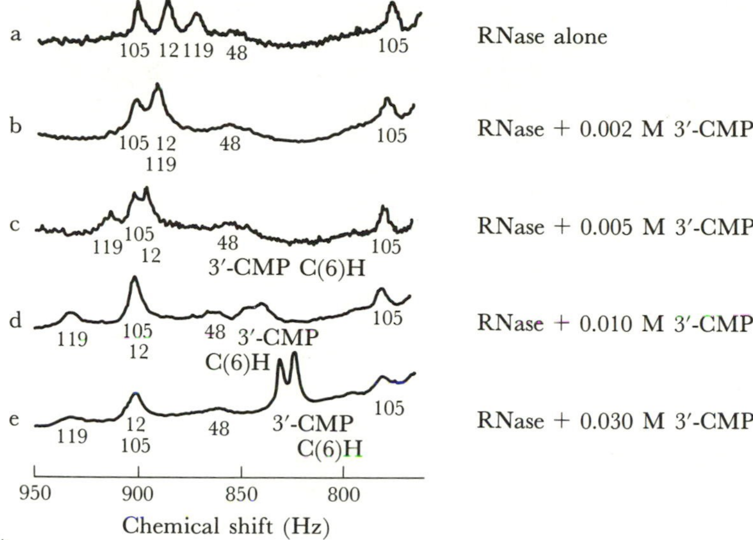

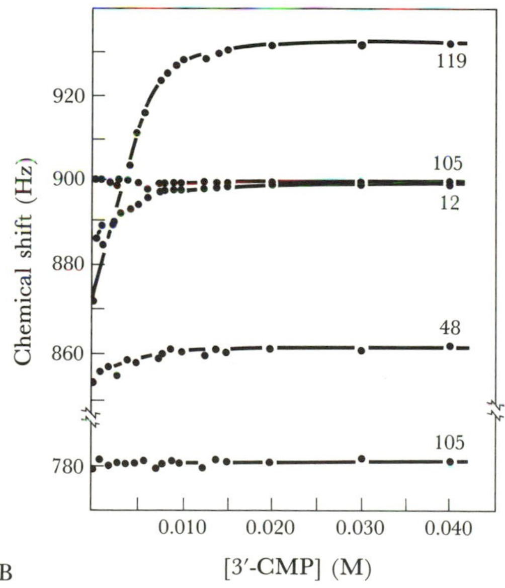

Figure 17-14
A. A part of the proton spectrum showing peaks corresponding to various histidines in RNase as a function of the amount of added 3′-CMP. Numbers refer to the particular histidine. B. Plot of the data in part A. [Data from D. H. Meadows and O. Jardetzky, *Proc. Natl Acad. Sci. U.S.A.* 61(1968):406–413.]

shows by far the greatest. Clearly, each histidine is in a new environment when 3′-CMP is added. The problem is to determine which histidines participate in direct bonding. A study of the chemical shift of the C-2 proton peak of imidazole (the aromatic part of histidine) as a function of phosphate concentration (i.e., using imidazole instead of histidine and phosphate instead of 3′-CMP) shows that the curve relating chemical shift and concentration matches that for histidine-119. Hence, the chemical shift for histidine-119 *may* be a result of binding to the phosphate group of 3′-CMP (Table 17-2, rule 7). This can be verified by further studies with the inhibitors 2′-CMP and 5′-CMP, whose binding constants to ribonuclease differ. The chemical shift of histidine-12, as well as the pH dependence of the shift, is the same for all three inhibitors even though their binding differs. In other words, histidine-12 is not responsive to factors that affect the binding of the inhibitors, suggesting that histidine-12 is not the site of binding of the phosphate. On the other hand, both the chemical shifts and the pK of binding for histidine-119 differ for the three inhibitors—that is, histidine-119 is sensitive to the position of the phosphate and therefore probably forms a direct bond with it. However, the changes in histidine-12 are sufficiently great that it is probably in or near the active site. The small shift in histidine-48 is probably a result of a conformational change in the protein induced by the binding.

Adjacent to histidine-119 is phenylalanine-120. The lines produced by this residue have not been unambiguously identified; however, because there are five protons in the ring and because there is a peak in the spectrum in the range of chemical shift corresponding to aromatic amino acids, the peak can be presumed to correspond to a phenylalanine. This peak shows a large change in position when 3′-CMP is added, but none when phosphate alone is added. This suggests that binding also includes an interaction between the pyrimidine ring of 3′-CMP and phenylalanine. Note that this is just suggestive because it is not clear that the line is actually due to phenylalanine-120.

Further evidence for the participation of the pyrimidine part can be derived from an examination of the spectrum of 3′-CMP (by rule 7, Table 17-2, these regions of the ligand that take part in binding should also show spectral changes). Indeed, it is found that the ring-proton lines are affected by binding and the shifts are the same for 2′-, 3′-, and 5′-CMP. This agrees with the idea that the pyrimidine ring participates in binding and, because its binding is insensitive to the position of the phosphate, it does not bind to histidine-119.

Figure 17-15 shows a postulated structure for the 3′-CMP complex. Although the details might not be correct, it is shown to give an idea of the sophisticated conclusions that may be drawn from NMR data.

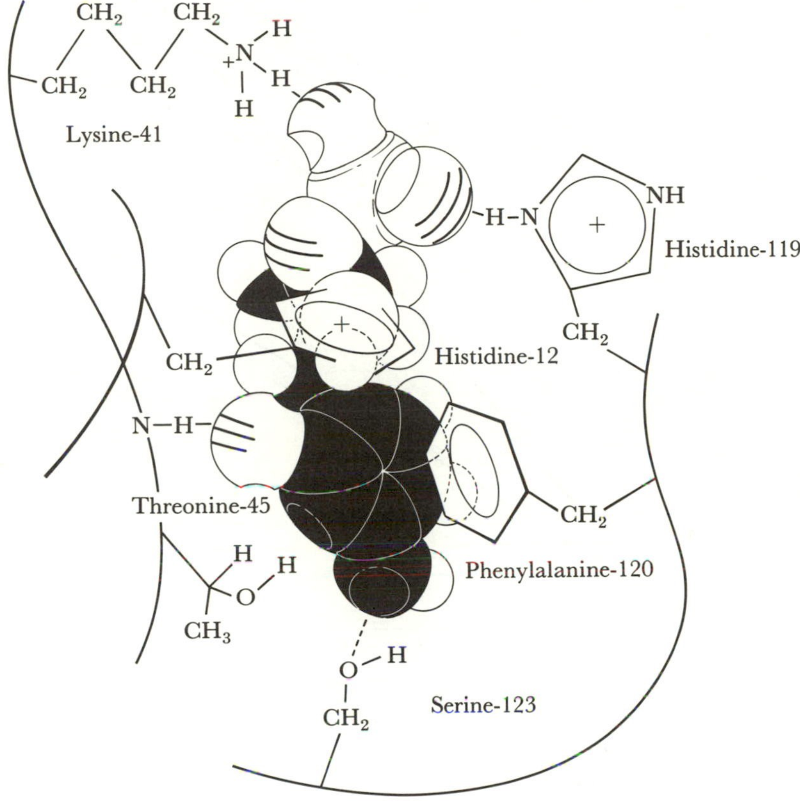

Figure 17-15
Postulated structure of the 3′-CMP–ribonuclease complex. [From D. H. Meadows, G. C. K. Roberts, and O. Jardetzky, *J. Mol. Biol.* 45(1969):491–511.]

Example 17-D

☐ Mode of binding of sulfacetamide to the protein bovine serum albumin, assayed from the spectrum of the ligand.

The spectrum of sulfacetamide consists of a single large peak due to the methyl group and a cluster of four peaks due to the aromatic ring. When bovine serum albumin is added, the line widths of the methyl and the aromatic protons increase substantially (Figure 17-16), which indicates a decreased mobility of all of these protons (Table 17-2, rule 6). A trivial explanation of the change in line width is that the increased viscosity of the solution by the protein decreases the proton mobility. However, this explanation cannot be correct because, as shown in the figure, the line width of the aromatic protons increases

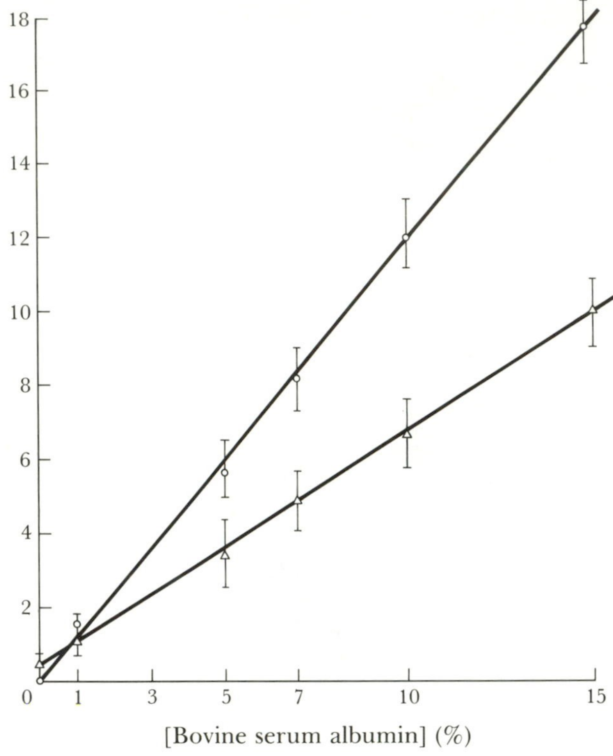

Figure 17-16
Relaxation rate ($1/T_1$) of 0.1 M sulfacetamide as a function of bovine serum albumin concentration in D_2O: circles represent *p*-aminobenzene sulfonamide protons; triangles represent methyl protons. [From O. Jardetzky and M. G. Wade-Jardetzky, *Mol. Pharmacol.* 1(1965):214–230.]

more rapidly than that of the methyl protons. If the broadening were due solely to the increase in solution viscosity caused by the addition of the bovine serum albumin, all lines would broaden by the same factor; indeed, this is the case if immunoglobulin (which does not bind sulfacetamide) is added. Hence, the change in line width probably results from binding; because the change is greater for the aromatic protons, the aromatic ring is probably the principal site of interaction. Note also that the change in line width can serve as an assay for binding; indeed, further studies of the effects of concentration of both bovine serum albumin and sulfacetamide, of pH, and of ionic strength on line width allows the calculation of the dissociation constant, and thus the determination of the role of ionic forces in binding.

☐ Effect of a cofactor on enzyme-substrate binding. **Example 17-E**

Yeast alcohol dehydrogenase converts ethanol into acetaldehyde but only if the enzyme cofactor nicotinamide-adenine dinucleotide (NAD) is present. A hint about the role of NAD in this reaction can come from NMR studies. The nicotinamide proton lines are broadened when the enzyme and the NAD are mixed. This broadening is specific to the enzyme (i.e., other proteins do not do this) and is therefore due to enzyme-cofactor binding and not to the viscosity effect mentioned in Example 17-D. If ethanol (the substrate) or acetaldehyde (the product) is added to the enzyme–NAD mixture, the methyl proton lines of both substrate and product are broadened. No broadening of any lines occurs if the enzyme is absent, so that both substrate and product must bind to the enzyme-cofactor complex. If enzyme, substrate, and product are mixed *in the absence of NAD*, there is no broadening of the methyl lines of either ethanol or acetaldehyde. Hence, the NAD must be bound to the enzyme in order for the substrate to bind.

☐ Binding of a polymer to a protein—the DNA-histone complex. **Example 17-F**

DNA binds tightly to histone to form nucleohistone; this is of great interest because of its role in chromosome structure.

From studies of amino acid sequences, it is known that most histones have an asymmetric distribution of amino acids (i.e., the amino acids are arranged in three clusters—a basic region, an acidic region, and a nonpolar region). The amino acid protons of these three regions have distinguishable resonance lines. If histones and DNA are mixed, the lines representing the basic region are broadened. This cannot be due to the effect of DNA viscosity because, if so, all lines would be broadened equally. This suggests that the DNA is bound to the basic region of the histones.

Studies of the effect of salt concentration on histones in the absence of DNA show that the lines of the nonpolar amino acids are broadened with increasing salt concentration. This is taken to mean that the histones self-associate by means of the nonpolar regions to generate large structures (Table 17-2, rules 5 and 6). From this, one may reasonably propose that the histone-DNA gel (called chromatin) consists of histones self-associated by means of the nonpolar groups with DNA bound to the basic groups.

The Nuclear Overhauser Effect

In Chapter 15, energy transfer between a fluorescent donor and an acceptor molecule was examined. It was observed that the efficiency of energy transfer could be used as a means of measuring intramolecular distances. The nuclear Overhauser effect (NOE) is the NMR analogue of this process. Consider a macromolecule that is folded so that two amino acids, which are well separated in the polypeptide chain, are very near one another. If their separation is less than 4 Å, there is a significant interaction of the magnetic fields of particular nuclei in the amino acids. For example, when a proton in amino acid A is irradiated, the area of the absorption line depends on the relative number of spins in the upper and lower spin-energy states; this is because radiation stimulates both emission and absorption equally. Consider now another amino acid B that is very near A. The proton B is then irradiated with such a high intensity of radiation that it becomes *saturated.* This means that the number of protons in the upper and lower spin states are equal; since the probabilities of absorption and emission are equal, there is no net absorption and the line of B vanishes. If energy is transferred to A, more A protons are raised from the lower to the higher spin state. This increases the degree of emission at the expense of the number of absorbing centers so that the A line *decreases* in area. This decrease is called the nuclear Overhauser effect. The magnitude of the effect can be related to the distance between A and B and the accuracy of the measurement is ± 0.1 Å.

Example 17-G ☐ Indication that one of two histidines is near a particular alanine in an enzyme.

An enzyme contains two histidines at positions 62 and 140 in the polypeptide chain; histidine-62 is thought to be in the active site because binding of the substrate causes a slight frequency shift of the histidine-62 line. However, this is not a strong conclusion because the shift could result from a small conformational change of the entire protein molecule that occurs when the substrate is bound. Alanine is known to be in the active site also. The sample is then irradiated to the point of saturation at the frequency corresponding to the histidine-62 line and then to the histidine-140 line. When histidine-62 is saturated, the area of the alanine line decreases significantly. Such a decrease does not occur when histidine-140 is irradiated. The nuclear Overhauser effect thus suggests that the histidine-62 is in the active site.

Use of NMR to Study Polynucleotide Structure

To date, NMR has had little use in the study of polynucleotides, partly because (1) DNA is so rigid and has such a low diffusion coefficient that the lines are extremely broad and in fact barely observed (its molecular weight is far above the 25,000 limit for proteins); and (2) most of the structural features of polyribonucleotides were fairly well worked out by the time NMR workers turned their attention to nucleic acids. An exception is in determining the structure of the smaller polynucleotide transfer RNA (tRNA), an example of which follows.

☐ Base-pairing structure of transfer RNA. **Example 17-H**

For studying the base pairing of tRNA, it is possible to use an H_2O solution because there are two clusters of lines far from the H_2O lines that give a great deal of information. One cluster consists of guanine N-1 protons of the guanine·cytosine (G·C) pairs and the other, uracil N-3 protons of the adenine·uracil (A·U) pairs. Because the intensity of a line is proportional to the number of protons present, the areas of the lines in the two clusters are indicative of the number of hydrogen bonds in the molecule. Furthermore, the ratio of the areas of the two clusters indicates the ratio of G·C to A·U pairs. To obtain information about the distribution of G·C and A·U pairs in the molecule, one can make use of the fact that tRNA can be enzymatically cleaved at particular sites to generate a discrete set of fragments, each of which apparently maintains the original hydrogen bonds. By examining the spectra of fragments, it is possible to determine for each fragment the number of G·C and A·U base pairs. Therefore, whenever the base sequence of a particular tRNA is known, it is possible to construct a three-dimensional model of the tRNA, at least insofar as hydrogen bonding is concerned. One of the most important conclusions drawn from these studies is that the so-called cloverleaf model (Figure 17-17) is the correct description of tRNA in solution. Since these experiments were done, the complete three-dimensional structure has been worked out by X-ray diffraction.

☐ Indication of particular base pairs in tRNA. **Example 17-I**

Hydrogen-bonded base pairs are sufficiently near to exhibit the nuclear Overhauser effect. Thus, NMR can be used to confirm base-pairing assignment as well as determining whether particular bases are stacked one above another. This has been done for a large number of bases in elegant experiments performed in the laboratory of Alfred Redfield.

Spin Labeling

An immediate aim of NMR spectroscopy is to determine distances—either between residues of a macromolecule or between a ligand and a binding site. The technique of spin labeling, first developed by Harden McConnell and his coworkers, allows fairly precise measurements to be made. The principle underlying this method is that a paramagnetic center (i.e., an unpaired electron), has an associated, large, fluctuating magnetic field generated by the unpaired electron spin, which produces a substantial broadening of the lines of any nucleus that is exposed to that fluctuating field. The amount of broadening decreases as the distance between the unpaired electrons and the nucleus of interest increases

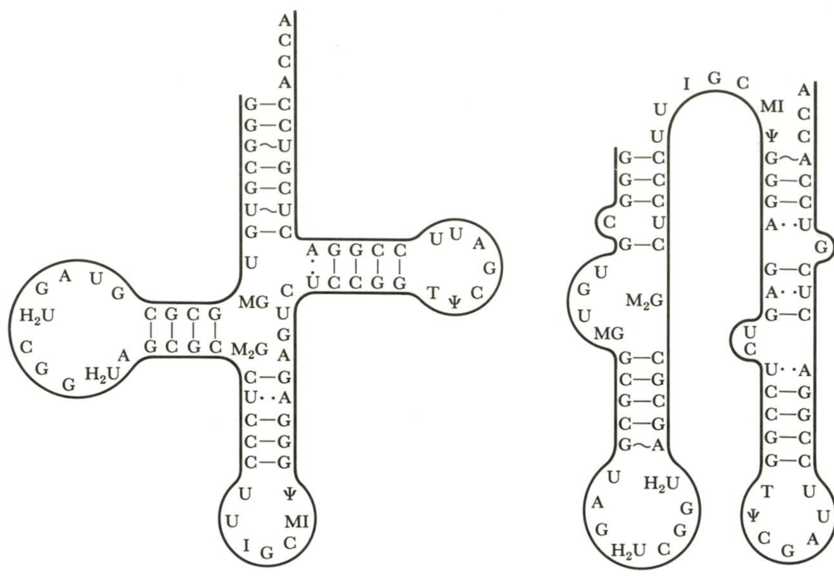

Figure 17-17
Two of several possible structures of yeast alanine tRNA arranged so that a large fraction of the bases participate in hydrogen bonding. The hydrogen bond of a G·C pair is indicated by —; that of an A·U pair is indicated by ··; the weaker hydrogen bonds found in other base pairs are indicated by ~. An NMR analysis yielding the number of base pairs of various types indicates that the cloverleaf structure (left) is the correct one. The symbols are: G, guanosine; C, cytidine; A, adenosine; U, uridine; I, inosine; H_2U, dihydrouridine; T, ribosylthymine; MG, methylguanosine; M_2G, dimethylguanosine; Ψ, pseudouridine; MI, methylinosine. [The base sequence is that reported by R. W. Holley, J. Apgar, G. A. Everett, J. T. Madison, M. Marquisee, S. H. Merrill, J. R. Fenswick, and A. Zamer, *Science* (Washington, D.C.) 147(1965): 1462–1465.]

or as the time of exposure of the nucleus to the paramagnetic center is decreased. With protons, an effect is seen with distances of as much as 40 Å; for ^{13}C nuclei (which have not yet been studied appreciably by the spin-labeling technique), the distance is much greater. In practice, a substance called a spin label (i.e., one that contains a free radical) is introduced into the system by reacting it either with a particular residue of a protein or with a ligand that is bound by the protein.

The most common spin labels are those containing the nitroxide group, namely,

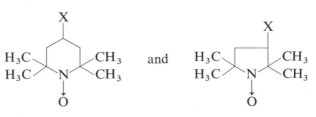

X; H_3C, H_3C, N, CH_3, CH_3, ↓ O   and   X; H_3C, H_3C, N, CH_3, CH_3, ↓ O

This group has an unpaired electron that is located primarily in the π orbital of the nitrogen. By appropriate choice of the group X, it is possible to couple these compounds at the desired positions on various biological molecules. The nitroxide group is an especially useful spin label because it is stable up to 80°C and from pH 3 to pH 10 and is generally unreactive. To obtain meaningful information from the spectra of a spin-labeled system, it is necessary to assume that the introduction of the spin label itself does not affect the conformation of the macromolecule. Usually, this is verified either by demonstrating identical enzymatic activity under a variety of conditions or by the identity of physical properties. These are not powerful criteria and constitute the only weakness of the method. The use of spin labels is best seen by an example.

Example 17-J

☐ Distance between binding site and an amino acid residue of a protein.

If histidine-15 of the enzyme lysozyme is covalently spin-labeled with nitroxyl nitrogen and the enzyme substrate *N*-acetylglucosamine is added, the resonance of the acetamido methyl proton of *N*-acetylglucosamine is broader than if the spin label were absent (Figure 17-18). By means of a simple equation, the distance between the unpaired electron in the nitroxyl group and the proton responsible for the resonance can be calculated from the line width. After the length of the spin-labeled compound itself has been corrected for, the true distance between the proton and histidine-15 can be calculated. Distances determined in this way agree with values determined by X-ray diffraction. The difference in the values is rarely more than 10–15%.

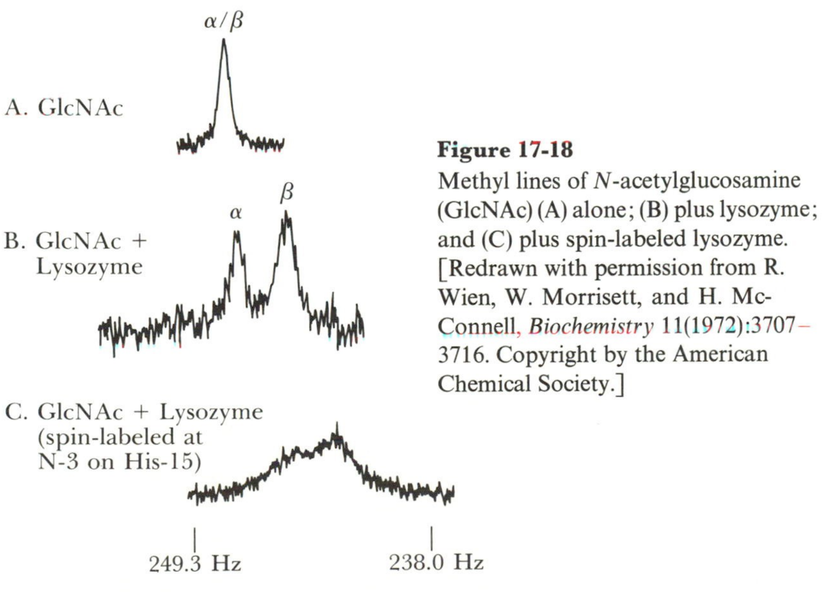

Figure 17-18
Methyl lines of *N*-acetylglucosamine (GlcNAc) (A) alone; (B) plus lysozyme; and (C) plus spin-labeled lysozyme. [Redrawn with permission from R. Wien, W. Morrisett, and H. McConnell, *Biochemistry* 11(1972):3707–3716. Copyright by the American Chemical Society.]

Fourier Transform NMR

As explained earlier, traditional NMR instruments generate spectra either by choosing a fixed frequency of the electromagnetic field or by fixing the field and varying the frequency. To obtain a good signal-to-noise ratio, the scan must be repeated many times and the data averaged. This is a slow and inefficient method because only a narrow region of the spectrum is being recorded at any one time. Consider the effect of using a fixed field and exciting with a short but very intense radio-frequency *pulse*. Since a pulse consists of a *superposition of a spectrum of frequencies* (the mathematics behind this statement—i.e., the concept of the Fourier transform—can be found in almost any text on wave theory), it can simultaneously excite the entire range of resonance frequencies. This is the principle of Fourier transform NMR spectroscopy and constitutes a great technical advance. This manner of excitation requires substantial changes in instrumentation because, to obtain a spectrum, the signal received (Figure 17-19) must be unraveled. This is

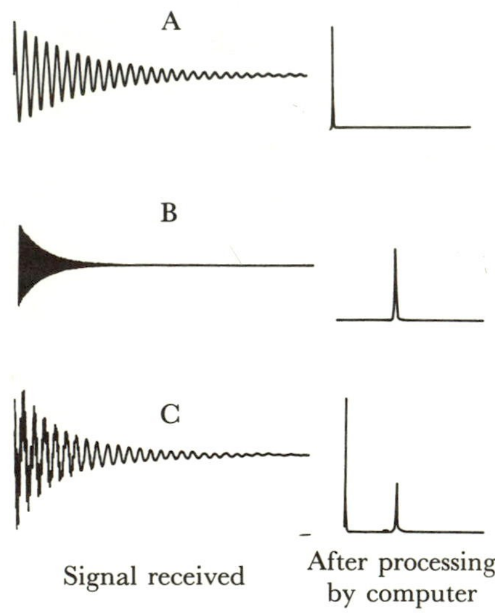

Figure 17-19
Signal received in Fourier transform NMR: the left-hand side shows the signals received by the computer; the right-hand side shows the frequency spectrum after computer processing. The parameters to notice are the frequency and the decay time. The frequency determines the position on the frequency scale at the right, and the decay time determines T_2, which is inversely proportional to band width: (A) low frequency, slow decay; (B) high frequency, rapid decay; and (C) a mixture of A and B. [From R. Dwek, *Nuclear Magnetic Resonance in Biochemistry*, Clarendon Press, 1973.]

done by an on-line digital computer. The details of the receiving and processing system are too complex to be described here but are handled by commercially available instrumentation. Again, repeat scans are made, and the data are averaged to improve the signal-to-noise ratio, as in conventional NMR work; however, the advantage of the Fourier transform method is that it takes roughly one-tenth the time to obtain a given signal-to-noise ratio, thus providing a tenfold increase in sensitivity. What this means is that (1), at concentrations used in conventional NMR, lines are detected that are normally obscured by noise and resolution is improved and (2), to obtain standard sensitivity, one-tenth the concentration is required. The ability to use lower concentrations has two important consequences: (1) other nuclei such as ^{13}C, for which the natural abundance and sensitivity is low (Table 17-1), can be studied; and (2) systems that are highly concentration dependent, such as enzyme reactions or associating polymers, can be studied, using a range of concentrations. In the future, it is likely that all NMR work will use Fourier transform instruments.

Use of Nuclei Other Than the Proton

To produce a resonance line, a nucleus must have nonzero spin. This, unfortunately, eliminates the common isotopes ^{12}C and ^{16}O (an empirical rule is that, if both the atomic number and mass number are even, $I = 0$). A second requirement is that the natural abundance of the isotope and the magnetic moment be high enough that a signal can be detected (the relative sensitivities of nuclei are roughly proportional to the cube of the ratio of the magnetic moments). It is also advantageous that the spin is not $>\frac{1}{2}$ because, for such nuclei, the lines are very broad owing to a phenomenon called quadrupole relaxation (which will not be discussed).

Three nuclei that satisfy these criteria are ^{13}C, ^{19}F, and ^{31}P. They are of special interest because the large polarizability of their electron clouds results in chemical shift ranges of approximately 350, 500, and 700 ppm, respectively. Because these shifts are about fifty times as great as those for the proton, resolution is much greater and it is possible to look at more subtle environmental differences.

The isotope ^{13}C has tremendous potential because of its wide range of chemical shifts, its greater sensitivity to structural changes, and its simple spectra (i.e., because of its low abundance, it is rarely adjacent to another ^{13}C so that the splitting of lines is unusual). Furthermore, with ^{13}C, information is obtained about the environment of carbon atoms in addition to that of protons. Furthermore, D_2O, which may sometimes have an effect on structure, is not needed. However, because of its low natural abundance (1%), it has been very difficult to detect by conventional means. This problem has to some extent been alleviated by Fourier transform NMR, although the sensitivity is still lower than with protons. The ultimate solution to the sensitivity problem will be the development of techniques for enriching with ^{13}C, although along with the obvious advantage of enrichment will come the disadvantage of splitting. However, even with the current problem of low sensitivity, ^{13}C NMR has already made important contributions to biochemistry.

The isotope ^{19}F has the disadvantage that it is not normally present in biological materials. It has the advantage that it can be introduced at defined sites in a molecule and therefore serves as an external probe. If the number of sites is small, the spectra consist of a small number of lines. In proteins, ^{19}F is used in two ways: (1) by introducing a ^{19}F at a site on the protein and observing the ^{19}F resonances as a function of various agents—pH, temperature, ligands, and so forth; (2) by using a fluorinated ligand and observing the signal from both a bound and an unbound ligand. This can be used to study chemical exchange, determine various parameters of binding, and learn something about the structure of the binding site.

The isotope ^{31}P has recently become important in studies with nucleotides, membranes, phospholipids, and other phosphorylated compounds. Of particular interest is the fact that the intracellular concentrations of many of these compounds (e.g., ATP, ADP, and inorganic phosphate) are so high that NMR studies can be carried out with living cells. For example, ^{31}P NMR has been used to determine the intracellular concentrations of various components of the glycolytic pathway in various growth conditions. An unusual example is the determination of intracellular pH. This is possible because the chemical shifts of the phosphorus nuclei of 2,3-bisphosphoglycerate depend on their ionization states and the ratio of the concentrations of these states depends on pH.

Electron Spin Resonance (ESR) Spectroscopy

Origin and Properties of ESR Spectra

An electron also possesses a spin magnetic moment that can assume two orientations in space and thus should interact with a magnetic field in the way just described for nuclei. In most chemical compounds, electrons are paired so that the net magnetic moment is zero and there is no condition of resonance for absorbance of radiation. However, there do exist substances that possess an unpaired electron—for example, free radicals, certain ions of transition elements, excited photochemical intermediates in the triplet state, and molecules such as O_2 that have a triplet ground state. Such substances are called *paramagnetic* and many can be detected by electron spin resonance (ESR) spectroscopy. (In some references the technique is called electron paramagnetic resonance or EPR spectroscopy.)

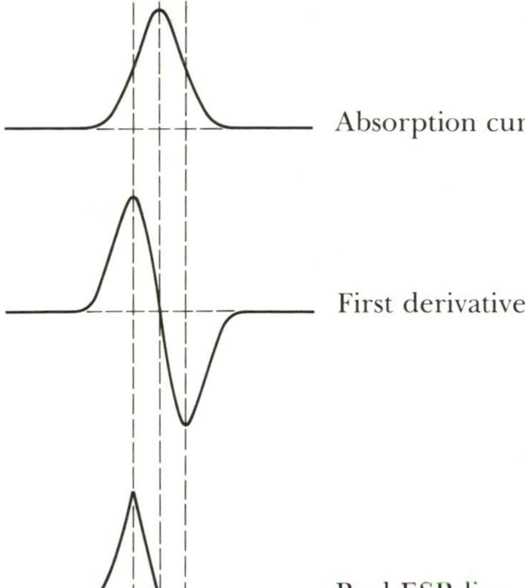

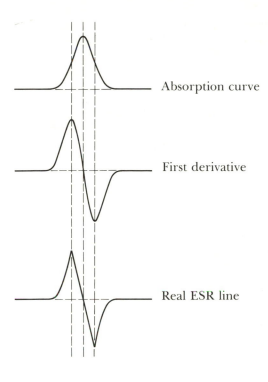

Figure 17-20
An absorption curve, as seen in NMR, and its derivative, as seen in ESR. For reasons discussed later in the text a real ESR line is sharper than the curve in the middle panel.

The electron spin magnetic moment is about 10^3 times greater than a typical nuclear magnetic moment so that the energy of interaction with an applied field is much greater in ESR than NMR. This means that for a particular magnetic field strength the frequency of the radiation absorbed is much higher in ESR than in NMR and, in fact, in the microwave region (rather than the radiofrequency region encountered in NMR). This difference has two important consequences. First, ESR is much more sensitive than NMR and can be used with substances in solution at a concentration of 10^{-6} M rather than the higher concentration of 10^{-3} M needed for NMR. Second, the technology for detecting electron spin resonance differs and yields a spectrum that is not an absorption spectrum but the first derivative of the absorption spectrum. What is meant by that is shown in Figure 17-20; instead of a symmetric peak, a peak–trough configuration is observed. This peak–trough pair is, nonetheless, called a *line.* It might be thought that the peak–trough configuration would have a disastrous effect on the resolution of the spectra, as would be the case in the NMR spectrum shown in Figure 17-11 if every peak became a peak–trough. However, this is not the case because a typical ESR spectrum does not contain tens or hundreds of lines but usually fewer than ten lines which are not very closely spaced. This is because there is rarely more than one unpaired electron in a compound.

The frequency of ESR lines is determined by the energy difference between the two spin states of the electron when in a magnetic field H. This difference ΔE is

$$\Delta E = g(eh/4\pi m)H$$

in which e and m are the charge and mass of the electron, h is the Planck constant, and g is a universal constant called the g factor, with the value 2.00232 for a free electron. The expression $eh/4\pi m$ is called the Bohr magneton and is usually denoted β. The frequency ν of the radiation corresponding to the transition is defined by $\Delta E = h\nu$ so that this frequency is

$$\nu = g\beta H/h.$$

In ESR just as in NMR spectroscopy, it is not the strength of the applied field but the strength of the field seen by the unpaired electron that determines the resonance condition (that is, the frequency). This change in frequency from that of a free electron occurs in two parts. First, there is a local magnetic field produced by the molecule that contains the unpaired electron. Second, there are fields produced by other nearby molecules. The first effect is the larger one and is usually expressed as a change in the value of the g- factor (that is, an effective g- factor). The measurement of g for an unknown line is a valuable aid in identifying the source of the line. (This is of course equivalent to contains the unpaired electron. Second, there are fields produced by other nearby molecules. The first effect is the larger one and is usually expressed as a change in the value of the g- factor (that is, an effective g- factor). The measurement of g for an unknown line is a valuable aid in identifying the source of the line. (This is of course equivalent to identification by the value of the chemical shift in NMR.) The second effect provides the information about molecules that is desired; how this is done is described later.

There is another consequence of the local magnetic field on the energy of the unpaired electron. The magnetic moment of the electron also interacts with the magnetic fields of nearby spinning nuclei (usually the nucleus of the atom carrying the electron). This interaction gives rise to splitting so that a cluster of lines, rather than a single line, results. This is known as *hyperfine structure*. The number of lines per cluster is determined by the number of possible orientations of the magnetic moment of the particular nucleus with which the electron interacts. In general, if the spin quantum number of the nucleus is I, the ESR spectrum has a cluster of $2I + 1$ lines of equal intensity. Thus if an unpaired electron approaches a ^{14}N atom, whose nucleus has a spin of 1, there will be $2(1) + 1 = 3$ lines in the cluster. This is shown in the spectrum of the free-radical ion $(SO_3)_2NO^-$ (Figure 17-21). If the electron is near a proton, for which $I = \frac{1}{2}$, there are $2(\frac{1}{2}) + 1 = 2$ lines. The spacing between adjacent lines of a cluster is constant for a particular cluster but differs for each molecular type. The magnitude of the spacing depends on several factors that are beyond the scope of this book.

An electron may interact with several nuclei. Since the spacing produced by the interaction with each nucleus will differ, the total number of lines is the product of the number produced by each nuclear interaction. Thus if an unpaired electron is very near both a ^{14}N atom (giving three lines per cluster) and a proton (giving two lines per cluster) and if the proton-induced spacing is less than the ^{14}N-induced spacing, the spectrum will consist of three *pairs* or six lines, as shown in Figure 17-22).

Often an electron interacts with several protons that are magnetically equivalent. An example is the interaction between the free electron of the carbon of the ethyl radical $CH_3\dot{C}H_2$ and the two methylene protons on the *same* carbon atom. Each proton produces a pair of lines but the higher-frequency line of one pair has the same interaction energy as the lower-frequency line of the other pair, a phenomenon already seen in the case of NMR, as shown in Figures 17-6 and 17-7. Thus three lines are seen and the central line has twice the intensity of each of the outer lines. In general if there are n equivalent protons, the spectrum consists of

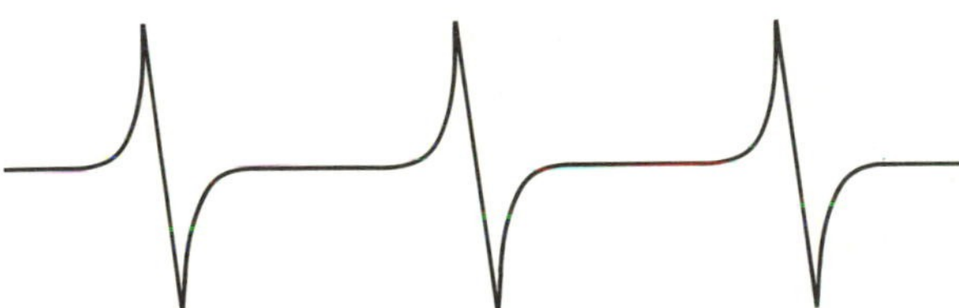

Figure 17-21
The ESR spectrum of the radical ion $(SO_3)_2NO^-$. The free electron is on the N atom and thus interacts with the N nucleus. The nuclear spin of ^{14}N is 1 so that hyperfine splitting to yield three lines occurs.

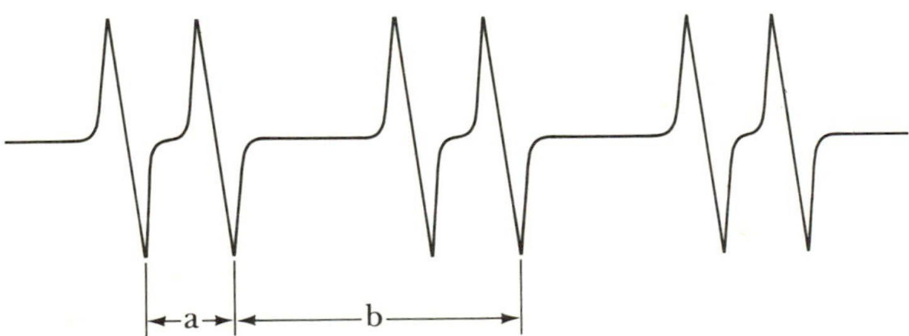

Figure 17-22
An ESR spectrum showing splitting caused by two different nuclei. The distances a and b are the separations induced by the proton and ^{14}N nucleus, respectively.

$n + 1$ lines, whose intensity distribution is given by the coefficients of the binomial distribution, namely,

n	Intensity distribution
0	1
1	1 1
2	1 2 1
3	1 3 3 1
4	1 4 6 4 1

This is shown in the seven-line spectrum of the benzene radical anion, as shown in Figure 17-23. This spectrum demonstrates one kind of information obtainable from ESR spectroscopy. If the free electron were localized on one of the six carbon atoms of this radical anion, the six protons would not be equivalent; instead the interaction would be with the single proton on that carbon atom and there would be two lines. (There is no interaction with the two adjacent carbon atoms because the nuclear spin of ^{12}C is zero.) However, the seven lines indicate that the free electron interacts with *each* of the six protons. Hence the free electron cannot be localized on a particular carbon atom but must move throughout the benzene ring, interacting equally with each proton.

Uses of ESR Spectroscopy

The existence of an ESR spectrum and the pattern of splitting can both give information in biochemical analysis. This is shown in the following examples.

Example 17-K □ Elucidation of the mechanism of an enzymatic reaction.

The reaction $A + B \rightarrow C + D$ has been shown by chemical analysis to proceed by one of the following mechanisms:

I.	II.
$Enz + A \rightarrow (EnzA)$	$Enz + A \rightarrow (EnzA)$
$EnzA + B \rightarrow (EnzAB)$	$EnzA \rightarrow 2C^{\cdot}$
$(EnzAB) \rightarrow (AB) + Enz$	$C^{\cdot} + B \rightarrow C + B^{\cdot}$
$(AB) \rightarrow C + D$	$B^{\cdot} + C^{\cdot} \rightarrow D$

There are no free-radical intermediates in scheme 1 but there are two such intermediates in scheme II. If lines are observed in an ESR spectrum of the reaction mixture, scheme I is unlikely. From the change in the spectrum with time it may be possible to determine whether $B^{\cdot}$ and $C^{\cdot}$ are intermediates. In scheme II, there are two different compounds containing free radicals; thus there should

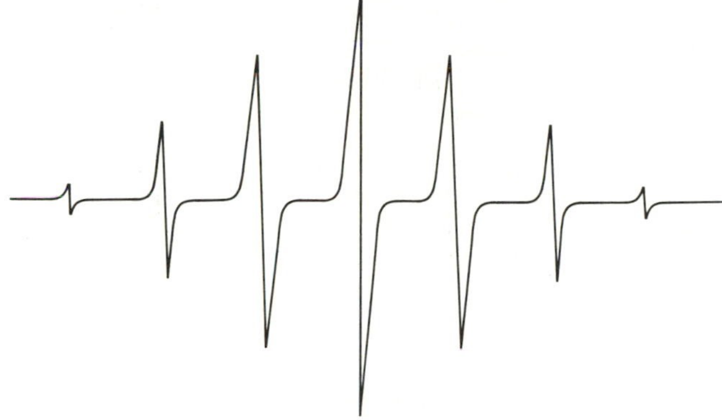

Figure 17-23
The ESR spectrum of the benzene radical anion in solution. [From *Physical Chemistry* by P. W. Atkins. Oxford University Press and W. H. Freeman and Company. Copyright © 1978.]

be two sets of lines (which may overlap). Furthermore, from the proposed structures of $B^\cdot$ and $C^\cdot$ the splitting pattern of each set can be predicted. If the spectrum does not support the presence of $B^\cdot$ and $C^\cdot$, another scheme having free-radical intermediates would have to be proposed. In no case would a radical-free mechanism be possible. If, on the contrary, no ESR spectral lines are seen, one may not conclude that the mechanism does not include a free-radical intermediate because the lifetime of the intermediate may be so short that its concentration is too low to be detectable. Observations of this sort have been extremely valuable in understanding the mechanism of molecular damage by ionizing radiation such as X-rays because well known radicals such as $OH^\cdot$ have been detected when water is irradiated.

☐ The site of binding of the Cu^{2+} ion by a protein. **Example 17-L**

An enzyme binds the Cu^{2+} ion but the binding site is unknown. It has been suggested that the ion binds to an imidazole N of histidine. This would produce splitting into three lines because the spin of ^{14}N is 1. Other known sites of binding of the Cu^{2+} ion are oxygen and sulfur; however, ^{16}O and ^{32}S both have zero spin and there would be no splitting. Three lines are observed, which suggests that binding to nitrogen is possible. An ESR spectrum was also obtained from a ^{15}N-substituted protein. This isotope has a spin of $\frac{1}{2}$. Confirming the first conclusion, only two lines are observed. Thus, the Cu^{2+} ion is bound to a nitrogen atom although there is no proof from this experiment that it is a histidine nitrogen.

☐ Binding of the Ca^{2+} ion to a bacterial enzyme. **Example 17-M**

Extensive studies of Ca^{2+} ion binding to a particular bacterial enzyme have suggested that three ions bind to three different imidazole nitrogens. The Ca^{2+} ion is not paramagnetic so that there is no ESR spectrum of the system. The Mn^{2+} ion, which is paramagnetic, also binds to the enzyme. Studies in which both Mn^{2+} and Ca^{2+} ions are present indicate that the Ca^{2+} ion competitively inhibits binding of the Mn^{2+} ion. Furthermore, the number of Mn^{2+} and Ca^{2+} binding sites are the same. Thus it is likely that the two ions bind to the same site (although the possibility of distinct but overlapping sites is not ruled out). The ESR spectrum of the Mn^{2+} complex shows three lines, which is consistent with binding to a nitrogen atom. To ascertain that ^{14}N nuclei are responsible for the splitting, bacteria are grown in medium containing ^{15}N instead of ^{14}N. The spin of the ^{15}N nucleus is $\frac{1}{2}$ so that splitting induced by ^{15}N would yield two lines rather than three. Two lines were observed when Mn^{2+} was added to the ^{15}N-labeled enzyme, confirming the binding to nitrogen. It should be noted that if the ^{15}N substitution were not complete so that the enzyme sample contained both ^{14}N- and ^{15}N-labeled molecules, $3 + 2 = 5$ lines would have been observed.

The interaction energy between an unpaired electron and an external magnetic field depends on the relative orientation of the electron spin vector and the magnetic field vector. Thus with a particular value of the external field the microwave frequency at which resonance occurs (which is proportional to the interaction energy) also depends on this relative orientation. The frequency is at a maximum when the spin and the field vectors are parallel. In most experimental arrangements, all molecules are in motion and hence do not have a preferred orientation. However, with a rigid system, such as a crystal, ESR can be used to determine the orientation of the molecules that contain the unpaired electron. ESR studies of crystals have been valuable in understanding the structure of the proteins myoglobin and hemoglobin. Both of these proteins contain heme, a planar molecule having a paramagnetic iron atom. ESR spectra of crystals of myoglobin and hemoglobin have been obtained with the crystals oriented at various angles with respect to the external magnetic field. By studying the variation of the resonance signal with crystal orientation, the angle of the heme group with respect to the crystal axes has been determined. This was important information that aided in the elucidation of the three-dimensional structure of these proteins by X-ray diffraction analysis of the crystals.

The orientation effect just described also gives information about the motion of molecules in solution. The theory, which relates the rate of tumbling with line width, is complex and beyond the scope of this book. The result is that *as the rate of tumbling increases, the lines become narrower*.* In addition to interpreting crystal data, this fact provides a

* One way to think about this is in terms of the Heisenberg Uncertainty Principle. This principle states that as the position of a particle becomes better defined, its energy (which can be measured as a frequency) is less well-defined—that is, the measured frequency covers a broad range and the spectral line is wide. As tumbling increases, position becomes poorly known so that the energy is defined more precisely—that is, the frequency range or line width becomes smaller.

powerful technique both for measuring molecular motion and for determining certain features of the structure of macromolecules when applied to spin labels like the nitroxide discussed earlier.

The nitroxides are stable molecules having an unpaired 2p electron. This electron produces a strong ESR spectrum, which consists of three lines because the electron is on a ^{14}N nucleus. The chemical structure of one of the more commonly used nitroxides in ESR studies, 2,2,6,6,-tetramethylpiperidinol-N-oxyl (called TEMPOL) is shown below.

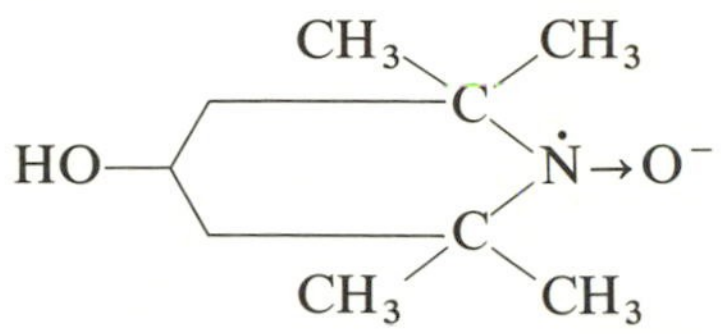

$$HO - C_5H_{?} \text{ ring: } (CH_3)_2C - \dot{N} \rightarrow O^- - C(CH_3)_2$$

A simple way to determine the effect of mobility on an ESR spectrum is shown in Figure 17-24. First, at a single temperature the mobility of TEMPOL in aqueous solution is decreased by addition of increasing amounts of the highly viscous substance glycerol. Second, the viscosity is increased and thermal motion is decreased by lowering the temperature. Third, the molecules are locked in the solvent by freezing. Notice how the three lines broaden and deform as the mobility decreases.

Nitroxide spin labels have been used in many biochemical studies. The technique is to select a biochemical that can interact with a biological macromolecule or system and form a covalent bond between the nitroxide and the biochemical. The coupling can sometimes be done at several points in the molecule without affecting the ability of the molecule to interact with the system of interest. One then uses ESR spectroscopy to answer questions about the mobility and the environment of the biochemical. Several kinds of experiments that have been done are described in the examples that follow.

Example 17-N

☐ The fluidity of the interior of the erythrocyte membrane.

When red blood cells (erythrocytes) are placed in a buffer of low osmotic strength, the cells swell and become porous. Hemoglobin and other cell contents leak out leaving a pure cell membrane. The fatty acid stearic acid is easily incorporated into this cell membrane and its carboxyl group may be located at either the inner or outer surfaces of the membrane while the organic chain remains inside. Two stearates containing a nitroxide have been synthesized with the spin label either near the carboxyl group or at the other end (see structures I and II, Figure 17-25). These molecules have been separately incorporated into two membrane preparations and ESR spectra of each were obtained, as shown in the figure. The spectrum obtained when structure I is present is much broader than that seen when structure II is present. By comparison to standards such as in Figure 17-24, the

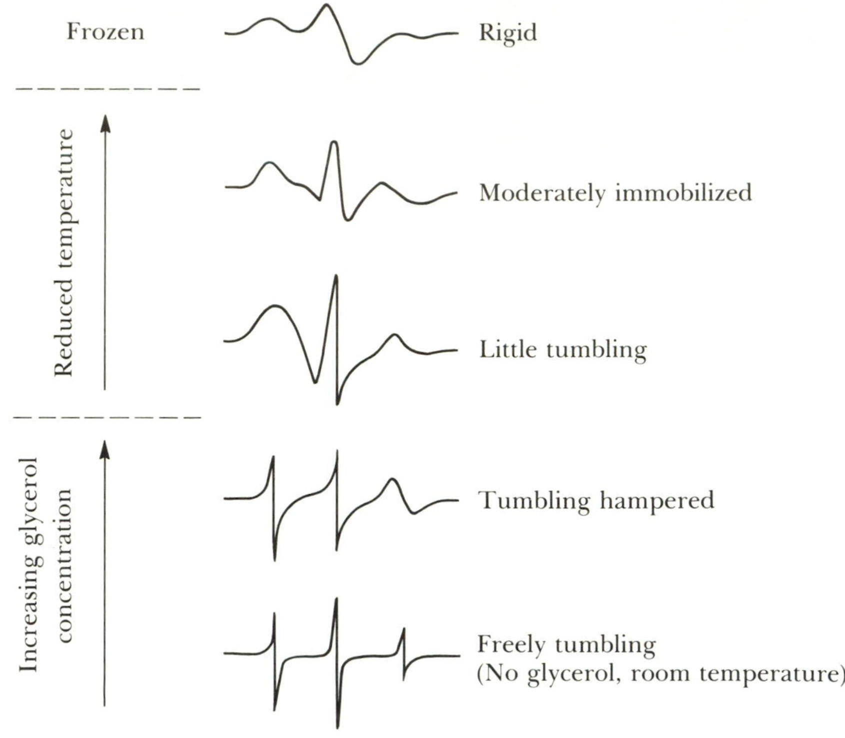

Figure 17-24
Typical ESR spectra showing line broadening as the mobility of the spin label is reduced. The lines are narrowest in aqueous solution and at room temperature. They broaden if glycerol is added to increase the viscosity. Broadening continues as the temperature is lower. When the sample is frozen, the line at the far right is so broad that it is difficult to discern.

data suggest that the carboxyl end of stearic acid is not free to move and the organic end moves freely—that is, the interior of the membrane is sufficiently fluid to allow substantial rotational freedom of incorporated fatty acid molecules. Note that a rigid-type spectrum can be seen even though the membranes themselves are tumbling in solution; this is because the membranes move only very slightly during the very small time interval required to absorb high-frequency microwave radiation. Studies of this sort have been used to understand membrane function. For instance, the addition of other substances can increase or decrease fluidity and at the same time affect the rate of transport of various molecules. Furthermore by lowering the temperature it is possible to determine the temperature at which the internal region of the membrane becomes solidlike. This has been correlated with transmembrane transport rates.

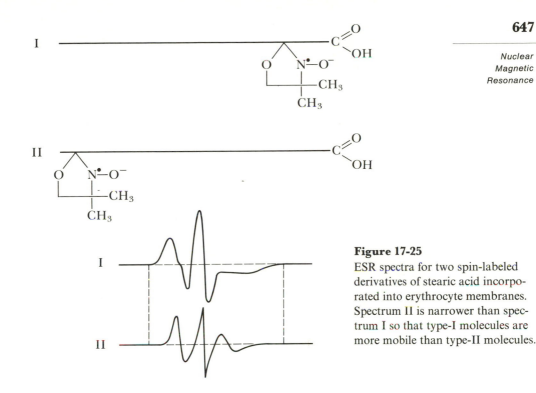

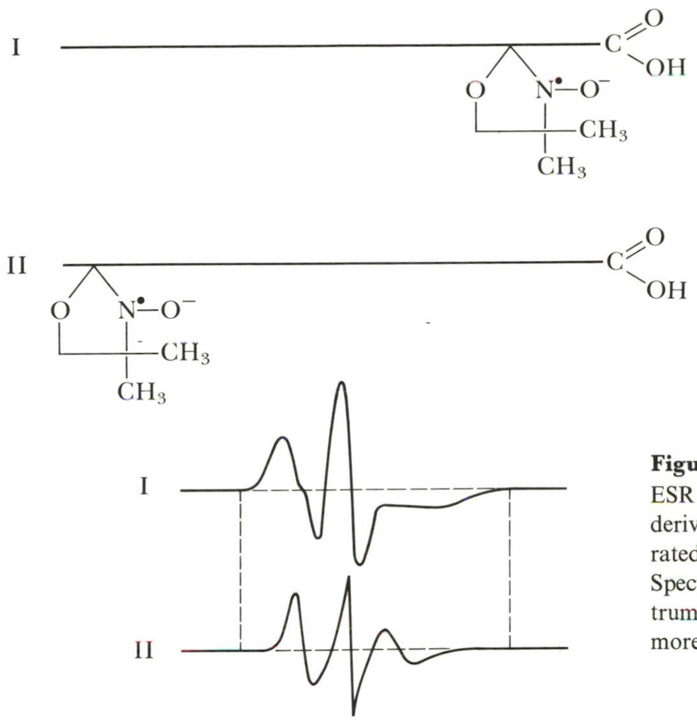

Figure 17-25
ESR spectra for two spin-labeled derivatives of stearic acid incorporated into erythrocyte membranes. Spectrum II is narrower than spectrum I so that type-I molecules are more mobile than type-II molecules.

Example 17-O

☐ The dynamic state of the membrane.

Roger Kornberg and Harden McConnell [*Biochemistry* 10(1971): 1111–1120] performed the following experiments that demonstrated that in a synthetic closed lipid bilayer (a *vesicle*) internal and external lipids exchange positions. They prepared vesicles consisting only of phosphatidylcholine and a spin-labeled analogue in which the spin label is at the polar end (Figure 17-26). Thus, the spin label is localized only at the inner and outer surfaces of the membrane. An ESR spectrum was measured. The vesicles were then treated with ascorbic acid, which does not penetrate the membranes but reduces the external nitroxide groups. Another ESR spectrum was obtained; this had only half the intensity (peak height) of the first spectrum because half of the nitroxide groups had been destroyed by the ascorbic acid. The peak height was continually measured for many hours and it was observed to decrease with a half-life of 6.5 hours. The explanation for this decrease is that internal spin-labeled phosphatidylcholine molecules come to the surface where the ascorbic acid immediately eliminates the spin label. The interpretation of this fact is that *pairs of phosphatidylcholine molecules exchange*. This process has been termed phospholipid flip-flop and may be important for the following reason. Many substances are able to move through the membrane without an energy supply (passive transport). The rate of transport

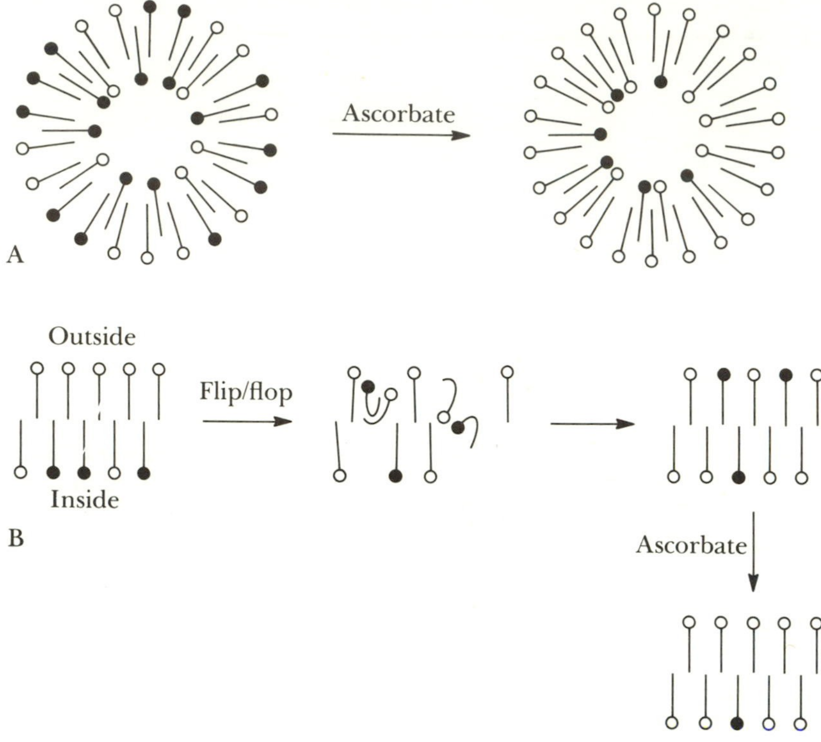

Figure 17-26
The flip-flop mechanisms. Panel A shows a bilayer vesicle consisting of phosphatidylcholine (open circles) and spin-labeled phosphatidylcholine (solid circles). Ascorbate, which cannot penetrate the vesicles, chemically reduces the external spin-labeled phosphatidylcholine. Panel B shows a portion of the vesicle in which flip-flop has occurred. Since ascorbate remains in contact with the external surface, spin-labeled phosphatidylcholine molecules that are brought to the surface by the flip-flop mechanism are immediately reduced. Thus, the total number of spin-labeled molecules decreases continually.

of the Cl^- ion, for instance, is consistent with binding of Cl^- to an external phosphatidylcholine molecule followed by flip-flop to the other side, where the ion is discharged. This has suggested to membrane biologists that the flip-flop may be a mechanism for the passive transport of some ions across naturally occurring biological membranes.

Example 17-P ☐ The depth of a crevice surrounding the binding site of a protein.

If a spin label were rigidly bound to the binding site of a protein, the ESR spectrum would be of the immobilized type. If instead it were attached to the binding site by a flexible connector, it would have the tumbling type of spectrum. However, if the binding site were in a deep crevice, the flexible connector would have to be long enough that the spin label is not in the crevice in order to obtain a tumbling spectrum. This reasoning has been used in studying the dimensions of the active site of the enzyme carbonic anhydrase [R. H. Ehrlich, D. K. Starkweather, and C. F. Chignell. *Mol. Pharmacol.* 9(1973):61–69], an enzyme that possesses, at the end of a crevice, a zinc atom to which a sulfonamide can bind. A spin label was linked to the sulfonamide by connectors of various lengths and ESR spectra were obtained for each compound. The data showed that the zinc atom is about 14 Å from the surface of the protein, which provides the depth of the crevice.

SELECTED REFERENCES

Anet, F. A. L., and G. C. Levy. 1973. "Carbon-13 Nuclear Magnetic Resonance Spectroscopy." *Science (Wash., DC)* 180:141–145.

Atherton, N. M. 1979. *Electron Spin Resonance: Theory and Applications.* Halsted Press.

Becker, E. D. 1969. *High Resolution Nuclear Magnetic Resonance.* Academic Press.

Bovey, F. A. 1969. *Nuclear Magnetic Resonance Spectroscopy.* Academic Press.

Bovey, F. A. 1972. *High Resolution Nuclear Magnetic Resonance of Macromolecules.* Academic Press.

Carrington, A., and A. D. McLachlan. 1979. *Introduction to Magnetic Resonance.* Halsted Press.

Dugas, H. 1977. "Spin-Labeled Nucleic Acids." *Acc. Chem. Res.* 10:47–54.

Dwek, R. A. 1973. *Nuclear Magnetic Resonance in Biochemistry.* Clarendon Press.

Dwek, R. A., I. D. Campbell, R. E. Richards, and R. J. P. Williams. 1977. *Nuclear Magnetic Resonance in Biology.* Academic Press.

Feher, G. 1970. *Electron Paramagnetic Resonance with Applications to Selected Problems in Biology.* Gordon and Breach.

Hamilton, C. L., and H. M. McConnell. 1968. "Spin Labels," in *Structural Chemistry and Molecular Biology,* edited by A. Rich and N. Davidson, pp. 115–149. W. H. Freeman.

James, T. 1975. *Nuclear Magnetic Resonance in Biochemistry.* Academic Press.

Jost, P. C., and O. H. Griffith. 1978. "The Spin-Label Technique," in *Methods in Enzymology,* vol. 49, edited by C. W. L. Hirs and S. N. Timasheff, pp. 369–418. Academic Press.

Kearns, D. R., and R. G. Schulman. 1974. "High Resolution NMR Studies of the Structure of Transfer RNA and Other Polynucleotides in Solution." *Accounts Chem. Res.* 7:33–39.

Knowles, P. F., D. Marsh, and H. W. E. Rattle. 1976. *Magnetic Resonance of Biomolecules.* Wiley. Levy, G., ed. 1976. *Topics in Carbon-13 NMR Spectroscopy.* Wiley.

McDonald, C. C., and W. D. Phillips. 1970. "Proton Magnetic Resonance Spectroscopy," in *Fine Structure of Proteins and Nucleic Acids,* vol. 4, edited by G. D. Fasman and S. N. Timasheff, pp. 1–48. Dekker.

McDonald, C. C., W. D. Phillips, and J. D. Glickson. 1971. "Nuclear Magnetic Resonance Study of the Mechanism of Reversible Denaturation of Lysozyme." *J. Am. Chem. Soc.* 93:235–246.

Redfield, A. G. 1978. "Proton Nuclear Magnetic Resonance in Aqueous Solutions," in *Methods in Enzymology*, vol. 49, edited by C. W. L. Hirs and S. N. Timasheff, pp. 253–269. Academic Press.

Roberts, G. C. K., and O. Jardetzky. 1970. "Nuclear Magnetic Resonance Spectroscopy of Amino Acids, Peptides, and Proteins." *Adv. Protein Chem.* 24:448–545.

Sheard, B., and E. M. Bradbury. 1970. "Nuclear Magnetic Resonance in the Study of Biopolymers and Their Interaction with Ions and Small Molecules." *Prog. Biophys. Mol. Biol.* 20:187–246.

Swartz, H. M., J. R. Bolton, and D. C. Borg. 1972. *Biological Applications of Electron Spin Resonance.* Wiley-Interscience.

Wertz, J. 1972. *Electron Spin Resonance: Elementary Theory and Practical Applications.* McGraw-Hill.

Wertz, J., and J. R. Bolton. 1974. *Electron Spin Resonance: Elementary Theory and Applications.* McGraw-Hill.

Wuthrich, K. 1976. *NMR in Biological Research: Peptides and Proteins.* American Elsevier.

PROBLEMS

17–1. How many proton resonances would you expect to see with (a) liquified methane, (b) liquified ethane, (c) chloroform, (d) formaldehyde, (e) acetone, (f) methanol, (g) ethanol in CCl_4?

17–2. Would you expect the proton spectrum of a mixture of phenylalanine and glycine to be the sum of the spectra of each? Would the total concentration of each be important?

17–3. Explain why ^{13}C spectra contain fewer lines than proton spectra and why the number of lines would increase as the abundance of ^{13}C is increased.

17–4. Suppose that you are trying to determine the mechanism of action of a certain enzyme. You hypothesize that the substrate passes through two intermediates in being converted into the product. The two intermediates are known compounds but extremely difficult to identify in the reaction mixture by standard chemical and physical tests. Explain how NMR might help.

17–5. Suppose that an enzyme has a single histidine. On the addition of the substrate, each histidine line moves from 2 ppm to 4 ppm. Can you unambiguously state that histidine is in the binding site?

17–6. Proton spectra of proteins invariably show pronounced changes on denaturation. The usual changes are line positions and widths. Would you expect the appearance or disappearance of splitting to be one of the changes?

17–7. A particular enzyme reaction involves an enzyme, a cofactor, and a substrate molecule. Design an experimental protocol using NMR to distinguish the following possibilities: (a) the enzyme binds the cofactor before the substrate; (b) the substrate binds the cofactor and the complex binds to the enzyme; (c) the enzyme binds the substrate but there is no reaction until the cofactor is added.

17–8. What differences might be expected between the spectra of water and ice? Explain.

17–9. What kinds of spectral changes might accompany dimerization of a protein?

17–10. A proton spectrum is obtained for an enzyme. About fifty lines show chemical shifts about twice those normally found for proteins. Furthermore, the lines are broader than those normally encountered. What might be the cause, and how could you test your hypothesis?

17–11. Would you expect there to be observable differences in the proton spectrum of a linear and a circular DNA?

17–12. Two guanines (bases 14 and 32) in the polynucleotide chain of a tRNA molecule are being studied to determine whether they are engaged in hydrogen bonds. To do this, a concentrated solution of the tRNA is diluted tenfold in D_2O and NMR spectra are observed as a function of time. No guanine-14 resonance is seen. The guanine-32 line gradually disappears over the period of a few hours. What can be said about hydrogen bonding of these guanines?

17–13. If dry proteins are X-irradiated and then dissolved in water, an ESR line is observed that gradually disappears over a period of several hours. What can be concluded?

17–14. You wish to study the binding of the Mg^{2+} ion to DNA. Can this be done by ESR spectroscopy?

17–15. The Mn^{2+} ion binds to a protein and produces six lines arranged in three pairs. What information does this give you about the Mn^{2+} binding site?

17–16. You are given two solutions. Solution A contains a DNA molecule whose molecular weight is 5×10^6 and solution B contains a double-stranded decanucleotide (five base pairs). $MnCl_2$ is added to each solution and then the solutions are mixed up accidentally. To distinguish them, ESR spectra are obtained. Tube 1 gives a line that is much narrower than that of tube 2. Which tube contains solution A?

17–17. A strange protein-digesting bacterium is isolated. This bacterium is unable to synthesize amino acids and cannot grow if free amino acids are put in the growth medium. It has an absolute requirement for proteins, which apparently adsorb to the cell wall and are degraded to amino acids on the cell surface. This bacterium fails to grow at 39°C. To investigate this temperature dependence the following experiment is done. The bacteria are fed a protein to which TEMPOL has been coupled. If the bacteria are collected from the TEMPOL-protein growth medium or resuspended in phosphate buffer, this suspension is found to give an ESR signal. If these bacteria are treated with ascorbate, the ESR signal disappears in several minutes. If this experiment is done at 40°C, the collected bacteria do not yield an ESR signal. What is the probable defect above 39°C?

17–18. Spin-labeled morphine has been used to study binding of morphine to various receptors. Explain how examining the band width of the ESR lines could be used to study binding.

Six

MISCELLANEOUS METHODS

18

Ligand Binding

A common event in living systems is the non-covalent binding of one or more molecules to a single macromolecule. A molecule bound to a macromolecule is called a *ligand*. In the cases usually studied, the ligand is a small molecule such as the *substrate* or a *cofactor* of an enzyme. The ligand may also be a macromolecule, as in the case of a regulatory molecule, such as a repressor protein binding to DNA. The study of ligand binding is important for the following reasons: (1) to understand the mechanisms of action of enzymes and of regulatory systems often requires knowledge of the number of binding sites and the strength of the binding, and (2) there are probably as many intracellular reactions in which binding is an integral part as there are chemical reactions.

There are many types of binding. For example, a molecule may have only a single binding site or there may be many sites. In the latter case, the sites may be either identical or different and they may be for only one kind of molecule or several distinct molecules. Furthermore, multiple sites may be interdependent in the sense that occupation of one site by a ligand may affect the affinity of other sites for their ligands. In order to understand biochemical systems in which binding plays a role, it is necessary to determine the strength and number of binding sites and the magnitude of the interaction between multiple sites.

Experimentally the measurement of binding is straightforward. However, interpretation of the data requires different modes of analysis that depend on various features of the binding process. These features can be incorporated into a mathematical theory of binding that is quite useful.

In this book, mathematical analyses have been avoided because they tend to detract from our understanding the experimental techniques. However, it is not possible to describe binding equilibria in any useful fashion without resorting to equations. Fortunately, the mathematics does not got beyond simple algebra. The analysis begins by first writing an equation for the dissociation constant of a binding reaction. The equation is then restated in terms of a particularly useful parameter, namely, the average number of ligand molecules bound to each macromolecule in a solution having a particular concentration of macromolecule and of added ligand. (For some types of binding the equation is very complex so that simplifying assumptions are made that apply to that mode of binding.) For convenience, the resulting equation is then rearranged to a form that enables the data to be presented usually in a straight-line plot; from this plot the number of binding sites and the dissociation constants can be obtained by measuring the slope and the intercepts. In the following sections, we proceed from the simple case of a single binding site to the more complex cases of multiple, dependent binding sites. It is important to recognize, though, that a linear plot will occur only when the data are plotted using the equation that actually describes the mode of binding and that the correct equation is not known in advance. Thus, the usual procedure is to plot the data using various equations that are available; *the equation that is found to yield a linear plot is assumed to be a correct one.*

In the following sections, the more common modes of binding and the relevant equations will be described.

Binding by Molecules with One Binding Site

Consider a macromolecule P that has a single binding site for a molecule A. At saturation, one mole of P combines with one mole of A to form a complex PA.

$$\mathrm{P} + \mathrm{A} \rightleftarrows \mathrm{PA} \tag{1}$$

The dissociation constant K_d for this association reaction is defined as

$$K_d = \frac{[\mathrm{P}][\mathrm{A}]}{[\mathrm{PA}]} \tag{2}$$

in which the brackets denote concentrations. (In chemistry, the equilibrium constant $K = 1/K_d$ is usually used whereas biochemists and molecular biologists prefer K_d. It is important to be aware of this difference when consulting other references.)

Equation (2) indicates that the fraction of ligands bound is affected by the value of [P] and [A]. A useful parameter for plotting data is r, the number of moles of ligand bound to one mole of macromolecules; it is defined as follows:

$$r = \frac{[\text{A}]_{\text{bound}}}{[\text{P}]_{\text{total}}} = \frac{[\text{PA}]}{[\text{P}] + [\text{PA}]} \qquad (3)$$

Most equations that describe binding are expressed in terms of K_d and r. To evaluate r, a particular concentration [A′] of ligand is added to a known concentration [P′] of the macromolecule and then either the concentration of bound ligand [PA] or of unbound ligand [A] is measured. Only one of these need to be measured because [A′] is known at the outset and [A] + [PA] = [A′]. The value of [P] also need not be measured since [P] + [PA] = [P′]. Methods for measuring [A] and [PA] will be described in a later section.

Combining equations (2) and (3) yields

$$r = \frac{[\text{A}]}{K_d + [\text{A}]} \qquad (4)$$

which is known as the *Langmuir isotherm*. This equation is usually rewritten in one of two forms that allow the data to be plotted on a straight-line graph. One form is

$$\frac{1}{r} = \frac{K_d}{[\text{A}]} + 1 \qquad (5)$$

Using this equation, a plot of $1/r$ versus $1/[\text{A}]$ (which is called a *double-reciprocal* plot) gives a straight line, whose slope is K_d. Another useful form is

$$\frac{r}{[\text{A}]} = \frac{1}{K_d} - \frac{r}{K_d} \qquad (6)$$

in which a plot of $r/[\text{A}]$ versus r gives a straight line having a slope of $-1/K_d$.* Equations (5) and (6) are both useful in handling data; however, in the example that follows we will see, that depending on the design of a particular experiment used to measure binding, one equation is usually preferable to the other.

Example 18-A ☐ Determination of K_d for binding of the Mg^{2+} ion to an enzyme having a single binding site.

* There is an advantage in using either equation (5) or (6) rather than equation (4) to determine K_d although, in both cases, the same parameters are measured. This will become clear when we consider a macromolecule with multiple binding sites; in this case, the shape of the curves obtained by using equation (5) or (6) provides, by simple inspection, certain information about the binding reaction and the effect of binding on the macromolecule itself.

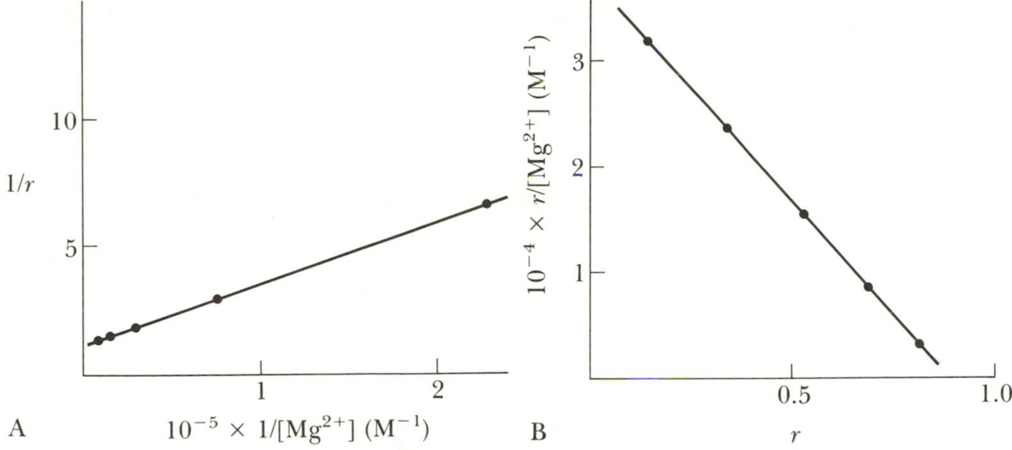

Figure 18-1
Binding data plotted according to equation (5) (panel A) and equation (6) (panel B).

In this experiment, various amounts of Mg^{2+} are added to a 100 μM solution of an enzyme and the amount of bound Mg^{2+} is measured. The data obtained are plotted first according to equation (5) yielding the graph shown in Figure 18-1A. K_d is the slope of the straight line drawn through the points. Note that if the experimental points were more evenly spaced, the line could be drawn more accurately. If the same data are plotted using equation (6), the graph shown in Figure 18-1B is obtained; here the points are more evenly spaced and the line can be drawn more precisely. Thus in this experiment, equation (6) is superior to equation (5). This will not always be the case because the spacing between the points is determined both by the values chosen by the experimenter for the amount of added Mg^{2+} and by K_d. Thus, one experimental design might give rise to clustered points when equation (6) is used and evenly spaced points with equation (5), and another experimental design might give the converse. However, the values of added ligand are not always freely chosen by the experimenter. For example, measurements at low concentration of added ligand (that is, at the right-hand end of the graph in Figure 18-1A) are usually more difficult to make and have a larger experimental error than those at higher concentration; thus, the experimental points are frequently clustered as in Figure 18-1A.

The preceding statements are not meant to imply that the means of displaying the data determines the accuracy of evaluating K_d or any other binding parameter. Clearly the information content of a particular set of binding data is independent of the method used to interpret the data. The plots are visually different but a statistical treatment to obtain the best line through the points must yield the same value of K_d and the same error in its determination, regardless of how the data are plotted. The point is that a statistical analysis is not usually carried out so that some graphical methods show some aspects of the information in the data more clearly.

Macromolecules with Several Binding Sites

Many biological macromolecules have more than one binding site. For example, a typical nucleic acid molecule can bind hundreds or even thousands of metal ions and a typical protein molecule can sometimes bind several hundred H^+ ions. Furthermore, the binding sites are not always identical; for example, an enzyme usually has distinct binding sites for a cofactor and a substrate molecule.

Multiple binding sites may either be *independent* of one another or *interactive*, in which case occupation of one site influences binding at a second site. We will discuss independent binding sites first, distinguishing between the cases of tight binding and weak binding.

Tight Binding

In the case of tight binding, it is assumed that every ligand molecule is bound if the number of ligand molecules added to a solution of macromolecules is less than the total number of binding sites *in the solution*. Also, if the amount of added ligand exceeds the number n of the binding sites in the solution, each of the n sites is occupied and additional ligand molecules remain unbound. From the concentration of added ligand at which some ligand remains unbound, the number of binding sites can be measured, as this is the concentration at which the amount of bound ligand reaches a maximum. This simple method of determining n when binding is very tight is shown in the following example.

Example 18-B ☐ Determination of the number of binding sites on a macromolecule.

Consider a protein with an unknown number of binding sites for some molecule X. To a solution of this protein at a concentration of 2 μM are added aliquots of X and a plot is made of free versus added ligand. The data which are obtained are shown in Figure 18-2. We can see that binding continues linearly until the concentration of added ligand is 6 μM; at this point the concentration of free ligand increases markedly (but still linearly) so that saturation of the sites occurs at 6 μM. Thus, the number of binding sites is 6 μM/2 μM $= 3$.

Weak Binding

The general equation that describes weak binding is rarely useful except when applied to two simple cases, namely, identical independent

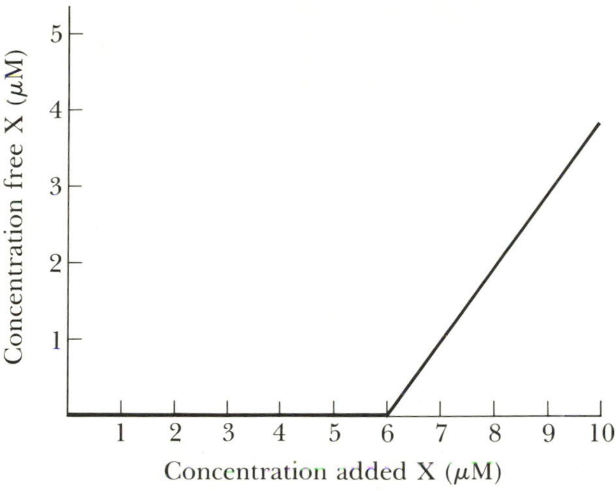

Figure 18-2
A binding curve illustrating tight binding. The initial protein concentration is 2 μM. There is no free X until the amount of added X exceeds 6 μM. Therefore 2 μmol of protein bind 6 μmol of X and there must be three binding sites for X per protein molecule.

sites and non-identical independent sites for the same ligand. When weak binding is cooperative, a separate theory is required; cooperative binding is discussed in a later section.

1. Identical and independent weak binding sites

If all binding sites are identical and independent, the relevant equations analogous to equations (5) and (6) are the following:

$$\frac{1}{r} = \frac{1}{n} + \frac{K_d}{n[A]} \tag{7}$$

$$\frac{r}{[A]} = \frac{n}{K_d} - \frac{r}{K_d} \tag{8}$$

in which K_d is an average dissociation constant. When equation (7) is used, a plot of $1/r$ versus $1/[A]$ gives a straight line having a slope of K_d/n and a y-intercept of $1/n$. With equation (8) (often called the *Scatchard equation*) a plot of $r/[A]$ versus r gives a line having a slope of $-1/K_d$ and an x-intercept of n/K_d. [Note that equations (5) and (6) are special cases of equations (7) and (8) with $n = 1$.]

A valuable aspect of this method of data analysis is that if the binding sites are *not* both identical and independent, neither of the plots, $1/r$ versus $1/[A]$ or $r/[A]$ versus r, will give straight lines. This is shown in

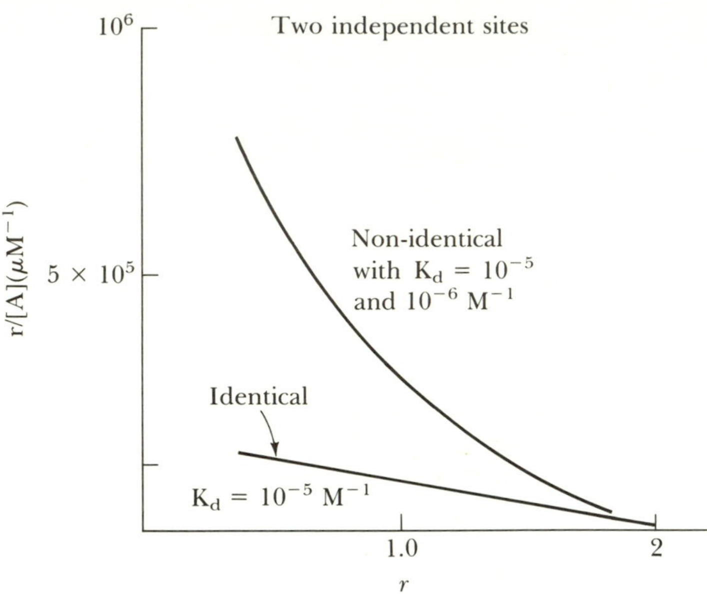

Figure 18-3
Curves resulting from use of equation (8) when there are two binding sites. A straight line only occurs if the binding sites have the same value of K_d. A curved line is a clear indication that there is more than one binding site.

Figure 18-3, which compares two macromolecules having, in one case, two identical sites and, in the other, two non-identical sites. In a typical experiment, a smooth curve will result. This is treated in the following section. If the sites are identical but dependent, that is, if binding is cooperative, a curved line will also result. This will be discussed later.

In the analysis of binding data, it is always a reasonable procedure to try first to apply equations (7) or (8). If a linear graph results, there is no reason to test any other equations; if a curved graph results, other equations must be tried.

2. Macromolecules with different but independent binding sites for the same ligand

Many macromolecules have distinct types of sites for the same ligand and often each has a different binding constant. For instance, in the binding of H^+ ions by a protein the sites of binding are chemically different groups such as amino, carboxyl, and hydroxyl groups. Other examples are the various binding sites on DNA for RNA polymerase and the different ribosome-binding sites on messenger RNA molecules. If binding sites are independent, the theoretical treatment is simple; however, extraction of information from experiments is difficult unless the number of classes of binding sites is small and the values of K_d differ greatly for each class. This is shown in the following discussion.

Consider a molecule having several different classes of binding sites denoted by 1, 2, 3, . . . When a ligand is added, binding can occur to any site, and those sites with greater affinity for the ligand will be filled first. For any particular concentration of ligand, each class will have a particular value of r (for example, $r_1, r_2, \ldots$) that will be determined by the dissociation constant for that class. Experimentally r, the *average* number of ligand molecules per macromolecule, is measured and this equals $r_1 + r_2 + r_3 + \cdots r_n$.

If there are n_x sites of class x, and so forth, the relevant equation is

$$r = \frac{n_1[A]}{K_{d,1} + [A]} + \frac{n_2[A]}{K_{d,2} + [A]} + \ldots \quad (9)$$

A plot of r versus $[A]/K_d$ for two classes of sites having several ratios of K_d is shown in Figure 18-4. Note that only when the values of K_d differ greatly is it clear that there is more than a single class. In order to determine the number of classes unambiguously, it is necessary that the value of K_d be such that nearly all sites of one class are filled before there is appreciable filling of sites in a second class.

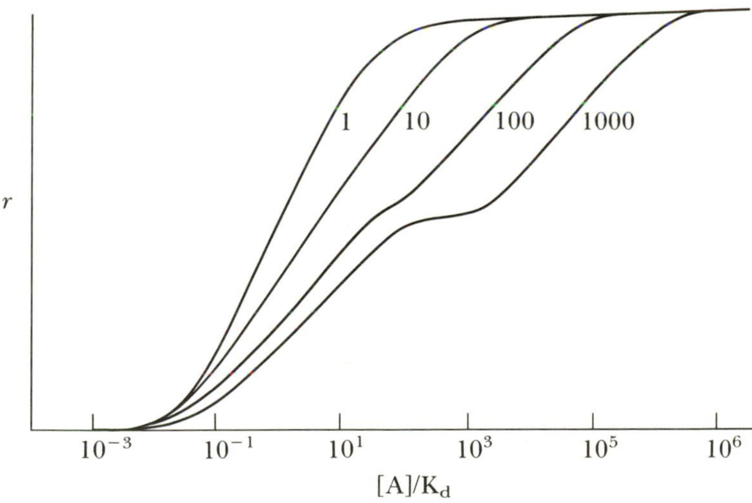

Figure 18-4
Binding data plotted according to equation (9). This type of plot is prepared when the use of equation (8) gives a curved line (as in Figure 18-3). In this figure, the molecule has two classes of sites. The numbers next to each curve is the ratio of the values of K_d for each class. Note that the two values of K_d must differ substantially before there is a clear indication that there are only two classes; if there were more than two sites, the difference in the dissociation constants would have to be even greater.

The basis of the theory of cooperative binding is the following. A molecule binds to one site of a macromolecule that has many weak binding sites. In response to this binding, one of two events occurs—either the affinity of all binding sites changes abruptly to a new value or the affinity of one or more of the sites changes gradually. In the second case, the affinity continues to change as more ligand molecules bind. The affinity can either increase (*positive cooperativity*) or decrease (*negative cooperativity*). The simplest case to analyze is that of *highly cooperative binding* of a single type of ligand.

Highly cooperative binding of a single ligand

In highly cooperative binding of a ligand at one site in a macromolecule the affinity of the other binding sites is altered so that each site is immediately filled if ligand molecules are available. Thus, if there are n binding sites per macromolecule, we need only consider the equation $\mathrm{P} + n\mathrm{A} \rightleftarrows \mathrm{PA}_n$, for which $K_\mathrm{d} = ([\mathrm{P}][\mathrm{A}]^n/[\mathrm{PA}_n])$ and

$$r = \frac{[\mathrm{A}]_{\mathrm{bound}}}{[\mathrm{P}]_{\mathrm{total}}} = \frac{n[\mathrm{PA}_n]}{[\mathrm{P}] + [\mathrm{PA}_n]} = \frac{n[\mathrm{A}]^n}{K_\mathrm{d} + [\mathrm{A}]^n} \tag{10}$$

which is commonly rewritten as

$$\frac{r}{n - r} = \frac{[\mathrm{A}]^n}{K_\mathrm{d}} \tag{11}$$

This is known as the *Hill equation*. Another common form of this equation uses a parameter Y—the *fraction* of binding sites that are filled. Since $Y = r/n$,

$$\frac{r}{n - r} = \frac{Y}{1 - Y} = \frac{\text{fraction of sites filled}}{\text{fraction of sites not filled}}$$

and equation (11) becomes

$$\frac{Y}{1 - Y} = \frac{[\mathrm{A}]^n}{K_\mathrm{d}} \tag{12}$$

In order to use equation (12) to evaluate K_d or n, Y must be measured. Since Y is the fraction filled, it is necessary to determine the maximum amount of A that can be bound, the *saturation value*. Thus Y is sometimes called the *fractional saturation of sites*. Equation (12) should be tried whenever a plot according to equation (8) gives a curved line.

Highly cooperative binding is easily detected by a *Hill plot*, namely a graph of $\log(Y/1 - Y)$ versus $\log[\mathrm{A}]$. According to equation (12), this plot should yield a straight line with slope n and intercept $-\log K_\mathrm{d}$.

Equation (12) is a statement of an idealized situation, when cooperativity is so great that there is immediate filling of each site after the first site is filled. This is of course not a real situation and experimentally a value of n equal to the number of binding sites is never observed (when $n > 1$). For example, hemoglobin has four binding sites for O_2 yet n is observed to be 2.8. Thus, the observed value of the slope of a Hill plot, which is called the Hill coefficient, n_H, is a qualitative indicator of the degree of cooperativity. The value of n_H is always less than the actual number of binding sites. Note that a Hill plot is not as powerful as a Scatchard plot in indicating whether cooperativity is present. A Scatchard plot indicates cooperativity merely by being curved whereas if one wishes to draw the same conclusion from a Hill plot the total number of binding sites must be known, which usually requires an independent determination. However, the Hill plot is useful because it clearly indicates when there is *no* cooperativity. For example, if the binding sites are independent, binding is described by equation (8), which can be rearranged to

$$\frac{r}{n - r} = \frac{1}{K_d}[A] \tag{13}$$

A Hill plot of this equation yields a slope of 1.

When cooperativity is not complete, i.e., when $n_H < n$, the Hill plot is not linear because equation (12) does not describe the real situation. Thus, if a Hill plot is made over a very wide range of values of [A], it is found that the curve is not a straight line for the entire range of values (Figure 18-5)—at the extremes of [A] the line has a slope of approximately one. This has the following explanation. When the ligand concentration is very low, there is no cooperativity. Thus, a Hill plot should represent single-site binding—that is, the binding of the first ligand

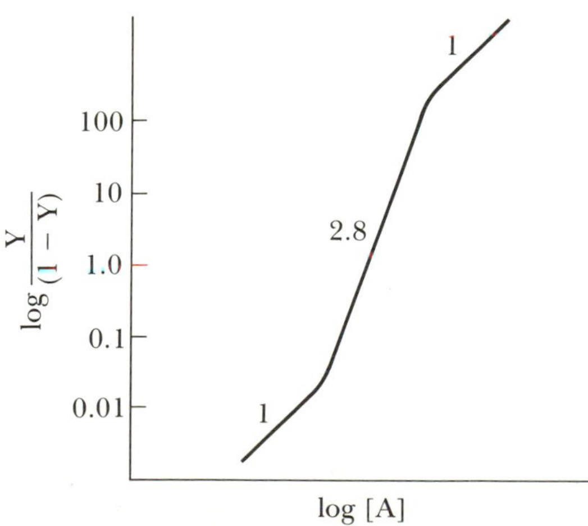

Figure 18-5
A Hill plot for a hypothetical protein having three cooperative binding sites. The numbers indicate the slopes of various parts of the curve.

molecule and the value of K_d for the first ligand should, in theory at least, be obtainable from extrapolation of this part of the curve. At high ligand concentration all sites are filled but one. Thus, this region of the Hill plot should also indicate single-site binding with a value of K_d for the last ligand. As mentioned above, an approximate slope of 1 is seen in the extremes of the Hill plot and the values of $\log(1/K_d)$ should be obtainable by extrapolation to the y-axis. However, binding date at very low and very high ligand concentrations are usually not very accurate so that, in practice, these two values of K_d cannot be derived from a Hill plot.

Parameters used to describe cooperativity

In the preceding section, the situation in which binding of the first ligand facilitates subsequent binding, that is, *positive cooperativity*, has been considered. *Negative cooperativity* occurs when a bound ligand *reduces* the binding of subsequent ligand molecules, especially in ion and proton binding to proteins. Whenever cooperativity is suspected, that is, when the use of equations (7) or (8) yields a curve, a plot of r versus [A] should be made and equation (4) should be used. This equation is not useful in determining n or K_d but gives a clear indication of cooperative binding. Examples of binding curves for positive and negative cooperativity obtained in such a plot are shown in Figure 18-6. Note that, in positive cooperativity, a plot of r versus [A] is sigmoidal; that is, as [A] increases, r rises slowly at first, then rapidly, and finally levels off, becoming asymptotic to $r = n$. This sigmoidal shape is the most important criterion of positive cooperativity. Negative cooperativity is also made evident by a Hill plot because n_H is observed to be less than 1.

Three parameters are often used to discuss a cooperative binding curve. These are the values of r at 10%, 50%, and 90% of the maximum (the saturation value) and are written r_{10}, r_{50}, and r_{90}, respectively. A curve can be described by stating $[A]_{10}$, $[A]_{50}$, and $[A]_{90}$, the values of [A] corresponding to r_{10}, r_{50}, and r_{90}, respectively. The value of $[A]_{50}$ is a measure of the binding constant; the larger $[A]_{50}$ is, the less tight is the binding. The difference $[A]_{90} - [A]_{10}$ is a measure of the degree of cooperativity; its value is small when the degree of cooperativity is great (that is, the curve rises steeply). This is because with strong cooperativity, after the first site is filled, there is a great acceleration in the rate at which subsequent sites are filled. This point was also made in Chapter 1 when it was indicated that the sharpness of DNA melting curves indicates the cooperative interactions that stabilize the DNA double helix.

Biological significance of cooperativity

The curves in Figure 18-6 give some indication of the utility of cooperativity in biological systems. For instance, consider a macromolecule that has no biological activity when no ligand is bound and maximum

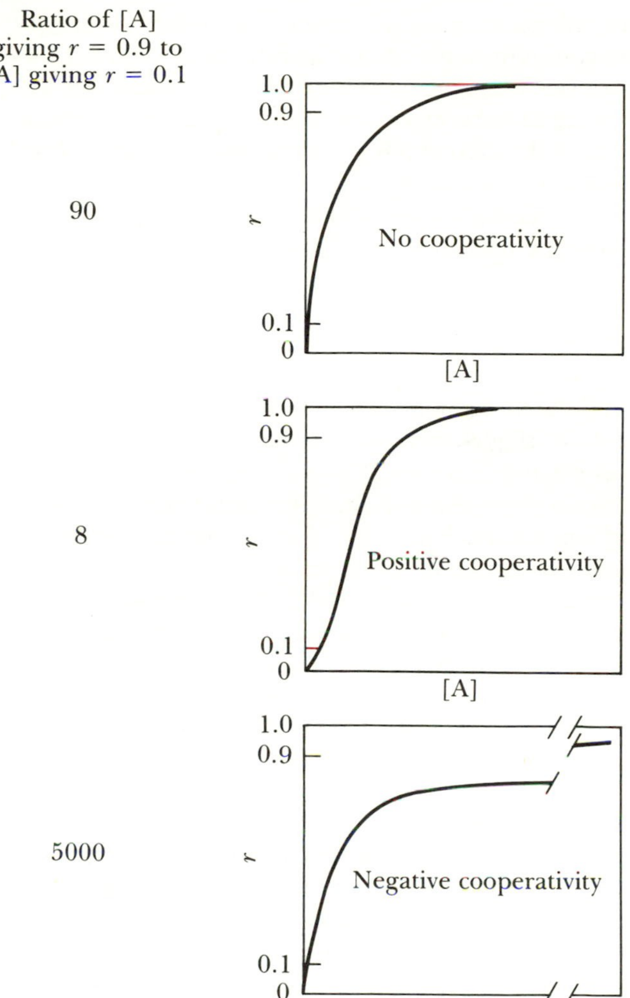

Figure 18-6
The types of binding curves obtained when equation (4) is used, showing the two types of cooperativity.

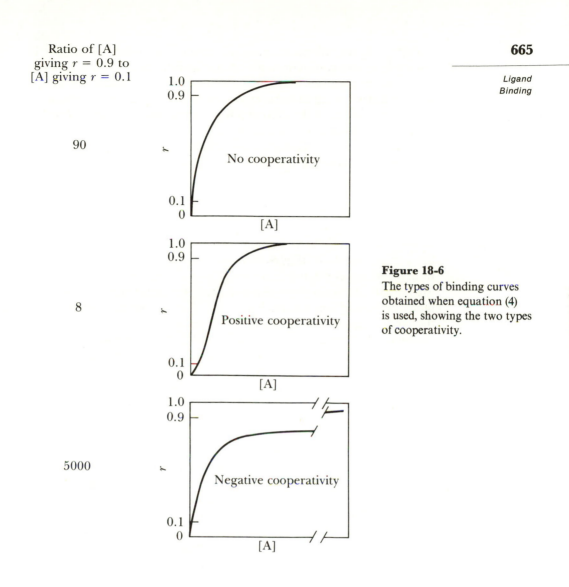

activity only when all binding sites are filled. In this case, the y-axis of Figure 18-6 could be labeled the fraction of molecules having biological activity. Cooperativity then becomes a means of controlling activity of the protein by the concentration of ligand. Thus, if the system is positively cooperative, there is little activity at low ligand concentration and this activity will increase rapidly in the low midrange of ligand concentration. If negative cooperativity is present, some activity exists at low ligand concentrations and the magnitude of this activity rapidly becomes insensitive to concentration changes as concentration increases. This modulating effect is even greater if a macromolecule binds two ligands A and B and either (1) B can be bound only if A is bound first (positive cooperativity) or (2) B cannot be bound easily if A is bound first (negative cooperativity). If the function of the macromolecule is to bind B (for example, an enzyme binding its substrate), the biological activity is regulated by the concentration of A. Cooperativity is one of the most important mechanisms for regulating biological activity and it has been suggested that the ability of a macromolecule to bind ligands cooperatively may have been the driving force causing living systems to evolve with macromolecular components.

In the following section, the most common type of cooperative binding, namely, allostery, is described.

Allostery

Many functional proteins are not single monomeric polypeptide chains but are aggregates of single-chain *subunits* (usually ranging from two to six) that may or may not be identical. In the case of a tetramer in which each subunit has a single binding site, the tetramer is in essence a macromolecule having four identical sites and, if these sites are independent, the binding properties can be analyzed by use of equations (7) or (8). However, binding is often cooperative and the cooperativity is thought to result from structural changes in the polypeptide chains caused by or stabilized by binding of the first ligand molecule. This mode of cooperative binding requires a mathematical formalism that differs from that which has already been presented. In the case of a single macromolecule, it is easy to see how binding of one ligand molecule can affect the binding of subsequent ligand molecules, because binding of the first can induce a change in the shape of the macromolecule which alters the shape of the remaining binding sites. The mechanism of cooperativity in a multisubunit protein is more complex because it is necessary to assume that a shape change in one subunit is somehow responsible for altering the shape of a binding site in other subunits. This can happen, though, because the subunits are in contact. Thus, the following sequence of events is envisioned. Binding of the first ligand molecule alters the shape of the subunit to which it is bound, and this results in changes in the site on this subunit which interacts with other subunits (Figure 18-7). If the subunits remain in contact, each subunit adjoining the first will suffer a shape change at its respective subunit-interaction site, and this in turn alters the ligand-binding site of each subunit. This may increase or decrease the affinity for the ligand molecules. Multisubunit proteins undergoing such modifications are called *allosteric proteins*.*

Two models have been presented to describe the mechanism by

---

* The term *allosteric* (Greek) means *other site* or *other structure*. Originally the terms was used to refer to an enzyme having two distinct binding sites, one for the substrate and the other for an effector molecule, which could be either an activator or an inhibitor. Later it was found that for some enzymes there are multiple copies of the active site and the substrate itself is an effector; such molecules were said to be *homotropic* in contrast with *heterotropic* enzymes, which have more than one class of binding site.

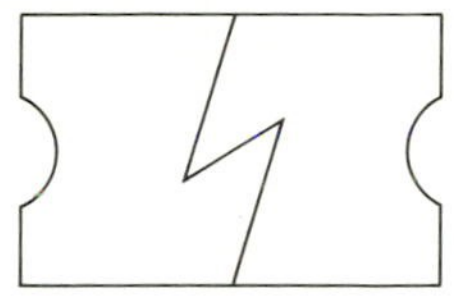

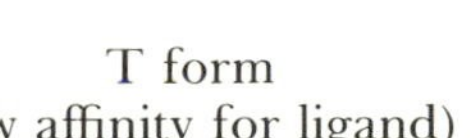

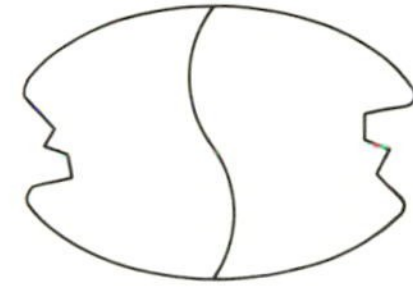

R form
(High affinity for ligand)

Figure 18-7
A representation of the R and T forms of an allosteric protein. [From *Biochemistry*, 2nd edition, by Lubert Stryer. W. H. Freeman and Company. Copyright © 1981.]

which the initial shape change occurs and how this results in modification of the protein aggregate. These are called the *symmetry or concerted model* and the *sequential model*. In both models the following conditions are assumed.

1. Each subunit exists in two forms, T and R, which bind the ligand with low and high affinity, respectively,
2. If a ligand molecule is firmly bound to a subunit, that subunit must be in the R form.

A few assumptions are not common to the two models. In the concerted model, it is assumed that the T and R forms are in equilibrium, that significant binding *only* occurs to the R form, and that this binding shifts the equilibrium strongly in the R direction. In contrast, the sequential model assumes that binding tends to occur first to the T form and this binding *induces* a transition to the R state. A second major difference between the two models concerns the number of possible forms of an allosteric protein whose binding sites are not all filled. Thus in the concerted model, it is assumed that the symmetry in arrangement of the subunits is an essential component of the forces stabilizing the macromolecular aggregate. Thus, an R subunit can interact stably with another R form and two T subunits can aggregate, but an R and a T form cannot form a stable pair. Thus, as long as the subunits remain in contact, the stabilization of one R subunit causes conversion of *all* remaining T subunits to R subunits—that is, the symmetry of the protein aggregate is preserved. For instance, for a tetramer, only TTTT and RRRR are allowable forms of the complex; that is, TTTR, TTRR, and TRRR do not exist. In the sequential model, this notion of symmetry is discarded and it is assumed instead that the conversion of TTTT to RRRR occurs through steps such as TTTT → TTTR → TTRR → TRRR → RRRR, in which each step occurs as another ligand molecule is bound. These two models are shown diagrammatically in Figure 18-8.

A

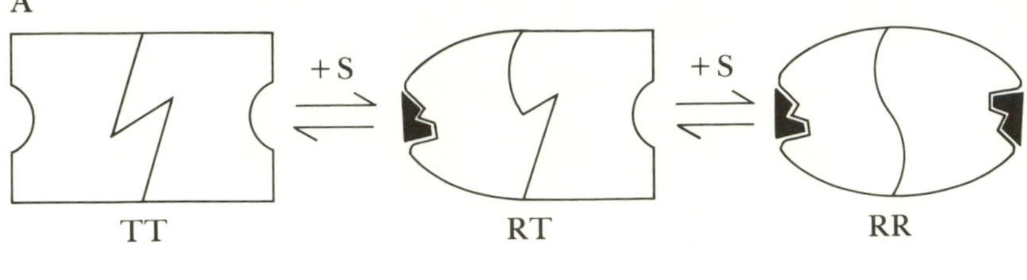

B

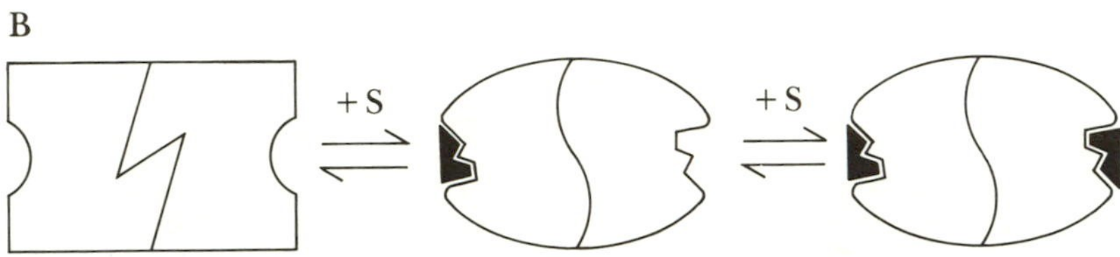

Figure 18-8
A. Sequential model for the cooperative binding of ligand in an allosteric protein. The empty active site in RT has a higher affinity for the ligand than the sites in TT. B. Concerted model for the cooperative binding of ligand in an allosteric protein. The low-affinity form, TT, switches to the high-affinity form, RR, upon binding the first ligand molecule. [From *Biochemistry*, 2nd edition, by Lubert Stryer, W. H. Freeman and Company. Copyright © 1981.]

The concerted model can be described by a simple formalism, which extends the analysis shown in Figure 18-6. We consider a protein consisting of several subunits, each having one binding site; the high- and low-affinity forms of a subunit *having no bound ligand* are designated as R_0 and T_0. These two states are in equilibrium and are related by a dissociation constant, L, called the *allosteric constant*:

$$L = \frac{[T_0]}{[R_0]} \tag{14}$$

The binding of a ligand molecule A to an R- or T-type subunit is described by the equilibrium constants K_R and K_T respectively. The relevant equation (whose derivation can be found in the Selected References near the end of this chapter) is

$$Y = \frac{Lc\alpha(1 + c\alpha)^{n-1} + \alpha(1 + \alpha)^{n-1}}{L(1 + c\alpha)^n + (1 + \alpha)^n} \tag{15}$$

in which

$$\alpha = [A]/K_R \text{ and } c = K_R/K_T$$

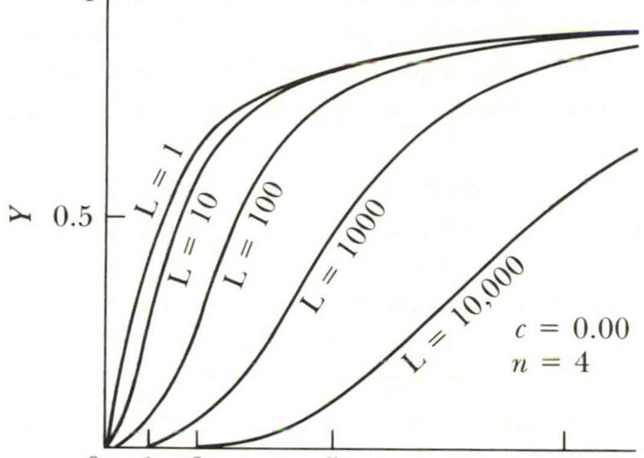

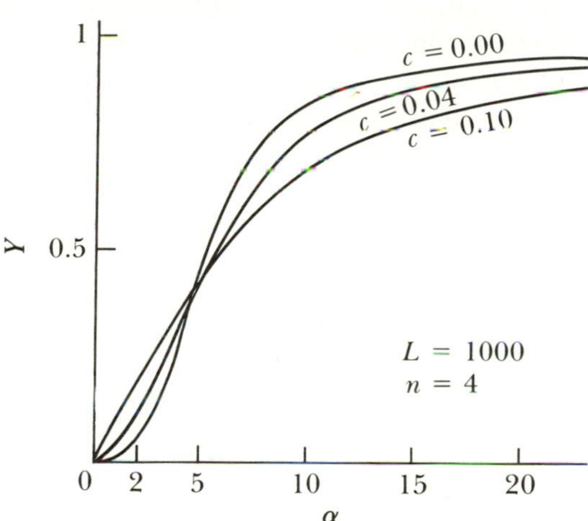

Figure 18-9
Theoretical curves of the saturation function Y drawn to various values of the constants L and c, with $n = 4$ (i.e., for a tetramer). [From J. Monod, J. Wyman, and Jean-Pierre Changeux, *J. Mol. Biol.* 12(1965):88.]

A plot of Y versus [A] or α (Figure 18-9) has the sigmoid shape characteristic of cooperativity, As the value of L increases, the curve broadens and shifts to higher values of α. Thus, cooperativity increases when the $R_0 \rightleftarrows T_0$ equilibrium leans heavily toward T_0. Note also the effect of the value of c. As c becomes very small, the sigmoid shape of the curve becomes more obvious. Thus, cooperativity is more significant when binding to the R form is much greater than to the T form. It should also be noted that when c approaches zero, equation (15) simplifies to

$$Y = \frac{\alpha(1 + \alpha)^{n-1}}{L + (1 + \alpha)^n} \qquad (16)$$

This is a useful equation and data should be plotted with this equation whenever cooperativity is suspected.

Equations (15) and (16) enable one to understand the molecular basis of the cooperative binding of ligands to macromolecules consisting of subunits. For instance, the variation in the rate of reaction of some enzymes with substrate concentration can often be explained by the concerted model.

The sequential model is much more complicated. In this model, it is assumed that when a single T subunit binds a ligand molecule, the T subunit is, with high probability, converted to the R form. As each additional ligand is bound, the T forms are sequentially converted to the R state. This process for a tetramer in a square array is depicted in Figure 18-10, which points out the major complication in the theory. Both the ligand-protein interactions and the subunit interactions are governed by equilibria, as shown in the figure; thus the dissociation constants for these interactions must be included in the derivation. Furthermore, if a second ligand binding to the tetramer has the same probability of binding as the first ligand, there would be no cooperativity. Thus, it is assumed that the existence of a single R form increases the probability that other subunits *in contact* with that R will switch to an R form. This

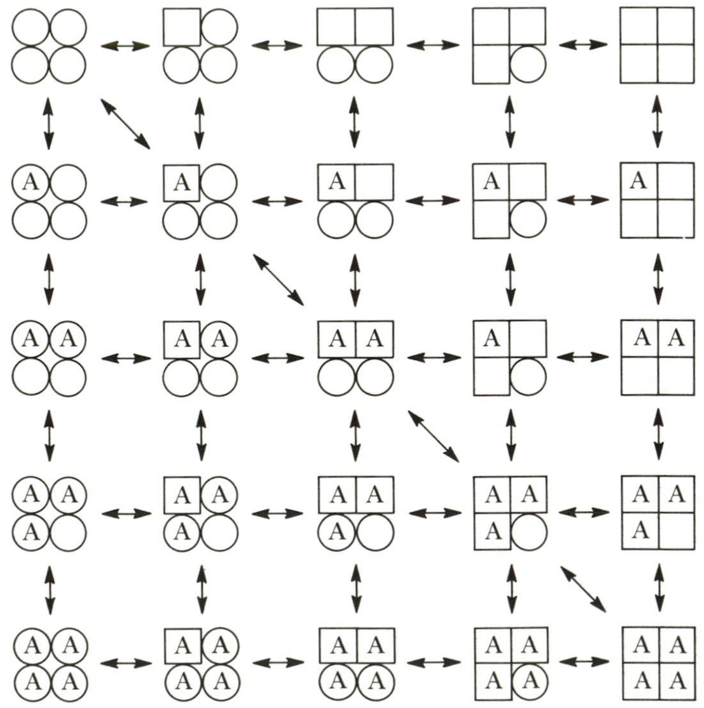

Figure 18-10
The possible intermediates in sequential binding of A to a tetramer. The circles and squares are the T and R forms respectively.

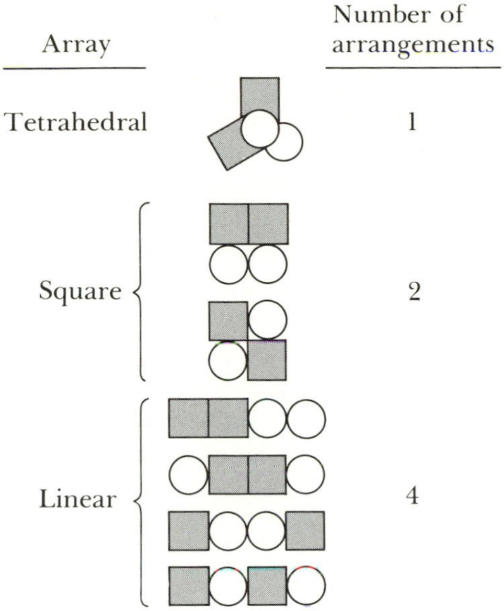

Array	Number of arrangements
Tetrahedral	1
Square	2
Linear	4

Figure 18-11
Possible arrangements of two bound and two unbound subunits in a tetramer for three different arrays.

creates the next complexity because the geometric arrangement of the subunits determines how many points of contact there are, as shown in Figure 18-11. With a tetramer the array of subunits can be linear, rectangular, tetrahedral, or more complex in form. Note in the figure that when two subunits are in the R form, there are one, two, and four possible arrangements for the tetrahedral, rectangular, and linear arrays, respectively.

The derivation of the equations describing sequential binding of the ligand A is fairly involved. The final equation for the rectangular case is

$$Y = \frac{K_{TR}^2 K_{TT}^4 Q + K_{TR}^4(1 + 2K_{TT}^4) + 3K_{TR}^2 K_{TT}^2 Q^3 + K_{TR}^4 Q^4}{K_{TR}^4 K_{TT}^4 + K_{TR}^2 K_{TT}^4 Q + 2K_{TT}^3(K_{TT} + 2K_{TR}^2)Q^2 + 4K_{TR}^2 K_{TT}^2 Q^3 + K_{TR}^4 Q^4} \quad (17)$$

in which $Q = L[\mathrm{A}]/K_d$, $K_d = [\mathrm{T}][\mathrm{A}]/[\mathrm{A}]$, $K_{TR} = [\mathrm{RR}][\mathrm{T}]/[\mathrm{TR}][\mathrm{R}]$, $K_{TT} = [\mathrm{R}][\mathrm{T}]^2/[\mathrm{TT}][\mathrm{R}]^2$; and [RR], [TT], and [TR] are the concentrations of the interacting subunits. (These concentrations are usually not easily measurable so that K_{TT} and K_{TR} are rarely known.) This is clearly a very complex equation and, in a binding experiment, only the values of Y and [A] are available so that the determination of the values of the other parameters is a formidable task. One possible method is to select arbitrary values for L, K_d, K_{TR}, and K_{TT} and calculate curves of Y versus [A] and compare these to the experimental curves. This is not profitable since with four varying parameters the number of possibilities is too great and often different combinations of values produce theoretical curves that are experimentally indistinguishable. Therefore, a different approach is taken. It can be shown that the basic shape of the saturation curves for the rectangular, tetrahedral, and linear cases are different; frequently one or two of these shapes can be eliminated by comparison to experiment. Alternatively, the arrangement of the subunits might be known from other data, for example X-ray diffraction analysis. At present, the use of the formalism of the sequential model is limited by the inability to measure or calculate the parameters needed to determine whether a particular experimental binding curve can be explained by sequential binding. For some macromolecules the sequential model may be ruled out. For instance, the binding curve may only be consistent with sequential binding to a linear array but it might be known from other techniques that the array is tetrahedral; in this case, it would be unlikely that binding would be sequential.

Allosteric Activation and Inhibition in Terms of the Concerted Model

Many multisubunit macromolecules have two distinct binding sites for two different substances A and B; filling one site with ligand A can affect the binding of B either positively or negatively. That is, the binding of A to an A site might either facilitate (activate) binding to B to its site or reduce (inhibit) the binding. The terms *allosteric activation* (an example is the action of a cofactor on an enzyme) and *allosteric inhibition* (for example, end-product inhibition of enzymes) are used for these processes.

Allosteric activation and inhibition can easily be understood in terms of the concerted model. In activation, the $R_0 \rightleftarrows T_0$ equilibrium shifts to the left since the activator increases the probability that a subunit is in the R form; similarly, the inhibitor shifts the equilibrium to the right by stabilizing the T form. Thus, activation and inhibition merely represent a decrease and increase, respectively, of the value of L. It is important to appreciate two points. First, the binding of an activator need not *always* result in conversion of a T subunit to an R subunit, but may only increase the concentration of the R form. Second, the T and R terminology refers to the binding of the second ligand of B. Thus, the protein is allosteric only in the binding of B and the activator merely decreases the value of L in equations (15) and (16). The consequence of the changes in L is shown in the following example.

Example 18-C ☐ Activation and inhibition of a dimer for the binding of B.

Let us consider an allosteric dimer ($n = 2$) for which $L = 10^4$ and $K_R = 10^{-5}$M. This value of L implies that at any time there is, on the average, 1 R-type molecule for every 10^4 T-type molecules. Let us also assume that binding to the T form is very poor ($c = 0$) so that equation (16) may be used. Since the dimer is allosteric, the the addition of the ligand B increases the number of R forms since, after it has bound to the R form, the complex cannot shift to the T form (Figure 18-12). Thus, the fraction of R-type molecules increases as more ligand is added. When all sites are occupied (that is, when the sites are saturated), all molecules are R forms, as shown in Figure 8-12. Applying equation (16), with $n = 2$, $K_R = 10^{-5}$ M, and $L = 10^4$ as the state for which neither activator nor inhibitor is present, we can see in Figure 18-13 the effect of a 10-fold activation ($L = 10^3$) or inhibition ($L = 10^5$) on the plot of Y versus [R]. At any value of [B], Y is decreased by the inhibitor and increased by the activator.

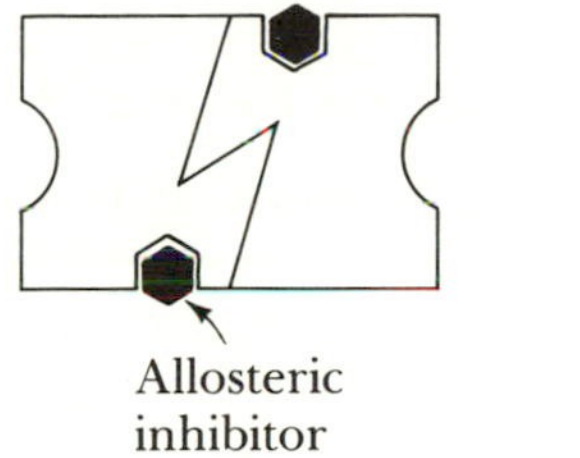

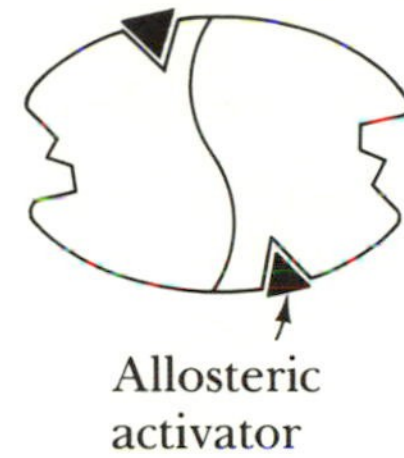

Figure 18-12
In the concerted model, an allosteric inhibitor (represented by a hexagon) stabilizes the T state, whereas an allosteric activator (represented by a triangle) stabilizes the R state. [From *Biochemistry*, 2nd edition, by Lubert Stryer. W. H. Freeman and Company. Copyright © 1981.]

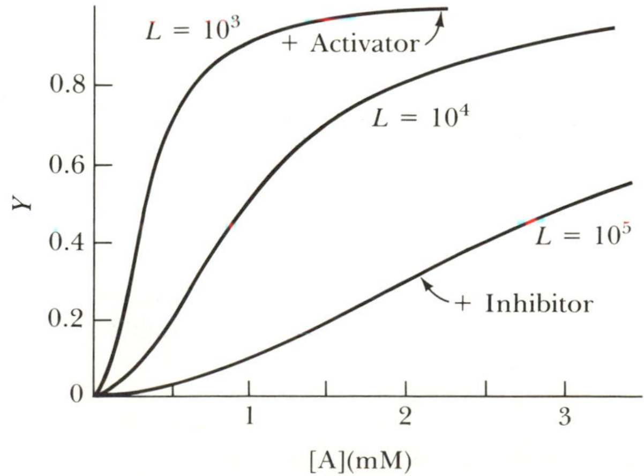

Figure 18-13
Saturation function Y, as a function of ligand concentration, [A], according to the concerted model. The effects of an allosteric activator and inhibitor are also shown. [From *Biochemistry*, 2nd edition, by Lubert Stryer. W. H. Freeman and Company. Copyright © 1981.]

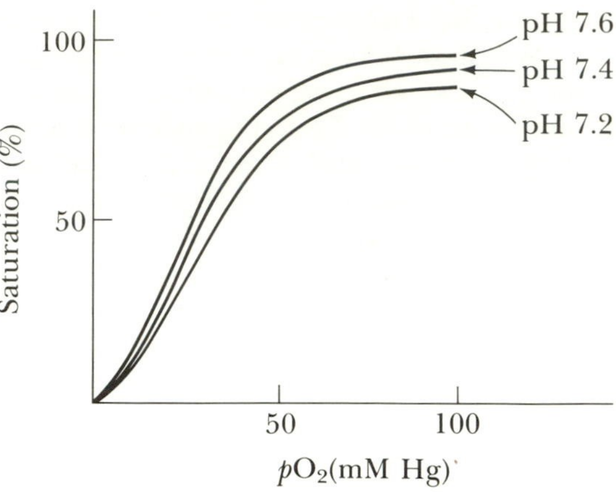

Figure 18-14
O_2 binding by hemoglobin at various pH values.

Hemoglobin—An Example of an Allosteric Protein

Hemoglobin is the O_2-carrying protein of blood. It is a tetrameric protein containing two α and two β subunits, each of which binds O_2. A plot of the percentage saturation (r) against the partial pressure of O_2 (pO_2), which by Henry's law is proportional to the concentration of dissolved O_2, shows a sigmoidal curve characteristic of positive cooperativity (Figure 18-14). As always, the cooperativity means that the affinity of the hemoglobin for the first O_2 molecule is low but, once it is bound, the affinity increases. There is a tendency to interpret Y versus [A] curves in terms of starting with [A] = 0 and examining the results of increasing [A]. It is just as reasonable to do the reverse, that is, begin with the sites saturated and study the effect of decreasing the ligand concentration. With this line of thought, cooperativity implies that once a single ligand molecule is removed, the remaining ones will follow quickly. Therefore, in the case of saturated hemoglobin, we may say that as pO_2 drops, the loss of one O_2 molecule from saturated hemoglobin causes rapid dissociation of the remaining ones.

We have mentioned that regulator molecules can affect cooperativity by altering the $R_0 \rightleftarrows T_0$ equilibrium. Although not usually thought of as a regulator, the H^+ ion can often cause very great shape changes by changing the interactions between individual amino acids. The pH can also regulate in another way if binding of the ligand involves a simultaneous release or adsorption of a proton. This latter mechanism occurs with hemoglobin. The effect of pH on the O_2-binding power of hemoglobin (a phenomenon known as the Bohr effect) is also shown in Figure 18-14. That is, the higher the pH, the greater is Y, and at lower pH values Y decreases. This is a result of the following equilibrium:

$$Hb^+ + O_2 \rightleftarrows HbO_2 + H^+$$

in which Hb^+ is a protonated subunit of hemoglobin (that is, the low-affinity or T form). The O_2-binding properties of hemoglobin can be successfully analyzed by assuming that release of the H^+ ion results in a transition to the R form. An X-ray diffraction study of the structure of hemoglobin, to which various amounts of O_2 are bound at several pH values, confirms this assumption in that the study indicates that the first two O_2 molecules bind to the two α subunits of hemoglobin and release two H^+ ions. This results in a conversion of the two α subunits to R forms. The transition changes the interaction with the two β subunits, which then also undergo a transition to the R form. These subunits then bind O_2 with high affinity and release the remaining two H^+ ions. Thus, the H^+ ion can also be though of as an allosteric inhibitor that decreases binding at low pH.

The Bohr effect explains how the pO_2 and the pH regulate the oxygen-carrying property of hemoglobin in circulating blood. In the lungs, pO_2 is about 100 mm Hg, the pH is relatively high, and hemoglobin is 96% saturated with O_2. Within body tissues, because of O_2 consumption, pO_2 is lower (approximately 45 mm Hg); furthermore, owing to the CO_2 formed as the end product of respiration, the pH is relatively low. Thus, hemoglobin binds O_2 less strongly and 96% saturated hemoglobin releases O_2 to the tissue until the hemoglobin is 65% saturated. In this way, hemoglobin cycles between 65% and 96% saturation with O_2 and continually oxygenates the body tissues through which blood flows.

Experimental Methods to Measure Binding

There are two principal means to measure binding. In the direct methods, the macromolecules and the ligands are first mixed and then the macromolecules to which ligand molecules have bound are physically separated from the free ligand molecules. This is commonly done by *equilibrium dialysis*, *membrane filtration*, and centrifugation. In the indirect methods, binding is measured by observing a change in a physical property of either the macromolecule or ligand that is induced by the binding. Common observations are spectral changes or changes in sedimentation properties.

Equilibrium Dialysis

Equilibrium dialysis is a convenient way to measure the binding of a small molecule to a macromolecule, as long as the small molecule is dialyzable and an assay for the small molecule exists. The principle of equilibrium dialysis follows. A solution of macromolecules is placed inside a dialysis bag (Chapter 7), as shown in Figure 18-15. The bag is suspended in a medium containing a particular concentration of a small molecule that binds to the macromolecule. The small molecules then diffuse into the bag. If no macromolecules were present, the concentrations inside and outside the bag would be the same. However,

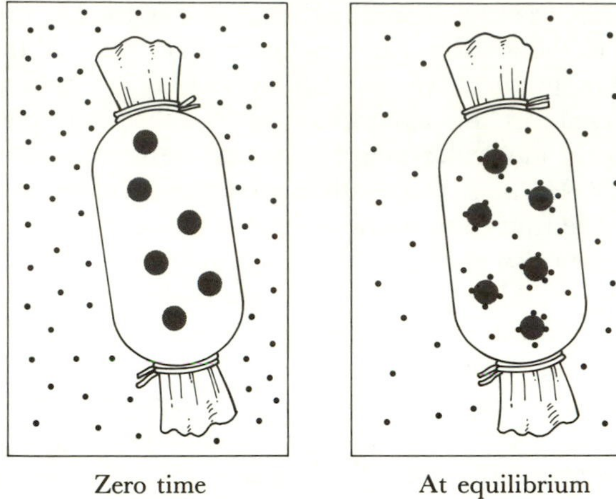

Figure 18-15
Equilibrium dialysis. A dialysis bag filled with macromolecules (shaded circles) is placed in a solution containing dialyzable small molecules (solid circles) that can bind to the macromolecules. At equilibrium, the concentration of *free* small molecules is the same inside and outside the bag. Because the macromolecules bind some of the small molecules, the *total* concentration of small molecules is greater inside the bag than outside.

because the macromolecules are present, the concentration inside the bag is greater by virtue of the number of bound molecules, because the concentration of unbound molecules within the bag will always equal the concentration outside the bag.

The values needed to determine the concentration [A] of unbound ligands and the concentration [PA] of the complex are the following:

$$[\text{Unbound}] = [\text{A}] = [\text{A}]_{\text{outside}}$$
$$[\text{Complex}] = [\text{PA}] = [\text{A}]_{\text{inside}} - [\text{A}]_{\text{outside}}$$

The concentration [P] of the unbound macromolecule is

$$([\text{P}]_{\text{initial}} - [\text{PA}]) \quad \text{or} \quad ([\text{P}]_{\text{initial}} - [\text{A}]_{\text{inside}} + [\text{A}]_{\text{outside}})$$

Thus, for the simple case of binding with a single site, described by the reaction of P + A $\rightleftarrows$ PA (equations 1 and 2),

$$K_{\text{d}} = \frac{[\text{A}]_{\text{inside}} - [\text{A}]_{\text{outside}}}{[\text{A}]_{\text{outside}}([\text{P}]_{\text{initial}} - [\text{A}]_{\text{inside}} + [\text{A}]_{\text{outside}})}$$

To obtain reliable data, certain conditions must be met.

1. The nonspecific binding of ligand and macromolecule to the dialysis bag itself must be small and preferably measurable.
2. Binding must be strong because of the following effect, which tends to indicate an artifactual negative binding (which, of course, is meaningless). That is, if the macromolecule is very large and does not bind the small molecule at all, its great size takes up so much volume that it excludes the small molecule from the solution within the dialysis bag. Hence, binding must always be great enough that this negative effect is negligible.
3. If the macromolecule and the ligand are charged, which is often the case, there will be a Donnan effect. This effect, which is a result of the inability of the charged macromolecule to traverse the membrane and the requirement for electrical neutrality throughout the solution, causes an inequality in the concentrations of charged ligand molecules across the membrane. Since this inequality may have nothing to do with binding of the ligand to the macromolecule, the data can be misleading. To avoid the Donnan effect, one need only increase the salt concentration of the solution so that the charges causing the Donnan effect are shielded.

The use of equilibrium dialysis to evaluate K_d is shown by the following example:

Example 18-D

☐ Determination of K_d for the binding of Mg^{2+} to a protein having a single binding site.

Five milliliters of a solution containing a protein at 5×10^{-3} M is placed in a dialysis bag containing 10^{-3} M radioactive $^{24}MgCl_2$. At equilibrium the bag contains 5.5 ml and 1653 cpm/ml. The external fluid contains 1555 cpm/ml. One milliliter of 10^{-4} M $^{24}MgCl_2$ contains 1565 cpm. The bag is washed out and counted; only 35 cpm above background are found and this is negligible.

The internal and external concentrations of $^{24}MgCl_2$ are $1653/1565 = 1.056 \times 10^{-4}$ M and $1555/1565 = 9.936 \times 10^{-5}$ M, respectively. The volume in the bag has increased from 5.0 and to 5.5 ml so that the total protein concentration is $(5.0/5.5)(0.005) = 4.545 \times 10^{-3}$ M. Thus

$$K_d = \frac{(1.056 \times 10^{-4} - 9.936 \times 10^{-5})}{(1.056 \times 10^{-3})(4.545 \times 10^{-3} - 1.056 \times 10^{-4} + 9.936 \times 10^{-5})}$$

$$= 1.3\ \text{M}^{-1}$$

Equilibrium analysis has also been used as a means of detecting macromolecules that are identifiable only by their ability to bind a particular ligand. An example of this procedure is the following.

Example 18-E □ Measurement of the binding of the *E. coli* repressor for the *lac* operator.

The lactose (*lac*) operon of the bacterium *E. coli* is regulated by a repressor protein that binds to the operator base sequence of the operon. When the repressor is bound to the operator, RNA polymerase is incapable of binding to the DNA and thus fails to initiate synthesis of *lac* mRNA. Several substances are known that are inducers of *lac* mRNA synthesis. Prior to purification of the *lac* repressor, genetic experiments had been performed that led to the hypothesis that an inducer binds to the repressor causing a conformational change of the repressor that prevents the repressor from binding to the operator. Walter Gilbert developed a procedure for purifying the *lac* repressor, detecting it by means of repressor-inducer binding. Cell extracts were fractionated by chromatography and each fraction was tested by equilibrium dialysis for binding of the radioactive inducer isopropylthiomethyl galactoside. That is, extracts containing the ^{14}C-labeled galactoside were dialyzed and the ^{14}C concentrations inside and outside the bag were determined. The fraction having an internal ^{14}C concentration greater than the outside concentration contained the repressor. By using separation procedures that continually maximized the extent of binding, the *lac* repressor was ultimately purified. The relative values of $^{14}C_{inside}$ and $^{14}C_{outside}$ were also used to determine the dissociation constant for isopropylthiomethyl-galactoside–repressor binding.

Centrifugation

Centrifugation is a useful procedure for detecting binding as long as two conditions are met. (1) The macromolecule or binding structure must have a sufficiently high sedimentation coefficient that it can be pelleted. (2) Dissociation is sufficiently slow that the ligand is not lost when the pellet is washed to free it of extraneous, unbound ligand. The procedure is to mix the macromolecule and the ligand and then sediment the macromolecule to the bottom of the centrifuge tube. The ligand in the pellet represents bound ligand while that in the supernatant is the unbound ligand. It is necessary for pelleting to occur very rapidly so that the amount of ligand bound is not changed by the high concentration of the macromolecule in the pellet. The Beckman Airfuge (page 441) is an excellent tool because of its high centrifugal force and its ability to reach speed and to come to a stop rapidly. The use of centrifugation is shown in the following example.

Example 18-F □ Detection of binding of a drug Q to a drug receptor.

Ten milliliters of solution contain 50 mg of a drug receptor and 10^{-4} M [^{3}H]Q (3.5 × 10^4 cpm/μmol). After incubation the drug receptor is collected by centrifugation. The pellet weighs 138 mg and contains 660 cpm of [^{3}H]Q. The supernatant contains 3468 cpm/ml. Assuming that all of the 50 mg of the drug receptor are collected, the mass of liquid in the pellet is $138 - 50 = 88$ mg. Assuming a solvent density of 1 g/ml, the volume of liquid in the pellet is 0.088 ml and there are $0.088 \times 3468 = 305$ cpm of free [^{3}H]Q. Thus there are $660 - 305 = 355$ cpm of [^{3}H]Q in the pellet. This represents $355/3.5 \times 10^4 = 1.01 \times 10^{-2}$ μmol bound. Thus $r = 1.01 \times 10^{-2}/50 = 2.02 \times 10^{-4}$ mol Q/mg drug receptor.

Membrane Filtration

If the macromolecule can be collected on a filter through which unbound ligand molecules freely pass, binding can be measured by filtering a solution containing both macromolecules and ligands and measuring the amount of ligand retained on the filter. This is a rapid and sensitive method that is applicable to the binding of ligand molecules to nucleic acids, proteins, and large particles such as membrane fragments. The filters most commonly used are the nitrocellulose or nitrocellulose–cellulose-acetate membranes (Chapter 7). Usually conditions can be found under which the macromolecule binds tightly to the filter but in which a ligand molecule that is not bound to the macromolecule passes through the pores unhindered. The calculation of bound ligand is carried out as in Example 18-F.

Spectral Changes

Spectroscopic procedures are easily performed whenever the spectrum of either the macromolecule or the ligand changes significantly. The spectral change might be a shift in the absorption maximum, an increase or decrease of the absorbance, a change in the intensity of fluorescence, or other more complex changes such as the band width in NMR spectra. If the spectrum in the absence of binding and the spectrum at saturation are known, and if the spectral change is linear with ligand concentration, the degree of binding at intermediate stages can be measured. An example is given in Figure 18-16, which shows the changes in the spectrum of the dye proflavin that accompany its binding to DNA. When unbound, proflavin has an absorption maximum at 450 nm; when bound, the maximum is at 468 nm. Note that the absorbance at 475 nm is 0 when no DNA is present but substantial in the DNA-proflavin complex. Thus the absorbance at 475 nm can be used to measure binding.

Sedimentation Changes

An elegant method for detecting binding makes use of the shape changes of a macromolecule that sometimes occur when a ligand is bound. The shape change can be measured by observing changes in the sedimentation coefficient when a solution containing macromolecules and ligand are centrifuged. How this is done is shown in the example that follows.

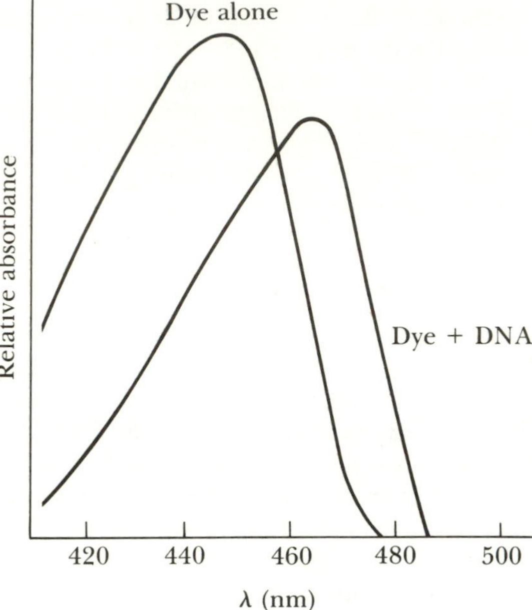

Figure 18-16
Absorption spectra of proflavin and proflavin saturated with DNA.

Example 18-G □ Measurement of the binding of a ligand A by a protein P by sedimentation analysis.

Suppose a protein P has a sedimentation coefficient, s, of 7.2 S and this value changes to 6.8 S when a ligand A is bound; that is, the complex, PA, has an s value of 6.8 S. If a solution containing known concentrations of both P and A is centrifuged, the ratio f of the amount of material having s values of 6.8 S and 7.2 S is $f = [PA]/[P]$. If $[P']$ and $[A']$ are the starting concentrations, $[P] = [P']/(1 + f)$ and $[PA] = f[P']/(1 + f)$.

In the case of a macromolecule that binds several ligand molecules at saturation, it is sometimes possible, by the sedimentation method, to determine the number of binding sites. This can be done if the addition of each ligand molecule produces a change in s so that as $1, 2. \ldots, n$ molecules are bound, $1, 2, \ldots, n$ distinct sedimenting species having different s values are observed. The number of binding sites determined in this way is always a minimum value because it is possible that, at some stage in the binding process, there is either no change in s or the change is too small to be observed. In an allosteric protein having several subunits, the existence of many sedimenting forms might be taken as evidence against the concerted model and should suggest analysis in terms of the sequential model.

In a particularly interesting variant of the sedimentation method, one observes a change in the s value of a ligand when a molecule is added.

Usually a fraction, f, of the total ligand is observed to sediment at the same or nearly the same rate as the macromolecule. For example, there are many enzymes that use nicotinamide-adenine nucleotide (NAD) as a cofactor. NAD absorbs ultraviolet light at wavelengths not absorbed by most protein molecules so that the binding of the NAD to the protein can be detected by the appearance of ultraviolet-absorbing material sedimenting with the s value of the protein (s = 2–10 S, depending on the protein) because NAD is too small to be sedimentable. When this occurs, $[PA] = f[A']$ and $[A] = (1 - f)[A']$.

SELECTED REFERENCES

Dahlquist, F. W. 1978. "The Meaning of Scatchard and Hill Plots," in *Methods in Enzymology*, vol. 48, edited by C. W. H. Hirs and S. N. Timasheff, pp. 270–299. Academic Press.

Koshland, D. E. 1970. "The Molecular Basis for Enzyme Regulation," in *The Enzymes*, 3rd edn, vol. 1, edited by P. Boyer, pp. 341–396. Academic Press.

Koshland, D. E. 1973. "Protein Shape and Biological Control." *Sci. Am.* 229:52–65.

Monod, J., J. Wyman, and J. P. Changeux. 1965. "On the Nature of Allosteric Transitions: A Plausible Model." *J. Mol. Biol.* 12:88–118. This is the initial statement of the theory.

Price, N. C., and R. A. Dwek. 1974. *Principles and Problems in Physical Chemistry for Biochemists*. Clarendon Press. This is a particularly clear description of how binding data are used.

Van Holde, K. E. 1971. *Physical Biochemistry*. Prentice-Hall. A good description of the theory of ligand binding.

PROBLEMS

18–1. The binding of a small molecule X to a protein is studied by holding the protein concentration constant and varying the concentration of the small molecule. The data obtained are the following:

Added X (mM)	20	50	100	150	200	400
Bound X (mM)	11.5	26.1	42.7	52.9	58.8	69.3

Which method of plotting the data will give a more reliable value for K_d, equation (6) or equation (7)?

18–2. The binding of a substrate to an enzyme increases if the ionic strength of the solution increases. Give two explanations for this increase.

18–3. The binding of a ligand A to a protein P is greater in a 50% methanol: 50% water mixture than in water. Propose two explanations for the difference.

18–4. A protein at a concentration of 10^{-3} M is placed on one side (I) of a semipermeable membrane. An excess volume of a 10^{-4} M solution of a small molecule, X, is placed on the other side (II) of the membrane. The concentrations of X in compartment II have the values 10^{-4}, 9.2×10^{-5}, 8×10^{-5}, 6.25×10^{-5}, 6.12×10^{-5}, and 6.12×10^{-5} M at times 0, 1, 3, 7, 15, and 24 hours, respectively. After 24 hours the concentration of X in compartment I is 6.54×10^{-5} M.

a. Does X bind to the protein and, if so, what is the dissociation constant?
b. A second molecule Y is added to compartment II. At equilibrium, the the two concentrations of X are 10^{-4} M (side I) and 10^{-4} M (side II) and the concentrations of Y are 10^{-3} M (side I) and 5×10^{-6} M (side II). What can you say about the binding sites for X and Y?
c. Now 10^{-3} M Z is added instead of Y. At equilibrium, the concentration of Z is the same on both sides of the membrane but the concentrations of X are 9.72×10^{-5} M (side II) and 9.78×10^{-5} M (side I). What can you say about the interaction of Z and X?

18–5. Cytochrome *c* is a red protein that, under certain conditions, can bind to DNA molecules. The sedimentation coefficient, *s*, of cytochrome *c* is less than 5 S in a wide variety of aqueous solvents having different salt concentrations. The *s* value of the DNA of bacteriophage T7 is 32 S and is constant throughout the range of NaCl concentrations from 0.01 M to 2 M. If particular concentrations of DNA and cytochrome *c* are mixed together in 0.01 M NaCl and the solution is centrifuged, the *s* value of 10% of the cytochrome *c* is 32 S and the remainder is less than 5 S. In 0.5 M NaCl, no red material sediments with an *s* value of 32 S. What information do these experiments give you?

18–6. The antibody(Ab)-antigen(Ag) reaction is an example of very tight binding. The following data were obtained in a binding experiment in which [Ab] = 1 μM.

Ag added (μM)	0.50	1.0	1.5	2.0	2.5	3.0
Free Ag (μM)	0.015	0.025	0.035	0.05	0.5	1.0

How many binding sites are there per antibody molecule?

18–7. A protein P is at a concentration of 10^{-4} M and a substance X that binds to the protein is added at a concentration of 10^{-5} M. After ten minutes it is found that there is, on the average, 1 X bound per 100 molecules of P. If [P] = 10^{-3} M and [X] = 10^{-4} M, after ten minutes there is 1 molecule of X bound per 10 molecules of P. If this more concentrated sample, in which binding has already occurred, is diluted 10-fold (that is, back to 10^{-4} M P and 10^{-5} M X), it is found even after several hours that there is no unbound X—there is still 1 molecule of X per 10 molecules of P. What does this tell you about the binding of X to P?

18–8. A protein P has two identical binding sites for a molecule A. The shape of the protein is dependent on the pH. If the pH is varied, the affinity of P for A changes but the number of potential binding sites remains constant. How would the Scatchard plots obtained at various values of the pH differ from one another?

18–9. An enzyme E is suspected of consisting of several subunits but no one has measured the number or determined whether they are identical or not. In the course of studying the binding of the substrate S to the enzyme, you obtain the data shown in the following Scatchard plot, in which $r = [S]_{bound}/[E]_{total}$.

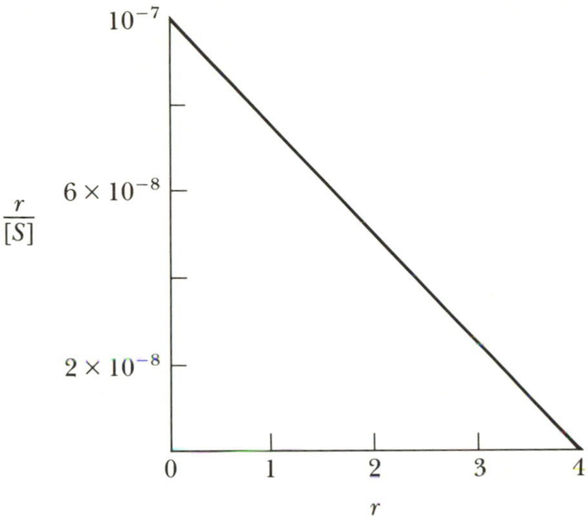

a. What can you say about the structure of the protein?
b. What is K_d for the binding?

18–10. A titration curve of a weak acid is a binding curve in the sense that as the pH is decreased by the addition of a strong acid like HCl, the dissociation of the weak acid will decrease. For such a system r = [dissociated]/[undissociated] so that a Scatchard plot of $r/[H^+]$ versus r can be made. Describe qualitatively the differences between the Scatchard plot of a monoprotic acid and a triprotic acid.

18–11. A particular protein, called the T4 gene-32 protein, binds tightly to DNA. Its curved Scatchard plot suggests that the binding is cooperative. The gene-32 protein is readily visualized by electron microscopy when it is bound to a DNA molecule. Which of the following drawings depict what you would expect to see? The dots represent the protein molecules.

18–12. The binding of a ligand B to a protein is studied. The protein concentration is 10^{-3} M. The amounts bound, b, for each value of added B (that is, [B]) are the following:

[B](M)	0.01	0.02	0.05	0.07
(b)(M)	5×10^{-6}	3.3×10^{-3}	3.78×10^{-3}	4.72×10^{-3}
[B](M)	0.1	0.2	0.5	
(b)(M)	4.95×10^{-3}	5×10^{-3}	5×10^{-3}	

Prepare a Hill plot and determine if the binding is cooperative, the number of binding sites, and the value of K_d.

18–13. A macromolecule, H, consists of four identical subunits, M. The binding of a substance X for various values of [X] is shown in the figure below.

a. Explain the different saturation concentrations for M and H.
b. Interpret the different shapes of the curves.
c. What kind of plot could be made to test the interpretation in part (b)? If your interpretation is correct, what would the plot show?

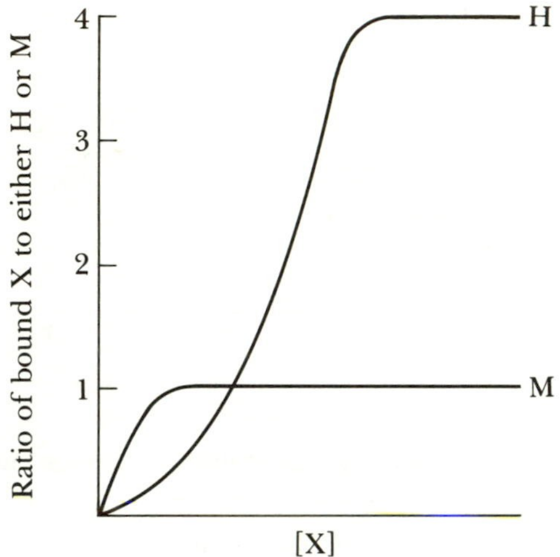

18–14. A Hill plot typically consists of three regions as shown below.

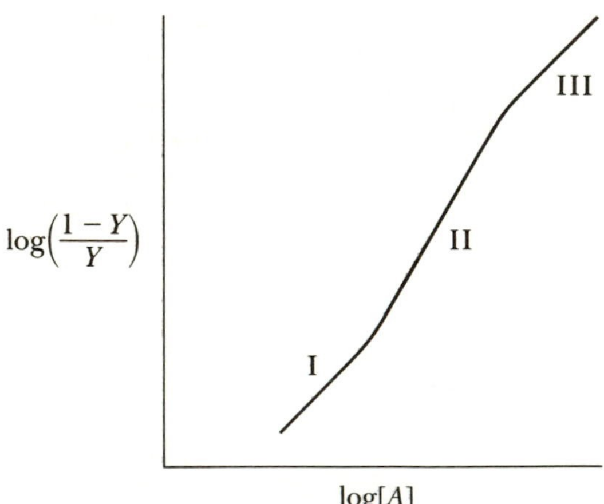

The regions I and III invariably have a slope of 1. K_d can be determined for each of these regions by extrapolating to the *y*-axis (the intercept is log K_d). If a Hill plot is made for an allosteric protein, what constants are determined that are necessary to perform an analysis of allostery?

19

Miscellaneous Methods

This chapter comprises brief descriptions of miscellaneous, but useful, techniques that do not belong with the topics treated in the other chapters.

Hydrodynamic Shear and Sonic Degradation

Both the viscosity and the sedimentation coefficient of a DNA solution decrease when the solution flows rapidly through a narrow orifice (e.g., a hypodermic needle, a fine-tipped pipette, or a narrow-bore viscometer) and when it is stirred or shaken too vigorously. These effects are due to breakage of the molecule by hydrodynamic shear forces.

As explained in Chapter 13, if a liquid containing a rodlike molecule is flowing through a tube, the shear force will tend both to orient the rod and to stretch it. If the molecule is stretched to its limit, it will break, *somewhere near its middle*. Similar shear forces can be generated by the violence of turbulent agitation or violent stirring. By controlling the flow rate of solutions through narrow tubes or the speed of rotation of the blades of a stirrer, it is possible to break a linear molecule such as DNA successively into halves, quarters, eighths, and so forth. As might be expected, the critical shear force (i.e., the smallest shear stress that will break the molecule) decreases with increasing molecular weight.

In general, shearing is avoided in the laboratory in order to isolate DNA of maximum molecular weight and to maintain the size of isolated DNA. Table 19-1 gives the effects of various laboratory practices on DNA of various molecular weights. Note that the higher the molecular weight, the more care is required.

Determination of critical shear values has been used as a semiquantitative analytical tool. For example, if a molecule is known to be linear, it is possible to estimate its molecular weight from the shear stress required to halve the molecule. It is important to understand, though, that the actual shear stress is never measured. What is done is to take a series of DNA molecules whose molecular weights are known, stir solutions of them at various stirring speeds, and plot the lowest speed necessary to produce halving versus molecular weight. This provides a standard calibration that can be used to estimate M for a sample by measuring the critical stirring speed. Figure 19-1 gives an example of such an experiment. A calibration curve could also be prepared by

Table 19-1

Extent of Breakage of Double- and Single-Stranded DNA of Various Molecular Weights by Common Laboratory Operations.

	Double-stranded DNA		Single-stranded DNA	
Operation	$M = 25 \times 10^6$	$M = 106 \times 10^6$	$M = 12.5 \times 10^6$	$M = 53.0 \times 10^6$
Pipetting				
with 1-ml serological pipettes	Safe	Safe	Safe	Safe
with 0.1-ml pipettes	Safe	Some breakage	Safe	Breaks
Filling centrifuge cells				
with number 22 needles	Safe	Safe	Safe	Breaks
with number 24 needles	Safe	Some breakage	Some breakage	Breaks
Pouring	Safe	Safe	Safe	Breakage with splashing
Mixing with another solution	Safe	Safe	Safe	Breakage if densities differ
Shaking	Safe	Breakage with turbulence	Breakage with turbulence	Breaks
Bubbling air	Safe	Safe	Safe if slow	Breaks

General trend

1. Sensitivity to shear increases with molecular weight.
2. A single-stranded molecule is more sensitive than a double-stranded molecule having the same length.

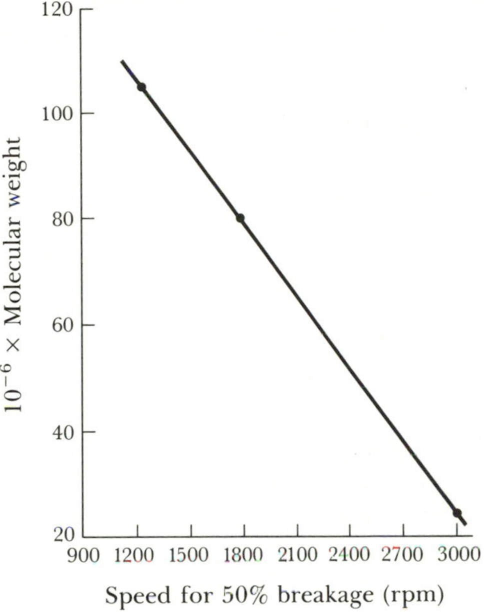

Figure 19-1
Typical calibration curve, relating stirring speed that produces 50% breakage for DNA having various molecular weights. DNA solutions are 20 μg/ml in 1 M NaCl. These curves are strongly concentration dependent, the critical speed decreasing markedly with lower DNA concentration. To determine molecular weight, it is necessary to use the same DNA concentration and the same stirring system.

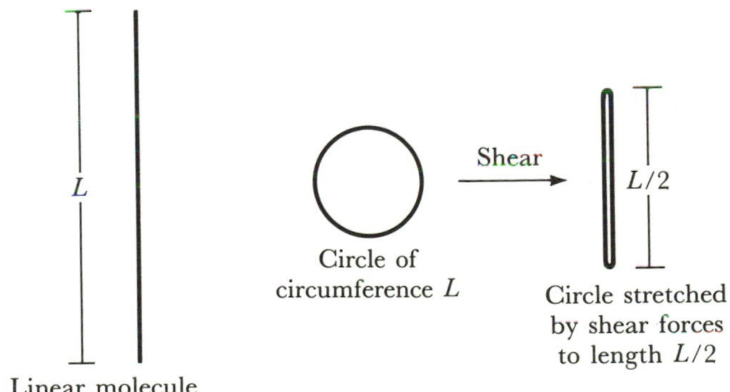

Figure 19-2
Shearing of a circle. A circle of circumference L is stretched to a nearly linear structure of length $L/2$, having twice the number of strands.

flowing the sample through narrow tubes and plotting critical flow rate versus molecular weight; however, the stirring procedure is easier.

If the molecular weight is known, one can tell if the molecule is nonlinear because an anomalous shear sensitivity results. For example, if a circular molecule of molecular weight M were stretched, it would form a loop having length equal to that of a molecule of $M/2$ and would also be twice as thick. Hence, its shear sensitivity would be one-quarter that expected from its molecular weight (Figure 19-2).

To make an estimate of molecular weight from shearing studies, it is necessary to maintain a constant DNA concentration because the resistance to shear increases substantially with concentration ("self protection").

Preparation of half-molecules, quarter-molecules, and smaller fractions is also of great value in isolating specific regions of a DNA molecule because frequently the fragments are separable by their buoyant density in solutions of cesium salts (see Chapter 11).

Hydrodynamic shear can also be used to disrupt large structures, even whole cells. For example, one of the most common methods of breaking cells is to expose them to intense sonic fields, although it is not commonly understood that sonication causes breakage by hydrodynamic shear. When a liquid is placed in a high-intensity sonic field (either by putting it in a resonating cavity or, more commonly, by inserting a vibrating probe into the liquid), the liquid *cavitates*—that is, microbubbles rapidly form and collapse. Cavitation is caused by the dissolved gases coming out of solution as bubbles, and the disruption of cells, aggregates, or macromolecules is caused by the violent collapse of these bubbles. (In extremely intense sonic fields, some chemicals decompose to produce free radicals, which can cause other types of molecular damage.) It is not clear whether the collapse of the bubble produces turbulence or local laminar flow, but the mechanism of breakage is certainly by means of hydrodynamic shear; in fact, by controlling the sonic intensity, DNA molecules have been successively broken into halves, quarters, and so forth. Controlled sonication has rarely been used as an analytical tool; its principal use is to rupture cells in preparation for the isolation of internal components. In general, the technique is simple and the only precaution necessary is to avoid heating. Because sonication pours a considerable amount of vibrational energy into water, the temperature rises rapidly. The usual protocol is to chill samples to ice temperature, sonicate for 30 seconds, recool, and repeat. In this way, a low temperature can be maintained.

Hydrogen Exchange

Many substances dissolved in deuterated (^2H_2O) or tritiated (^3H_2O) water will exchange some of their hydrogen atoms with ^2H or ^3H. Amino acids, nucleosides, short-chain polypeptides, proteins in the random-coil configuration, and single-stranded nucleic acids rapidly exchange hydrogen atoms bound to nitrogen, oxygen, and sulfur atoms; carbon-bound hydrogen exchanges much more slowly. In proteins, for chemical reasons, exchangeable protons on some amino acid side chains (e.g., serine OH and the NH_2 of lysine and arginine) exchange much more rapidly than those in the peptide bond and in the amides of glutamine and asparagine. These two classes are distinguished by their pH dependence—the first class having a minimum at pH 7, the second at pH 3. Each class may be subdivided according to whether the protons participate in the formation of hydrogen bonds. Because the chemical rate of exchange is usually much slower than the making and breaking of hydrogen bonds (which controls the access of the solvent to the groups that are hydrogen-bonded), the rate of exchange of any group is the product of the chemical exchange rate and the fraction of time exposed to the solvent. Thus, the primary effect of a structure such as a hydrogen bond is to reduce the fractional exposure of the group to the solvent because the group is exposed only when local opening takes place. Hence, by measuring the chemical exchange rate for free groups and the rate when in a macromolecule, the fraction open at any given time can be determined.

In double-stranded nucleic acids, the exchange of a hydrogen-bonded proton is also greatly facilitated by an agent that breaks the hydrogen bonds. In summary, a potentially rapidly exchangeable proton exchanges poorly if it is hydrogen-bonded and this can be used to determine the number of hydrogen bonds, the effectiveness of various agents as denaturants, or the dynamic state of a macromolecule. When the technique was first employed, only deuterium exchange was followed because ^3H_2O was not available. Detection of the deuterium was fairly difficult and was done either by density determination in a Linderstrøm-Lang density-gradient column (see Chapter 12), or by infrared spectrophotometry (Chapter 14). However, since ^3H_2O has become available, tritium exchange has totally replaced deuterium exchange because ^3H is easily detected by its radioactivity.

The tritium-exchange technique, which employs gel chromatography with various molecular sieves (Chapter 8, Figure 8-21), is performed by means of the following procedure. Tritiated water is added to a protein or nucleic acid in H_2O solution. At various times, samples are removed and applied to the column. The ^3H_2O is strongly retarded by the column, whereas proteins and nucleic acids pass through rapidly. Using a column whose length is such that the protein or nucleic acid emerges after 10 seconds, the concentration of unbound ^3H is reduced approximately 10^8-fold. Fractions are taken from the column and the radioactivity in each fraction is determined by scintillation counting; the concentration of protein or nucleic acid is determined by spectrophotometry (Chapter 14) or, in some cases, by color tests for proteins. Hence, the number of exchanged hydrogens per unit weight or per molecule (if the molecular weight is known) can be measured as a function of time.

Several important results have been obtained from the ^3H-exchange method. For example, the protons in the hydrogen bonds of double-stranded DNA exchange. From the pH effect, kinetics, and equilibria, it has been shown that the hydrogen bonds in double-stranded DNA are continually breaking and reforming; this process is called *breathing* and provides a potential mechanism for such processes as the initiation of DNA synthesis and DNA-DNA pairing in genetic recombination. A second example is the determination of the number of hydrogen bonds in transfer RNA from a measurement of the fraction of exchangeable hydrogens that exchange rapidly; these data have been used to substantiate one of the proposed structures of tRNA. A third example is a major effort that is now in progress to study the kinetics of protein folding by determining the number of hydrogen bonds as a function of time following the transfer of a denatured protein to conditions resulting in reformation of the native structure.

In the applications just described, rapidly exchanging protons were examined. Useful information can also be obtained by studying protons that exchange very slowly. Tritiated water is used and acquisition of radioactivity is measured. This technique is particularly useful in probing the structure and interactions of tRNA. The proton that is studied is the hydrogen on C-8 of purines.

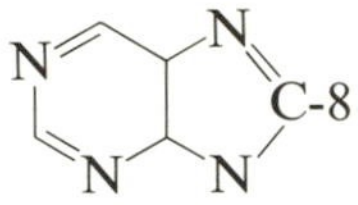

N N C-8 N N

This proton can be examined specifically by incubating tRNA in ^3H_2O for a few hours. About one percent of the C-8 protons and virtually all protons in hydrogen-bonded groups exchange. However, if the tRNA is returned to ^1H_2O, the rapidly exchanging protons are restored to the non-radioactive form but H-8 remains radioactive for a very long time.

The following kind of experiment has been done with tRNA. The molecules are incubated in ^3H_2O, transferred to ^1H_2O and then digested with an enzyme that makes a cut after each guanine. This generates a unique set of fragments that can be separated by homochromatography (Chapter 8). Each fragment is collected and digested with other enzymes that release each guanine (G) and each adenine (A) and the radioactivity in each sample of G and A is determined. Since the base sequence of the tRNA is known, the amount of radioactivity acquired by each G and A in the tRNA molecule can be measured. It is found that the exchange rate varies strikingly from one base to the next. Bases that exchange very slowly are presumed to be engaged in intramolecular interactions. With yeast phenylalanyl-tRNA there is close agreement between these findings and X-ray crystallographic data. This suggests that it may be possible to determine the structure of other tRNA molecules by this technique.

The ^3H-exchange method has also been used to determine the points of contact between tRNA and an aminoacyl-tRNA synthetase, the enzyme that covalently links the amino acid to each tRNA type. Bases that exchange in the free tRNA molecule and fail to exchange when the aminoacyl-tRNA synthetase is bound are either in contact with the enzyme or are in regions that undergo conformational changes resulting from enzyme binding. Since other data show that there is no significant conformational change, these bases are presumably points of contact with the enzyme.

If a tRNA proton that exchanges poorly is made to exchange more rapidly by some interaction, this can be taken as evidence for a change in conformation. Experiments of this sort have recently been performed in the laboratory of Charles Cantor to determine whether tRNA undergoes a shape change when it binds to a ribosome. Whether there are such changes has been a major question in molecular biology. Tritium-exchange experiments show that the guanine at position 57 in yeast phenylalanyl-tRNA exchanges at double the rate when the tRNA is bound to the ribosome. Other changes are also observed that support the view that there is such a change in conformation.

Light Scattering

A measurement of the angular dependence of the light scattered by a solution of macromolecules yields the molecular weight and the radius of gyration of the molecule. At one time, light scattering was a major technique for characterizing macromolecules. However, scattering techniques are not widely used today for the following reasons. (1) Technically light scattering is difficult to measure; (2) the molecular weight of proteins is much more rapidly obtained by gel electrophoresis and uses considerably less material than does light scattering, and the values obtained by the two procedures are equally accurate; and (3) the radius of gyration is not a particularly informative number and can be estimated more easily from a comparison of the sedimentation coefficient and the molecular weight. Nonetheless, a brief description of light scattering is given here in order that the reader may be able to read and understand many of the classic experiments of the 1950s. It is also possible that with the availability of laser light sources that allow the use of small volumes, dusty solutions, and measurement at very low angles, there will be a resurgence of the technique.

The physical basis of scattering is the following. When a light wave falls on a molecule, its oscillating electric field sets up an oscillating dipole, which is itself a source of radiation. The ability of the dipole to oscillate is determined by how tightly the molecule holds on to its electrons or alternatively by the force required to move an electron; this is measured by the polarizability α. Since the dipole has an orientation, the direction of emission of the wave or the photon is related to the component of the dipole vector seen by an observer at a particular observation angle with respect to the direction of the incident light. If the molecules were rotating rapidly, the scattered intensity averaged over many rotations would be the same at all angles. With a collection of molecules, other effects occur and an angular dependence is found. If the molecules are well separated (this constraint will be discussed in a moment), the intensity of the light scattered at a particular angle by a collection of molecules is proportional to the number of scattering centers (that is, the number of the molecules). It is this point that enables one to determine the molecular weight. The equation that describes the angular dependence of the scattered intensity is

$$\frac{I_\theta}{I_0} = \frac{8\pi^2 a\alpha^2}{\lambda^4 r^2}(1 + \cos^2\theta) \qquad (1)$$

in which I_0 and I_θ are the intensities of the incident light and the light seen at an angle θ with respect to the beam direction at a distance r from the molecules, λ is the wavelength of the light, α is the polarizability, and a is the number of particles per cubic centimeter (Figure 19-3). If the particles are in solution, the equation must be modified. As written, the equation describes scattering by a sample consisting of identical particles or, at least, particles having the same value of α. In the study of macromolecules, solutions are used and in solution solvent molecules and macromolecules will rarely have the same polarizability; in fact if they do, the intensity of the scattered light by the solution is identical to that of the pure solvent. In order to detect the scattering by the macromolecule, the solvent molecules and the macromolecules must have different polarizabilities and the values of the polarizabilities of each must appear in the equation needed to describe the phenomenon. Polarizability is not easily measured but is related in a straightforward way to the index of refraction; thus, the required equation can be written in terms of the indices of refraction of the solvent and of the macromolecule. The value for the pure solvent is easy to obtain with a refractometer but the index of refraction of a macromolecule is not easily measurable unless a dried sample can somehow be made transparent. However, it can be expressed as the product of the concentration c (in g/cm^3) of the macromolecule and the *change* in the refractive index n of the solution with concentration, namely dn/dc, which can be mea-

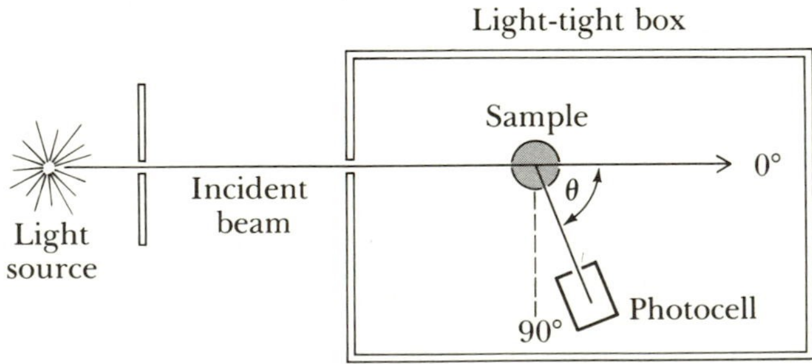

Figure 19-3
Diagram of a light-scattering photometer showing the scattering angle θ. The photocell is mounted on a rotating arm centered on the sample so that the intensity can be measured at various values of θ.

sured in a straightforward way. When these factors are all taken into account, there results the equation

$$\frac{I_\theta}{I_0} = \frac{2\pi^2 n_0^2 (\mathrm{d}n/\mathrm{d}c)^2}{N_\mathrm{A}\lambda^4 r^2}(1 + \cos^2\theta)Mc \tag{2}$$

in which M is the molecular weight of the macromolecule, N_A is Avogadro's number, and n_0 is the index of refraction of the pure solvent. This equation indicates that the *intensity of the scattered light is proportional to the molecular weight* and this is the basis for the determination of M by the light-scattering method. In an effort to simplify this equation the substitutions

$$R_\theta = \frac{r^2}{(1 + \cos^2\theta)}\frac{I_\theta}{I_0} \quad \text{and} \quad K = \frac{2\pi^2 n_0^2 (\mathrm{d}n/\mathrm{d}c)^2}{N_\mathrm{A}\lambda^4} \tag{3}$$

are usually made so that equation (2) becomes

$$R_0 = KMc \tag{4}$$

This equation describes an ideal situation, which is not always approached in practice. In the derivation (which is not given) of equation (2), an assumption is made that the scattering by a particles is a times the scattering by a single particle. This is not the case with samples of an ordinary concentration because as the concentration increases, there are various contributions that break down the proportionality. This has the effect (with which the reader should now be familiar) that *the observed value of* M *calculated from equation* (4) *is concentration-dependent.* This dependence is expressed by the equation

$$\frac{Kc}{R_\theta} = \frac{1}{M} + 2Bc + \cdots \tag{5}$$

in which B is a constant. Thus, to determine the true value of M, Kc/R_θ is plotted against c and $1/M$ is obtained by extrapolating the resulting curve to $c = 0$.

Equations (4) and (5) are applicable to molecules whose dimensions are small compared to the wavelength of the incident light because a complication occurs with large molecules. The limit for the validity of the equations is reached when the size of the molecule is about $\frac{1}{50}\lambda$. For a typical wavelength of visible light (450 nm), this limit is about 9 nm, which allows one to use these equations for all but the largest globular proteins and for small fibrous proteins. For most nucleic acids, however, and for macromolecular aggregates such as viruses, the equations are not valid.

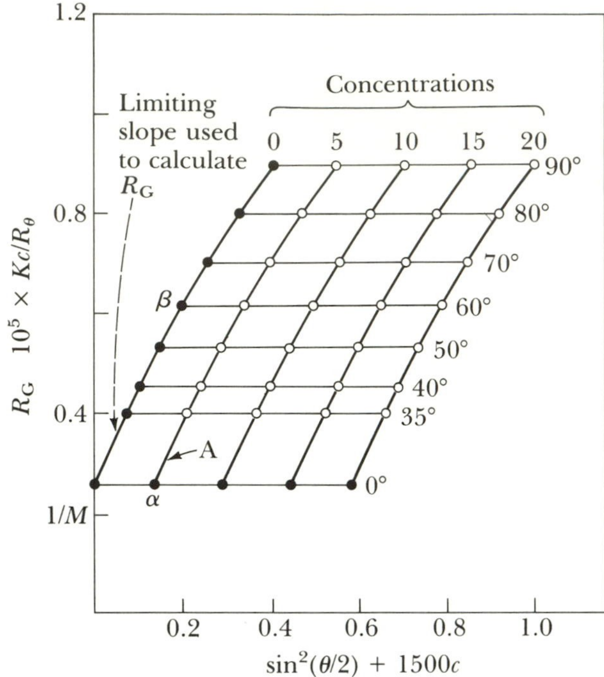

Figure 19-4
A Zimm plot. The open circles are the experimental points. The solid circles are obtained as described in the text. The y-intercept of the 0° line is $1/M$, which in this case is 3×10^5. The limiting slope of the dashed line as it reaches the y-axis is used to calculate R_G.

The problem with a large molecule is that different regions of the same molecule serve as scattering centers and the scattered waves can destructively or constructively interfere. When this interference is incorporated into the theory, the resulting equation is

$$\frac{Kc}{R_\theta} = \left[1 + \frac{16\pi^2 R_G^2}{3\lambda^2}\sin^2\frac{\theta}{2}\right]\left[\frac{1}{M} + 2Bc\right] \tag{6}$$

in which R_G is the radius of gyration. Now to obtain M, one must perform an extrapolation to zero concentration and to zero angle. This extrapolation is carried out by an elegant procedure called a Zimm plot (Figure 19-4). In a Zimm plot Kc/R_θ is the ordinate and $\sin^2(\theta/2)$ + kc is the abscissa, in which k is an arbitrary constant that enables the graph to have a convenient scale. Usually k is selected to be between 1000 and 5000; the actual value of k is irrelevant because, in the extrapolation to $c = 0$, $\sin^2(\theta/2) + Kc$ becomes $\sin^2(\theta/2)$. Measurements of

R_θ are made for a series of concentrations and for each concentration the values of R_θ are graphed for different values of θ; the points (open circles) lie on a vertical, slightly curved, line such as the line A in the figure. The line terminates at the bottom at a point α having the abscissa value $\sin^2(\theta/2) + kc$. This point α corresponds to the value of Kc/R_θ obtained by extrapolating the data to $\theta = 0$. A set of vertical lines is obtained, one line for each concentration used. The set defines a series of points (lower solid circles) such as α, that lie on a horizontal line. For these points $\theta = 0$ so that the equation of this horizontal line is simply equation (5). Thus 1/M is obtained from the y-intercept of this line. The experimental points also lie on horizontal lines; each horizontal line consists of a series of points obtained at a particular angle but at different concentrations. These lines can be extrapolated to $c = 0$, namely, to the point at which $\sin^2(\theta/2) + kc = \sin^2(\theta/2)$. For the curve corresponding to $\theta = 50°$, this point is denoted β in the figure. This process can be carried out for each horizontal line to generate a set of points (left solid circles) such as β. These points are then connected to form the vertical dashed line in the figure. The equation of this line is simply equation (6) with the $2Bc$ term equal to 0. The y-intercept of this line is also $1/M$ and the slope is $(1/M)(16\pi^2R_G^2/3\lambda^2)$ from which R_G can be calculated.

The use of light scattering to determine the molecular weight of a protein is shown in the following example.

Example 19-A

☐ Determination of the molecular weight of a protein Q by light scattering.

A series of solutions, having different concentrations of a small protein Q in water, is prepared. The index of refraction at each concentration is measured and a plot of n versus c is drawn. A straight line results, whose slope, which is dn/dc, is 0.172. The index of refraction n_0 of water is 1.333. Light scattering is performed at a wavelength of 436 nm so that the constant K in equation (4) is 4.65×10^{-7} cm^5 g^{-2} mol. Samples at protein concentrations of 0.025, 0.050, and 0.075 g/cm^3 are separately placed in the sample vessel and illuminated with light of intensity I_0. A photocell is placed 7 cm away from the sample; thus $r = 7$ cm. The light intensity is measured at an angle of 90° with respect to the incident light beam. The observed intensities for the three solutions are: 0.025 g/cm^3, $I_{90} = 5 \times 10^{-6} I_0$; 0.050 g/cm^3, $I_{90} = 10^{-5} I_0$; 0.075 g/cm^3, $I_{90} = 1.49 \times 10^{-5} I_0$. The values of $R_{90} = [(7^2/1 + \cos^2 90)]/(I_{90}/I_0) = 49(I_{90}/I_0)$ are 5.34×10^{-5}, 1.08×10^{-4}, and 1.61×10^{-4} for concentrations of 0.025, 0.050, and 0.075 g/cm^3 respectively. From equation (4), $R_{90}/Kc = 4600$. The same value is obtained for all three concentrations, which indicates that the protein molecule is sufficiently small that equations (5) and (6) are unnecessary. Thus, the molecular weight of the protein is 4600.

There are several experimental problems with the light-scattering technique. The principal one is the presence of dust in the solutions because a few dust particles may scatter light considerably more than all of the dissolved macromolecules. A second problem arises with larger molecules, when it is necessary to measure I_θ at very small values of θ in order to perform a reliable extrapolation to zero angle. However, it is not possible to measure I_θ when θ is nearly zero because the scattered light is superimposed on the transmitted light beam (refer again to Figure 19-3). The use of light-scattering photometers with laser light sources seems to minimize this problem.

The scattering of X-rays, for which $\lambda = 0.1\text{–}0.3$ nm, has also been used to determine M and R_G for small molecules. Traditionally a different equation, known as the Law of Guinier, is used in the treatment of X-ray scattering data. This equation, which is valid only for very low angles, is

$$i(s) = i(0) \exp[-4\pi^2 R_G^2 s^2/3]$$

in which $i(s)$ is the scattered intensity at an angle θ, $i(0)$ is the scattered intensity extrapolated to $\theta = 0$ (not the intensity of its incident beam, as appeared in the light-scattering equations) and $s = (2/\lambda)\sin(\theta/2)$. Thus, a plot of $\log[i(s)]$ against s^2 yields a straight line whose slope is $4\pi^2 R_G^2/3$ and whose intercept is $i(0)$. The value of $i(0)$ can be used to calculate M in a somewhat complex way. Low-angle X-ray scattering can also be used to measure the mass per unit length and the cross-sectional area of rodlike molecules although this is rarely done at the present time.

Further information about low-angle X-ray scattering can be found in the Selected References near the end of the chapter.

Light scattering has also been used to monitor changes in shape. This is a valuable procedure and has been used in studying the permeability of membranes. The principle of the measurement is the following. Small, nearly spherical units, called vesicles, are either isolated from tissue or prepared from synthetic membranes. These are hollow structures bounded by a membrane. They are incubated in a particular solution for a long time so that the internal fluid has the same composition as the solution in which they are suspended. The vesicles are then transferred to a solution in which one of the solutes is at a higher concentration than that of the internal fluid. In response to the osmotic pressure created across the membrane, water rapidly flows out of the vesicle causing it to shrink. This shrinkage results in production of vesicles with highly irregular shapes, which scatter light at 90° more effectively than the original unshrunk vesicles. The ions in the external fluid are still at a higher concentration than those within the vesicle so that ions flow into the shrunken vesicles. This flow is accompanied by a flow of water so that the vesicles gradually return to their normal size. The intensity, I_{90}, of light scattered at 90° also decreases and at a rate determined by the permeability of the membrane to the ions. This can be used to

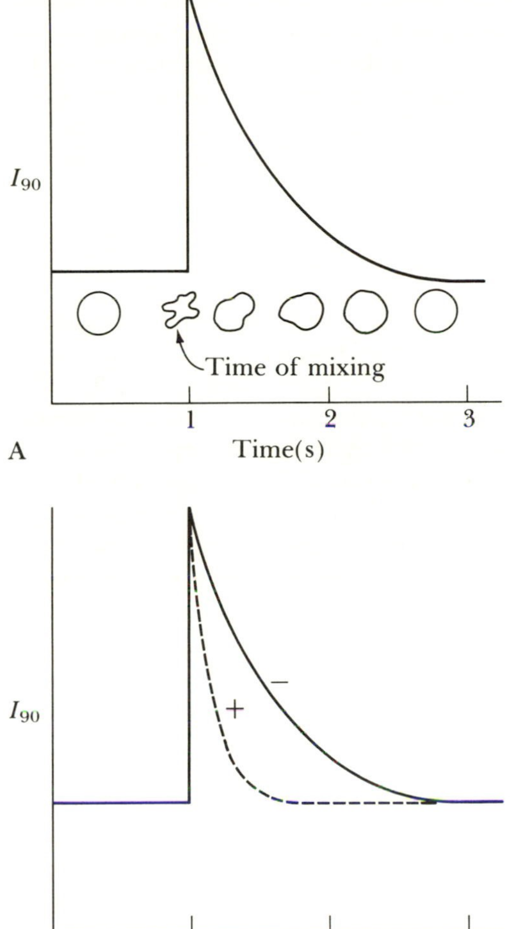

Figure 19-5
The use of light scattering to monitor permeability of vesicles to ions. Panel A shows the scattering intensity at 90° (I_{90}) as a function of time and a representation of the shapes of the vesicles at various times. Panel B shows a measurement of the flux of KCl in the presence (+) and absence (−) of the drug valinomycin, which accelerates the flow of the K^+ ion.

measure rapid fluxes of ions if mixing is rapid and if I_{90} can be monitored continuously. The kind of data obtained is shown in Figure 19-5A. An example of its use is described in the following.

Example 19-B

☐ Measurement of the relative rates of movement of the K^+ and Cl^- ions across a membrane.

Vesicles are incubated overnight in 0.1 M KCl, then diluted into 1 M KCl, and I_{90} is measured. The results are shown in Figure 19-5B. In order to maintain electroneutrality both K^+ and Cl^- ions must flow through the membrane. Thus, the solid curve in the figure indicates the flow of the ion that moves more slowly. The antibiotic valinomycin increases the flow of K^+ ions tremendously. Thus if valinomycin is added, the decrease in I_{90} should occur more rapidly if the solid curve represents the K^+ flux; there is no change in the curve if it represents the Cl^- flux. In the experiment shown in the figure, the addition of valinomycin results in the dashed line, which shows a more rapid decrease. This again is the slower of the two fluxes and is the Cl^- flux.

Fluctuation Spectroscopy

Fluctuation spectroscopy (Flusy) is a recently developed technique for measuring the molecular weight of large DNA molecules [M. Weissman, H. Schindler, and G. Feher. *Proc. Natl. Acad. Sci. U.S.A.* 73(1976): 2776–2780]. It has had little use but yields values that may be as accurate or more accurate than those obtained by viscoelastic measurements.

The technique is based upon a statistical principle, namely, that if a large volume of a solution is divided up into a large number of equal subvolumes, the number of solute molecules in each subvolume will not be exactly the same—that is, the number of molecules per unit volume, or the concentration, fluctuates. An equation from statistical mechanics states that

$$\overline{\left(\frac{\delta N}{\bar{N}}\right)^2} = \overline{\left(\frac{\delta C}{\bar{C}}\right)^2} = \frac{1}{\bar{N}} \tag{7}$$

in which δ represents a fluctuation, and $\bar{N}$ and $\bar{C}$ are the equilibrium values of the number and concentration (in g/cm^3) of the molecules around which the fluctuations occur. By measuring the fluctuations of the concentration of molecules in a given volume V (in cm^3), in which the fluctuations are detected, the molecular weight M of the solute molecules can be calculated. The relevant equation is

$$M = \bar{C}\left(\frac{\delta C}{\bar{C}}\right)^2 VN_A \tag{8}$$

in which N_A is Avogadro's number. The use of this equation is shown in the following hypothetical example.

Example 19-C ☐ Determination of M by measuring the concentration of 10^4 aliquots of a DNA solution initially at a concentration of 1.3×10^{-6} g/cm^3. The size of each aliquot is 1 cm^3 and the standard deviation of the 10^4 concentration measurements is 2.5×10^{-4}.

The value of $\bar{C}$ is 1.3×10^{-6} g/cm^3 and the square of the standard deviation, $(2.5 \times 10^{-4})^2 = 6.25 \times 10^{-8}$, is equivalent to $(\delta C/\bar{C})^2$. Thus $M = (1.3 \times 10^{-6})(6.25 \times 10^{-8})(1)(6 \times 10^{23}) = 4.9 \times 10^{10}$.

The procedure just described in the example is inconvenient because of the necessity of making so many independent measurements. Thus, the following technique has been developed for DNA solutions. First, ethidium bromide is added to the DNA solution. The weak fluorescence of ethidium bromide is enhanced enormously when the molecules are bound to DNA molecules. Thus, the DNA concentration can be measured at very low concentrations simply by monitoring the fluorescence of the ethidium bromide. Second, the DNA–ethidium-bromide solution is placed in a rotating cylinder and the ethidium bromide is excited by a very narrow beam of light from a laser. (Figure 19-6A). The fluorescence is measured at right angles to the laser beam, using an automatic recording device that indicates fluorescent intensity I as a function of time.

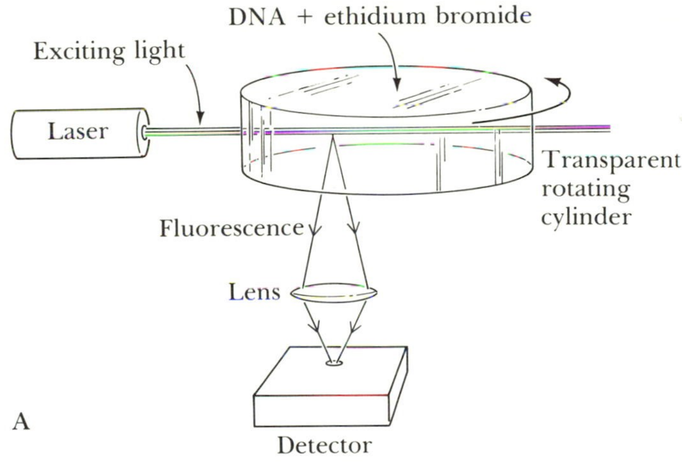

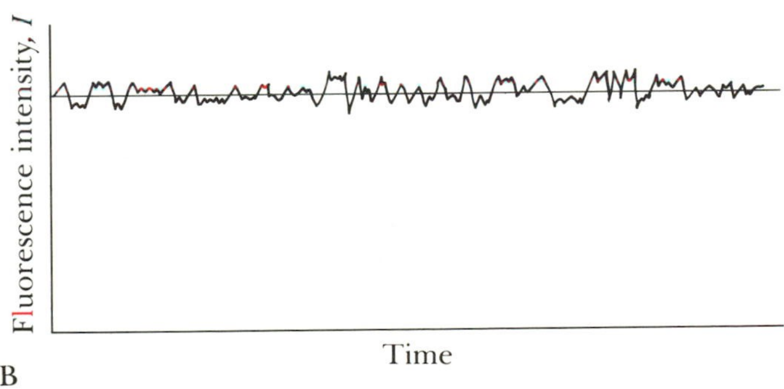

Figure 19-6
A. A schematic representation of the apparatus used in fluctuation spectroscopy. B. A typical plot of the data obtained.

The kind of data obtained are shown in Figure 19-6B. An electronic processing system is used to convert the values of $\delta I/\bar{I}$ to $\overline{(\delta I/\bar{I})^2}$. The only parameter in equation (2) whose value is not readily apparent is V, the volume of solution from which I is measured. To obtain the value of V the apparatus is filled with an aqueous suspension of synthetic polystyrene spheres, whose concentration is precisely known by direct microscopic counts of an accurately measured volume. The fluctuations in the intensity of the *scattered* light are measured (there is no fluorescence in this experiment) and V can be determined from equation (2) written as $M = \bar{I}_s\overline{(\delta I_s/\bar{I}_s)^2}VN_A$ in which I_s is the intensity of the scattered light.

This method has been used successfully to determine M in DNA molecules in the range of 10^8 to 10^{11} and should be particularly valuable in measuring M for chromosomal DNA from eukaryotic cells.

Hybridization Between Single-Stranded Polynucleotides

If double-stranded DNA is denatured, the individual polynucleotide strands become physically separated (Chapter 1). If the ionic strength is low (e.g., 0.01 M), the negative charge of the phosphate groups of the sugar-phosphate chain repel one another. This has two effects: (1) two different strands cannot come into contact because of mutual electrostatic repulsion and (2) an individual strand tends to be highly extended and relatively rigid, because all parts of the backbone repel one another. Hence, at low ionic strength, no two nucleotide bases ever come near enough to form hydrogen bonds. If the ionic strength were suddenly increased (e.g., to 0.5 M), the charge on the phosphates would be shielded and the strands would rapidly approach the random-coil configuration. However, in the absence of charge repulsion, hydrogen bonds would reform, that is, between guanine and cytosine and between adenine and thymine (for DNA) or adenine and uracil (for RNA). Because of the flexibility of these long polynucleotide strands, the probability of forming intrastrand (same strand) hydrogen bonds is greater than that of forming interstrand (different strand) hydrogen bonds, unless the concentration is very high. The intrastrand bonds will tend to be between very short, complementary tracts of bases (Figure 19-7), because long, complementary sequences will not frequently be found in a single strand. For steric reasons, not all bases will form hydrogen bonds, and this is reflected by the fact that the OD_{260} of the DNA is not restored to the original value for native DNA (Figure 19-8A). Because the hydrogen-bonded regions are very short, they have low thermal stability and can be disrupted at temperatures well below that needed to denature native DNA (Figure 19-8B). Therefore, the intrastrand-hydrogen-bonded DNA can be converted to single-stranded random coils at temperatures and ionic conditions in which native DNA would be stable, and it is expected that, given sufficient time for the complementary single strands to find one another, native DNA will reform. This is indeed the case, and

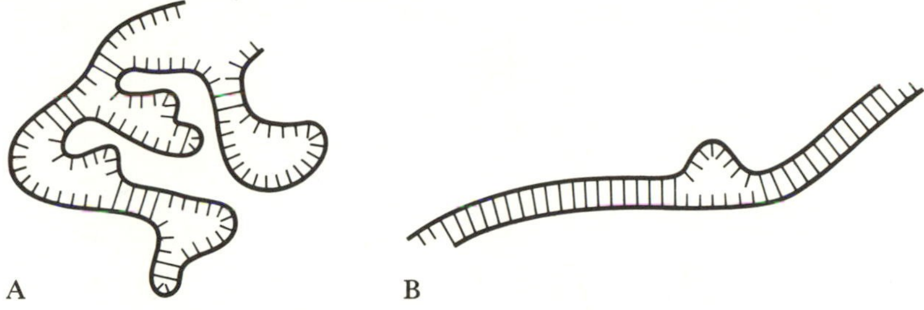

Figure 19-7
A. Internally aggregated single-stranded DNA. Base pairs are formed between different parts of the strand, thus making a very compact structure. B. Renatured DNA, showing one short region of noncomplementarity.

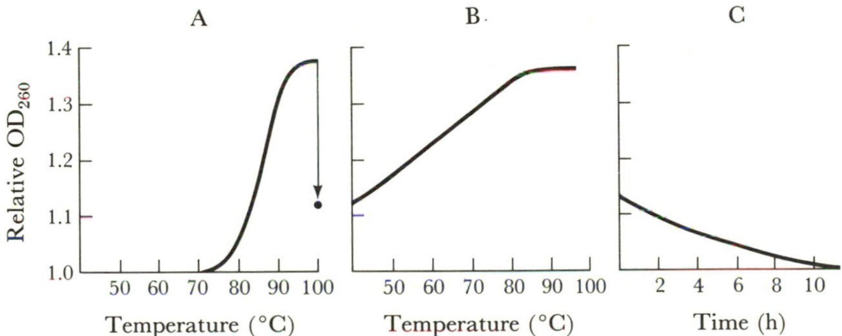

Figure 19-8
A. The OD_{260} of DNA in high ionic strength as a function of temperature. After reaching 100°C, the temperature has been lowered to 20°C. Thus OD_{260} drops, as indicated, to a value of 1.12. B. The DNA that has been heated to 100°C and lowered to 20°C is reheated to give a broad curve of OD_{260} versus temperature. C. The DNA used in part B is heated to 88°C instead, for the times indicated. The OD_{260} drops, indicating reformation of hydrogen bonds (renaturation).

this process is called *renaturation* or *annealing* (Figure 19-8C). The two-strand molecule need not consist of two identically complementary strands because one strand could be DNA and the other RNA and some regions (ranging from a single base pair to an extended tract) might be noncomplementary (Figure 19-7B). The process by which a two-strand molecule is formed from unlike strands is usually called *hybridization*, although the terms renaturation and hybridization are used interchangeably. A description of the techniques for detecting hybridization, together with several examples of its use, follows.

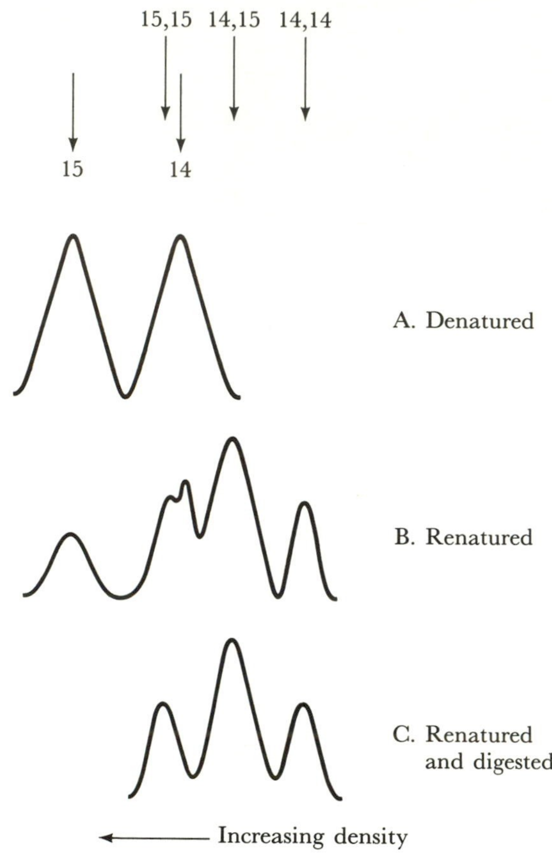

Figure 19-9

Detection of hybridization by equilibrium centrifugation in CsCl (Chapter 11). The curves are photometric traces of ultraviolet-absorption photographs, showing the DNA concentration distribution in the centrifuge cell: (A) mixture of denatured ^{14}N- and ^{15}N-labeled phage T7 DNA; (B) after renaturation of the sample in part A (note that there are five peaks—three of renatured DNA and two of unrenatured DNA); (C) result of treatment of the sample in part B with an exonuclease specific for single-stranded DNA—only the renatured DNA remains. The positions of single-stranded ^{14}N- and ^{15}N-labeled DNA are indicated by the arrows numbered 14 and 15, respectively; the positions of the double-stranded [^{14}N]DNA, hybrid DNA and [^{15}N]DNA are indicated by the arrows labeled 14, 14; 14,15; and 15, 15, respectively.

Detection by Equilibrium Centrifugation in Cesium Chloride Solution

This technique is rarely used now but was so common at one time that it deserves brief mention. For the preparation of DNA·RNA hybrids, use was made of the fact that, for polynucleotides of roughly 50% G + C content, the densities in CsCl of DNA and RNA differ by approximately 0.1 g/cm^3 (a very large difference). Hence, a denatured and a renatured mixture could be centrifuged to equilibrium in a CsCl solution and hybridization was detectable by the appearance of a band at the appropriate density. For DNA·DNA hybrids of DNA from the same species of organism, usually one DNA was density labeled with ^{15}N, ^{2}H, or ^{13}C and the other with ^{14}N, ^{1}H, and ^{12}C. If the DNAs were from different organisms having different G + C content, the intrinsic densities of the DNAs would not be the same and the hybrid would be detected by the appearance of a band indicating DNA of intermediate density (Figure 19-9). Note that in this technique hybridization is performed between DNAs in solution.

Hybridization on Nitrocellulose Filters

If single-stranded DNA in solutions of high ionic strength is passed through nitrocellulose filters, the DNA is bound by the filters (Chapter 7). If the filters are then dried in a vacuum, the DNA is bound so tightly that it cannot be washed off even at high temperature and low ionic strength. Binding occurs by means of the sugar-phosphate chains and leaves the bases available for hydrogen bonding to complementary bases. Thus complementary single-stranded DNA can be hybridized to these bound DNA molecules. An immediate problem arises in that added single strands will, of course, bind to the filters whether or not DNA is already present on the filters. To avoid this problem, after the DNA is bound to the filters, the filters can be treated in various ways so that no further single-stranded polynucleotide can be bound. One way to do this is to wash the filter with a solution of bovine serum albumin, which seems to saturate the DNA-binding sites. After the filters have been prepared and treated, a small volume of a solution of radioactive single-stranded DNA or RNA is added and the mixture is incubated under conditions appropriate for hybridization—that is, high ionic strength and a temperature between 65°C and 70°C. Complementary radioactive DNA or RNA binds to the DNA on the filter (see Figure 7-4). The filter is then washed to remove unhybridized material; after it has dried, the radioactivity is counted.

Examples of the Use of Hybridization

The following examples illustrate the use of hybridization both as an analytic tool and as a preparative one.

Example 19-D ☐ Measurement of genetic relatedness between organisms.

Genetic relatedness between two different organisms implies that there should be regions of the DNA that show partial or complete homology—that is, some extended base sequences should be identical or nearly so. To test this, nitrocellulose filters containing single-stranded DNA of organism A are prepared. Then radioactive, single-stranded DNA of very low molecular weight (M is usually reduced by sonication) from organism B is added, and hybridization is allowed to occur. The molecular weight of the B DNA is so low that a large piece of DNA will not be bound by virtue of a small region having homology. The fraction of the added B DNA that binds to the filter is a measure of the fraction of the sequences that are common of the two organisms. The observed fraction must be corrected for the efficiency of hybridization. This is usually done by adding single-stranded A DNA (radioactively labeled with another isotope) and determining the fraction bound. Hence, the fraction of B DNA that is homologous to A DNA is (fraction B bound)/(fraction A bound). The completeness of homology can be estimated from the thermal stability of the hybridized DNA, using certain empirical relations.

This method can also be used to measure the fraction of a DNA sample that is of a particular type. For example, suppose that a bacteriophage infects a host bacterium in a radioactive growth medium such that both newly synthesized phage DNA and bacterial DNA are labeled. The fraction of the label that is phage DNA can be determined by hybridization of the mixture to filters containing phage DNA. As in Example 19-D, a correction must be made for hybridization efficiency.

Example 19-E ☐ Purification of messenger RNA.

If a bacteriophage infects a bacterium in growth medium containing [^{3}H]uridine, radioactive phage and bacterial mRNA are synthesized. If the RNA is isolated and hybridized to filters containing single-stranded phage DNA, only phage mRNA is bound to the filters. Washing the filters will free them of all bacterial mRNA, because it is unbound. By placing the washed filter in a buffer and heating to the appropriate temperature for denaturing an RNA·DNA hybrid, phage mRNA will be dissociated from the filter and will be in the buffer. The filter can be removed and the phage mRNA is thereby purified.

Example 19-F ☐ Identification of the DNA template for mRNA.

If mRNA, purified as in Example 19-E, is hybridized with DNA that is genetically deleted for certain sequences and then treated with RNase to digest unbound RNA, the amount of bound RNA will decrease if the mRNA is made partly from the deleted region. Thus, DNA sequences corresponding to an mRNA can be identified.

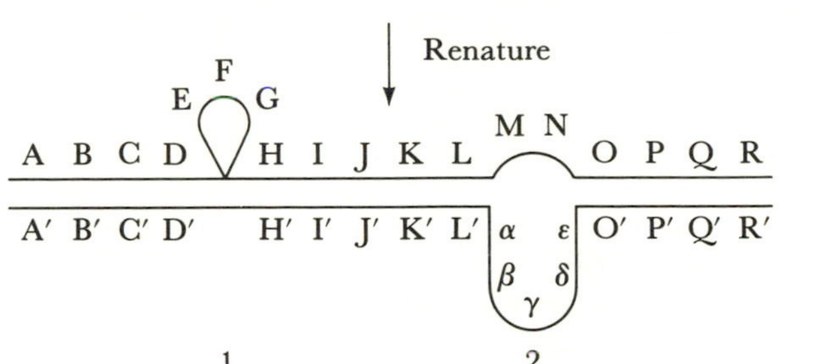

Figure 19-10
Diagram of heteroduplex analysis. Two different DNA strands are renatured. Complementary bases (e.g., AA′) reform hydrogen bonds. Regions of DNA for which no complementary regions exist (e.g., EFT, MN, and $\alpha\beta\gamma\delta\varepsilon$) form loops. Note that there are two different types of loops. Type 1 contains one piece of single-stranded DNA and occurs with a true deletion. Type 2 contains two pieces of single-stranded DNA and occurs when each strand contains a different sequence in the same region.

Example 19-G

☐ Physical mapping of genetic deletions in DNA—an electron microscopic analysis.

If two single-stranded DNA molecules, one of which is deleted for certain sequences, are hybridized, hydrogen-bond formation will be complete except that the sequences in the normal DNA corresponding to the deletion will have nothing with which to form hydrogen bonds. Therefore, a double strand will result of the same length as that of the deleted DNA but with a single-strand loop at the site of the deletion whose length corresponds to the length of the deleted sequence. Such molecules, which are prepared by hybridization in liquid, can be examined with the electron microscope. The position and length of the loop allows the deletion to be mapped physically with respect to the length and ends of the normal DNA (Figure 19-10). This is described in greater detail in Chapter 3 (Figure 3-17).

C_0t Analysis

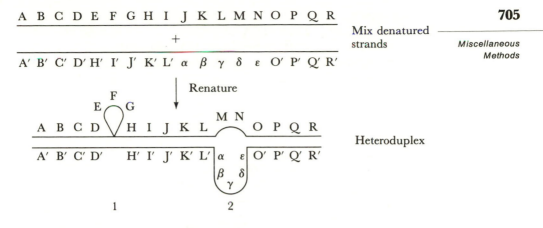

A study of the dependence of the rate of renaturation of denatured DNA, a procedure called c_0t analysis, can yield the number of copies of each base sequence in the DNA complement of a particular cell type. This is because a repeated sequence is present in a higher molar concentration than a sequence for which there is only a single copy. Thus, a repeated sequence has more potentially complementary strands than a single-copy sequence and hence renatures faster. This technique is an important tool in the study of the DNA of eukaryotic cells because these cells contain numerous copies of particular DNA sequences.

Renaturation is both a bimolecular and second-order reaction and is described by the rate law

$$\frac{d[A]}{dt} = -k[A]^2 \tag{9}$$

in which [A] is the concentration of the single strands. If $[A]_0$ is the initial concentration, the rate law can be integrated to yield

$$\frac{1}{[A]} = \frac{1}{[A]_0} + kt \tag{10}$$

Experimentally it is convenient to measure the fraction f of the strands that remain dissociated. This is defined as

$$f = \frac{[A]}{[A]_0} \tag{11}$$

Combining this with equation (10) yields

$$f = \frac{1}{1 + [A]_0 kt} \tag{12}$$

This equation tells us two things. (1) After a very long time, the fraction that is dissociated approaches zero—that is, ultimately, renaturation is complete. (2) The rate of renaturation increases as the initial DNA concentration is increased, as expected of a bimolecular reaction.

Concentration has a special meaning in the analysis of renaturation and requires some discussion. In ordinary chemical kinetics, concentration is expressed as the number of molecules per unit volume (or some form of this ratio). This is a useful measure because, in general, each molecule contains only a single site at which a particular reaction can occur. This is not the case for renaturation because any sequence of bases can appear several times in a particular DNA chain. For the purposes of studying renaturation the best measure of concentration is actually the number of copies of each sequence per unit volume. The effect of multiple copies of a particular sequence can be seen by examination of three simple types of DNA molecules: types I, II and III (Figure 19-11).

Consider a type I DNA molecule, whose base sequence is ABC · · · IJ in one strand and A′B′C′ · · · I′J′ in the other strand. The prime denotes a

Type I:	A	B	C	D	E	F	G	H	I	J	
	A′	B′	C′	D′	E′	F′	G′	H′	I′	J′	
Type II:	A	B	C	D	E	F	G	H	I	J	L
	A′	B′	C′	D′	E′	F′	G′	H′	I′	J′	L′
	M	N	O	P	Q	R	S	T			
	M′	N′	O′	P′	Q′	R′	S′	T′			
Type III:	A	B	C	D	E	A	B	C	D	E	
	A′	B′	C′	D′	E′	A′	B′	C′	D′	E′	

Figure 19-11
Three types of DNA molecules described in the text in discussing c_0t curves.

complementary base. That is, according to the standard base-pairing scheme G′ = C, A′ = T, C′ = G, and T′ = A; thus, AA′ is a proper base pair, as are the pairs BB′, CC′, and so forth. In order that a proper double helix forms, a base *X* in one strand must collide with the complementary base X′ in another strand. Since collisions are random, most of the collisions will not lead to base pairing. If the rate of renaturation of this type I single-stranded DNA, which has 10 bases, is compared to that of another molecule (a type II molecule which has 20 base pairs and whose base sequences are ABC ··· ST and A′B′C′ ··· S′T′, the type II molecules having more bases and hence a higher molecular weight will renature more slowly for the same amount of DNA per unit volume because so many more *in*effective collisions occur per unit time. Thus the initial concentration $[A]_0$ of DNA that appears in equations (10) and (12) must somehow take into account the molecular weight or the number of base pairs in a complementary base sequence. If the type I molecules are cut in half, so that a sample contains molecules having either the base sequence ABCDE or the sequence FGHIJ, the rate of renaturation will *not* be altered. (Be sure you understand this point.) This is because the number of ineffective collisions is unchanged—that is, B can initiate renaturation only by colliding with B′ and failure results whether B collides with an A′ on one strand or a G′ on the other strand. Let us now consider the rate of renaturation of the single strands of a type III molecule, that is, one whose sequences are ABCDEABCDE and A′B′C′D′E′A′B′C′D′E′. Such a molecule is said to have two copies of the *unique sequence* ABCDE. For a particular concentration (expressed as g/cm^3), this molecule will renature faster than a type I molecule because in a collection of type III molecules more of the collisions with base B are effective than with a collection of type I molecules. Again, cutting the type III molecules into fragments does not affect the renaturation rate. Thus if c_0 is the initial concentration, now expressed in base pairs/cm^3,* and x is the number of base pairs in each unique sequence, then

$$[A]_0 = \frac{c_0}{x} \tag{13}$$

and equation (12) becomes

$$f = \frac{1}{1 + c_0 tk/x} \tag{14}$$

Let us now see how k can be evaluated. Consider a molecule containing only bases of type A in one strand and only of type A′ in the other. An example of this is poly(dA)·poly(dU), which consists of one strand of polydeoxyadenylic acid and one strand of polydeoxyuridylic acid. If this DNA is denatured, *all* collisions between complementary strands are effective so that $x = 1$. In general, for the time interval $t_{\frac{1}{2}}$ when $f = \frac{1}{2}$, it is the case that $1 + (c_0 t_{\frac{1}{2}}/x) = 2$ so that

$$c_0 t_{\frac{1}{2}} = \frac{x}{k} \tag{15}$$

Hence for the DNA just described, for which $x = 1$, $k = 1/c_0 t_{\frac{1}{2}}$ and a measurement of the time required for 50% renaturation for any initial concentration c_0 yields the value of k. It is important to understand that k *is a constant characteristic of the physical process of renaturation and is independent of the molecular weight of the DNA and the value of* x. Furthermore, since x is also a constant, $c_0 t_{\frac{1}{2}}$ is a constant although neither c_0 nor $t_{\frac{1}{2}}$ is a constant. Thus, for any particular DNA concentration c_0 that is chosen, $t_{\frac{1}{2}} = x/kc_0$. The important point is that *the product* $c_0 t_{\frac{1}{2}}$ *is proportional to x, the number of base pairs per unique sequence.*

The measurement of renaturation rates to obtain the value $c_0 t_{\frac{1}{2}}$ is called $c_0 t$ (or *cot*) analysis. In order to understand the value of such an analysis, the result of two more experiments will be considered.

In the first experiment, we compare the curves of f versus $c_0 t$ for a variety of DNA molecules whose molecular weights differ. It should be noted that in such an experiment, DNA of any concentration can be chosen, that is, any (or several) values of c_0 can be selected because the curve of f versus $c_0 t$ is independent of c_0 (that is, $c_0 t$ is a constant). If

* Note that by expressing c as base pairs/cm^3, we are ignoring the molecular weight of the single strand and using instead the molarity of the bases. Thus, if a DNA molecule whose molecular weight is 10^7 is at a concentration of 10 mg/l, the molecular-weight-dependent molar concentration is $0.01/10^7 = 10^{-9}$ M, whereas expressed as base pairs (molecular weight = 660), the molecular-weight-independent concentration is $0.01/660 = 1.5 \times 10^{-5}$ M.

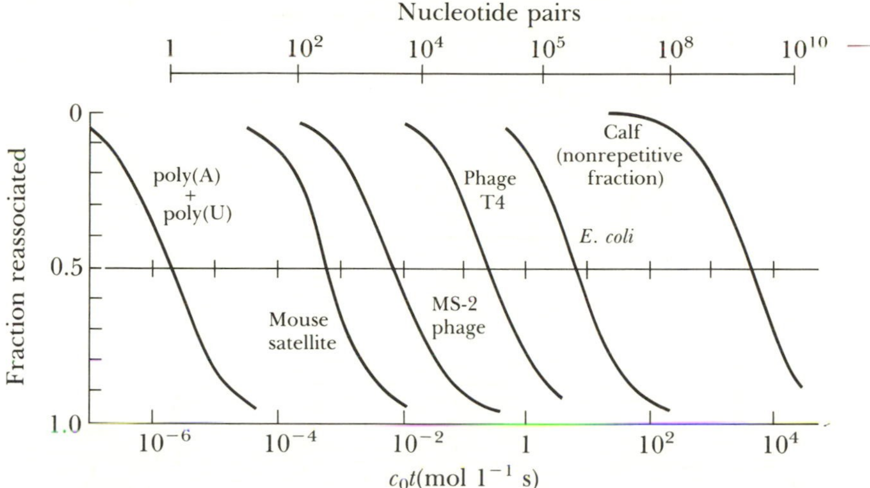

Figure 19-12
Reassociation of denatured DNA of various types as a function of time and concentration. [After E. J. Britten and D. E. Kohne, *Science* 161(1968):529.]

these DNA molecules each contains a single base sequence (that is, they have no repeated sequences), x is simply the number of base pairs in the unbroken DNA molecule. That is, $c_0t_{\frac{1}{2}}$ *(not c_0t) is proportional to x and hence to the molecular weight of the unbroken DNA.** This is the fundamental rule used to analyze experimental c_0t curves. The result of such an experiment is shown in Figure 19-12, in which it can be seen that $c_0t_{\frac{1}{2}}$ is proportional to the number of base pairs per molecule. This means that the measured value of $c_0t_{\frac{1}{2}}$ can be used to obtain the size of a particular sequence and the number of copies of the sequence in the organism from which the DNA was isolated. Before proceeding to the second experiment, an important point must be made: the value of c_0 used in a c_0t curve is the *total* DNA concentration in the renaturing sample, rather than the concentration of each component. Thus, when analyzing multicomponent curves, the observed $c_0t_{\frac{1}{2}}$ is the true $c_0t_{\frac{1}{2}}$ value (i.e., the value if only that component were present) divided by the fraction of the total DNA represented by that component. Since the scale in Figure 19-12 relates size to $(c_0t_{\frac{1}{2}})_{\text{true}}$, each $(c_0t_{\frac{1}{2}})_{\text{obs}}$ must be converted to $(c_0t_{\frac{1}{2}})_{\text{true}}$ before using such a scale.

In the second experiment, the molecule shown in Figure 19-13A is examined. This molecule contains four copies of a sequence that is 5% the length of the intact molecule. If a c_0t curve is prepared and if x is $(c_0t_{\frac{1}{2}})_{\text{true}}$ for the unique sequence, $(c_0t_{\frac{1}{2}})_{\text{obs}}$ will be $x/0.8 = 1.25 \times$ (80%

* For technical reasons the DNA molecules are broken into small fragments but, as we saw when discussing type I, II and III molecules, such breakage does not alter the value of c_0t.

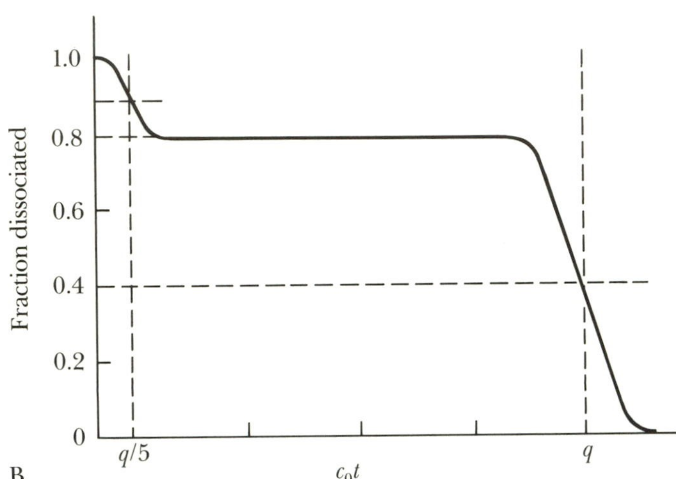

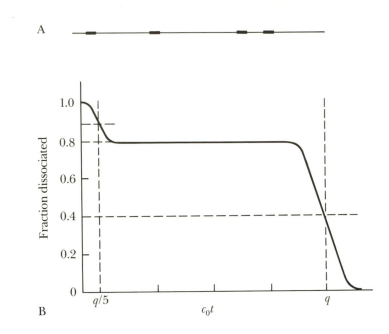

Figure 19-13
A DNA molecule (panel A) that has four repetitive sequences (heavy line) each 5% of the total molecule and the c_0t curve (panel B) for this DNA. See text for details.

of the total sequence is unique). For the redundant sequence $(c_0t_{\frac{1}{2}})_{\text{true}}$ will be $0.05x$ and $(c_0t_{\frac{1}{2}})_{\text{obs}}$ will be $(0.05) \times (0.20) = 0.25x$. This is shown in Figure 19-13—20% has $(c_0t_{\frac{1}{2}})_{\text{obs}}$ of $q/5$ and 80% has $(c_0t_{\frac{1}{2}})_{\text{obs}}$ of q. Notice in the figure how the $c_0t_{\frac{1}{2}}$ values are identified—namely, as an inflection point in the descending part of the c_0t curve or as the value of c_0t at which each transition is 50% complete. The fraction of material having a particular c_0t value is the fraction of the vertical distance in the figure.

Three techniques are used in c_0t analysis, namely, absorbance measurements, membrane filtration, and hydroxyapatite chromatography. The absorbance measurement is based on the greater absorbance of single-stranded DNA compared to double-stranded DNA. A sample of denatured DNA is adjusted to a high salt concentration and elevated temperature, that is, the conditions that induce renaturation. The sample is then placed in an automatic recording spectrophotometer and the absorbance is plotted as a function of time. The resulting curve is a c_0t curve. Sometimes the time required for complete renaturation is many days, a long time to occupy a recording spectrophotometer (a very expensive instrument) for a single experiment. In this case, either membrane filtration or hydroxyapatite chromatography is used. When membrane filtration is used, samples are withdrawn from the renaturing sample and filtered through a nitrocellulose filter. Single-stranded but not double-stranded DNA is retained on the filter. When hydroxyapatite is used, the phosphate concentration is adjusted so that single- but not double-stranded DNA is retained on the column (Chapter 8, page 268). The double-stranded DNA is then eluted by applying a higher phosphate concentration. If radioactive DNA is available, the relative amounts of radioactivity on the filter and in the filtrate, or in the void volume and the eluate, are measured. If the DNA cannot be made radioactive, the amount of DNA in the two fractions is measured by its absorbance. These two techniques have the disadvantage that continuous monitoring is not done and that sometimes it may be necessary to take samples at night.

c_0t analysis is used as a means of detecting repetitive sequences in DNA, as seen in the previous experiment. This is of great interest because in animal cells there are short sequences repeated up to one million times. The reason for this is unknown but is a topic of current research interest. An example of how one draws conclusions about the complexity of a DNA sample is given in the following.

Example 19-H

☐ Determination of the size and number of copies of various sequences in a DNA sample whose c_0t curve is shown in Figure 19-14.

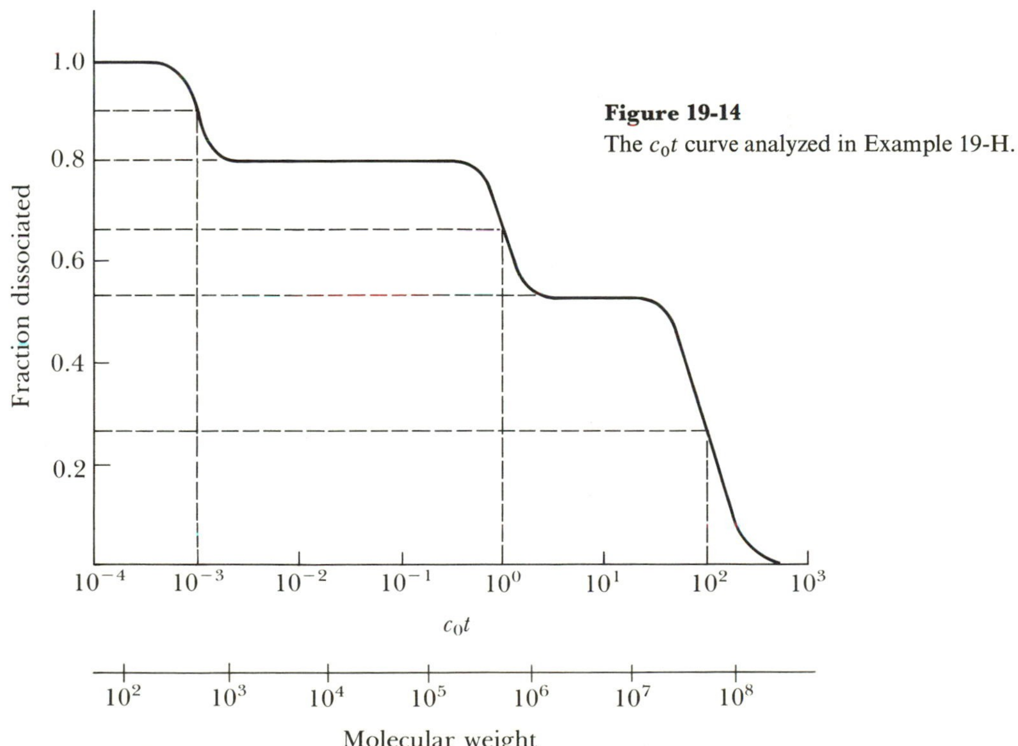

Figure 19-14
The c_0t curve analyzed in Example 19-H.

The following $c_0t_{\frac{1}{2}}$ are observed for a DNA sample: 10^2, 53%; 1, 27%; 10^{-3}, 20%. The number of copies of each sequence is determined from the size scale in Figure 19-12. We first calculate $(c_0t_{\frac{1}{2}})_{true}$ for each component by multiplying each $(c_0t_{\frac{1}{2}})_{obs}$ by the fraction of the total DNA that it represents. Thus, the $(c_0t_{\frac{1}{2}})_{true}$ are 0.53×10^2, 0.27×1, and 0.20×10^{-3}, which correspond to 4.2×10^7, 2.2×10^5, and 160 base pairs, respectively, for each component. The number of copies of each sequence is inversely proportional to $(c_0t_{\frac{1}{2}})_{obs}$; thus, if we assume that the genome contains one copy of the longest sequence, then the cell from which the DNA has been isolated would contain 1, 10^2, and 10^5 copies of sequences having 4.2×10^7, 2.2×10^5, and 160 base pairs, respectively. The total number of base pairs per genome would be $4.2 \times 10^7 + 100(2.2 \times 10^5) + 10^5(160) = 8 \times 10^7$, and the molecular weight of the total cellular DNA would be $(8 \times 10^7)(660) = 5.3 \times 10^{10}$. Note that if an independent measurement of the total molecular weight of the DNA were to yield the value $1 \times 10^{11} = 2(5.3 \times 10^{10})$, then there would be 2200, and 2×10^5 copies of these sequences, respectively.

Concentration of Macromolecules

In the course of isolating and purifying macromolecules, solutions often become very dilute and must be concentrated. A straightforward procedure is to concentrate the molecules by centrifugation, but often the sedimentation coefficient (Chapter 11) is too small for this to be effective. There are many other successful procedures of which six are described here.

Salting Out with Ammonium Sulfate

Charged solute macromolecules (e.g., macromolecules in polar solvents) are solvated and are thereby rendered soluble. If high concentrations of electrolytes are added, the solvent molecules are bound so tightly by the ions that they are unable to solvate the solute molecules. Hence, the solute molecules come out of solution. This is called salting out and is a useful technique for precipitating macromolecules such as proteins.

In particular, for most, if not all, proteins there is a concentration of ammonium sulfate (a highly soluble, highly pure, inexpensive reagent) above which the protein precipitates.* Hence, solid $(NH_4)_2SO_4$ can be added to a protein solution until precipitation occurs. The precipitate is collected by centrifugation and then redissolved in a small volume of

* The precise amount differs for each protein so that, by varying the amount, a protein mixture can be fractionated.

whatever buffer is required. This rather general method sometimes fails if the total protein concentration of the starting solution is too low.

For complex reasons DNA is not salted out and thus cannot be concentrated in this way. In fact, salting out is an effective way to remove proteins from a DNA sample.

Both salting out and centrifugation are based on the removal of the sample from the solution. The following methods are based on the removal of water.

Flash Evaporation

In this technique, the solution is placed in a rapidly spinning flask, which is then evacuated (Figure 19-15). Because of the spinning, the liquid forms a relatively thin film, which significantly increases the surface-to-volume ratio. The vacuum serves to reduce the boiling point of water to room

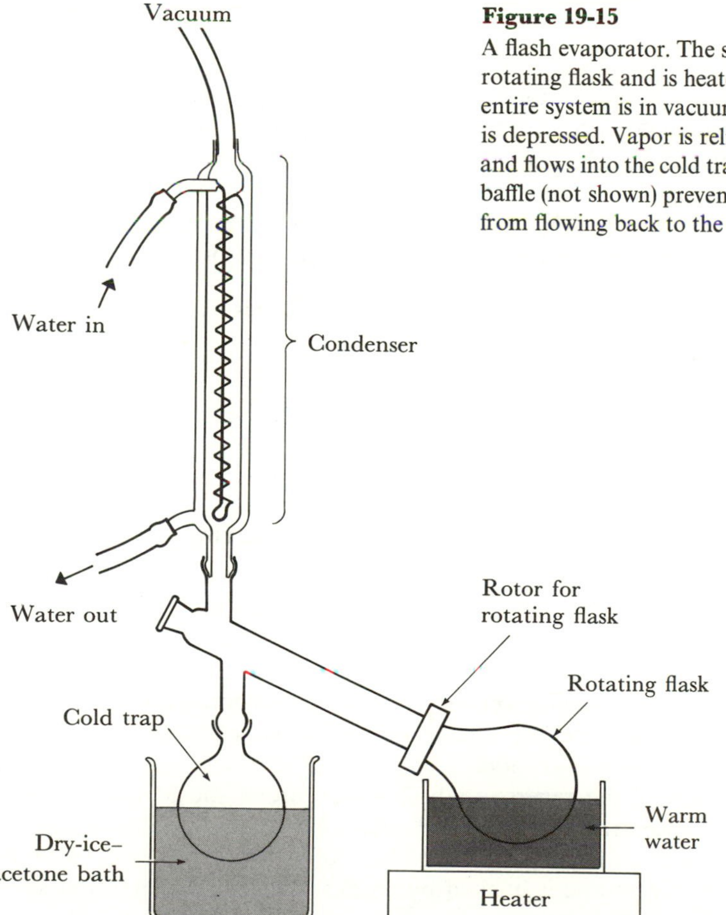

Figure 19-15
A flash evaporator. The sample is contained in the rotating flask and is heated by the water bath. The entire system is in vacuum so that the boiling point is depressed. Vapor is reliquified in the condenser and flows into the cold trap where it freezes. A special baffle (not shown) prevents the condensed liquid from flowing back to the rotating flask.

temperature or below. In this way, water can be removed without subjecting the molecules to temperatures that might cause denaturation or dissociation.

This method is not effective for very small amounts of material because a large fraction of the solute may remain stuck to the wall of the rotating flask and be difficult to remove.

Lyophilization

This technique is based on the fact that, at sufficiently low pressure, ice sublimes. Hence, the sample is frozen and subjected to high vacuum. Note that both lyophilization and flash evaporation can be used to concentrate small molecules as long as they are nonvolatile.

With proteins lyophilization yields a light, fluffy precipitate that is easy to handle and to redissolve. Lyophilized DNA, however, dissolves very slowly; in solutions of low ionic strength (less than 0.01 M), the DNA will dissolve in 12–24 hours. If the ionic strength is 0.1 M or higher, complete solution may require a week.

Pressure Dialysis

Semipermeable membranes (e.g., Diaflo and Pellicon) that pass small molecules but not macromolecules were described in Chapter 7. If a solution is placed in a chamber, one wall of which is such a membrane, the rate of passage of the small molecules (including water) through the membrane increases with increasing pressure. Hence, when pressure is applied to the chamber, water is forced through and the macromolecules are concentrated (Figure 19-16).

Reverse Dialysis

If a solution is placed in a dialysis bag and the bag is allowed to remain in air, water will evaporate from the bag, thus concentrating the solution. If the bag is surrounded by a dry, highly soluble polymer that cannot pass through the membrane, water will leave the bag to dissolve the dry polymer. This tends to be faster than air-drying. The material most commonly used is polyethylene glycol.

In this and the three preceding methods, either water is removed and salts are not or the salts are removed more slowly than the water. In any case, there is an increase in the salt concentration. This may or may not cause concern, depending on the particular macromolecule. In the following two procedures, the macromolecule is concentrated without an increase in salt concentration.

A variant of the reverse dialysis procedure is the Minicon multipurpose concentrator made by Amicon Corporation. The sample is placed in a chamber, which is schematically shown in Figure 19-17. A permeable membrane is backed by an absorbent pad. Water and salts

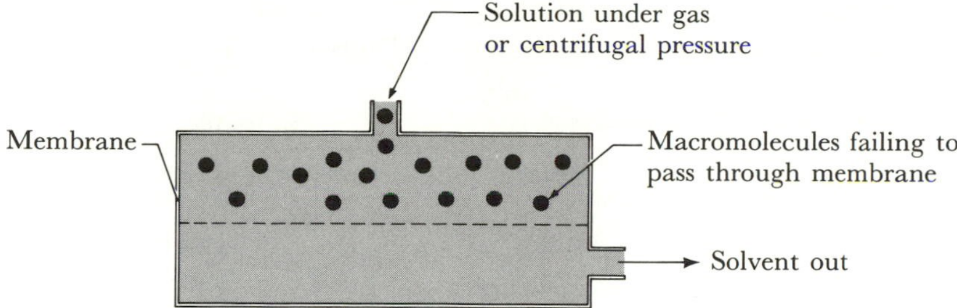

Figure 19-16
Concentration of a solution of macromolecules, using a Diaflo or Pellicon membrane. Under pressure, solvent and small solute molecules pass through the membrane. Macromolecules do not and are thereby concentrated.

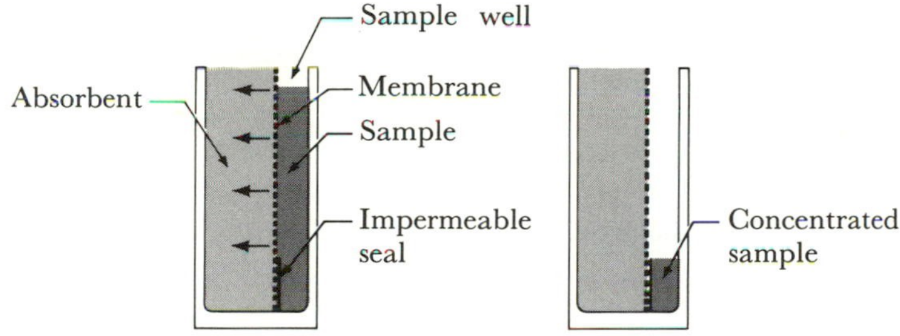

Figure 19-17
A Minicon concentrator. The sample is placed in the right-hand compartment. Liquid passes through the membrane and is absorbed by the material in the left-hand compartment. When the level of the liquid reaches the plastic barrier, concentration is complete.

pass through the membrane into the pad until the sample volume is such that it is in contact only with an impermeable seal. This is usually set for twentyfold concentration.

Hollow Fiber Membranes

The rate of liquid removal by pressure dialysis (see above) increases with the surface-to-volume ratio. If the solution is contained in hollow porous fibers (see Chapter 7, Figure 7-10), the surface-to-volume ratio becomes huge. These devices (commercially available as Bio-Fiber, made by Bio-Rad Laboratories) consist of bundles of semipermeable, hollow fibers attached to filling and emptying ports. Pressure is established across the fiber wall and solvent passes through the fiber wall, leaving behind a more concentrated solution.

Table 19-2
Examples of Separation and Concentration by Partitioning.

Common solvent systems	Factors affecting separation
Dextran and polyethylene glycol	Ionic strength and pH
Dextran and hydroxypropyl dextran	Ionic strength
Dextran and methyl cellulose	pH
Dextran sulfate and polyethylene glycol	Particular ions
Dextran sulfate and methyl cellulose	Temperature
Substances separated	
Native and denatured DNA	
Covalently closed circular DNA and open circles or linear DNA	
Native and denatured proteins	
Various proteins (useful in enzyme purification)	
Protein and nucleic acid	
DNA and RNA	
Various polynucleotides	
Virus and virus-antibody complex	
Various viruses and phages	
Various microorganisms	
Male and female *E. coli*	
Different species of *Chlorella*	
Different poliovirus strains	
Intact and broken chloroplasts	
Erythrocytes from various species of animals	
Lymphocytes, leucocytes, and platelets	
Spores and vegetative cells	
Insulin-secreting granules and acid phosphatase particles from pancreatic β cells	

SOURCE: P. A. Albertsson, *Partition of Cell Particles and Macromolecules*, Wiley-Interscience, 1971, and P. A. Albertsson, *Adv. Protein Chem.* 24(1970): 309–341.

Separation by Partitioning Between Dextran and Polyethylene Glycol Solutions

Concentrated aqueous solutions of the polysaccharides dextran and polyethylene glycol are immiscible. Many biological polymers, cellular components, and even cells show markedly different solubility in these two solutions and will therefore separate by partitioning. The standard procedure is to add the sample to either the dextran or the polyethylene glycol solution, add the other solution, shake for complete mixing, and then allow the two phases to separate. Table 19-2 lists several of the numerous materials that have been separated in this way.

Flow Birefringence

As explained in Chapter 13, elongated macromolecules can be oriented by flow under the influence of a shear gradient; the greater the axial ratio (ratio of length to width), the more easily the molecules are oriented. As explained in Chapter 2, oriented molecules exhibit birefringence—that is, they are capable of polarizing light. When birefringence is produced by a shear gradient, it is usually called flow birefringence. The amount of birefringence as a function of the shear gradient can be measured and this yields the rotational diffusional coefficient, a parameter from which the axial ratio can be calculated.

Consider a Couette viscometer (Chapter 13) with a stationary inner cylinder and a rotating outer cylinder. Between these cylinders is an annulus, which is observed by looking down the axis of the instrument. If there is a polarizer between the light source and the instrument and the emerging light passes through a second polarizer (the analyzer) whose axis is perpendicular to the first, no light reaches the observer. If the annulus contains a solution of elongated molecules and the outer cylinder is rotated, these molecules are oriented by the shear gradient and they resolve the light transmitted by the polarizer, so that some light is passed by the analyzer. However, because the system has cylindrical symmetry, there are two regions in which the average orientation is parallel to the direction of the polarizer (see Figure 19-18). At these points (which are opposite one another), there is no resolution of the polarized light and no light is transmitted by the analyzer. Similarly, there are two regions in which the molecules are oriented in the direction of the analyzer and again no light is transmitted. Hence, there are four regions that appear dark against the light background, as shown in Figure 19-18. This is called the *cross* and the angle χ is the *extinction angle*. It can be shown that

$$\chi = 45^\circ - \frac{15G}{\pi\Theta} + \cdots \qquad (16)$$

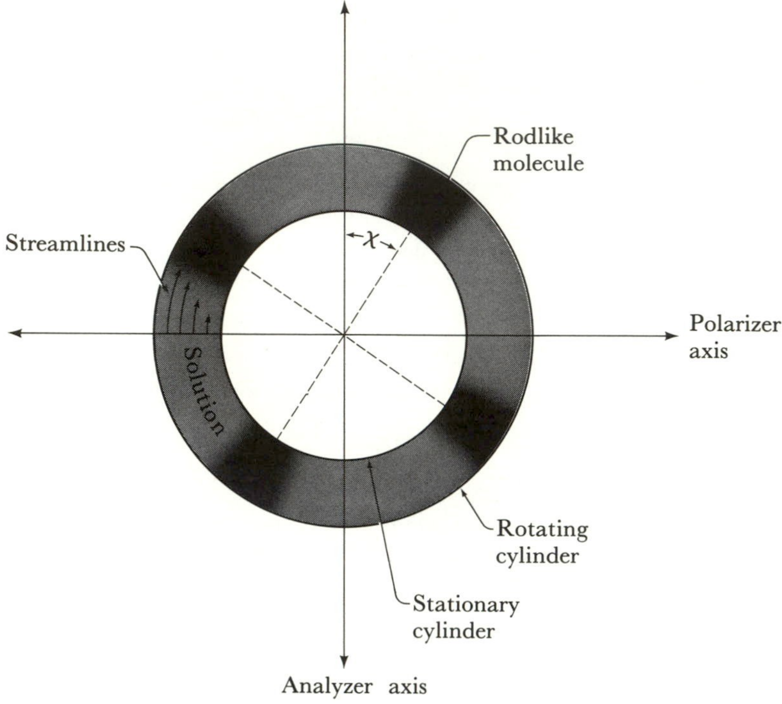

Figure 19-18
View down the axis of a flow birefringence instrument. The solution is contained between the inner and outer cylinders. When the outer cylinder rotates, the molecular rods are oriented by the streamlines. At two points, they are parallel to the analyzer; at two other points, they are parallel to the polarizer. No light is transmitted at these four points; the pattern of four dark regions is known as the *cross*. The angle χ formed by the axes of the cross is called the extinction angle.

in which G is the shear gradient and Θ is the rotational diffusion coefficient. Hence, by plotting χ versus G, a straight line with slope $-15/(\pi\Theta)$ is obtained. Equation (16) is valid only for low G and a small axial ratio.

This method has been used primarily to determine axial ratio and, in some cases, to monitor helix-coil transitions. In the latter case, when the molecule becomes disordered, it can be less easily oriented and the birefringence decreases.

The flow birefringence apparatus can be used to measure intrinsic birefringence or dichroism (Chapter 14) of molecules that can be oriented almost totally at low shear force (e.g., DNA). Most molecules cannot be completely aligned because the required shear gradient would be so great that turbulence would set in and disrupt the orientation. However, this is not the case for very elongated molecules.

Two simple examples serve to indicate how orientation within a molecule can be measured.

☐ Orientation of nucleotide pairs in DNA. **Example 19-I**

The bases of DNA are dichroic—that is, they have a preferred direction of absorption of ultraviolet light with respect to their axes (Chapter 14). Hence, if DNA is totally oriented in a flow-birefringence instrument and illuminated with polarized ultraviolet light, the oriented molecules absorb only if they are situated at certain angles with respect to the plane of polarization. Hence, the orientation of the bases with respect to the DNA axis can be determined.

☐ Orientation of proflavin molecules bound to DNA. **Example 19-J**

Proflavin also has a preferred axis for the absorption of blue light. By choosing a wavelength absorbed by proflavin but not by DNA, the orientation of the bound proflavin can be determined in the same way as in example 19-I. Such a measurement has shown that the proflavin molecules are parallel to the base pairs.

SELECTED REFERENCES

Albertsson, P. A. 1971. *Partition of Cell Particles and Macromolecules.* Wiley-Interscience. The best book on the subject.

Britten, R. J., and D. E. Kohne. 1968. "Repeated Sequences in DNA." *Science* (*Wash., DC*) 161:529–540. The original paper on c_0t analysis.

Cerf, R., and H. A. Scheraga. 1952. "Flow Birefringence in Solutions of Macromolecules." *Chem. Rev.* 51:185–261. A good review of fiow birefringence.

Davis, R. W., M. Simon, and N. G. Davidson. 1971. "Electron Microscopic Heteroduplex Methods for Mapping Base Sequence Homology in Nucleic Acids," in *Methods in Enzymology*, vol. 21, edited by L. Grossman and K. Moldave, pp. 413–328. Academic Press.

Englander, W. S. 1967. "Measurement of Nucleic Acid Hydrogen Exchange," in *Methods in Enzymology*, vol. 12B, edited by L. Grossman and K. Moldave, pp. 379–386. Academic Press.

Englander, W. S., and J. J. Englander. 1978. "Hydrogen-Tritium Exchange," in *Methods in Enzymology*, vol. 49, edited by C. H. W. Hirs and S. N. Timasheff, pp. 24–38. Academic Press.

Kasai, M., and F. Osawa. 1972. "Flow Birefringence," in *Methods in Enzymology*, vol. 26, edited by C. H. W. Hirs and S. N. Timasheff, pp. 289–323. Academic Press.

Printz, M., and P. H. von Hippel. 1965. "Hydrogen Exchange Studies of DNA Structure." *Proc. Natl. Acad. Sci. U.S.A.* 53:363–370. Experiments on "breathing" of DNA.

Weissman, M., H. Schindler, and G. Feher. 1976. "Determination of Molecular Weights by Fluctuation Spectroscopy: Application to DNA," *Proc. Natl. Acad. Sci. U.S.A.* 73:2776–2780.

Miscellaneous Methods

19–1. A substance partitions so that 98% appears in 20% dextran and 2% in 20% polyethylene glycol. If solid polyethylene glycol were added to a solution of the substance to make a final concentration of 30%, what would probably happen to the substance?

19–2. Which of the following would show flow birefringence: a sphere, a prolate ellipsoid, a rigid rod, a flexible rod, a chain of beads, a circular DNA molecule?

19–3. Why do you think the sample vessel in a flash evaporator is made to rotate?

19–4. Actively growing bacilli are rod-shaped, typically with a ratio of length to width of about 4. A stationary culture consists of bacteria with a length-to-width ratio of about 2. If you were going to prepare a cell-free extract by sonication, would you choose a growing or stationary culture? Explain.

19–5. What factors should be considered in selecting a method for concentrating?

19–6. Suppose that, in doing a flow birefringence measurement, you are measuring the position of the cross at various speeds. In the course of increasing the speed, the cross is suddenly observed to fade and move to a larger angle. What has probably happened?

19–7. The ribosomes of a protozoan contain two RNA molecules whose molecular weights are 5×10^5 and 10^6. The haploid DNA content of each cell is 2×10^{10}. A c_0t analysis of the DNA yields the curve shown below. How many copies of the DNA sequence corresponding to these RNA molecules exist in each cell?

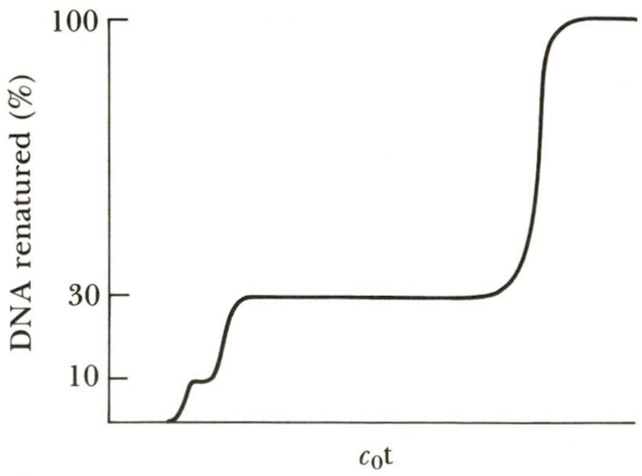

19–8. DNA molecule A is 100 units long. Variant B is 90 units long lacking the region from 40 to 50. Variant C is 105 units long; the region from 0 to 10 has been replaced by a nonmatching segment 15 units long and the region from 80 to 90 is replaced by another nonmatching segment 10 units long. Draw the two heteroduplex molecules resulting from denaturing and then renaturing (a) A and B, and (b) A and C. Label all lengths that you know.

1–6. The DNA would denature because the bases would interact with the solvent rather than with themselves.

1–7. There are six possible linear types:

However, if the subunits were either isotropic or symmetric, structure 1 would be identical with structure 3 and structure 4 would be identical with structure 6; then there would be four types.

There are two possible triangular types:

1–8. Eleven.

1–9. There are several possibilities. Each subunit might have a positively charged and a negatively charged region. These regions could interact by a simple charge interaction. If they have hydrophobic regions on the surface, they might interact so that water can be excluded.

1–10. In the first case,

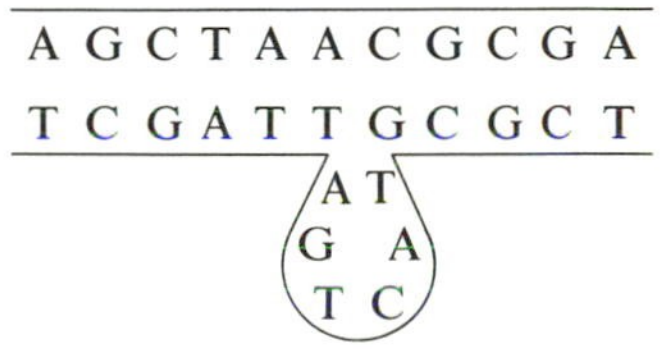

In the second case,

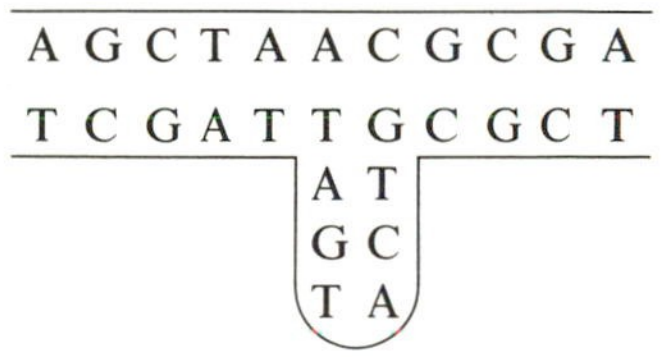

1–11. $\frac{3}{8}$ hybrid, $\frac{9}{16}$ heavy, and $\frac{1}{16}$ light.

1–12. There are only four (↑) peptide bonds because the C—N bond (⇣) to the left of proline is not a peptide bond.

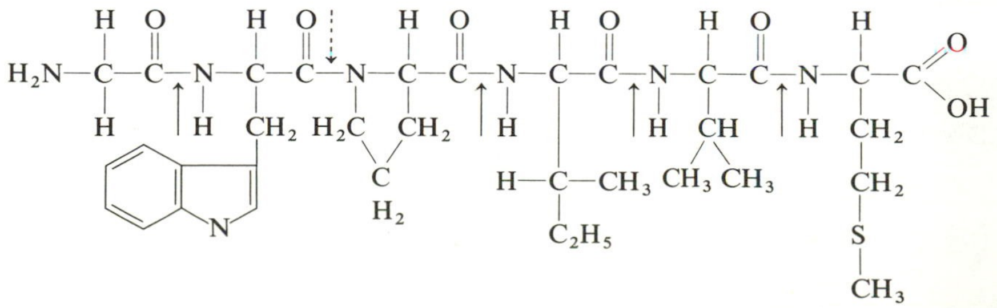

19–9. A phage DNA has a molecular weight of 30×10^6. Two deletion mutants A and B are selected and each is found to have a molecular weight of 25×10^6.

a. The DNA of each mutant is mixed with an equal number of molecules of the original, nonmutant DNA and then denatured and renatured. These heteroduplexes are examined by electron microscopy and the molecules shown below are seen. What can be said about the deletions (i.e., the size and number) in A and B?

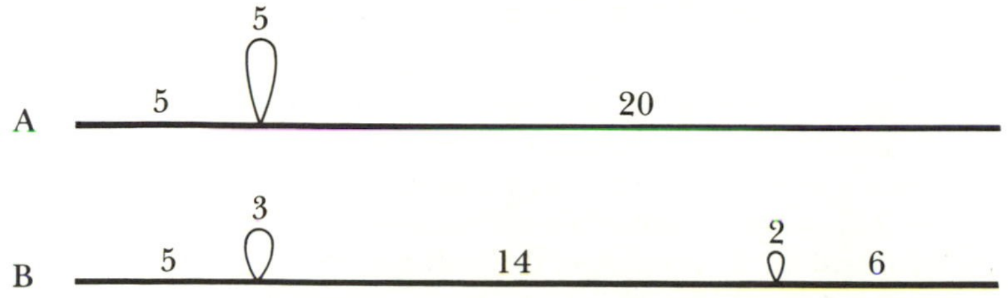

b. Both A and B are mutagenized in a way that leads to genetic duplications (that is, a base sequence is repeated in tandem) and two new mutants A′ and B′ are isolated. The DNA of both A′ and B′ has a molecular weight of 30×10^6. To confirm that A′ and B′ have a new piece of DNA at the site of the deletion, each DNA is used to form a heteroduplex with the original, nonmutant DNA. The molecules shown below are observed. Is the insertion at the site of the deletion in each case? If there is a duplication, is the sequence that is duplicated the sequence at the right or at the left of the deletion, assuming that all molecules that have been shown have the same left-to-right orientation?

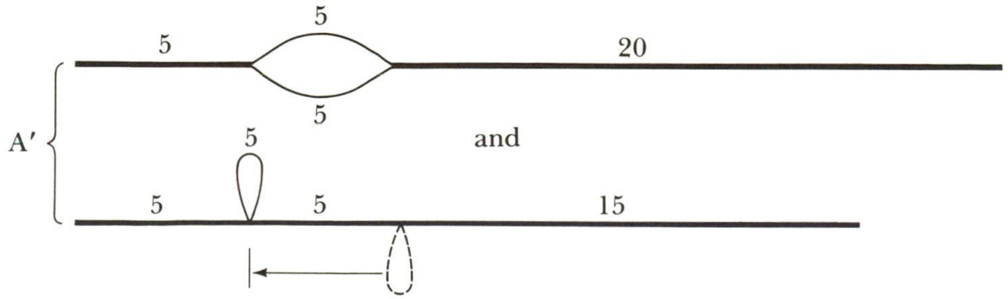

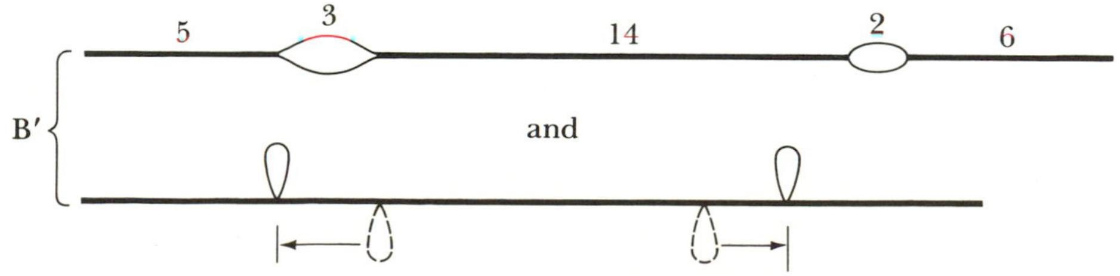

The dashed loops can be in any of the positions indicated by the arrows.

c. A c_0t analysis is carried out for both A′ and B′. Draw the expected c_0t curve assuming that each deletion has been filled in by a duplication.

19–10. How does the y-intercept of a Zimm plot change if a protein containing four subunits is subjected to a treatment that dissociates the subunits?

19–11. How does the y-intercept of a Zimm plot change if a very compact protein, containing only a single polypeptide chain, is treated in a way that destroys all attractive interactions between the amino acids? How does the dashed line of Figure 19-4 change?

19–12. A particular protein is a double-stranded, fairly rigid rod. The molecule contains many positively and negatively charged amino acids that should be able to attract one another. It is believed that heating a solution of the protein to 90°C causes complete separation of the strands and that the strands do not reform when the solution is cooled. Two models are proposed for the structure of the protein. In model I, it is assumed that one strand contains only the positively charged amino acid and the other contains only the negatively charged ones. In model II, it is assumed that each strand contains amino acids of both charges. To distinguish the models, light scattering is studied both before and after heating in 0.01 M NaCl, a salt concentration in which there is no shielding of charges. The Zimm plots for each solution are shown below. Which model is probably correct?

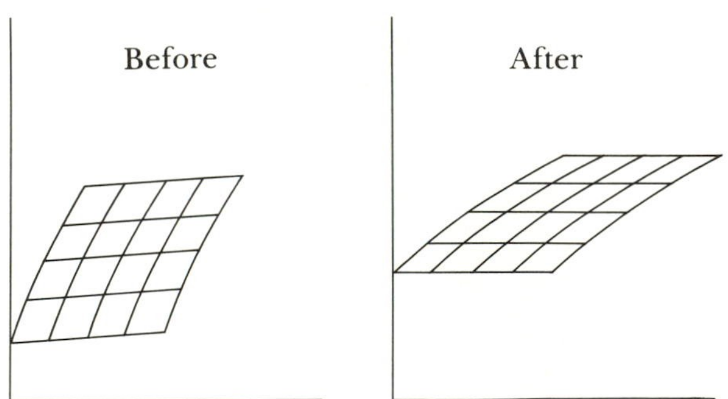

19–13. How would the rate of ^{3}H exchange in a protein be affected by increasing temperature?

19–14. A DNA solution increases in OD_{260} when a particular DNA-binding protein is added. However, the rate of ^{3}H exchange of the protein-DNA complex is less than that of the free protein, a little less than that of double-stranded DNA, and very much less than the rate for single-stranded DNA. Explain these results.

Answers to Problems

Chapter 1

1–1. Because of the alternating sequence of A and T, any region of the chain can form hydrogen bonds with any other region. Hence, the structures could range from a hairpin

```
A T A T A T A
T A T A T A T
```

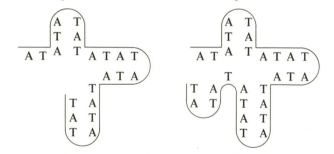

to various complex branched structures; for example,

```
      A T                       A T
      T A                       T A
  A T A T A T A T           A T A T A T A T
            A T A               T     A T A
            T               T A   A T
        T A                 A T   T A
        A T                       A T
        T A                       T A
```

1–2. Because of the alternating base sequence, each molecule forms *intra*molecular hydrogen bonds, as shown in the answer to problem 1–1, leaving very few bases available for *inter*molecular hydrogen bonds.

1–3. Fifteen.

1–4. Linear and circular dimers, trimers, and so forth, because at high concentration the probability of two different molecules interacting with one another approaches that of the two ends finding one another.

1–5. Polyvaline, because the charged amino group of lysine causes mutual repulsion of the monomers. Polylysine shows the greater effect of pH on shape because, in a certain pH range, the net charge varies.

1–13. This strongly hydrophobic patch on the surface of the protein would tend to interact with a similar patch. Thus, the polypeptide chain would probably dimerize.

1–14. The disulfide bonds force certain regions of the polypeptide chain to be near one another. This stabilizes many interactions that are so weak that they could not form without such a constraint.

1–15. $M_n = 38{,}050$. $M_w = 39{,}524$.

1–16. Seventy percent of the weight is dimers. Thus the weight ratio of monomers to dimers is 30:70. Since a dimer weighs twice as much as a monomer, the number ratio of monomers to dimers is $30{:}\frac{1}{2}(70) = 30{:}35 = 6{:}7$. Thus, there are six monomers for every seven dimers.

1–17. $M_n = 47{,}778$. $M_w = 49{,}500$.

1–18. $M_n = 15.7 \times 10^6$. $M_w = 21.6 \times 10^{-6}$.

1–19. If the molecule is spherical, $R_G = 14.8$ Å. If it is a random coil, R_G, calculated from the end-to-end distance, is 14.7 Å. These values agree so that the molecule is probably spherical.

1–20. If the molecule is spherical, $R_G = 17$ Å. The observed value is 3.81 times larger so the molecule is ellipsoidal.

1–21. The center of mass is at the center of the square. Each sphere is $\sqrt{50}$ cm from the center of mass. Thus $R_G^2 = 50$ and $R_G = 5\sqrt{2} = 7.07$ cm.

1–22. For a random coil, it would increase because the greater number of collisions with solvent molecules would increase the average size of the molecule. If the rod were rigid, there would be no change.

1–23. In 1 M NaCl, the molecule becomes flexible so that h and R_G decrease. In 0.01 M NaCl, the flexibility is unchanged so that the decrease in h and R_G is very slight.

1–24. The two interactions are independent and presumably are disrupted at two distinct temperatures. The transition to the random-coil configuration occurs at the higher temperature.

1–25. The transition is cooperative for the molecule in which the lysines and aspartates are both adjacent because two adjacent hydrogen bonds are more stable than two separated hydrogen bonds.

1–26. The one that is not looped out is more stable because, in looping out, hydrogen bonds are broken. The adjacent hydrogen bonds have less stability than they would have if all possible hydrogen bonds were present.

Chapter 2

2–1. The limit of resolution is the minimum distance between two particles that allows the observer to see that there are two particles. In this case, this limit is $5.4 \times 10^{-5}/1.32 = 0.41$ μm. Since the holes are 20 μm apart, they are visible as separate particles. They appear as points of light but the observer cannot measure their size because to do so is equivalent to distinguishing opposite sides of a single hole. These sides are 0.01 μm apart, which is less than the limit of resolution.

2–2. $453.7 \times 10^{-9}/1.28 = 3.54 \times 10^{-7}$ cm $= 0.354$ μm.

2–3. With increasing magnification, an image of an object is spread over a larger viewing area. The number of photons per second is constant so that the number of photons per second per unit viewing area decreases; that is, the brightness decreases.

2–4. If the virus takes up acridine orange, it can be visualized as a point of light against a black background. The fact that the size of the point is below the limit of resolution is irrelevant because resolution refers to the ability to separate nearby objects. Hence, the virus should not aggregate or the observed count will be too low. A DNA virus is easier to see than an RNA virus because the fluorescence of acridine orange is greater if bound to DNA than if bound to RNA.

2–5. If viewed from the side, they will appear the same.

2–6. Contrast is produced by shifts in phase, which are converted into intensity differences. If the object has regions that differ in absorbance, the contrast effects caused by phase shifts may be either weakened, cancelled, or enhanced by intensity differences.

2–7. Either the polarizer or the analyzer is removed and the sample is rotated. A dichroic object will vary in brightness as it is rotated.

2–8. If a blue filter is placed anywhere in the light path, contrast should be enhanced because red or pink light is usually absorbed by a blue filter.

2–9. It probably consists of either partially oriented fibers or parallel fibers, each of which is internally collapsed.

2–10. Fluorescence microscopy because contrast increases with increasing intensity.

2–11. First, the sample could dry out; second, the sample could curl; and, third, if the sample were in liquid, the liquid surface could act as a lens.

2–12. The cell wall would not be visible. By phase microscopy, the cell itself would be visible if the index of refraction of the cell contents differed from that of the suspending medium. If the cell wall were very thin, its absence might be unnoticed because of the phase halo.

2–13.
a. They have a higher index of refraction than the solvent and would be light against a dark background.
b. They will be invisible because there is no difference in the indices of refraction.
c. Four

2–14.
a. 45°. The fiber and parallel disc regions would be the brightest.
b. Remove the analyzer and align the polarizer with the axis of the cell. The disc region will be black. Then, align the polarizer to be perpendicular to the axis of the cell and the disc region will be bright.

2–15. A 30% albumin solution has the same index of refraction as the bacteria so they are not visible.

2–16.
a. The tail contains oriented fibers whereas the molecules in the head do not have a preferred orientation.
b. They are both at an angle of 45° with respect to the axes of the polarizer and the analyzer.
c. The tails form a 90° angle. Thus any light that passes through one tail cannot pass through the other.

2–17. The object either must consist of linear units emanating radially from a single point or a small region or it must consist of concentric rings, each consisting of parallel fibers. If the analyzer were removed, two arms of the cross would disappear.

2–18.

	Nucleus	*Cytoplasm*
a.	Green	Colorless
b.	Bright green	Dull green
c.	Bright green	Orange

2–19. The first observation shows that the ultraviolet light has not broken the chromosome. The second shows that material has been either added or lost; loss is more probable. The third observation confirms the second. The fourth and fifth show that DNA has been lost. Therefore, the ultraviolet irradiation has resulted in the depletion of DNA. Other components must be present in the chromosome that are capable of maintaining chromosome structure even when the DNA chain is broken.

Chapter 3

3–1. Those with dark centers are empty particles.

3–2. 750 × (820/1250) = 492 Å. The viruses with short shadows have probably collapsed.

3–3. If single- and double-stranded DNAs are denoted by thin and heavy lines, respectively, the structure will be:

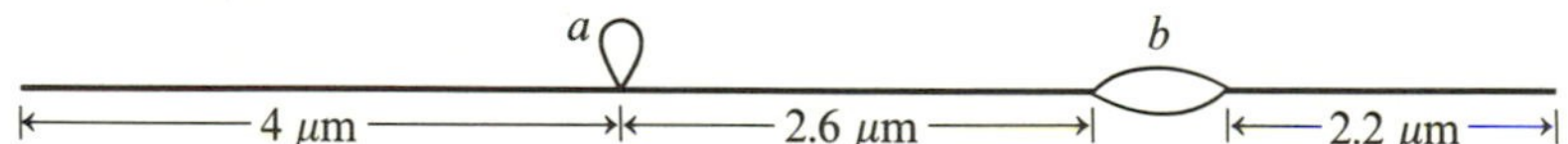

Loop *a* will have a length of 0.5 μm; each strand of loop *b* will be 0.7 μm long.

3–4. The molecule is circular, and one part of it consists of a double-stranded region equal in length to 1% of the total length flanked by two single-stranded regions of equal length that is also approximately 1% of the total. The remainder of the molecule is double-stranded. The short double-stranded region is the terminally redundant region. The total length of the molecule is 99% of the original length.

3–6. No, because the metal film would be less transparent to electrons than the sample, which would therefore be invisible.

3–7. The molecules will overlap so that it will be difficult to follow a single molecule from end to end.

3–8. There are approximately (100/400) × 22 = 5.5 viruses per 100-Å layer. The cell contains roughly 5000 such layers, assuming that its shape has not been altered by the processes required for preparing thin sections. Hence, there are approximately 27,500 viruses per cell. The most important assumption in this calculation is that the viruses are uniformly distributed throughout the cell (such an assumption would rarely be satisfied). Another assumption, made for arithmetical simplicity, is that all layers have the same diameter; although this is not possible in a sphere, it introduces a smaller error than the assumption about homogeneity.

3–9. If the slices are perpendicular to the long axis of the cylinder, circles will be observed. If they are parallel to the long axis, stacked discs will be seen. The width of the disc will vary according to the placement of the cut and will be greatest if the cut is on the long axis. If the cut is at an angle, the cylinder will appear short, but discs will still be seen.

3–10. a. It decreases because the wavelength decreases with increasing voltage.

b. The kinetic energy E is eV, in which $e = 1.6 \times 10^{-19}$ coulomb = the charge of the electron, so that $E = (1.6 \times 10^{-19})(2 \times 10^5) = 3.2 \times 10^{-14}$ joule (J). The wavelength $\lambda = hc/E$, in which h is Planck's constant (6.63×10^{-34} J s) and c is the velocity of light (2.8×10^8 m s^{-1}), so that $\lambda = 5.8 \times 10^{-12}$ m $= 0.0058$ nm.

c. The principal advantage is greater resolution; however, sample destruction would occur very rapidly so that the minimal-beam-exposure technique would be needed.

3–11. As film thickness increases, contrast decreases. Therefore the thinnest support that can resist destruction by the electron beam is used. If the sample has a very high electron density, a thick film can be used.

3–12. The fixative ensures that cell molecules and structures in the sample have the same relative positions they had in the living cell or tissue. The criteria for good and bad fixation are often subjective. However, broken vesicles, non-parallel membrane components, broken nuclei or cell walls, and disoriented fibers are usually taken to indicate that fixation is bad.

3–13. Negative staining: ribosomes, a large protein, arrangement of subunits, empty viruses. Replica: arrangement of subunits, surface of a membrane. Sectioning: inner structure of a bacterium. Freeze-fracture: inner regions of a membrane.

3–14. The height is $0.022 \tan 30° = 0.013$ μm.

3–15. The height is $(2.3/8.4)(0.069) = 0.019$ μm.

3–16. a. The entire molecule would be seen but the metal shadow would often switch from one side of the molecule to another. Also the thickness of the shadow would vary.

b. In both cases, contrast would be reduced and the molecule would be difficult to see.

c. Silver absorbs fewer electrons per unit weight than platinum. Therefore, silver would give poorer contrast.

3–17. It is possible that all SV40 molecules do not have the same length.

3–18. The spherical aberration is not corrected in the electron microscope.

3–19. The bacteria are thick and thus are very electron-dense. Thus, they appear as opaque rods so that a phage is visible only if it happens to be situated at the periphery of the opaque area. When the multiplicity of infection is 500, the probability that one will be at the edge is greater than when there are 10 phages per bacterium.

3–20. Each end must have a region that has mirror-image symmetry in the base sequence—that is, it is a palindrome. The length is 5% of the total length.

3–21. a. In spreading 1, the DNA concentration is too high and there are many overlapping monomers that are scored as catenanes.

b. The most dilute and thus most reliable spreading is number 3 in which 28/279 or 10% of the molecules are scored as catenanes.

3–22. The range is ±7%, which is not very large. This could be the sum of the following: instrumental errors such as fluctuations in voltage leading to changes in magnification; variations in the degree of stretching or contraction of the molecule during the spreading operation; difficulty in determining precisely where the ends of the molecules are; and errors in following the track of the molecule from one end to the other.

3–23. Heads lacking DNA, which are filled with phosphotungstic acid.

3–24. No, because the error in the diameter is about 5%, which leads to an error in volume of about 14%.

Chapter 4

4–1. 10^{-2} M and 10^{-10} M.

4–2. pH 7.2.

4–3. a. The concentration of the HCl within the glass electrode has become small.

b. The regeneration is faster in the higher concentration of HCl.

4–4. The pK, ultraviolet absorbance, chelating ability, interaction with or even precipitation of components in the dissolved sample.

4–5. One possibility is to prepare it at 80°C. Alternately it can be prepared at room temperature and the pH checked at 80°C; if the pH is incorrect, acid or base can be added at room temperature as long as the pH is measured at 80°C.

4–6. The barbiturate buffer absorbs ultraviolet light and thereby is acting as a screen.

4–7. First, the enzyme might bind phosphate and be inactivated. This is not likely, though, because few proteins bind phosphate. Second, the enzyme might require a cofactor such as the Ca^{2+} ion, which is precipitated by phosphate. Third, the enzyme may contain a bound metal ion such as Zn^{2+}, which is removed because of stronger binding to the phosphate ion.

Chapter 5

5–1. One millimole contains 6×10^{20} methyl groups. If all were labeled with one tritium atom, half or 3×10^{20} would decay in twelve years. Hence, the activity is 2.5×10^{19} decays per year or 8×10^{11} decays per second, or 22 curies. Hence, $\frac{6}{22} = 27\%$ of the thymidine molecules are radioactive.

A molecule of DNA having a molecular weight of 25×10^6 contains 75,700 bases or 18,940 thymidines of which $0.27 \times 18{,}940 = 5100$ are radioactive.

If a sample registers 1000 cpm on a counter with 52% efficiency, it contains an amount of radioactivity corresponding to $1000/0.52 = 1923$ cpm $= 8.7 \times 10^{-10}$ curie. This would be contained in $(\frac{1}{6} \times 8.7 \times 10^{-10} = 1.45 \times 10^{-10}$ millimoles of thymidine $= 1.45 \times 10^{-13}$ mole. If 50% of the base pairs in a molecule of DNA are A·T, for every mole of thymidine nucleotide, there are three moles of other nucleotides. Hence, there are $4 \times 1.45 \times 10^{-13} = 5.8 \times 10^{-13}$ mole of nucleotides. The average weight of one mole of nucleotide is 330 grams. Therefore the weight of DNA corresponding to 1000 cpm is $330 \times 5.8 \times 10^{-13} = 1.9 \times 10^{-10}$ grams.

5–2. Rewriting equation (2) as $N = N_0 \exp(-0.693t/\tau_{\frac{1}{2}})$ in which $\tau_{\frac{1}{2}} =$ 14.2 days, $t = 26$ days (January 23 to February 18), $N_0 = 5.3$ Ci/mmol, then $N = 1.49$ Ci/mmol.

5–3. Rewriting equation (2) as $t = -(\tau_{\frac{1}{2}}/0.693)\ln(N/N_0)$ in which $\tau_{\frac{1}{2}} =$ 14.2 days and $N/N_0 = 100(2 \times 10^6)$, then $t = 203$ days. No, because the half-life of tritium is so long that nearly 200 years would be required.

5–4. ^3H and ^{32}P would be the pair of choice because the levels of a scintillation counter can be adjusted so that only a small fraction of the ^{32}P activity is in the ^3H channel.

5–5. [^3H]Thymidine would be the better choice because it is incorporated only in DNA, whereas ^{32}P is also incorporated in RNA, phosphoprotein, and phospholipid.

5–6. There is no ^3H activity in the ^{14}C channel and 20% of the ^{14}C activity is in the ^3H channel. Hence, the 1620 cpm in channel B represents ^{14}C only. However, $0.2 \times 1620 = 324$ cpm of ^{14}C is in channel A. Therefore, the total ^{14}C activity is $1620 + 324 = 1944$ and the ^3H activity is $1450 - 324 = 1126$. The $^3H/^{14}C$ ratio is $1126/1944 = 0.58$.

5–7. Correcting for counting efficiencies, the ^3H activity is 4504 dpm and the ^{14}C activity is 2371 dpm. The $^3H/^{14}C$ ratio is $4504/2371 = 1.90$. The molar ratio of thymidine to uridine is then $1.90/(6/0.5) = 0.16$.

The number of curies of thymidine is $4504/(2.2 \times 10^{12}) = 2.1 \times 10^{-9}$, which equals $(1/6000) \times 2.1 \times 10^{-9} = 3.5 \times 10^{-13}$ mole. In a DNA that is 43% G + C, 0.285 of the nucleotides are thymidine. Therefore, there are $(1/0.285) \times 3.5 \times 10^{-13}$ mole, or 4×10^{-10} gram, of DNA. The number of curies of uridine is $2371/(2.2 \times 10^{12}) = 1.08 \times 10^{-9}$ or 2.16×10^{-12} mole. The number of moles of RNA nucleotides is $(1/0.28) \times 2.16 \times 10^{-12} = 7.7 \times 10^{-12}$ or 2.54×10^{-9} gram of RNA.

5–8. The DNA with a molecular weight of 20×10^6 contains 6×10^4 P atoms; therefore the ratio $^{32}P/^{31}P$ is $1/(6 \times 10^4)$ and the concentration of ^{32}P in the medium must be $10^{-3}/(6 \times 10^4) = 1.6 \times 10^{-8}$ M. This is 1.6×10^{-11} mol/ml.

5–9. A 1% probable error means roughly that $N^{-1/2}$, in which N is the total number of counts, should be 0.01. Hence, one must count 10^4 counts and the sample should be counted for at least $(1/752) \times 10^4 = 13.3$ minutes.

19–9. A phage DNA has a molecular weight of 30×10^6. Two deletion mutants A and B are selected and each is found to have a molecular weight of 25×10^6.

a. The DNA of each mutant is mixed with an equal number of molecules of the original, nonmutant DNA and then denatured and renatured. These heteroduplexes are examined by electron microscopy and the molecules shown below are seen. What can be said about the deletions (i.e., the size and number) in A and B?

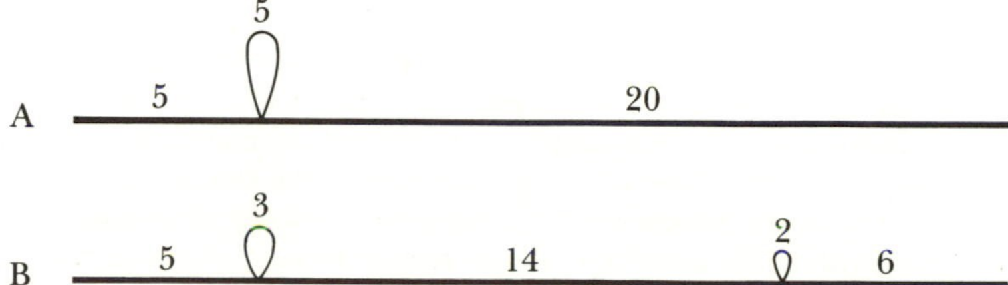

b. Both A and B are mutagenized in a way that leads to genetic duplications (that is, a base sequence is repeated in tandem) and two new mutants A′ and B′ are isolated. The DNA of both A′ and B′ has a molecular weight of 30×10^6. To confirm that A′ and B′ have a new piece of DNA at the site of the deletion, each DNA is used to form a heteroduplex with the original, nonmutant DNA. The molecules shown below are observed. Is the insertion at the site of the deletion in each case? If there is a duplication, is the sequence that is duplicated the sequence at the right or at the left of the deletion, assuming that all molecules that have been shown have the same left-to-right orientation?

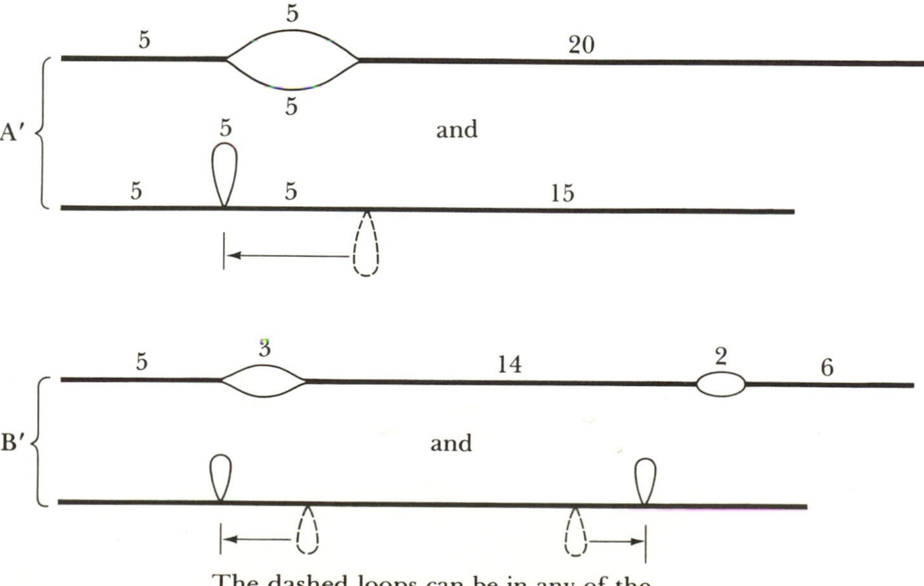

The dashed loops can be in any of the positions indicated by the arrows.

c. A c_0t analysis is carried out for both A′ and B′. Draw the expected c_0t curve assuming that each deletion has been filled in by a duplication.

19–10. How does the y-intercept of a Zimm plot change if a protein containing four subunits is subjected to a treatment that dissociates the subunits?

19–11. How does the y-intercept of a Zimm plot change if a very compact protein, containing only a single polypeptide chain, is treated in a way that destroys all attractive interactions between the amino acids? How does the dashed line of Figure 19-4 change?

19–12. A particular protein is a double-stranded, fairly rigid rod. The molecule contains many positively and negatively charged amino acids that should be able to attract one another. It is believed that heating a solution of the protein to 90°C causes complete separation of the strands and that the strands do not reform when the solution is cooled. Two models are proposed for the structure of the protein. In model I, it is assumed that one strand contains only the positively charged amino acid and the other contains only the negatively charged ones. In model II, it is assumed that each strand contains amino acids of both charges. To distinguish the models, light scattering is studied both before and after heating in 0.01 M NaCl, a salt concentration in which there is no shielding of charges. The Zimm plots for each solution are shown below. Which model is probably correct?

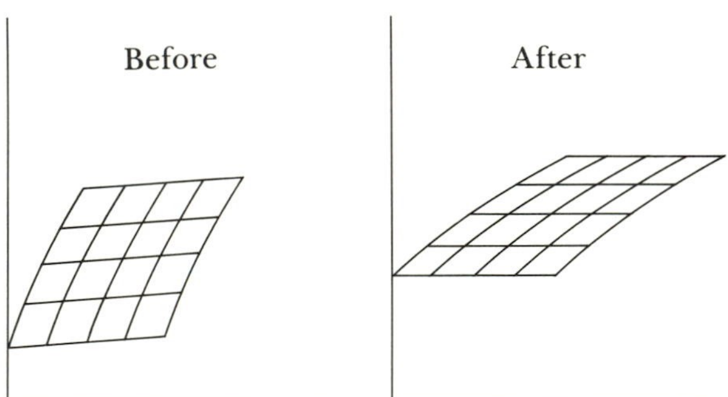

19–13. How would the rate of ^{3}H exchange in a protein be affected by increasing temperature?

19–14. A DNA solution increases in OD_{260} when a particular DNA-binding protein is added. However, the rate of ^{3}H exchange of the protein-DNA complex is less than that of the free protein, a little less than that of double-stranded DNA, and very much less than the rate for single-stranded DNA. Explain these results.

Answers to Problems

Chapter 1

1–1. Because of the alternating sequence of A and T, any region of the chain can form hydrogen bonds with any other region. Hence, the structures could range from a hairpin

A T A T A T A
T A T A T A T

to various complex branched structures; for example,

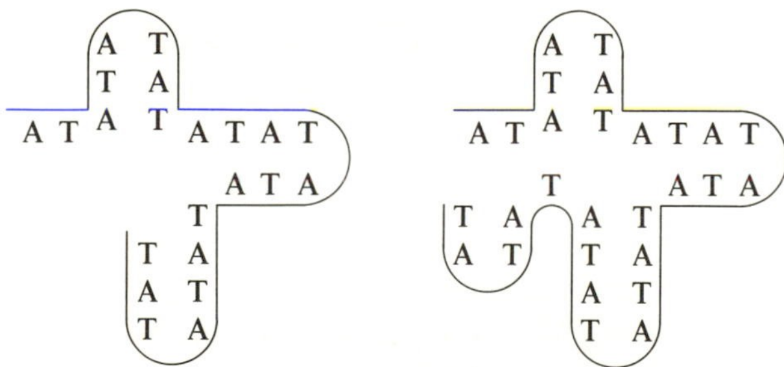

1–2. Because of the alternating base sequence, each molecule forms *intra*molecular hydrogen bonds, as shown in the answer to problem 1–1, leaving very few bases available for *inter*molecular hydrogen bonds.

1–3. Fifteen.

1–4. Linear and circular dimers, trimers, and so forth, because at high concentration the probability of two different molecules interacting with one another approaches that of the two ends finding one another.

1–5. Polyvaline, because the charged amino group of lysine causes mutual repulsion of the monomers. Polylysine shows the greater effect of pH on shape because, in a certain pH range, the net charge varies.

1–6. The DNA would denature because the bases would interact with the solvent rather than with themselves.

1–7. There are six possible linear types:

However, if the subunits were either isotropic or symmetric, structure 1 would be identical with structure 3 and structure 4 would be identical with structure 6; then there would be four types.

There are two possible triangular types:

1–8. Eleven.

1–9. There are several possibilities. Each subunit might have a positively charged and a negatively charged region. These regions could interact by a simple charge interaction. If they have hydrophobic regions on the surface, they might interact so that water can be excluded.

1–10. In the first case,

A G C T A A C G C G A
T C G A T T G C G C T
A T
G A
T C

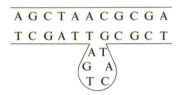

In the second case,

A G C T A A C G C G A
T C G A T T G C G C T
A T
G C
T A

1–11. $\frac{3}{8}$ hybrid, $\frac{9}{16}$ heavy, and $\frac{1}{16}$ light.

1–12. There are only four (↑) peptide bonds because the C—N bond (⇣) to the left of proline is not a peptide bond.

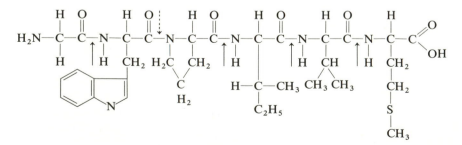

H O H O H O H O H O H
H_2N—C—C—N—C—C—N—C—C—N—C—C—N—C—C—N—C—C(=O)OH
H H CH_2 H_2C CH_2 H H CH H CH_2
C H—C—CH_3 CH_3 CH_3 CH_2
N H_2 C_2H_5 S
CH_3

1–13. This strongly hydrophobic patch on the surface of the protein would tend to interact with a similar patch. Thus, the polypeptide chain would probably dimerize.

1–14. The disulfide bonds force certain regions of the polypeptide chain to be near one another. This stabilizes many interactions that are so weak that they could not form without such a constraint.

1–15. $M_n = 38{,}050$. $M_w = 39{,}524$.

1–16. Seventy percent of the weight is dimers. Thus the weight ratio of monomers to dimers is 30:70. Since a dimer weighs twice as much as a monomer, the number ratio of monomers to dimers is $30{:}\frac{1}{2}(70) = 30{:}35 = 6{:}7$. Thus, there are six monomers for every seven dimers.

1–17. $M_n = 47{,}778$. $M_w = 49{,}500$.

1–18. $M_n = 15.7 \times 10^6$. $M_w = 21.6 \times 10^{-6}$.

1–19. If the molecule is spherical, $R_G = 14.8$ Å. If it is a random coil, R_G, calculated from the end-to-end distance, is 14.7 Å. These values agree so that the molecule is probably spherical.

1–20. If the molecule is spherical, $R_G = 17$ Å. The observed value is 3.81 times larger so the molecule is ellipsoidal.

1–21. The center of mass is at the center of the square. Each sphere is $\sqrt{50}$ cm from the center of mass. Thus $R_G^2 = 50$ and $R_G = 5\sqrt{2} = 7.07$ cm.

1–22. For a random coil, it would increase because the greater number of collisions with solvent molecules would increase the average size of the molecule. If the rod were rigid, there would be no change.

1–23. In 1 M NaCl, the molecule becomes flexible so that h and R_G decrease. In 0.01 M NaCl, the flexibility is unchanged so that the decrease in h and R_G is very slight.

1–24. The two interactions are independent and presumably are disrupted at two distinct temperatures. The transition to the random-coil configuration occurs at the higher temperature.

1–25. The transition is cooperative for the molecule in which the lysines and aspartates are both adjacent because two adjacent hydrogen bonds are more stable than two separated hydrogen bonds.

1–26. The one that is not looped out is more stable because, in looping out, hydrogen bonds are broken. The adjacent hydrogen bonds have less stability than they would have if all possible hydrogen bonds were present.

Chapter 2

2–1. The limit of resolution is the minimum distance between two particles that allows the observer to see that there are two particles. In this case, this limit is $5.4 \times 10^{-5}/1.32 = 0.41$ μm. Since the holes are 20 μm apart, they are visible as separate particles. They appear as points of light but the observer cannot measure their size because to do so is equivalent to distinguishing opposite sides of a single hole. These sides are 0.01 μm apart, which is less than the limit of resolution.

2–2. $453.7 \times 10^{-9}/1.28 = 3.54 \times 10^{-7}$ cm $= 0.354$ μm.

2–3. With increasing magnification, an image of an object is spread over a larger viewing area. The number of photons per second is constant so that the number of photons per second per unit viewing area decreases; that is, the brightness decreases.

2–4. If the virus takes up acridine orange, it can be visualized as a point of light against a black background. The fact that the size of the point is below the limit of resolution is irrelevant because resolution refers to the ability to separate nearby objects. Hence, the virus should not aggregate or the observed count will be too low. A DNA virus is easier to see than an RNA virus because the fluorescence of acridine orange is greater if bound to DNA than if bound to RNA.

2–5. If viewed from the side, they will appear the same.

2–6. Contrast is produced by shifts in phase, which are converted into intensity differences. If the object has regions that differ in absorbance, the contrast effects caused by phase shifts may be either weakened, cancelled, or enhanced by intensity differences.

2–7. Either the polarizer or the analyzer is removed and the sample is rotated. A dichroic object will vary in brightness as it is rotated.

2–8. If a blue filter is placed anywhere in the light path, contrast should be enhanced because red or pink light is usually absorbed by a blue filter.

2–9. It probably consists of either partially oriented fibers or parallel fibers, each of which is internally collapsed.

2–10. Fluorescence microscopy because contrast increases with increasing intensity.

2–11. First, the sample could dry out; second, the sample could curl; and, third, if the sample were in liquid, the liquid surface could act as a lens.

2–12. The cell wall would not be visible. By phase microscopy, the cell itself would be visible if the index of refraction of the cell contents differed from that of the suspending medium. If the cell wall were very thin, its absence might be unnoticed because of the phase halo.

2–13.
a. They have a higher index of refraction than the solvent and would be light against a dark background.
b. They will be invisible because there is no difference in the indices of refraction.
c. Four

2–14.
a. 45°. The fiber and parallel disc regions would be the brightest.
b. Remove the analyzer and align the polarizer with the axis of the cell. The disc region will be black. Then, align the polarizer to be perpendicular to the axis of the cell and the disc region will be bright.

2–15. A 30% albumin solution has the same index of refraction as the bacteria so they are not visible.

2–16.
a. The tail contains oriented fibers whereas the molecules in the head do not have a preferred orientation.
b. They are both at an angle of 45° with respect to the axes of the polarizer and the analyzer.
c. The tails form a 90° angle. Thus any light that passes through one tail cannot pass through the other.

2–17. The object either must consist of linear units emanating radially from a single point or a small region or it must consist of concentric rings, each consisting of parallel fibers. If the analyzer were removed, two arms of the cross would disappear.

2–18.

	Nucleus	*Cytoplasm*
a.	Green	Colorless
b.	Bright green	Dull green
c.	Bright green	Orange

2–19. The first observation shows that the ultraviolet light has not broken the chromosome. The second shows that material has been either added or lost; loss is more probable. The third observation confirms the second. The fourth and fifth show that DNA has been lost. Therefore, the ultraviolet irradiation has resulted in the depletion of DNA. Other components must be present in the chromosome that are capable of maintaining chromosome structure even when the DNA chain is broken.

Chapter 3

3–1. Those with dark centers are empty particles.

3–2. 750 × (820/1250) = 492 Å. The viruses with short shadows have probably collapsed.

3–3. If single- and double-stranded DNAs are denoted by thin and heavy lines, respectively, the structure will be:

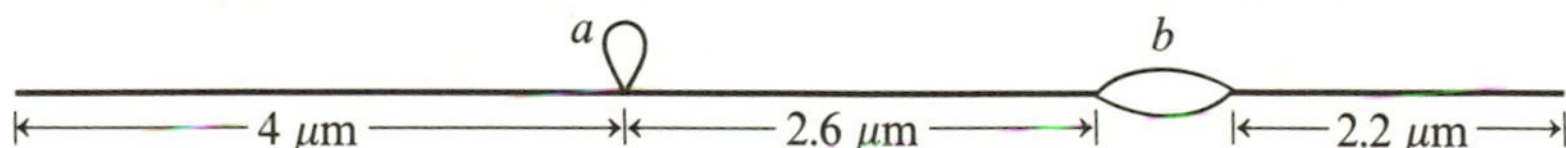

Loop *a* will have a length of 0.5 μm; each strand of loop *b* will be 0.7 μm long.

3–4. The molecule is circular, and one part of it consists of a double-stranded region equal in length to 1% of the total length flanked by two single-stranded regions of equal length that is also approximately 1% of the total. The remainder of the molecule is double-stranded. The short double-stranded region is the terminally redundant region. The total length of the molecule is 99% of the original length.

3–6. No, because the metal film would be less transparent to electrons than the sample, which would therefore be invisible.

3–7. The molecules will overlap so that it will be difficult to follow a single molecule from end to end.

3–8. There are approximately (100/400) × 22 = 5.5 viruses per 100-Å layer. The cell contains roughly 5000 such layers, assuming that its shape has not been altered by the processes required for preparing thin sections. Hence, there are approximately 27,500 viruses per cell. The most important assumption in this calculation is that the viruses are uniformly distributed throughout the cell (such an assumption would rarely be satisfied). Another assumption, made for arithmetical simplicity, is that all layers have the same diameter; although this is not possible in a sphere, it introduces a smaller error than the assumption about homogeneity.

3–9. If the slices are perpendicular to the long axis of the cylinder, circles will be observed. If they are parallel to the long axis, stacked discs will be seen. The width of the disc will vary according to the placement of the cut and will be greatest if the cut is on the long axis. If the cut is at an angle, the cylinder will appear short, but discs will still be seen.

3–10. a. It decreases because the wavelength decreases with increasing voltage.

b. The kinetic energy E is eV, in which $e = 1.6 \times 10^{-19}$ coulomb = the charge of the electron, so that $E = (1.6 \times 10^{-19})(2 \times 10^5) = 3.2 \times 10^{-14}$ joule (J). The wavelength $\lambda = hc/E$, in which h is Planck's constant (6.63×10^{-34} J s) and c is the velocity of light (2.8×10^8 m s^{-1}), so that $\lambda = 5.8 \times 10^{-12}$ m = 0.0058 nm.

c. The principal advantage is greater resolution; however, sample destruction would occur very rapidly so that the minimal-beam-exposure technique would be needed.

3–11. As film thickness increases, contrast decreases. Therefore the thinnest support that can resist destruction by the electron beam is used. If the sample has a very high electron density, a thick film can be used.

3–12. The fixative ensures that cell molecules and structures in the sample have the same relative positions they had in the living cell or tissue. The criteria for good and bad fixation are often subjective. However, broken vesicles, non-parallel membrane components, broken nuclei or cell walls, and disoriented fibers are usually taken to indicate that fixation is bad.

3–13. Negative staining: ribosomes, a large protein, arrangement of subunits, empty viruses. Replica: arrangement of subunits, surface of a membrane. Sectioning: inner structure of a bacterium. Freeze-fracture: inner regions of a membrane.

3–14. The height is 0.022 tan 30° = 0.013 μm.

3–15. The height is (2.3/8.4)(0.069) = 0.019 μm.

3–16. a. The entire molecule would be seen but the metal shadow would often switch from one side of the molecule to another. Also the thickness of the shadow would vary.

b. In both cases, contrast would be reduced and the molecule would be difficult to see.

c. Silver absorbs fewer electrons per unit weight than platinum. Therefore, silver would give poorer contrast.

3–17. It is possible that all SV40 molecules do not have the same length.

3–18. The spherical aberration is not corrected in the electron microscope.

3–19. The bacteria are thick and thus are very electron-dense. Thus, they appear as opaque rods so that a phage is visible only if it happens to be situated at the periphery of the opaque area. When the multiplicity of infection is 500, the probability that one will be at the edge is greater than when there are 10 phages per bacterium.

3–20. Each end must have a region that has mirror-image symmetry in the base sequence—that is, it is a palindrome. The length is 5% of the total length.

3–21. a. In spreading 1, the DNA concentration is too high and there are many overlapping monomers that are scored as catenanes.

b. The most dilute and thus most reliable spreading is number 3 in which 28/279 or 10% of the molecules are scored as catenanes.

3–22. The range is ±7%, which is not very large. This could be the sum of the following: instrumental errors such as fluctuations in voltage leading to changes in magnification; variations in the degree of stretching or contraction of the molecule during the spreading operation; difficulty in determining precisely where the ends of the molecules are; and errors in following the track of the molecule from one end to the other.

3–23. Heads lacking DNA, which are filled with phosphotungstic acid.

3–24. No, because the error in the diameter is about 5%, which leads to an error in volume of about 14%.

Chapter 4

4–1. 10^{-2} M and 10^{-10} M.

4–2. pH 7.2.

4–3. a. The concentration of the HCl within the glass electrode has become small.

b. The regeneration is faster in the higher concentration of HCl.

4–4. The p*K*, ultraviolet absorbance, chelating ability, interaction with or even precipitation of components in the dissolved sample.

4–5. One possibility is to prepare it at 80°C. Alternately it can be prepared at room temperature and the pH checked at 80°C; if the pH is incorrect, acid or base can be added at room temperature as long as the pH is measured at 80°C.

4–6. The barbiturate buffer absorbs ultraviolet light and thereby is acting as a screen.

4–7. First, the enzyme might bind phosphate and be inactivated. This is not likely, though, because few proteins bind phosphate. Second, the enzyme might require a cofactor such as the Ca^{2+} ion, which is precipitated by phosphate. Third, the enzyme may contain a bound metal ion such as Zn^{2+}, which is removed because of stronger binding to the phosphate ion.

Chapter 5

5–1. One millimole contains 6×10^{20} methyl groups. If all were labeled with one tritium atom, half or 3×10^{20} would decay in twelve years. Hence, the activity is 2.5×10^{19} decays per year or 8×10^{11} decays per second, or 22 curies. Hence, $\frac{6}{22} = 27\%$ of the thymidine molecules are radioactive.

A molecule of DNA having a molecular weight of 25×10^6 contains 75,700 bases or 18,940 thymidines of which $0.27 \times 18{,}940 = 5100$ are radioactive.

If a sample registers 1000 cpm on a counter with 52% efficiency, it contains an amount of radioactivity corresponding to $1000/0.52 = 1923$ cpm $= 8.7 \times 10^{-10}$ curie. This would be contained in $(\frac{1}{6} \times 8.7 \times 10^{-10} = 1.45 \times 10^{-10}$ millimoles of thymidine $= 1.45 \times 10^{-13}$ mole. If 50% of the base pairs in a molecule of DNA are A·T, for every mole of thymidine nucleotide, there are three moles of other nucleotides. Hence, there are $4 \times 1.45 \times 10^{-13} = 5.8 \times 10^{-13}$ mole of nucleotides. The average weight of one mole of nucleotide is 330 grams. Therefore the weight of DNA corresponding to 1000 cpm is $330 \times 5.8 \times 10^{-13} = 1.9 \times 10^{-10}$ grams.

5–2. Rewriting equation (2) as $N = N_0 \exp(-0.693t/\tau_{\frac{1}{2}})$ in which $\tau_{\frac{1}{2}} = 14.2$ days, $t = 26$ days (January 23 to February 18), $N_0 = 5.3$ Ci/mmol, then $N = 1.49$ Ci/mmol.

5–3. Rewriting equation (2) as $t = -(\tau_{\frac{1}{2}}/0.693)\ln(N/N_0)$ in which $\tau_{\frac{1}{2}} = 14.2$ days and $N/N_0 = 100(2 \times 10^6)$, then $t = 203$ days. No, because the half-life of tritium is so long that nearly 200 years would be required.

5–4. ^3H and ^{32}P would be the pair of choice because the levels of a scintillation counter can be adjusted so that only a small fraction of the ^{32}P activity is in the ^3H channel.

5–5. [^3H]Thymidine would be the better choice because it is incorporated only in DNA, whereas ^{32}P is also incorporated in RNA, phosphoprotein, and phospholipid.

5–6. There is no ^3H activity in the ^{14}C channel and 20% of the ^{14}C activity is in the ^3H channel. Hence, the 1620 cpm in channel B represents ^{14}C only. However, $0.2 \times 1620 = 324$ cpm of ^{14}C is in channel A. Therefore, the total ^{14}C activity is $1620 + 324 = 1944$ and the ^3H activity is $1450 - 324 = 1126$. The $^3H/^{14}C$ ratio is $1126/1944 = 0.58$.

5–7. Correcting for counting efficiencies, the ^3H activity is 4504 dpm and the ^{14}C activity is 2371 dpm. The $^3H/^{14}C$ ratio is $4504/2371 = 1.90$. The molar ratio of thymidine to uridine is then $1.90/(6/0.5) = 0.16$.

The number of curies of thymidine is $4504/(2.2 \times 10^{12}) = 2.1 \times 10^{-9}$, which equals $(1/6000) \times 2.1 \times 10^{-9} = 3.5 \times 10^{-13}$ mole. In a DNA that is 43% G + C, 0.285 of the nucleotides are thymidine. Therefore, there are $(1/0.285) \times 3.5 \times 10^{-13}$ mole, or 4×10^{-10} gram, of DNA. The number of curies of uridine is $2371/(2.2 \times 10^{12}) = 1.08 \times 10^{-9}$ or 2.16×10^{-12} mole. The number of moles of RNA nucleotides is $(1/0.28) \times 2.16 \times 10^{-12} = 7.7 \times 10^{-12}$ or 2.54×10^{-9} gram of RNA.

5–8. The DNA with a molecular weight of 20×10^6 contains 6×10^4 P atoms; therefore the ratio $^{32}P/^{31}P$ is $1/(6 \times 10^4)$ and the concentration of ^{32}P in the medium must be $10^{-3}/(6 \times 10^4) = 1.6 \times 10^{-8}$ M. This is 1.6×10^{-11} mol/ml.

5–9. A 1% probable error means roughly that $N^{-1/2}$, in which N is the total number of counts, should be 0.01. Hence, one must count 10^4 counts and the sample should be counted for at least $(1/752) \times 10^4 = 13.3$ minutes.

5–10. The count rate should be unaffected by volume as long as the two photomultipliers observe the entire sample. The background due to environmental radioactivity will decrease as the volume decreases.

5–11. The channel ratio (B/A) for each quenched standard must be calculated and plotted against the efficiency of detection—for example, (A + B)/1000. This graph has the following equation: efficiency of detection = channel ratio. Because the observed channel ratio for the sample is 1211/1822 = 0.66, the actual count rate is (1211 + 1822)/0.66 = 4595 cpm.

5–12. a. To get the same factor for different isotopes there would have to be color quenching.

b. Chemical quenching would produce less quenching of a ^{14}C-containing sample than of a ^{3}H-containing sample.

5–13. The 0.1-ml sample contains $0.04 \times 25 = 1\ \mu g$ of leucine. This has $1251/0.70 = 1787$ cpm $= 8.1 \times 10^{-4}\ \mu Ci$. The specific activity is $8.1 \times 10^{-4}\ \mu Ci/\mu g$. In the original culture, the concentration of radioactivity was $(0.1 \times 50 \times 25)/10 = 12.5\ \mu Ci/ml$ and the concentration of added leucine was $(0.1 \times 25)/10 = 0.25\ \mu g/ml$. Because the specific activity in the protein is much lower than the specific activity of the added [^{14}C]leucine, there must be [^{12}C]leucine in the medium at a concentration of $1.54 \times 10^{4}\ \mu g/ml$. This is so high that the increase in this concentration by the added leucine is negligible. Therefore, to increase the count rate from 1251 to 5000 cpm, a fourfold increase, simply requires the addition of four times as much [^{14}C]leucine, or 0.4 ml.

5–14. The ratio of count rate to background is 3.7 for the Geiger counter and 1.9 for the scintillation counter. Because the count rate is expected to be small, the reliability of the measurement will be better if the background is the smaller fraction of the total. The Geiger counter is preferable. Of course, the sample must be counted for a longer period of time than with the scintillation counter to achieve the same percentage of error.

5–15. There is no quenching when 10^{8} cells or fewer are counted. Thus, there are 998 + 4001 or 5000 cpm per 10^{8} cells or 5×10^{4} cpm/mg protein = 0.023 μCi/mg protein.

5–16. a. After subtracting the large value of the ^{14}C in channel A from the large value of total activity in channel B, the difference is so small that it has no statistical reliability.

b. Precipitate the sample with HCl instead of trichloroacetic acid. Little, if any, [^{14}C]protein will be precipitated so that the ^{3}H values in channel A will be reliable, though still small.

5–17. The [^{3}H]oleic acid has been radiochemically altered during storage and has become some substance that is incorporated into non-lipid material.

5–18. a. When subtracting the ^{14}C in channel A from the total in channel A, the ^{3}H, which represents DNA, is much greater than the ^{14}C, which represents RNA.

b. Pure DNA and RNA have been used to prepare the standards yet the sample consists of whole cells, in which there is probably quenching. Thus, the actual distribution of ^{14}C between the two channels is not that which is indicated by the standard.

c. Prepare a standard that is identical to the samples.

5–19. If the molecule has a molecular weight of 100 million, 1 μg contains 6×10^9 molecules or 1.2×10^{10} ^{32}P atoms or 2×10^{-14} mole. A specific activity of 500 cpm per 2×10^{-14} mole is equivalent to 1.13×10^4 curies per mole.

5–20. Glycine is converted to other substances that can find their way into DNA and RNA.

5–21. Chloroform is a well-known chemical quencher. The problem could be solved by preparing samples containing [^3H]toluene and various amounts of chloroform and obtain a channels-ratio correction curve. If the channels ratio is obtained for each experimental sample, corrected values can be calculated.

Chapter 6

6–1. Grow phages in medium containing [^3H]thymidine. X-irradiate with various doses (including zero). Let the phages adsorb to the bacteria and then inject DNA. Remove the phages from the bacteria by violent agitation. Separate the bacteria, which now contain injected DNA, from the phages. Place the bacteria on a glass slide, dry, and coat with either stripping film or a thin layer made by the dipping method. After exposure and development, count the number (n) of grains over the cells and the number (N) of cells with grains. The value of n obtained for zero dose describes the number of grains corresponding to a whole DNA molecule. If an X-irradiated phage injects all of its DNA, n and N will be independent of dose. If only a fragment is injected, n will decrease, but N will remain nearly the same.

6–2. With 94% efficiency, only $0.94 \times 42 = 39$ rays will be detectable. Because of logarithmic decay, the equation describing the time t to get N rays from $N_0 = 39$ detectable ^{32}P atoms is:

$$39 - N = 39e^{-0.693t/\tau_{\frac{1}{2}}} = 39e^{-0.693t/14.2} = 39e^{-0.049t}$$

Then, to get $N = 12$, $t = 7.5$ days.

6–3. If a ^3H-labeled compound is used and an autoradiogram is prepared using either stripping film or the dipping method, the grain density will be maximal near the observed cell boundary but spread across the entire cell area because the cell wall is both above and below the cell when observed. If the compound is contained only in the cytoplasm, the grain density will be either uniform across the cell area or maximal in regions where the cell is thick. If an autoradiogram is prepared of a thin section of the cell, all grains will be peripheral if the compound is localized in the cell wall.

6–4. Cells are grown to stationary phase (i.e., no more cell division) for long periods of time in [^3H]thymidine. The cells are spread on an agar surface so that individual cells are well separated. The cells are then allowed to grow in the absence of [^3H]thymidine to form a microcolony consisting of several hundred cells and this is overlaid with stripping film. Because of the semiconservative nature of DNA replication, if there is a single chromosome, two cells in the microcolony will be labeled. If the cell contains four chromosomes, eight cells in the microcolony will be labeled.

6–5. The short range of ^3H would reduce efficiency so that ^{14}C is preferable. The range of ^{14}C is not so great that resolution would be decreased because the track length of ^{14}C is very short compared with the size of spots on chromatograms.

6–6. A ^3H-labeled amino acid should be used, together with either stripping film or dipping. Cells are grown first at 37°C in nonradioactive medium. Labeled amino acid is added, and the culture is divided into two parts, one at 37°C and one at 42°C. After some time has elapsed, the cells are autoradiographed. If the rate at 42°C reflects the fact that 10% of the cells incorporate normally and 90% fail to incorporate, then grains will be found above only 10% of the cells. If the rate per cell is reduced tenfold, all cells will be labeled, but the average number of grains per cell grown at 42°C will be 10% that of a cell grown at 37°C. To count grains, different exposure times are necessary so that the number of grains is small enough that individual grains can be unambiguously counted.

6–7. If cells are prelabeled by growth in [^3H]thymidine and then autoradiographed, grains will be formed over all cells. If the small cells contain no DNA, no grains will be found over these cells. A reduced number of grains indicates a lower amount of DNA.

RNA can be studied by growth in a medium containing [^3H]uridine and a large amount of unlabeled thymidine. Uridine is a precursor of both RNA and DNA. The thymidine prevents the appearance of ^3H in DNA by competing with the thymidine synthesized by means of the uridine–deoxyuridylate–thymidylate pathway.

6–8. This cannot be done by growth in [^3H]thymidine because too many grains will result from the huge amount of chromosomal DNA in which the nucleoli are immersed.

6–9. Male and female cells are usually not distinguishable by light microscopy. However, if one type is grown in medium containing [^3H]thymidine and the other in nonradioactive thymidine, the two types can be distinguished by autoradiography. Hence, if ^3H-labeled males are mixed with ^1H-labeled females, pairs will be observed consisting of one labeled and one unlabeled cell. If one culture of ^3H-labeled males is mixed with another culture of unlabeled *males* and if there is homosexual pairing, pairs will be found consisting of one labeled and one unlabeled cell.

Chapter 7

7–1. Nitrocellulose is soluble in acetone. Hence, fiberglass is better.

7–2. Filtration is preferable if the particle sediments very slowly and if the pellet is hard to resuspend. It is a poor method if the amount of precipitate is so great that the filter would clog or if the precipitate binds to the filter material.

7–3. Study the retention as a function of pore size. If adsorption is occuring, retention is usually independent of pore size.

7–4. At high flow rate, the linear molecules are oriented and pass through the pores end-on, whereas circular molecules cannot become thin enough and are retained. At low flow rates, both are retained.

7–5. Sterilize filter, filter holder, and collection vessel.

7–6. The sample is collected on a fiberglass filter and incubated in alkali. The filter is removed and the liquid made acidic. The resulting precipitate is collected on a separate filter. The background should be 0.0001 × 0.01% = 0.000001%. Nitrocellulose filters are alkali soluble and therefore not usable.

7–7. Note that between 10^7 and 2×10^7 cells remain on the filter. Therefore, if 10^7, 10^6, or 10^5 cells were filtered, none could be removed.

7–8. Use radioactive protein or nucleic acid. Dialyze, empty, and dry tubing. Then count tubing to see if radioactivity is bound to the membrane.

Chapter 8

8–1. Triadenylic acid, because it has a smaller charge than penta-adenylic acid.

8–2. The three fractions are the fragments and the intact protein. The order, in terms of molecular weights, is 65,000; 43,000; 22,000, with the intact protein eluting first.

8–3. The linear form elutes first.

8–4. B moves twice as far as A at all times.

8–5. a. This is just like dialysis. The NaCl will enter the gel and water will enter the solution. Thus, both the NaCl and the DNA concentrations decrease.

b. There will be two concentric circles. The smaller RNA molecule will move further.

8–6. Gel chromatography.

8–7. By gel chromatography.

8–8. Chromatography on hydroxyapatite.

8–9. The linear molecule would elute first because the circular one occupies a smaller spherical domain and would more easily penetrate the agarose pores. Denatured ribosomal RNA would elute first for the same reason.

8–10. At high concentration, the protein probably aggregated. This larger structure could not penetrate the pores and would pass through with the void volume.

Increasing the diameter is like increasing the number of columns and should be sufficient.

8–11. Two-dimensional chromatography using isopropanol–HCl and acetic acid.

8–12. Because only 5% is lost and enzyme X is 30% of the total, it could not be the case that X was not eluted. Therefore, the activity of X has been lost. There are many possible explanations of which two are the dissociation of subunits and denaturation caused by binding to the charged DEAE-cellulose.

8–13. In the single-strand region because single-stranded DNA binds more tightly than double-stranded DNA. If the phosphate concentration is sufficient to dissociate the double-stranded material from the hydroxyapatite, the molecule will remain bound by the single-stranded piece.

8–14. In the ion-exchange procedure, the enzyme and the Mg^{2+} ions are probably eluted at different ionic strengths. With the gel, they probably separate because of their different sizes. In both procedures, the enzyme will be found in a solution lacking Mg^{2+} and this may cause denaturation. This can be avoided in gel chromatography if the gel has been equilibrated with Mg^{2+} beforehand and if the eluting buffer contains Mg^{2+}.

8–15. a. The protein contains two different polypeptide chains. It has been dissociated by the ion exchanger and the two chains have been separated into fractions 8 and 12. Only a small portion remains undissociated; it is in fraction 3.

b. The crude cell extract contains an inhibitor that is not present in fractions 8 and 12.

c. No, because dissociation would occur again.

d. The protein would not dissociate and activity would be retained.

e. If the inhibitor has not eluted from the column, there will be 250 units. If it has eluted, there will be 100 units unless the lower concentration of the inhibitor makes it less effective.

8–16. The protein contains subunits that dissociate. As the protein concentration decreases, the number of monomers increases. The monomers elute later than the intact protein.

8–17. Morphine could be coupled to a matrix and affinity chromatography could be performed.

8–18. Gel chromatography at pH 8.5 should be used. Each fraction should be readjusted to pH 5.5 before assaying the protein.

Chapter 9

9–1. At high ionic strength, a solution would be too conductive, the current would be too high, and the solution would overheat.

9–2. No.

9–3. pH usually has the greatest effect because the total charge and the sign of the charge can be varied. Reducing ionic strength increases the availability of charged groups to the solvent, although it can have the reverse effect by changing protein conformation. Temperature also can affect conformation but is not always a useful parameter because the alterations are frequently reversible.

9–4. No, because they may have different shapes and hence different frictional coefficients.

9–5. No, because two proteins could easily have the same mobility—for example, if their charge-to-mass ratios were very nearly the same.

9–6. The protein appears to consist of two kinds of subunits, one having twice the molecular weight of the other. If the SDS-treated protein were fractionated by gel chromatography, the relative values of the molecular weights could be ascertained from the relation between elution volumes and molecular weight. After the removal of the SDS from the fractionated subunits, they could be chromatographed one by one with the untreated protein to determine the relative molecular weight of each subunit compared with the intact protein; in this way, it could be known whether the protein consisted of one subunit of each type, two of each type, and so forth.

9–7. Using the equation $D = a - b\log(M)$, calculate a and b from the distances moved by each molecule. This yields $a = -40.84$ and $b = -4.34$. From these values and $D = 2.8$ cm the molecular weight of protein C is 23,278.

9–8. The relative distances migrated are the ratio log 26,000/log 1800 = 1.36. The relative areas reflect the total mass ratio, or (192 × 26,000)/(64 × 1800) = 43.3.

9–9. It would be most useful to vary the pore size of the gel (by varying the concentration or the degree of cross-linking), because in a particular range it might be expected that two proteins having the same mobility but different molecular weights will separate by the molecular-sieve effect. To aid in ascertaining homogeneity, the protein could be analyzed by gel chromatography because two proteins having different molecular weights can have the same molecular volume. Homogeneity can also be checked by sedimentation (see Chapter 11).

9–10. Cuts have not been made at sites 35 and 83. Thus, the fragments of length 10, 17, 20, and 38 are replaced by the fragments 45 − 15 = 30 and 100 − 45 = 55.

9–11. The two terminal fragments made by cuts at positions 20 and 90 are absent and a fragment 20 + 10 = 30 units long, which is the sum of the terminal fragments, is present. Thus the molecule is circular.

9–12. With the circular molecule, the 0.10 and 0.16 fragments are missing. Thus they are terminal and have summed to make 0.26, a size that is already present. In the short reaction, the 0.10 and 0.30 fragments are replaced by an 0.40 fragment. Thus the 0.10 and 0.30 fragments must be adjacent. Thus the order of the fragments is either 0.16, 0.18, 0.26, 0.30, 0.10 or 0.16, 0.26, 0.18, 0.30, 0.10. The 10% deletion has lost the 0.26 fragment, which has become replaced by an 0.26 − 0.10 = 0.16, a size that already exists. It is stated that the deletion is very near the center so that the 0.26 fragment must be central. Thus the cuts are made at positions 0.16, 0.34, 0.60, and 0.90.

9–13. 5′XYGGTACTAGGGTATCAATGGATCGTCAGATC.

Chapter 10

10–1. Because the OD increases with dilution, the initial concentration must be between 0.00 and 0.05 μg. Hence, the concentration is approximately 0.03 μg.

10–2. The position of the band is determined both by the diffusion coefficient of each reactant and by the relative concentration of each at equivalence. Because the equivalence point for precipitin formation is not necessarily the same for two proteins having the same diffusion coefficient, two bands will result. One band would appear only if the two proteins had precipitin curves with the same equivalence point. Similarly, two proteins with different diffusion coefficients might not result if the equivalence points were different but such that a balance was achieved.

10–3. The reaction would probably be very different because the three-dimensional structure of the antigenic site of the native protein would probably be greatly changed on denaturation.

10–4. If the amino acid change is in a region of the protein that is not the antigenic site and if it does not affect the structure of this site by virtue of an overall change in the structure of the protein, there will be no effect on antigenicity. If it does these things, antigenicity can be either lost if the changes are great or altered if they are small.

10–5. Use the radioimmunoassay. Couple the drug to bovine serum albumin and immunize a rabbit to prepare antidrug antibody. Prepare the drug in radioactive form. Obtain a standard curve by adding various amounts of the nonradioactive drug to a constant amount of a mixture of the radioactive drug and antidrug. Then add various aliquots of blood to the mixture of the radioactive drug and the antibody and determine the amount of drug from the standard curve.

10–6. a. The concentration is considerably higher than 0.010 μg/0.1 ml because the amount of radioactivity is so small. The value of 15 cpm is too low to be reliable.

b. Dilute the blood sample and measure the concentration of X in each dilution.

c. 0.003 μg/0.1 ml or 0.03 μg/ml.

10–7. The number must be greater than seventeen because some substances may not be antigenic and some precipitin lines may be too faint to be seen.

Chapter 11

11–1. The flexible rod would have the higher *s* because its frictional coefficient would be lower. The hollow and the solid spheres would have the same *s*.

11–2. It undergoes a shape change that increases its functional coefficient.

11–3. One reason is that the molecule has so few bound counterions that its molecular weight is slightly lower. The main reason is that it carries bound charges of only one sign along the tube leaving behind nonsedimenting ions having the opposite charge. This creates a retarding electrical force.

11–4. The speed is doubled so that the centrifugal force is quadrupled. Thus it would take $\frac{1}{4} \times 20 = 5$ min.

11–5. When it loses structure, its *s* value increases; before denaturation it moves more slowly than a random coil so that it is likely to be rodlike.

11–6. The *s* value is higher in 1 M NaCl because there is no interstrand charge repulsion owing to the phosphates.

11–7. Velocity = $s \times$ centrifugal force. To calculate centrifugal force, the rotational velocity must be expressed in radians per second. Thus the linear velocity is $(22 \times 10^{-13})[(40{,}000)(2\pi)/60]^2(6) = 2.32 \times 10^{-4}$ cm/sec. In 20 min (=1200 sec), the molecule will move approximately 0.28 cm.

11–8. a. No, because two molecules having different values of *M* can also have different functional coefficients and the values can balance so that both molecules have the same *s*.

b. No, because two molecules can have the same value of M and different frictional coefficients.

11–9.

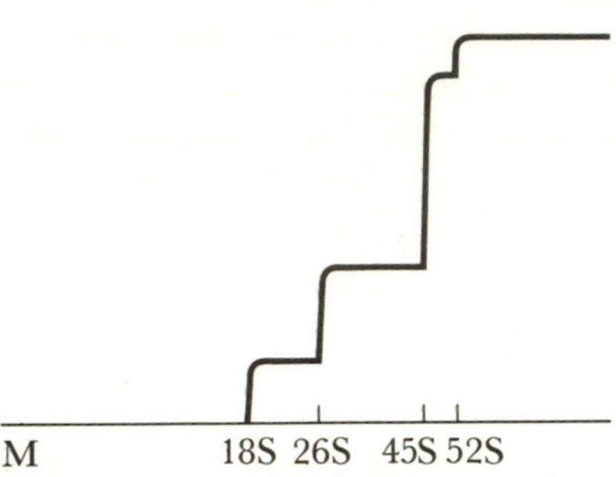

11–10. B.

11–11. Half of the molecules contain one or more strand breaks in both strands or every double-stranded molecule contains breaks in only one single strand. If C had been observed, the breaks would be in a unique position.

11–12. Diagram A is for the sample in which the breaks are randomly arranged.

11–13. The number of subunits is six; four have a molecular weight of 5000 and two have a molecular weight of 15,000.

11–14. As the number of bound ethidium bromide molecules increases, the supercoiled DNA unwinds until it is no longer supercoiled and then winds up again in the opposite direction. In the form that is not supercoiled, the molecule is least compact and therefore has the minimum value of s.

11–15. The lipoprotein must have a lower density than 1 M NaCl and then sediments in a direction opposite to the centrifugal force.

11–16. In part B, the reagent has precipitated the DNA so that it has formed a pellet before reaching speed. In part C, the reagent has hydrolyzed the DNA to either mononucleotides or small oligonucleotides, neither of which sediments at the centrifugal forces normally encountered.

11–17. At 25 Ci/mmol, almost every thymidine contains one tritium atom. This increases the molecular weight by two atomic mass units. For a molecule of DNA for which 50% of the base pairs are A·T, this would be two atomic mass units per mass of two base-paired nucleotides having a molecular weight of approximately 1320. Therefore, if the density were originally 1.698 g/cm^3, it would increase to $1.698 \times (1322/1320) = 1.701$ g/cm^3, an increase of 0.003 g/cm^3.

^{14}C has a half-life of roughly 5000 years or 400 times that of ^{3}H. Therefore, one ^{14}C per thymidine would have a specific activity of $25/400 = 0.62$ Ci/mmol. Therefore, at 1 mCi/mmol, there would be $1/0.62 = 1.6$ ^{14}C/thymidine. Compared with ^{12}C, this would increase the mass by 3.2 atomic mass units. By a calculation similar to that given for ^{3}H, the density of a 50% A·T DNA would be 1.702 g/cm^3.

For ^{32}P, a similar calculation yields one ^{32}P per eight nucleotides, which would increase the density by 0.0006 g/cm^3.

11–18. There could be a shape change, sufficient binding of the iodide ion to affect the molecular weight, or a dimerization accompanied by a shape change. A change in molecular weight could be detected by sedimentation equilibrium.

11–19. DNA can be detected by absorption optics. The absorption coefficient of DNA at the absorption maximum (i.e., 260 nm) is much higher than that of protein at 280 nm. For proteins with strongly absorbing groups (e.g., hemoglobin or cytochrome *c*), this is not necessarily the case.

11–20. It is more likely that the material that had pelleted on the tube bottom did not resuspend. The *s* obtained in the second centrifugation is too high because the concentration is lower than that initially placed in the cell.

11–21. The combination of acridine orange and visible light alters the DNA in such a way that a bond in the sugar-phosphate backbone is cleaved at alkaline pH.

11–22. Sediment for different times and show that the distance moved is proportional to the time of sedimentation. A density gradient that is isokinetic for DNA would not be isokinetic for protein because DNA and protein have different densities.

11–23. The $s_{20,w}$ of linear DNA should decrease because the molecule increases in length and becomes more rigid. This effect is greater than that produced by the greater mass of the complex.

11–24. For native protein, the effect, if detectable, would be small because the compactness of a protein is usually not an effect of the presence of disulfide bonds.

11–25. In B, the molecules are both extended and have the same shape. Therefore the 5S one has a lower molecular weight than the 8S one. Since the area of the 5S is twice that of the 8S, it suggests that there are at least two 5S RNAs per 8S molecule. In A, there is a 12S peak. Its area is twice that of the 8S material. Thus the 12S and 5S substances are probably the same molecules except that the 12S one is more compact. The 8S molecules do not change in *s* value so that the molecules are probably already extended almost completely (Actually 9S would have been a more realistic value since it should be somewhat more extended in low salt.) Thus the particle contains at least 2 compact RNAs and 1 extended molecule whose molecular weight is at least twice as high as the compact molecule. Another possibility is that the 12S molecule is a double-stranded RNA molecule, whose strands separate in the denaturant.

11–26. The protein is hitting the walls of the tube and sticking. These molecules are not recovered during the fractionation. The longer the run, the greater is the probability of hitting the wall and more is lost.

11–27. If the boundary of a homogeneous substance becomes narrower, this usually indicates that the diffusion coefficient has decreased. A decrease in the diffusion coefficient usually means that the molecule has expanded. Such expansion also causes a decrease in the *s* value. Hence, in the first case, the reagent has disrupted internal bonds (e.g., disulfide bonds) so that the molecule is less compact. If the boundary broadens, the diffusion coefficient has increased and the molecule becomes more compact. This increases the *s* value. Hence, a decrease in the *s* value with a broader boundary is not due simply to a shape change. The molecular weight has probably decreased and the reagent probably dissociates subunits.

11–28. No. The bands would be too wide.

11–29. The density difference between [^{14}N]DNA and [^{15}N]DNA = 0.015 g/ml. Therefore, the density gradient is 0.015/1.32 g cm^{-3} $mm.^{-1}$ Because *E. coli* is 50% G+C, the difference in density of each from that of *E. coli* DNA is 0.098 × 0.20 = 0.0196 g/cm^3. This corresponds to a distance of (0.0196/0.015) × 1.32 = 1.72 mm from *E. coli* DNA.

11–30. Mg^{2+} must be bound to the DNA in CsCl and MgDNA is less dense than CsDNA.

11–31. When the molecule is 10% replicated, it contains 90 units of density equal 1.75 g/cm^3 and 20 of density = $\frac{1}{2}$(1.75 + 1.7) = 1.725 g/cm^3. Therefore the density of the partially replicated molecule is (1/110)[(90)(1.75) + (20)(1.725)] = 1.745 g/cm^3.

11–32. The density, ρ, of ^{14}N- and ^{15}N-labeled DNA is 1.700 g/cm^3 and 1.714 g/cm^3, respectively. The density of the linked DNA is

$$\frac{M_1 + M_2}{(M_1/\rho_1) + (M_2/\rho_2)}$$

where the subscripts 1 and 2 refer to the normal and deleted DNA, respectively. Using 30.24 × 10^6 instead of 30 × 10^6 for the value of M_1 for [^{15}N]DNA, the value of M_2 of the molecule with the deletion is 22.5 × 10^6.

11–33. ($\frac{1}{3}$)(1.700) + ($\frac{2}{3}$)(1.300) = 1.433 g/cm^3.

11–34. Since the speed and temperature are the same when both DNA and transfer RNA are studied, the relation between M and band width d can be written $M = k\rho/d^2$, in which k is a constant including all of the necessary factors and ρ is the density. Thus, for two different molecules 1 and 2, $d_2 = d_1\sqrt{\rho_2 M_1/\rho_1 M_2}$. The densities of DNA and RNA are 1.7 and 1.8 g/cm^3, respectively, so that the band width of the tRNA is 42.3 mm, which is larger than the length of an analytical centrifuge cell.

11–35. All fragments do not have exactly the same G+C content so that there is density heterogeneity, which spreads the band thus yielding an artificially low value of the molecular weight.

11–36. a. They are identical.
b. With saturating amounts of ethidium bromide, the nicked circle, like a linear molecule, binds more ethidium bromide than a covalent circle. Therefore the density of a nicked circle is less than that of a covalent circle.

11–37. The sample contains fragments. These produce a broader band that is superimposed on the narrower main band, which makes the observed band broader than a band without fragments.

11–38. In CsCl, one obtains the molecular weight of CsDNA. In NaCl it is for NaDNA. Since the nucleotide molecular weight is 330 and $M_{NaCl} = 23$ and $M_{CsCl} = 137$, if the $M_{NaDNA} = 25$ million, then M_{CsDNA} is [(137 + 330)/(23 + 330)] × 25 = 33 million.

11–39. Use equation (13) with $a = 6.425$ cm, $c_a = 1.012$ mg/cm^3, $r = 6.703$ cm, and $C_r = 6.905$ mg/cm^3 and remember to express angular velocity in radians per second [20,000 rpm = (2000)(2π)($\frac{1}{60}$) = 2094 radians per second]. The value of R in the appropriate units is 8.314×10^7 erg deg^{-1} mol^{-1} and $T = 293$ K. Thus, $M = 20{,}600$.

11–40. The molecule consists of two linked covalent circles; each has the same molecular weight. One nick yields a covalent circle linked to a nicked circle; the density of this structure is an average of 1.55 and 1.60 or 1.575 g/cm^3. In alkali, the nicked strand falls off (this is C) leaving behind a denatured double-stranded covalent circle linked to a single circle (B). This structure contains three units of molecular weight compared to one in C; this accounts for the 3:1 ratio of areas of B and C.

11–41. The fast and slow material consist of circles and linears respectively. A survival of $1/e = 0.37$ is obtained after irradiation with 2000 rads. This is the dose that produces, on the average, one break per circle.

11–42. Two-thirds of the molecules in the DNA sample are covalent circles; one-third are open circles or linear molecules. If there were an average of one break per molecule, $1/e = 0.37$ of the covalent circles would not receive a break. Therefore, $0.37 \times (\frac{2}{3}) = 0.246$ of the molecules would remain covalent circles and 0.754 would not. The ratio of the area of the denser band to that of the lighter band would be $0.246/0.754 = 0.326$.

Chapter 12

12–1. The effect would be quite small for proteins because the cause of the effect for amino acids is that Li^+ and K^+ can bind to the free carboxyl groups and alter the density of the molecules. Proteins have few free carboxyl groups.

12–2. Yes, because of the relative mass increases due to the binding of Na^+ and Mg^+ to the phosphates.

12–3. The pycnometer holds 9.9249 g of H_2O and therefore has a volume of $9.9249/0.9982 = 9.9428$ cm^3. The weight of the solution in the pycnometer is 11.3251 g and the density is $11.3251/9.9428 = 1.1390$ g/cm^3. The solution consisting of 3.5921 g of solute and 9.9413 g of H_2O has a total weight of 13.5334 g and a volume of $13.5334/1.1390 = 11.8818$ cm^3. The volume of water used to prepare the solution is $9.9413/0.9982 = 9.9592$ cm^3. Hence, if the increase in mass is 3.5921 g, the change in volume is $11.8818 - 9.9592 = 1.9226$ cm^3. Therefore, $\bar{v} = 1.9226/3.5921 = 0.535$ cm^3/g.

12–4. Binding of the Hg^{2+} ion increases the density and thus decreases $\bar{v}$.

12–5. Curvature usually means that the sample is not homogeneous but contains many components. However, thermal convection can cause curvature. If the curve has two components, the sample contains two types of molecules.

12–6.
$$M = \frac{s^\circ_{20,w} RT}{D^\circ(1 - \bar{v}\rho)}$$
$$= \frac{14.2 \times 10^{-13}}{5.82 \times 10^{-6}} \cdot \frac{(8.3100 \times 10^7)293}{1 - 0.74(0.9982)}$$
$$= 2.275 \times 10^5.$$

12–7. Lower speed is preferable because this would allow a longer time for diffusional spreading to occur. A mixture of two components could be measured as long as the two boundaries are totally resolved.

12–8. Decrease due to increased friction. The flexible rod has greater D because it encounters less friction.

12–9. a. $D = kT/f$ and for a sphere, $f = 6\pi r\eta$, in which r is the radius. The mass m of the sphere $= \frac{4}{3}\pi r^3\rho$, in which ρ is the density, which is 1.5 g for the virus. The mass must be expressed in grams so that $m = (5 \times 10^7)/(6 \times 10^{23}) = 8.3 \times 10^{-17}$ g. Substituting all values, $D = 9.1 \times 10^{-8}$ cm^2/sec.

b. Smaller, because the asymmetry of the particle will increase friction.

12–10. Repeating the calculation of problem 12-9a, D would be 1.26×10^{-7} cm^2/sec if the molecule were spherical. The value of f/f_0 is 5.7 so that the molecule is very long and thin.

12–11. The molecule has six subunits. Repeating the calculation of problem 12–9, the values of D for the protein and one subunit, if each were spherical, would be 4.45×10^{-7} cm^2/sec and 8.1×10^{-7} cm^2/sec, respectively. The values of f/f_0 for the subunit and the protein are 1.30 and 1.03 respectively. Thus, the axial ratio of the subunit is 6 and that of the protein is 2 and the six subunits could not be arranged end-to-end but could be in a hexagonal array.

12–12. 2.28×10^{-19} or a molecular weight of 1.37×10^4.

12–13. 6×10^{-19} or a molecular weight of 3.6×10^4.

Chapter 13

13–1. At the higher NaCl concentration, intrastrand hydrogen bonds form and the single-stranded DNA is more compact; hence $[\eta]$ is lower. In formaldehyde, the amino groups of the bases are titrated and intrastrand hydrogen bonds cannot form. There would be no effect in 0.01 M NaCl because there are no hydrogen bonds in the absence of formaldehyde. In 0.5 M NaCl, the single-stranded DNA would be less compact if formaldehyde were present so that $[\eta]$ would be greater than 3 ml/g. However, $[\eta]$ would be less than 20 ml/g because, in 0.5 M NaCl, the negatively charged phosphates would be neutralized and a random coil could be assumed.

13–2. The bacteria have become permeable to proteins and the intracellular proteins have come out of the cells.

13–3. (a) The sphere with larger radius; (b) the solid sphere; (c) the rigid rod; (d) the lollipop.

13–4. Both will be the same; see equation (2).

13–5. (a) The circle because in a shear field it will be deformed to a structure of half length and double thickness; (b) the denatured DNA because it will be very compact owing to intramolecular hydrogen bonds; (c) the DNA in 1 M NaCl because it will be more compact; (d) they will be nearly the same, but the twisted circle will be slightly more resistant because it will be shorter.

13–6. If $[\eta]$ increases, it is probably compact. If $[\eta]$ decreases, it is probably an extended fiber. With a decrease of $[\eta]$ uncertainty arises because the protein might consist of subunits that dissociate in 6 M guanidine chloride.

13–7. The ions might disrupt the quasi-crystalline water lattice (i.e., decrease the interaction between water molecules).

13–8. The polynucleotide has become more extended. If the ionic strength is low, the pH change might break intrastrand base pairing with the result that charge repulsion between the phosphates would produce an extended single-stranded molecule. At high ionic strength, a likely possibility is that the polynucleotide would undergo a transition from being single-stranded to being double-stranded.

13–9. The DNA sample probably contains basic proteins. At low ionic strength, the proteins bind to the DNA and cause intramolecular aggregation, which decreases viscosity. If the DNA is diluted before decreasing the ionic strength, the concentration of DNA and protein can be sufficiently low that aggregation does not occur.

13–10.

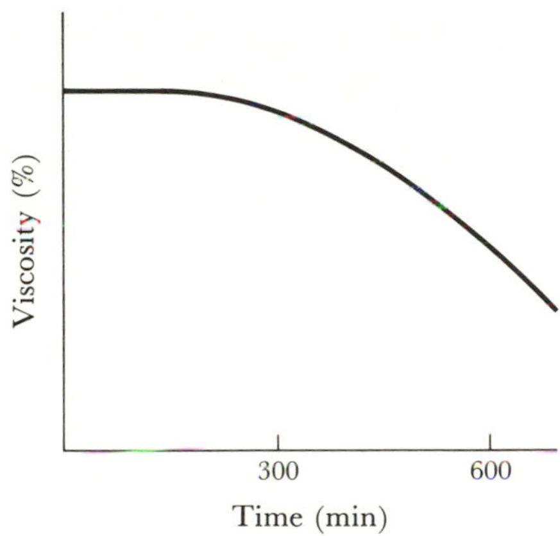

13–11. Both η/r and $[\eta]$ would be higher after joining.

13–12. The viscosity should decrease as the optical density increases.

Chapter 14

14–1. The frequency is $2989 \times c = 2989(3 \times 10^{10}) = 8.97 \times 10^{13}\ s^{-1}$. The wavelength = 3345.6 nm. It is in the infrared part of the spectrum.

14–2. Tyrosine absorbs ultraviolet light; isoleucine does not.

14–3.
a. 1/100 = 0.01 M.
b. Use equation (3). The concentration is 0.02 M.
c. The molecular weight is 78 so that the concentration of benzene is 10.26 M. Thus, 0.001 cm would have an absorbance of 1.

14–4. 32 μg/ml = 0.032 g/1 = 0.032/423 = 0.000076 M. Therefore, ε = 0.27/0.000076 = 3552 $M^{-1}\ cm^{-1}$.

14–5. By mutual dilution of both A and B, the optical densities of the mixture should be OD_{260} = 0.301 and OD_{450} = 0.27. There appears to be an interaction between A and B. It has been assumed that Beer's law is obeyed during the dilution.

14–6. Beer's law is not obeyed until the solution has been diluted 1:2. The 1:4 dilution is definitely in the linear range, and its molarity is 0.84/348 = 0.0024. The molarity of the original solution is fivefold greater or 0.12 M.

14–7. The ratio OD_{260}/OD_{280} of the isolated substance A is less than that of pure A so that some B must be present. If for every part of A there are x parts of B, the following equation may be written:

$$\frac{5248 + 311x}{3150 + 350x} = \frac{2.50}{2.00}.$$

Therefore, $x = 10.5$. Then the fraction of the observed OD_{260} due to A is $5248/[5248 + (10.5)311] = 0.616$. Hence, the OD_{260} due to A only is $0.616 \times 2.50 = 1.54$, and the concentration of A is $1.54/5248 = 0.00019$ M.

14–8. One solves the pair of equations

$$11{,}300\,[\mathrm{Tyr}] + 1960\,[\mathrm{Trp}] = 0.717$$

$$1500\,[\mathrm{Tyr}] + 5380\,[\mathrm{Trp}] = 0.239$$

obtaining $[\mathrm{Tyr}] = 58.6\ \mu M$ and $[\mathrm{Trp}] = 28.1\ \mu M$. The solution is at a concentration of 0.1 g/l. Thus, the content of tryptophan and tyrosine is 281 and 586 μmol/g protein, respectively.

14–9. Native and denatured DNA can be distinguished by the hyperchromicity at 260 nm resulting from boiling or from the addition of alkali. An alternative is to measure the spectrum because the absorption maxima of native and denatured DNA differ.

14–10. Measure the optical density at the absorption maximum of cytochrome *c*. However, the bacteria scatter strongly so that it is necessary to determine the spectrum at long wavelengths in order that a scattering correction can be carried out.

14–11. The DNA is partially denatured at 20°C.

14–12. Protein A has increased the OD by 20%. Boiling produces a final increase of 30%, which is about the expected increase for denaturation. Thus A is probably a protein that binds to DNA and produces single-stranded regions. Protein B produces a 60% increase in OD and there is no further increase on boiling. Clearly there is no double-stranded DNA left. Furthermore a 60% increase is expected if free bases were produced. Thus, protein B is probably a nuclease.

14–13. In late stages of growth of the culture, growth has stopped but cell division continues.

14–14. All internal amino acids become exposed to the polar solvent. Thus according to rule 1a (Table 14-2) λ_{max} and ε will decrease on complete hydrolysis.

14–15. The tryptophan could be on the surface or be in an internal, highly polar region.

14–16. When the protein is in 0.1 M NaCl and at high concentration, the tryptophan and tyrosine are no longer exposed to the solvent and seem to be in a nonpolar region. The simplest explanation is that the protein has aggregated to form a multisubunit protein and that these two amino acids are in the regions that have joined together. These amino acids must also be surrounded by nonpolar amino acids.

14–17. All of the absorbing groups must be internal or at least in crevices inaccessible to ethylene glycol.

The protein has eight tryptophans. None is on the surface. Four are probably in crevices narrower than 9 Å and larger than 4 Å because they are accessible to dimethylsulfoxide but not to sucrose. Four are either internal or in very narrow crevices. Alternately, they may all be internal if dimethylsulfoxide causes partial unfolding.

14–18. Six tyrosines are on the surface. Complete unfolding occurs at pH 11.7. When the protein is restored to pH 6, it refolds but in such a way that only four tyrosines are on the surface.

14–19. There are two classes of DNA: 20% of the DNA has a G + C content less than 50%, and 80% has a G + C content greater than 50%. This frequently occurs if the bacterium contains an accessory DNA molecule called a plasmid. Alternately, the culture from which the DNA is isolated might be contaminated with another bacterium.

Ten percent of the DNA was denatured by the heating at 75°C.

Chapter 15

15–1. I, absorption; II, fluorescence.

15–2. The protein dimerizes at high concentration and either the region of contact is the region binding ANS or dimerization causes a conformational change, which increases the polarity of the surface.

15–3. The tryptophan could be internal, in a crevice too narrow for iodide to enter, or adjacent to negatively charged amino acids.

All tryptophans may not have the same quantum yield, and some of the surface tryptophans may be in negatively charged regions.

15–4. No, because the absorption spectrum of a dimer usually differs from that of a monomer.

15–5. No.

15–6. The complex may have different energy levels or the probability of quenching may be reduced. Spectral shifts in either direction are possible because of the different energy levels.

15–7. No, because some of the transitions may not lead to fluorescence.

15–8. The tryptophans may already be quenched. The ligand may remove a quenching factor in the protein yet introduce another. Binding of the ligand may produce a conformational change in the entire molecule so that quenching of the fluorescence of those in the binding site is counteracted by enhancement of the fluorescence of tryptophans that are elsewhere in the molecule and were partially quenched before ligand binding.

15–9. By an energy transfer process, O_2 could be excited. Alternately, the DNA bases could be excited and more susceptible to oxidation by unexcited O_2. The acridine orange and the O_2 could form a complex that, if excited, would be highly reactive.

15–10. As ionic strength increases, the protein unfolds so that the region containing the fluor is more flexible. Alternately, the protein might become very compact or very symmetric, thus allowing more rapid rotation.

15–11. For the first observation, the fluor is sufficiently near the tryptophan that energy transfer occurs. The fluor is on the surface. Therefore, the tryptophan is on the surface. The tryptophan is probably either in a negatively charged region or is protected by the fluor from quenching by iodide.

The explanation for the second observation is the same as that for the first except that the fluor does not protect the tryptophan from collisional quenching.

For the third observation, there is a conformational change at pH 9 that moves F away from the tryptophan.

For the fourth observation, the tryptophan and the fluor are very near one another at pH 9.

15–12. When I^- quenches, λ_{max} shifts to shorter wavelengths; this means that the unquenched tryptophan is in a nonpolar region. Since only half of the fluorescence is quenched by the I^- or Cs^+ ions, both tryptophans cannot be on the surface. Thus, the second person is correct.

15–13. a. It will be brighter for A-(proline)$_8$-B because A and B are nearer.

b. The fluorescence is more intense for the alanine compound. A and B are nearer in the alanine compound because (alanine)$_7$ is flexible, whereas the (proline)$_7$ chain, which has no peptide bonds, is not flexible.

Chapter 16

16–1. The wave that lags by 90° should be advanced by 90°. For elliptically polarized light, the maximum length of the **E** vectors should be different.

16–2. They are probably mirror images.

16–3. The conformation in solution probably differs from that in the crystal. Because the change is very great, there is probably a region that is either extremely hydrophilic or extremely hydrophobic.

16–4.

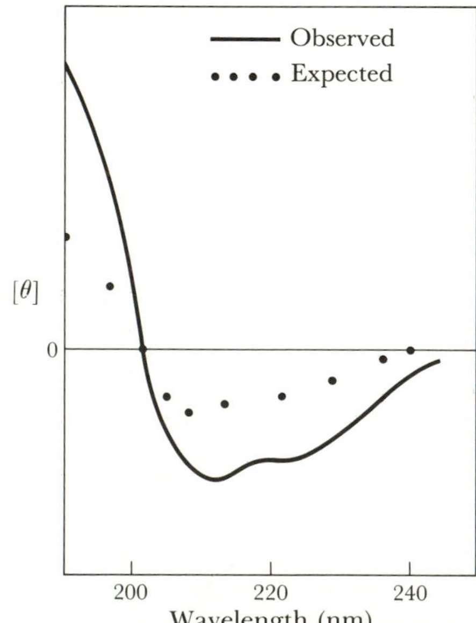

16–5. They can be seen against a baseline that is nearly zero.

16–6. By dilution.

16–7. In the range of the absorption peak.

16–8. The DNA is unwound and the ethidium bromide assumes the helical arrangement of the DNA.

16–9. The decrease would be greater in 0.5 M NaCl because, in 0.01 M NaCl, the molecule would be extended and the bases would all be stacked.

16–10. DNA is tightly folded in the phage head. The manner of folding differs from one phage to the next.

16–11. The fraction of unstacked bases (i.e., those at the ends) becomes small.

16–12. Some form α helices; others do not. In some, there are no side-chain interactions; in others, there are and these can be attractive or repulsive, depending on the degree of polarity. Polyproline does not have peptide bonds.

16–13. The protein binds the small molecule and, in so doing, undergoes a conformational change.

16–14. The protein binds to the DNA in 0.01 M NaCl but not in 1 M NaCl. Furthermore, since the CD spectrum of DNA is entirely due to helicity, the protein must induce a change in the helicity of the DNA. Possible functions of the protein are denaturing DNA, supercoiling, or coiling, such as in chromatin.

16–15. No. The effect of the Ca^{2+} ion seems to be a result of the ionic strength rather than its chemical identity.

Chapter 17

17–1. (a) 1; (b) 1; (c) 1; (d) 1; (e) 1; (f) 2, one for the hydroxyl and one for the methyl hydrogens; (g) 3, one for the hydroxyl, one a quartet for the CH_2 group, and a triplet for the methyl group.

17–2. At all concentrations, the spectra of phenylalanine and glycine would be additive because it is unlikely that the concentrations could ever be high enough for the phenylalanine to induce a ring-current shift in the glycine. However, with phenylalanine and tyrosine, there would be a change at high concentration because phenylalanine and tyrosine have a tendency to stack. Hence, there would be mutual ring-current effects.

17–3. Due to the low abundance of ^{13}C, it would be very rare that two ^{13}C nuclei would be adjacent. Therefore, splitting would not occur. With enrichment, the probability of adjacent nuclei would increase and this advantage would be lost.

17–4. Lines corresponding to the intermediates should appear while the reaction is in progress. The area of the lines with respect to the lines of the initial reactants indicates the relative concentrations of the intermediates and the reactants as a function of time.

17–5. No, because binding of the substrate might introduce a conformational change in the protein that could cause the shift. The shift in line position would indicate that further investigation of the histidine is worthwhile, especially because the shift is fairly large. If there were no shift, it would be very likely that the histidine is not in the binding site.

17–6. No, because splitting requires that the interacting nuclei are covalently coupled and denaturation does not involve changes in covalent bonds.

17–7. First, mix enzyme and cofactor in the absence of substrate and look for displacement of peaks. Second, mix substrate and cofactor and look for displacement. Then add enzyme and look for further displacement or shifts in enzyme peaks. Third, mix enzyme and substrate. If no displacement, add cofactor and look for displacement.

17–8. The line width would differ because the protons in ice are less mobile than in water. Changes in line position could also result if freezing imposed a well-defined orientation between H_2O molecules.

17–9. Lines derived from protons in the region of mutual binding would shift. Most of the line widths would increase owing to the reduced mobility of the dimer compared with that of the monomer.

17–10. It is likely that the protein contains at least one paramagnetic center—probably a transition metal ion. The addition of a chelating agent should reduce the size of the chemical shift and cause narrowing. This test might not work if the metal is covalently bound. Chemical tests for the presence of the metal would be called for.

17–11. Not for open circles. If the circle were supercoiled, there might be differences because, with a high degree of supercoiling, many of the base pairs would be disrupted, leaving single-stranded regions in the DNA.

17–12. When a compound is deuterated, proton spectral lines vanish. Thus, guanine-14, which is rapidly deuterated, is not hydrogen-bonded. Guanine-32 exchanges slowly, presumably because it is hydrogen-bonded.

17–13. X irradiation produces a free-radical intermediate which in water is gradually converted to a substance that is not paramagnetic.

17–14. Not directly, because the spin is zero. If it can be replaced by the Mn^{2+}, which is often the case, study by ESR is possible.

17–15. It is bound to two atoms, one having spin $\frac{1}{2}$ (possibly a proton) and the other having a spin of 1.

17–16. The narrower line (tube 1) is produced by the larger molecule, which tumbles more slowly than the decanucleotide.

17–17. The bacterium fails to adsorb the protein above 39°C.

17–18. The lines become narrower when the morphine is bound.

Chapter 18

18–1. Equation (7) gives points that are more equally spaced.

18–2. The enzyme may undergo a change in shape so that its binding site has greater affinity for the substrate. Alternately, the binding might be inhibited by charge repulsion between two like charges, one near the binding and one on the substrate. At high ionic strength these are shielded from one another.

18–3. The protein may undergo a shape change that increases the affinity of the binding site for A. Alternately, the binding may be a result of electrostatic attraction and decreasing the dielectric constant by the addition of methanol will decrease the shielding and increase the attraction.

18–4. a. Since the concentration is constant after fifteen hours, the system has come to equilibrium. There is binding because the concentration on side I is greater than that on side II. According to equation (2), $K_d = (10^{-3})(6.12 \times 10^{-5})/(6.54 \times 10^{-5} - 6.12 \times 10^{-5}) = 1.46 \times 10^{-2}$ M.

b. X and Y bind at the same site. Y binds much more tightly than X. Since no X is bound when [Y] = [protein], it suggests that there is only one binding site for X. There are not enough data to say whether there are additional binding sites for Y to which X cannot bind.

c. Z does not bind to the protein because [Z] is the same on both sides of the membrane. The binding of X is markedly decreased. Z certainly inhibits the binding of X. Since it does not bind to the protein, Z must act directly on X. One possible explanation is that Z binds to X and alters X so that X binds less strongly. Another possibility is an electrostatic effect on the system. For instance, if the protein-X binding is the result of an attraction between positive and negative charges and Z is strongly dipolar, it would produce electrostatic shielding, as would high ionic strength.

18–5. Cytochrome *c* binds to DNA in 0.01 M NaCl but not in 0.3 M NaCl.

18–6. Two.

18–7. Binding is irreversible.

18–8. The *x*-intercept would be constant but the slope would change.

18–9. a. The *x*-intercept is four and the curve is linear. Thus there are four independent binding sites. This suggests (but of course does not prove) that the enzyme contains four identical subunits. Other subunits that do not bind S might also be part of the enzyme.

b. The *y*-intercept is $10^{-7} = 1/K_d$ so that $K_d = 10^7$ M.

18–10. For a monoprotic acid a straight line will result whose slope is $-K$. For a triprotic acid the *x*-intercept will be three; since the equilibrium constants will rarely be the same, the plot will be curved.

18–11. If the binding is cooperative, the protein molecules will be clustered as shown in panel B.

18–12. The binding is cooperative with five binding sites. $K_d = 10^{-7}$ M.

18–13. a. Each subunit binds one molecule of X so that the tetramer binds four times as much per molecule as the monomer does.

b. The sigmoid shape of the H curve suggests cooperative binding.

c. A Scatchard plot would show curvature. However, a Hill plot would be more informative. If binding is cooperative, a slope of four will be found in the Hill plot.

18–14. K_d for region I is K_T; K_d for region II is K_R.

Chapter 19

19–1. It would probably precipitate.

19–2. All but the sphere, except that, for the circular DNA, high shear stress would be necessary.

19–3. Because a film of liquid is carried upward from the main body of solution, the surface-to-volume ratio is increased, thus increasing the rate of evaporation.

19–4. A growing culture, because the longer cells would be more susceptible to breakage by shear forces.

19–5. Thermal stability, sensitivity to low or high ionic strength, and the materials to which the substance of interest can irreversibly adsorb.

19–6. The molecules have probably been broken in half by the shear stress.

19–7. Since there are two DNA single strands corresponding to each RNA molecule, the effective molecular weight of the RNA must be doubled before performing the calculation. Thus, there are $0.20(2 \times 10^{10})/(2 \times 10^{6}) = 0.10(2 \times 10^{10})/(10^{6}) = 2000$ copies each of the two rRNA molecules per DNA molecule whose molecular weight is 2×10^{10}. Actually, the number is smaller because all eukaryotic rRNA genes contain spacers that are part of the repeating sequence.

19–8.

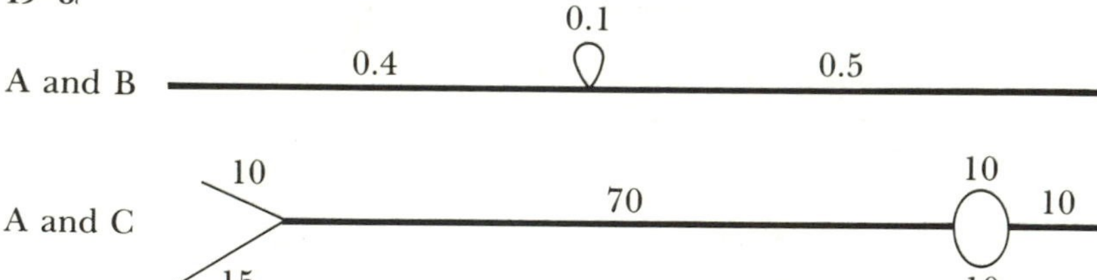

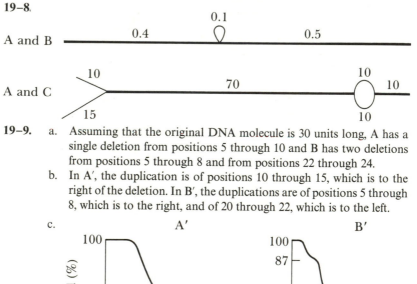

19–9. a. Assuming that the original DNA molecule is 30 units long, A has a single deletion from positions 5 through 10 and B has two deletions from positions 5 through 8 and from positions 22 through 24.

b. In A′, the duplication is of positions 10 through 15, which is to the right of the deletion. In B′, the duplications are of positions 5 through 8, which is to the right, and of 20 through 22, which is to the left.

c.

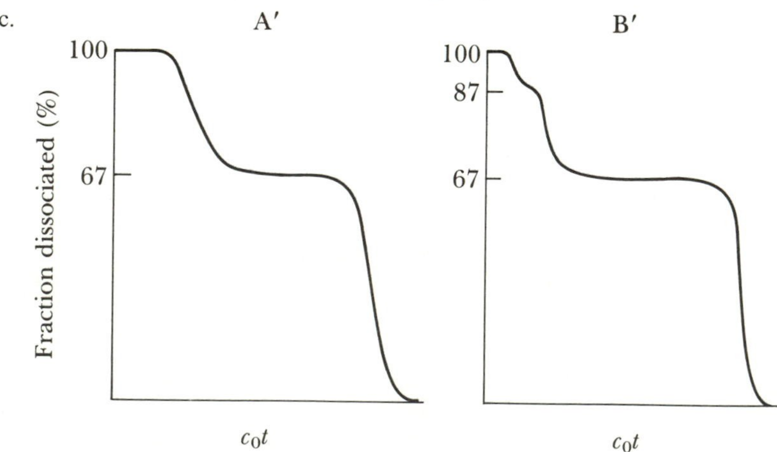

19–10. It becomes four times as large.

19–11. The y-intercept does not change. The slope of the vertical line increases because R_G increases.

19–12. R_G has decreased after heating so that the molecule is more compact than the rod. This would not happen in model I because charge repulsion would maintain rigidity. In model II, there is intrastrand charge attraction making each strand very compact. Thus, model II is correct.

19–13. It would increase as interactions are destroyed, making all groups available for exchange.

19–14. The protein probably forms hydrogen bonds with the bases of single-stranded regions thus reducing the availability of exchangeable protons.

Index

Aberration
  chromatic, 42–43
  spherical 42, 75
Absorbance, 497
Absorption of light by macromolecules
  binding and, 516–517
  DNA, 508–510
  factors affecting, 500–503
  orientation effects, 502
  pH effects, 501
  proteins, 511–517
  solvent effects, 501
Absorption optics in ultracentrifuge, 374, 379–380
Absorption spectroscopy, 496–518
  rules for interpretation, 503
  uses of, 504–517
Acridine orange, 548, 550–551, 563–565, 567
Active site, 163, 516, 528, 545, 556, 626–628
  fluorescence for studying, 545, 556
Adsorption chromatography, 220–232
  theory of, 220–221
Adsorption to purify antibodies, 325
Affinity chromatography, 257–265
  applications, 260–262
  concanavalin A in, 260
  cyanogen bromide-Sepharose in, 258
  DNA-cellulose in, 265–266
Affinity elution, 264
Agarose, 240, 258, 292
Agarose-gel electrophoresis, 281, 286
Agglutination, passive, 339
Airfuge, 441–443
Airy disc, 43–44
Allostery, 666–680
α helix, 7, 9, 10, 521, 524, 528, 589–590, 622, 624
α particles, 173–174
Alumina, 221, 229
Amino acids, 3–5
Ammonium sulfate, precipitation of proteins with, 271, 712
Ampholine, 310
Analyzer, 58
Anastigmat, 43
Angle of rotation, 577
Anion exchanger, 249
ANS, 548–549
Antibody
  definition, 323
  ferritin-conjugated, 352–353
  fluorescent, 64–65, 567–568
  preparation of, 324–325
  purification by adsorption, 325
  purification by affinity chromatography, 260
Antibody-antigen reaction, 326–334
Antigen
  definition, 323
  preparation of labeled, 336–337
Antiserum, 325
Aplanatic lens, 43
Apochromat, 47
Approach-to-equilibrium method, 423–424
Archibald method, 423–424

Argon ionization detector, 233
Associating systems, 394–396, 422, 563
Astigmatism, 43
Autoradiography
  α particles in, 171, 173
  background in, 117–118
  β particles in, 170, 173–174
  contact between source and emulsion, 179–180
  dipping method, 172, 182–183
  of DNA, 184–185
  electron microscopic, 190–191
  emulsion choice of, 175–177
  exposure time, 178, 185–187
  fading, 178
  fog, 178
  of gels, 289, 300
  isotope, choice of, 177
  molecular, 187–188
  mounting of emulsion, 179–185
  resolution, 175–176
  sensitization, 185–187
  source-emulsion relation, 174–175, 179–185
  stripping film method, 172, 183–185
  tracks, 174

Background
  autoradiography, 177
  scintillation counting, 140–141
Band sedimentation in self-generating density gradients, 405–411
Base stacking, 22, 595–597
Bed volume, 221
Beer's law, 497–498
Benzalkonium chloride technique, 93–94
Beta decay, 129–130, 134
β particles, 129–130, 173–174
  in autoradiography, 170, 173–174
β sheet, 7, 10–12, 521
β structure, 521, 528, 589–590
β turn, 590
Binding
  assay by sedimentation, 391–392
  detection by change in fluorescence, 543
  fluorescence depolarization and, 562
  general properties, 655–680
  nitrocellulose filters and, 198–203
  use of spin labeling to study, 635
Birefringence
  flow, 717–719
  form, 57–58
  intrinsic, 57–59
  in polarization microscopy, 57, 60
Bohr effect, 674–675
Bohr magneton, 640
Boundary cell, synthetic, 463
Breathing of DNA, 689
Buffers, 122–125
  exchange of, 244
  rule of, 252
Buried groups, 513–515, 523

Calcium phosphate gel, 221
Carbon, activated, 221
Carbon-13
  NMR and, 637–638
Catenane, 30
  detection of, 414–416
Cation exchanger, 248–249
Cations, divalent, removal of, 255
Cavitation, 688
CD, 579–600
  absorbance and, 581
  advantage over ORD, 584–586
  conformation and, 589–590
  measurement of, 586–587
  relation to ORD, 584–586
  rules for interpreting, 587–588
  theory, 574–582
Cells
  internal structure determined by immunological methods, 352–353
  separation by partitioning, 716–717
Cellulose acetate electrophoresis, 280–281
Centrifugation. *See* Sedimentation.
Cesium salts, in sedimentation equilibrium, 424–440
Channeling, 221
Chemical exchange, 610–611, 619
Chemical shift, 607–611, 617–620
Chromatic aberration, 42–43
Chromatography
  column, 221–222
  column gel, 282–283
  covalent, 263
  DNA-cellulose, 265–266
  electrophoresis and, 312–316
  elution in, 221–224
  gas, *see* Gas chromatography
  gel, 238–248, 689
  high-performance liquid, 255–257
  HPLC, 255–257
  hydrophobic interaction, 262
  hydroxyapatite, 268–270, 710–712
  malachite green, 267
  metal chelate, 262
  paper, *see* Paper chromatography
  partition, *see* Partition chromatography
  reverse phase, 255
  thin-layer, *see* Thin-layer chromatography
Chromophore, 494
Chromosomes, visualization by fluorescence, 64, 66, 67
Cibacron blue agarose, 261
Circular DNA, 29, 438
Circularly polarized light, 573–577
Circular proteins, 483
Closed circle, 29
Coefficient of viscosity, 471
Cohesive ends of DNA, 30
Coil
  near-random, 9
  random, 8, 9

Collagen, 32–33, 88–89
Column chromatography, 221–222
Columns
  chromatographic, 221–222
  density gradient, 457
Coma, 43
Compact proteins, 9
Compensator, 59–60
Complement, 329
Complement fixation, 329–333, 341–342
Con A, 260
Concanavalin A, 260
Concatemer, 97
Concentrating, methods for, 712–715
Concentration, measurement of, 504
Concentration gradient, 233, 371, 396–397
Conformational change, 592–593
Contour length, 16
Contrast
  in electron microscopy, 75–90, 106–109
  in light microscopy, 51
  phase, 53–56, 61, 66–67
Convection, 398
Cooperative transition, 20–21, 24
Copolymer, 2
Cotton effect, 581–586
Counter
  end-window, 135–136
  Geiger-Müller, 134–137
  scintillation, 137–155
Coupling constant, 614
Covalent circle, 29, 418
Crewe electron microscope, 107–109
Critical point technique, 80–81
Crossed immunoelectrophoresis, 355–356
Cross-link, 12
Cross-reaction, 324
  examples of, 344–346
C_0t analysis, 705–712
Curie, 133
Curvature of field, 47
Cyanogen bromide-Sepharose, 258

Dansyl chloride, 548–549, 555
Dark-field electron microscopy, 106–108
Dark-field light microscope, 51–52
Degradation
  shear, 474–475, 685–688
  sonic, 685–688
Degree of polymerization, 2
Dehydration, 78
Denaturation, 9, 19–27, 515–516, 545, 563, 592, 623–626
Denaturation mapping, 93–99
Density, 427–428, 456–459
Density gradient centrifugation, 424–440
  use of Cs_2SO_4, 434–435
  examples of use of, 420–429, 431–443
Density gradient, preparation of, 223
Density gradient column, 457
Depth filter, 196
Desalting, 244
Dextran, 239–241, 249, 717
Diaflo membranes, 209
Dialysis, 205–212
  equilibrium, 657–678
  glass fiber, 210–212
  molecular filters for, 208–209
  reverse, 208–209, 714
Dichroism, 518–525
  measurement with polarizing microscope, 62
Difference spectra, 513–515
Diffusion, 419–420, 460–467, 623, 632
Diffusion coefficient, 419–420, 460–467, 623, 632
  measurement of, 460–467
Diffusional spreading, 219, 225, 231, 250, 255
Disc electrophoresis, 287–290
Dissociation constant, relation to *K*, 655
Disulfide bond, 3, 9, 283, 482–483, 529, 568
  detection by viscometry, 482–483
  fluorescence and, 568
Divalent cations, removal of, 255
DNA
  absorption spectra, 503
  alkaline sedimentation, 411–419
  antiparallel, 7
  autoradiography, 183–185, 188–189
  base stacking, 22
  bases, 3, 6
  binding to nitrocellulose filters, 198–203
  breathing, 24, 689
  catenanes, 414
  chromatography of, 265–270
  circular, 29, 199–200, 297–299, 414–418, 436–440, 483
  covalent circles, 29, 418
  degradation by shear, 686–687
  denatured, 19, 416, 503–504, 508–510
  density, 427–440
  depurination of, 412
  effective segment length of, 16
  electrophoresis of, 268–308
  fluorescence polarization of, 562
  fractionating by partitioning, 716–717
  G + C content, 430
  glucosylated, 435
  helix-coil transition, 342–343, 503, 508–510
  heteroduplexing of, 92–93
  hybridization of, 700–705
  hydrogen bonds, 21, 520–522
  hydrophobic interactions, 21
  infrared spectra, 521–522
  intercalation of dyes into, 484–485, 564
  melting temperature, 22
  molecular weight by end-group labeling, 158
  molecular weight by viscoelasticity, 488
  nicks, 412

DNA *continued*
orientation of bases by flow birefringence analysis, 719
partially denatured, 23
phosphodiester bonds, 8
protein-binding assay, 202
renaturation, 26, 508–510
replicating, 428–429
restriction mapping, 294–297
satellite bands, 429–430
sedimentation in alkali, 411–419
separation by density gradient centrifugation, 424–440
sequencing, 303–308
single-strand breaks, 412, 415
strand separation, 387, 435–436
structure, 6
supercoiled, 30–32, 268–270, 415, 417
T_m, 22–23
viscoelasticity, 486–489
visualization by electron microscopy, 89–102
DNA-protein complex, assay on nitrocellulose filter, 198–203
DNS, 548, 555
DNA·RNA hybrid, 200–201
Dose-response curve, 445–447
Double-labeling, 143–145
Double reciprocal plot, 655–656
Dowex, 250
Drude equation, 583

EDTA, 123
Effective segment length, 16
Electrode, 118–121, 125–127
glass, 118–119
ion-selective, 126
pH, 125–127
Electromagnetic lens, 73–74
Electromagnetic spectrum, 497
Electron microscope
autoradiography, 190–191
contrast, 75–78
Crewe, 107–109
dark-field, 106–108
image formation, 73
sample preparation, 78–93
scanning, 110
specimen support, 75–78
Electron spin resonance, 638–649
origin of signal, 639
line width, 644–646
spin labels for, 645–649
Electronic absorption band, 496
Electrophoresis
agarose-gel, 292–299
cellulose acetate, 280–281
chromatography and, 312–316
crossed, 355–356
disc, 287–290
of DNA, 268–308
electrical parameters, 290–291
gel, 281–299
gradient, 309
high voltage, 279–280
immunoelectrophoresis, 318, 353–357
moving boundary, 277–278
PAGE, 283–286, 309
paper, 278–280
polyacrylamide, 283–289
preparative, 317
rocket, 356–357
starch-gel, 281–282
submarine gels, 293
theory, 276–277
two-dimensional, 280
wicks, effects of, 290–291
Electrostatic interaction, 28
Elliptically polarized light, 571–576
Ellipticity, 575, 580
Elution, 221–224
End-group labeling, 158
End-to-end distance, 15
Energy level diagram, 495–496
Energy transfer, 540, 551–559
Enhance, 187
Enzyme
active site, fluorescence for studying, 545, 556
assay by absorbance, 506–507
assay by fluorescence, 568–569
cofactor binding, detection of, 391–392
purification of, 270–271
substrate binding to, detection of, 629–631
Enzymobeads, 337–338
Equilibrium dialysis, 675–678
Ethidium bromide, 293–294, 463–469, 569, 699
Exchange
hydrogen, 403, 407, 688–691
reaction, 161–162
tritium, 246, 689–691
Excimer, 559
Excited state, 496
Excluded volume, 15
Extrinsic fluorescence, 547–551
Eyepiece, 47

Ferritin-conjugated antibody, 352–353
Ferrous sulfate dosimetry, 507
Fibrous proteins, 9
Films, support, 75–78
Filters, depth, 196
Filters, fiberglass, 196–198
Filters, nitrocellulose, 194–205
binding assays with, 198–202
cellulose acetate in, 196
counting of radioactivity with, 197–198
detergents, 196
DNA-binding, 199
DNA-protein complex, assay for, 202–203

glycerol in, 196
hybridization, measurement of, 703–704
Filters, screen, 196
Fingerprinting, 314–316
Fixative, 78
Flame ionization detector, 234
Flash evaporation, 713
Flipping of nuclei, 605–617
Flow birefringence, 717–719
Fluctuation spectroscopy, 698–700
Fluorescein, 548
Fluorescence
detection of binding by, 543
extrinsic, 63, 547–551
instrumentation for measuring, 540–541
intrinsic, 542–547
polarization, 559–567
quenching, 540
theory of, 538–540
tryptophan, 542, 544
tyrosine, 542, 544
uncorrected spectra, 541
Fluorescence microscope, 62, 567
Fluorescence spectroscopy, 537–570
Fluorescent antibody, 64–65, 352, 567–568
Fluorescent immunoassay, 350–351
Fluors
extrinsic, 547–548
in gamma ray counting, 155
in liquid scintillation counting, 138
Formamide spreading, 92–93, 95
Förster theory, 553
Fourier transform NMR, 618, 620, 636–637
Fraction collector, 223
Free radicals, 642
Freeze etching, 80–81
Freeze fracture, 81–82
Frictional coefficient, 276, 363, 461, 465–467
relation to diffusion coefficient, 465–467
Fringes, 275–276

g factor, 640
γ rays 131, 154–156
detection of, 155–156
Gap, 30
Gas chromatography, 232–238
detectors for, 232–234
identification of components in, 235–237
uses of, 237–238
Geiger-Müller counter, 134–137
Gel chromatography, 238–248, 689
advantages, 243
applications, 243–246
materials, 239–241
molecular weights by, 242, 248
theory of, 238–239
thin-layer, 247–248
Glass fiber dialysis, 210–212
Glass fiber filters. *See* Filters, fiberglass
Glow discharge procedure, 76, 87
Glutaraldehyde technique, 93
Glycoproteins, 2
Good's buffers, 124–125
Gradient maker, 224
Grid, sample, for electron microscopy, 75–76
Ground state, 495

Half-life of common isotopes, 132
Hapten, 324
Helix, properties of, 9–11
Helix-coil transition, 20–27, 508–510, 563
Helix destabilizing protein, 24
Hemoglobin, 674–675
Hershey circle, 30
Heteroduplex methods in electron microscopy of DNA, 93, 705
High-performance liquid chromatography, 255–257
High-pressure liquid chromatography, 255–257
High-voltage electrophoresis, 280
Hill equation, 662–664
Hollow fiber membranes, 715
Homochromatography, 316–317, 690
Homopolymer, 2
HPLC, 255–257
Hybrid DNA, 28
Hybridization, 28, 200–201, 299–300
detection of, on nitrocellulose filters, 200–201
Hydrogen bonds, 5, 9–11
study by tritium exchange, 688–691
Hydrogen exchange
study by gel chromatography, 688–691
study by infrared spectroscopy, 523
study by Raman spectroscopy, 529
Hydrophilic, definition, 3
Hydrophobic, definition, 3
Hydrophobic forces, 595
Hydrophobic interaction, 9, 17–18
Hydrophobic interaction chromatography, 262
Hydroxyapatite chromatography, 268–270, 710–711
Hyperchromicity, 508–509
Hyperfine structure, 640–641
Hypochromicity, 502–503, 594

Illumination, 29–30
Image reconstruction, 104–105
Immunoassays
examples of, 350–357
fluorescent, 350
protein A, 350–352
Immunodiffusion, 328–330
Immunoelectrophoresis, 353–357
Immunoglobulin, 323, 335
Immunology, 323–357
Immunoprecipitation, 159

Immunoradiometric assay, 337–338
Index of refraction, 41–42, 576–578, 580, 692
Infrared spectroscopy, 519–524, 531
  determination of α, β, and random coil structure by, 521. 524
Inhibition assays, immunological, 329–339
Intensifying screen, 187
Intercalation, 484–485
Interference microscope, 56–57
Interference optics in the ultracentrifuge, 423
Intramolecular shielding, 609
Intrinsic viscosity, 473, 481
Ion-exchange chromatography, 248–257
  applications, 254–255
  ionic conditions, 252
  rule of buffers, 252
  starting conditions procedure, 253
  techniques, 252–253
  types of exchangers, 248–250
Ion exchangers, 248–252
Ion-retardation resin, 254
Ion-selective electrode, 126
Isoelectric focusing, 309–312
Isoelectric point, 309
Isotopes, 132

Johnston-Ogston effect, 382

Karplus equations, 614–615
Kleinschmidt technique, 89–101, 245

λ_{max}
  definition, 499–500
  table of values, 501
Langmuir isotherm, 655
Larmor frequency, 606
Lectins, 260–261
Lens
  anachromatic, 42
  aplanatic, 43
  apochromat, 46–47
  electromagnetic, 73
  image formation by, 41–42
  objective, 45–46
  oil-immersion, 46–47
Ligand, definition, 654
Light, energy of, 494–495
Light scattering, 505–506, 691–698
Light wave, 495
Linderstrøm-Lang density gradient column, 457
Line splitting, 611–613, 617
Lipopolysaccharide, 2
Lipoprotein, 2
Liquid scintillation counter. *See* Scintillation counting
Lyophilization, 714

Macromolecule
  definition, 2
  precipitation with acids, 149–151, 158, 198

Magnetic moment, 604–606, 612, 638
MAK chromatography, 266–267
Malachite green chromatography, 267
Mass spectroscopy, 236–237
Mean residue ellipticity, 580
Mean residue rotation, 579
Melting temperature, 20, 22, 508–510
Membrane filtration, 149–150, 158, 193–205, 679, 710
Mesh size, 225
Messenger RNA, assay by filter binding, 199–201
Metal chelate chromatography, 262
Methylated albumin-kieselgur chromatography, 266–267
Micelles, 208
Microscope
  adjustment of, 49–51
  dark-field, 51–52
  fluorescence, 62–66, 567–568
  interference, 56–57
  parts of, 44–49
  phase contrast, 53–56, 61, 66–67
  polarization, 57–62
  resolution limit, 43–44
Minicon concentrator, 714–715
Mixed bed resin, 255
Mobility, electrophoretic, 276–277
Molar absorption coefficient, 378, 497, 576, 579, 580
Molar ellipticity, 580
Molar rotation, 579
Molecular autoradiography, 187–188
Molecular filters, 208–211
Molecular sieve chromatography. *See* Gel chromatography
Molecular weight
  number average, 13–14
  weight average, 13–14
Molecular weight, determination by
  airfuge, 442–443
  approach-to-equilibrium, 423–424
  end-group labeling, 158
  gel chromatography, 242–243
  gel electrophoresis, 284–286
  intrinsic viscosity, 481
  SDS-gel electrophoresis, 285
  sedimentation-diffusion, 419–420
  sedimentation equilibrium, 420–423
  sedimentation equilibrium in a density gradient, 426–427
  thin-layer gel chromatography, 248
  viscoelasticity, 488
Monodisperase, 370
Motion of molecules, measurement of, 562, 615–616, 645–647

NA, 43, 46, 48
Native, definition, 19
Near-random coil, 19
Negative contrast technique, 86
Neutralization by antiserum, 324, 343–344

Nick
definition, 30
detection of, 411–417
Nicked circle, 29–30
Nitrocellulose filters. *See* Filters, nitrocellulose
NMR. *See* Nuclear magnetic resonance
Noncooperative transition, 20–21
Northern transfer, 302–303
Nuclear emulsion
background in, 177–178
efficiency of, 177
tracks in, 171, 173, 176–177, 181
types of, 171
Nuclear magnetic resonance, 603–649
carbon-13 in, 637
fluorine-19 in, 638
Fourier transform, 636–637
identification of lines, 613, 621–622, 632
instrumentation, 616
rules for interpretation of spectra, 619
saturation and, 632
theory, 604–607
Nuclear magnetic resonance spectrometer, 616–617
Nuclear Overhauser effect, 631
Nuclear spin, 605–606
Number-average molecular weight, 13–14
Numerical aperture, 43, 46, 48

Open circle, 29–30
Optical activity, 576–577, 580
Optical density
concentration and, 505–506
definition, 497
Optical mixing spectroscopy, 464
Optical rotatory dispersion. *See* Circular dichroism
ORD. *See* Circular dichroism
Ouchterlony double-diffusion technique, 329, 330, 342–344

Palindrome, 37
Paper chromatography
ascending, 228–229
comparison with thin-layer chromatography, 231
descending, 228–229
detection of spots in, 228–230
methodology, 227
R_f, 227
Paper electrophoresis, 278–280
Paramagnetism, 609, 619, 633, 638, 643
Partial specific volume, 422, 455–460
Partition chromatography, 217–219, 238
Partitioning, separation by, 717
Pellicon, 209
Peptide bond, 3, 8
Peroxidase–anti-peroxidase assay, 352–354
pH, 118–127
complications in measuring, 121–122
definition, 118
ionic concentration and, 121
meter, 119–121
sodium error, 122
theory and measurement, 120
Phase, 53
Phase plate, 53–54
Phenyl neutral red chromatography, 267
Phosphodiester bond, 8
Plain curve, 582–583
Plane of polarization, 574–578
Plasmid, 297–301, 417–418, 431, 437
Pleated sheet, 10
Polarization of fluorescence, 559–567
Polarization microscopy. *See* Microscope, polarization
Polarized light, 57–60, 518–519, 573–580
absorption of, 518–519
circular, 573–578
elliptically, 575–576, 580
plane of, 574–578
rotation of, 575–577
Polyacrylamide, 240, 243, 258
Polycarbonate filter, 203–204
Polydisperse, 13
Polyethylene glycol, partitioning with, 717
Polylysine film procedure, 77–78, 101–102
Polymerization, degree of, 2
Polynucleotide
base stacking, 595–597
CD spectra, 594–600
circular, 229–232
definition, 6
hydrophobic forces in, 596
linear, 29–30
structure by NMR, 632–633
Polypeptide chain, 6
Ppm (parts per million), 607
Precipitation, acid, of macromolecules, 149–151, 158, 195
Precipitin, 326–329
Preflashing, 186–187
Pressure dialysis, 714–715
Primary structure, 7
Proflavin, 548
Prosthetic group, 591
Protein A, 261
Protein A immunoassay, 350–352
Proteins
association and dissociation, 563
binding of small molecules, detection, 516–517, 523, 562
buried groups in, 514–515
denaturation, detection of, 515, 545, 563, 592, 623–626
dissociation of subunits, 517, 563
DNA-binding, purification with DNA-cellulose, 265
external amino acids, 515
fluorescence used to determine structure, 543–547, 549–550
helix-coil transition, 515
internal amino acids, 525

Proteins *continued*
  molecular weight, determination by gel chromatography, 242, 248
  NMR used to determine structure, 621–630
  ORD and CD used to determine structure, 588–591
  purification, 270–271
  purification by affinity chromatography, 260
  radius of gyration, 17
  Raman spectra used to determine structure, 528–529
  shape by diffusion, 465–466
  shape by viscometry, 482
  solvent-perturbation method used to study, 510, 513–515
  spectrophotometric titration, 512–515
  structure determined by fluorescence, 543–547, 549–550
  structure determined by NMR, 621–630
  structure determined by ORD and CD, 588–591
  structure determined by Raman spectra, 528–529
Pulse-height counting, 142–144, 151–155
Pycnometry, 456–458

Quantum yield
  definition, 540
  factors affecting, 540–542
  measurement of, 544
Quaternary structure, 7
Quenching
  chemical, 141, 152–154
  collisional, 540
  color, 142, 149–150, 152, 198
  dilution, 142, 154
  fluorescent, 540, 544, 547
  in Geiger counting, 136
  in scintillation counting, 141–154
Quinacrine staining, 66–67

R_f, 227
Radial migration separator, 489
Radioactive decay, theory of, 131–132
Radiochemical purity, 133
Radioimmunoassay, 333–337, 346–350
Radioisotope half-lives, 132
Radius of gyration, 8, 15, 17, 691, 694–695
Raman spectroscopy, 524–532
Random coil
  definition of, 8–9
  detection of, 521, 528, 623–624
Random walk, 15
Rayleigh scattering, 504–505
Reactigel, 259
Refractive index. *See* Index of refraction
Relative viscosity, 472, 482, 486
Relaxation in NMR, 615–616, 630
Renaturation
  definition, 28
  of DNA, 510
Replica formation, 80–81
Reporter groups, 517–518
Resolution, 43
Resonance energy transfer, 540, 551–559
Retention time, 235–236
Reverse dialysis, 208–209, 714
Reverse phase chromatography, 255
Rheopexy, 474
Rhodamine B, 548
RNA
  messenger, 199–201
  separation by gel electrophoresis, 287
Rocket electrophoresis, 356–357
Rotary shadowing, 86
Rotation of polarized light, 579–581
Rotational filtering, 105–106
Rotational levels, 495
Rotational strength, 582, 597–598

$s_{20,w}$. *See* Sedimentation coefficient
Scanning electron microscope, 109
Scatchard plot, 659, 663
Scattering correction, 504–505
Schlieren optics, 374–375
Scintillation cocktail, 148
Scintillation counting
  background noise in, 140–141
  channel ratio method in, 152–153
  external standards ratio, 152–154
  fluors for, 138
  internal standard ratio in, 152–154
  quench correction in, 152–154
  quenchers in, 141–142
  sample preparation for, 145–151
Screen filters, 196
SDS-gel electrophoresis, 283–286
Secondary structure, 7
Sedimentation
  boundary, 370, 396
  of DNA in alkali, 411–419
  through preformed sucrose gradients, 396–405
  velocity, 363–364, 370, 380–386
Sedimentation coefficient
  concentration dependence of, 380–382
  definition, 364, 443–445
  measurement of, 389–390
  molecular weight, effect on, 387–388, 401
  shape effect on, 386–387
  speed dependence on, 383–385
  standard, 384–389
  from zonal centrifugation, 400–401
Sedimentation-diffusion for determining molecular weights, 419–420
Sedimentation equilibrium
  in a density gradient, 424–440
  molecular weight from, 420–424
  partial specific volume from, 459–460
  theory of, 420–421
Sedimentation velocity
  concentration dependence in, 380–382
  determination of, 363, 374–380, 384–385
  examples of the use of, 390–394

Sedimentation velocity *continued*
Johnston-Ogston effect in, 382
rules of, 363
theory of, 363
Self-absorption, 145–146, 150, 154–155, 177
Sensitization in autoradiography, 185–187
Sephacryl, 239
Sephadex, 219, 239–243, 249, 251, 258–259, 271
Sepharose, 240–241, 245, 258–262
Sequencing of nucleic acids, 303–308
Serum blocking power, 340–341
Shadow casting, 80, 83–86
Shear
degradation, 475, 685–688
gradient, 473–474
stress, 473–474
Side-chain, amino acid, 3, 9
Slab gel, 282–284, 294
s-M relation, 388, 401
Sodium error, 122
Solubilization, 148–149
Solvation, 17
Solvent-perturbation method, 510, 513–515
Sonic degradation, 270, 688, 704
Sorbent, 217
Southern transfer, 299–302
Specific activity, 133
Specific rotation, 579
Spectrophotometer, 499
Spectrophotometric titration, 512, 523
Spectropor tubing, 209
Spin, nuclear, 604–605, 611–612, 637, 640
Spin labeling, 633–635, 645–649
Spin-lattice relaxation, 615
Spin-spin
decoupling, 613–614, 622
interaction, 611–615
relaxation, 615–616
Splitting, 611–613, 617
Staining 51, 78, 86–89
negative, 86–88
positive, 88–89
Starch-gel electrophoresis, 281–282
Star experiment, 181
Streptomycin to precipitate nucleic acids, 270
Stripping film, 172, 178, 183–185
Subunits, 19, 467, 666
Sucrose gradients, sedimentation through preformed, 396–405
Suction filtration, 195
Supercoil, 30–31, 268–270
Superhelix, 30–32
Support film, 76–77
Survival curve, 445–446
Synthetic boundary cell, 423, 463

T_m, 22
Tautomerism, detection by infrared spectroscopy, 522
Temperature, melting, 22
Thermopile, 541
Theoretical plates, 217–218
Thin-layer chromatography, 229–232
Thin-layer gel chromatography, 247–248
TNS, 547–551
Tracers, 158
Transmission, 497
Tris buffer, problems with, 122–124
Tritium exchange, 246, 689, 691
tRNA, structure of, 511, 523–524, 632–634
Tryptophan
difference spectrum of, 514
fluorescence of, 539, 544
Twisted circle, 30
Two-dimensional chromatography-electrophoresis, 227–228, 312–316
Tyrosine
absorption spectrum of, 502
difference spectrum of, 514
titration of, 512

Ultracentrifuge, 364–368
optical systems in, 374–380
rotors and cells, 366–369

$\bar{v}$, 363–364, 455–456, 459–460
Van der Waals interaction, 18–19
Velocity sedimentation. *See* Sedimentation velocity
Vesicles, 617–618
Vibrational levels, 495–496
Viscoelasticity, 486–489
Viscometers, 476–480
Viscosity
effect of macromolecules on, 472–474
fluorescence polarization and, 432–433, 437
intrinsic, 473–475, 481–483
measurement of, 470–481
relative, 472, 482, 486
specific, 472
theory of, 470–472
Void volume, 221

Weight-average molecular weight, 13–14
Western transfer, 302–303
Windowless flow counter, 137

Zimm plot, 694
Zonal centrifugation, 396–419
examples of use of, 402–405, 413, 417–419
sedimentation coefficient by, 400–401
Zone electrophoresis. See Electrophoresis, zone